LARGE SCALE INTEGRATED TECHNOLOGY:
STATE OF THE ART AND PROSPECTS

NATO ADVANCED STUDY INSTITUTES SERIES

Proceedings of the Advanced Study Institute Programme, which aims at the dissemination of advanced knowledge and the formation of contacts among scientists from different countries.

The series is published by an international board of publishers in conjunction with NATO Scientific Affairs Division

A	Life Sciences	Plenum Publishing Corporation
B	Physics	London and New York
C	Mathematical and Physical Sciences	D. Reidel Publishing Company Dordrecht and Boston
D	Behavioural and Social Sciences	Martinus Nijhoff Publishers The Hague, Boston and London
E	Applied Sciences	

Series E: Applied Sciences — No. 55

LARGE SCALE INTEGRATED CIRCUITS TECHNOLOGY: STATE OF THE ART AND PROSPECTS

Proceedings of the NATO Advanced Study Institute on "Large Scale Integrated Circuits Technology: State of the Art and Prospects",
Erice, Italy, July 15–27, 1981

edited by

Leo Esaki
IBM T.J. Watson Research Center
Yorktown Heights, N.Y. 10598
U.S.A.

and

Giovanni Soncini
CNR-Lamel
40126 Bologna
Italy

1982

Martinus Nijhoff Publishers
The Hague / Boston / London

Distributors:

for the United States and Canada
Kluwer Boston, Inc.
190 Old Derby Street
Hingham, MA 02043
USA

for all other countries
Kluwer Academic Publishers Group
Distribution Center
P.O.Box 322
3300 AH Dordrecht
The Netherlands

Library of Congress Cataloging in Publication Data

NATO Advanced Study Institute on "Large Scale
    Integrated Circuits Technology: State of the
    Art and Prospects" (1981 : Erice, Italy)
    Large scale integrated circuits technology.

    (NATO advanced study institutes series. Series E,
Applied sciences ; no. 55)
    1. Integrated circuits--Large scale integration--
Congresses. I. Esaki, Leo 1925-   . II. Soncini,
Giovanni. III. North Atlantic Treaty Organization.
IV. NATO Advanced Study Institute. V. Title.
VI. Series.
TK7874.N337 1981   621.381'73   82-14229
ISBN-13:978-94-009-7647-4

ISBN-13:978-94-009-7647-4   e-ISBN-13:978-94-009-7645-0
DOI: 10.1007/978-94-009-7645-0

PREFACE

A NATO Advanced Study Institute on "Large Scale Integrated Circuits Technology: State of the Art and Prospects" was held at Ettore Majorana Centre for Scientific Culture, Erice (Italy) on July 15-27, 1981, the first course of the International School of Solid-State Device Research.

This volume contains the School Proceedings: fundamentals as well as up-to-date information on each subject presented by qualified authors. The material covered in this volume has been arranged in self-consistent chapters. Therefore, the Proceedings may be used as a suitable textbook or authoritative review for research workers and advanced students in the relevant field.

The nascent information society is based on advanced technologies which will revolutionize human abilities to manipulate and communicate information. One of the most important underpinnings for developing such an information society lies in innovations in semiconductor microelectronics. Such innovations, indeed, are dramatically reducing the cost of transmitting, storing, and processing information with improved performance, ushering in an era characterized by large scale integration - the subject of this book.

L. Esaki

G. Soncini

March 1982

Contributors

D. A. Antoniądis

G. Baccarani

C. M. Bailey

F. Bertotti

A. N. Broers

G. J. Declerck

R. H. Dennard

C. Donolato

R. C. Eden

L. Esaki

B. Hoefflinger

W. Holt

R. W. Keyes

K. Kimura

B. O. Kolbesen

M. Montier

B. Murari

F. H. Reynolds

H. Ryssel

G. Soncini

T. Takahashi

E. J. VanDerveer

B. M. Welch

CONTENTS

These Proceedings were prepared by Peggy Powers and Maryann Pulice.

CHAPTER I

OVERVIEW

SEMICONDUCTOR DEVICES
AND THE ROLE OF PHYSICS IN THEIR DEVELOPMENT

L. Esaki

IBM Thomas J. Watson Research Center
Yorktown Heights, New York 10598

ABSTRACT: Following a brief description of early semiconductor history, important events in device development are presented in perspective, with emphasis on the role of semiconductor physics.

I. INTRODUCTION

New scientific knowledge, arising from great inventions or discoveries, often leads to a large-scale engineering effort which eventually has far-reaching consequences in our society. The invention of the transistor by three solid state physicists, Shockley, Bardeen, and Brattain, is one such example. The development of the transistor began in 1947 through interdisciplinary cooperation with physicists, chemists, metallurgists and electronic engineers at Bell Laboratories. A large-scale developmental effort for a variety of semiconductor devices followed in a number of institutes throughout the world. Particularly, integrated circuits (ICs) have made an impressive evolution toward higher levels of integration during the past twenty years. Semiconductor know-how, thus established, has revolutionized the whole world of electronics --- data processing, telecommunications, industrial process control, military gears, scientific instruments and consumer products.

Solid-state or semiconductor physics undoubtedly has given an impetus in creating semiconductor technology. Semiconductor physics involves experimental investigation as well as theoretical understanding of the physical properties, including electrical, optical, and thermal properties and interactions with all forms of radiation in semiconductors. Many of these have been of interest since the 19th century, partly because of their practical applications and partly because of the richness of intriguing phenomena that semiconductor materials present.

Point-contact rectifiers made of a variety of natural crystals found practical applications as detectors of high-frequency signals in radio telegraphy in the early part of this century.[1] The natural crystals employed were lead sulphide (galena), ferrous sulphide, silicon carbide, etc. Plate rectifiers made of cuprous oxide or selenium were developed for handling large power output. The selenium photocell was also found useful in the measuring of light intensity because of its photo-sensitivity.

In the late 1920's and during the 1930's, the new technique of quantum mechanics was applied to develop electronic energy band structure,[2] whereby a modern picture of the elementary excitations in semiconductors was obtained. Of course, this modern study has its roots in the discovery of x-ray diffraction by von Laue in 1912, which provided quantitative information on the arrangements of atoms in semiconductor crystals. Within this framework, attempts were made to obtain a better understanding of semiconductor materials and quantitative or semiquantitative interpretation of their transport and optical properties, such as rectification, photoconductivity, electrical breakdown, etc.

During this course of investigation on semiconductors, it was recognized in the 1930's that the phenomena of semiconductors should be analyzed in terms of two separate parts: surface phenomena and bulk effects. Rectification and photovoltaic effects appeared to be surface or interface phenomena, while ohmic electrical resistance with a negative temperature coefficient and ohmic photocurrent appeared to belong to bulk effects in homogeneous semiconductor materials. Relatively thick depletion layers near the surface are formed because of the existence of surface states which trap electrons and, also because of long screening lengths in semiconductors arising from much lower carrier concentrations than in metals. Thus the depletion of carriers creates potential barriers on the semiconductor surface or at the interface between a semiconductor and a metal contact, or between two semiconductors. The early recognition of the importance of surface physics was one of the significant aspects in semiconductor physics.

II. TRANSISTORS

Since the rectification in semiconductor diodes is analogous to that obtained in a vacuum diode tube, a number of attempts had been made to build a solid-state triode by inserting a "grid" into semiconductors or ionic crystals - a solid-state analog of the triode tube amplifier.[3] Because of relatively low densities of carriers in semiconductors, Shockley thought, control of the density of carriers near the semiconductor surface should be possible by means of an externally applied electric field between the surface and a metal electrode insulated from the surface -- the field effect device. The observed effect, however, was much less than predicted.[4] In 1947, in the course of trying to make a good "field effect" device with two gold contacts less than fifty microns apart on the germanium surface, Bardeen and Brattain made the first point-contact transistor where they discovered a phenomenon -- minority carrier injection into a semiconductor.[5] The importance of this phenomenon was soon recognized and led to the invention of the junction transistor by Shockley. The realization of this junction device, which did not occur until 1950,[6] was far more significant than its precursor.

The development of such junction transistors, as well as the progress in semiconductor physics on Ge and Si, would not have been accomplished without the key contribution of materials preparation techniques. Soon after Teal and Little prepared large Ge single crystals, Sparks successfully made a grown junction transistor.[7] The subsequent development was Pfann's zone refining and then Theuerer's floating zone method for silicon processing. These developments made it possible to make Ge and Si crystals of controlled purities and unprecedented perfection.

The early Ge junction transistors had poor frequency response and relatively low reliability. In fabricating these transistors, the grown-junction technique, or the alloying technique, was used to form p-n junctions. Then a procedure for forming p-n junctions by thermal diffusion of impurities was explored in order to obtain better reproducibility and tighter dimensional tolerances. This technique, indeed, enabled bringing forth the double diffused transistor with desirable impurity distribution, which was the prototype of the contemporary transistor.[8] Attention was also turned toward Si because of its expected high reliability and improved temperature capability.

In the 1940's, a team at Bell Laboratories selected elemental semiconductors, Ge and Si, for their solid-state amplifier project, primarily because of the possible simplicity in understanding and material preparation, in comparison with oxide or compound semiconductors. This not only was a foresighted selection but also had important implications: Ge and Si single crystals exhibited long diffusion lengths of hundreds of microns at room temperature, arising from both high carrier mobilities and long trapless lifetimes of minority carriers, which were a prerequisite to the desirable operation of the transistor. The long lifetimes may arise from the indirect energy-gap in these elemental semiconductors in contrast with the direct energy-gap in some III-V compound semiconductors which exhibit high rates of radiative recombination of electrons and holes.

The exploration of the III-V compound semiconductors was initiated through Welker's ingenuity and imagination, in the early 1950's, to produce semiconductor materials even more desirable for transistors than Ge or Si.[9] Although this initial expectation was not easily met, III-V compound semiconductors found their most important applications in LED, injection lasers, Gunn microwave devices, etc.; this could not have been achieved through elemental semiconductors. One might add that some III-V compound semiconductors demonstrate very intriguing characteristics: small effective masses for electrons (GaAs:$0.067m_o$ and InAs:$0.023m_o$), hence high electron mobilities (more than $100,000$ cm^2/volt•sec at low temperatures) desirable for high-speed FETs; semi-insulating materials suitable for IC substrates; nearly perfect heteroepitaxy between two compounds enabling us to fabricate novel structures, etc. Such characteristics obviously will be exploited further in device development of the future.

III. IMPORTANT DEVICES

Now, in order to reach a perspective in semiconductor device development, it may be worthwhile to comment on some selected semiconductor devices:

1) Solar Cells - In 1940, Ohl observed a photovoltage as high as 0.5V by flashlight illumination in "naturally" grown Si p-n junctions.[10] The modern Si solar cell, however, was created by bringing together the seemingly unrelated activites, namely, large area p-n junctions by Fuller's diffusion method, Pearson's effort for power rectifiers, and Chapin's search for power sources of communication systems in remote locations.[11] This cell showed a conversion efficiency from solar energy to electrical energy of 4%. Low as this efficiency may seem today, in 1953 it was very exciting, improving on selenium by a factor of five. Development and production of solar cells were stimulated by the needs of the space program.

In 1972, heterojunction solar cells consisting of $pGa_{1-x}Al_xAs-pGaAs-nGaAs$, exhibiting power conversion efficiency of 16-20%, were reported by Woodall and Hovel.[12] The improved efficiencies were attributed to the reduction of both series resistance and surface recombination losses resulting from the presence of the heavily-doped $Ga_{1-x}Al_xAs$ layer. The recent advent of the energy crisis, however, generated a renewed interest in research and development for such solar cells as are economically viable for terrestrial applications. The development of amorphous Si solar cells is one such example.[13]

2) Tunneling Devices - Interest in the tunnel effect goes back to the early years of quantum mechanics. In the early 1930's, attempts were made to explain phenomena in solids such as rectification, contact resistance, etc., in terms of electron tunneling across the insulating barriers. However, since theories and experiments often gave conflicting results, not much progress was made at that time. Around 1950, semiconductor p-n junctions generated a renewed interest in the tunneling process. Experiments to observe such a process in the reverse breakdown of the junctions, however, were again inconclusive.

In 1957, Esaki demonstrated convincing experimental evidence for tunneling in his heavily-doped (narrow) p-n junction -- the tunnel diode.[14] This diode found usefulness in microwave applications because of its differential negative resistance being responsive to high frequencies. The discovery of the tunnel diode not only generated an interest in heavily-doped semiconductors but also helped to open a new research field on tunneling in semiconductors as well as in superconductors. The 1973 Nobel prize citation states, "....the pioneering work by Esaki provided the foundation and direct impetus for Giaever's discovery and Giaever's work in turn provides the stimulus which led to Josephson's theoretical predictions." The Josephson tunnel-junction devices, operated at superconducting temperatures, now find usefulness in rather unique applications The attempt has been made to use such quantum-mechanical devices as the basic switching elements for ultra high-speed computers, although their development is still at the early stage.

3) Integrated Circuits and MOS Devices - In 1958, Kilby initiated the fabrication of a circuit which included a number of transistors, diodes, resistors, and capacitors, all residing on one semiconductor chip.[15] This structure is called the (monolithic) integrated circuit. Around the same time, Noyce and Moore introduced improved fabrication techniques called the "planar" process which enabled the birth of the first modern transistor - a landmark in semiconductor history. It was soon realized that this transistor with dished junctions (extending to the surface) and oxide passivation (protecting the junctions), was most suited for assembling integrated circuits, because metal stripes evaporated over the surface oxide layer could be readily used for interconnections.[16]

As mentioned earlier, the transistor was invented while searching for a field-effect device. The field-effect concept originated as early as the 1920's, and yet no successful device was made in spite of a number of attempts because of lack of adequate technology. However, thermally-grown SiO_2 on Si single crystal surfaces, which was originally developed for the above-mentioned oxide passivation of junctions in the later 1950's, was found to be a most ideal insulator for such a field effect device by Kahng and Atalla.[17] This insulator, indeed, had relatively low loss and high dielectric strength, enabling the application of high gate fields. More importantly, the density of surface states at the Si-SiO_2 interface was kept so low that the band bending in Si near the interface was readily controllable with externally applied gate fields. Thus, a simple, yet most practical, Si MOS field-effect transistor (FET) was created whereby the surface inversion layer conductance ("channel") was modulated by gate voltages. This transistor is called a unipolar device because of no minority carrier involvement; it needs fewer processes in fabrication than the bipolar transistor because of its structural two-dimensionality, and is especially adaptable for large-scale integrated circuits.

Integrated circuits of digital as well as linear types have had one of the largest impacts on electronics; they are now the main building block in computers, instrumentation, control systems and consumer products. This is quite evident in the fact that, for instance, the generation of the computer always has been identified by the stage of the progress of integrated circuits used, since computer cost and performance have been critically dependent on the IC technology. Presently, integrated circuits, consisting of MOS FET or MOS based components such as dynamic memory cells,[18] charge-coupled devices,[19] nonvolatile memory cells, etc., are even more extensively used than bipolar transistors, in computer memories, microprocessors, calculators, digital watches, etc., while being challenged by advances in bipolar-based devices such as $I^2 L$ (Integrated Injection Logic).

In the early 1960's the era of integrated circuits began with a small scale integration (SSI) such as several bipolar logic gates per chip. The development of ICs progressed through medium scale integration (MSI) and large scale integration (LSI). The 1 K-bit, 4 K-bit, and 16 K-bit MOS random access memory (RAM) chips made in the 1970's can be considered to be examples of LSI. Now, with further reduction of the individual device size, the techniques of very large scale integration (VLSI) have been established: the 64 K-bit MOS RAM, introduced in the late 1970's, is generally accepted to be the first commercial chip in the VLSI class. Thus, the complexity of ICs has almost doubled each year, now approaching one million components on a single Si chip of, say, a half centimeter square, and yet the cost per function has decreased over ten thousandfold since their introduction. Meanwhile, system performance and reliability have been tremendously improved.

There are three main reasons for the achievement of such high level of integration, as were pointed out by Noyce:[20]

1. By decreasing minimum dimensions with the advance of the photolithographic techniques, resulting in higher density circuit elements.

2. By decreasing defect density through improved processing techniques, allowing the practical production of circuits of larger area.

3. Through innovations in circuit forms, allowing higher functional density.

By scaling device dimensions down, keeping the electric fields in the devices constant, device parameters such as speed and power consumption move in a favorable direction. As the size of individual FETs continues to decrease for large integration with the application of advanced processing techniques, the "channel" distance is shortened to the submicron range and the oxide thickness is thinned to a few hundred angstroms. However, if one pushed this to the extreme, then new physical problems arise from excessively high fields across thin oxide films as well as in the "channel" direction. There has been some discussion on physical limits in digital electronics.[21]

While taking measurements of Si surface transport properties at low temperatures, Fang and Howard discovered that electrons in the "channel" were two-dimensional,[22] which provided a unique opportunity for studies of quantum effects.[23]

4) GaAs Devices - We will discuss three types of devices, namely, injection lasers, microwave oscillators and GaAs FETs, in this order:

Since the early part of this century, the phenomenon of light emission from SiC diodes was recognized and studied, although a practical light emitting diode had not materialized until the development of efficient p-n junctions made of III-V compound semiconductors.[24] Apparently, reports of high-efficiency radiation in GaAs stimulated a few groups to engage in a serious experimental effort to find lasing action in semiconductors: These possibilities were previously discussed.[25] In 1962, the announcement of the successful achievement of lasing action in GaAs came on the same date, independently, from two groups: Hall et al. at

General Electric; and Nathan et al. at IBM; and a month later from Quist et al. at Lincoln Laboratory.[26] All of them observed a pulsed coherent radiation of 8400Å from liquid nitrogen-cooled, forward-bias GaAs p-n junctions. This happening is not surprising in the present competitive environment of the technical community where new scientific information is rapidly disseminated and digested, and new ideas are quickly implemented. There was a two-year interval between the first reports of the Ruby and He-Ne lasers and the announcement of the injection laser.

The performance of the device was improved with incorporation of heterojunctions by Alferov et al.[27] With double-heterostructure the threshold current density for lasing was substantially reduced by confinement of both carriers and photons between two heterojunctions.[28] Finally, in 1970, Hayashi et al.[29] succeeded in operating the device continuously at room temperature. Because of the compactness and the high efficiency of this laser, the achievement paved the way towards many practical applications such as optical (light-wave) communication, signal processing, display and printing. There is a developmental effort in integrated optics to mount miniaturized optical components, including injection lasers and waveguides, on a common substrate using heterojunction structures of III-V compound semiconductors, analogous to the integrated circuit, for improved signal processing.

In 1962, Gunn discovered that, when the applied field across a short bar of reasonably pure n-type GaAs exceeded a threshold voltage of several thousand volts per cm, coherent microwave oscillations could be extracted by synchronizing the random current fluctuations with a resonator.[30] Furthermore, by his ingenious probe technique, he was able to show that the oscillations were related to the periodic formation and propagation of a narrow region of very high field -- "domain." It took two years to confirm that Gunn's experimental discovery of oscillations was indeed due to the Ridley-Watkins-Hilsum transferred electron effect, proposed in 1961 and 1962.[31]

As is true of any important discovery, Gunn's work triggered a wide spectrum of experimental and theoretical activity from device physics to microwave engineering: Apparently this achievement rejuvenated the work of microwave semiconductor devices in general, and, in 1964, IMPATT (IMPact ionization Avalanche Transit Time) diodes finally started to oscillate - which was rather overdue since Read's proposal in 1958.[32] The operation of the device was explained on the basis of dynamics of electrons involving the transit time and avalanche. IMPATT and Gunn devices are now used in many microwave gears.

In addition to the development of discrete microwave GaAs FETs, GaAs integrated circuits have been recently offered for high-speed digital applications as an alternative to Si devices, where planar GaAs Schottky-Gate FETs (MESFET) are the basic circuit elements.[33] It has been demonstrated that the high mobility and electron velocities in short channel ($\sim 1\mu m$) give transconductances or drain currents several times those for Si n-channel FETs at equivalent gate biases. More recently, attempts have been made to further improve high-speed performance of GaAs MESFET by applying selectively doped GaAs-Ga$_{1-x}$Al$_x$As heterostructures prepared by molecular beam epitaxy.[34] It is certainly not unreasonable to postulate that, in the future, integrated circuits made of devices with narrow-gap semiconductors such as InAs-based FETs, operated at $77°K$, show performance far superior to those made of GaAs devices, simply because of smaller electron effective masses.[35]

IV. SUMMARY

Figure 1 schematically illustrates the development path of a variety of semiconductor devices. It should be noted that the development path of each device appears to have had its own sequence of conception (theory) and observation (experiment): Typically, the theoretical prediction was later confirmed by the experiment, but, in many instances, the experimental discovery came first, followed by the theory and yet, in other instances, the initial idea which led to the discovery was irrelvant to its consequence. Obviously, this article cannot possibly cover all landmarks and indispensable innovations, not to mention a great number of wonderful, but nonworkable ideas.

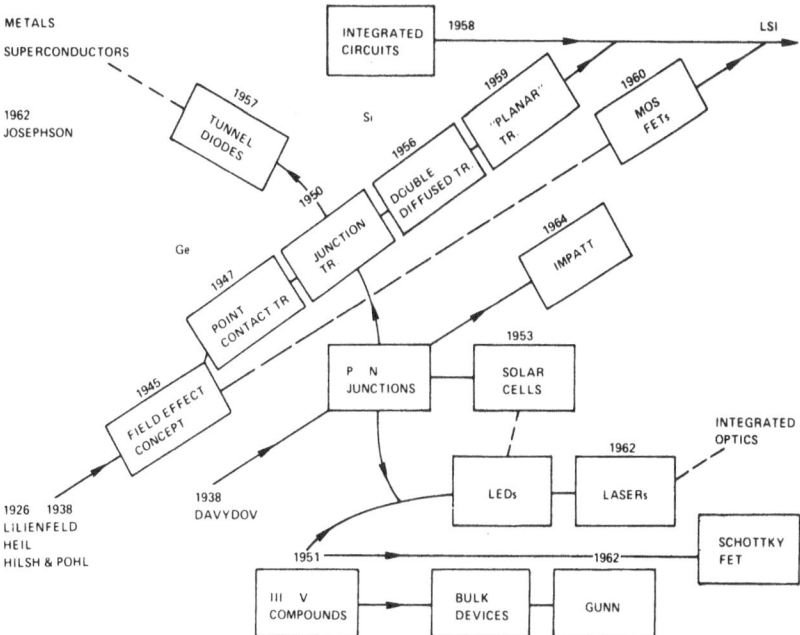

Fig. 1 - Schematic illustration of the development path of a variety of semi-conductor devices.

Not all of this progress arises from engineering ingenuity and advanced material technology; pioneering research in semiconductor physics which often tries to break new ground has also contributed to each significant development, exploring intriguing phenomena such as electron-hole multiplication (avalanche), tunneling, hot electrons, lasing by high carrier injection, field-effect transport, two-dimensional electrons on surfaces or in semiconductor superlattices, etc. For a qualitative understanding, involved materials, crystalline or amorphous, as well as surfaces, have been extensively investigated - often under extreme conditions with advanced instruments; measurements at high pressure or at high electric or magnetic fields, or in ultrahigh vacuum, or under the synchrotron radiation fall into this category.

In the device development path, we have witnessed that good communications have existed between scientists whose main interests are research for new knowledge and engineers whose main interests are the innovative applications of scientific knowledge. The mutual stimulation is obvious in industrial research laboratories, where new technology and new science often come from the same building, and, sometimes, from the same heads. Particularly, in semiconductors, the separation between basic discoveries and applications is far less distinct than that in some other fields.

Following the view expressed by the Physics Survey Committee,[36] some of the most startling technological advances in our time are closely associated with basic research. As compared with thirty years ago, the highest vacuum readily available with commercial systems has improved more than a thousandfold; materials can be manufactured that are one hundred times purer; the submicroscopic world can be seen at hundreds of times higher magnification; the detection of trace impurities is hundreds of times more sensitive either in bulk or on surfaces. These examples are only a small sample. On the other hand, scientific research is crucially dependent on advanced technology. Among numerous such examples, one can cite digital scientific instrumentation with computer-assisted data acquisition and analysis capabilities, which has been widely employed in a variety of investigations. Such useful instruments have not been brought forth without the advance of semiconductor IC technology. We can see here an intricate interdependency between science and technology.

Reviewing the semiconductor device development over the past thirty years, we believe, it has been proven that the interaction between scientists and engineers is indeed the most effective means for creating new technology.

REFERENCES

1. H. K. Henisch, Rectifying Semiconductor Contacts, Clarendon Press, Oxford 1957.

2. A. H. Wilson, Proc. Roy. Soc. A, $\underline{133}$, 458 (1931).

3. J. E. Lilienfeld, U.S. Patents 1,745,175 (1926); 1,877,140 (1928); 1,900,018 (1928); O. Heil, British Patent 439457 (1935); R. Hilsh and R. W. Pohl, Zeits. f. Phys. $\underline{111}$ 399 (1938).

4. W. Shockley and G. L. Pearson, Phys. Rev. $\underline{74}$, 233 (1948).

5. J. Bardeen and W. H. Brattain, Phys. Rev. $\underline{74}$, 230 (1948); Phys. Rev. $\underline{74}$, 231 (1948); Phys. Rev. $\underline{75}$, 1208 (1949).

6. W. Shockley, IEEE Trans. Electron Devices $\underline{ED-23}$, 597 (1976).

7. G. K. Teal, IEEE Trans. Electron Devices $\underline{ED-23}$, 621 (1976).

8.   M. Tanenbaum and D. E. Thomas, B.S.T.J. 35, 1 (1956).

9.   H. J. Welker, IEEE Trans. Electron Devices ED-23, 664 (1976).

10.   R. S. Ohl, U.S. Patent 2,402,662 (1941).

11.   F. M. Smits, IEEE Trans. Electron Devices ED-23, 640 (1976).

12.   J. M. Woodall and H. J. Hovel, Appl. Phys. Lett. 21, 379 (1972).

13.   D. E. Carlson and C. R. Wronski in Amorphous Semiconductors edited by M. H. Brodsky (Springer-Heidelberg 1979) Chapter 10.

14.   L. Esaki, Nobel Lecture, Dec. 11, 1973, Pub. by Les Priz Nobel, p. 66 (1974); IEEE Trans. Electron Devices ED-23, 644 (1976).

15.   J. S. Kilby, IEEE Trans. Electron Devices ED-23, 648 (1976); U. S. Patent 3,138,743 (1959).

16.   R. N. Noyce, U. S. Patent 2,981,877 (1959).

17.   D. Kahng, IEEE Trans. Electron Devices ED-23, 655 (1976).

18.   R. H. Dennard, U. S. Patent 3,387,286 (1968).

19.   W. S. Boyle and G. E. Smith, IEEE Trans. Electron Devices ED-23, 661 (1976).

20.   R. N. Noyce, Computer Age, edited by M. L. Dertouzos and J. Moses, MIT Press, Cambridge, Mass. 1979, page 321.

21.   R. W. Keyes, Proc. IEEE 63, 740 (1975).

22.   F. F. Fang and W. E. Howard, Phys. Rev. Lett. 16, 797 (1966).

23.   A. B. Fowler, et al., Phys. Rev. Lett. 16, 901 (1966).

24.   E. E. Loebner, IEEE Trans. Electron Devices, ED-23, 675 (1976).

25.   M. G. A. Bernard and G. Duraffourg, Physica Status Solidi 1, 699 (1961); W. P. Dumke, Phys. Rev. 127, 1559 (1952).

26.   R. N. Hall, IEEE Trans. Electron Devices, ED-23, 700 (1976).

27.   Zh. I. Alferov et al., Fiz. Tekh. Poluprov. 2, 1545 (1968), and Fiz. Tekh. Poluprov. 3, 1328 (1969).

28.   H. Kressel and F. Z. Hawrylo, Appl. Phys. Lett. 17, 169 (1970).

29.   I. Hayashi, M. B. Panish, et al., Appl. Phys. Lett. 17, 109 (1970).

30.   J. B. Gunn, ED-23, 705 (1976).

31.   B. K. Ridley and T. B. Watkins, Proc. Phys. Soc. (London) 74, 293 (1961); C. Hilsum, Proc. IRE 50, 185 (1962).

32.   B. C. DeLoach, Jr., IEEE Trans. Electron Devices, ED-23, 657 (1976).

33.   R. C. Eden, B. M. Welch, R. Zucca and S. I. Long, IEEE Trans. Electron Devices, ED-26, 299 (1979).

34.   T. Mimura, S. Hiyamizu, K. Joshin and K. Hikosaka, Jpn. J. Appl. Phys. 20, L317 (1981).

35.   L. Esaki, L. L. Chang and E. E. Mendez, Jpn. J. Appl. Phys. 20, L529 (1981).

36.   Physics in Perspective, Vol. I and II, National Academy of Sciences, Washington, D. C., 1972.

LSI: PROSPECTS AND PROBLEMS

Robert W. Keyes

Thomas J. Watson Research Center
Yorktown Heights, NY   10598

ABSTRACT: VLSI technology is aimed at maximizing the utilization of silicon area, electrical power, and costly production facilities, while providing high functional capability. Integration and miniaturization are the routes by which this is accomplished. Progress in integration and miniaturization involves solving a continual series of technological problems in lithography and chemical process technology.

I. INTRODUCTION

Modern solid state electronics has demonstrated a persisting capacity to provide increasing functional capability while decreasing the cost of components and the systems built from them. The technological themes that have made continuing advances possible are miniaturization and integration. Integration is the key to low cost because it permits many devices to be fabricated by handling and processing one object. Integration also leads to improvements in reliability because the interconnections formed by integrated semiconductor technology, in addition to being less costly, have turned out to be far more reliable than solder joints and pluggable connections.

Miniaturization, of course, is one of the ingredients that makes high levels of integration possible. In addition, however, it leads to increased speed of operation because electrical signals have shorter distances to travel and because capacitances are smaller. The lower capacitances also reduce the power dissipation of circuits. Thus, the benefits accruing from miniaturization and integration are to be found in cost, speed, power dissipation and reliability. It must be observed that low cost, low power dissipation and high reliability are the things that make it possible to construct and operate large systems containing a great many components.

This paper is an attempt to glimpse the future of solid state electronics. Increasing levels of integration are driven by strong economic forces. We will extrapolate the physical characteristics of semiconductor technology through the remaining two decades of this century[1]. Clearly, new problems will be encountered and new technologies will be needed to continue the present rate of progress. The extrapolations will provide an idea of the magnitudes involved and the time scale on which problems must be faced. (It should be noted that the dates quoted are those of commercial introduction and follow laboratory demonstration by many years.) Various aspects of the VLSI technology future will be discussed within this framework.

II. EXTRAPOLATIONS

Extrapolations are broadly based on the assumptions that present levels of development effort will continue and produce results comparable to those achieved in the past. They also assume that such results will be economically desirable, that they will enable a user to excerise data processing functions at reduced cost. Extrapolation through the earlier part of the period covered has been substantially guided by available projections of others. It is influenced by judgements as to where limits will be found in the latter part of the period.

Several factors have combined to make the rapid growth in level of integration that has occurred during the last twenty years possible. These are increasing chip size, decreasing dimensions of structures, and compaction of devices that reduces their size with a fixed dimensional constraint. They will now be considered in turn.

A frequently used measure of the status of semiconductor technology is the minimum dimension of structures fabricated on a chip. Although minimum dimension is a useful way to characterize a technology, it must be used with some caution. In the first place, commercial practice usually lags far beyond the best results obtainable in research and development laboratories. Many years may elapse between the first demonstration of a method to produce small dimensions and the refinement of the method to the point at which routine, economical use of it with high yield is possible. Thus, the projections will seem unduly conservative to a laboratory scientist. In the second place, great effort may be devoted to a single step that determines a dimension crucial to the performance of a device, such as the smallest dimension of the gate of a field effect transistor, but this dimension may not be available anywhere else on the chip.

The size of chips has grown with time. The growth is often somewhat erratic; the introduction of a new kind of chip with an increased number of components generally requires an increase in chip size. Experience with processing the new chip makes redesigns that utilize less area possible and chip size decreases slowly until a new product is introduced. Chip size is limited by the size of the field that can be exposed by lithorgraphic tools and by the number of devices and the amount of interconnection wiring that can be processed reliably. The extrapolation of chip size and minimum dimension is shown in Fig. 1.[1-3]

The third source of increased integration mentioned above, compaction of devices, is the development of new device structures that require less area within a fixed dimensional constraint. For example, a thinner dielectric layer could allow a certain amount of capacitance to occupy less area. An additional layer of wiring can reduce the size of a cell that contains a logic gate and its interconnections. A new method of fabricating metal-semiconductor contacts might allow a contact of fixed ohmic resistance to occupy less area. Such compaction usually involves additional process steps or improved processes or both.

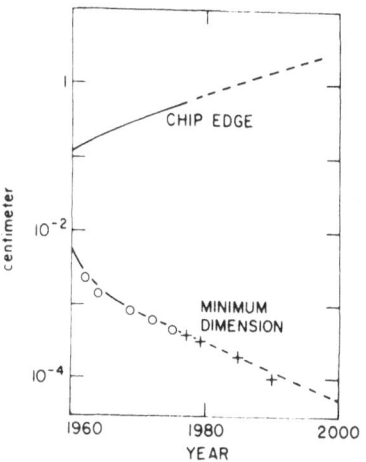

Fig. 1 - History and projection of the minimum lithographic dimension and the maximum technologically feasbile chips size. [1] (Copyright 1971 IEEE).

Compaction can be quantitatively characterized in the following way. The minimum area that one can hope to control independently of other areas on a chip is the square of the minimum dimension mentioned above. Call this area a resolvable element. Then the state of the compaction of a structure is described by the number of resolvable elements that it

occupies.   Contemporary chips contain 10^6 to 10^7 resolvable elements.   A memory bit

(including associated wiring and circuitry) may occupy 100 resolvable elements, while a logic

gate requires several thousand.

The number of resolvable elements contained on a chip can be calculated from Fig. 1 and

this number is plotted in Fig. 2.   There has

been a steady decrease in the number of

elements needed to make a circuit that

stores one bit of information in memory, as

shown in Fig. 2.   It is assumed that the

number of elements used to construct a log-

ic gate will decrease at a rate similar to the

historic rate of memory compaction and an

extrapolation based on this premise is also

given in Fig. 2.

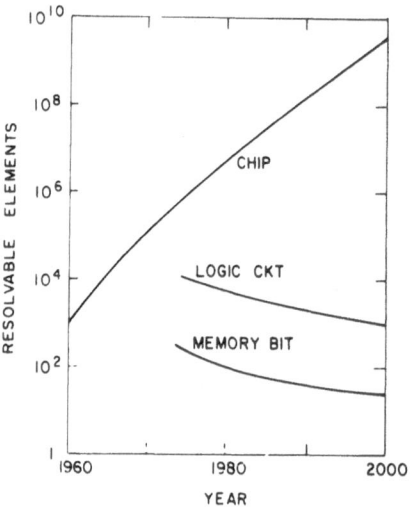

Fig. 2 - The number of resolvable elements on a chip of maximum dimension, calculated
from Fig. 1, and estimates of
the number of such elements
needed to construct a logic circuit and a memory bit plus the
other elements, such as wires,
decoders, and drivers that ac-
company them on the chip [1]
(Copyright 1979 IEEE).

III. VOLTAGE-A CONSTRAINT

Operating voltage is an important fac-

tor in all aspects of integrated circuit tech-

nology.   Low voltage is desirable to mini-

mize power dissipated on a chip and to a-

void problems associated with high electric fields, such as dielectric breakdown.   Voltage,

however, cannot be decreased without limit.   A minimum voltage amplitude of electrical

signals is required to insure correct operation for several reasons.[4] The most important of

these, at least in field-effect devices, are the various differences between nominally identical

devices.   An electrical signal must control a device whose characteristics are not precisely

known.   The signal voltage must be large enough to control all possible recipient devices.

Differences among devices have a number of sources.   One is process variability; the threshold voltage of a field effect transistor depends on substrate doping level, surface charge, insulator thickness, and other geometrical parameters that can only be controlled within a certain tolerance during manufacture.[5] Even if perfect control could be achieved, there would be some variability arising from the randomness of the location of donor and acceptor atoms on an atomic scale.[6] The average number of atoms in significant regions of a device may be only a few hundred, and the actual number varies around the average in a statistical way with a standard deviation equal to the square root of the average number. Still another source of differences arises from the dependence of device characteristics on temperature and the variability of temperature within a system.

Voltage is also needed to achieve high gain and non-linearity in semiconductor logic circuits. As can be seen from the ideal p-n junction characteristic,

$$i = i_o \left[\exp (qV/kT) - 1 \right]$$

semiconductors respond in only a linear, resistive way to voltage less than the thermal voltage, (kT/q).

IV. THE COMPONENTS OF AN INTEGRATED CIRCUIT

Of all of the components of integrated circuit technology, devices attract the most attention. Their complex structure offers many opportunities for invention. Three types of devices compete for use in VLSI, the bipolar transistor, the metal-oxide-semiconductor field effect transistor (MOSFET), and the Schottky gate field effect transistor (MESFET).

The bipolar transistor was the first transistor to find wide application in electronics and thus may serve as a standard against which to compare other types. The early success of the bipolar transistor is undoubtedly a result of the insensitivity of its characteristics to process variations; the exponential dependence of current through a p-n junction on voltage means that there is only a logarithmic dependence of threshold voltages on parameters such as doping levels and junction areas. The bipolar transistor is well-adapted to high performance VLSI because a small device can control a large current. Its transconductance is high.

The MOSFET has come to play a very important role in large scale integration. A simpler process, as compared to bipolar transistor technology, has made the high yields per device that make VLSI possible feasible. Nevertheless, the introduction of the MOSFET into electronics was delayed by the need for high quality oxide insulation between the gate and substrate and rigorous control of the SiO_2-Si interface properties. The MOSFET is well-suited to high impedance devices and has made "dynamic" circuits, in which information is stored on a capacitor that is only occasionally refreshed, possible. The transconductance (for a device that occupies a fixed area of a chip) of a MOSFET is less than that of a bipolar transistor because the current is controlled by the gate through a capacitive effect rather than through the exponential characteristic of a p-n junction. Electrical control is exercised, in effect, through a capacitive voltage divider.[7] The threshold voltage of a MOSFET depends on parameters of the device structure, such as oxide thickness and doping level, that cannot be controlled perfectly during manufacture. The differences among devices must be overcome by large signal swings, and MOSFET circuits therefore operate at higher voltages than bipolar circuits.

MESFETs are similar to MOSFETs in that the current is controlled by varying the number of current carrying electrons by a capacitive effect. MESFETs are free of the problem of producing a very thin high quality oxide. The threshold voltage of MESFETs depends on the thickness and the doping level of a thin conductive layer. However, in contrast to the case of MOSFETs, in which voltage is applied across a high breakdown field insulator, the variability among MESFETs is not so easily surmounted by high voltage signals; voltage amplitudes are limited by the properties of the Schottky barriers. Thus, excellent control of the parameters of the conductive layer are required for the successful application of MESFETs.

MESFETs have attracted interest for another reason: they permit transistors to be made in high electron mobility III-V compounds, in which techniques for creating high quality oxide layers comparable to SiO_2 on silicon have not been developed. GaAs MESFETs have had great success in filling requirements for applications in the high frequency microwave transistor field. However, further development is needed to demonstrate that III-V transistors can meet the additional requirements of good reproducibility in a production environment and low cost that characterize VLSI.

In fact, the wiring on a chip occupies more space on modern highly integrated logic chips than devices. Miniaturization of wire cannot keep pace with the miniaturzation of devices. Electromigration and ohmic resistance limit the miniaturization of wires. Maintaining the width of wires while increasing their density will require more layers of wiring and, consequently, the development of the technology to provide larger numbers of wire layers on a chip and the vias, layer-to-layer connections, that interconnect them. The electrical characteristics of the wires from the viewpoint of circuit design may also require modification in that longer wires may have to be regarded as resistive transmission lines while short wires are treated as capacitors.

Miniaturization of device isolation must be considered seperately from miniaturization of devices. As voltages will not decrease in proportion to dimension, electric fields in the isolation structures will increase, requiring more attention to breakdown and leakage. The processing of oxide isolation must be developed essentially independently of device development. It is probable that interest in semiconductor-on-insulator configurations, such as the well-studied silicon-on-sapphire technique, will continue, as they offer the possibility of relatively simple isolation. A reduced sensitivity to soft errors caused by ionizing radiation is an additional benefit of such methods.

Capacitance is found almost everywhere in an integrated circuit chip. Often it is undesirable, requiring time and energy to charge and discharge it. Sometimes, however, as in "dynamic" memory cells, it is essential to the functioning of a circuit. Memory depends on the ability of a capacitor to retain a charge for a significant period of time. When all dimensions of a structure are scaled downwards capacitance diminishes in proportion to dimension and the charge stored is correspondingly decreased, which is desirable from the point of view of power dissipation if leakage can be controlled. However, when the 16K bit and 64K bit memory chips were introduced it was found that ionizing radiation, most notably α-radiation from the radioactive decay of heavy elements, can excite enough charge collectable by the memory capacitors to upset the stored bit.[8,9] This problem can be attacked by error-correcting coding of stored information, modifying the structure of memory devices to reduce thin collection efficiency, and purification of packaging materials, which are the chief source of radiotive impurities. Nevertheless, one is forced to ask how far the trend to less stored charge in dynamic circuits can continue. Since the decrease in stored charge is a direct result of miniaturization, which is driven by strong economic forces, there may be a need to stem the decrease of capacitance by raising voltage or by invention of means for providing more capacitance per unit area. The demands on thin insulating films may be more severe than those arising from simple scaling of dimensions.

V. FABRICATION-LITHOGRAPHY AND PROCESSES

Miniaturization is accomplished by improvements in lithography, which is here regarded as the exposure of films of radiation-sensitive material to patterns of radiation, and processing, which is conversion of the patterns of exposure to physical structures that implement devices and circuits, and also encompasses a few unpatterned operations.

Lithography has two aspects, pattern generation and pattern replication. Most often a patterned mask is created either by mechanical motion of a beam of light or a cutting tool or by deflection of an electron beam. The pattern is then reproduced by exposure to radiation that has passed through the mask. The mask may be placed in contact with the substrate or be some distance from the substrate and exposed with well-collimated radiation. The pattern may also be written directly on the wafer by a deflected electron beam without the intermediary of a mask.

The original creation of patterns with very small minimum dimension, smaller than can be attained by mechanical means, requires the use of focusing optical elements. The word "optical" here must be extended to include electron optics.

The resolution of optical lithography, exposure by visible or ultraviolet radiation, is limited by diffraction effects on the scale of the wavelength of the exposing light. High quality lenses are available in the spectral regions with wavelengths longer than about 2000A, making exposure by projection demagnification of a mask possible. Light with wavelengths in the infrared is unsuitable, both because of the degradation of resolution by diffraction and because of the lower sensitivity of photosensitive materials. At high photon energies the lack of optical elements means that radiation can be used only for contact or near-contact printing. The higher the photon energy, the greater the variety of effects that can be produced in a radiation-sensitive material. In particular, photons in the x-ray region can cause the creation of an energetic electron that can expose resist at an appreciable distance from the point of photon absorbtion.

Focused electron beams can be used to produce very small structures. The wavelength of the electrons used for lithography is negligible and the limits to resolution must be sought in phenomena other than diffraction. In fact, they are found in the aberrations of electron lenses and in the fact, mentioned above, that energetic electrons can travel through solids and expose resists at significant distances from the location they were intended to affect. The exposure of resist at undesired locations in the vicinity of the target is known as the "proximity effect." Compensation for the proximity effect can be included in the computer programs that control the exposure of resist patterns by electron beams. The exposure of a resolvable element is made to depend on the nature of the pattern in its neighborhood.

The proximity effect can be greatly reduced by using ion beams instead of electron beams to expose resists; the ions have much shorter ranges in solids. In addition to reducing the proximity effect, the short range of ions means that more of their energy is deposited in the resist. Electrons penetrate the resists and lose a large part of their energy in the underlying semiconductor. The use of ion beams in lithography is hindered by the lack of bright, convenient ion sources, however.

A large amount of processing is needed to translate the lithographic images, however produced, into functional electronic circuits. Many processes may be used. Epitaxial deposition of silicon, doping by diffusion or ion implantation, etching, oxidation, deposition of films of metals, insulators, and polycrystalline silicon, and lift-off are some of the process steps that may be employed. Here only a few general features that affect the application of the available techniques can be mentioned.

The thrust toward miniaturization means that all dimensions of a structure must be reduced. In particular, the widths of depletion layers must be reduced, which implies that doping levels must be increased. Heavier doping also enables doped regions, such as the base regions of bipolar transistors, to withstand the necessary voltage while their thickness is reduced.

Reduction of dimensions calls for modification of processes to produce and control thinner layers. As the rate of chemical and diffusive processes increases rapidly with increasing temperature, lower temperature processing is indicated. Ion implantation also permits improved control of doping depths, although the annealing process allows further movement of dopant atoms. Ion implantation, in addition, simplifies processing in that the masking layers used do not have to stand high temperature, as they do in diffusive processes. Laser processing shortens the time at high temperature and frequently offers an alternative way of processing thin layers.

As structures become smaller surface effects become relatively more important compared to volume effects. Aspects of processing that involve surface phenomena will change. For example, problems caused by adhesion of one layer to another or by surface diffusion may increase in importance. The size of grain boundaries in metal lines will bear a different relation to the size of the line. The scale of irregularities at an alloyed metal-silicon contact will not automatically change with the thickness of the layer to be contacted. Clearly, continual process development and refinement will be essential.

VI. SYSTEM ASPECTS

VLSI circuitry is constrained by the need to interact with its external environment. For example, chips produce heat that must be efficiently removed to keep the chip temperatures to a tolerable level. The permissible investment in cooling means and, therefore, the amount of power that can be dissipated on the chip, depend on the application in which the chip is to be used. High-speed expensive computers devote substantial resources to heat removal while small, lower cost systems devote very little. Extensive discussion of this question is beyond the scope of this school, but it constrains various aspects of VLSI chip technology.

The supply of power to the chip is intimately connected to the question of cooling; the heat removed must be supplied as electrical power. Power supply has a more direct effect on chip design, because connections to power supply leads must be provided and as the various chip functions are exercised the power drain fluctuates and the changes in the current drawn from the supply are translated by the inductance of the connections into variations of the on-chip power supply. Circuits must endure these fluctuations.

Chips must also have connections that make the transmission of signals to other entities possible. The number of external connections, called pins, varies with the nature of the chip; chips with relatively self-contained functions such as large memory chips and microprocessors need fewer pins than chips in large computers, where the results of operations may be required quickly in another part of the machine. As levels of integration and the length of words used to process information grow it is almost inevitable that more external connections per chip will be desired. This will affect chip technology in that the circuits that transmit signals on the external interconnections use space on the chip and use more power than internal logic circuits. In addition, the miniaturization of external interconnections has not kept pace with the miniaturization of features on chips.

Although the parameters that describe integrated circuits have been discussed as though they were continuous variables, such is often not the case. Systems may be assembled from components acquired from different manufacturers and compatibility requires standardization of certain specifications. For example, a 5V power supply standard will endure even though it may not be the optimum for chip technology because high-quality low-cost 5V power supplies become available. Capital equipment designed to handle wafers of a certain size and thickness that has been standardized by agreement with silicon suppliers will not be replaced until some very major economic advantage can be foreseen. Chip manufacturers may be restricted by standarization of interconnection patterns such as the 16 pin DIP. Thus the necessity for interaction with the rest of the electronic world often requires that parameters charge in rather large, abrupt steps, awaiting the acceptance of new standards rather than being the subject of continuous optimization.

VII. ECOMOMICS OF VLSI

Factors other than level of integration affect the economics of LSI. Progress to larger wafer sizes has also helped to lower the cost of silicon electronics. Much of the processing is carried out on wafers and its cost is divided among more chips as wafer size increases. Silicon wafers are cut from single crystal ingots grown by the Czochralski method. Freedom from defect and uniformity and control of impurity content are essential. The development of techniques for growing silicon ingots of increased diameter is the gating factor in progress to layer wafer sizes. Change also comes slowly because of large investments in tooling designed for a particular wafer size.

Yield is another ingredient of chip economics. Yield means the fraction of the chips fabricated on a wafer that are acceptable. Yield depends on the complexity of the chip, that is, on the number of devices and the length of wire. The complexity at any time is limited to that which can be produced with sufficient yield; for example, the 16K bit memory chip could not replace 4K bit chips until it could be produced with reasonable yield.

The dimension of structures also influence yield; higher yields can be obtained with larger dimensions. Larger dimensions, however, mean larger chips and fewer chips per wafer. The product of yield and chips per wafer, the number of good chips per wafer, may have a maximum as a function of dimension.

It is common experience in semiconductor manufacturing that yield increases with time as a given product is produced. This phenomenon is easily understood as an increase in the skill of operators with practice and the identification and elimination of yield detractors. It is known as the "learning curve" and is taken into account in projecting the economics of a semiconductor product. There is a continuing conflict in LSI production between the profitability of continuing the manufacture of a product that is far down on the learning curve and the competitive advantage of introducing a new advanced product.

The very large number of circuits that can be included on a VLSI chip can lead to great complexity in their logical design, layout, and interconnection. Design becomes a major factor in the cost of a product. Therefore, memory chips with their simpler structure lead the advance of integration. The cost of the design of a specialized complex logic chip can only be supported when there is a large market for the chip. Several methods for making the advantages of VLSI available to small volume users have evolved. The microprocessor chip can be programmed to perform a variety of functions in the same way as a computer. Gate arrays or masterslices are chips that take advantage of volume production by manufacturing an array of standardized logic gates whose function is defined in the final stages of production by one or two layers of wiring. Programmable arrays are similar in concept but are designed to be programmed in a simpler way, namely, by deleting certain connections.

Chip manufacturing facilities require a large capital investment. The rapid advance of technology can make the facilities obsolete in a few years. Their cost must be debited to the chips they produce. Low cost per chip implies that the rate of chip processing must be maximized. The term "throughput" is used to describe the quantification of this objective in modern semiconductor technology. It is frequently measured in "wafer starts per week", but, of course, the economics also depends on chips per wafer and yield.

VIII. CONCLUSION

VLSI technology must be aimed at optimization of the utilization of various resources. Although not mutually independent, these may be identified as captial investment in production facilities, silicon area, power, and time spent in processing signals. Such optimization will pose a continuing series of technological questions as the level of integration increases, but history gives confidence that the problems will be mastered and continued progress towards VLSI can be anticipated.

REFERENCES

1.   R. W. Keyes, IEEE J. Solid State Circuits 14, 193-201 (1979).

2.   G. Moore, Technical Digest 1975 International Electron Device Meeting (IEEE, New York, 1975) pp. 11-13.

3.   G. Marr, Fall Compcon 77 Digest (IEEE, New York, 1977) pp. 242-244.

4.   R. W. Keyes, Proc. IEEE 69, 267-278 (1981).

5.   E. Demoulin, *et al.*, Technical Digest 1979 International Electron Devices Meeting (IEEE, New York, 1979) pp. 34-37.

6.   R. W. Keyes, Appl. Phys. 8, 251-159 (1975).

7.   E. O. Johnson, IEEE Trans. Electr. Dev. 22, 1044-1045 (1975).

8.   T. C. May, Proc. 29th Electronic Components Conference (IEEE, New York, 1979) pp. 247-256.

9.   J. F. Ziegler and W. A. Lanford, Science 206, 776-788 (1979).

CHAPTER II

SILICON TECHNOLOGY

SILICON CRYSTALS FOR LARGE SCALE INTEGRATED CIRCUITS

B. O. Kolbesen

Siemens AG, Research Laboratories
Otto-Hahn-Ring 6, D-8000
Munich 83, Fed. Rep. Germany

1. INTRODUCTION

Silicon is the basic material of the electronic industry. The considerable continuous growth of this industry brings about a steadily increasing demand for electronic-grade poly-crystalline silicon[1], which in 1980 has already exceeded 2000 tons/year (Fig. 1). The major driving force

DOE/JPL PUBL. 79-110 NOV. 1979

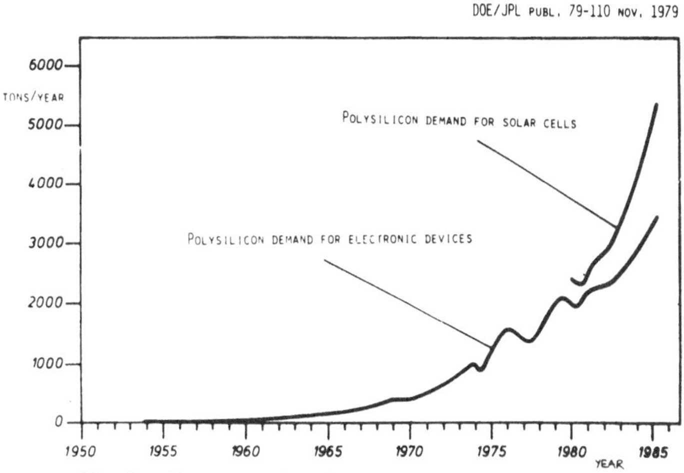

Fig. 1 - Current and projected demand for polycrystalline silicon for electronic devices (Ref. 1).

of the electronic industry is integrated circuits, which require high quality large diameter (75-125 mm) single crystalline wafers and ingots.

Currently these large diameter silicon single crystals are grown by the Czochralski (CZ) method (Fig. 2a) from quartz crucibles or by the crucible-free floating zone (FZ) method (Fig. 2b). The most important features of these two different growth methods are listed in Table I, the essential physical properties of CZ and FZ crystals are contrasted in Table II.[2,3] The CZ method on the whole is more easily handled on a production scale than the FZ method, e.g. in CZ crystal growth larger diameters and weights of crystals are achieved routinely.[2,3] On the

TABLE I

Czochralski and Floating-Zone Silicon
Comparison of Growth Methods after Keller and Mühlbauer (1981)

	CZ	FZ
Purity	Poor (Oxygen) $\rho<100\Omega cm$ one pass	High $\rho>100\Omega cm$ several passes poss.
Crystal Size Diameter Weight	Large Easy 1981: 150mm 20-60 kg	Large Possible 1981: 100-125mm 20-40 kg
Mech. Growth Conditions	Gravity Stabilizes Melt and Crystal	Gravity Destabilizes Melt and Crystal
Thermal Growth Conditions	Small Gradients	Large Gradients
Economy	Large Weights Pull Rate<2mm/min Simple Power System if Resistance Heated Overall Costs About the Same	Reasonable Weights Pull Rate>3mm/min RF-heating Necessary No Crucible
Application	Low Voltage Devices IC's	High Voltage Devices Power Devices
Market Share - Total IC's	80% >95%	20% < 5%

TABLE II

Czochralski and Floating-Zone Silicon
Comparison of Important Features

		CZ	FZ
Resistivity Range	P-type N-type	0.005-50 ohm.cm	0.1 -> 1000 ohm.cm
Dopants		B,P,As,Sb	B,P
Life Time		10-50μsec	100-3000μsec
Defect Densities		High	Low
Mech. Behavior Yield Point Slippage		High Highly Resistance to	Low Little Resistant to
Capability for Internal Gettering		Yes	No
Oxygen Range Conc. Average		2×10^{17}-2×10^{18} cm^{-3} 8×10^{17} cm^{-3}	<5×10^{14}-2×10^{16} cm^{-3} 5×10^{15} cm^{-3}
Carbon Range Conc. Average		<5×10^{15}-5×10^{17} cm^{-3} 4×10^{16} cm^{-3}	<5×10^{15}-3×10^{17} cm^{-3} 2×10^{16} cm^{-3}

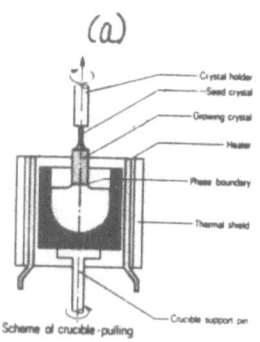

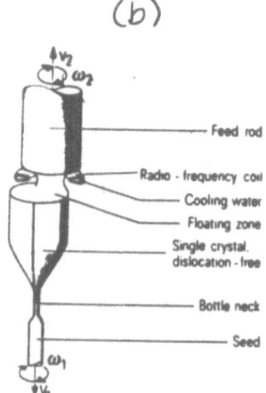

Fig. 2 - Growth methods for silicon single crystals (schematic)
a)   Czochralski (CZ)
b)   Floating-zone (FZ)

other hand, FZ crystals are characterized by a low impurity content. The
typical ranges of oxygen and carbon
contents[1] of CZ and FZ crystals are
shown in Fig. 3.[4] The considerable difference in oxygen content between FZ
and CZ silicon is obvious. This difference in oxygen content is the most important reason for the difference in the
properties and behavior of FZ and CZ
crystals. Oxygen up till now has been
known rather for its harmful effects

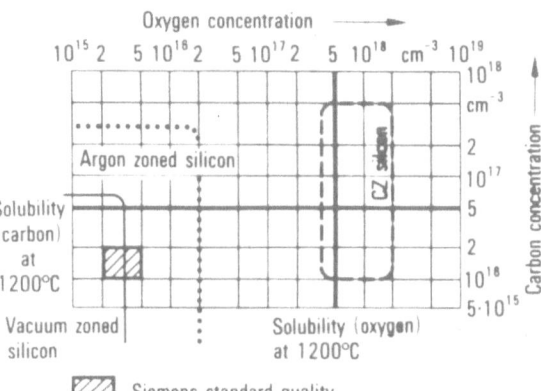

Fig. 3 - Oxygen [0] and carbon [C] concentrations of FZ and CZ silicon crystals of various
suppliers determined by infrared spectroscopy
according to DIN (Ref. 4).
$$[0] = 2.45 \times 10^{17} \text{ cm}^{-2} \cdot \alpha \text{ (cm}^{-1})$$
$$[C] = 1.0 \times 10^{17} \text{ cm}^{-2} \cdot \alpha \text{ (cm}^{-1}) \quad \alpha =$$
absorption coefficient

on the electrical characteristics of devices, like the formation of unwanted donors and

crystalline defects during annealing processes.

In recent years, however, it has been realized that oxygen also has beneficial effects on the

material properties and on the devices fabricated on CZ substrates. One desired feature is the

hardening effect of interstitially dissolved oxygen, which increases the resistance of CZ wafers

to slippage during device processing as compared to oxygen-poor FZ-wafers. On the other

hand high oxygen concentrations in CZ wafers can cause warpage, i.e. bowing of the wafers.

This may lead to severe problems in photolithography, in particular, due to the ever-increasing

wafer diameters and ever-decreasing feature sizes.

Another beneficial effect of oxygen can be summarized under the heading "internal

gettering" which in the meantime is utilized by many device manufacturers to improve the

yield of devices, in particular LSI circuits. Internal gettering is based on the formation of

---

1)   The oxygen and carbon concentrations of Fig. 3 were determined by infrared spectros-
      copy applying the procedure of DIN.[5] The DIN calibration factor for oxygen differs
      from the ASTM[5] calibration factor: $[0]_{DIN} = 0.51 \times [0]_{ASTM}$. In the following DIN
      values are used, if not otherwise noted.

oxygen precipitates in the bulk of CZ wafers. These oxygen precipitates and the crystal

defects caused by them act as very effective sinks for metallic impurities.

The technically simpler production and the beneficial effects of oxygen constitute some of

the reasons why today worldwide more than 90% of large diameter crystals are produced by

the CZ method, and why in the fabrication of IC's more than 95% of the material used is CZ

silicon.

The high quality of silicon starting material regarding crystal perfection and impurity

content cannot be maintained during device processing. Frequently, process-induced defects

like dislocations, stacking faults and precipitates are formed due to process-induced mechanical

stresses and agglomeration of intrinsic point defects and impurities originating from the

starting material or introduced during processing.

In the major part of the lecture we will compare the chemical and physical properties of

CZ and FZ crystals and we will discuss important aspects of oxygen, the dominating impurity

in CZ crystals, for circuit manufacturing. Finally we will deal briefly with process-induced

defects in silicon which can reduce the yield of LSI circuits considerably.

2.   DOPANT DISTRIBUTION

The improvement of the axial and radial dopant uniformity is still one of the major goals

of large diameter crystal growth, in particular for the production of CZ crystals from large

melts.[6]

The incorporation of impurities in melt growth in general is described by the theory of Burton, Prim and Slichter (BPS).[7] The BPS theory is based on the concept of a diffusion boundary layer at the moving solid-liquid interface (Fig. 4). For steady state growth conditions it yields an effective distribution or segregation coefficient

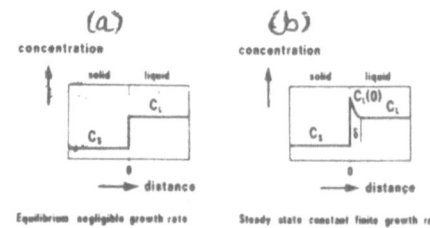

Fig. 4 - Impurity concentration near the growing interface in melt growth according to Burton, Prim and Slichter.[7]
a)   in equilibrium (negligible growth rate)
b)   at steady state: constant finite growth rate

$$k = \frac{C_S}{C_L} = \frac{k_o}{k_o + (1 - k_o) \exp\left(-\frac{\delta}{D} \cdot v\right)} \tag{1}$$

where C_S and C_L are the dopant concentrations in the solid and the liquid, respectively, k_o is the equilibrium distribution coefficient (negligible growth rate), δ is the width of the diffusion boundary layer, D is the diffusion coefficient of the dopant impurity in the liquid and v is the growth rate. Equation (1) contains the assumption that the interface distribution coefficient k_i = $C_S/C_L(o)$ equals the equilibrium coefficient K_0; ($C_L(o)$=dopant concentration at the interface in Fig. 4). In real crystal growth a of parameters, which in part are interdependent, such as crystal orientation, interface shape, rotation rate and melt flow, thermal or forced, exert a strong influence on the effective distribution coefficient and therefore on the axial and radial macroscopic and microscopic dopant distribution.[2,3,6,8-10]

2.1 CZ Silicon Crystals

In CZ crystal growth the dopant is already contained in the polycrystalline starting material or is added to the silicon melt. The dopant concentration varies from the seed to the bottom end due to segregation of the dopant. Because the effective segregation coefficient k of B,P,As and Sb, in general used as dopants in silicon crystal production is smaller than unity, the dopant concentration in the

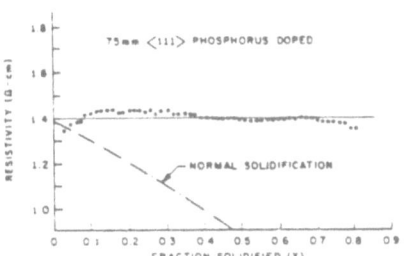

Fig. 5 - Axial resistivity profile of a double crucible grown crystal and the profile calculated for a conventional growth according to eq. (2) (Ref. 8).

melt, and consequently also in the growing crystal, increases during growth. Assuming complete mixing in the melt, the axial dopant profile according to Pfann[11] is given by

$$C_S = k\, C_{LO}(1 - g)^{k-1} \tag{2}$$

where C_{LO} is the initial dopant concentration in the melt and g the solidified fraction of the melt. Equation (2) is consistent with experimental data.[9] An example of the axial resistivity distribution of a P-doped CZ crystal calculated according to Eq. (2) is included in Fig. 5. For boron, due to a k-value of about 0.8, the segregation effect and thus the decrease from seed to bottom end is of the order of 25% depending on the length of the crystal and the volume of melt. Thus the stringent demands for many NMOS devices ($\frac{\Delta\rho}{\rho} \leq 0.3$) can be met fairly well. For phosphorus, due to k = 0.35, the decrease from seed to bottom end can amount to a factor of up to 2.5. Thus only small portions of a crystal are within the specification, in particular of PMOS or CMOS devices.

The radial variations depend on the parameters mentioned in Sec. 2, and are typically around 20% for phosphorus doped crystals.

In order to increase the production yield of crystals, the following methods are used or are in development.[6]

i) The crystals are divided into portions which meet the specifications of different types of devices.

ii) Many short crystals are grown by repeated recharging of polycrystalline silicon.

iii) Polysilicon is fed continuously into the crucible during growth.

iv) Growth with an effective segregation coefficient, k, nearly to unity by making the diffusion boundary layer very thick according to Eq. (1) by controlling the melt flow. This has recently been achieved by a novel magnetic CZ growth method (MCZ) which applies magnetic fields for controlled melt flow.[12,13] MCZ grown crystals show axial and radial uniformities better than 10%[12,13]

v) Double crucible method:[8] A small diameter crucible with an opening in the bottom is submerged in the melt contained in a larger crucible (Fig. 7). If the target dopant concentration in the crystal is C_O, the dopant concentration in the inner crucible should be C_O/k to account for the segregation. The consumed melt in the inner crucible is replenished by melt of dopant concentration C_O of the outer crucible through the opening (Fig. 7). The axial profile of a phosphorus doped crystal grown by the double crucible method is depicted in Fig. 5. The radial variations are below 20%.[8]

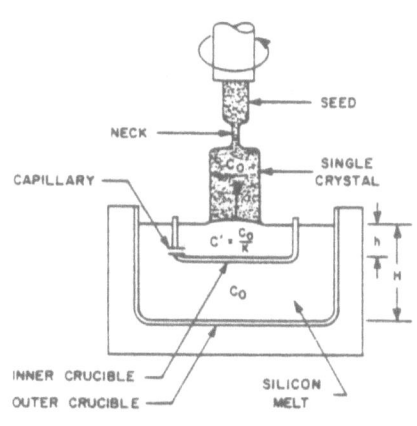

Fig. 7 - Scheme of double crucible . growth arrangement (Ref. 8.)

2.2 FZ Silicon Crystals

In FZ growth of silicon a narrow liquid zone generated by rf heating is passed vertically through a silicon rod. According to Pfann[11] the segregation process, for impurities, except for the last zone length, is described by

$$C_S = C_O[1-(1-k) \exp (-kx/h)]$$ (3)

where C_O is the concentration of an initially homogeneously distributed impurity, h is the height or length of the liquid zone and x is the distance. The distribution of an impurity after one zone pass is depicted if Fig. 6.

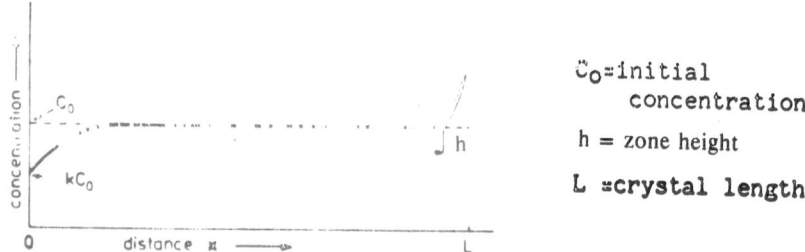

Fig. 6 - Concentration profile of an impurity in a silicon rod after one pass of a floating zone. C_0 is the initially homogeneous concentration of the impurity, h is the height or length of the floating zone (Ref. 11).

A comprehensive description of doping in FZ growth has recently been given by Keller Mühlbauer.[2]

In the case of doping by predoped polycrystalline feed rods, the segregation effect leads to large deviations of the ρ-target value near the seed end of the crystal. This effect can be avoided by gas doping. Here the dopant is added to the liquid zone during growth of the single crystal in form of a gaseous compound like phosphine, PH_3, or phosphornitrilochloride, $(PNCl_2)_3$. The axial resistivity fluctuations can be kept below $\pm 10\%$. Extremely homogeneous phosphorus doping, especially for resistivity levels above $10\Omega cm$, can be accomplished by neutron transmutation doping (NTD) of the completed single crystal. The axial and radial dopant uniformity achieved by this method is better than $\pm 5\%$.

Compared to the CZ method, the FZ method provides considerably better axial ρ-uniformity, whereas the radial ρ-uniformity of FZ crystals is not as good, except for NTD doped material. Since the trend in MOS devices moves to higher and tighter resistivity demands, the FZ material may gain more significance for some applications in the future.

3. OXYGEN INCORPORATION AND DISTRIBUTION IN CZ CRYSTALS

In CZ crystal growth, the silicon melt reacts with the quartz crucible resulting in the dissolution of SiO_2 at the crucible surface.[8] As a consequence of this reaction, large amounts of oxygen are dissolved in the melt. As illustrated in Fig. 8 a certain fraction of the oxygen leaves the melt via evaporation as SiO from the melt surface. Part of the oxygen is incorporated into the growing silicon crystal. The precise mechanism by which oxygen is incorporated is not well understood up to now.[8] In general the concentration of interstitially incorporated oxygen $[O_i]$ is

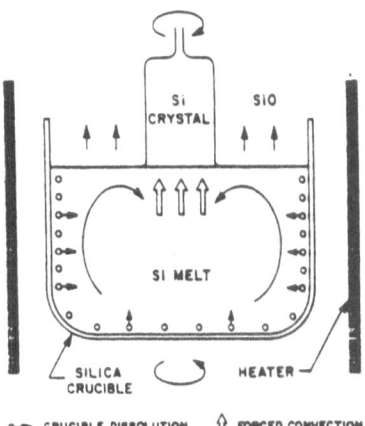

Fig. 8 - Scheme of a CZ growth system including relationship between oxygen controlling functions (Ref. 8.)

not uniform axially and radially. Usually $[O_i]$ decreases from seed end to bottom end (Fig. 9). As can also be seen in Fig. 9, this effect increases with increasing crystal diameter and varies according to the system utilized for growing the crystals (Fig. 10). Under reduced pressure in

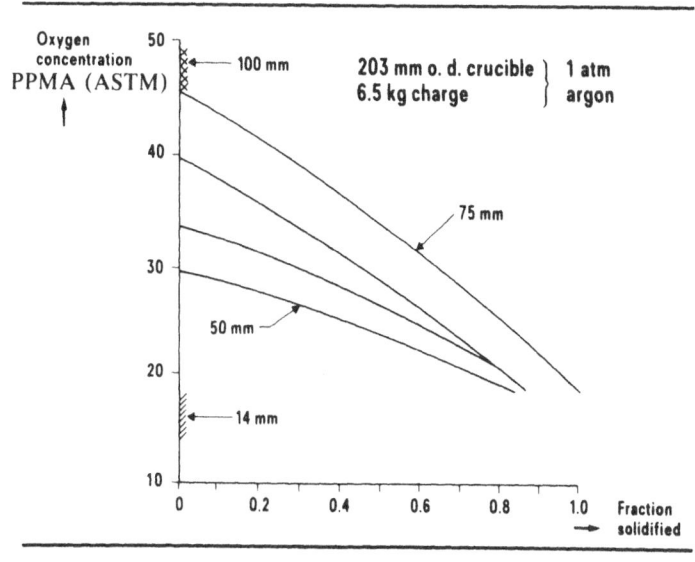

Fig. 9 - Axial oxygen distributions for different diameter CZ crystals grown under the same conditions (Ref. 8).

the growing system, the slope of $[O_i]$ decreases in conjunction with a reduction of $[O_i]$ at the seed end whereas at the bottom end $[O_i]$ remains relatively unchanged (Fig. 11).

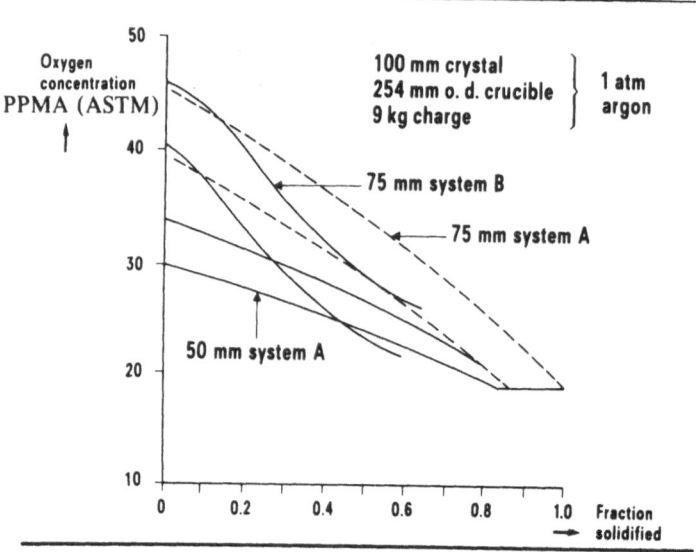

Fig. 10 - Axial oxygen distributions for CZ crystals grown using different systems and parameters (Ref. 8).

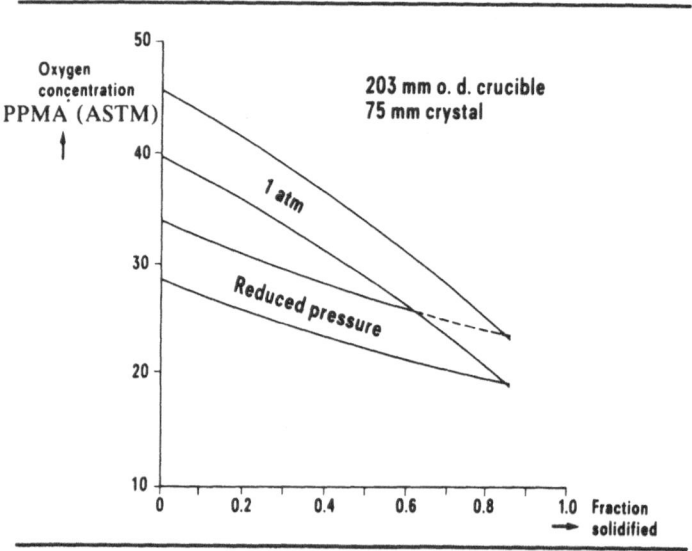

Fig. 11 - Effect of reduced pressure (14-30 Torr) on axial oxygen distribution of CZ crystals (Ref. 8).

According to a recent review by Benson *et al.*[8] it is now well established that the most significant parameters determining the average value of $[O_i]$ and its axial and radial distribution are: crystal diameter, volume of melt, size of the crucible area being in contact with the melt, melt convection, thermal and/or forced (rotation rate), evaporation rate of SiO from the free melt surface.

The experimental data available from the literature,[8] in part shown in Figs. 9 to 11, can qualitatively be explained as follows:[8] At the beginning of crystal growth the crucible dissolution, i.e. the oxygen flux into the melt, is maximum. Thus the oxygen concentration in the melt $[O_m]$ and the oxygen concentration $[O_i]$ incorporated into the growing crystal is maximum. The oxygen transport in the melt is dominated by thermal convection. The oxygen concentration near the crystal-melt interface is not uniform; it increases by forced convection due to high rotation rates of crystal and crucible. Reduced pressure increases the SiO evaporation rate and thus lowers $[O_m]$ and consequently $[O_i]$. Larger crystal diameters decrease the evaporation rate by reducing the free melt surface of a given growth system. Thus $[O_i]$ increases. As the melt level gradually drops the crucible wall dissolution decreases and the thermal convection is reduced, hence $[O_m]$ and $[O_i]$ decreases. At low melt levels the oxygen flux from the crucible wall becomes negligible; that from the bottom then becomes dominant. The thermal convection is minimum and the ratio of the free melt surface / wetted crucible wall area becomes maximum. Thus $[O_i]$ approaches minimum values at the bottom end of the crystal. These are little influenced by forced convection and evaporation rate since the oxygen distribution in the melt seems to be more uniform.

Murgai[14] recently reported a quantitative approach to describe the axial $[O_i]$ distribution. He found that the crucible dissolution is a diffusion limited process and that the axial $[O_i]$ profile is characterized by a non-steady state condition of the net rate of oxygen flux into the melt during growth. He also showed that the radial $[O_i]$ profiles are controlled primarily by melt hydrodynamics at the crystal-melt interface. At fast crystal-crucible rotation rates the $[O_i]$ variations across the major part of the crystal diameter are minimized.

Since for intrinsic gettering by oxygen precipitates (see Sec. 6) a uniform axial and radial oxygen distribution in the crystal is highly desired, efforts have been undertaken to achieve this goal.

Recently Lin and Pearce[15] applied a double cruci-ble system as described in Sec. 2 and shown in Fig. 7. The inner crucible serves as a baffle, which suppresses thermal convection in a large melt. A balance between thermal convection, SiO evaporation from the surface and crucible dissolution is established resulting in me-dium level $[O_i]$ with good axial and radial uniformity (Fig. 12).

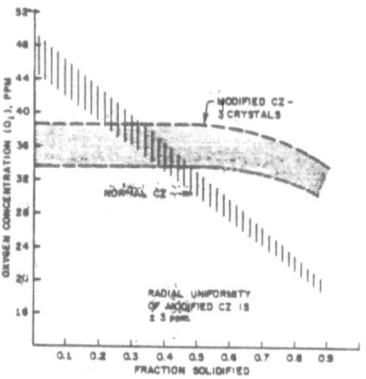

Fig. 12 - Axial oxygen distribu-tions of double crucible grown and normal CZ silicon crystals (Ref. 15).

Alternative approaches to achieve controlled and uniform oxygen incorporation are the application of a magnetic field[12] and of a rotating magnetic field.[13] In both cases, thermal convection can be suppressed. $[O_i]$ down to $1 \times 10^{17} cm^{-3}$ can be achieved as well as good axial and radial uniformity.[12,13]

In case of the oxygen incorporation the BPS-theory[7] is not valid and the definition of a segregation coefficient is not meaningful.[8] For the existence of oxygen striations, microscopic concentration fluctuations, experimental evidence has been obtained by Abe *et al.*[16] However, there is still a discrepancy in experimental results: In earlier experiments,[16] oxygen striations did not coincide with phosphorus and carbon striations but in recently published experiments they do.[6]

4. MECHANICAL PROPERTIES

The prevention of slippage and warpage of silicon wafers during furnace processing in device fabrication is of tremendous practical importance for achieving high yields. Empirically it has been known for many years that the mechanical behavior of CZ silicon is superior to FZ silicon.[17] However, reliable experimental data and the understanding of the physical reasons of this phenomenon were rather limited in the past. In recent years more systematic work on the mechanical behavior of silicon, and especially on the role of oxygen therein, has been performed[18-21] and recently reviewed by Sumino.[18]

4.1 Generation, Mobility and Multiplication of Dislocations in FZ and CZ Silicon

The mechanical behavior of semiconductor crystals is linked closely to the conditions under which dislocations can be generated, moved and multiplied. The generation, mobility and multiplication of dislocations is in turn strongly influenced by the temperature, the nature and magnitude of the stresses acting, the availability of dislocation sources and the concentrations of impurities incorporated in the crystal.

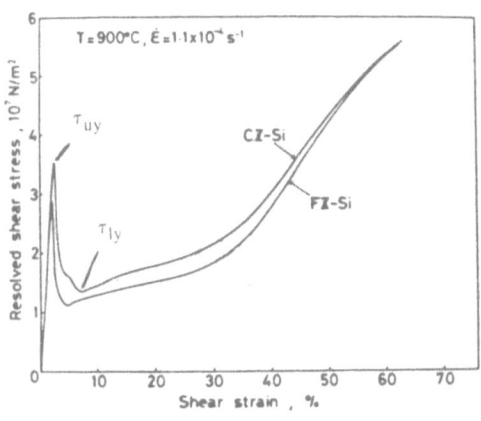

Fig. 13 - Stress-strain curves of undoped FZ and CZ silicon crystals deformed in [123] direction at 900°C at a (tensile) strain rate of 1.1×10^{-4} sec^{-1} (τ_{uy} and τ_{ly} are upper and lower yield stress) according to Sumino[18].
a) dislocation-free crystals

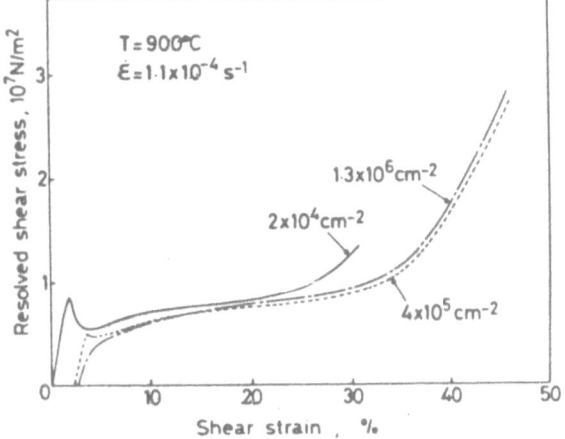

Fig. 13 - b) FZ crystals of various dislocation densities as indicated in the diagrams

Remarkable differences in the mechanical behavior between CZ and FZ silicon have been found.[18] In Fig. 13 stress-strain curves of FZ and CZ crystals of various dislocation densities are shown. The oxygen content of the CZ crystals was about 7×10^{17}cm^{-3}. Whereas the difference between the dislocation-free FZ and CZ crystals regarding the upper and lower yield stress is not very significant (Fig. 13(a)), a drastic difference is obvious for the FZ and CZ crystals containing dislocations (Fig. 13(b,c)). For FZ crystals the stress drop between upper and lower yield point decreases with increasing dislocation density and finally vanishes (Fig. 13(b)). CZ crystals still exhibit a distinct stress drop at high dislocation densities (2×10^6 cm^{-12}). The FZ crystals doped with oxygen to about 1.5×10^{17} cm^{-3} exhibited a behavior between that of the oxygen-poor FZ and the oxygen-rich CZ crystals.[18]

Whereas different dislocation mobilities were observed in un-doped FZ and CZ crystals the dislocation velocities were the same.[18] For CZ crystals, a critical stress of about 3×10^6 N/m^2 was required to move dislocations, whereas for FZ crystals the critical stress was found to be below 1×10^{16} N/m^2 (lowest stress applicable in the experimental

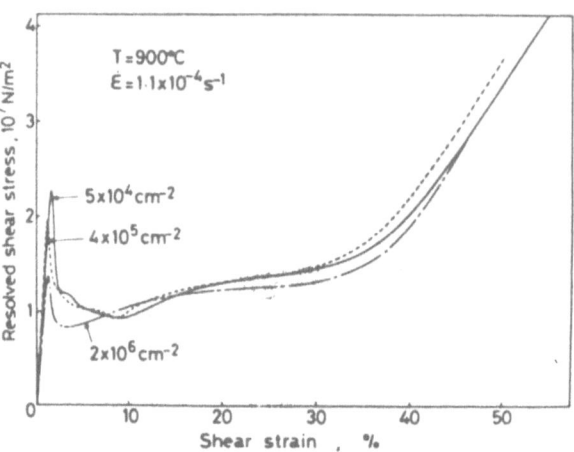

Fig. 13 - Stress-strain curves of undoped FZ and CZ silicon crystals deformed in [12$\bar{3}$] direction at 900°C at a (tensile) strain rate of 1.1×10^{-4} sec^{-1}: (τ_{uy} and τ_{ly} are upper and lower yield stress).
c) CZ crystals of various dislocation densities as indicated in the diagrams (after Sumino).[18]

setup used).[18] In FZ crystals carbon concentrations up to 1×10^{17} cm^{-3} did not exert any noticeable effect on the dislocation velocity and mobility.

The mechanical strength of CZ crystals degrades drastically if heat-treatments in the temperature range from about 800 to 1100°C are performed which lead to the precipitation of oxygen.[18,22] Figure 14 shows the decrease of the upper and lower yield stress with increasing concentration of precipitated oxygen. The oxygen precipitates can be dissolved again by heat-treatments above 1200°C . Thereby the mechanical strength of CZ silicon is restored (Fig. 14).

Kondo[21] showed that for CZ silicon

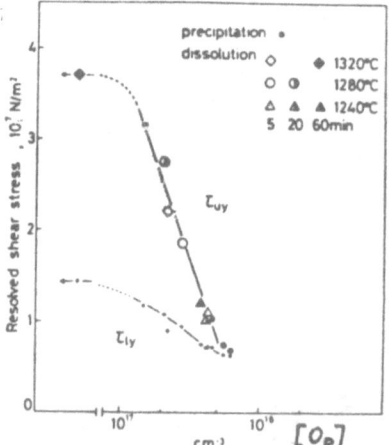

Fig. 14 - Upper and lower yield stress τ_{uy} and τ_{ly}, respectively, in dislocation-free CZ crystals vs. concentration of oxygen precipitated [O$_p$]. Deformation parameters are the same as in Fig. 13 (Ref. 18).

($[O_i]=9.5\times10^{17}$ cm^{-3}) which was not heat-treated the plastic deformation temperature for a given applied stress (e.g., 5×10^7N/m^2) was about 50°C higher than for FZ silicon. After heat treatment at 1050°C it dropped to about 50°C below the value of FZ silicon due to the precipitation of oxygen.

For the interpretation of the differences in the mechanical behavior of FZ and CZ crystals the high oxygen concentrations in the CZ silicon play a key role:[18] It is supposed that oxygen atoms create atmospheres around stationary or slowly moving dislocations by agglom-erating upon the latter and locking them firmly. Therefore, in contrast to FZ silicon, where this effect is absent because of the low oxygen content, in CZ silicon a certain (critical) stress is necessary to render the dislocations mobile again. The existence of a critical shear stress for the generation of dislocations in CZ silicon can be attributed to a deactivation of dislocation sources, such as surface flaws and grown-in microdefects, caused also by the aforementioned effect of oxygen. This is the reason why CZ silicon is more resistant to the generation of slip in furnace processing than FZ silicon. However, this beneficial effect of the oxygen is lost as soon as the oxygen precipitates due to heat-treatment (see Sec. 6). Oxygen precipitates, due to their large strain field, punch out series of dislocation loops which act as very effective dislocation sources. Since the region around a precipitate is depleted of oxygen, locking of the punched out dislocations does not take place effectively.

The dislocation hardening effect of oxygen can best be maintained in silicon containing oxygen in the concentration range from 1 to 5×10^{17} cm^{-3}.[18,21] Such oxygen contents can be realized in CZ crystals grown by the recently developed techniques which apply magnetic fields.[12,13] However, at these oxygen levels the beneficial effect of internal gettering due to oxygen precipitation in the bulk of CZ wafers cannot be achieved (see Sec. 6).

4.2 Warpage of Silicon Wafers

Furnace processing of silicon wafers can produce temperature gradients (ΔT wafer center-to-edge>150°C) resulting in mechanical stresses high enough ($>1\times10^7$N/m^2) to cause generation and motion of dislocations from the edge toward the center of the wafer and/or in

the center of the wafer.[19] This phenomenon is generally called slip or slippage.
In FZ wafers, slippage usually occurs at
the wafer periphery only. In CZ wafers,
slippage is often observed in the center
of a wafer. Slippage can result in considerable warping of a wafer in particular if it takes place in the wafer center.
Warpage of wafers can cause tremendous difficulties in photolithography.
Moreover, the high dislocation densities
also cause a considerable deterioration

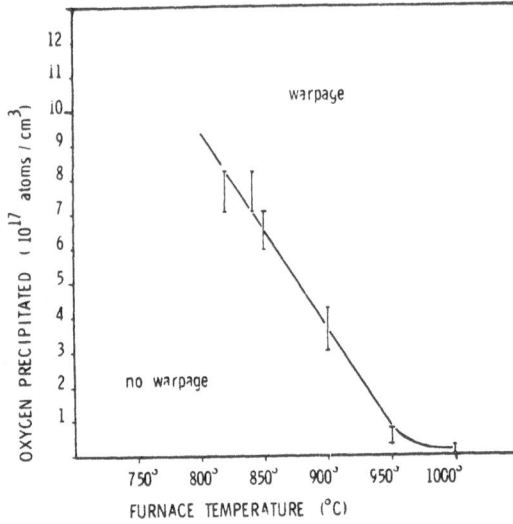

Fig. 15 - Critical furnace temperature as a
function of oxygen concentration precipitated
for generation of warpage (Ref. 19). Solid
line: calculated curve.

of the electrical characteristics of the devices.

Thus the prevention of warpage is imperative for
high device yields.

Moerschel *et al.*[23] found that the oxygen content $[O_i]$ of the silicon, the initial bow of the wafer and the furnace conditions constitute the most
important parameters influencing warpage. Leroy
and Plougonven[19] studied the warpage phenomenon in detail. They arrived at the following conclusions which are consistent with the results discussed in Sec. 4.1.

  i)   The susceptibility of wafers to
         warpage increases with increasing
         concentration of oxygen precipi-

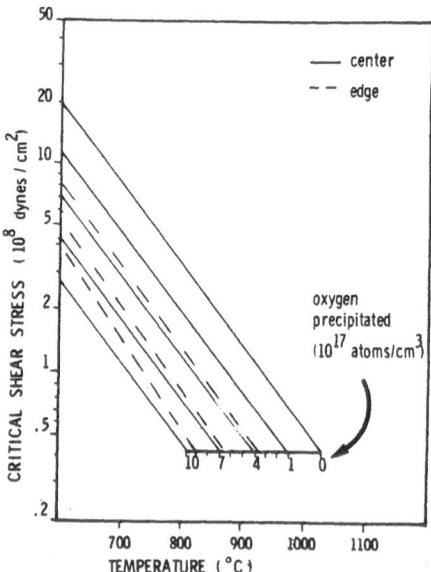

Fig. 16 - Critical shear stress at the
center and at the edge of the wafer as
a function of temperature and oxygen
precipitated (Ref. 19).

tated $[O_p]$ and diameter/thickness ratio of the wafers.

ii)   With increasing $[O_p]$ the critical furnace temperature at which warpage takes place decreases (Fig. 15) and so does the critical shear stress for plastic deformation (Fig. 16).

iii)   Removing the wafers from the furnace causes a tensile stress at the wafer edge and a (smaller) compressive stress at the wafer center. The latter adds to the compressive stress exerted onto the silicon lattice by the oxygen precipitates and favors the generation of dislocations in the center.

iv)   In case of initial bow of the wafers the dislocation generation effect is larger at the concave side than at the convex side of the wafer because the compressive stress is about 1.5 times larger at the concave side.

Measures for preventing warpage are:

a)   Reduction of $[O_i]$ below 5×10^{17} cm^{-3}.

b)   Reduction of temperature gradients e.g., by ramping.

c)   Furnace processing at temperatures $\leq 900°C$.

Measure b) is most appropriate in practice. In case a) the effect of internal gettering gets lost. Processing temperatures $< 900°C$ are far too low for many oxidations and diffusions in present device fabrication.

5. GROWN-IN MICRODEFECTS

At present large diameter (>50mm) silicon single crystals are free of line dislocations. As a matter of fact for the production of large crystals dislocation-free growth is imperative.[2] As soon as dislocations are present or generated, due to the high stresses caused by large temperature gradients high densities of dislocations are formed. These can even cause cracking of the crystals.[2]

Large diameter as-grown FZ crystals grown at rates >3mm/min are free of swirl defects[1)]

---

1)   Swirl defects are point-like microdefects arranged in a swirl-like pattern on slices cut out perpendicular to the crystal growth axis[25].

and contain rather low concentrations of microdefects ($<10^3$ cm^{-2}) giving rise to saucer pits (shallow etch pits) after preferential etching. Also, in large diameter as-grown CZ crystals, microdefects can hardly be revealed by etching methods.[24] However, due to thermodynamics, intrinsic atomic point defects (self-interstitials, vacancies) are contained inevitably in the bulk of the crystals. These point defects, in conjunction with impurities, in particular with oxygen and carbon, which are present in high concentrations in the crystals (see Fig. 33), create complexes and defect nuclei. Therefore at least after heat treatment, especially in CZ crystals, microdefects in a wide range of concentrations frequently arranged in a swirl-like pattern may appear and may be investigated more easily. They can be made visible and studied by a variety of diagnostic techniques, among others, by preferential etching, x-ray transmission topography with and without decoration and transmission electron microscopy (TEM). In particular high-voltage TEM (HVEM) provides results on the microscopic structure of the defects. Recently de Kock reviewed the present knowledge of defects in silicon crystals[25] and in particular that of swirl defects.[26]

In small diameter (25mm) CZ crystals swirl defects of the A-, B- and C-type were identified.[27] The A-defects in CZ crystals consist of interstitial-type dislocation loops[28] similar to those in FZ crystals.[29,30] The nature of the B-defects, as in the case of FZ crystals, and that of the C-defects is still unknown. In CZ crystals their concentrations strongly depend on the carbon content of the crystals[27] ranging from 10^7 cm^{-3} at carbon levels of about 10^{17} cm^{-3} to almost zero at carbon concentrations below 10^{16} cm^{-3}. Doping with donors above 10^{17} cm^{-3} suppresses the formation of A-defects; doping with acceptors above 10^{17} cm^{-3} prevents the formation of B and C defects.

In larger diameter (> 50mm) CZ crystals, hillocks are sometimes revealed by preferential etching.[24] Below such hillocks tiny precipitates, surrounded by a dislocation loop (20Å and 200Å, respectively, in diameter) were found by TEM.[24,31] The precipitates very likely consisted of silicon oxide. The concentration of the hillocks was independent of the oxygen and carbon content of the crystals, but was rather related to the growing conditions. In fact it has been observed that a curved interface gives rise to a higher density of microdefects than a flat interface in CZ crystal growth.[31] Swirl defects usually show the highest density at the seed end and decrease distinctly toward the bottom end.[31]

At present the knowledge about the nature and the formation mechanisms of microdefects in CZ silicon is rather limited. However, it is well established that the grown-in defect nuclei play an important role for the precipitation behavior of oxygen (see Sec. 6).

In addition they can provide nuclei for oxidation induced stacking faults[32] and are considered harmful when located in the active regions of devices.[32] But their presence in active regions can be prevented by proper device processing (see Sec.. 6).

6. OXYGEN PRECIPITATION AND INTERNAL GETTERING

6.1 Solubility and Diffusivity of Oxygen in Silicon

Oxygen can be contained in CZ crystals up to concentrations of about 2×10^{18} cm^{-3} (Fig. 3). At the usual furnace processing temperatures (800-1200°C) during IC wafer fabrication, the initial oxygen concentration $[O_i]$ exceeds the solubility and thus oxygen tends to precipi

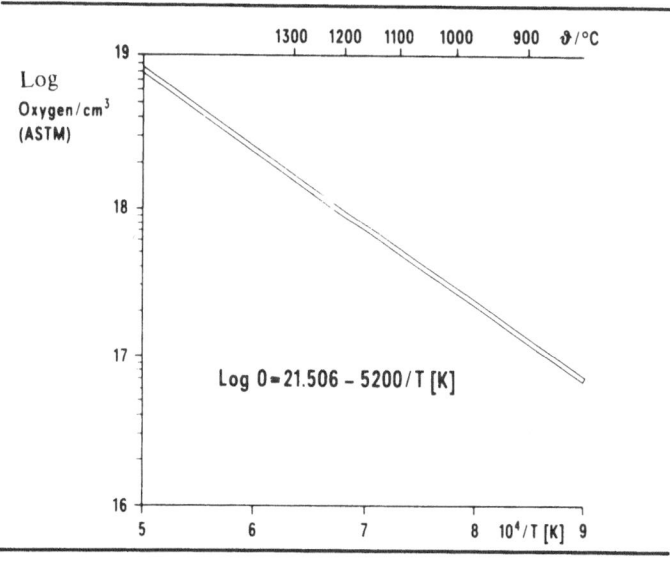

Fig. 17 - Solubility of oxygen in silicon vs. reciprocal temperature after Craven.[35]

-tate. There is some discrepancy regarding the solubility data for oxygen in silicon as recently discussed by Patel.[33,34] Figure 17 shows the curve recently determined by Craven[35] which seems to be the most reliable at present.[1] The soluble concentration of oxygen $[O_s]$ at a temperature T (K) is given by

$$\log [O_s] = 21.506 - 5200/T(K) \quad \text{(ASTM)} \tag{4}$$

$[O_s]$, using the DIN factor, at the melting point $T_m = 1420°C$ (1693K) is 1.38×10^{18} cm^{-3}, at 1000°C (1273K), a typical temperature of many furnace processes, it is 1.34×10^{17} cm^{-3}, at 800°C (1073K) it is 1.93×10^{16} cm^{-3}. Depending on $[O_i]$ of the crystal and the annealing temperature, very high supersaturations $[O_i]/[O_s]$ can result.

The diffusion coefficient of oxygen in silicon, D, according to Takano and Maki[36] is given by

$$D = 0.093 \exp (-2.4/kT). \tag{5}$$

6.2 Precipitation of Oxygen

In recent years extensive work has been performed on the precipitation of oxygen in CZ silicon. This work has been reviewed comprehensively by Patel.[33,34]

6.2.1 Nucleation, Growth and Shrinkage of Oxygen Precipitates

Oxygen can precipitate both homogeneously and heterogeneously in bulk silicon during heat treatment.[37] The terms homogeneous and heterogeneous nucleation are defined as follows:[37] Homogeneous nucleation means generation and growth of nuclei take place simultaneously during heat treatment. In heterogeneous nucleation, only growth of precipitates already nucleated in crystal growth, is supposed to take place. Both types of nucleation mechanisms seem to be important for the precipitation of oxygen. At annealing temperatures below 950°C the experimental data are consistent with a homogeneous nucleation model.[35,37]

---

1)   Craven[35] used the ASTM[5] calibration factor for evaluating $[O_i]$ from the infrared absorption measurement data. These differ from the data evaluated according to the now generally accepted DIN[5] calibration factor by a factor of 0.51.

At higher temperatures heterogeneous nucleation becomes dominant.[36,37] Grown-in defects and high carbon concentrations ($>5\text{x}10^{16}$ cm^{-3}) then play an important role.[35,37] The effect of the grown-in defects surpasses that by the carbon by far.[35] Recently, de Kock and van de Wijgert[27] demonstrated the role of grown-in defects and atomic point defects in the precipitation of oxygen.

The most important parameters determining the nucleation and precipitation behavior of oxygen are[35,27,38]

    i)   initial oxygen concentration $[O_i]$

    ii)   annealing temperature

    iii)   annealing ambient

    iv)   grown-in defect nuclei

Inoue and colleagues[37-39] studied the interdependence of nucleation rate J, nuclei concentration N, critical size r_c of nuclei, annealing temperature T_a and initial oxygen

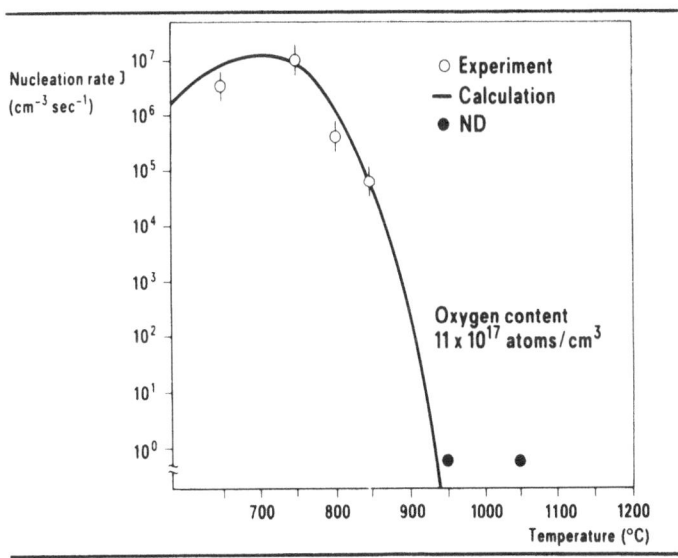

Fig. 18 - Precipitation of oxygen in silicon (Ref. 37):
a) Nucleation rate vs. annealing temperature. $[O_i]$ = $1.1\text{x}10^{18}$ cm^{-3}.

54   *Kolbeson*

concentration $[O_i]$. They utilized a two step process which consisted of an annealing step at low temperatures for generating nuclei, followed by a heat treatment at 1050°C causing the precipitates to grow to an observable size. Density and size of the precipitates were determined by preferential etching and TEM. They found:

a)   The nucleation rate J peaks at an annealing temperature (first step of 750°C for $[O_i]$ of about 1.1×10^{18} cm^{-3} (Fig. 18(a)).

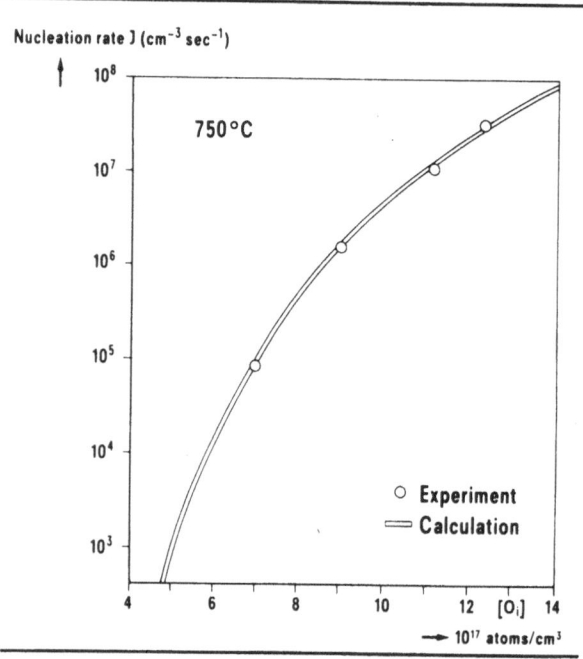

b)   J increases strongly with increasing $[O_i]$ (Fig. 18(b)).

c)   At low annealing temperatures the critical nucleus size r_c is small (it consists of a few

Fig. 18 - Precipitation of oxygen in silicon (Ref. 37):
b) Nucleation rate at 750°C vs. initial oxygen concentration

atoms) but the concentration of nuclei N is large, at higher temperatures r_c is large, but N is small.

d)   N increases with $[O_i]$.

e)   Saturation of nucleation occurs when the oxygen depletion layers around nuclei overlap.

f)   The precipitate densities observed at annealing temperatures above 950°C are much higher than expected from a homogeneous nucleation model. This indicates that grown-in defects contribute noticeably to N.

g)   The growth of oxygen precipitates is oxygen diffusion limited and is described

by a t^n (n=3/4) law similar to that observed for the growth of stacking faults

due to oxygen precipitation in silicon.[40] Shrinkage of precipitates takes place

if their size is below the critical size and if an oxygen undersaturation exists

around the precipitates. The latter is important for the formation of a denud-

ed zone (see below).

Inoue *et al.*[37] achieved good agreement between their experimental data and theoretical

curves (Fig. 18(a,b)) calculated according to their homogeneous nucleation model. Craven[35]

calculated the critical size of oxygen precipitates as a function of the temperature for various

$[O_i]$ (Fig. 19) according to the homogeneous nucleation model of Inoue *et al.* and found good

agreement with his own ex-

periments.   However, the

validity of the assumptions

made in this model has re-

cently been criticized by

Patel.[34] He noted that the

assumption of an equilibri-

um distribution of nuclei

sizes is not tenable under

their experimental condi-

tions.   According to Patel

the nuclei are supercritical

and their growth rate is lim-

ited by the diffusion of oxy-

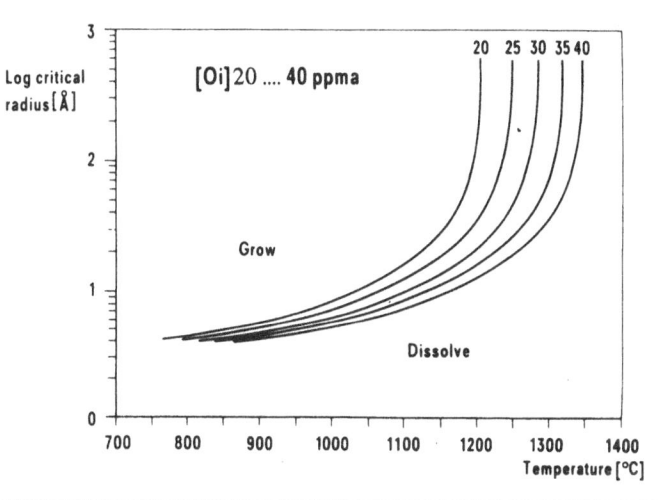

Fig. 19 - Relationship between critical radius and temperature for various oxygen concentrations (20-40ppma
ASTM) calculated by Craven[35] using the homogeneous
nucleation model of Ref. 37.

gen and not by nucleation, since the critical nucleus should contain only one or two atoms at

750°C.

6.2.2. Denuded Zone Formation

Since crystal defects degrade the electrical properties of devices, a defect-free zone (denuded zone), extending several microns below the wafer surface in which the active devices of IC's are located, has to be formed. During annealing, oxygen diffuses to the wafer surface. The resulting oxygen profile (Fig. 20) obeys the diffusion equation.[37] In a second heat treatment step, oxygen continues to diffuse on to the surface, but also precipitates in the bulk (Fig. 20). In a zone below the surface the oxygen concentration is below the equilibrium concentration. Within this zone nuclei generated in the first step dissolve.

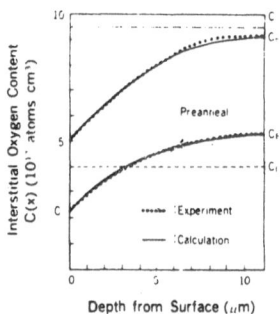

Oxygen Content Depth Profile

Fig. 20 - Depth profile of oxygen concentration after preannealing at 750°C (97 hours) and denuding at 1100°C (0.6 hours). C_i=initial [O], C_B=bulk [O], C_E=equilibrium [O] at 1100°C; [O] determined by spreading resistance via 450°C oxygen donors. (Ref. 37.)

Craven[35] investigated also the influence of the annealing ambient on the formation of the denuded zone. He found that the width of the denuded zone decreases in the following sequence of ambients: dry oxygen, nitrogen and a mixture of nitrogen, 3% oxygen and 3% HC1. A typical depth of the denuded zone was about 15μm for $[O_i] = 9 \times 10^{17}$ cm^{-3} (DIN), T=1100°C, 240 min in nitrogen. This agrees with results obtained earlier by Hu.[41]

The explanation of these results is a matter of speculation at present. In oxidizing ambient SiO$_2$ is formed at the surface, giving rise to the creation of excess silicon interstitials which are injected into the bulk.[42] Since the precipitation of oxygen creates silicon interstitials too, in the light of a mass action law the injection of interstitials from the Si/SiO$_2$ interface hinders the oxygen precipitating reaction. By contrast, it is believed that during HC1 oxidation, vacancies are generated at the interface or at least the creation of interstitials is reduced drastically. This would favor the precipitation of oxygen. According to a recent model by Hu[42] vacancy clusters are involved in oxygen precipitation.

6.3 Internal Gettering by Oxygen Precipitates

Internal or intrinsic gettering[43,44] has received increasing attention for several years. It is based on the precipitates of oxygen formed in the bulk of oxygen-rich CZ silicon (Sec. 6.2). These oxygen precipitates and the lattice defects like dislocation-loop arrays and stacking faults (Fig. 21(a,b)) caused by the oxygen precipitates act as very effective sinks for metallic

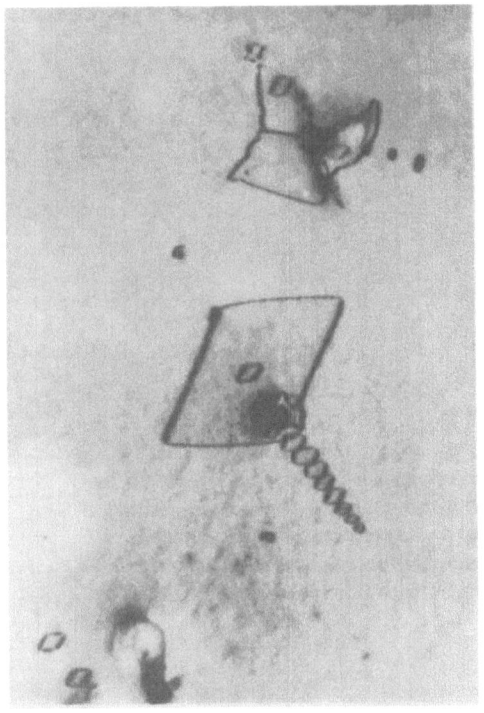
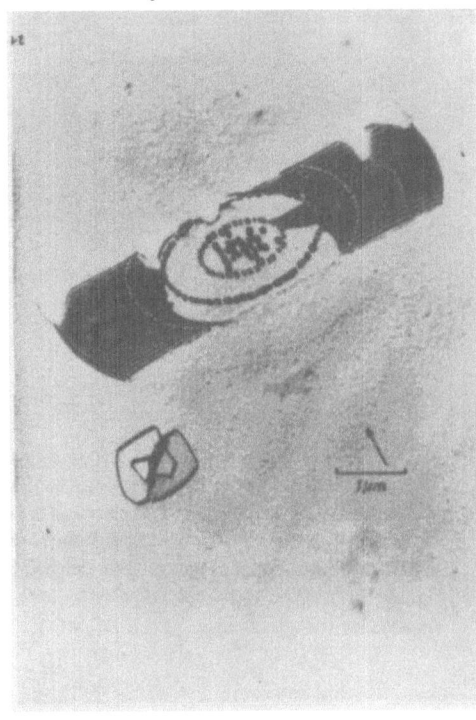

a)   1μm   b)

Fig. 21 - High-voltage electron micrographs of oxygen precipitation induced defects
a) Precipitate and series of punched out dislocation loops
b) Complicated arrangement of stacking faults and precipitates (multi-step annealing)

impurities. A qualitative example for internal gettering is presented in Fig. 22[45,46] Figure 22(a) displays the yield map of bipolar transistors fabricated in a 15μm thick epitaxial layer deposited on a CZ substrate. Devices rejected due to increased base-collector reverse current are indicated by the dark shading. Figure 22(b) shows the result of preferential etching of the substrate. The swirl-like pattern of "oxygen defects" (bright areas) in the substrate coincides excellently with the unshaded regions of the yield map. The metallic impurities have largely been removed from those regions of the epitaxial layer below which oxygen defects were

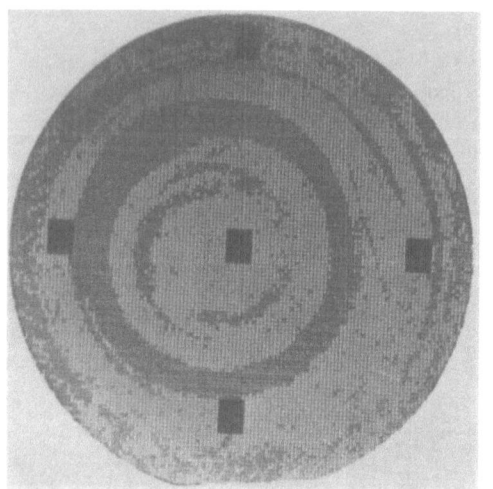 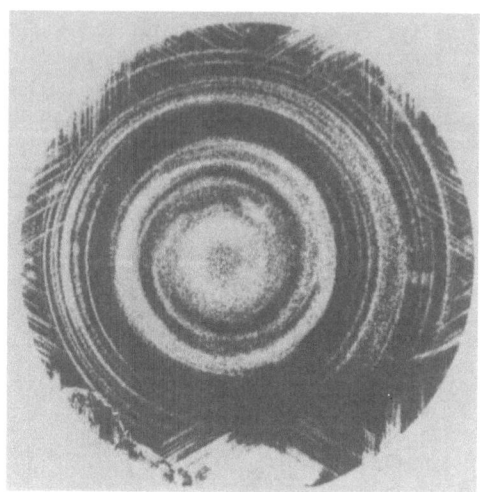

Fig. 22 - Effect of intrinsic gettering of metallic impurities by 'oxygen defects' in the substrate (2") on the yield of bipolar transistors in the epitaxial layer. a) Yield map: devices rejected due to increased base-collector reverse current I_{CB} are shaded. b) Swirl-like pattern of oxygen defects (bright) in the CZ substrate revealed by preferential etching. The defect pattern coincides excellently with the unshaded (good) device regions of the yield map. (Ref. 45, 46.).

located in the substrate. However, Fig. 22 is an example for a highly undesired inhomogeneous distribution of oxygen defects in a wafer and thus a locally non-uniform internal gettering effect. A similar non-uniformity results also from the non-uniform axial oxygen distributions in CZ crystals (Sec. 3). This gives rise to considerably differing yields on wafers from the seed end as compared to those from the bottom end of a crystal (Fig. 23).[47]

The effect of internal gettering regarding metallic impurities has recently been investigated in our laboratories[48] by etching, X-ray topography and neutron activation analysis (NAA). The parameters of the CZ silicon substrates were: (100), 100mm, $[O_i] = 7 \times 10^{17}$ cm^{-3} (DIN). The process sequence used to form oxygen precipitates in part corresponded to a bipolar technology and consisted of the following steps:

1)   Steam oxidation at $1020°C$ ($0.5 \, \mu m$ oxide)

2)   Annealing in nitrogen at $1220°C$

3)   Epitaxy at $1130°C$, $2 \mu m$ epi-layer

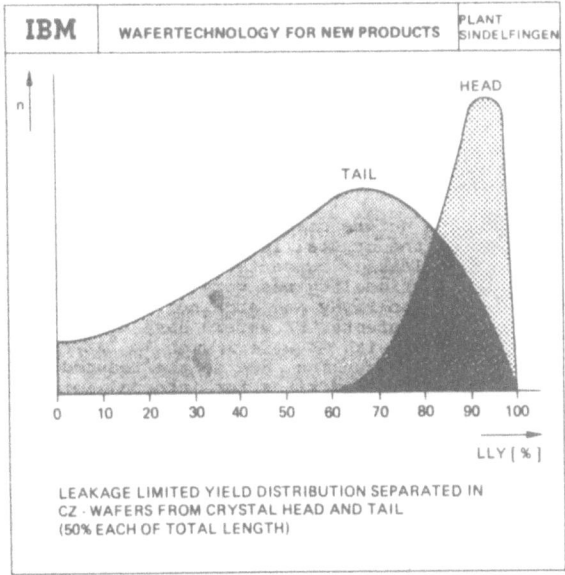

Fig. 23 - Distribution of yields of MOS memories fabricated on wafers taken out from the seed half and the tail (bottom) half of a CZ crystal. Yield was limited by leakage currents (Ref. 47).

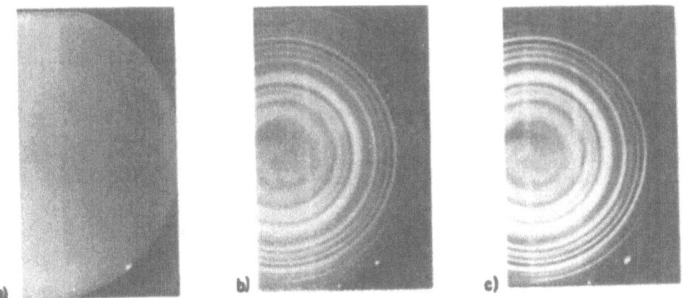

Fig 24 - X-ray transmission topographs of CZ substrates a) after oxidation b) after N_2 annealing c) after epitaxy. b) and c) display concentric ring patterns of bulk oxygen defects (Ref. 48).

Typical X-ray topographs of wafers of this experiment are shown in Fig. 24: a) after oxidation b) after N_2-annealing c) after epitaxy. b) and c) display concentric ring patterns of bulk oxygen precipitates. Figure 25 shows the depth profiles of Cu determined by NAA in wafers annealed in nitrogen at 1220°C for various times. Wafer M9 corresponds to the situation after the oxidation at 1020°C. It is obvious that the average Cu concentration, which increases by about 3 orders of magnitude from steps 1) to 2), does not rise significantly with annealing time. The shape of the depth profile changes drastically from step 1) to 2). Wafer

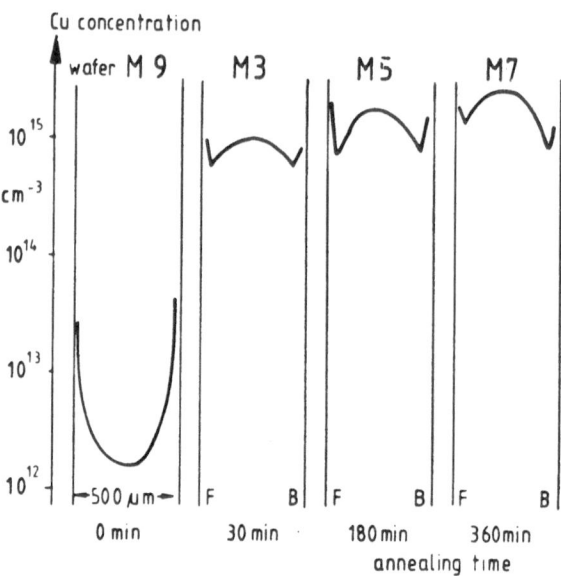

F = front surface
B = back surface

Cu-profiles after different furnace annealing times.

Fig. 25 - Depth profiles of Cu determined by neutron activation analysis (NAA) in wafers annealed for 0, 30, 180 and 360 min in nitrogen at 1220°C. Wafer M9 reflects the situation prior to annealing and contained no oxygen defects. M3, M5 and M9 contained oxygen defects as displayed in Fig. 24(b,c) (Ref. 48.)

M9 contains no oxygen defects visible by X-ray topography. Thus the Cu profile corresponds to the usual U-shaped profile obtained by diffusion, as it is also well known for Au.[49] Wafers M3, M5 and M7 contain oxygen defects and exhibit a distinct W-shaped Cu profile. The W-shape results from the agglomeration of Cu at oxygen defects in the bulk. In Fig. 26(a) the Cu profile of a wafer (W 44) is shown which contained no oxygen precipitates in the bulk. For comparison the profile of M5 from the same experiment is depicted. After the epitaxy the average Cu concentrations decrease by more than one order of magnitude but the profiles did not change very much (Fig. 26(b)). The decrease of the average Cu content is very likely explainable by the lower epitaxy temperature (1130°C) compared to the annealing temperature (1220°C) since the solubility of Cu decreases and thus the excess Cu diffuses out of the

wafer. A

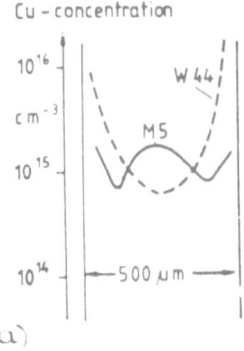

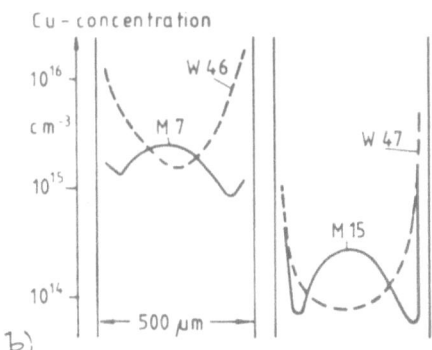

Fig. 26 - Cu profiles in CZ substrates after annealing steps (Ref. 48).
a) Wafer W 44 exhibits U-shaped profile due to absence of oxygen defects. For comparison a wafer (M 5 of Fig. 25) from the very same experiment is shown.
b) During the epitaxy at 1130°C, following the nitrogen annealing at 1220°C, the average Cu concentration of both types of wafers (M7, M15 with and W46, W47 without oxygen precipitates) decreases by more than one order of magnitude. The shape of profiles is nearly maintained.

comparison of the X-ray topograph and Cu autoradiograph of the very same wafer (Fig. 27) verifies that the Cu in fact is concentrated at the ring-like arranged oxygen defects. Yet the ring pattern of the topograph and the autoradiograph are shifted against each other because the topograph images the entire bulk of the wafer whereas the autoradiograph reveals the surface near zone. As can be seen in Fig. 28 on a preferentially etched cross-sectional specimen prepared from the wafer of Fig. 27, the oxygen precipitates are arranged in bands inclined to the wafer surface. They reflect the

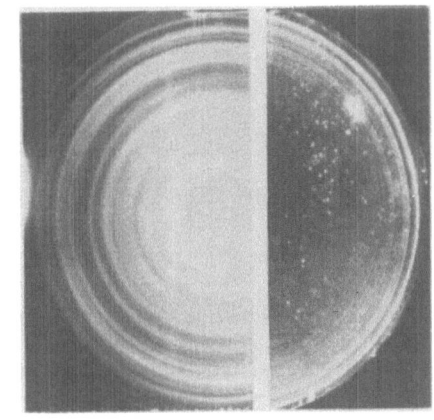

a) X-ray topograph   b) autoradiograph

Oxygen precipitation in Cz-Si

Fig. 27 - Comparison of a) X-ray topograph and b) Cu autoradiograph made from the very same wafer. Concentric ring pattern is shifted slightly between a) and b). For details see text (Ref. 48).

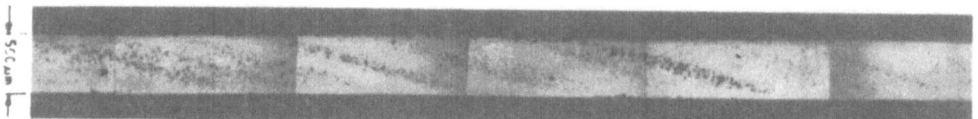

Fig 28 - Preferentially etched cross-sectional specimen prepared from the wafer of Fig. 27: **Oxygen defects are** revealed as band-like arrays reflecting the curved shape of the **crystal-melt interface during** crystal growth. (Ref. 48.)

curved shape of the crystal melt interface during growth.

From the above it is obvious that the gettering effect of the oxygen precipitates can reduce metallic impurities from the active regions of circuits at the wafer surface to some degree only.

A successful application of internal gettering by oxygen precipitates in device fabrication has to meet the following demands:

    i)   Macroscopically uniform distribution of bulk oxygen precipitates in proper density across the wafer diameter: The oxygen precipitate distributions of CZ wafers displayed in Fig. 29(a,c) fulfill this requirement fairly well; distributions as shown in Fig. 29(b,d) are not suitable.[48] Non-uniformity on a microscopic scale as revealed by the concentric ring pattern in Fig. 29(a) and on the micrograph of the cross-section of Fig. 28 can be tolerated.

    ii)   Denuded zone (Sec. 6.2.2.): During formation of the bulk oxygen precipitates, creation of defects in a zone extending several microns below the wafer surface has to be prevented. This includes oxygen precipitates and dislocation loops emitted by them as well as oxidation induced surface stacking faults nucleated at oxygen precipitate embryos.

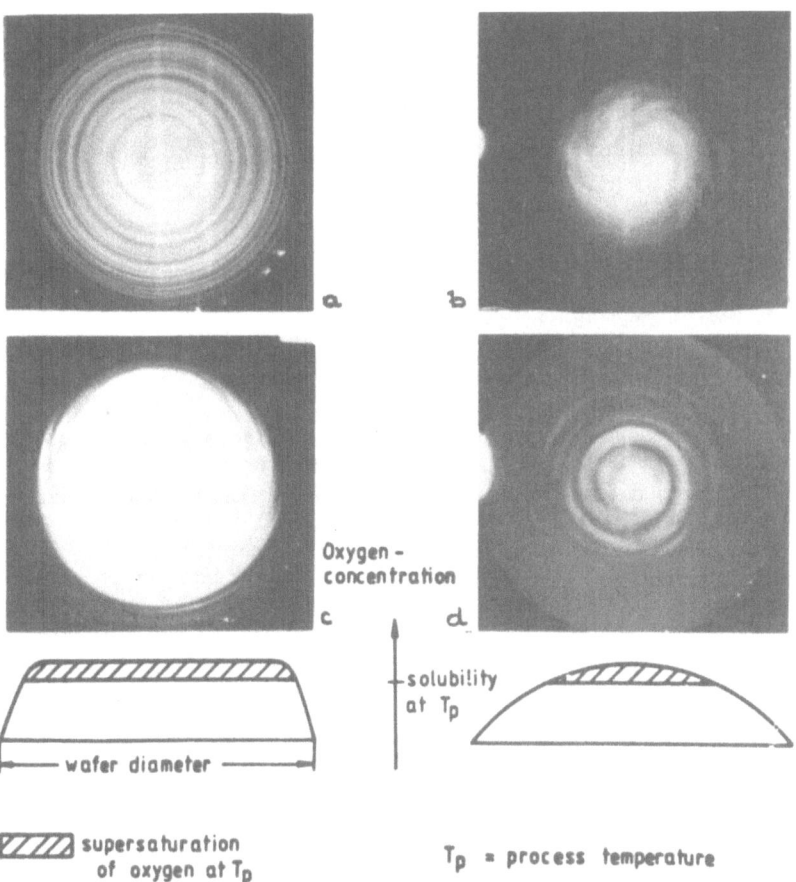

Fig. 29 - X-ray transmission topographs of CZ substrates which have received an oxygen precipitation treatment as described in the text.
a) concentric ring pattern and c) homogeneous pattern of oxygen defects across large part of wafers; b) and d) contain oxygen defects in the center only.
The initial radial oxygen distributions and the regions where supersaturation exists at the annealing (process) temperature of both groups of wafers are depicted schematically below (Ref. 48).

iii)   Optimization of the entire wafer processing: The application of internal

gettering by bulk oxygen precipitates requires proper tailoring of starting

material and wafer processing for each specific circuit technology.[50,51] For

example, in bipolar technology with its higher furnace processing temperatures

and larger number of temperature steps major attention has to be paid to the

problem of wafer warpage (Sec. 4.2).

Item i) and ii) can be achieved by using proper CZ material and suitable precipitation

procedures, both of which have been developed recently.

As shown in Fig. 12, CZ crystals with an $[O_i]$ of about 9×10^{17} cm^{-3}, and restricting axial

and radial $[O_i]$ variations restricted to within $\pm 10\%$, can be grown by the double crucible

method.[8,15] CZ crystals with similar or even better specifications can now be obtained also by

commercial suppliers (see e.g.[35]).

Various procedures for the precipitation of oxygen and the formation of a denuded zone

have recently been published.[35,37,51] Many of them consist of at least two steps. Inoue *et*

al.[37] suggest a first annealing step at $750°C \leq T \leq 950°C$, followed by a second annealing step

at about $1000°-1100°C$. For $[O_i] > 7 \times 10^{17}$ cm^{-3} (DIN) a precipitate concentration $\geq 10^{10}$

cm^{-3} is created in this procedure, which is sufficient for a proper internal gettering effect.

Some authors[35,51] suggest implementation of a preannealing step at higher temperatures

($\cong 1100°C$) in dry oxygen or inert ambient (nitrogen or argon) prior to the low temperature

nucleating step. Thereby the out-diffusion of oxygen is enhanced and a wider denuded zone

achieved.

The oxygen precipitation process can also be included into the processing sequence of a

particular circuit technology. This is highly desirable from an economical point of view.

However, it should be kept in mind that internal gettering is no panacea. It should not

deter efforts to steadily improve the cleanliness in wafer processing of advanced LSI and VLSI

circuits.

7. PROCESS-INDUCED DEFECTS IN SILICON

The author has reviewed this subject recently.[45] Therefore, in the following, only a summary of the content of that paper is given. For a more detailed description and for references, the reader is referred to Ref. 45.

Crystal defects and impurities originating from the starting material or induced by the processing can be important yield detractors of integrated circuits. Their harmful effect is intensified with increasing complexity and decreasing feature sizes of IC's. Impurities, especially metals, not only play a major role in the electrical activity of the defects, but also provide effective sources in the generation of dislocations as well as nuclei for the formation of stacking faults in epitaxy and in oxidation processes.

In high-speed bipolar integrated circuits with oxide-isolated transistor elements, process-induced dislocations can cause emitter-collector shorts due to enhanced emitter dopant diffusion along these dislocations. Such dislocations are formed in a selective oxidation process, where nitride masking is applied, at the perimeter of the nitride pads. The driving force in the generation of these oxide-nitride edge dislocation is primarily an intrinsic tensile stress of the nitride film and a stress caused by volume expansion of the growing oxide. Surface damage, metal precipitates, glide dislocations, stacking faults, etc., act as dislocation sources in the dislocation generation process. The dislocations are mainly of 60° type with Burgers vectors 45° inclined to the (001) wafer surface. Their formation can be drastically reduced by:

i)   elimination of dislocation sources

ii)   reduction of the stresses by optimization of the nitride-buffer oxide layer and of the isolation oxidation temperature, taking advantage of the visco-elastic behavior of oxides.

Stacking faults found in bipolar devices are generally more complicated than those observed in MOS devices. These complicated defects, like microtwins, sailboat stacking faults and multiple stacking faults are most harmful to bipolar circuits with narrow base width

($\leq 0.3 \mu m$). They cause diffusion pipes between emitter and collector. Simple stacking faults are detrimental only in about 50% of all cases investigated. In MOS devices, the electrical activity of stacking faults is strongly linked to the degree of decoration of their partial dislocations by impurity precipitates. Decorated stacking faults cause reduced relaxation time of MOS capacitors anf refresh failures of dynamic MOS memories. Metals like Cu and Fe constitute the main contaminants of wafer processing. Metals also provide abundant nuclei at which stacking faults start to grow during epitaxy and oxidation processes. The reduction of the impurity level can be accomplished by improved cleaning procedures and/or gettering techniques.

8. FINAL REMARKS AND CONCLUSIONS

At present, an overall comparison of the properties of large diameter CZ anf FZ silicon crystals and substrates yields advantages for CZ silicon with regard to the application for LSI circuits: The discussed physical-technological aspects such as the superior mechanical behavior and the capability of internal gettering as well as a number of practical-economical aspects such as the easier handling and automatization capability of CZ silicon on a production scale, distinctly are in favor of CZ silicon. In addition, the large market share of CZ silicon in the IC production for so many years resulted, consequently, in the optimization of IC wafer processing with CZ silicon. In view of this striking advantage of CZ silicon, there is at present no important reason on the horizon why the producers of LSI circuits should switch over to FZ silicon.

Nevertheless, inherent to FZ silicon is a higher purity and lower defect density which prevents some harmful effects of the oxygen in CZ silicon: in particular one gets ride of the creation of oxygen precipitation induced detrimental defects in the electrically active zone of devices. On the long term range this property of FZ silicon might gain more importance. Internal gettering based on oxygen precipitates can be abandoned in advanced LSI and VLSI circuit fabrication if the cleanliness of the wafer processing, in particular with regard to metal contamination, is improved. This, in part, is achieved by using lower processing temperatures, required to obtain the small lateral and vertical dimensions of the devices, and by applying

further in-process gettering procedures such as back side damage or phosphorus gettering. In

this case, the device technologist may prefer silicon substrates with low oxygen content ($<$

5×10^{17} cm^{-3}), such as FZ silicon or CZ silicon grown in a magnetic field (Sec. 3), in order to

safely avoid the aforementioned harmful effects of oxygen such as the formation of donors or

detrimental crystal defects.

ACKNOWLEDGEMENT

The author is grateful to Dr. G. Franz, Dr. W. Keller, Dr. H. Strack, and Dr. P. and Mrs.

M. Voss for many discussions and valuable comments on the manuscript. The careful

photographic work of Mrs. L. Bernewitz, Mrs. H. Mylonas, Mr. Riedl and Mrs. S. Tugrul is

gratefully acknowledged.

REFERENCES

1. E. Costogue, R. Ferber, W. Hasbach, R. Pellin and C. Yaws in Silicon Materials Outlook Study for 1980-85 US Department of Energy (DOE)/Jet Propulsion Laboratory (JPL) JPL Publication 79-110, Nov. 1979.

2. W. Keller and A. Mühlbauer, Floating-Zone Silicon, Preparation and Properties of Solid State Materials, ed., W. R. Wilcox, Marcel Dekker Inc., New York, Basel 1981.

3. H. Herrmann, H. Herzer and E. Sirtl, Advances in Solid State Physics XV, p. 279, ed. H. J. Queisser, Pergamon/Vieweg, Braunschweig (1975).

4. B. O. Kolbesen and A. Mühlbauer, Solid State Electronics, in print.

5(a). ASTM (American Society for Testing and Materials) Designation F 120, F 121 (Oxygen) and F 123 (Carbon) - 70T, Book of ASTM Standards, Part 8 (1970).

5(b). DIN 50 438 Teil 1 (Oxygen) and Teil 2 (Carbon); Deutsches Institut für Normung 1980.

6. T. Abe, K. Kikuchi, S. Shirai and S. Muraoka, Semiconductor Silicon 1981, ed. H. R. Huff, R. J. Kriegler and Y. Takeishi, The Electrochem. Soc., Pennington, N.J. p. 54.

7. J. A. Burton, R. C. Prim and W. P. Slichter, J. Chem. Phys. 21, 1987 (1953).

8. K. E. Benson, W. Lin and E. P. Martin, Ref. 6, p. 33.

9. J. R. Carruthers, A. F. Witt and R. E. Reusser, Semiconductor Silicon 1977, ed. H. R. Huff and E. Sirtl, The Electrochem. Soc., Princeton, N.J., p. 61.

10. J. Burtscher in Scientific Principles of Semicond. Tech., Proc. European Summer School, ed. H. Weiss, 1974, p. 63.

11. W. G. Pfann, Zone Melting, second edition Wiley, New York (1965).

12. T. Suzuki, N. Isawa, Y. Okubo and K. Hoshi, Ref. 6, p. 90.

13. K. Hoshikawa, H. Hirata, H. Nakanishi and K. Ikuta, Ref. 6, p. 101.

14. A. Murgai, Ref. 6, p. 113.

15. W. Lin and C. W. Pearce, J. Appl. Phys. $\underline{51}$, 5540 (1980)

16. T. Abe, K. Kikuchi and S. Shirai, Ref. 9, p. 95.

17. S. M. Hu and W. J. Patrick, J. Appl. Phys. $\underline{46}$, 1869, (1975).

18. K. Sumino, Ref. 6, p. 208.

19. B. Leroy and C. Plougonven, J. Electrochem. Soc., $\underline{127}$, 961 (1980).

20. A. George and G. Champier, Phys. Stat. Solidi(a) $\underline{53}$, 529 (1979).

21. Y. Kondo, Ref. 6, p. 220.

22. J. R. Patel, Discuss. Faraday Soc. $\underline{38}$, 201 (1964).

23. K. G. Moerschel, C. W. Pearce and R. E. Reusser, Ref. 9 p. 170.

24. R. W. Series, K. G. Barraclough and W. Bardsley, Ref. 6 p. 304.

25. A. J. R. de Kock in Handbook on Semiconductors Vol. 3, ed. S. P. Keller, North Holland Publishing Comp. 1980, p. 247.

26. A. J. R. de Kock in Defects in Semiconductors, ed. J. Narayan and T. Y. Tan, North Holland, 1981 p. 309.

27. A. J. R. de Kock and W. M. van de Wijgert, J. Cryst. Growth $\underline{49}$, 718 (1980).

28. A. J. R. de Kock, W. T. Stacy and W. M. van de Wijgert, Appl. Phys. Lett. $\underline{34}$, 611 (1979).

29. H. Föll and B. O. Kolbesen, Appl. Phys. $\underline{8}$, 319 (1975).

30. P. M. Petroff and A. J. R. de Kock, J. Cryst. Growth $\underline{30}$, 117 (1975).

31. K. Daido, S. Shinoyama and N. Inoue, Rev. Electrical Comm. Lab. $\underline{27}$, No. 1-2 (1979).

32. C. J. Varker and K. V. Ravi, Semiconductor Silicon 1973, Ed. H. R. Huff and R. R. Burgess, The Electrochem. Soc., Princeton, N.J., p. 670.

33. J. R. Patel, Ref. 9, p. 521.

34. J. R. Patel, Ref. 6, p. 189.

35. R. A. Craven, Ref. 6, p. 254.

36. Y. Takano and M. Maki, Ref. 32, p. 469.

37. N. Inoue, K. Wada and J. Osaka, Ref. 6, p. 282.

38. J. Osaka, N. Inoue and K. Wada, Appl. Phys. Lett. 36, 288 (1980).

39. K. Wada, N. Inoue and K. Kohra, J. Cryst. Growth 49, 749 (1980)

40. J. R. Patel, K. A. Jackson and H. Reiss, J. Appl. Phys. 48, 5279 (1977).

41. S. M. Hu, Appl. Phys. Lett. 36, 561 (1980).

42. S. M. Hu, Defects in Semiconductors, ed. J. Narayan and T. Y. Tan, North Holland, 1981, p. 333.

43. W. K. Tice and T. Y. Tan, Appl. Phys. Lett. 28, 564 (1976).

44. T. Y. Tan and W. K. Tice, Phil. Mag., 34, 615 (1976).

45. B. O. Kolbesen and H. Strunk, Inst. Phys. Conf. Ser. No. 57, 21, (1981).

46. B. O. Kolbesen and K. R. Mayer, Res. Rept. No. T79-156 Fed. Dept. Res. and Technol. FRG, 1979.

47. H. H. Steinbeck, Abs. 530 Vol. 80-2. The Electrochem. Soc. Fall Meeting, Hollywood, Fla. 1980.

48. G. Franz and B. O Kolbesen, Res. Rept. No. T81-NT 947, Fed. Dept. Res. and Technol. FRG, 1981.

49. W. R. Wilcox and T. J. La Chapelle, J. Appl. Phys. 35, 240 (1964).

50. D. Huber, P. Stallhofer and M. Blätte, Ref. 6, p. 756.

51. K. Kugimiya, S. Akiyama and S. Nakamura, Ref. 6, p. 249.

STRUCTURAL TECHNIQUES
FOR BULK DEFECTS CHARACTERIZATION

C. Donolato

CNR - Istituto LAMEL
Via Castagnoli 1
40126 Bologna, Italy

ABSTRACT: This lecture considers briefly the most widely applied structural techniques for the characterization of crystal defects in silicon, like X-ray topography, transmission and scanning electron microscopy. Examples of defects analysis are reported concerning the characterization of the defects introduced by the phosphorous predeposition in Si and the study of dark current sources in charge-coupled devices. A final section is dedicated to a more detailed discussion of the assessment of the electrical activity of semiconductor defects by the scanning electron microscope in the electron beam induced current mode.

1. INTRODUCTION

Crystal defects in silicon single crystals have been studied intensively, because they are known to reduce the yield in the fabrication and to affect the performance of Si-based devices.[1,2] The most widely applied techniques for the assessment of crystal defects in semiconductors are:[2]

a)   Chemical etching and optical microscopy

b)   X-ray topography

c)   Transmission electron microscopy (TEM)

d)   Scanning electron microscopy (SEM) on the electron beam induced current (EBIC) or cathodoluminescence (CL) mode.

Although quick and useful, chemical etching cannot be regarded as a structural technique, so it will not be included in the discussion. Since each technique is sensitive to different properties of a defect, the combined use of different methods allows a more complete characterization.

However, while XRT and SEM observations are non-destructive, TEM analysis requires thinned samples and is therefore employed after all others.

The combined application of some of the b), c), d) techniques is discussed briefly in the first part of this paper on the basis of selected examples of defect characterization performed at the LAMEL Institute.

In the second part the attention will be focussed on SEM techniques, in particular on the problem of relating the SEM-EBIC images of a defect to its configuration and electrical activity.

2. CHARACTERIZATION OF THE DEFECTS INTRODUCED BY THE PHOSPHORUS PREDEPOSITION IN SILICON

Phosphorus predeposition in silicon usually causes the formation of precipitates in the crystal. Here a short description is given of the crystallographic characterization of these defects in especially diffused wafers; then an example is shown of the influence of P precipitates on the electrical properties of a simple device.

2.1 Structure

The results reported here are taken from the work of Servidori and Armigliato.[3] CZ grown p-Si (111) slices were P-predeposited at 1000°C for 12 min in a 94% N_2 and 6% O_2 atmosphere, with $POCl_3$ as a source.

The presence of defects induced by this process was first established by X-ray topography in the Lang transmission mode. This method is non-destructive and yields an image of a wide area (a few centimeters square) of the sample. Figure 1 shows a typical X-ray topograph of a predeposited sample: the defects appear as black and white dots $30 \div 40 \mu m$ in diameter. Analysis of the contrast of these dots has shown that they correspond to precipitates, but the resolution of the technique does not allow more detailed statements.

Further information can be obtained by transmission electron microscopy. This techniqe requires thinning of the sample down to a few thousand angstroms (for observations at ~100 kV), but offers a much higher resolving power. With the usual jet-thinning technique, there

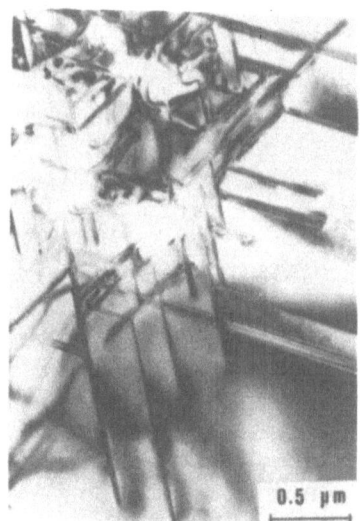

Fig. 1 - 220 X-ray topo-
graph showing precipitates
in P-diffused (111) Si.
Cu $K_{\alpha 1}$ radiation.
(Courtesy of M. Servidori)

Fig. 2 - TEM micrograph of
an aggregate of precipi-
tates, corresponding to a
dot of Fig. 1. (Courtesy of
M. Servidori)

can be some difficulty in getting the sample thin just in the region of interest; however, large

area (a few mm^2) preparation techniques have been developed.[4]

Figure 2 is the TEM image of a region corresponding to a black and white dot and reveals

that the defect consists of an aggregate of flat rod-like precipitates. These particles were

found to be aligned with the <110> directions of the silicon crystal parallel to the foil surface

and to lie on inclined {111} planes.

By using the TEM in the selected-area diffraction mode, the diffraction pattern of a single

rod can be obtained (Fig. 3.). The information that is achieved from such a pattern is

twofold: first we can deduce the lattice spacing and the angles between the various crystal

planes of the precipitate. This analysis led to the conclusion that the defects consist of

orthorhombic SiP crystals. Second, the diffraction pattern gives the crystallographic relation-

ship between the precipitate and the Si matrix; for example, evidence has been obtained that the precipitates grow so as to relieve the tensile stress which is known to develop in the sample during the phosphorus predeposition.

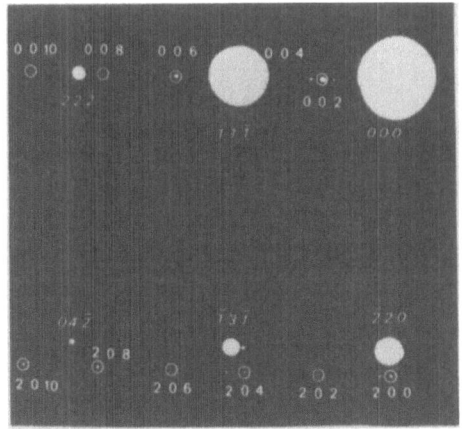

2.2 Eletrical Effects

The phosphorus predeposition process (without subsequent drive-in diffusion) produces shallow p-n junctions, which are usually required in photo-sensitive devices.

We consider here the application of this process to the fabrication of avalanche photodiodes. A major problem in

Fig. 3 - Selected area diffraction pattern of a precipitate, taken in the (112) projection of Si. The larger spots arise from the matrix. The encircled spots are due to the precipitate and were used for structure characterization. (Courtesy of M. Servidori)

the production of such devices is the occurrence of localized sites of premature breakdown (microplasmas), which reduce the current gain and limit the possibility of obtaining good large-area devices.[5] In the example discussed here,[6] microplasmas are found to be associated with P precipitates introduced by the predeposition process.

The structure of the diodes employed in this study is illustrated in Fig. 4. The active area was a circle $35 \div 80 \mu m$ in diameter, surrounded by a guard ring $8 \mu m$ deep. The predepositon process occurred at $920°C$ for 7 min, resulting in a junction depth of

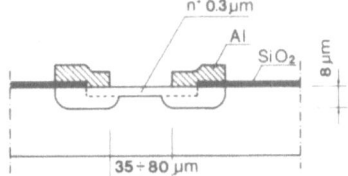

Fig. 4 - Structure of the avalanche photodiodes used in the study of microplasmas.

$0.3 \ \mu m$ with an integrated P concentration $Q = 5.7 \ 10^{15} \ cm^{-2}$.

The reverse characteristic of a defective diode is shown in Fig. 5; two kinks are recognizable in the breakdown region, which actually correspond to two microplasmas. This is demonstrated by the optical micrograph of the active area of the diode under high reverse bias (Fig. 6), which shows the presence of two bright spots.

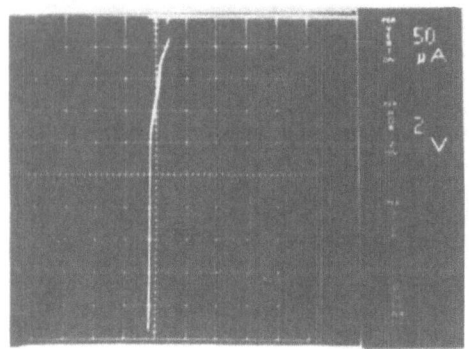

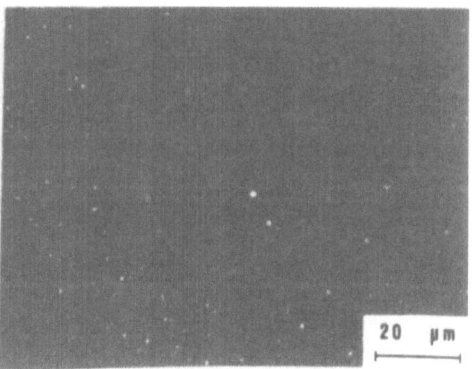

Fig. 5 - I-V characteristic in the breakdown region of a defective photo
diode.

Fig. 6 - Optical image of the active
area (inner circle) of the photodiode of
Fig. 6, showing two microplasmas (V
= V_B, I = 2 mA).

The current multiplication at a microplasma can be studied with higher resolution by the

SEM in the EBIC mode. Figure 7 shows the SEM/EBIC image of a microplasma at a reverse

voltage near its turn-on voltage. We see that the current enhancement is not uniform across

the defect; this is related to the internal structure of the defect, but the SEM resolution in this

mode does not allow more detailed statements.

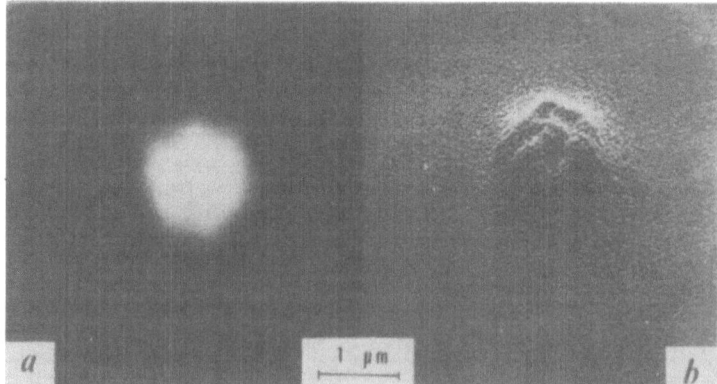

**Fig. 7 - SEM-EBIC image of a microplasma at high reverse
bias: a) normal image; b) Y-modulation image**

The sample has then been thinned and the microplasma region has been observed in the

TEM at 200 kV. The TEM image of a typical microplasma is given in Fig. 8 and shows that the defect consists of an aggregate of precipitates of SiP surrounded by a network of dislocations. Therefore these observations indicate that the formation of microplasmas is related to the phosphorus diffusion and that this processing step needs to be improved.

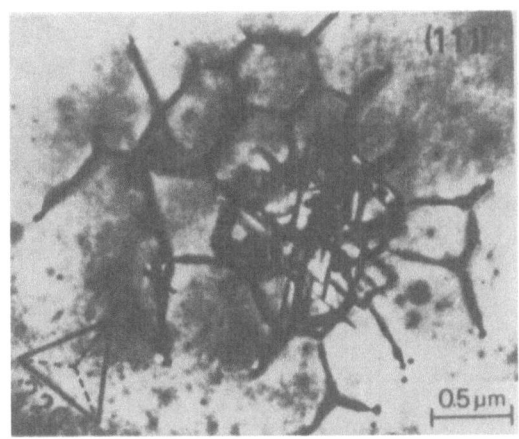

Fig. 8 - TEM micrograph of a typical defect corresponding to a microplasma, showing an aggregrate of precipitates surrounded by dislocations.

3. STUDY OF DARK CURRENT SOURCES IN CCD's

Defects in charge-coupled devices (CCD) give rise to anomalous charge generation and affect the information stored in the devices. Charge generation in CCD's is usually studied by stopping the transfer process for a given integration time and then shifting to the output the charge packets collected.[7] The improved method used here[8] consists in repeating the usual measurement for various configurations of the clocks, differing for the number and the position of the potential wells opened (Fig. 9(a)).

Figure 9(b) shows the charge collected in a definite cell of a defective device after a fixed integration time, for different stop configurations. It appears that the generation is reduced only when there is no potential well under the gate ϕ_4; this suggests the presence of a bulk defect there. For defects smaller than the gate

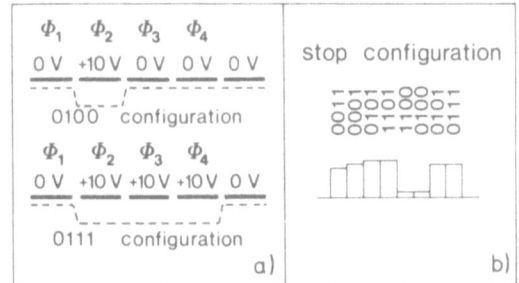

Fig. 9 - a) Examples of stop configuration of the clocks; b) Charge collected in a defective cell after a fixed integration time for various stop configurations.

width, this procedure indicates by purely electrical means under which gate the defect lies.

A more precise localization and some further insight into the nature of the defect can be obtained by X-ray topography. Figure 10 is a (220) Lang topograph of a complete CCD,

Fig. 10 - X-ray topograph of a CCD (Courtesy of M. Servidori)

showing that many defects are present in the gate region, the most extended features being dislocations.   However, not all these defects give rise to anomalous charge generation: generally, electrically active defects appear in the X-ray topograph as short segments parallel to the long side of the gates. Figure 11(a) shows a detail of the Lang topograph of the device which has been characterized by the electrical technique: a crystal defect is actually observed in the lower part of the expected gate. X-ray diffraction contrast experiments suggested the presence of precipitates or tangle of defects aligned along a preferential crystallographic direction.

After removing all surface layers by chemical etching, the device was observed by SEM in the secondary electron mode. The resulting micrograph of Fig. 11(b) confirms the presence of the defect with higher resolution. An attempt was made of analyz-

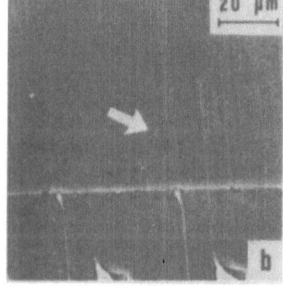

Fig. 11 - a) X-ray topograph showing a defect in the faulty cell: b) SEM micrograph showing a defect in the position indicated by X-ray topography.

ing by an energy-dispersive X-ray spectrometer the defective region, but only the emission lines of Si were observed.

TEM observations were also tried on some devices, but the depth of the defects was probably too high to be observed in a 125 kV electron microscope. High voltage TEM (0.5 ÷ 1 MV) could be useful in this kind of studies, since specimen thicknesses up to some microns can be observed, and the defects are less likely to be destroyed by the thinning procedure. In addition, at high voltages the electron transparent area is much larger, even if the sample is thinned by usual methods, and the correlation is therefore made easier.

4. STRUCTURE AND ELECTRICAL ACTIVITY OF STACKING FAULTS IN SILICON

Typical semiconductor processes, like epitaxial deposition or oxidation, can lead to the formation of stacking faults (SF) in the material. As illustrated in a previous lecture (B. O. Kolbesen: Silicon Single Crystals for LSI), these defects can have a harmful influence on the performance of the devices and for this reason have been intensively investigated.

A relevant result of such studies is that not all SF's, as evidenced for instance by etching, have detrimental effects; the discriminating factor appears to be the degree of decoration by impurities.

The structure of these defects is usually investigated by TEM, while their electrical activity is conveniently probed by SEM in the EBIC mode.[9] Figure 12 is a TEM micrograph of a (small) SF in (100) Si; the sequence of Fig. 13 gives the SEM-EBIC images of an oxidation-induced stacking fault at different beam energies. In the TEM micrograph the SF shows up as a defective area delimited by a nearly semi-circular partial dislocation. In the SEM-EBIC images,

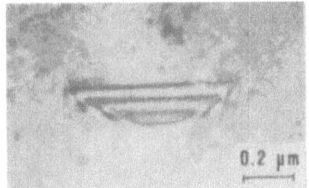

Fig. 12 - TEM micrograph of a stacking fault in (100) silicon. (Courtesy of A. Armigliato)

however, the contrast appears to originate from the boundary only: this shows that the electrical activity of the defect is essentially due to the bounding dislocation.

The TEM image of the SF exhibits a more complicated structure than that of SEM-EBIC images. However, there is a well developed contrast theory for interpreting TEM images, describing the diffraction (elastic scattering) of the electron waves from imperfect crystals;[10] a similar theory exists for XRT images as well.[11]

The situation is quite different for SEM-EBIC images, in spite of their simple appearance; in fact, only recently has the problem of the formation of EBIC images of defects received some attention (for a review on this subject see Ref. 13). An account is given here of the contrast model recently proposed by the author.[13,14]

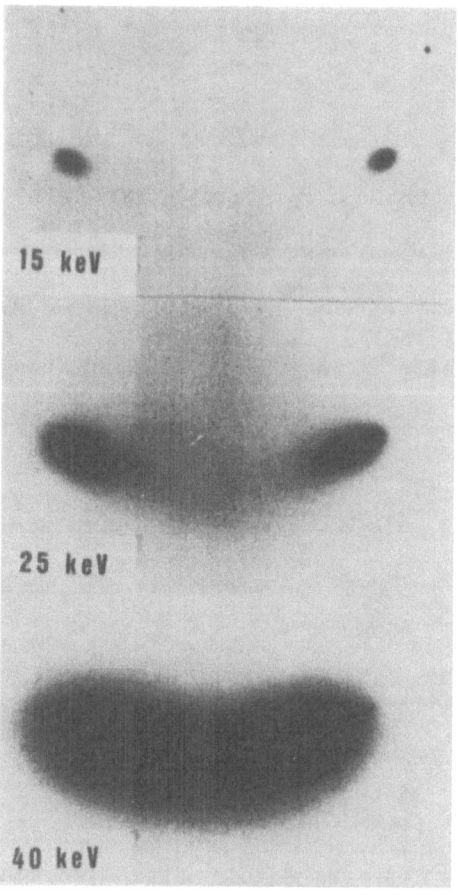

10 µm

Fig. 13 - SEM-EBIC images of an oxidation-induced stacking fault in (100) Si at different beam energies, as obtained by the Schottky barrier technique /12/. (Courtesy of S. Kawado)

5. CONTRAST FORMATION IN SEM-EBIC IMAGES OF SEMICONDUCTOR DEFECTS

The observation of semiconductor defects in the SEM by the EBIC method relies upon the energy dissipation (inelastic scattering) of the electron of the beam in the sample by production of electron-hole pairs. Beam-generated carriers are collected through a p-n junction or a Schottky diode (Fig. 14) and the resulting current is used as video signal of the SEM. Therefore, the EBIC image represents a map of the collection efficiency of the sample; defects appear as regions of reduced collection efficiency, i.e., when the beam is near a defect less current is induced in the device due to enhanced carrier recombination at the defect.

In order to model these observations, the following processes are to be described:

i)   the generation of carriers by the electron beam;

ii)   the transport and collection of beam-generated carriers;

iii) their recombination at a defect.

A number of simplifying assumptions will be made to obtain a simple mathematical formulation of the above processes. However, because of the localized character of the electron beam excitation, it is essential to discuss the problem in three dimensions. We consider preliminarly the very schematic case of a point generation

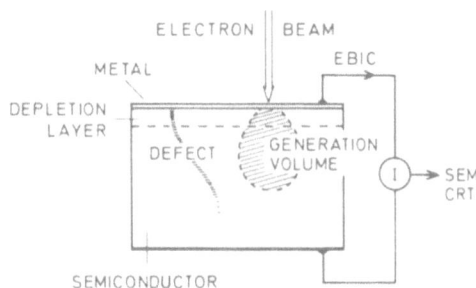

Fig. 14 - Schematic illustration of the experimental arrangement used for imaging semiconductor defects by the Schottky barrier EBIC technique.

in an infinite homogeneous semiconductor and the related diffusion problem, since this will

evidence some peculiar properties of the three-dimensional case.

5.1   Point Generation in an Infinite Semiconductor

Suppose that an infinite homogeneous semiconductor, for instance of n-type, contains a

point source of minority carriers (holes). If the holes are generated at a rate G (s^{-1}) and

recombine with a lifetime τ, their density p($\underline{r}$) obeys the diffusion equation

$$D\nabla^2 p(\underline{r}) \; - \; \frac{1}{\tau}\, p(\underline{r}) \; = \; - \; G\, \delta(\underline{r}) \tag{1}$$

where D is the hole diffusion coefficient and $\delta(\underline{r})$ is the Dirac delta function. It is easy to see

that the solution of Eq. 1 is:

$$p(r) \; = \; \frac{G}{4\pi D} \; \frac{e^{-r/L}}{r} \tag{2}$$

where r is the distance from the source and L $= \sqrt{D\tau}$ is the hole diffusion length. The

expression (2) shows that the excess hole concentration decreases with the distance from the source as a result of two factors: a) the exponential term exp(-r/L) due to the recombination, and b) the geometrical factor 1/r due to the three-dimensional character of the diffusion.[+]

At distances from the source small in comparison to L (i.e., for r/L << 1) we may write:

$$p(r) \cong \frac{G}{4\pi D} \; \frac{1}{r} \tag{3}$$

Equation (3) shows that near the source the hole density is practically the same as in a semiconductor having L = ∞; this conclusion also holds for an extended source, if the source dimensions are << L.

This latter condition is usually verified in EBIC experiments in silicon. In fact, the extension of the generation volume (~ R_p = R_p (E), the primary electron range) in Si at E=40 keV, a typical maximum energy for an SEM, is ~10 μm; values of L much greater than 10μm are common in Si. In conclusion, to study phenomena occurring near the generation region in Si, we may neglect the minority carrier recombination in the bulk, i.e., we may assume L = ∞ (or, equivalently, τ = ∞). It is worth pointing out that for τ = ∞ Eq. 1 reduces to the well-known Poisson's equation of electrostatics, for which well-established methods of solution exist.

5.2   The Uniform Generation Sphere in a Semi-infinite Semiconductor

The experimental situation of Fig. 14 will be described in a simplified manner by use of the following assumptions:

i)   The electron beam generates e-h pairs uniformly over a sphere tangent to the surface and having radius R = R_p/2 (the uniform generation sphere); outside, the generation rate is zero;

[+] The solution of the corresponding problem in one dimension is ~ exp(-x/L).

ii)   beam-injected minority carriers move by diffusion only and do not recombine

in the bulk (see Sec. 5.1); their collection by the surface barrier is described

by a condition of infinite surface recombination velocity;

iii)   recombination at a defect produces only a small change in the original carrier

distribution.

It is not difficult to give an explicit solution
of the resulting diffusion equation under
these hypotheses, by using the above-
mentioned analogy with electrostatics and
the well-known method of images.[15]   The
calculated minority carrier distribution is
shown in Fig. 15. It appears that the volume over which the minority carrier density
is relevant is about the same as the generation volume. The interaction of the minority carrier cloud of Fig. 15 with a localized

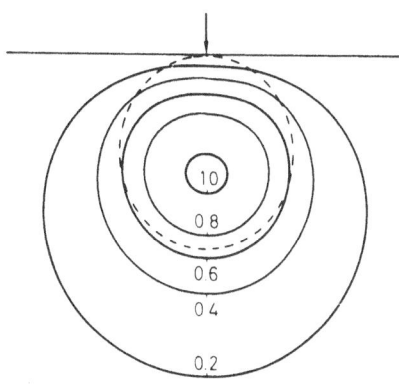

Fig. 15 - Calculated distribution of electron-
beam-generated minority carriers in a semi-
infinite semiconductor for L=∞, displayed as
contours of equal carrier density. The dashed
circle is the trace of the generation sphere.

defect gives rise to the EBIC contrast and is discussed in the next section.

5.3   The Collected Current at a Defect

We consider first the elementary case of a small localized defect acting as recombination

center. When the beam is far away from the defect, under the assumptions of Sec. 5.2, all

generated holes are collected and the induced current I_0 is simply equal to the total generation

rate. When the beam approaches the defect, a fraction of the injected holes is lost by

recombination; if I^* is the number of holes subtracted by the defect per unit time, the

collected current becomes

$$I = I_0 - I^*$$

(4)

The current term I^* is dependent on the position of the beam relative to the defect, and corresponds to the signal by which the defect is imaged. The rate of hole capture by the defect can be assumed to be proportional to the density of the hole cloud at the point where the defect lies, thus

$$I^* = k\, p(\underline{r}) \qquad (5)$$

where k is a proportionality factor and $\underline{r}$ is the position of the point-like defect. To a first approximation, according to the assumption iii) of Sec. 5.2, we may use for $p(\underline{r})$ the hole density in absence of the defect, which can be calculated according to the procedure indicated in that section.

It is useful to introduce the contrast profile

$$i^* = I^*/I_o = k\frac{p(\underline{r})}{I_o} \qquad (6)$$

which describes the contrast distribution of the image independently of the beam current. The EBIC contrast profile of the point-like defect turns out to be:[13]

$$i^* = \frac{k}{4\pi D}\begin{cases}1/d - 1/f & d \geq R \\[2mm] \frac{1}{2R}\left[3 - (d/R)^2\right] - 1/f & d \leq R\end{cases} \qquad (7)$$

where d and f are the distances from the center of the generation sphere of the defect and its image in the surface plane, respectively (Fig. 16).

An extended defect can be considered as a continuous distribution of point-like defects over a suitable region V (line, surface or volume). The corre-

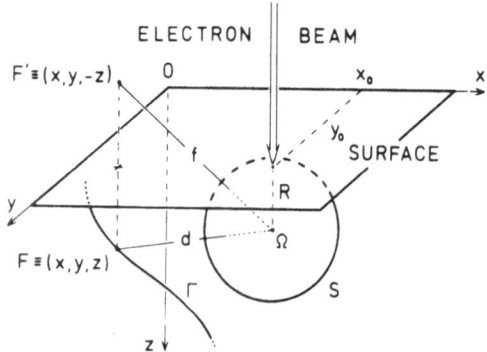

Fig. 16 - Geometrical relations of the model used for calculating the EBIC image of a defect.

sponding contrast profile is found by integrating (7) over V

$$i^* \text{ (extended defect)} = \int_V i^* \text{ (point} - \text{like defect) } dV \qquad (8)$$

Equations (7) and (8) allow the image of a defect having known geometry to be calculated for any value of the beam energy (E appears in Eq. 7 through the sphere radius $R=R(E)$).

A comparison with the experimental EBIC images of known defects will verify the accuracy of the contrast model; on the other hand, the model is expected to be useful for deducing the geometry and recombination properties of a defect from its EBIC images.

5.4   Computer-simulated EBIC Images of Stacking Faults

It has been established that oxidation-induced stacking faults in (100) Si lay on {111} planes and are bounded by a nearly semi-circular $1/3 <111>$ Frank partial dislocation (see Fig. 17). Since only the fault boundary appears to be electrically active (see Sec. 4), the fault will be represented as a recombination line coinciding with its boundary. The corresponding EBIC image is calculated by integrating numerically Eq. 7 along this line.

The results are most conveniently displayed as continuous tone pictures, which can be directly compared to experimental images. This was obtained by connecting a minicomputer to a CRT display through suitable D/A converters. About 50x50 picture elements were computed for each image; this number proved to be adequate, since EBIC images do not show usually many details.

Computer-simulated EBIC images of the stacking fault at increasing beam energies are shown in Fig. 17.[14]   A comparison with the corresponding experimental images of Fig. 13 shows that the representation used is essentially correct. Residual differences are most probably due to the very schematic nature of the uniform generation sphere model.

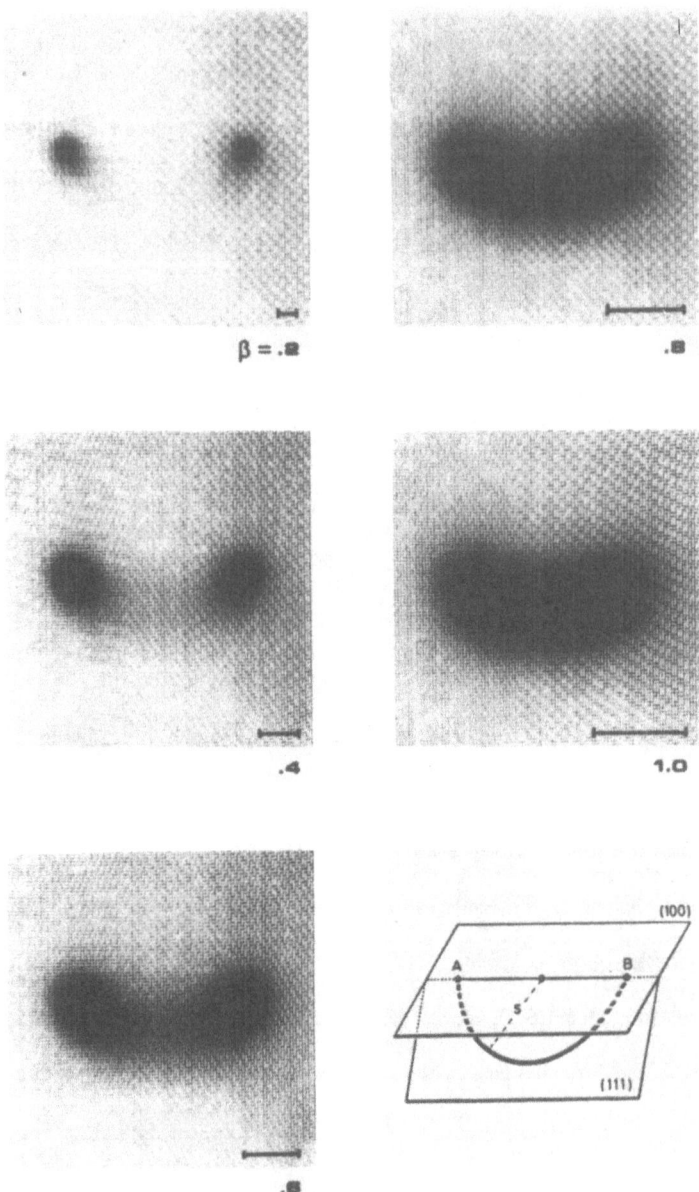

Fig. 17 - Simulated EBIC images of a stacking fault in (100) Si
for increasing values of the ratio $\beta = 2R/s$. Higher values of
β correspond to higher beam energies. On each image, the
segment gives the diameter 2R of the generation sphere.

6. CORRELATION BETWEEN SEM-EBIC AND TEM IMAGES OF DEFECTS

Figure 18 shows one of the best published examples[16] of correlation between SEM-EBIC (a) and TEM (b) images of defects in a p-n junction, showing single dislocations (A,C), a dislocation tangle (B) and a stacking fault (D). This clear correlation has been achieved by the use of a high voltage (1 MV) TEM, whose advantages have been briefly discussed in Sec. 3.

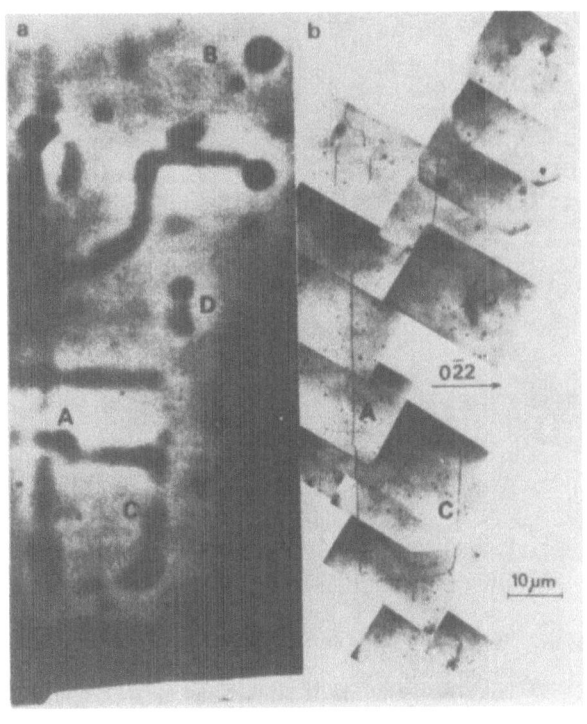

Fig. 18 - Correlation between electrical activity and structure of crystal defects in a Si p-n junction; a) SEM-EBIC image at 15 kV; b) High voltage (1 MV) TEM diffraction contrast image. (After Blumtritt and Gleichmann, Ref. 16)

The correspondence between the features of the two images is clear, although there are remarkable differences between them. First, the EBIC contrast shows strong variations along a dislocation, which are chiefly connected to variations of the depth of the defect. Second, the resolution of the EBIC micrograph is much poorer than that of the TEM image, being chiefly related to the extension of the generation volume ($\cong 2 \ \mu$m at 15 kV).

The model of the EBIC contrast described previously has proved suitable for predicting the appearance of the image, starting from the geometry and the recombination properties of a defect. Therefore the model could also be useful for solving the inverse problem, i.e., reconstructing the "electrical shape" of a defect from its EBIC images.

The expected result of this reconstruction for the dislocations of Fig. 18(a), for instance, is an EBIC micrograph with greater resolution completed with information about the depth of the dislocation lines. This should offer the possibility of making more detailed comparison with the corresponding TEM image and consequently allow more detailed statements about the relationship between electrical and structural properties of these defects.

ACKNOWLEDGEMENTS

The author is grateful to M. Servidori for useful discussions on TEM and XRT observations.

REFERENCES

1.   H. R. Huff and E. Sirtl, Editors, Semiconductor Silicon 1977, (The Electrochemical Society, Princeton, 1977).

2.   P. A. Barnes and G. A. Rozgonyi, Editors, Semiconductor Characterization Techniques, (The Electrochemical Society, Princeton, 1978).

3.   M. Servidori and A. Armigliato, J. Mater. Sci. 10, 306 (1975).

4.   B. O. Kolbesen, K. R. Meyer and G. E. Schuh, J. Phys. E 8, 197 (1975).

5.   H. Melchior and W. T. Lynch, IEEE Trans. Electron Devices, ED-13, 829 (1966).

6.   C. Donolato, P. G. Merli and I. Vecchi, J. Electrochem. Soc. 124, 473 (1966).

7.   D. G. Ong and R. F. Pierret, IEEE Trans. Electron Devices, ED-22, 593 (1975).

8.   C. Donolato, P. Gargini, C. Morandi and M. Servidori, Proc. 8th ESSDERC, Montpellier, 1978, Abstracts p. 162.

9.   K. V. Ravi, C. J. Varker and C. E. Volk, J. Electrochem. Soc. 120, 533 (1973).

10. M. J. Whelan, in Modern Diffraction and Imaging Techniques in Materials Science, ed. by S. Amelinckx *et al.* (North-Holland, Amsterdam, 1970) p.35.

11. A. Authier, *ibid.*, p. 481.

12. S. Kawado, Y. Hayafuji and T. Adachi, Jpn. J. Appl. Phys. 14, 407 (1975).

13. C. Donolato, in Proc. 12th Annual SEM Symposium, ed. by O. Johari (SEM Inc., AMF O'Hare, Illinois, 1979) p. 257.

14. C. Donolato and H. Klann, J. Appl. Phys. 51, 1624 (1980).

15. W. K. Panofsky and M. Phillips, Classical Electricity and Magnetism (Addison-Wesley, Reading, Mass., 1962).

16. H. Blumtritt and R. Gleichmann, Ultramicroscopy 2, 405 (1977).

CHAPTER III

PROCESSING

PLANAR PROCESSING: SILICON OXIDATION

G. J. Declerck

ESAT Laboratory
Katholieke Universiteit Leuven
Heverlee, Belgium

ABSTRACT: Thermally grown SiO_2-layers and CVD-deposited oxide films are widely used throughout silicon integrated circuit processing as gate oxide, implantation or doping mask, field oxide or protective glass layer.

This paper will focus on the requirements for thin gate oxides as used in LSI and VLSI-technology. Some problems concerning the properties and the growth of field oxides will also be addressed. Finally, advanced oxidation techniques such as HCl-oxidation, and high pressure oxidation will be briefly discussed.

I. REQUIREMENTS FOR HIGH QUALITY OXIDES

A. Oxide Thickness

In a classical $5\mu m$ nMOS or CMOS process the gate oxide thickness lies between 70 and 100 nm and field oxides range between 0.7 and $1.0\mu m$. As a result of the scaling principle the gate oxide thickness for VLSI is reduced to the 20-50 nm range and field oxide thickness to 300-500 nm. Important processing requirements in this respect are:

a) An excellent uniformity over large diameter wafers and a good reproducibility from batch to batch.

b) A thorough understanding of the oxidation rate dependence on substratae properties such as orientation, type of doping and doping level. A considerable oxidation rate difference can exist between polysilicon and low-doped substrates. It should be noted here that the oxide grown on top of the poly serves as isolation oxide in double poly processes.

The kinetics of the thermal oxidation of silicon can be characterized quite satisfactorily by the linear-parabolic growth model derived by Deal and Grove[1]:

$$d^2 + A d = B(t + \tau) \tag{1}$$

B is the parabolic rate constant, B/A is the linear rate constant, d is the oxide thickness, t is the oxidation time and τ accounts for the initial rapid oxidation phase.

This equation can also be expressed as follows:

$$\frac{(d^2 - d_o^2)}{B} + \frac{(d - d_o)}{B/A} = t - t_o \tag{2}$$

where d_o and t_o represent a minimum oxide thickness and oxidation time below which the linear-parabolic law no longer holds. This initial oxidation regime has been extensively studied by Irene who clearly demonstrated that deviations from the classical theory occur for oxides thinner than 20 nm and grown in dry O_2.[2] This effect has been attributed to the presence of micropores in the thin SiO_2 films, allowing rapid diffusion of the oxidant (dry O_2). These micropores are also believed to be responsible for the premature dielectric breakdown observed for those oxides. The role of micropores in the initial oxidation regime has been confirmed recently by A. G. Revesz.[3]

The oxidation kinetics of thin oxides grown in H_2O-containing ambients can be fitted to the classical linear-parabolic growth model with $d_o = o$.[2] Those oxides also show better dielectric breakdown behavior. This illustrates the close correlation between the physical properties of thin SiO_2-layers and their electrical characteristics. The growth of very thin oxides (3-14 nm) using low pressure oxygen (0.25-2.0 Torr) has been studied by A. C. Adams et al..[4] A parabolic or linear-parabolic growth model was used to fit their experiments.

The oxidation kinetics discussed so far are only valid for low doped substrates, this is for substrates that are intrinsic at oxidation temperatures. For higher substrate doping of boron, phosphorus or arsenic a strong doping enhanced oxidation effect is observed.[5-9] This DEO-effect is illustrated in Fig. 1, showing the oxide thickness versus oxidation time in dry O_2 at 900°C and 1100°C for various phosphorus doping levels.[6]

It has been demonstrated by Ho *et al.* that this DEO-effect is mainly caused by a substantial increase of the linear rate constant B/A at higher doping levels, whereas the parabolic rate constant B is only slightly affected, as

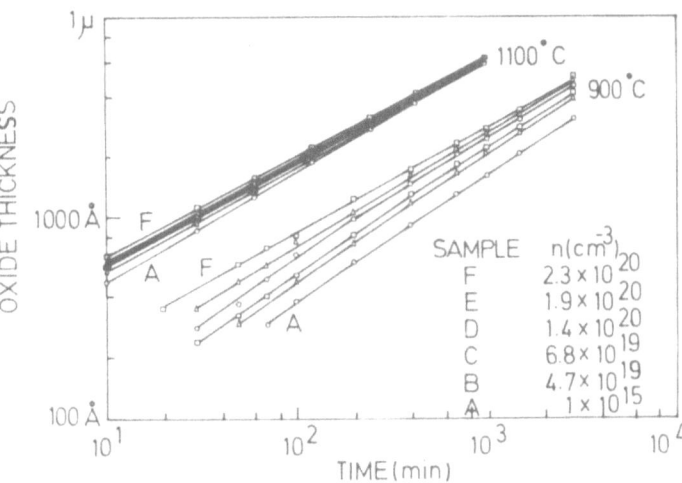

Fig. 1 - Oxide thickness vs. oxidation time in dry O_2 at 900°C and 1100°C with substrate phosphorus doping level as parameter. After Ho *et al.*[6]

shown in Fig. 2. The temperature dependence of B/A is plotted in Fig. 3 with substrate doping as a parameter. No significant effect on the activation energy is observed.

In order to get a better insight in the origin of the DEO-effect one has to concentrate on the oxidation mechanism itself going on at the oxide-silicon interface.

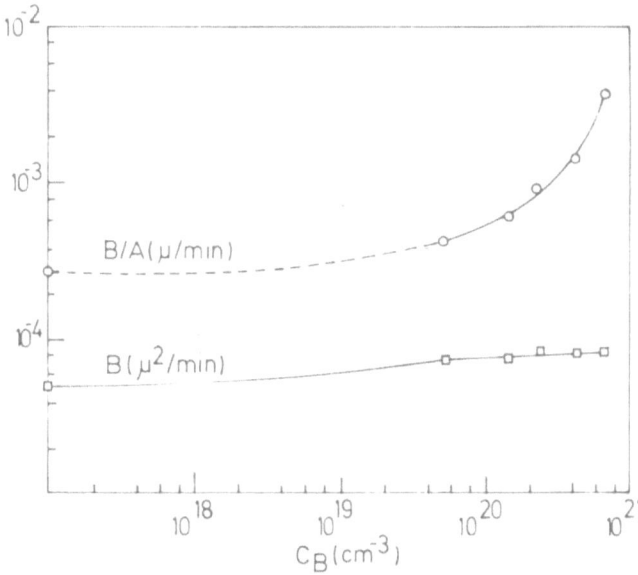

Fig. 2 - Linear and parabolic oxidation rate constants at 900°C vs. initial substrate phosphorus doping level. After Ho *et al.*[6]

As pointed out by Dobson, the thermal oxidation of silicon leads to the creation of excess silicon atoms (silicon interstitials) or to the absorption of silicon vacancies.[10] In the former case the oxygen atom occupies a silicon lattice position and creates a silicon interstitial; in the latter case the oxygen atom fills a vacancy in the silicon lattice. It is obvious that any disturbance of the interstitial or vacancy equilibrium at the interface will affect the oxidation rate.

The nature and the properties of the point defects in silicon have received considerable attention over the last decade. However, so far only very little quantitative data on interstitials has been reported. The silicon vacancies are believed to exist in a neutral state V^o, in a positively charged state V^+, and in negatively charged states V^- and $V^=$ (double charged). The position of the energy levels of these states as a function of temperature is shown in Fig. 4.[11] It has to be emphasized that these charged vacancies have well-defined energy levels in the bandgap and that, as a consequence, their concentration depends on the Fermi level.

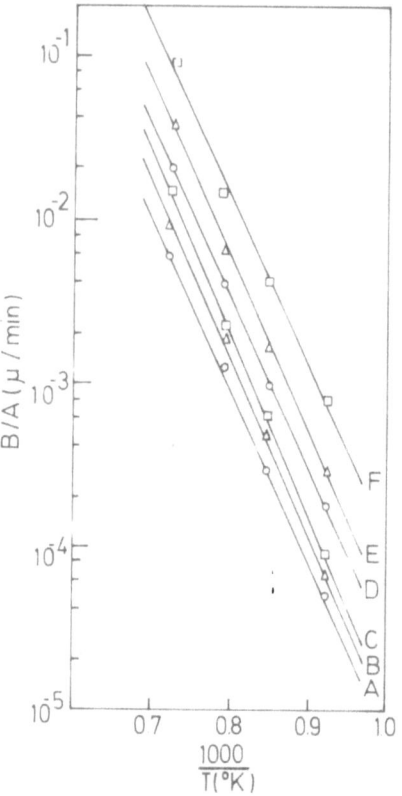

Fig. 3 - Linear rate constant vs. oxidation temperature with substrate phosphorus doping level as parameter. The doping levels are the same as in Fig.1. After Ho *et al.*[6]

This forms the basis of the DEO-model proposed by the Stanford-group.[6-8] At sufficiently high doping level the material is no longer intrinsic at the oxidation temperature. This will shift the Fermi level, resulting in a variation of the total number of vacancies. The relative concentration of the charged vacancies as a function of the Fermi level at 750°C is shown in Fig. 5.[12] It can be seen that the number of charged vacancies considerably increases for heavy n^+-material and that a somewhat smaller increase is expected for p^+-substrates.

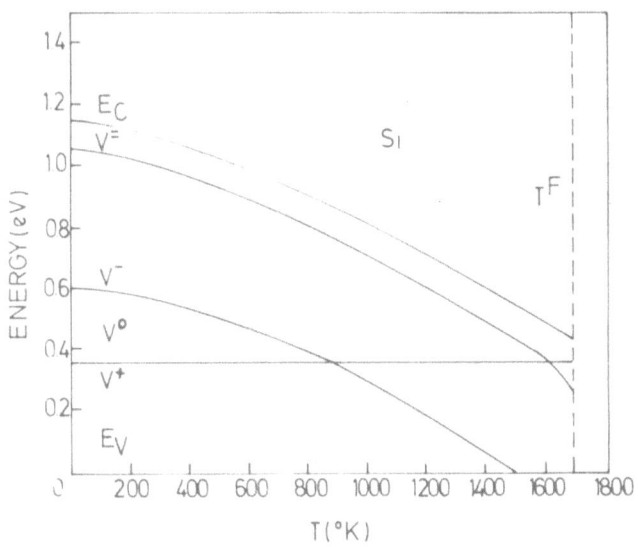

Fig. 4 - Position of the energy levels for vacancies in silicon as a function of temperature. After Van Vechten and Thurmond.[11]

Based on the previous model, Ho and Plummer express the linear oxidation rate as follows:

$$B/A = R_1 + K C_{VT} \quad (3)$$

R_1 accounts for the reaction rate for low-doped material; K is a proportionality factor; C_{VT} is the total concentration of vacancies at the interface, given by:

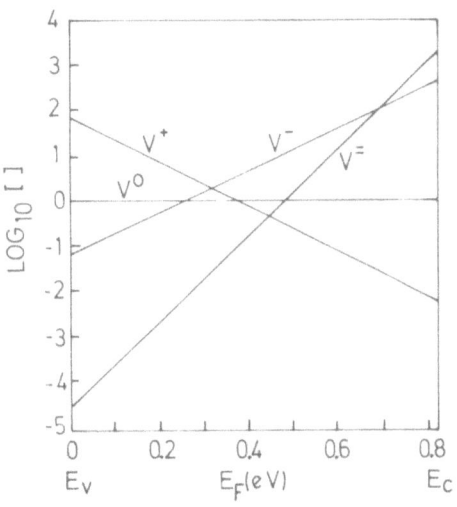

Fig. 5 - Relative concentration of the charged vacancies V^0, V^+, V^-, and $V^=$ for the single vacancy in silicon as a function of the Fermi level E_F at T = 750°C. After Van Vechten.[12]

$$C_{VT} = C_{V^o} + \frac{n_i}{n} C_{V+}^i + \frac{n}{n_i} C_{V-}^i + \left(\frac{n}{n_i}\right)^2 C_{V=}^i \qquad (4)$$

C_{V^o} is the concentration of neutral states; C_{V+}^i is the concentration of positively charged vacancies for an intrinsic substrate; n_i is the intrinsic concentration at oxidation temperature; n is the electron concentration.

The excellent agreement between experiment and theory is illustrated in Fig. 6. For the sake of completeness it should be mentioned that a very pronounced DEO-effect has been observed for arsenic-doped layers by Ohkawa and Nakajima, as shown in

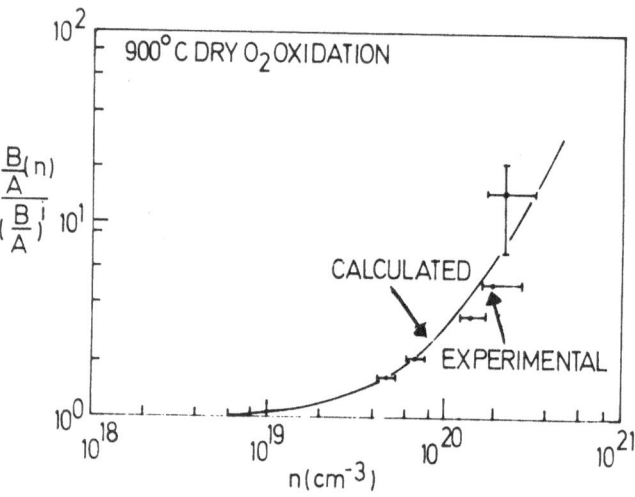

Fig. 6 - Comparison of experimental linear rate constant for 900°C dry O_2 oxidation (111) phosphorus-doped substrates with calculated B/A based on vacancy contribution model. After Ho and Plummer.[7]

Fig. 7.[9] These authors attributed the effect to a catalytic action of arsenic. Ho and Plummer, however, have analyzed these arsenic-data by means of the vacancy-model and have found an excellent agreement with theory.[8]

It should be clear from previous discussion that a very good understanding of the basic oxidation mechanisms is needed in order to model the oxide thicknesses over various parts of a VLSI-chip where different doping levels, crystal orientations (V-grooves) or masking conditions (nitride covering) can exist. Furthermore, the exact shape of two-dimensional profiles or geometries such as the birds beak in nitride-covered selective oxidation techniques is becoming of great importance for precise device modeling.

Application of the DEO-effect is found in the oxidation of heavily doped source and drain regions for reduced parasitic capacitance,[13] and in the growth of isolation oxides on top of doped polysilicon layers.

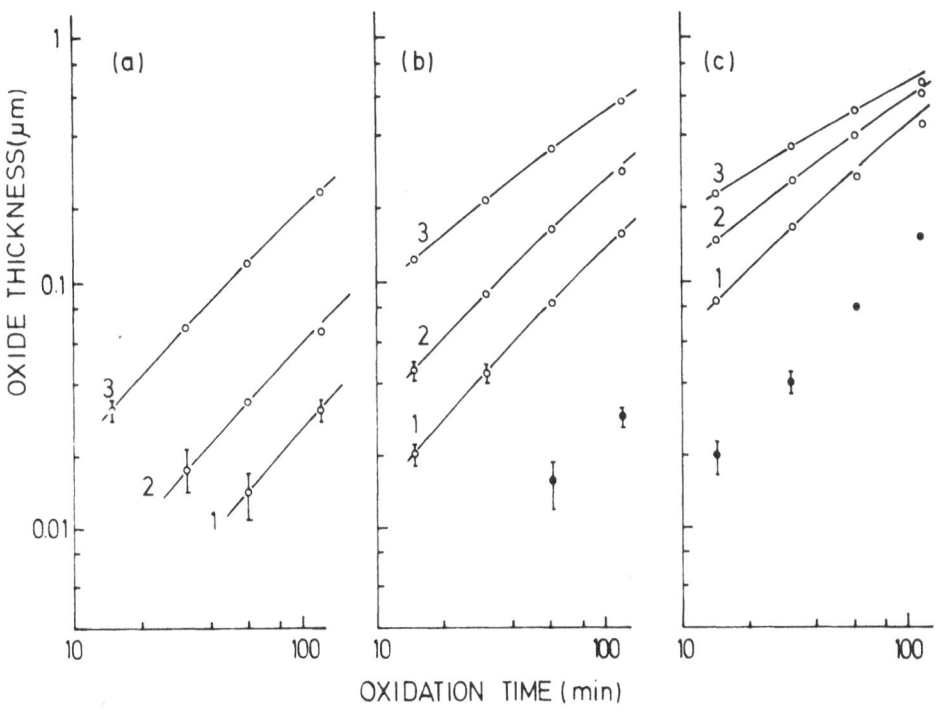

Fig. 7 - Oxide thickness vs. oxidation time at (a) 650; (b) 750 and (c) 850°C. Curves 1,2 and 3 are calculated values for surface concentration C_s of 2.5×10^{20}, 5.5×10^{20} and 2.2×10^{21} cm^{-3}, respectively. Experimental values: o, arsenic-diffused; •, undiffused silicon. After Ohkawa and Makajima.[9]

B. Oxide Charge and Interface Traps

Four different kinds of charges can be distinguished as associated with the oxide and the oxide-silicon interface: (Q refers to C/cm^2; N to elementary charges per cm^2)

1. Mobile ionic charge - Q_{mi}, N_{mi}

2. Oxide fixed charge - Q_{of}, N_{of}

3. Interface trap charge - Q_{it}, N_{it}

   The interface trap charge density as a function of energy is denoted by D_{it}.

4. Oxide traps induced by processing - Q_{ot}, N_{ot}.

Mobile Ionic Charge

The presence of mobile ions (mostly Na, K) leads to an instability of the threshold voltages and to a deterioration of the wear-out properties of the SiO$_2$-layer. The mobile ion density should be smaller than 10^9 cm^{-2}. In modern VLSI processing the contamination itself by mobile ions should be avoided. Passivation of the oxide by means of a PSG-layer or neutralization of the mobile ions by means of HCl-oxidation techniques are not preferable. Precleaning of the furnace tubes in a chlorine containing ambient and the use of polysilicon gates and of sodium-free metallization systems are mandatory.

Oxide Fixed Charge and Interface Traps

The production of VLSI circuits with high performance and excellent reliability puts very stringent demands on the quality of the silicon-SiO$_2$ interface. Both oxide fixed charge and interface trap density should be low and reproducible. For (100)-material, N_{of} should be smaller than 5×10^{10} cm^{-2} and D_{it} should be smaller than 1×10^{10} cm^{-2} eV^{-1}.

Interface traps affect the surface mobility of minority or majority carriers; they have a strong effect on 1/f-noise in MOS-circuits and they deteriorate the transfer efficiency of surface channel CCD's. The most important effect, however, for VLSI-circuits is probably the direct relation between interface traps and surface generation or recombination rates. It is well known that in CCD's, when care is taken to minimize bulk generation by means of special gettering techniques, the dominant dark current component will be caused by surface generation at a depleted surface

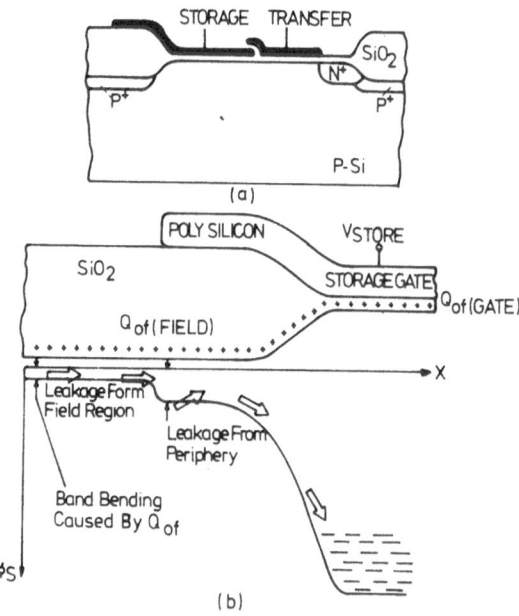

Fig. 8 - (a) Basic structure of a one transistor dynamic RAM cell, (b) Magnified view of the periphery and gate overlap region show the potential distribution that causes leakage charge to be collected in the storage region. After Chatterjee *et al.*[14]

(denoted by s_0). A study by Chatterjee *et al.*[14] shows that the most significant contribution to leakage rate in dynamic MOS memories originates from sources outside the storage gate region. They found an effective surface generation rate s_0 in the overlap region which was 20 times higher than under the gate or field regions (Fig. 8). This observation was explained by the higher probability of having oxidation induced stacking faults in the tapered transition region between the field and the active region due to the local oxidation technique. Similar effects will be of great importance for a large variety of VLSI circuits when leakage currents are to be further reduced.

The dependence of Q_{of} and N_{it} on several processes and material parameters has recently been reviewed by Razouk and Deal.[15] They came to the following conclusions:

1. A common origin of oxide fixed charge and the so-called structural type of interface traps can be deduced. This is illustrated in Fig. 9 where the midgap interface trap density is plotted versus the oxide fixed

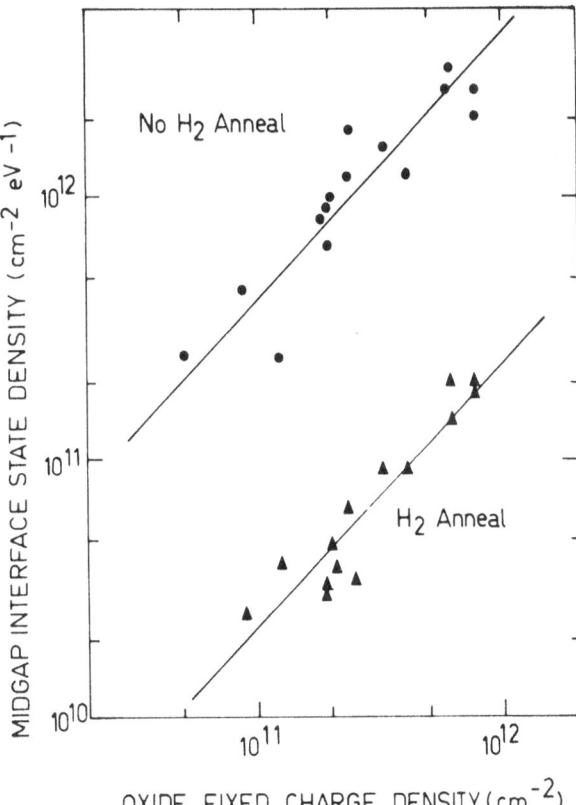

Fig. 9 - Midgap interface state density vs. oxide fixed charge for unannealed and H_2-annealed n-type and p-type silicon samples, (100) and (111) orientation, oxidized in dry O_2 at 1000° and 1200°C (Hydrogen anneal: 2 liters/min, 10% H_2 in N_2, 10 min, 500°C). After Razouk and Deal.[15]

charge for different oxidation and anneal conditions.

2. Small differences in processing techniques, especially during sample cooling (cooling rate and ambient) can significantly vary the values of N_{it}. The presence of moisture in the oxidation or cooling ambient is particularly impor-tant.

3. After the oxide growth, densities of N_{it} and Q_{of} can be independently altered by some particular treatments. A low temperature forming gas anneal, e.g., can reduce N_{it} without af-fecting Q_{of}.

4. The lowest D_{it}-values $(1 \times 10^{10}$ cm^{-2} $eV^{-1})$ were found after a low temperature forming gas anneal of wafers oxidized in HCl-containing ambients and cooled in N_2.

The interface trap density as a

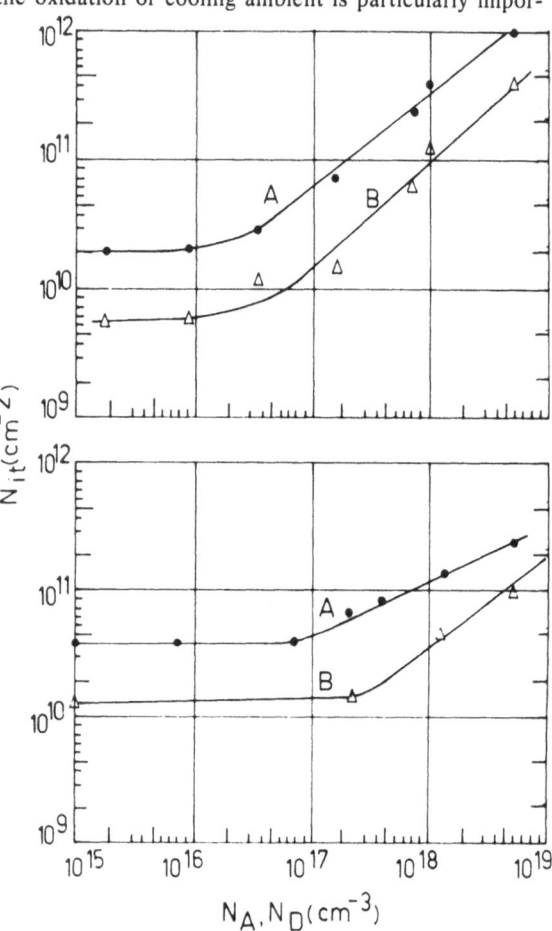

Fig. 10 - Number of interface states N_{it} as a function of (a) the interfacial boron concentra-tion N_A and (b) the interfacial phosphorus concentration N_D before (curve A) and after (curve B) low-temperature annealing at 450°C in N_2/H_2O. The Si-SiO$_2$ interface was made by thermal oxidation (1100°C dry O$_2$, fol-lowed by N$_2$ annealing; oxide thickness 100 nm) of epitaxial layers doped during epitaxial growth. After Snel.[16]

function of substrate doping level has been studied by Snel.[16] His results, shown in Fig. 10, clearly indicate that N_{it} increases for higher doping levels. The effect is more pronounced for p-type substrates where it is apparent for concentrations above 10^{16} cm^{-3} whereas for n-type

substrates a concentration of at least 10^{17} cm^{-3} is needed to make the effect measurable. It

was pointed out by Snel that these results are independent of the doping process used as

classical predeposition, ion implantation and doping during epitaxial growth all lead to the

same conclusion. The driving force behind this phenomenon, according to Snel,[17] is a

decrease of solid solubility of substitutional acceptor-type metal impurities in the p$^+$-doped

substrates. These metals will diffuse interstitially to the interface where they precipitate and

enhance the density of extrinsic interface traps.

It should be remarked that this effect could be of great importance in view of the

enhancement of surface generation rate observed at the edges of the storage cells in dynamic

RAM's.

Several experiments have recently been described to reduce the interface trap density by

means of a very active hydrogen species. Risch *et al.*[18] reported very low densities (5×10^8

cm^{-2} eV^{-1} - 1×10^9 cm^{-2} eV^{-1}) by depositing a H$_2$-containing layer at the end of the conven-

tional double poly process, followed by a low temperature annealing step. The use of a

hydrogen implantation step before the final low temperature anneal has been reported by

Kellner and
Goetzberger[19] to im-
prove the low-current β
of bipolar transistors
having a Si$_3$N$_4$ layer on
top of the SiO$_2$ as protection against ionic
contamination. High
temperature H$_2$-anneals
(800°C-1000°C) are
widely used to reduce

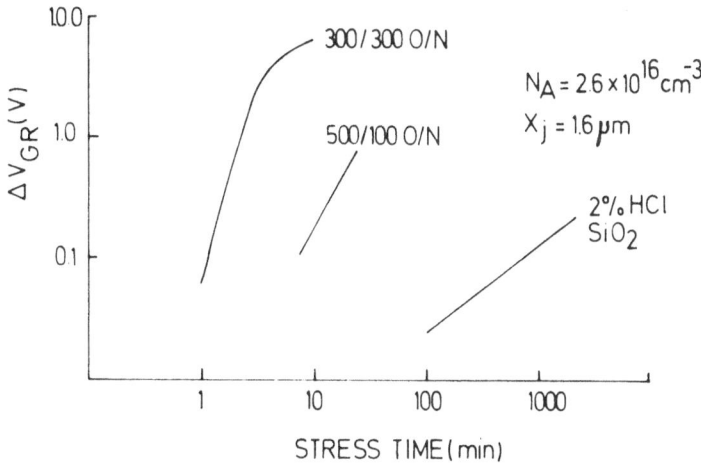

Fig. 11 - Threshold voltage shift for a 15μm-device stressed at $V_G = V_D = 15V$ and with a substrate bias of 3V. After Dockerty and Abbas.[21]

the N_{it} density in MNOS-circuits.   Viktorovitch and Pananakakis[20] obtained very low

N_{it}-values in MNOS and MNS structures using atomic hydrogenation in an RF plasma.

Oxide Traps Induced By Processing

As device dimensions shrink, the electric fields in the silicon tend to increase leading to

the generation of channel and substrate hot carriers which gain sufficient energy to surmount

the energy barrier at the oxide-silicon interface.   Part of these injected carriers will be trapped

in the gate insulator and this results in device instabilities such as threshold shifts and junction

breakdown walk-out phenomena.

Abbas and Dockerty have shown that the trapping of hot electrons produces large threshold shifts, especially when a double dielectric SiO_2/Si_3N_4 layer is used.[21] This effect is illustrated in Fig. 11.

The electron trapping efficiency of SiO_2-layers strongly depends on the processing techniques used. Gdula investigated the role of the oxidation technique. Some of his results are shown in Fig. 12.[22] The smallest trapping rate is observed for the dry thermal

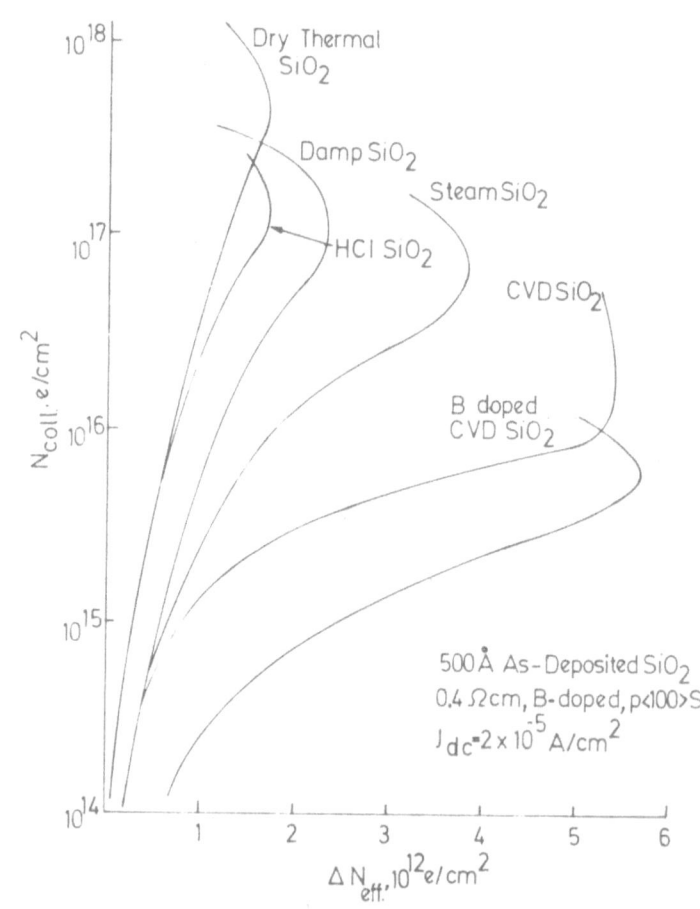

Fig. 12 - Trapping behavior of SiO_2 formed by different processes; no high temperature anneal. After Gdula.[22]

oxide, immediately followed by the HCl-oxide, whereas wet oxides, and to a much larger extent CVD-oxides, show increased electron trapping. The influence of various annealing treatments has been extensively studied by Young *et al.*.[23]

It has to be remarked that the use of advanced processing techniques such as direct e-beam writing, x-ray lithography, plasma based etching and deposition, reactive ion etching, electron gun evaporation and magnetron sputtering can generate charged and neutral traps in the gate dielectric material. This is

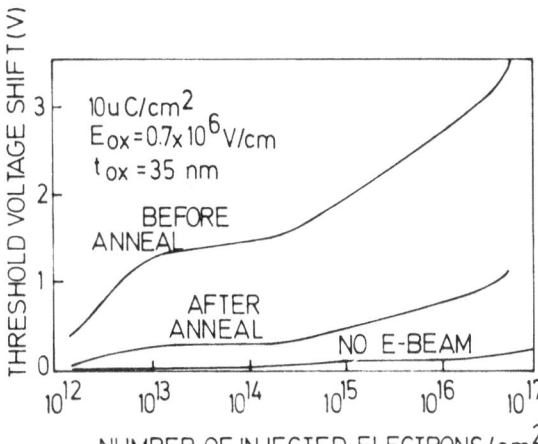

illustrated in Fig. 13 where the threshold voltage shift is plotted versus the number of injected electrons for both an irradiated (before and after anneal) and an unirradiated sample.[24] It should be noted that the radiation dosage used is compa-

Fig. 13 - Shift in threshold voltage in optically defined, irradiated polysilicon MOSFET's as a function of the number of electrons injected per square centimeter of gate area. The trapping in an irradiated sample before anneal is shown in the upper curve and that for an irradiated sample after a 30 min anneal at 400°C is shown in the middle curve. The trapping in a sample which was not irradiated is shown in the bottom curve for comparison. After Aitken.[24]

rable to the dosage needed for currently available e-beam resists.

The presence of neutral traps forms a severe reliability hazard when hot electron injection in the oxide can occur. This problem can be attacked by proper design (smaller fields) and by appropriate process modifications (e.g., the development of contact metallurgy enabling higher temperature anneals).[24]

C. Oxide Reliability

The presence of weak spots or pinholes in the oxide, the intrinsic gate-oxide breakdown and the long-term dielectric strength are of major importance for improved VLSI reliability.

Two different dielectric breakdown phenomena have to be discussed:

1. The breakdown under voltage ramp conditions, characterized by a maximum breakdown field and by a defect density which is the density per unit area of weak spots having a reduced breakdown strength.[25]

2. The <u>wear-out</u> or time-dependent breakdown under accelerated bias-temperature conditions.[26]

Several models exist for the oxide breakdown mechanism:

1. Early breakdown events may be caused by the presence of <u>weak</u> <u>spots</u> <u>or</u> <u>pinholes</u> in the oxide.

2. <u>Barrier</u> <u>height</u> <u>lowering</u> due to mobile contaminants such as sodium present close to the injecting electrode.[27]

3. <u>Impact</u> <u>ionization</u> giving rise to the build-up of positive charge within the oxide which leads to field enhancement and catastrophic failure.[28]

4. <u>Creation</u> <u>of</u> <u>deep</u> <u>electron</u> <u>traps</u> close to the injecting electrode by the combined effect of a high stress field and a high injection current.[29] These traps will then increase the internal field to destructive breakdown. It has been pointed out by

Harari[30] that the oxide breakdown strength for very thin oxides increases with decreasing thickness and approaches an upper limit of approximately 30 MV/cm (Fig. 14).

The experimental

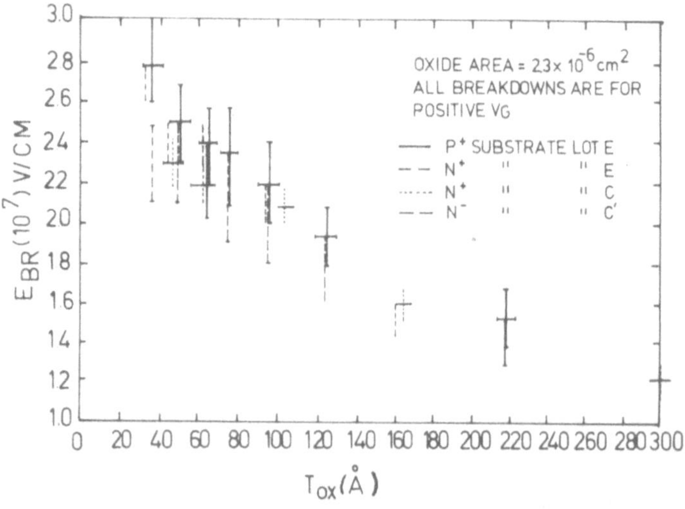

Fig. 14 - Oxide breakdown strength for ultrathin oxides as a function of oxide thickness. All data shown for positive V_G. Negative V_G breakdowns are lower by 5-10%. After Harari.[30]

observation that thicker oxides break down at lower external fields (8-9 MV/cm) is
explained by the very high density of negatively charged traps (up to 10^{19} cm^{-3}) which
may enhance the local internal fields to the level seen in the thinnest oxides.[30] It
should be remarked however that these extremely high breakdown fields for very thin
oxides have not been confirmed by Adams *et al.*.[4]

A typical set of wear-out curves is represented in Fig. 15.[31] The cumulative percentage of failures is plotted as a function of the stress time at 300°C with positive electrode bias for both polysilicon and aluminum electrodes. It is clear from the figure that the capacitors with polysilicon electrodes can withstand a given applied field about 3 orders of magnitude longer than do the capacitors with aluminum electrodes. It should be remarked, however, that histograms of breakdown voltages using the voltage ramp technique were the same for both electrode materials.[31] The superior dielectric reliability becomes even more evident from Fig. 16 showing the wear out

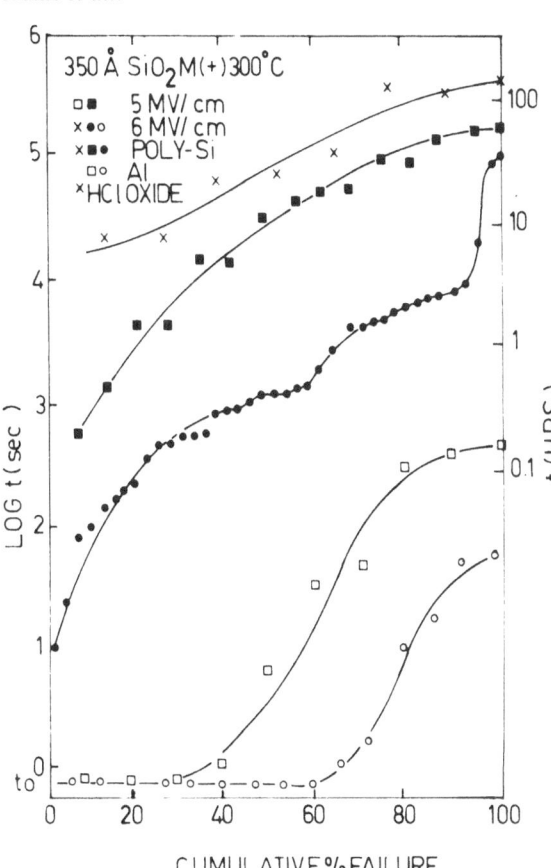

Fig. 15 - Statistical failure times of 35 nm SiO$_2$ films having either poly-Si or Al electrodes as measured at 300°C with positive electrode bias. After Osburn and Bassous.[31]

performance as a function of oxide thickness. The maximum time to failure is nearly independent of oxide thickness when poly electrodes are used, whereas for aluminum electrodes the wear-out behavior is much worse for thinner oxides. It should also be noticed that the

superior behavior of poly-
electrodes has only been observed for positive applied voltages. Under negative applied
fields both electrode materials
behave nearly the same as is illustrated in Fig. 17.

D. Influence On Silicon Bulk

Properties

So far we have confined
our discussion to the properties
of the oxide layer and of the
oxide-silicon interface. However, in modern silicon oxidation
technology one should not only
care about the oxide quality
but at the same time one
should also keep a proper control on the silicon bulk proper-

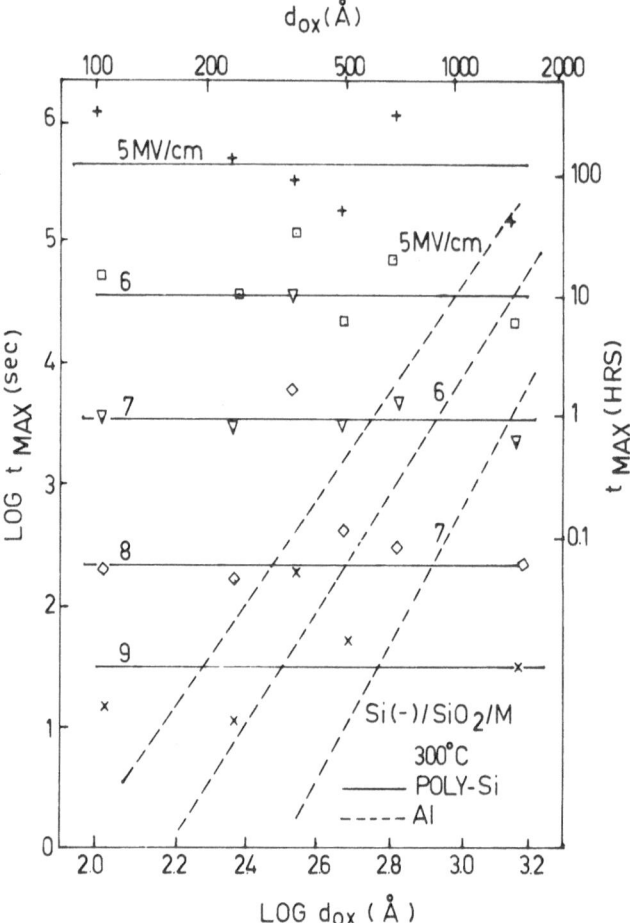

Fig. 16 - Maximum time to failure as a function of oxide thickness for poly-Si electrodes biased positively at 300°C for different applied fields. The dotted lines represent the Al electrode data of Osburn and Chou. After Osburn and Bassous.[31]

ties. It has become clear during recent years that oxidation of the silicon strongly affects
several important silicon bulk characteristics. The presence of oxidation-induced stacking
faults, dislocations or dislocation loops can drastically reduce the carrier lifetime in the silicon,
especially when metal precipitates are formed. They can also cause severe leakage current
problems in CCD's, dynamic RAM's or CMOS-circuits.[32-34]

The point defect concentration (silicon vacancies or silicon self-interstitials) at the silicon
surface is considerably altered by the thermal oxidation mechanism. It has been mentioned
before in this paper that at normal temperatures thermal oxidation results in a deficiency of

vacancies and/or an excess of interstitials at the interface[10]. This
can lead to the growth of OSF's, if
stacking fault nuclei are present, or
to the formation of other crystalline
defects.   Experiments   such   as
oxidation-enhanced   diffusion[35-37]
and the influence of impurity diffusion   on   the   OSF-behavior,[38-40]
strongly support the hypothesis that
silicon interstitials play a dominant
role in these phenomena.   Furthermore,   the   addition   of   a   chlorine
containing compound to the oxidation ambient strongly affects the Si-
interstitial distribution at the interface. The stacking fault length ver-

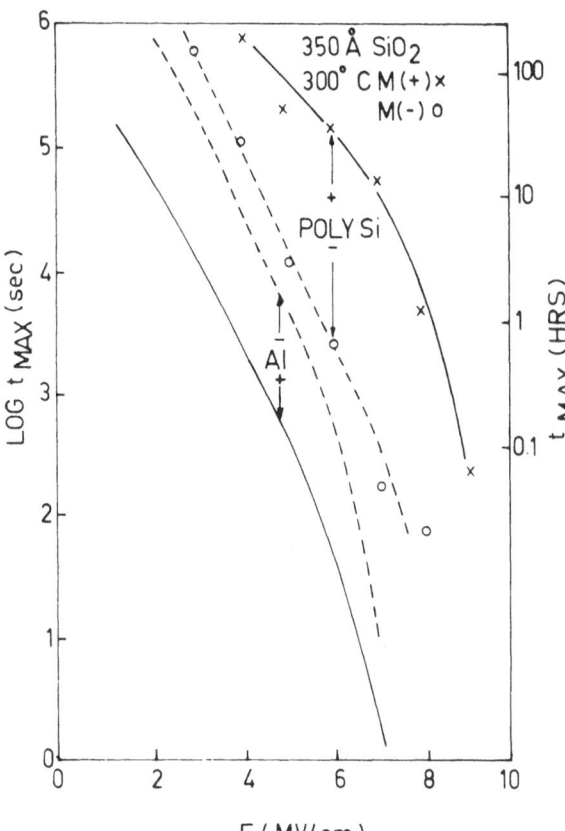

Fig. 17 - Maximum time to failure as a function of applied field at 300°C for poly-Si and
Al electrodes. After Osburn and Bassous.[31]

sus oxidation time at 1150°C for various concentrations of 111-trichloroethane (C_{33}) is given

in Fig. 18.[33] The slower growth or shrinkage of the OSF's in the chlorine containing ambient

is attributed to the chlorination reaction going on at the interface. This reaction results in a

lowering of the concentration of Si-interstitials due to the formation of Si-Cl bonds.[33]

The point defect concentration also affects the diffusion of most common dopants such as

boron, phosphorus or arsenic. The oxidation enhanced diffusion effect has to be accounted

for when a precise control of doping profiles is necessary. This will be discussed in the lecture

on diffusion.

In local oxidation techniques care should be taken to avoid excessive stresses at the edge of the nitride oxidation mask.[41] Optimization of the birds beak geometry and minimization of crystal defects due to the local oxidation is one of the major tasks of the process engineer.[42]

II. OXIDATION TECHNIQUES

It is not within the scope of this paper to give an extended review on the large variety of oxidation techniques available. In the following, a few headlines of current research in the oxidation technology will be briefly discussed.

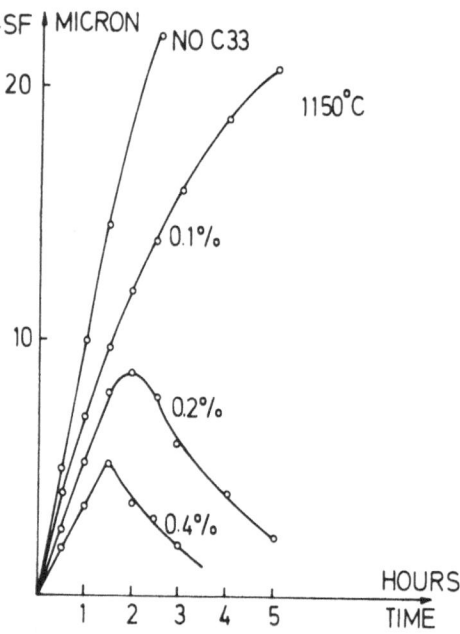

Fig. 18 - Length of OSF's vs. oxidation time at 1150°C for various concentrations of C_{33}. After Claeys *et al.*[33]

A. Chlorine-Oxidation

The beneficial effects of the addition of small amounts of a chlorine containing compound to the oxidizing atmosphere on device characteristics and on integrated circuit performance have been described in the literature over the last ten years. Oxidation processes using, respectively, HCl-gas,[43,44] trichloroethylene (TCE),[45,46] and 111-trichloroethane[47] have been proposed and their properties have been described in several review papers.[48,49]

The superior electrical behaviour of these oxides is mainly due to a reduction of both the mobile ionic charges and the density of fast interface traps, and to improved SiO_2 breakdown characteristics and a smaller defect density. Pre-oxidation cleaning of furnace tubes in a chlorine containing ambient has found widespread use. As already discussed before, one of the main advantages of the chlorine-oxidation is its great influence on the structure and electrical activity of crystalline defects and impurities in the silicon substrate.[50,51]

B. High Pressure Oxidation

For the historical background of pressure oxidation and for a detailed discussion on the technique with a description of available production equipment, we refer to the review paper by Zeto *et al.*.[52]

The use of high pressure oxidation should be seen in view of the general trend towards lower temperature processing, which is required to reduce dopant diffusion and to maintain the submicron dimensions needed in VLSI circuits. Figures 19 and 20 compare the oxidation time to grow, respectively, 100 nm gate oxide or $1\mu m$ field oxide in different oxidation conditions. It can be seen, for instance, that a 100 nm thick oxide can be grown in 90 min. at:

- 1050°C in dry oxygen, 1 atm.

- 850°C in wet oxygen, 1 atm.

- 750°C in wet oxygen, 5 atm.

- 730°C in dry oxygen, 500 atm.

The oxidation rate in high pressure oxidation systems, used at lower temperatures, can be described by the classical expression:

$$d^2 + Ad = B (t + \tau) \quad (1)$$

where both the parabolic rate constant B and the linear rate constant B/A are proportional to the partial pressure of the oxidizing species in the gaseous ambient. The rate controlling step in the high pressure - low temperature oxidation process is the reaction at the SiO_2/Si-interface.

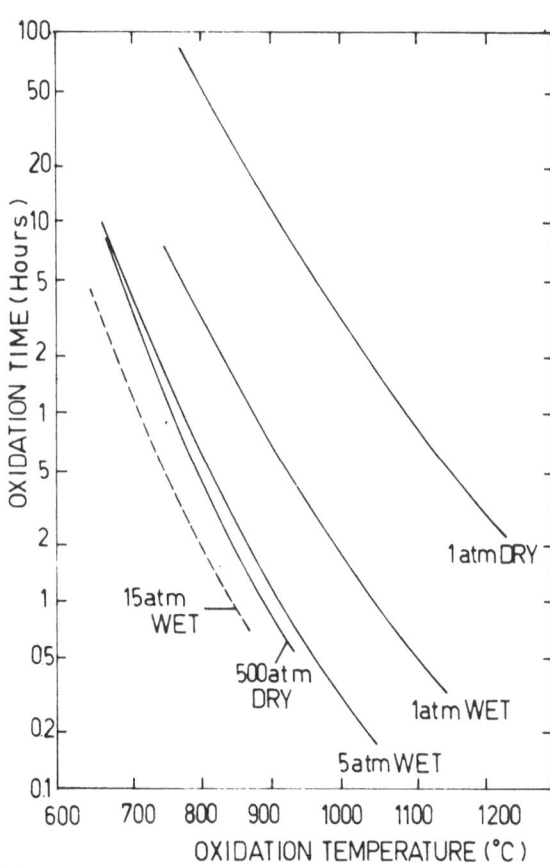

Fig. 19 - Oxidation time vs. temperature for the growth of $0.1\mu m$ of SiO_2 on (100) silicon under various conditions of pressure and ambient. After Zeto *et al.*[52]

Some important physical properties of

pressure oxides are:

1. The fixed charge

and the interface trap

density are compara-

ble to those of con-

ventional oxides.

2. The refractive in-

dex is somewhat high-

er (1.475 for a 500

atm., dry oxygen,

800°C oxide as com-

pared with 1.468 for

a 1 atm., dry O_2,

800°C. oxide).

3. The chemical etch

rate is about 10%

slower.

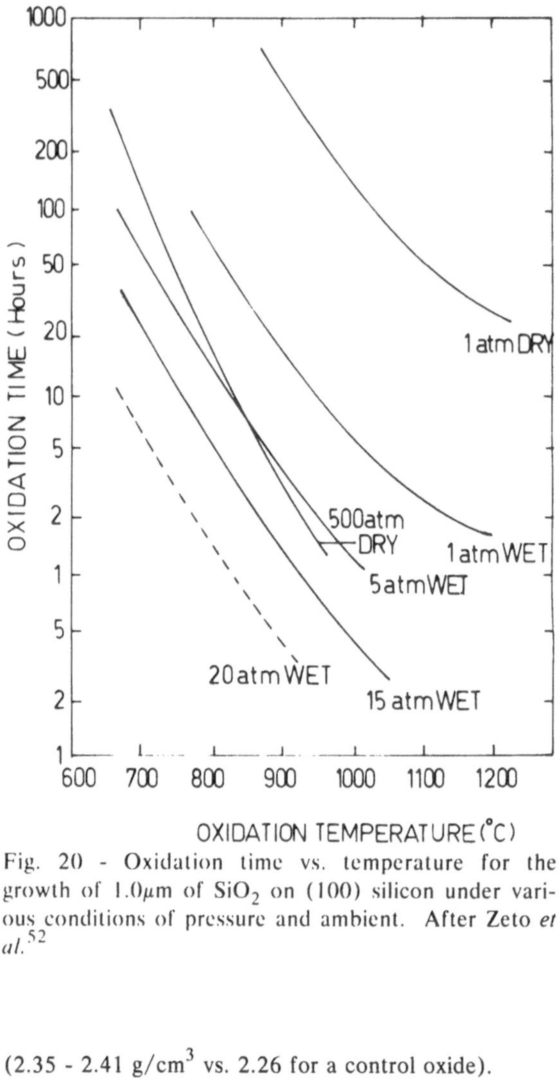

Fig. 20 - Oxidation time vs. temperature for the growth of 1.0μm of SiO_2 on (100) silicon under various conditions of pressure and ambient. After Zeto *et al.*[52]

4. The density is higher (2.35 - 2.41 g/cm^3 vs. 2.26 for a control oxide).

Silicon nitride has been proven to be a suitable oxidation mask for selective oxidation process-

es utilizing high pressure steam and high pressure dry oxygen.

The length and density of OSF's are significantly reduced due to the low temperature

processing.[53,54] However it should be remarked that, as the oxide growth is purely reaction-

limited up to the thickness of commonly used field oxides, very long bird's beaks are formed.[42]

At the same time, severe stresses can be developed at the oxide-nitride edges during local

oxidation, which leads to the generation of high densities of dislocations.[41] Because of these two disadvantages, the use of high pressure oxidation at lower temperatures ($<950°C$) for

nitride-masked selective field oxidation must be studied with very great care.

REFERENCES

1. B. E. Deal and A. S. Grove, J. Appl. Phys. 36, 3770 (1965).

2. E. A. Irene, J. Electrochem. Soc. 125, 1708 (1978).

3. A. G. Revesz, Electrochem. Soc. Meeting, Minneapolis 1981, abstract no. 281.

4. A. C. Adams, T. E. Smith and C. C. Chang, J. Electrochem. Soc. 127, 1787 (1980).

5. B. E. Deal and M.Sklar, J. Electrochem. Soc. 112, 430 (1965).

6. C. P. Ho, J. D. Plummer and J. D. Meindl, J. Electrochem. Soc. 125, 665 (1978).

7. C. P. Ho and J. D. Plummer, J. Electrochem. Soc. 126, 1516 (1979).

8. C. P. Ho and J. D. Plummer, J. Electrochem. Soc. 126, 1523 (1979).

9. S. Ohkawa and Y. Nakajima, J. Electrochem. Soc. 125, 1997 (1978).

10. P. S. Dobson, Phil. Mag. 26, 1301 (1972).

11. J. A. Van Vechten and C. D. Thurmond, Phys. Rev. B 14, 3539 (1976).

12. J. A. Van Vechten, in "Lattice Defects in Semiconductors," 1974, 212-220, Institute of Physics, London (1975).

13. C. P. Ho and J. D. Plummer, IEEE Trans. on Electron Devices, ED-26, 623 (1979).

14. P. K. Chatterjee, G. W. Taylor, A. F. Tasch and H. S. Fu, IEEE Trans. on Electron Devices, 26, 564 (1979).

15. R. R. Razouk and B. E. Deal, J. Electrochem. Soc. 126, 1573 (1979).

16. J. Snel, Inst. Phys. Conf. Ser., 50, 119 (1980).

17. J. Snel, Interface Specialist Conference, New Orleans 1979.

18. L. Risch, E. Pammer and K. Friedrich, Inst. Phys. Conf. Ser., 50, 114 (1980).

19. W. Kellner and A. Goetzberger, IEEE Trans. on Electron Devices 22, 531 (1975).

20. P. Viktorovitch, G. Pananakakis, A. Chenevas-Paule and V. Le Goascoz, Proc. ESSDERC 79, 157.

21. S. A. Abbas and R. C. Dockerty, Appl. Phys. Lett. 27, 147 (1975).

22.   R. A. Gdula, J. Electrochem. Soc. 123, 42 (1976).

23.   D. R. Young, E. A. Irene, D. J. DiMaria, R. F. De Keersmaecker and H. Z. Massoud, J. Appl. Phys. 50, 6366 (1979).

24.   J. M. Aitken, IEEE Trans. on Electron Devices ED-26, 372 (1979).

25.   C. M. Osburn and D.W. Ormond, J. Electrochem. Soc. 119, 591 (1972).

26.   C. M. Osburn and N. J. Chou, J. Electrochem. Soc. 120, 1377 (1973).

27.   T. H. Di Stefano, Appl. Phys. Lett. 19, 280 (1971).

28.   T. H. Di Stefano and M. Shatzkes, Appl. Phys. Lett. 25, 685 (1974).

29.   E. Harari, J. Appl. Phys. 49, 2478 (1978).

30.   E. Harari, Appl. Phys. Lett. 30, 601 (1977).

31.   C. M. Osburn and E. Bassous, J. Electrochem. Soc. 122, 89 (1975).

32.   K. Tanikawa, Y. Ito and H. Sei, Appl. Phys. Lett. 28, 285 (1976).

33.   C. L. Claeys, E. E. Laes, G. J. Declerck and R. J. Van Overstraeten in "Semiconductor Silicon" 1977, edited by the Electrochemical Society, Princeton, 773 (1977).

34.   S. P. Murarka, T. E. Seidel, J. V. Dalton, J. M. Dishman and M. H. Read, J. Electrochem. Soc. 127, 716 (1980).

35.   S. M. Hu, J. Appl. Phys. 45, 1567 (1974).

36.   S. M. Hu, J. Vac. Sci. Technol. 14, 17 (1977).

37.   D. Antoniadis, A. G. Gonzales and R. W. Dutton, J. Electrochem. Soc. 125, 813 (1978).

38.   C. L. Claeys, G. J. Declerck and R. J. Van Overstraeten in "Semiconductor Characterization Techniques", edited by the Electrochemical Society, Princeton 366, (1978).

39.   C. L. Claeys, G. J. Declerck and R. J. Van Overstraeten, Rev. Phys. Appl. 13, 797 (1978).

40.   R. Francis and P. S. Dobson, J. Appl. Phys. 30, 280 (1979).

41.   G. Franz, B. O. Kolbesen, R. Lemmer and H. Strunk, Semiconductor Silicon 1981, edited by the Electrochemical Society, Pennington 821, (1981).

42.   R. Lemme and H. Oppolzer, Semiconductor Silicon 1981, edited by the Electrochemical Society, Pennington 811, (1981).

43.   R. J. Kriegler, Y. C. Cheng and D. R. Colton, J. Electrochem. Soc. 119, 388 (1973).

44. R.J. Kriegler, Semiconductor Silicon 1973, edited by the Electrochemical Society, Princeton, 363 (1973).

45. M. C. Cheng and J. W. Hile, J. Electrochem. Soc. 119, 223 (1972).

46. G. J. Declerck, T. Hattori, G. A. May, J. Beaudouin and J. D. Meindl, J. Electrochem. Soc. 122, 436 (1975).

47. E. J. Janssens and G. J. Declerck, J. Electrochem Soc. 125, 1696 (1978).

48. B. R. Singh and P. Balk, J. Electrochem. Soc. 125, 453 (1978).

49. G. J. Declerck, Inst. Phys. Conf. Ser. 53, 133 (1980).

50. C. Claeys, G. J. Declerck, R. J. Van Overstraeten, H. Bender, J. Van Landuyt and S. Amelinckx, Semiconductor Silicon 1981, edited by the Electrochemical Society, Pennington, 730 (1981).

51. L. E. Katz, P. F. Schmidt and C. W. Pearce, J. Electrochem. Soc. 128, 620 (1981).

52. R. J. Zeto, N. O. Korolkoff and S. Marshall, Sol. St. Tech., p. 62, July (1979).

53. N. Tsubouchi, H. Miyoshi and H. Abe, The 9th Conference on Solid State Devices, Tokyo 1977, No. A-5-3.

54. L. E. Katz and L. C. Kimmerling, J. Electrochem. Soc. 125, 1680 (1978).

PLANAR PROCESSING: DIFFUSION

G. J. Declerck

ESAT Laboratory
Katholieke Universiteit Leuven
Heverlee, Belgium

ABSTRACT: As the dimensions of the devices used in VLSI-circuits shrink, the precise control of junction depths and of lateral diffusion, of implanted field regions and of threshold or barrier implants is of utmost importance.   In many of those cases the classical diffusion theory can no longer be applied because of the close interaction between the various diffusing dopants and also because of the pronounced influence of the annealing or oxidation ambient on the diffusion data.

The purpose of this lecture is to provide some insight in the physical mechanisms involved in solid state diffusion in silicon.   In the first part of the paper the diffusion equations and their classical solutions will be briefly reviewed; the interaction between the diffusing species and the silicon point defects will be addressed and some of the experimental techniques available for the study of the diffusion data will be described.

In the second part of the paper the deviations from the classical theory will be discussed. Field aided diffusion, point defect enhancement and impurity clustering at high concentration will be taken into account.   Finally, the oxidation enhanced diffusion effect and the relation between diffusion enhancement and stacking fault growth will be described.

I. GENERAL PRINCIPLES

The diffusion equations and their classical solutions will be described first.   It is important to remark that these solutions can only be employed for the so-called "intrinsic diffusion behavior" under the following conditions:

1.   There is only one dopant present.

2.   The concentration of the dopant is low compared with the intrinsic carrier concentration n_i at the diffusion temperature.

3.   There is no steep concentration gradient.

4.   There are no non-equilibrium effects present such as an oxidizing ambient. These ambients disturb the point defect equilibrium in the silicon and affect the diffusion behavior.

A. Diffusion Equations

The diffusive flux of a species along the concentration gradient $\frac{\partial C}{\partial x}$ is expressed by the first law of Fick:

$$F = -D\, \frac{\partial C}{\partial x} \tag{1}$$

where the diffusion coefficient D can be written as:

$$D = D_0\, \exp\, (-E_a/kT) \tag{2}$$

D_0 is a pre-exponential factor and E_a is the activation energy. Using the transport equation and assuming that no material is generated or lost, the second law of Fick is derived in one dimension:

$$\frac{\partial C}{\partial t} = -\frac{\partial F}{\partial x} = \frac{\partial}{\partial x}\left(D\, \frac{\partial C}{\partial x}\right) \tag{3}$$

In three dimensions Eq. (3) becomes:

$$\frac{\partial C}{\partial t} = -\, \text{div}\, F \tag{4}$$

B. Classical Solutions to the Diffusion Equation

If the diffusion coefficient D is not a function of the concentration C or of the distance x,

Eq. (3) can be written as:

$$\frac{\partial C}{\partial t} = D \frac{\partial^2 C}{\partial x^2} \tag{5}$$

The solution to this diffusion equation will be different for different boundary conditions. Two extreme cases will be considered below[1]:

1. Diffusion with constant surface concentration

During diffusion the surface concentration C_s of the dopant can be kept constant, e.g., by using a classical predeposition technique where the surface concentration is equal to the maximum solubility of the dopant. In this case the solution to the diffusion equation is given by the complementary error function:

$$C(x,t) = C_s \, \text{erfc} \, \frac{x}{2\sqrt{Dt}} \tag{6}$$

with

$$C_s = C(o,t) = \text{constant} \tag{7}$$

and

$$Q(t) = \frac{2}{\sqrt{\pi}} \sqrt{Dt} \, C_s \tag{8}$$

$Q(t)$ represents the total amount of dopants per unit area, incorporated in the silicon.

2. Diffusion from a limited amount of dopants

If during diffusion the total amount of impurities Q is constant, the diffusion profile will be given by the Gaussian distribution:

$$C(x,t) = C_s(t) \, . \, \exp \, (- x^2/\sqrt{Dt}) \tag{9}$$

$$C_s(t) = \frac{Q}{\sqrt{\pi Dt}} \tag{10}$$

$$Q = \text{constant} \qquad (11)$$

In practice, if D is not a function of the concentration C or of x, most experimental profiles can be fitted to one of the two solutions given above. The complementary error function and the Gaussian distribution are compared in Fig. 1. Deviations from this first order theory will be discussed in the second part of this paper.

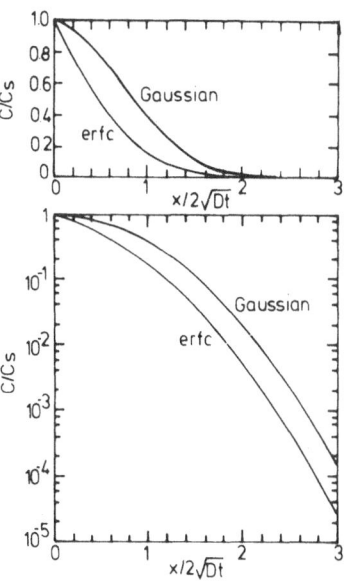

Fig. 1 - Comparison between the erfc-function and the Gaussian. After Grove.[1]

C. Diffusion Mechanisms

It is not within the scope of this paper to discuss in very much detail the physics of the various diffusion models. We prefer to focus our attention to a general description of the different mechanisms in order to provide a background for a good understanding of the engineering models used in process simulation programs such as SUPREM[2] or ICECREM.[3] For a more extended review on self-diffusion and impurity diffusion in semiconductors we refer to the excellent papers by Seeger and Chik,[4] Hu,[5] Shaw,[6] Willoughby,[7] and Gösele and Frank.[8]

All diffusion processes in silicon, i.e., self-diffusion, dopant or impurity diffusion, are governed by interactions between the diffusing species and point defects in the silicon lattice. It is not clear at all whether under thermal equilibrium the predominant intrinsic point defects in silicon are vacancies or self-interstitials. One school of thought argues in favor of the vacancy-model,[9,10] whereas the other group claims that the silicon self-interstitial is the native defect,[11,12] although some agreement exists that probably both types of point defects occur simultaneously.[10,12]

                                                                                    Declerck

It is also interesting to note that interstitials, or vacancies, or even both, are used to explain some of the diffusion related phenomena observed over the last decade. Oxidation-enhanced diffusion and stacking fault growth are generally attributed to an excess of silicon interstitials, whereas doping-enhanced oxidation is explained by an increase of the vacancy concentration.

Three models are widely used in silicon diffusion theory: the vacancy, the interstitial and the interstitialcy model. Each of these will now be briefly reviewed.

a. Vacancy Mechanism

When a diffusing atom jumps from one lattice site into an adjacent vacant lattice position, the atom is said to diffuse via a vacancy mechanism. This is illustrated in Fig. 2a. Since in thermal equilibrium the number of vacancies is rather low, diffusion via a vacancy mechanism is usually much slower than diffusion via an interstitial mechanism.

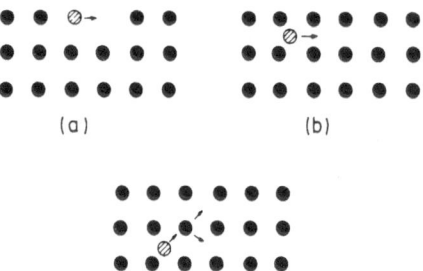

b. Interstitial Mechanism

Small atoms such as Ni do not occupy lattice sites in the silicon crystal but are located in the "interstices" or spaces between the

Fig. 2 - Schematic presentation of the vacancy diffusion mechanism (a), the interstitial mechanism, (b) and the interstitialcy mechanism.

silicon lattice positions. These atoms can diffuse through the crystal by jumping from one interstitial site to another, as illustrated in Fig. 2b. This interstitial mechanism is usually very fast.

Interstitialcy Mechanism

In this mechanism an interstitial atom first pushes one of the neighbouring lattice atoms into an interstitial position and then occupies this vacant site (Fig. 2c).

It has become clear during recent years that the interaction between the diffusing species

and the silicon point defects is strongly influenced by the charge state of the latter. This will

be addressed in the second part of this paper.

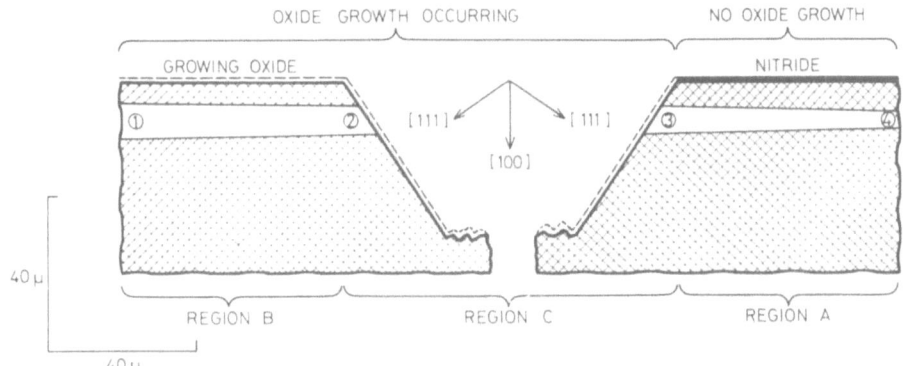

Fig. 3 - Marker layer structure developed to distinguish between the relative contributions of diffusion direction and orientation of the oxidizing surface to the apparent
anisotropy of dopant diffusion in oxidizing ambients. This figure shows a diagram of a
vertical cross section through the regions adjacent to the trench etched in region C.
After Hill.[13]

D. Marker Experiments and Intrinsic Diffusion Data

As mentioned above, the intrinsic diffusion data can only be obtained under well-

controlled experimental conditions where the point defect equilibrium in the silicon is not

disturbed by heavy doping, by the presence of a second dopant, or by the use of oxidizing

ambients. A detailed investigation of both the intrinsic diffusion behaviour of boron, phos-

phorus and arsenic, and of several diffusion anomalies, has recently been reported by Hill.[13]

His work is based on the use of a buried marker technique, which is illustrated in Fig. 3 and 4.

The buried marker has been fabricated by ion implantation in a low doped substrate $(2 \times 10^{14}$

$cm^{-3})$, followed by epitaxial growth of a $10\mu m$ thick layer with a doping level of 2×10^{15} cm^{-3}.

In this way a thin, well-defined layer of silicon doped with the marker dopant, and extending

laterally throughout the slice at a constant depth, is obtained, while the substrate and the

epitaxial layer are uniformly doped at a lower concentration with a dopant of opposite type.

Any heat treatment of the structure will cause the marker dopant layer to broaden by diffusion, and the extent of this broadening is used to calculate the diffusion coefficient of the marker dopant.

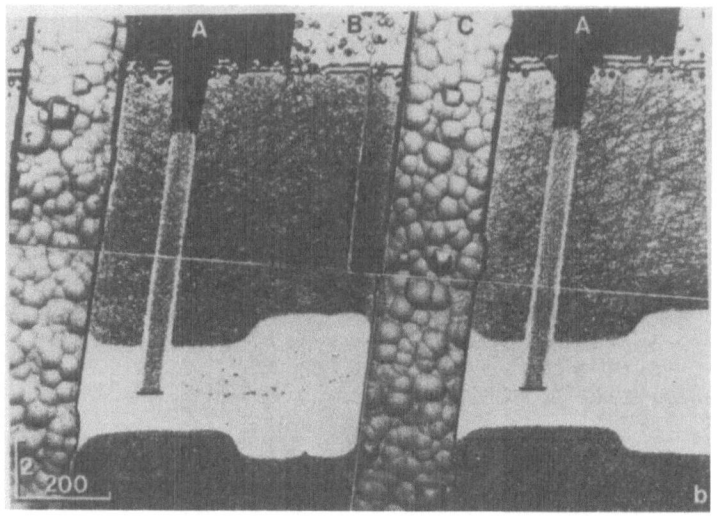

Fig. 4 - Bevelled and stained marker structure, after 16 hours at 977°C in dry oxygen. Prior to heat treatment, a trench bounded by (111)-planes was etched in region C (Fig. 3), using a selective etch. Substrate orientation in (100). Dimension bars in microns. After Hill.[13]

Furthermore, by means of deposition, photolithography and etching techniques parallel strips of bare silicon (region C in Fig. 3), deposited SiO_2 (region B) and deposited Si_3N_4 (region A) are defined, allowing the simultaneous study of diffusion under oxidizing and non-oxidizing conditions. On some wafers the unprotected region C was selectively etched to a depth of about $50\mu m$, resulting in a deep trench with sides of different crystallographic orientation. These test structures were used to study the orientation dependence of the oxidation enhanced diffusion effect. One of the very interesting conclusions of this work is that the diffusion behavior in silicon is isotropic and that the differences in diffusion coefficients observed under oxidizing conditions are entirely associated with the orientation of the adjacent crystal plane being oxidized as illustrated in Fig. 4.[13]

For a comparison of intrinsic diffusion data from different authors, we refer to Ref. 13.

II. DEVIATIONS FROM FIRST ORDER THEORY

In this part of the paper several diffusion anomalies which cannot be modeled adequately by means of the first order theory described in previous sections will be discussed.

A. Field Enhancement

Very early it was found that at high doping concentrations the diffusion behavior is considerably enhanced. In an attempt to explain this concentration effect Lehovec and Slobodskoy calculated the effect of the internal electric field on the motion of charged impurities.[14] The first law of Fick, in the case of a positively charged species, is then modified as:

$$F = -D_i \frac{\partial C}{\partial x} + \mu E C \qquad (12)$$

Where E is the electric field, D_i is the intrinsic diffusion coefficient and μ is the mobility of the dopant.

By solving the Poisson equation the electrostatic potential and the built-in electric field can be calculated, leading to an effective diffusion coefficient given by:

$$D = D_i \left[1 + \frac{C}{(C^2 + 4 n_i^2)^{1/2}} \right] \qquad (13)$$

The field enhancement effect is illustrated in Fig. 5 showing the effective diffusion coefficient as a function of doping concentration with the diffusion temperature as parameter. As can be seen, the maximum value of the enhancement is 2, which is clearly too small to explain the enhancement experimentally observed.

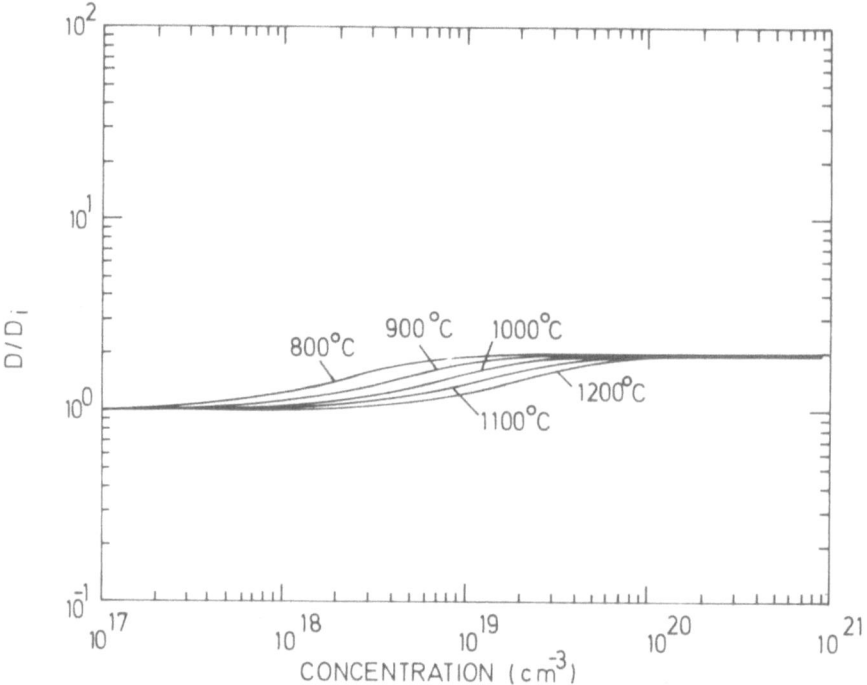

Fig. 5 - Field enhancement of the diffusion coefficient according to Eq. (13). After Ryssel *et al..*[3]

B. Point Defect Enhancement

One of the most interesting aspects of present diffusion theories is the strong interaction between the diffusing species and the various charge states of the point defects. Not so much is known about the energy levels induced by the presence of silicon interstitials. Vacancies, however, are known to occur in a neutral state (V^0), a singly negative (V^-), a doubly negative ($V^=$) and a positive charge state (V^+).[15] The diffusion coefficient of these different charge states may in general differ from each other but, and this should be emphasized, the relative densities of the various states depend on the position of the Fermi level in the bandgap and thus on the doping level of the silicon crystal. This so-called "Fermi-level" effect is of utmost importance for a good understanding of a variety of diffusion anomalies observed in silicon diffusion studies.

The transition from a neutral vacancy to, e.g., a singly negative charge state is modeled by the ionization reaction:

$$V^0 + e^- = V^-$$ (14)

The equilibrium for this reaction is given by

$$K(T) = \frac{[V^-]}{[V^0].n}$$ (15)

where n is the electron concentration and K(T) is the reaction constant which is a function of temperature only (and of the position of the vacancy energy level relative to the conduction band edge.) The concentration of neutral vacancies $[V^0]$ does not depend on the Fermi-level, but is only a function of temperature.[16] Equation (15) can be written for the intrinsic condition $n = n_i$:

$$K(T) = \frac{[V^-]_i}{[V^0].n_i}$$ (16)

Equating (15) and (16) leads to:

$$[V^-] = [V^-]_i \frac{n}{n_i}$$ (17)

It can be shown in a similar way that the concentration of a singly ionized donor vacancy $[V^+]$ is related to its concentration under intrinsic conditions by:

$$[V^+] = [V^+]_i \frac{n_i}{n}$$ (18)

For a doubly ionized acceptor state Eq. (15) becomes[6]:

$$[V^=] = [V^-]_i \left(\frac{n}{n_i}\right)^2$$ (19)

In a vacancy diffusion model the total diffusion is now considered as the simultaneous effect of the diffusivities associated with the various charge states of the vacancies. If D^0, D^-,

$D^=$ and D^+ represent the effective contribution under intrinsic conditions from the neutral, the singly and doubly negative and the positive charge state, respectively, the total diffusivity is written as:

$$D_i = D^o + D^- + D^= + D^+ \tag{20}$$

Each of these effective diffusivities is composed of the fractional concentration of that particular charge state, multiplied by the appropriate diffusion constant.[6] For the extrinsic condition the diffusivity is given by:

$$D = D^o + D^- \frac{[V^-]}{[V^-]_i} + D^= \frac{[V^=]}{[V^=]_i} + D^+ \frac{[V^+]}{[V^+]_i} \tag{21}$$

Using Eqs. (17), (18), and (19) this can be written as:

$$D = D^o + D^- \frac{n}{n_i} + D^= \left(\frac{n}{n_i}\right)^2 + D^+ \frac{n_i}{n} \tag{22}$$

It should be remarked that Eq. (22) can be extended to situations where interstitials and vacancies are present in different ionized states.[6]

Figure 6 shows the experimental diffusivity of arsenic as a function of arsenic concentration, at several temperatures and anneal conditions.[17] As indicated in the figure by the solid lines, the experimental points closely fit a diffusion behavior expressed by

$$D = D^o + \frac{n}{n_i} D^- \tag{23}$$

The vacancy enhancement of the diffusion coefficient discussed here gives a good explanation for the enhanced diffusion behavior observed at high concentrations of arsenic and phosphorus. It should be noticed that this effect enables the fabrication of shallow junctions having very steep gradients. This is illustrated in Fig. 7 where shallow ion-implanted arsenic profiles are shown after various heat treatments.[18] The agreement between experiment and computer simulation is excellent.

The intrinsic concentration n_i used in the previous equations is the intrinsic concentration at diffusion temperatures. Figure 8 shows n_i (T) as used in the SUPREM II program. It is worth noticing that at a diffusion temperature of 900°C the material behaves as extrinsic for doping levels higher than 5×10^{18} cm^{-3}.

Equations (22) and (23) serve as basis for the computer models

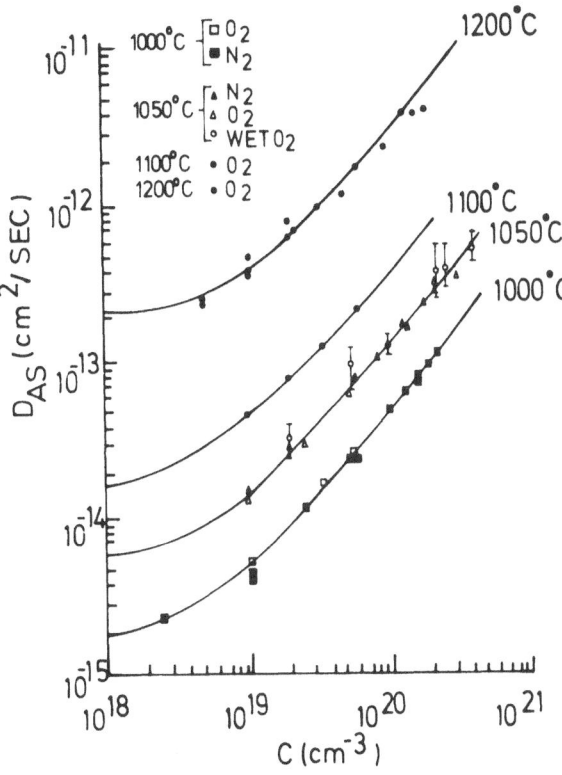

Fig. 6 - The diffusivity of arsenic in silicon as a function of arsenic concentration at several temperatures and annealing ambients. After Fair.[17]

used in process simulation.[2,3] If it is assumed that only the neutral and the singly charged states are involved in the diffusion mechanism, the diffusivity can be modeled by[2]:

$$D = D_i \frac{1 + \beta f}{1 + \beta} \tag{24}$$

where D_i is the intrinsic diffusion coefficient, measured at $n = n_i$. The parameter f is defined by

$$f = \frac{n}{n_i} \tag{25}$$

for a donor impurity, which is assumed to diffuse via neutral and singly negative charge state vacancies and by

$$f = \frac{n_i}{n} \quad (26)$$

for an acceptor impurity, diffusing via neutral and singly
positive charge state vacancies.

The parameter β ex-
presses the effectiveness of
diffusion via the charged
state as compared with diffusion via the neutral state:

$$\beta = \frac{D^-}{D^0} \text{ or } \beta = \frac{D^+}{D^0} \quad (27)$$

Equation (24), as used in
SUPREM, is illustrated in
Fig. 9. The β-values in SU-
PREM are $\beta = 3$ for boron
and $\beta = 100$ for arsenic.

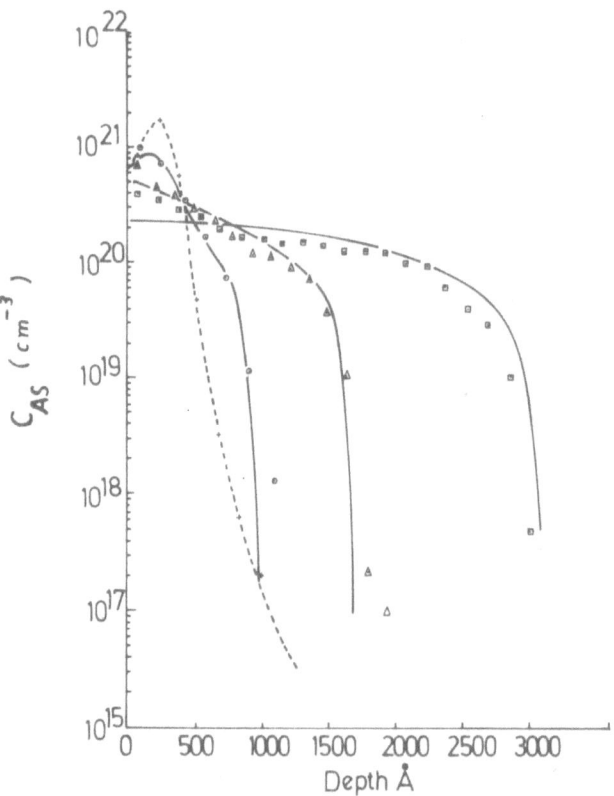

Fig. 7 - Comparison of experimental and computer calculated
arsenic diffusion profiles.
+++ : dose 5×10^{15} cm^{-2}, 30 mins in argon at 725°C.
ooo : dose 3.8×10^{15} cm^{-2}, 30 mins in argon at 900°C.
▲▲▲ : dose 3.9×10^{15} cm^{-2}, 30 mins in argon at 950°C.
⊡⊡⊡ : dose 4.5×10^{15} cm^{-2}, 30 mins in argon at 1000°C.

For phosphorus diffusion a more complicated model, based on the work by Fair and Tsai,
is used.[19] It is assumed that in the strongly doped surface region phosphorus forms
$P^+V^=$-pairs with doubly ionized vacancies. These $P^+V^=$-pairs dissociate at a depth where the
Fermi level drops below 0.11 eV from the conduction band, which is the energy level of the
doubly negatively charged vacancy. The $P^+V^=$-pair dissociation causes a flow of V^--vacancies
into the bulk leading to an enhanced phosphorus tail diffusion. A direct result of this diffusion
enhancement is the characteristic kink in the phosphorus profile, as illustrated in Fig. 10. The
generation of excess vacancies also enhances the diffusion of other dopants such as boron and

is believed to be responsible for the well-known emitter dip effect found in npn structures. For a detailed discussion on this model we refer to Ref. (19).

At phosphorus concentrations higher than $3 - 4 \times 10^{20}$ cm^{-3} the tail diffusivity decreases. According to Fair this effect may be attributed to the stress induced on the lattice by the high dopant concentration. An additional stress effect may be due to permanent lattice disorder produced by

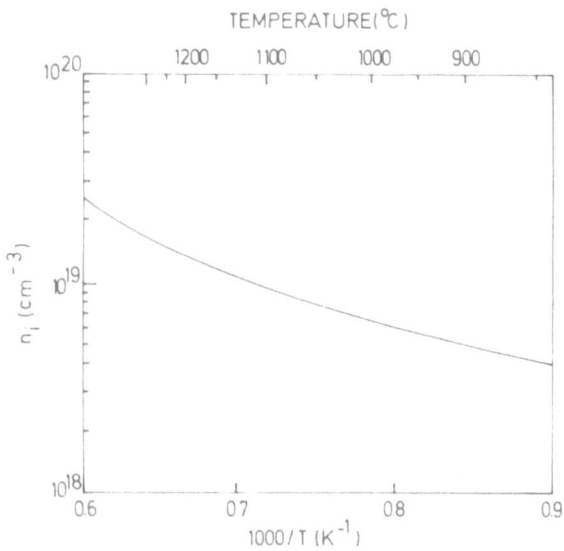

Fig. 8 - Intrinsic carrier concentration in silicon vs. temperature. After Antoniadis *et al.*[2]

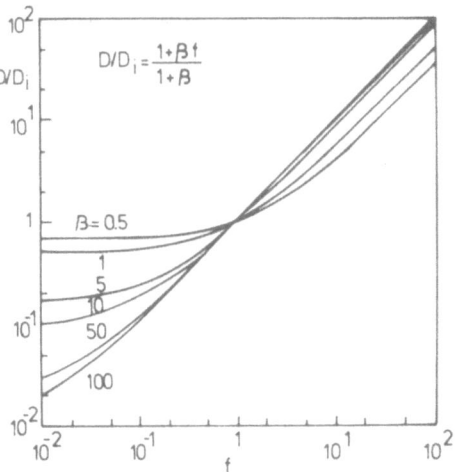

Fig. 9 - Normalized diffusivity vs. normalized carrier concentration for different values of β. After Antoniadis *et al.*[2]

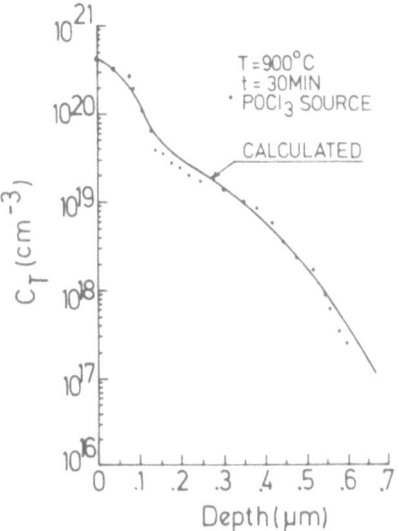

Fig. 10 - Calculated and measured total phosphorus profiles in silicon. After Fair *et al.*[19]

high dose implantations. Both stress effects can be modeled by a bandgap narrowing - ΔE_g, which effectively increases the intrinsic carrier concentration n_i^2:

$$n_i = n_i \text{ (unstressed)} \quad \exp . \; (-\Delta E_g/kT) \tag{28}$$

C. Clustering at High Dopant Concentrations

At high dopant concentrations clusters or precipitates of dopant atoms may form, resulting in a smaller number of substitutional dopants, active in the diffusion process. This effect is particularly important for arsenic[20,21] and for boron[22] at high doping levels. Modeling of the clustering mechanism is done by expressing the thermodynamic equilibrium of the clustering reaction, which for arsenic can be written as:

$$m \; A_s \; \overset{K_{eq}}{} \; As_m \tag{29}$$

m is the number of arsenic atoms in a cluster and K_{eq} is the arsenic equilibrium clustering coefficient. The cluster concentration $[As_m]$ is given by:

$$[As_m] = K_{eq} \, [As]^m \tag{30}$$

where $[As]$ is the concentration of substitution-al arsenic atoms, which is also the donor con-centration at room temperature. The total ar-senic concentration N_T is:

$$N_T = [As] + m \, [As_m] \tag{31}$$

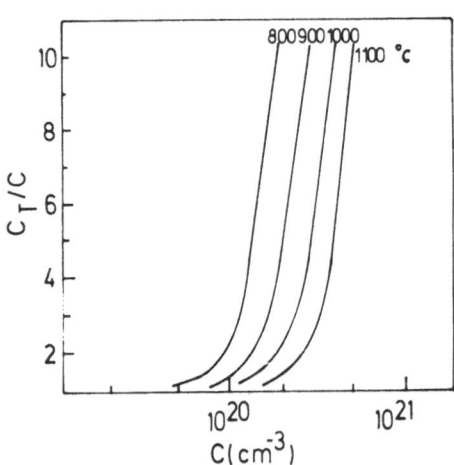

Fig. 11 - Normalized total arsenic con-centration vs. the substitutional arsenic concentration, resulting from clustering model, as a function of process temper-ature, under thermal equilibrium. Af-ter Antoniadis *et al.*[2]

or $N_T = [As] + m \, K_{eq} \, [As]^m \tag{32}$

Equation (32) is plotted in Fig. 11 for the case

of arsenic with a cluster size m equal to 4, as

used in SUPREM.[2] It is clear that the total number of substitutional arsenic atoms reaches a maximum as soon as the clustering effect starts. In ICECREM, a cluster size of 2 is used for arsenic and 12 for boron.[3]

It has to be recognized that arsenic clustering at lower temperatures (800°C) can considerably decrease the conductivity of shallow n^+-layers.[20,21] As the clustering and declustering rates at these low temperatures are quite slow, reaction (29) no longer reaches equilibrium and the kinetics of clustering and declustering have to be taken into account. Equation (29) has to be replaced by the kinetic equation of As clustering:

$$\frac{d}{dt} [As_m] = K_c [As]^m - K_d [As_m] \tag{33}$$

where K_c and K_d are the clustering and declustering coefficients, respectively.

D. Oxidation Enhanced Diffusion

It has been observed by several people that the diffusivity of boron, phosphorus and arsenic is considerably enhanced under oxidizing conditions.[23-27] The oxidation enhanced diffusion effect (OED) is more pronounced for (100)-oriented surfaces than for (111)-surfaces, and depends on temperature and oxidation time.

The OED-effect is generally attributed to the generation of excess silicon interstitials at the silicon-SiO_2 interface by the oxidation reaction, as suggested first by Hu.[24] Due to the incompleteness of oxidation, the silicon interface acts as a source of silicon interstitials which will diffuse into the silicon bulk material and enhance the diffusivity of most common dopants through an enhancement of the interstitialcy component of the diffusion mechanism.

It has been shown recently by Hill[13] that the activation energy of the diffusion enhancement is 2.31 ± 0.08 eV, which is exactly the same activation energy as observed for the growth of oxidation induced stacking faults (OFS's).[29] It should be noted that the linear oxidation rate (B/A) has a similar activation energy of 2.3 eV. All of this strongly indicates that both the OED-effect and the OSF-growth are controlled by the same point defects, injected into the silicon substrate by the oxidation mechanism. The correspondence between the OED-effect and the OSF-growth is even more striking since at high temperatures and for long oxidation times a diffusion retardation is observed, together with a stacking fault shrinkage.[18]

A numerical model for the OED-effect has recently been proposed.[27,28] If D_{ox} and D_i represent, respectively, the diffusivities in an oxidizing and neutral ambient, the diffusion enhancement can be written as[28]:

$$\frac{D_{ox}}{D_i} - 1 = G_E \cdot F(t) \qquad (34)$$

with G_E the diffusion enhancement factor, given in Fig. 12 as a function of temperature. $F(t)$ is a factor which expresses the reduction with time (Fig. 13). It should also be mentioned that Antoniadis *et al.* have shown that the diffusion coefficient can be expressed as:

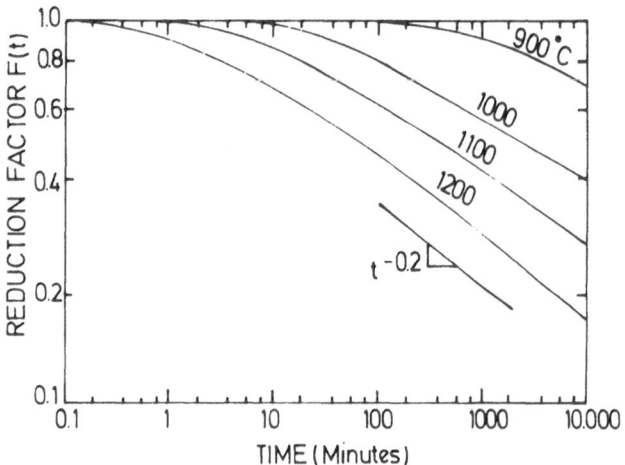

Fig. 12 - The reduction factor $F(t)$ is calculated as a function of oxidation time. Oxidations used in the calculations start without initial oxide. After Lin *et al.*[28]

$$D_{ox} = D_i + d_I \cdot K\left(\frac{dX}{dt}\right)^n \qquad (35)$$

with d_I a proportionality factor accounting for the diffusivity via the interstitialcy mechanism, K a reaction constant and $\frac{dX}{dt}$ the oxidation rate. The value for the power exponent n has been found to lie between 0.4 and 0.6, which is similar to the value found for the relation between OSF-growth and oxidation rate.[30]

Finally, it should be pointed out that in a chlorine-containing oxidation ambient the diffusion enhancement is reduced as shown in Fig. 14.[31,13] This is explained by a reduction of the concentration of silicon interstitials at the interface due to the formation of Si-Cl bonds, which is known to speed up the oxidation rate in chlorine-oxidations.

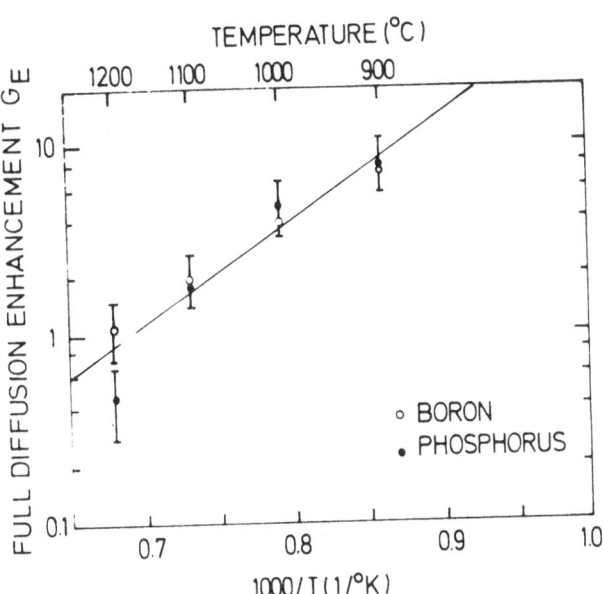

Fig. 13 - The maximum diffusion enhancement factors G_E for boron and phosphorus at different oxidation temperature. After Lin *et al.*[28]

CONCLUSION

The present diffusion models have been reviewed with emphasis on the role of the point defects in the silicon. In the future, a good understanding of the influence of the various processing steps on the silicon interstitial and silicon vacancy distribution will be necessary to enable an accurate modeling of the

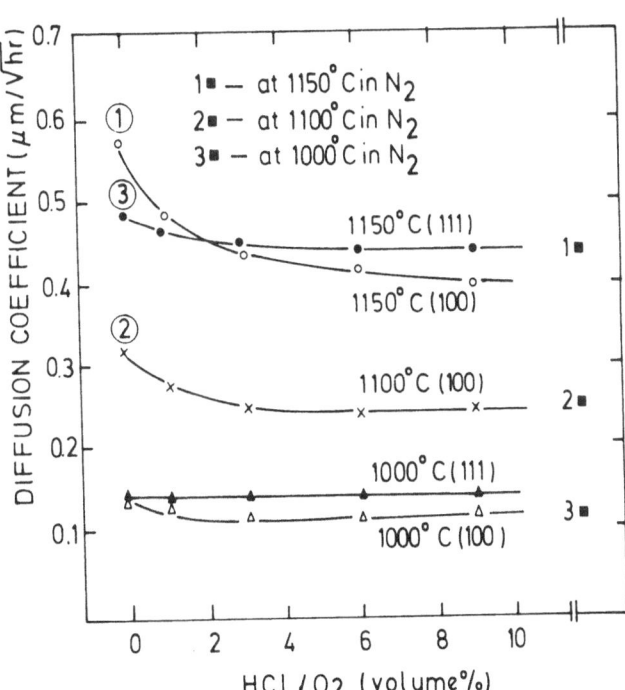

Fig. 14 - Diffusion coefficient of phosphorus in silicon at different temperatures and HCl volume concentrations. After Nabeta *et al.*[31]

structures needed in VLSI-technology.

REFERENCES

1.   A. S. Grove, John Wiley & Sons, 1967.

2.   D. A. Antoniadis and R. W. Dutton, IEEE Journal of S.S. Circuits 14, 412, (1979).

3.   H. Ryssel, K. Haberger, K. Hoffmann, G. Prinke, R. Dümcke and A. Sachs, IEEE Trans. on Electron Devices 27, 1484 (1980).

4.   A. Seeger and K. P. Chik, Phys. Status Solidi 29, 455, (1968).

5.   S. M. Hu in "Diffusion in Semiconductors," D. Shaw, editor (Plenum Press, London 1973) 217.

6.   D. Shaw, Phys. Status Solidi (b) 72, 11 (1975)

7.   A. F. W. Willoughy, Rep. Prog. Phys. 41, 1665 (1978).

8.   U. Gösele and W. Frank in "Defects in Semiconductors," J. Narayan and T. Y. Tan, editors (North-Holland, Amsterdam 1981).

9.   J. A. Van Vechten, Phys. Rev. B 17, 3197 (1978).

10.   R. B. Fair, J. Appl. Phys. 51, 5828 (1980).

11.   A. Seeger, W. Frank and U. Gösele in "Defects and Radiation Effects in Semiconductors," 1978 (Institute of Physics, London) p. 148.

12.   U. Gösele, F. Morehead, H. Föll, W. Frank and H. Strunk, Silicon Semiconductor 1981, edited by the Electrochem. Soc., Pennington 1981, p. 766.

13.   C. Hill, Silicon Semiconductor 1981, edited by the Electrochem. Soc., Pennington 1981, p. 988.

14.   K. Lehovec and A. Slobodskoy, Solid State Electrons 3, 45 (1961).

15.   J. A. Van Vechten and C. D. Thurnmond, Phys. Rev. B 14, 3539 (1976).

16.   W. Schockley and J. L. Moll, Phys. Rev. 119, 1480 (1960).

17.   R. B. Fair, Sensors and Actuators 1, 305 (1981).

18.   C. Hill, Lecture presented at the DINEMITE summer course, Leuven 1980.

19.   R. B. Fair and J. C. C. Tsai, J. Electrochem. Soc. 124, 1107 (1977).

20.   M. Y. Tsai, F. F. Morehead, J. E. E. Baglin and A. E. Michel, J. Appl. Phys. 51, 3230 (1980).

21. R. B. Fair, Silicon Semiconductor 1981, edited by the Electrochem. Soc., Pennington 1981, p. 963.

22. H. Ryssel, K. Müller, K. Haberger, R. Henkelman and F. Jalmel, Appl. Phys. 22, 35 (1980).

23. G. Masetti, S. Solmi and G. Soncini, Solid State Electron 16, 1419 (1973).

24. S. M. Hu, J. Appl. Phys. 45, 1567 (1974).

25. D. A. Antoniadis, A. G. Gonzales and R. W. Dutton, J. Electrochem. Soc. 125, 813 (1978).

26. D. A. Antoniadis, A. M. Lin and R. W. Dutton, Appl. Phys. Lett 33, 1030 (1978).

27. K. Taniguchi, K. Kurosawa and M. Kashiwagi, J. Electrochem. Soc. 127, 2243 (1980).

28. A. M. Lin, D. A. Antoniadis and R. W. Dutton, J. Electrochem. Soc. 128, 1131 (1981).

29. S. P. Murarka, Phys. Rev. B 16, 2849 (1977).

30. A. M. Lin, R. W. Dutton, D. A. Antoniadis and W. A. Tiller, J. Electrochem. Soc. 128, 1121 (1981).

31. Y. Nabeta, T. Uno, S. Kubo, and H. Tsukamoto, J. Electrochem. Soc. 123, 1416 (1976).

ION IMPLANTATION

Heiner Ryssel

Fraunhofer-Institut für Festkörpertechnologie
Paul-Gerhardt-Allee 42
8000 München 60
Germany

1. INTRODUCTION

Accelerated ions injected into a solid will lose their energy in a series of collisions with electrons and nuclei of the target. These ions are "implanted" into the solid, and the process is called ion implantation.

Ion implantation was invented very early after the discovery of the bipolar transistor. The basic patent of Shockley[1.1] already describes virtually all aspects of ion implantation. Nevertheless, it took a long time (until the end of the 60's) before this technique was introduced into device manufacturing. In 1962, the first real devices, nuclear detectors, were fabricated by phosphorus implantation,[1.2] and solar cells were also fabricated as early as 1963.[1.3] At the same time, the theoretical background for ion implantation, the so-called LSS theory, was developed by Lindhard, Scharff and Schiøtt.[1.4] From about 1970 on, the implantation technique was applied more and more in semiconductor technology. At first, threshold-adjust and self-aligned source and drain structures were fabricated, and later, bipolar transistors as well.[1.5-1.7]

In only about one decade, ion implantation has evolved from a laboratory curiosity to a standard production process for integrated circuits. The reasons for this are threefold:

1.   Ion implantation produces extremely homogeneous and reproducible doping concentrations through an on-line measuring of the implanted ion current.

2.   Ion implantation fits well into silicon planar technology. The oxide layers used for masking against diffusion can be used to mask against the ion beam. Furthermore, ion implantation can be performed through thin passivating layers (e.g., SiO_2, Si_3N_4), or using photoresist masks.

3.   Ion implantation is a low-temperature process (although an annealing step is usually required to recrystallize the damaged lattice).

The main problems of ion implantation - the range of the implanted ions depending on their mass and energy, the damage introduced into the crystal lattice during the implantation and the annealing process - will be treated in the following chapters together with applications of this technique.

2. RANGE DISTRIBUTIONS OF IMPLANTED IONS

The first calculations of the range of accelerated particles in solids were performed by Lenard, Rutherford and Bohr at the beginning of this century. If fast particles interact with solids, different processes take place which slow down the particles:

1.   inelastic impacts with bound electrons leading to ionization

2.   inelastic impacts with nuclei leading to bremsstrahlung, nuclear reactions or excitations

3.   elastic impacts with bound electrons

4.   elastic impacts with atoms leading to a partial transfer of kinetic energy

5.   emission of Cherenkov radiation

At high energies, only electronic stopping (inelastic impacts leading to ionization) is of any importance; at the low energies (up to several 100 keV) relevant for ion implantation, electronic stopping as well as nuclear stopping (elastic impacts with atoms) have to be taken into account.

In Fig. 2.1, the energy dependence of electronic and nuclear stopping is given schematically. The characteristic energies E_1, E_2, E_3 are mass dependent; values for the main doping elements in silicon are given in Table 2.1.

In the following, the LSS theory according to Lindhard, Scharff and Schiøtt[2.1] which is usually used for range calculations in the field of ion implantation, as well as a newly-developed theory,[2.2] will be treated briefly.

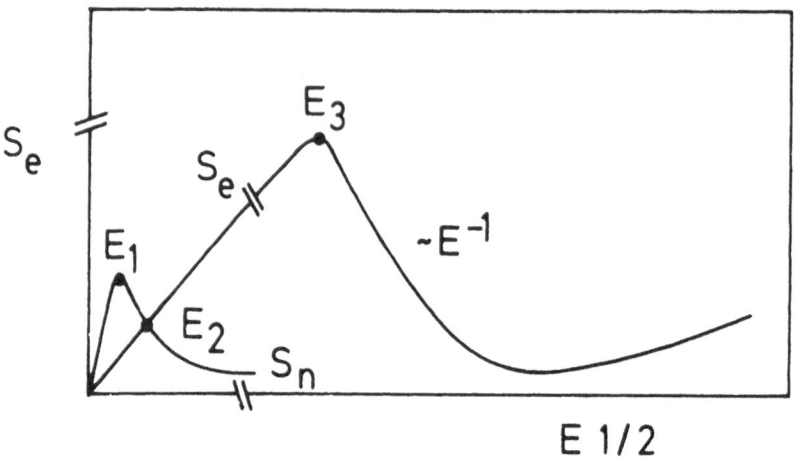

Fig. 2.1 - Energy dependence of electronic and nuclear stopping (schematic).

2.1 <u>LSS Theory</u>

Lindhard, Scharff and Schiøtt[2.1] assumed that the energy losses in electronic and nuclear colli-sions are separable.

TABLE 2.1   Characteristic energies E_1, E_2, E_3 according to Fig. 2.1

Ion	E_1 (keV)	E_2 (keV)	E_3 (keV)
B	3	17	3000
P	17	140	30000
As	73	800	300000
Sb	180	2000	300000

The calculation of the nuclear stopping can easily be performed assuming a sum of independent two-body elastic collisions. In this case, the problem is well known in classical mechanics. In the theory, the Thomas-Fermi potential function is used to describe the atomic collisions, although other potential functions, such as the Lens-Jensen or the Moliere poten-tials, seem to provide an improved description.

The stopping power is proportional to the atomic density and to the sum of the energy transferred in all single collisions:

$$S_n(E) = -\frac{1}{N} \left(\frac{dE}{dx} \right)_n = \int_0^{T_m} T_n d\sigma \qquad (2.1)$$

with $d\sigma$ the differential interaction cross section, N the atomic density, T_m the maximum transferable energy and T_n the transferred energy.

T_m is given by

$$T_m = 4 \frac{M_1 M_2}{(M_1 + M_2)^2} E \qquad (2.2)$$

where M_1 and M_2 are the masses of ion and target, respectively. The calculation of T_n is a classical problem of mechanics. It holds:

$$T_n = E \frac{2 M_1 M_2}{(M_1 + M_2)^2} (1 - \cos \phi) \qquad (2.3)$$

ϕ is the scattering angle in the center of the mass system, and can be calculated assuming a suited interaction potential. Usually, this is only possible numerically.

The electronic stopping is calculated assuming a free electron gas. In this case, the stopping cross section is proportional to the velocity of the ions, and is therefore proportional to the square root of the energy:

$$S_e(E) = -\frac{1}{N} \left(\frac{dE}{dx} \right)_e = k E^{1/2} \qquad (2.4)$$

The constant k depends on the atomic weights and number of ion and target, and is given by:

$$k = \xi_e \frac{a^{3/2} q Z_1 Z_2 N^2}{(Z_1^{2/3} + Z_2^{2/3})^{3/4} A_1^{3/2}} \qquad (2.5)$$

a is the screening parameter (of the order of the Bohr radius) and ξ_e is a dimensionless constant of the order of $Z_1^{1/6}$.

For numerical calculations, the formulations by Sanders[2.3] for the LSS theory are usually used. One obtains integro-differential equations for range and range straggling, which lead to recursive integral equations which can be solved for the range parameters.

Calculations for this theory have been performed by several authors, e.g. Gibbons *et al.*[2.4] and Smith,[2.5] and are available in the form of tables.

2.2 Biersack Theory

Other approaches for the calculation of range distributions are to solve Boltzmann transport equations, to use Monte Carlo techniques or to use the diffusion model of Biersack.[2.2] The latter leads to very simple expressions, and will therefore be treated here in addition to the well-known LSS theory.

In the Biersack theory, the mean directional cosine of the ion motion during the slowing-down process is calculated. With each collision, the ion loses energy and at the same time changes direction. The nuclear energy loss is directly related to the deflection angle of the ion (see Eq. 2.3). Consequently, the ion will deviate, in average, more and more from its original direction. One can represent the directions of ion motion by polar and azimuthal angles and depict them as points on a unit sphere. Since the direction of the motion changes at random with each collision, the stochastic motion on the unit sphere is governed by a diffusion process such as a Brownian motion.

These arguments lead to the diffusion equation:

$$\frac{dw}{d\tau} = \frac{\partial}{\partial \eta} \left[(1 - \eta^2) \frac{dw}{d\eta} \right]$$

(2.6)

where w is the distribution function, η is the directional cosine and τ is equivalent to Dt in ordinary diffusion.

For calculating ranges, it is not necessary to know w explicitly, but rather, to calculate an average value of the directional cosine $\bar{\eta}$, which is found to be:

$$\bar{\eta} = e^{-2\tau}$$

(2.7)

The relation between τ and the energy loss is obtained using the Einstein equation:

$$\tau = -\frac{M_2}{4M_1} \int_{E_o}^{E} \frac{S_n}{S_n + S_e} \frac{dE}{E}$$

(2.8)

leading to equations for the range R and the projected range R_p:

$$R = \int_{E_o}^{0} ds = \int_{0}^{E_o} \frac{dE}{S_e + S_n} \qquad (2.9)$$

$$R_p = \int_{E_o}^{0} \bar{\eta} ds = \int_{0}^{E_o} e^{-2\tau} \frac{dE}{S_e + S_n} \qquad (2.10)$$

Each pathlength of the ion trajectory is projected on the x-axis by multiplying with the directional cosine. The projected range can be obtained by first obtaining τ through Eq. (2.8), and inserting into Eq. (2.10). An algorithm, however, was developed by Biersack to give R_p, ΔR_p and the lateral spread $\Delta R_{p,L}$.

These practical equations are:

$$R_p (E + \Delta E) = R_p(E) \left(1 - \frac{M_2}{2M_1} \frac{S_n}{S_n + S_e} \frac{\Delta E}{E}\right) + \frac{\Delta E}{S_n + S_e} \qquad (2.11)$$

$$\xi(E + \Delta E) = \xi(E) + \frac{2R_p}{S_e + S_n} \Delta E$$

$$\Delta R_{p,L}^2 (E + \Delta E) = \Delta R_{p,L}^2(E) + \left[\xi(E) - 2\Delta R_{p,L}^2 (E)\right] \frac{M_2}{M_1} \frac{S_n}{S_n + S_e} \frac{\Delta E}{E}$$

with
$$\xi = R_p^2 + \Delta R_p^2 + \Delta R_{p,L}^2.$$

The Eqs. (2.11) can also be programmed easily on a pocket calculator for fast iterative calculations of R_p, ΔR_p and $\Delta R_{p,L}$. The theory is also extendable to give higher moments of ion distributions, but in this case, the expressions are far more complicated.

Calculations of range parameters using this theory are given for several important ions in silicon, SiO_2, Si_3N_4 and photoresist in the appendix.

2.3 Range Profiles of Implanted Ions

According to the classical LSS theory as well as the alternative Biersack approach,[2.1,2.2] the range profiles are given by Gaussian distributions. A Gaussian distribution is described by two moments, the projected range R_p and the projected standard deviation or straggling ΔR_p. Together with the implanted dose $N_\Box$ they describe the implanted profile by

$$C(x) = \frac{N_\Box}{\sqrt{2\pi}\ \Delta R_p}\ \exp\left[-(x-R_p)^2/(2\Delta R_p^2)\right] \qquad (2.12)$$

x is the distance measured along the axis of incidence. Gaussian distributions are very useful for fast estimation of the range distribution of implanted ions, or for calculating the thickness of masking layers. Many experimental investigations have shown, however, that this simple description is not adequate for most ions in silicon and other semiconductors. It has been argued that this might be due to channeling because of the crystalline structure of the usual semiconductors. It has been found, however, that the profiles of many ions are assymetrical, in amorphous targets as well, and higher moments have to be used to construct range distributions.

Theoretical results on higher moments can be extracted from both of the theories described, but these calculations are far too complicated to be treated here.

A particularly useful distribution is the Pearson type IV function[2.6] with four moments. The advantages of Pearson distributions in comparison to other distributions, e.g., joint half-Gaussian distributions[2.7] or Edgeworth distributions,[2.8] are that the Pearson distributions have no negative values and have a single maximum. Moreover, they can also model residual channeling tails, which are still present even if a proper misalignment has been used.[2.9]

The Pearson distribution of type IV centered around the projected range R_p is given by:

$$f(x) = K \left[b_2 (x-R_p)^2 + b_1 (x-R_p) + b_0\right]^{\frac{1}{2b_2}}. \qquad (2.13)$$

$$\exp \left[-\frac{\dfrac{b_1}{b_2} + 2a}{\sqrt{4b_2 b_0 - b_1^2}} \; \arctan \; \frac{2b_2 (x-R_p) + b_1/b_2}{\sqrt{4b_2 b_0 - b_1^2}} \right]$$

where K is a constant necessary for normalizing the distribution. The four other constants a, b_0, b_1 and b_2 are given by:

$$a = -\frac{\Delta R_p \, \gamma(\beta + 3)}{A}$$

$$b_0 = -\frac{\Delta R_p^2 \, (4\beta - 3\gamma^2)}{A}$$

$$b_1 = a \qquad (2.14)$$

$$b_2 = -\frac{2\beta - 3\gamma^2 - 6}{A}$$

where $A = 10\,\beta - 12\gamma^2 - 18$.

The four constants a, b_0, b_1 and b_2 can be expressed by four moments μ_1, μ_2, μ_3 and μ_4 of the distribution f(x). The first moment μ_1 is well known as the average projected range

$$\mu_1 = R_p = \int_{-\infty}^{\infty} x \, f(x) \, dx \qquad (2.15)$$

The three higher moments μ_i are given by

$$\mu_i = \int_{-\infty}^{\infty} (x-R_p)^i \, f(x) \, dx, \quad i = 2,3,4 \qquad (2.16)$$

It is customary to use the standard deviation ΔR_p, which is defined by the square root of the second moment μ_2, and dimensionless expressions for the higher moments:

standard deviation $\qquad\qquad\qquad \Delta R_p = \sqrt{\mu_2}$

skewness
$$\gamma = \frac{\mu_3}{\Delta R_p^3} \qquad (2.17)$$

and kurtosis
$$\beta = \frac{\mu_4}{\Delta R_p^4}$$

The skewness γ indicates the titling of the profile, and the kurtosis β indicates the flatness at the top of the profile. The relation between the third and fourth moments has to be chosen to satisfy

$$\beta \leq \beta_{min} = \frac{48 + 39\gamma^2 + 6(\gamma^2 + 4)^{3/2}}{32 - \gamma^2} \qquad (2.18)$$

in order to give a Pearson distribution of type IV.

The maximum of the Pearson distribution occurs at $x = R_p + a$ unless $\gamma = 0$. For a skewness of zero, a Gaussian profile results. For a negative skewness, the peak is deeper than R_p and the distribution falls off more rapidly for $x > a$ than for $x < a$. For a positive skewness, the opposite is true.

The kurtosis is 3 for a Gaussian distribution. The limit is represented by Eq. (2.18). A universal expression for the kurtosis is given by Gibbons.[2.10]

$$\beta = 2.8 + 2.4\gamma^2 \qquad (2.19)$$

In Figs. 2.2 to 2.10, some examples of experimentally-determined distributions, and the moments extracted from these distributions, are given for boron in silicon, SiO_2 and Si_3N_4, in comparison to theoretical calculations.[2.11] Figure 2.2 shows range profiles of boron in silicon obtained by the $^{10}B(n,\alpha)^7Li$ nuclear reaction. The tails are also perfectly fitted by the Pearson distributions. These tails could be caused by channeling (see Chapter 2.7). The moments extracted from these profiles are given in Figs. 2.3 and 2.4, in comparison to theoretical calculations of Gibbons[2.4,2.10] and Biersack.[2.2] The calculations for R_p agree within 2% for ^{11}B, the difference in R_p between ^{11}B and ^{10}B is less than 1.36% for energies between 30 and 300 keV. This is well below the experimental error. The experimental ranges are excellently fitted by

the calculations. For the range straggling, the calculations for ^{10}B and ^{11}B show a difference of about 2%, but between the calculations of Biersack and Gibbons there is a maximum difference of 13%. The experimentally-determined data show a much longer straggling of up to 45%. However, this is probably due to a residual amount of channeling.

Skewness and kurtosis are given in Fig. 2.4, where the theoretical estimate for the skewness according to Gibbons *et al.*,[2.4] and the kurtosis calculated with Eq.

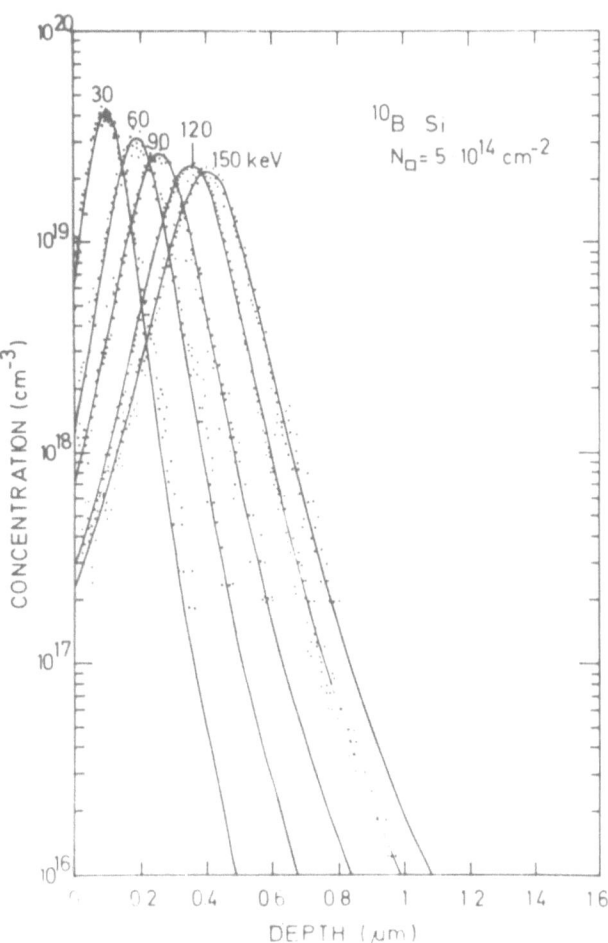

Fig. 2.2 - Comparison between measured boron profiles in silicon and Pearson IV distributions. No annealing was performed after implantation.[2.1]

(2.19), are also included. Both experimental and theoretical skewness are negative, but the experimental one is smaller by more than a factor of 2. At 30 keV, the depth resolution of the (n,α) method is not sufficient (400Å), resulting in a near-Gaussian profile which has a skewness of 0 and a kurtosis of 3. The negative skewness indicates that the profile is steeper towards the surface than towards the bulk.

In Fig. 2.5, a comparison between experimental boron profiles
in SiO_2, implanted at energies of
30 to 210 keV with doses of
5×10^{14} cm^{-2}, and Pearson-IV distributions, is given. A correspond-
ing set of curves is shown in Fig.
2.6 for boron in Si_3N_4. In both
cases, an excellent agreement is
again found between experiments
and fits over a concentration range
of more than two orders of magnitude.

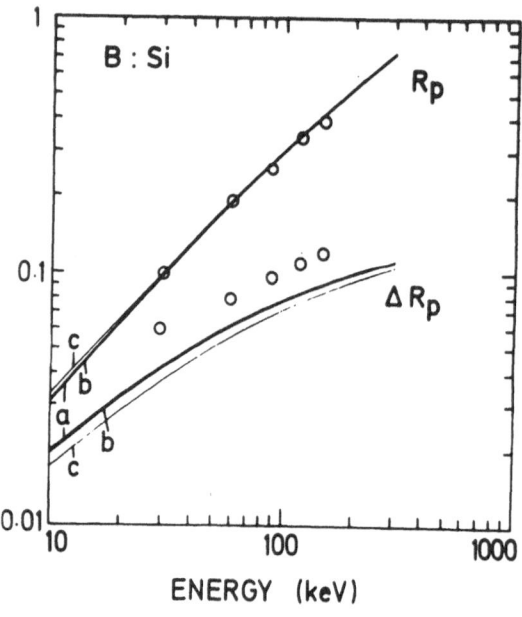

Fig. 2.3 - Range and range straggling of boron
in silicon. Experimental data, drawn line according to Gibbons (c), ^{11}B and Biersack (a)
^{10}B, (b), ^{11}B).[2.11]

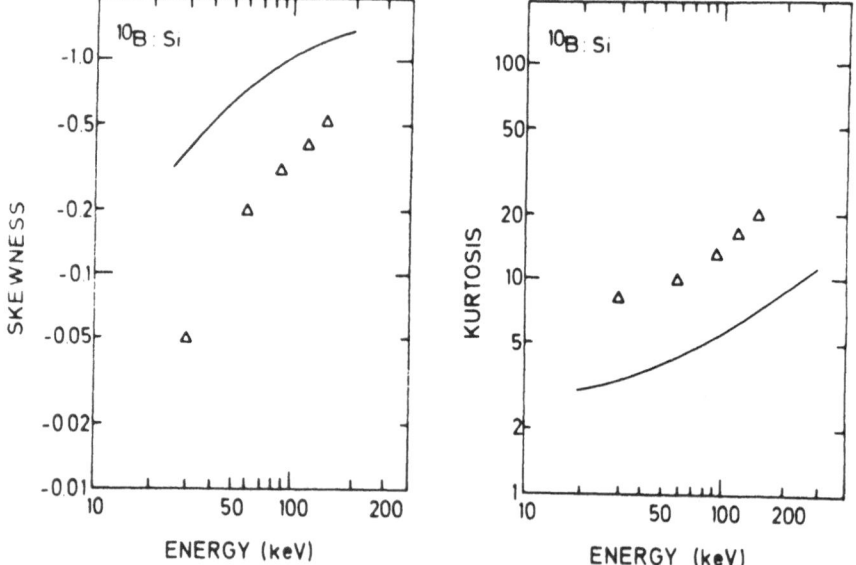

Fig. 2.4 - Skewness and kurtosis of boron in silicon. Solid line according to Gibbons.[2.4,2.10]

Range and range straggling for these cases, as extracted from the experiments, are given in Figs. 2.7 and 2.8. For comparison, the theoretical calculations for ^{11}B according to Gibbons *et al.*,[2.4] and for ^{11}B and ^{10}B according to Biersack,[2.2] are again included. R_p and ΔR_p for boron in SiO_2 are well-described by the calculation according to Biersack. The difference between ^{11}B and ^{10}B is below the experimental error, which

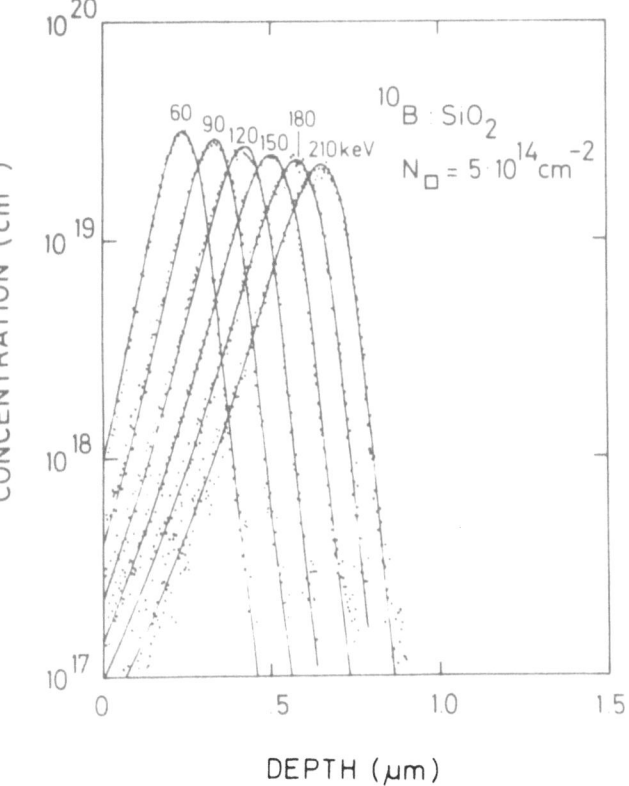

Fig. 2.5 - Comparison between measured boron profiles in SiO_2 and Pearson IV distributions. No annealing was performed after implantation.[2.11]

is estimated to be about 5% (due to uncertainties in energy calibration and in stopping power). For the range, the Gibbons data also fit well, whereas for the straggling, there is some discrepancy for lower energies. In Si_3N_4, the measured ranges are smaller by up to 26% than the theoretical calculations. In contrast, the straggling is well-described by theory at low energies, whereas at higher energies, the experimental straggling is smaller by 3 to 14% than the theoretical calculations.

Skewness and kurtosis are given in Figs. 2.9 and 2.10. The skewness of boron in SiO_2 is well-described by the theory, except for the lowest energy, where the resolution of the method is again insufficient. For Si_3N_4, the experimental data are larger than the theoretical calculations up to a factor of 2, but they show the same general tendency. Both show a negative value as in the case of silicon, i.e., the profiles are also steeper towards the surface. Similar measurements with arsenic are found in the same paper by Jahnel *et al.*[2.11] The results of all measurements in silicon, SiO_2 and Si_3N_4 show (provided that a proper misalignment was used to suppress channeling) that Pearson-IV distributions are well-suited to describe implantation profiles.

2.4 Lateral Spread of Implanted Ions

Ions implanted into a target through a thick mask will also penetrate laterally beneath

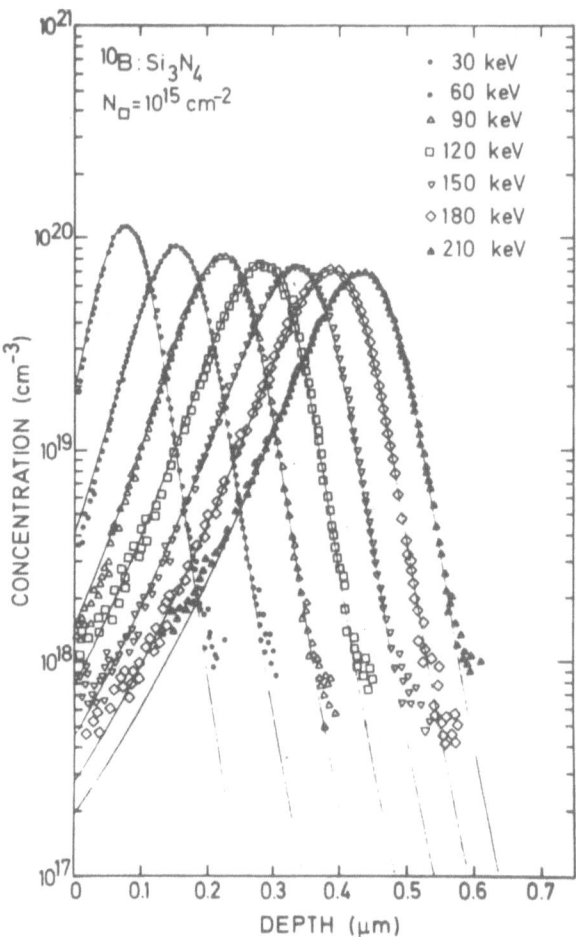

Fig. 2.6 - Profiles of boron implanted into Si_3N_4 fitted by Pearson distributions.[2.11]

the mask edge, as a result of the scattering of the ions. In the beginning of ion implantation, the lateral spread was thought to be negligible in comparison to the lateral dimensions of the devices and in comparison to the lateral diffusion. With VLSI circuits, dimensions are shrinking into the micron and submicron range and the lateral spread has to be taken into consideration. The lateral spread is of the order of the standard deviation. According to the calculations of Matsumura and Furukawa,[2.12] the two-dimensional profile is obtained by multiplying Eq. (2.1), in the case of an implantation through an ideal slit of width 2a and infinite length, with

$$\frac{1}{2}\left[\text{erfc}\ \frac{y-a}{\sqrt{2}\Delta R_{p,L}}\ -\ \text{erf}\ \frac{y+a}{\sqrt{2}\Delta R_{p,L}}\right] \tag{2.20}$$

where $\Delta R_{p,L}$ is the lateral spread of the implanted ions. For real masking layers with an arbitrary shape, Runge[2.13] found for Gaussian profiles:

$$C(x,y) = \frac{N_\square}{2\pi\ \Delta R_{p,L}\Delta R_p}\int_{-\infty}^{\infty}\left[\exp\ \left(-\ \frac{(x-\xi)^2}{2\Delta R_{p,L}^2}\ -\ \frac{(x-d_{ox}(\xi)-R_p)^2}{2\Delta R_p^2}\right)\right]\ d\xi \tag{2.21}$$

with d_{ox} (ξ) being the local thickness of the masking layer. In Fig. 2.11, two examples for such two-dimensional implantation profiles are given, for implantation through an infinitely steep mask and through a tapered mask. In these cases, Pearson distributions were used in the vertical direction.

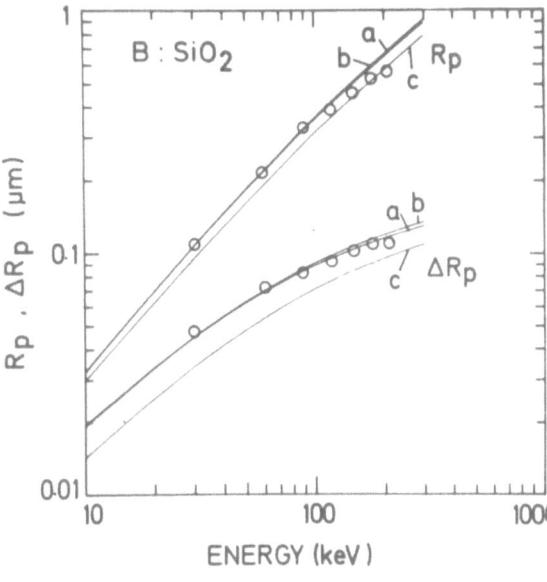

2.5 Two-Layer Structures

Very often, implantations are performed through thin masking layers, in order to avoid a contamination of the silicon, to provide for

Fig. 2.7 - Experimental range and range straggling data for boron in SiO_2: (a) [10]B, Biersack;[2.11] (b) [11]B, Biersack; (c) [11]B, Gibbons.

a scattering layer, or to adjust the depth distribution. Two important questions have to be answered: that of the depth distribution of the primary ions in the layer and the target, and that of the distribution of the atoms recoil-implanted from the surface coating into the target. The latter problem will be discussed in the next section.

Exact calculations require Monte Carlo simulations or the solution of Boltzmann equations.[2.14] Ishiwara and Furukawa, however, derived a simple model which is valid in materials whose average atomic numbers and masses are nearly equal.[2.15] This is true for the common combinations used with silicon, e.g., SiO_2, Si_3N_4, Al_2O_3. Moreover, only thin layers are used in such implants, further reducing the error. Assuming Gaussian profiles, the two parts of the profile are given by:

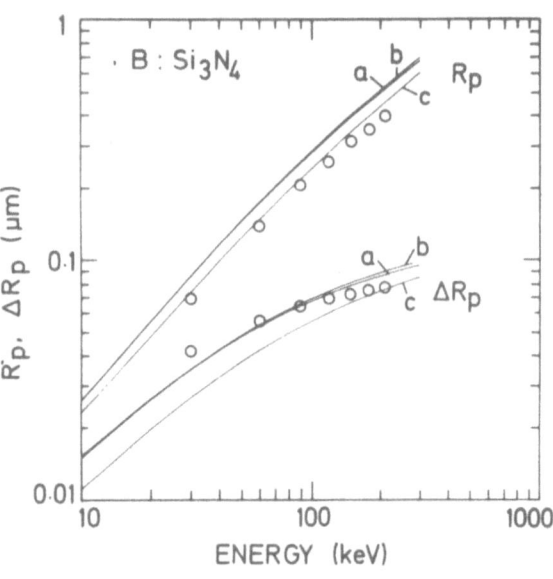

Fig. 2.8 - Experimental range and range straggling data for boron in Si_3N_4 (a) [10]B, Biersack; (b) [11]B, Biersack; (c)[11]B, Gibbons.

$$C_1(x) = \frac{N_\square}{\sqrt{2\pi}\,\Delta R_{p1}}\, \exp\left[-\frac{(R_{p1}-x)^2}{2\Delta R_{p1}^2}\right] \quad 0 \leq x \leq d \quad (2.22)$$

$$C_2(x) = \frac{N_\square}{\sqrt{2\pi}\,\Delta R_{p2}}\, \exp\left[-\frac{[d+(R_{p1}-d)\Delta R_{p2}/\Delta R_{p1}-x)]^2}{2\Delta R_{p2}^2}\right] \quad x \geq d$$

with d the thickness of the layer. The model can also be used for non-Gaussian profiles. In this case, the profile in material 1 is calculated, and the total number of atoms in this layer(N_1) is obtained by integration. Then, the profile is calculated in material 2, and the thickness d' is determined in which N_1 atoms are deposited. Finally, the profile is composed by joining profile 1 up to d with profile 2, starting from d'.

For very thin layers, it is usually sufficient to calculate the profile by reducing the energy, before looking up the range in the tables, by:

$$\Delta E = \frac{dE}{dx} d \tag{2.23}$$

where dE/dx is the total stopping power in layer 1 at the primary energy (which is tabulated in all range tables), to find the modified R'_{p2}. For the range straggling, no such reduction of the energy should be performed. In this case, $C_2(x)$ is given by

$$C_2(x) = \frac{N_\Box}{\sqrt{2\pi} \Delta R_{p2}} \exp \left[-\frac{(d + R'_{p2} - x)^2}{2\Delta R_{p2}^2} \right] \quad x \geq d \tag{2.24}$$

Examples of profiles obtained by these two models are given for a boron implant with 60 keV through 1000Å of Si_3N_4 in Fig. 2.12.

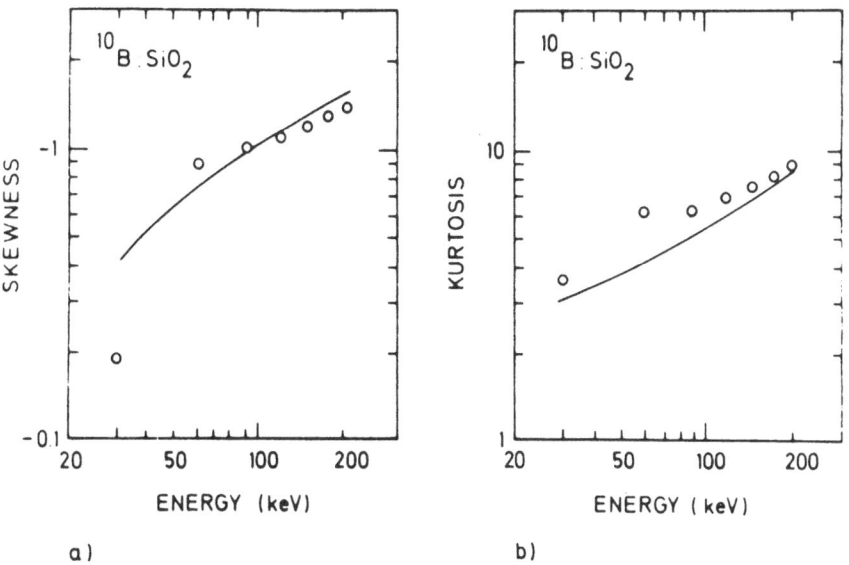

Fig. 2.9 - Skewness and kurtosis of boron in SiO_2. Solid line according to Gibbons.[2.11]

2.6 Knock-on Implantation and Atomic Mixing

The implantation of ions through a masking layer results in knock-on or recoil implantation of ions. If the mass of the implanted ions is not too different from the mass of the atoms of the masking layer, a large fraction of the energy of the primary particles can be transferred

to the atoms of the masking layer. These particles are then implanted into the substrate. The

energy which is transferred in an impact is given by Eq. (2.3). Profiles can then be construct-

ed, projecting the respective ranges into the x-direction. Theoretical calculations of the

distributions of recoil implants have been performed by Moline *et al.*,[2.16] Fischer *et al.*,[2.17] and

recently by Sigmund[2.18] and Hirao *et al.*.[2.19] The formalism is too complicated to deal with it

here in detail; no simple analytical profile description exists, and also no tabulated data.

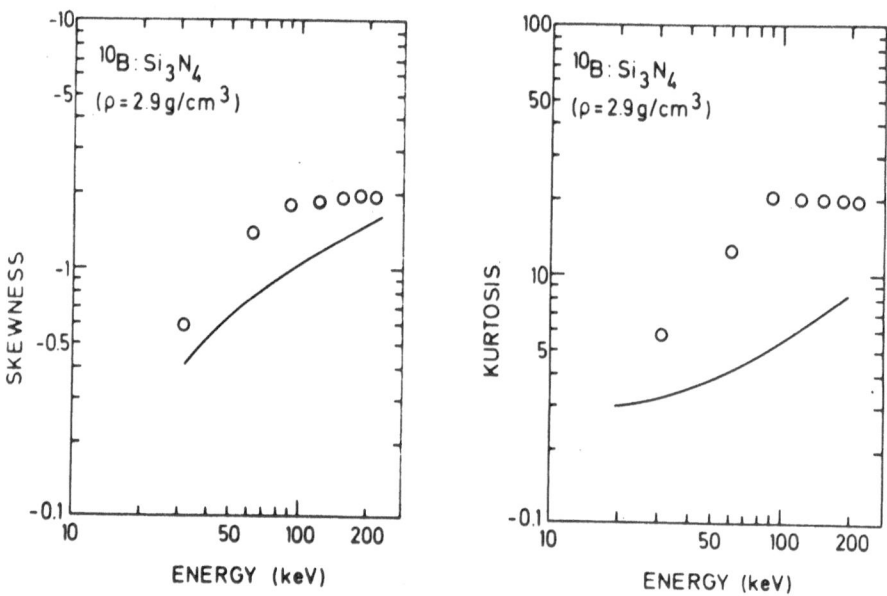

Fig. 2.10 - Skewness and kurtosis for boron in Si_3N_4. Solid line according to Gibbons.[2.11]

Many measurements have been reported on knock-on implants.[2.16, 2.19, 2.21] In Fig. 2.13,

a typical example is given for arsenic, implanted through a 650Å thick layer of Si_3N_4 with 335

keV to a dose of 10^{16} cm^{-2}.[2.19] The extremely shallow recoil profile with the maximum at the

surface is clearly seen. The primary ions come to rest deeper in the crystal than the recoil-

implanted atoms.

All experiments and theories show the same phenomena: a very shallow distribution of

the recoil-implanted ions and a deeper distribution of the primary atoms.

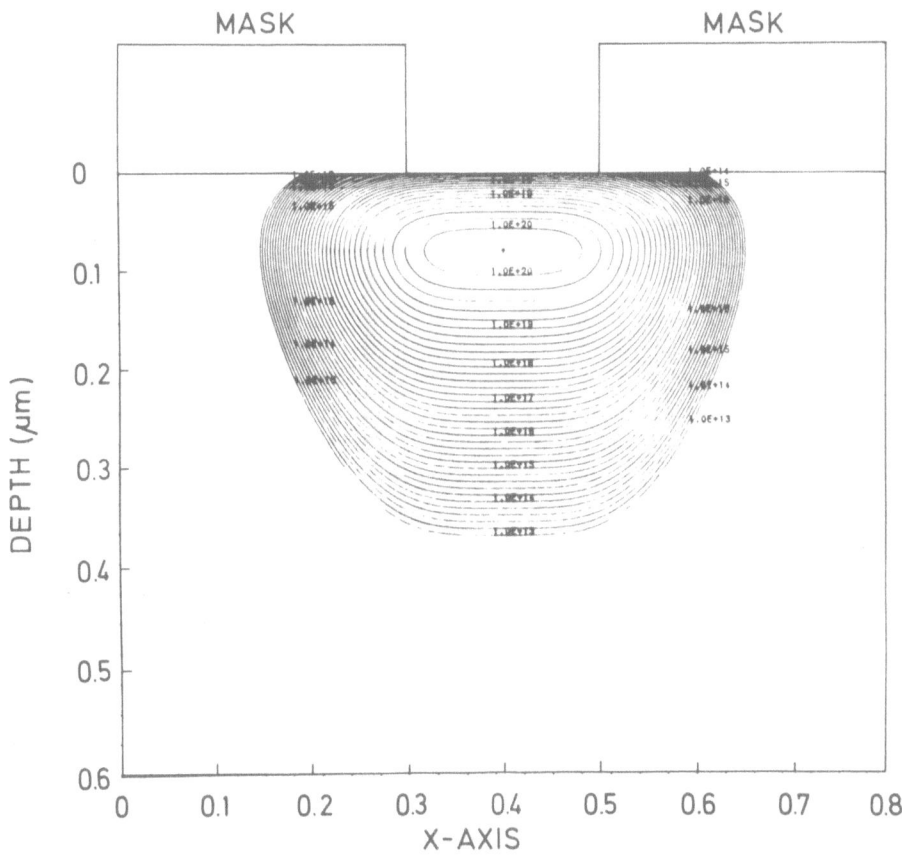

Fig. 2.11(a) - Two-dimensional simulation of an implantation profile (boron, 150 keV, 10^{15} cm^{-2}); infinite steep mask

Recently, some applications have also been found for knock-on implantations, e.g., for the control of the barrier height of Schottky diodes,[2.22] to produce non-volatile memories[2.23] as well as to manufacture solar cells.

The damage produced by the knocking ions, however, has to be considered. It will usually extend into the area of the p-n junction, and may severely degrade its properties. But there is undoubtedly no other implantation process capable of producing profiles as shallow as those obtainable using this process, with the maximum of the distribution at the surface.

Very recently, a new type of effect has been found to occur at interfaces: atomic mixing.[2.22-2.27] During the implantation of ions through an interface between a thin film and a substrate, not only are atoms from the thin film recoil-implanted into the substrate, but also, substrate atoms are transported into the thin film. This effect can be explained by a simple model.[2.26] Each incident ion initiates a collision cascade in a volume around the ion track. Within that volume, there is some atomic mobility for a very short time interval following the impact, resulting in an intermixing of the atoms near the interface. At the same time, reactions between the mixing atoms can also take place. Most of these studies have been made using metal-silicon reactions, e.g., platinum,[2.22, 2.27] molybdenum,[2.25] nickel and hafnium.[2.27] The formation of silicide induced by this effect at room temperature might offer a future method for production of low-temperature ohmic contacts or Schottky diodes.

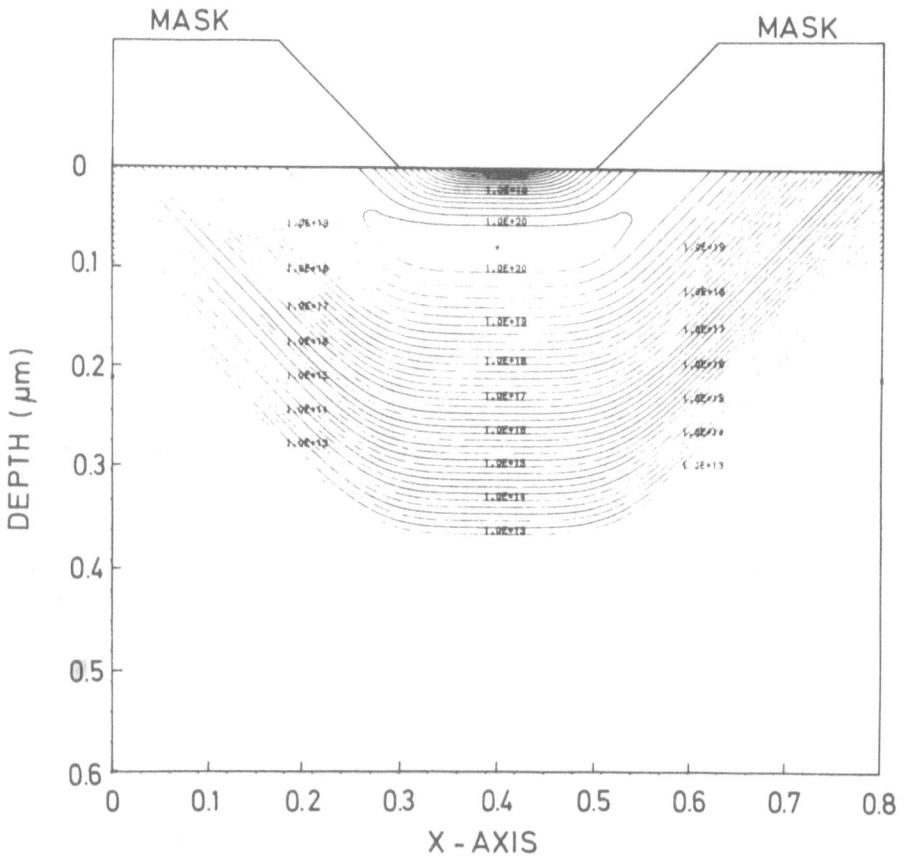

Fig. 2.11(b) - Two-dimensional simulation of an implantation profile (boron, 150 keV, 10^{15} cm^{-2}); tapered mask (45°).

2.7 Channeling

All range theories assume an amorphous target. Semiconductors such as silicon or GaAs are single crystals. Because of this crystalline structure, ions can penetrate much deeper into the crystals if they are implanted along a major crystal axis or plane, since they seldom come close enough to a target atom to lose significant energy by nuclear collisions. Therefore, electronic stopping mainly occurs, instead of a combination of electronic and nuclear stopping in a random direction.

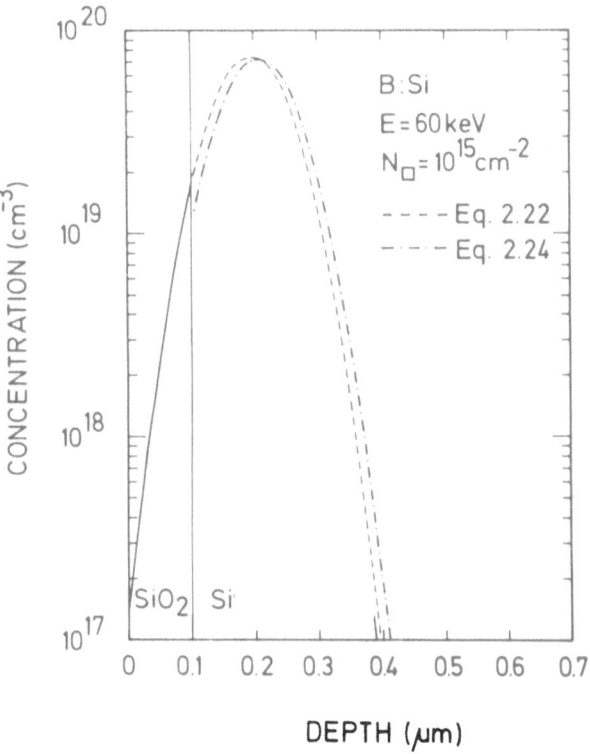

Fig. 2.12 - Comparison of range profiles in two-layer structures calculated by Eqs. (2.22) and (2.24) (B, 150 keV, 10^{15} cm^{-2}).

The basic principle of channeling is given in Fig. 2.14. Lindhard calculated the angle under which an ion is accepted into a channel.[2.28] For the energy range of ion implantation, it holds

$$\psi_{c2} = \left(\frac{a}{d\psi c_1} \right)^{1/2} \tag{2.25}$$

with

$$\psi_{c1} = \left(\frac{2Z_1 Z_2 q^2}{2\pi \epsilon_0 E d} \right)^{1/2} \tag{2.26}$$

where d is the distance between the atoms and

a is a screening parameter which is of the same

order of magnitude as the Bohr radius (a =

$0.8853 \, a_o \, (Z_1^{2/3} + Z_2^{2/3})^{-1/2}$; $a_o = 0.592$Å).

In Table 2.2, critical angles for the most im-

portant doping elements in silicon are given.

TABLE 2.2   Critical angles for channeling of various dopants in silicon

Ion	Energy (keV)	Critical Angle <100>	<110>	<111>
Boron	10	4.76	6.97	5.30
	100	2.67	3.47	2.98
	300	2.03	2.98	2.26
Phosphorus	10	5.79	7.51	6.45
	100	3.26	4.22	3.63
	300	2.47	3.21	2.76
Antimony	10	6.95	9.01	7.74
	100	3.91	5.07	4.35
	300	2.97	3.84	3.31

The range of channeled ions can be calculated approximately, assuming only electronic stopping of the implanted ions. Since no nuclear stopping takes place, channeling ranges are much larger than those in a random direction. The range is proportional to the velocity of the ions, which means a square-root dependence upon the energy. Presently, however, there exists no plausible theory describing channeling profiles depending

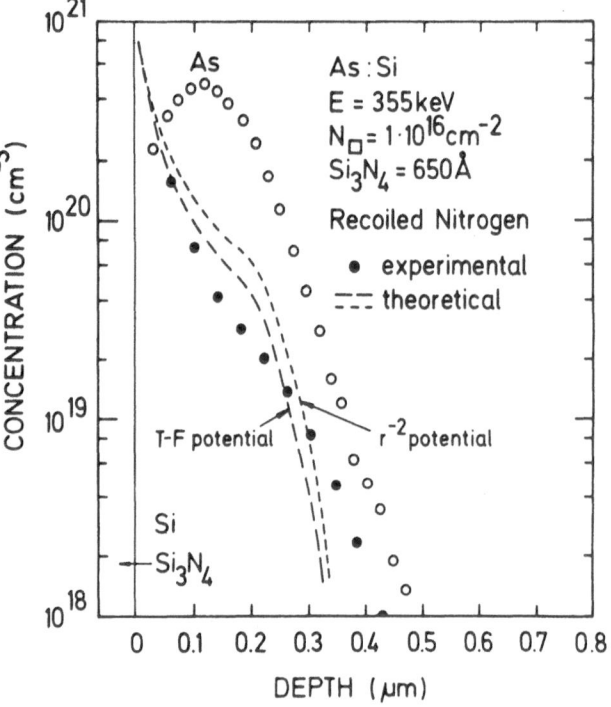

Fig. 2.13 - Comparison of experimental and theoretical profiles for arsenic implantation at 335 keV to a dose of 1×10^{16} cm^{-2} through Si$_3$N$_4$ of 650Å thickness.[2.19]

on crystal orientation, tilting, etc. Therefore, these profiles can only be described and

investigated experimentally.

In the following figures, examples of channeling profiles are given, showing their dependence on the orientation of the crystal, on the crystallographic perfection of the crystal, on the

temperature and on the scattering layer which is frequently used to suppress channeling.

Orientation   In Fig. 2.15, the dependence of channeling on the tilting angle for phosphorus

implants in silicon is shown.[2.29] In order to avoid channeling, one usually implants into tilted

wafers. Usually, $7°$ to $10°$ is used, but one has to be careful to avoid channeling into planar

channels with a too-strong tilting. This is illustrated in Fig. 2.16, where the influence of the

rotation of the wafer for a proper tilting of $7°$ is shown.

For an optimal suppression of the channeling effect, therefore, a proper tilting and

rotation of the wafers is an absolute necessity.

Another question is whether the channeling effect can be used to obtain deep distribu-

tions, e.g., for buried layers. For this, a parallel ion beam (either obtainable by mechanical

scanning of the wafers, or by an electrostatic double-deflection scanner) is required.

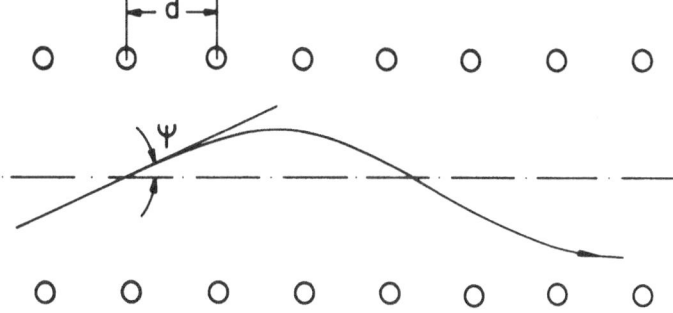

Fig. 2.14 - Schematic description of the channeling effect.

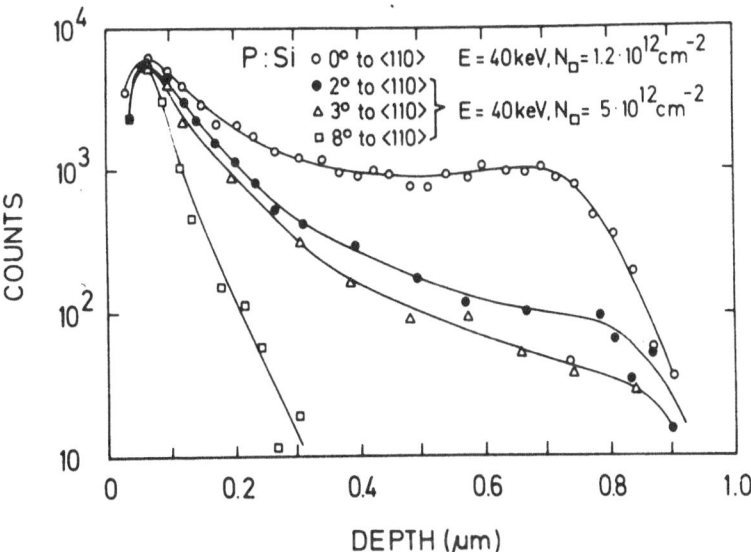

Fig. 2.15 - Dependence of channeling on the tilting angle.[2.29]

The crystal orientation, however, has to be exact within $\pm 0.1°$, which is very difficult to obtain. The influence which even very small deviations have on the profile shape is depicted in Fig. 2.17, for a phosphorus implant with 450 keV into <111> silicon.

Crystallinity With increasing ion dose, more and more damage is produced and the channels are closed. This does not reduce the channeling range, but rather the number of channeled ions, by scattering them into random directions. At high doses, the lattice is completely destroyed, an amorphous layer is formed and channeling is completely suppressed. Doses required to obtain amorphous layers are given in Fig. 2.18, for a 40 keV

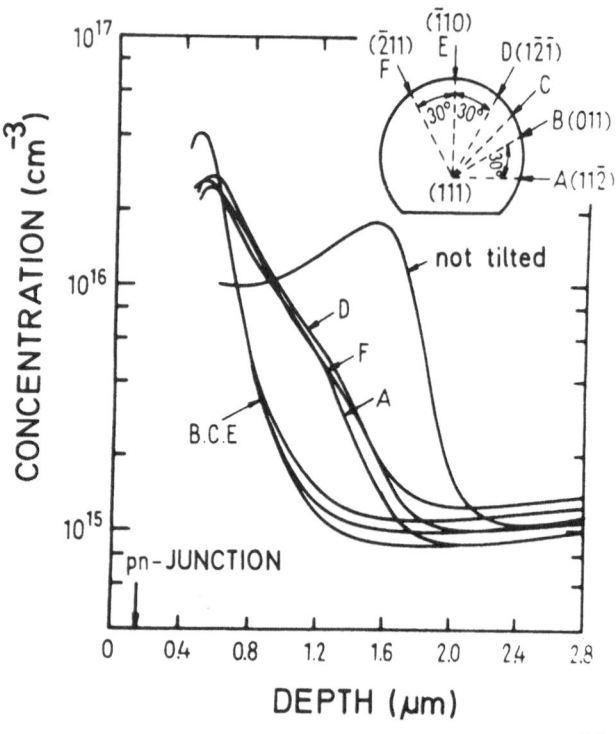

Fig. 2.16 - Dependence of phosphorus profiles on rotation.[2.9]

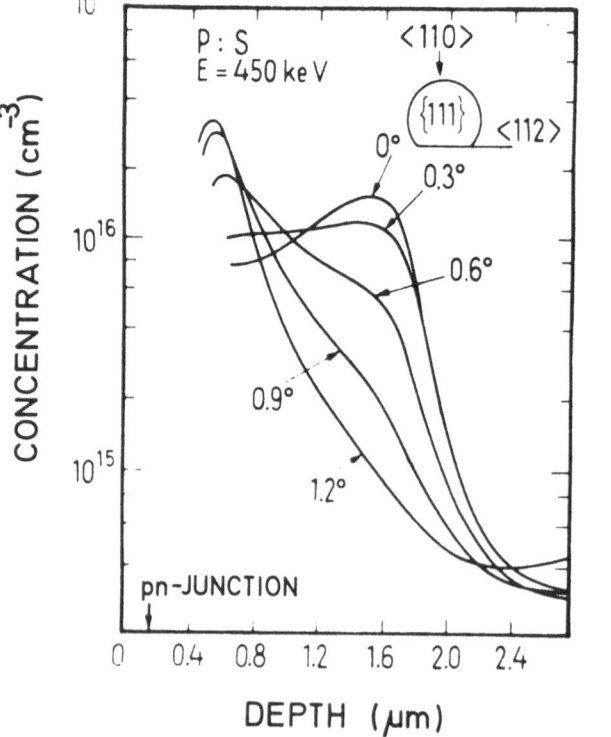

Fig. 2.17 - Dependence of phosphorus profiles on tilting.[2.9]

Ion Implantation

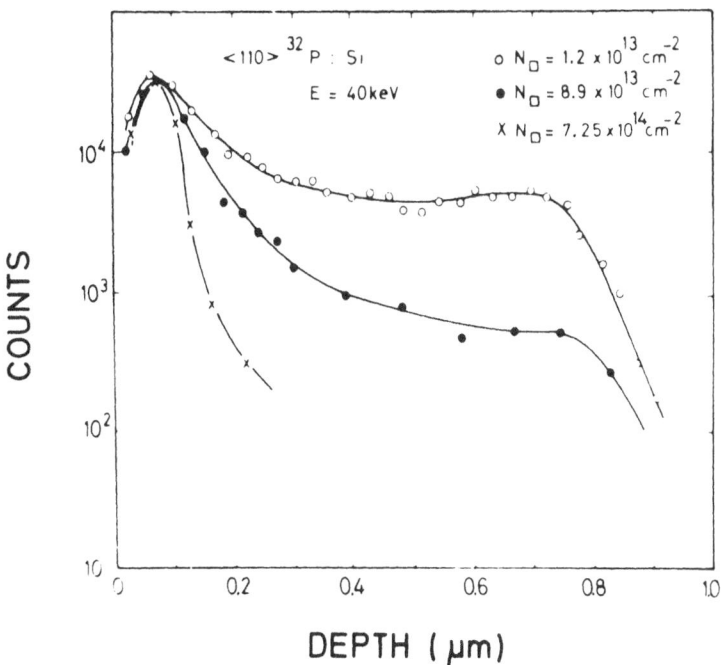

Fig. 2.18 - Dose dependence of channeling.[2.29]

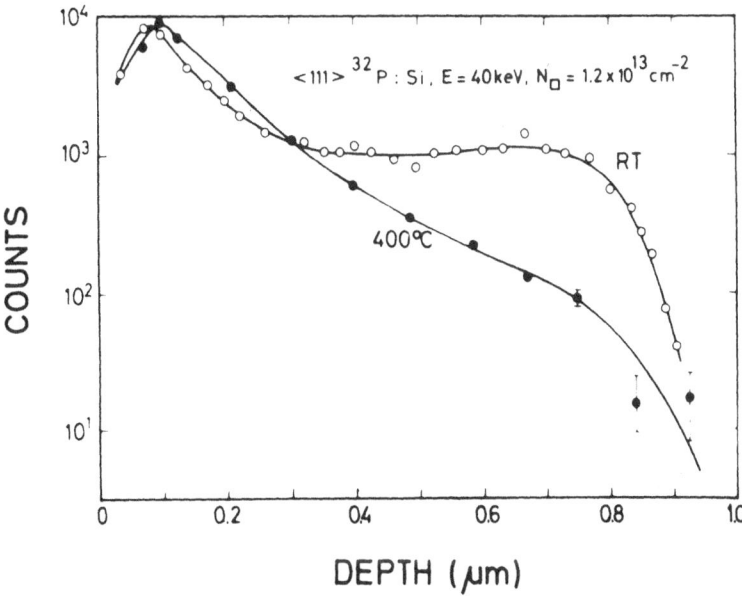

Fig. 2.19 - Temperature dependence of channeling.[2.29]

phosphorus implantation. The amorphous dose in this case is 5×10^{14} cm^{-2}. At doses producing an amorphous layer too, there is still a channeling tail visible, resulting from channeled ions before the amorphous layer was formed. A complete suppression of channeling is only possible if a thick amorphized layer is used, which can be produced using an additional implant with inert ions, such as argon or silicon.

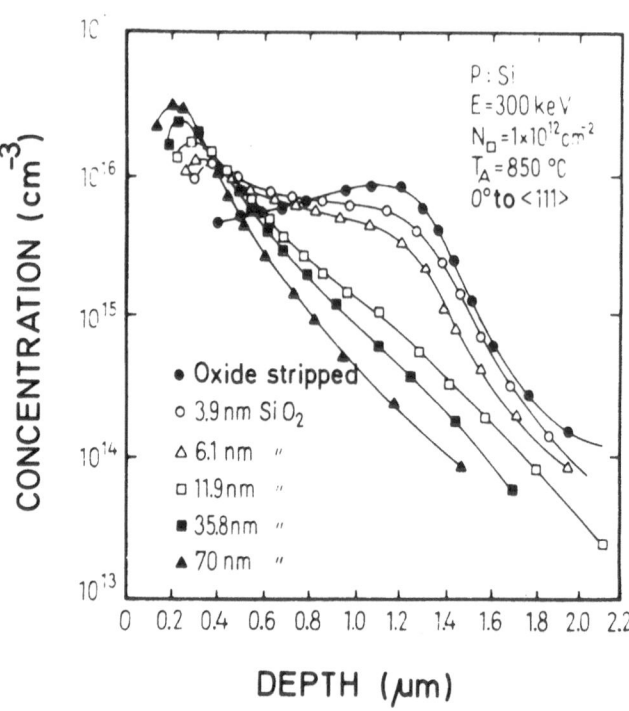

Fig. 2.20 - Dependence of channeling on oxide-layer thickness.[2.30]

Temperature   At high implantation temperatures which may result from beam heating, two effects occur: first, the damage produced during implantation partly anneals out, thus reducing the number of ions scattered out of the channels; more important, however, is the increase in the amplitude of lattice vibrations, thus reducing the effective channel diameter. This effect is shown in Fig. 2.19.

Scattering Oxide   If a crystal is covered with an amorphous layer (e.g., an oxide or nitride layer), the ions are scattered, depending on their velocity.   Thus, the number of ions which can penetrate channels is reduced.   In Fig. 2.20, the carrier concentrations of phosphorus implants through oxide layers with different thicknesses are shown.

Application of Channeling From the preceding examples of the different aspects of channeling, it is clear that a complete suppression of this effect is very difficult or impossible. The application of channeling to obtain deep impurity distributions, on the other hand, is also very limited. Due to the strong dose, orientation and temperature dependence, it seems very unlikely that it can be used for the mass production of integrated circuits.

3. DAMAGE INDUCED DURING IMPLANTATION

As an unwanted by-product of ion implantation, a damaging of the crystal lattice occurs which is detrimental to the desired electronic quality of the semiconductor and which, therefore, requires an annealing step (see Chap. 4).

Many theories have been developed to calculate the energy deposition in nuclear collisions and the damage produced by these collisions.[3.1-3.3] A great number of different defects are created, such as vacancies, interstitials, vacancy-impurity pairs, divacancies, spikes of heavy disorder and amorphous layers. The theories describing the energy deposition are well-developed; however, the interaction between the different defects created is not yet completely understood.

Typical depth distributions of the nuclear-collision energy loss are given in Fig. 3.1 for arsenic, and in Fig. 3.2 for boron, in silicon. The maximum

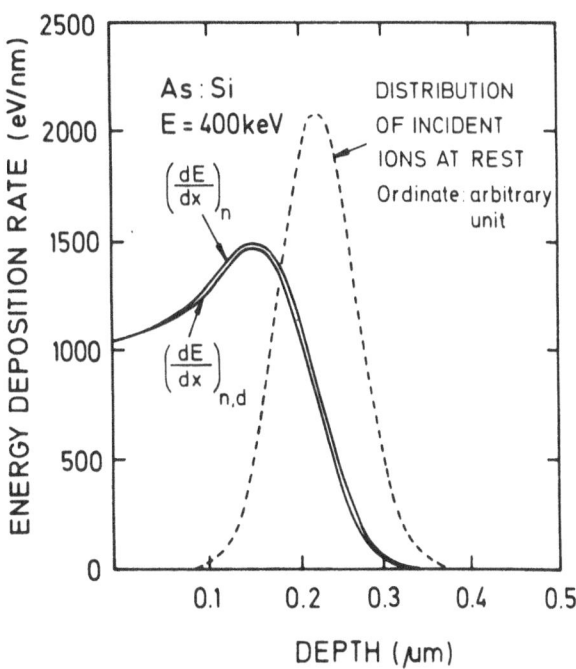

Fig. 3.1 - Depth distributions of the deposited energy in nuclear collisions (index n for total energy deposition; index n,d for energy deposition into displacements) for arsenic in silicon.[3.6]

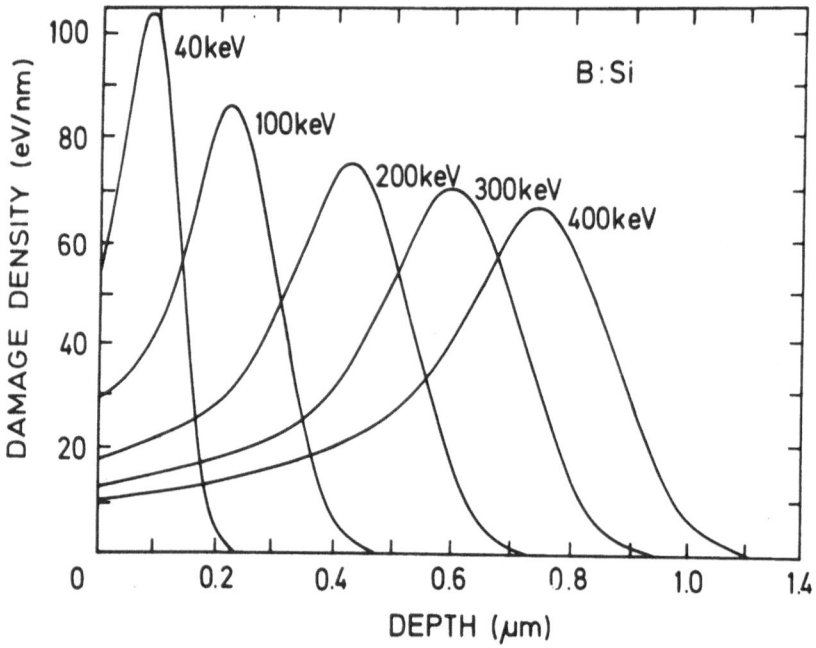

Fig. 3.2 - Depth distributions of the energy deposited into nuclear
collisions for boron in silicon.[3.3]

of the nuclear energy loss, and therefore of the damage distribution, is always closer to the

surface than the maximum of the impurity distribution. From such curves, the number of

displaced atoms can be calculated using the classical Kinchin-Pease formula:

$$\psi_{C2} = (\psi_{C1}\frac{d}{a})^{1/2} \tag{3.1}$$

where v is the energy deposited in nuclear collisions and E is the amount of energy required to

displace an atom (E_d = 15-25 eV in silicon). Tabulations of v are given by Brice[3.4] and

Winterborn[3.5].

For semiconductor applications, it is important to know whether or not an amorphous

layer has been formed by the implantation, since the annealing behavior is different. The

critical dose for producing an amorphous layer depends on the current density during implan-

tation since, for high current densities, the semiconductor is heated, causing an annealing

during implantation. At low current densities, the dose to produce an amorphous layer is

about

$$\psi_{C1} = \left(\frac{z_1 z_2 q^2}{2\pi \epsilon_0 Ed} \right)^{1/2} \tag{3.2}$$

In practice, however, the doses depend also on the ion species. In Table 2.3, data on measured amorphous doses for low current densities at 300 K are given. At lower temperatures, much lower doses are sometimes required. For boron, e.g., a dose of 10^{15} cm^{-2} is sufficient to induce an amorphous layer at 77 K. The width of the amorphous layer is dose-dependent, but usually smaller than R_p. In the amorphous layer, the nearest neighbors are covalently bonded, with no coordination to the next nearest neighbors.

TABLE 2.3 Amorphous dose for various elements at room temperature

Element	Mass	Dose (cm^{-2})
B	11	8×10^{16}
N	14	2×10^{15}
Ne	20	10^{14}
Al	27	5×10^{14}
P	31	6×10^{14}
Ar	40	4×10^{14}
Ga	70	2×10^{14}
As	75	2×10^{14}
Kr	84	2×10^{14}
Sb	122	10^{14}
In	204	10^{14}
Tl	204	5×10^{13}
Bi	209	5×10^{13}

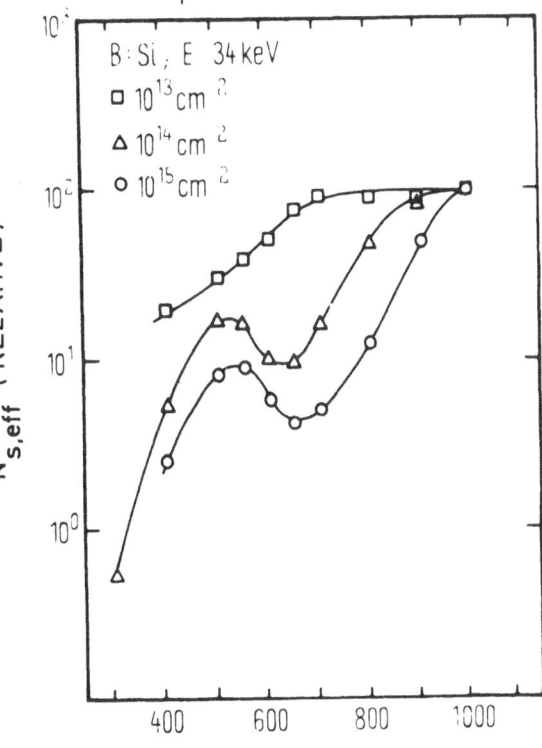

4. ANNEALING OF IMPLANTED LAYERS

The purpose of annealing is twofold, namely, to restore the crystal lattice which is damaged by the implantation process, and to render the implanted ions electrically active. To obtain electrical activity, the ions have to occupy a lattice site.

Fig. 4.1 - Annealing behavior of the sheet carrier concentration of boron implanted silicon.[4.2]

In this chapter, only thermal annealing in furnaces is treated; the new method of laser annealing is dealt with in a contribution by Soncini.[4.1]

In this chapter, only thermal annealing in furnaces is treated; the new method of laser

annealing is dealt with in a contribution by Soncini.[4.1]

4.1 Temperature Dependence of Annealing

At low doses, isolated

point defects, isolated damage

clusters and isolated amorphous

regions have to be annealed

out. Simple defects anneal out

at low temperatures, e.g., va-

cancies at 70 K to 150 K,

vacancy-group-V-element de-

fects between 400 and 500 K

and vacancy-group-III-element

defects at about 500 K. Dou-

ble vacancies anneal at 500 to

600 K. Between 700 and 800

K, clusters dissociate and form

dislocation loops, which partly

anneal out at temperatures

above 1300 K.

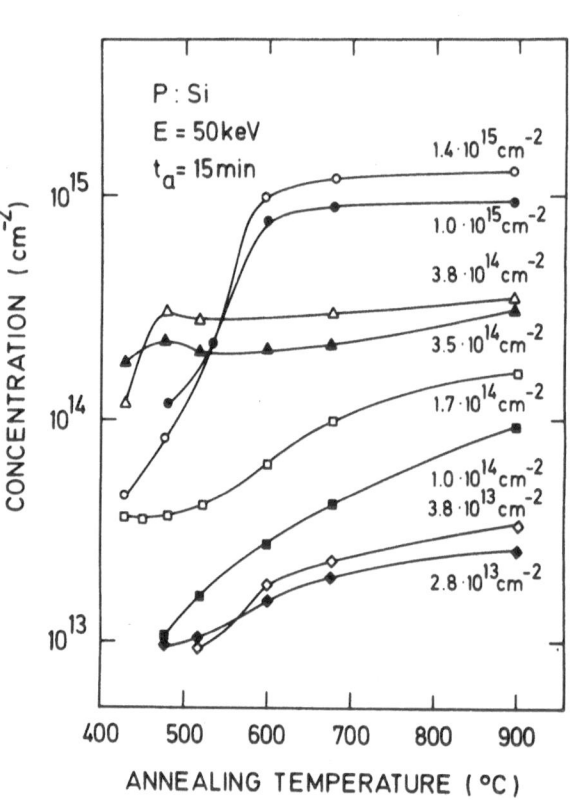

Fig. 4.2 - Change of surface carrier concentration due to annealing between 400 and 900°C for 15 min. Phosphorus ions with an energy of 50 keV and a dose of 2.8×10^{13} to 1.4×10^{15} cm^{-2} were implanted into a <111> p-type silicon substrate.

At high doses, amorphous layers have to be recrystallized. This takes place between 850

and 900 K, but some residual damage nonetheless remains, which partly anneals out at higher

temperatures. After annealing at 850 K, all standard doping elements of silicon will have

obtained their complete electrical activation, except for boron.

Boron shows a reverse annealing effect at doses or implantation temperatures which result

in no amorphous layer. This abnormal behavior is shown in Fig. 4.1. Only at very low doses is the effect not seen. If BF_2 is implanted, or if the implantation is done at liquid-nitrogen temperatures, an amorphous layer forms at a dose of 10^{15} cm^{-2}, and an activation behavior similar to the other dopant ions is found.

In Fig 4.2, the annealing behavior of phosphorus is depicted, showing a steep activation around 850 K and no reverse annealing.

For a complete recovery of the carrier mobility, slightly higher temperatures are required than for recrystallization; to restore a good carrier lifetime, temperatures of about 1200 to 1300 K are necessary. Especially, electrically-active disorder is responsible for reducing these parameters by producing deep levels or traps, thus reducing mobility and lifetime. Traps are especially harmful in space-charge regions of devices, by reducing the frequency response and by increasing reverse currents.

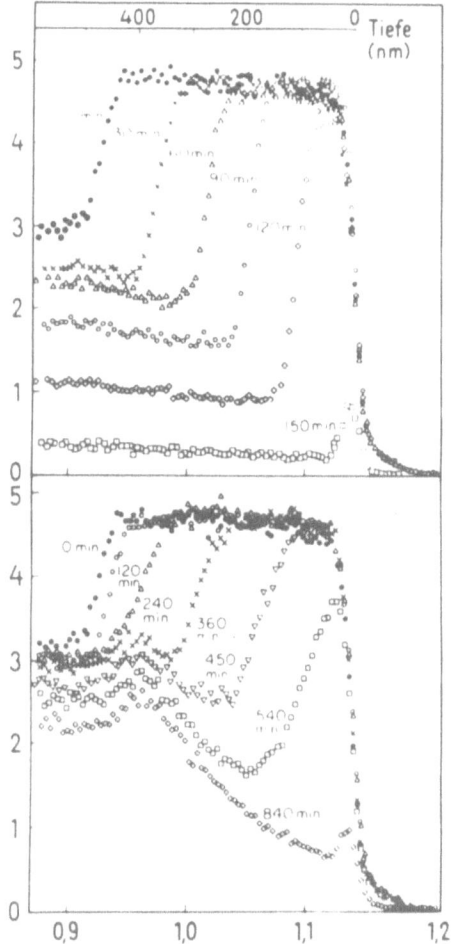

4.2   Orientation and Dopant Dependence of Annealing

The recrystallization of amorphous layers is strongly orientation-dependent. This effect was studied in detail by

Fig. 4.3 - Channeling spectra of self-implanted silicon (50 and 250 keV, 8×10^{15} cm^{-2}), preannealing at 400°C, annealing at 550°C. Upper curve <100>, lower curve <111> oriented sample.[4.4]

Csepregi.[4.3,4.4] <100>-oriented wafers recrystallize much faster than <111>-oriented wafers. This is shown in Fig. 4.3 for self-implanted silicon samples. Moreover, in the case of the <111> orientation, a second maximum develops at the border between the undamaged

substrate and the amorphous layer. The recrystallization rate is about 8 nm/min in <100>
and 2.5 nm/min in <110> material at 500°C; the activation energy is 2.3 eV in both cases.
For <111> material, the process is more complicated, and no activation energy can be given.

This orientation-dependent annealing depends strongly, in turn, on the thermal history of
the wafer. This is shown in Fig. 4.4. If the annealing is done immediately at high temperatures, many crystal defects remain (Fig. 4.4(a)). If the samples are annealed with steadily
increasing temperatures, a perfect annealing is obtained (Fig. 4.4(b)). A preannealing at

500°C with a subsequent high-temperature annealing step also results in a good annealing (Fig. 4.4(c)).

The orientation dependence of the regrowth rates can be explained by geometrical arguments, assuming that atoms can only be transferred from the amorphous to the crystalline phase at positions where at least two nearest neighbors at the interface are already in crystalline positions. On this basis, growth will not proceed along the <111> direction, because alternate planes have atoms with only one bond along this direction. Growth along

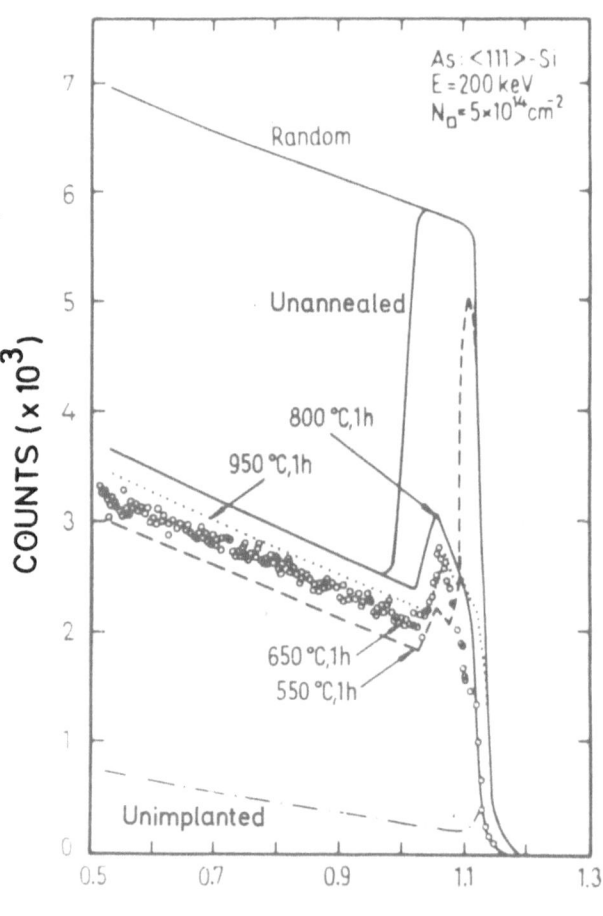

Fig. 4.4(a) - Random and channeling spectra of <111> oriented silicon implanted at 77 K with arsenic (200 keV, 5×10^{14} cm^{-2}),[4.13] one-step annealing

the <111> direction involves nucleation, and hence leads to non-uniform interface and twin formation.[4.3] A comparison between theory and experiment is given in Fig. 4.5. This simple model does not account for the non-zero <111> regrowth rate.

The annealing dependence on the dopant species was also first investigated by Csepregi.[4.5] For implanted concentrations of B, Ar, or P exceeding about 10^{20} cm^{-3}, there is an increase in the growth rate over that of only silicon-implanted amorphous layers. The increase is associated with the local impurity concentrations in the proximity of the amorphous-crystalline inter-face. The increase is by a factor of 6 to 25. For electrically-inactive impurities, the growth rate decreases by a comparable amount. This was measured for O, C, N, Ne, Ar and Kr.[4.6] Since the enhanced or retarded recrystallization rate depends on the impurity concentration, it is not constant during the re-growth process. The reasons for this behavior are not yet well-understood.

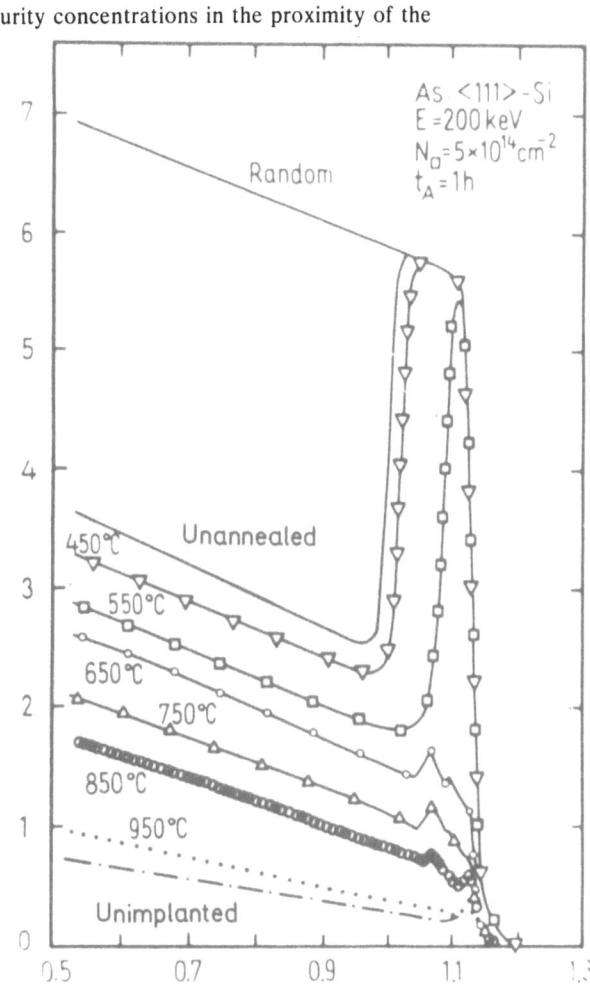

4.3 Dependence of Annealing on Atmosphere

In silicon technology, oxi-dizing treatments are very of-ten performed, either for driving-in a dopant or to obtain

Fig. 4.4(b) - Random and channeling spectra of <111> oriented silicon implanted at 77 K with arsenic (200 keV, 5×10^{14} cm^{-2}),[4.13] annealing in 100°C steps,

a masking layer for the next technological step. In the case of ion implantation, it was found that oxidizing annealing leads to the formation of many dislocations which grow into the bulk of the wafer. This is also the case if no amorphous layer was formed during the implant. Detailed studies of this effect have been performed by many groups.[4.7-4.12] If an inert preannealing is performed at 900 to 1000°C before the oxidation, virtually no dislocations remain. This can be done in two separate steps, or merely by changing the gas composition.

The importance of an inert annealing or preannealing for device characteristics was impressively shown by Seidel et al.[4.12] They studied the emitter-base currents and current gains as a function of the oxygen content in the annealing atmosphere of bipolar transistors, and found that the best results are obtained for an oxygen content below 0.1%, for through-oxide implants as well. This is especially true for <100> silicon, whereas <111> silicon is more tolerant to oxidizing treatments.

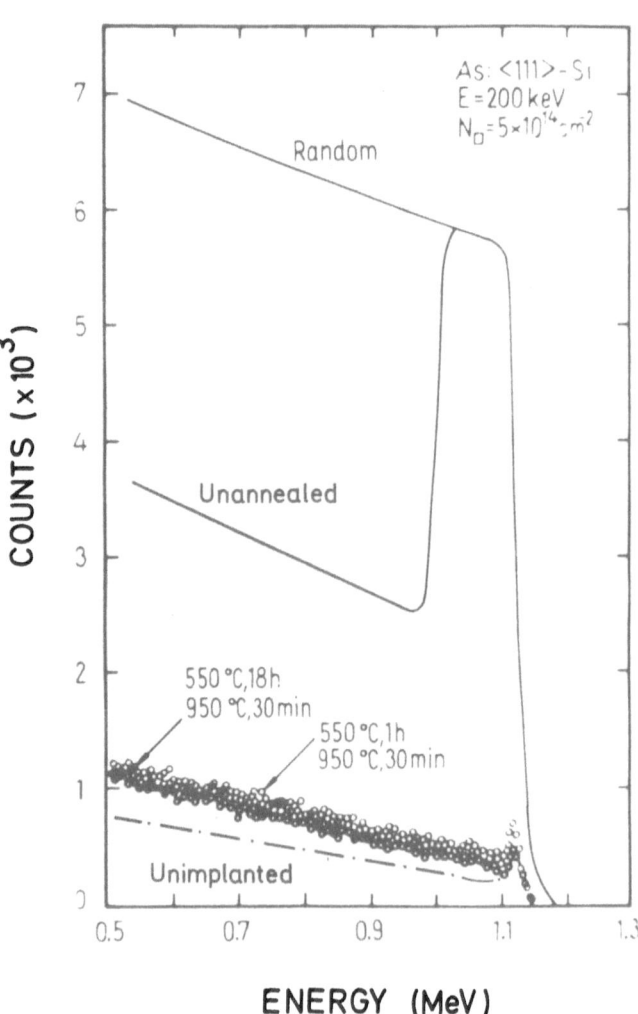

Fig. 4.4(c) - Random and channeling spectra of <111> oriented silicon implanted at 77 K with arsenic (200 keV, 5×10^{14} cm^{-2}),[4.13] 500°C preannealing.

4.4 Diffusion During Annealing

Although annealing at relatively low temperatures around 850 K usually produces a complete electrical activation of the implanted ions (provided that the solubility of the dopant was not exceeded by the implantation), the annealing is usually performed at higher temperatures ($\cong$ 1200 to 1300 K), in order to further reduce defects which influence mobility, lifetime and trap parameters. At such high temperatures, not only an annealing of defects and the

electrical activation of the implanted ions occur, but also profile modifications due to diffusion. Such a diffusion, in fact, is desirable in order to place the p-n junction behind the originally heavily-damaged layer, which still exhibits a certain amount of damage. Also in the case of amorphous layers, where the epitaxial recrystallization at low temperatures simultaneously incorporates the implanted ions (see Chap. 4.2), such a high-temperature annealing is used to diffuse the dopants

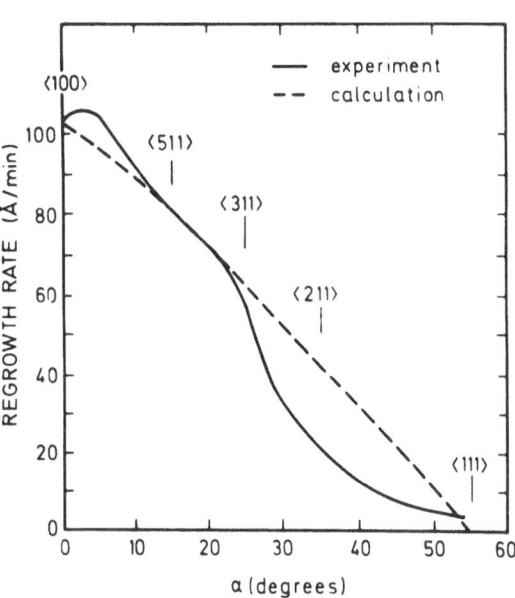

Fig. 4.5 - Plot of regrowth rate at 500°C vs. substrate orientation angle from the <100> direction.[4.3]

slightly and to reduce the amount of residual damage. Very often, a diffusion is also required to obtain a complete electrical activation, since the solubility has been exceeded. The problems of diffusion are treated in detail in the contributions of Antoniadis and Declerck;[4.14-4.15] here, only some implantation-relevant topics are dealt with. Diffusion coefficients depend on orientation, atmosphere and dopant concentration. To obtain shallow structures, relatively low-temperature anneals are required, but in this case, damage-enhanced diffusion still has to be considered. Also, it was found recently that at low temperatures, the diffusion coefficient might be larger than expected from an Arrhenius plot.[4.16]

This shows that even these basic parameters are not yet well-known. Ion implantation is a very good method for the study of these phenomena. For high concentrations, the diffusion coefficient becomes concentration-dependent, and complicates the profile control of implanted layers additionally. In Fig. 4.6, experimental profiles comparing the total and the electrically-active concentrations of a high-dose arsenic implant in silicon[4.17] are shown. Only about 50% of the arsenic atoms are electrically active, since the solubility at the annealing temperature of 950°C is only 3×20^{20} cm^{-2}. The concentration-dependent diffusion can also be seen from this figure.

The broken curve was calculated according to the process-modeling program ICECREM.[4.18]

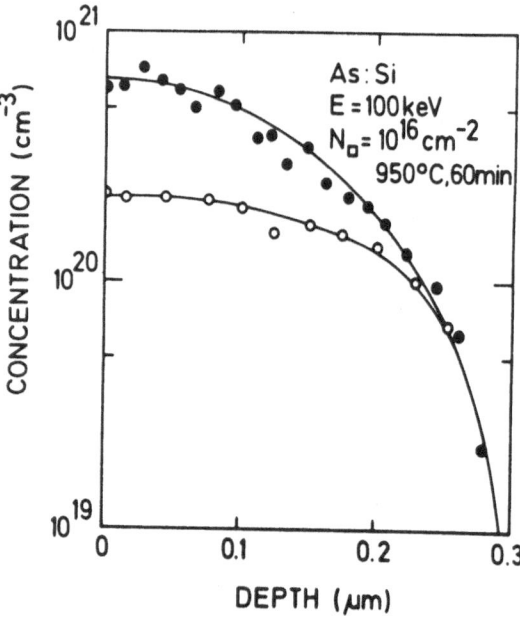

Fig. 4.6 - Total and electrical active arsenic concentration after annealing in comparison to a computer simulation.

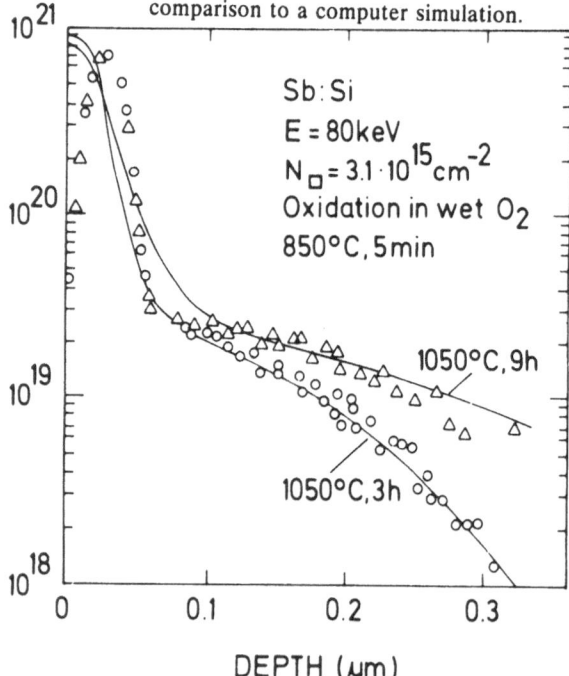

Fig. 4.7 - High-concentration diffusion behavior of antimony implanted into silicon; adapted from Ref. 4.19.

For antimony, the solubility is very low ($\cong 2 \times 10^{19}$ cm^{-3} at 1000°C), and therefore, doses as low as 10^{14} cm^{-2} are sufficient to exceed the solubility. In contrast to arsenic, the atoms above the solubility do not diffuse as shown in Fig. 4.7, whereas below the solubility, diffusion takes place.

Similar results are found for boron (see Fig. 4.8), where an immobile cluster also forms when the solubility is exceeded. During prolonged annealing, the boron profile broadens by diffusion and the excess boron can occupy lattice sites. The solubility found by such measure- ments is compared to other published data in Fig. 4.9.

For boron, another anomalous diffusion (besides the well-known vacancy and field enhancement found for all high-concentration diffusions with boron, arsenic, phosphorus, etc., which is treated by Antoniadis[4.14]) has been determined. In the first minutes of the annealing directly after implantation, a damage-enhanced diffusion takes place. The enhancement is

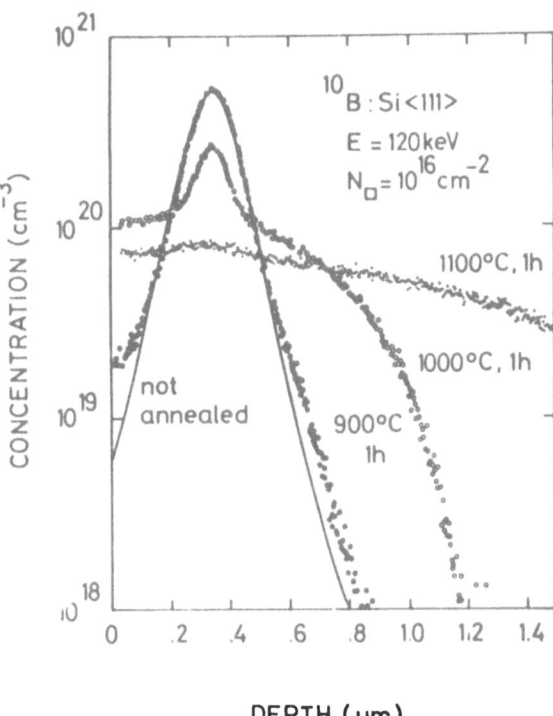

Fig. 4.8 - High-concentration diffusion behavior of boron in silicon.[4.20]

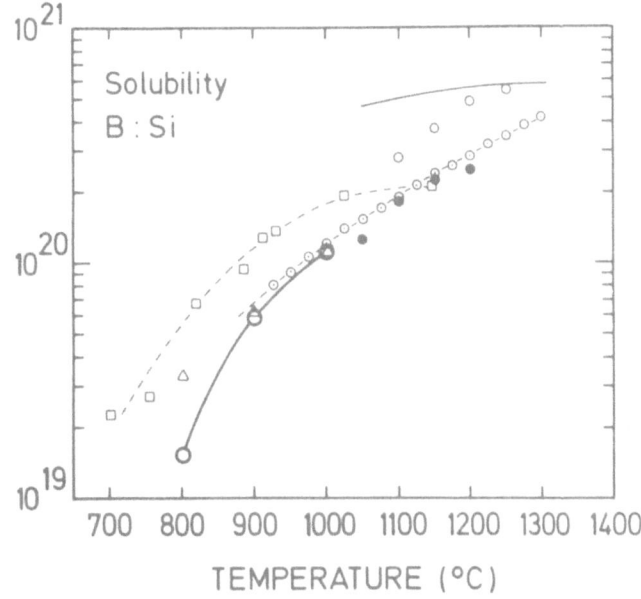

Fig. 4.9 - Comparison of boron solubility. o(n,α) nuclear reaction, ⊙ TEM, o(p,n) nuclear reaction, • □, △ electrical, - chemical.[4.20]

independent of the dose from 10^{13} to 10^{16} cm^{-2}, but is temperature-dependent. Because of the difficulty of measuring the diffusion coefficient after short-time heat treatments, we arbitrarily assume an enhanced diffusion in the first 10 minutes of the annealing. The experimentally-found results are given in Fig. 4.10. One can see that the enhancement is of more than a factor of 100 at low temperatures.

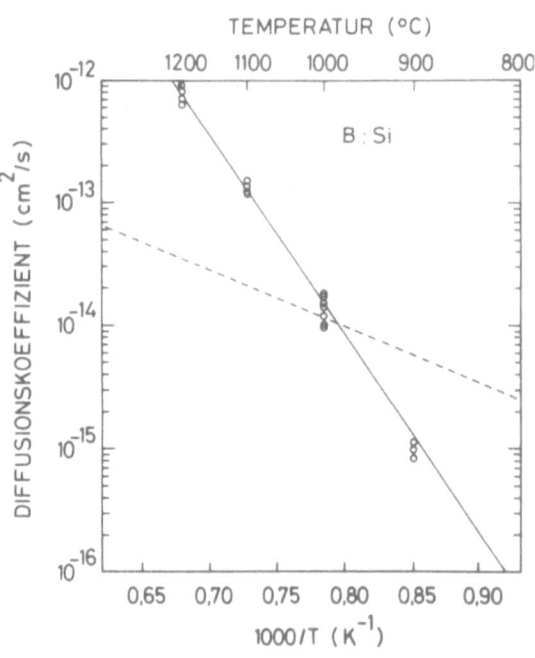

Fig. 4.10 - Damage-enhanced diffusion of boron; drawn line: intrinsic diffusion coefficient, broken line: additional diffusion coefficient during annealing of implantation damage (assumed to be effective for 1 h in this figure).

4.5 Residual Defects

After annealing, many defects remain; these are mainly dislocations. The best methods for determining the presence of such defects are electron transmission microscopy and defect etching (e.g. Sirtl etch, Secco etch). Backscattering is not sensitive enough to detect this low defect concentration.

In the following, some examples are given. At medium arsenic doses (5×10^{15} cm^{-2} at 80 keV), prismatic dislocation loops grow

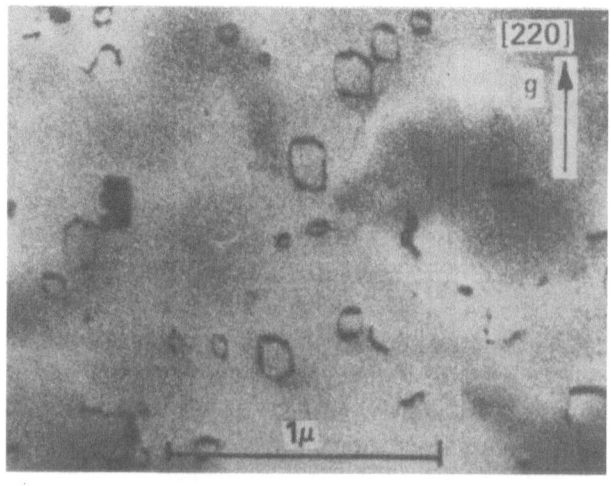

Fig. 4.11 - Prismatic dislocation loops in arsenic-implanted silicon (80 keV, 5×10^{15} cm^{-2}) annealed at 1000°C for 60 minutes.[4.22]

during annealing at 1000°C for 60 minutes (Fig. 4.11). At higher doses (10^{16} cm^{-2}), only large half-loops form. These dislocations are of the misfit type (Fig. 4.12). Implanting through thin SiO_2 layers also results in many defects, which are partly caused by knock-on particles. This knock-on effect is especially responsible for creating defects with doses in excess of 10^{16} cm^{-2}.

No amorphous layer is usually formed in the case of boron implantation. Defects nonetheless grow during annealing; however, the density after inert annealing at 900 to 1100°C is very low. Very detailed studies of the defects resulting from the annealing of phosphorus implants at a dose of 5×10^{14} cm^{-2} were made by Tamura.[4.21] He found a great variety of defects, depending on the implantation and annealing temperatures. A schematic description of his results is given in Fig. 4.13. After implantation at room temperature, point defects de-

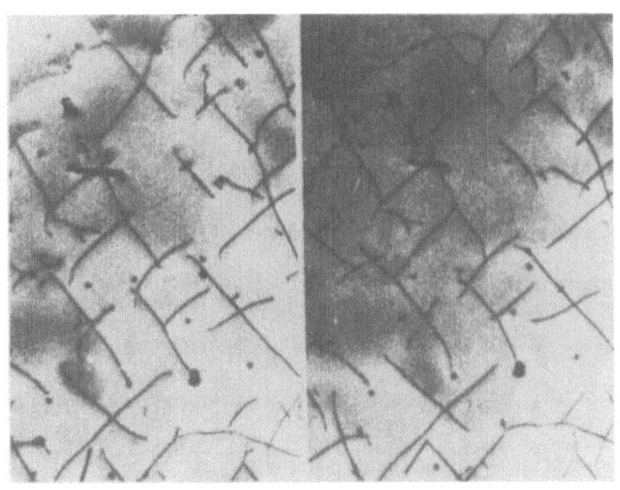

Fig. 4.12 - Stereo TEM photograph of dislocation lines after arsenic implantation (80 keV, 10^{16} cm^{-2}) and annealing at 1000°C for 60 minutes.[4.22]

T_A(°C) \ T_I(°C)	R.T.	200 400	500	600	700
500					
600					
700					
800					
900 -1000					
1100					
1200					

Fig. 4.13 - Defects in phosphorus-implanted silicon, depending on implantation and annealing temperatures.[4.21]

velop, which form dislocation loops at higher temperatures and dissolve at annealing temperatures above 1000°C. Implantations at elevated temperatures result in many different defects, which hardly anneal out. The appearance of rod-like defects can be correlated with a reverse annealing, similar to the case of boron.

The question of whether or not such defects are detrimental to device characteristics depends on the individual case. If an epitaxial layer is to be grown, defect-free material is required. This can be obtained by removing the layer containing defects by thermal oxidation, or by the usual etching step before deposition. In both cases, one has to diffuse the implanted ions before this treatment; otherwise, they are also removed. Very often, however, the defects have no effect on devices, especially if they do not grow during subsequent temperature steps, and do not penetrate junctions.

5. APPLICATION OF ION IMPLANTATION TO MOS DEVICES

The first applications of ion implantation to devices were in the field of MOS circuits. Ten years ago, the current delivered by an implanter was fairly low; therefore, applications requiring a low dose were favorable, and the requirements for the electrical properties of the implanted layers for MOS circuits were not any more stringent than those for bipolar circuits. In the meantime, however, device dimensions have been shrinking more, and more, and completely-implanted MOS circuits are becoming standard, which means increasing demands upon implantation technology.

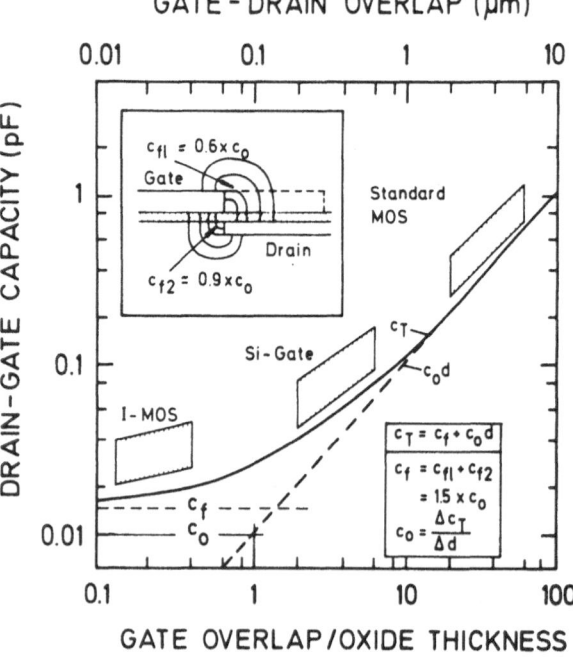

Fig. 5.1 - Miller capacitance of MOS transistors.[5.1]

5.1 Self-Aligned Gate

The oldest application of ion implantation in integrated circuit technology is the self-aligned or self-registered gate. Diffused MOS transistors always show a large gate-drain overlap, causing a large shunt capacitance (Miller capacitance).[5.1] In Fig. 5.1, the Miller capacitance is depicted, depending on the gate-drain overlap. An overlap is necessary due to alignment tolerances in lithographic processes. To reduce this effect, the source and drain are off-set from the gate. Using the gate as a mask, an implantation can be used to complete the structure with a minimum overlap. The different techniques are shown in Fig. 5.2. Figure 5.2(a) shows the conventional all-diffused technique; in Fig. 5.2(b) and (c), the

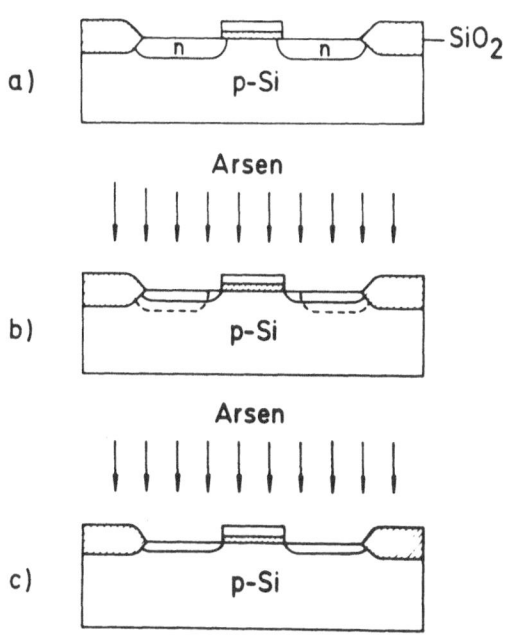

Fig. 5.2 - Implantation of source and drain; (a) all-diffused transistor, (b) implantation step used to self-align source/drain and gate, (c) fully implanted source and drain.

TABLE 5.1   Minimum Energy for depositing 50% of the implanted ions through SiO_2

Ion	Oxide Thickness (nm)				
	400	600	800	1000	1200
B	15	20	25	30	35
P	30	50	65	80	95
As	65	100	140	180	220

use of implantation to self-align source/drain and gate is depicted. If an aluminum-metallized gate is used as a mask, only temperatures up to the eutectic temperature (577°C) can be used for annealing. Usually, 450 to 500°C are applied. If polysilicon gates are used, higher annealing temperatures are possible, thus reducing 1/f noise and sheet resistivities. Modern MOS circuits make extensive use of this polysilicon gate technology. Advanced MOS devices use only implantation for source and drain doping. Together with the use of polysilicon as a gate material, high-temperature annealing (which is a prerequisite for obtaining low sheet resistivities) and self-aligned structures are possible. Usually, doses in excess of 10^{15} cm^{-2},

typically 3×10^{15} cm^{-2}, and annealing temperatures of about 950°C are used. For p-channel

transistors, boron is implanted; for n-channel transistors, phosphorus or arsenic is used. The

implantation energy depends on the thickness of the gate or screening oxide. In Table 5.1, the

implantation energies which are needed to implant 50% of the ions through the oxide are

given for boron, phosphorus and arsenic. The energies are rounded to the nearest 5 keV.

One can see that, for boron and phosphorus, energies deliverable by a standard implanter are

required, whereas with arsenic, only structures with thin oxides can be implanted.

In the near future, all diffusion steps in the manufacture of MOS-IC devices will be

eliminated, due to the greatly superior reproducibility and the better device parameters

obtainable by implantation.

5.2 Threshold Adjust

The threshold adjust is one of the most important applications of ion implantation. Here,

its exact dose control, as well as the possibility of implanting through the gate oxide, are the

great advantages which cannot be matched by any other method. The reduction of the

threshold voltage is necessary in order for integrated MOS circuits to be compatible with

bipolar circuits (e.g., the 5V level of TTL logic), and in order to reduce the power consump-

tion. This is very important for watch circuits and VLSI circuits, especially memories. The

threshold voltage is given by:

$$V_T = \phi_{MS} + 2\phi_F - \frac{Q_{ss} + Q_B}{C_{ox}} \qquad (5.1)$$

where ϕ_{MS} is the work-function difference of the gate and the silicon, ϕ_F is the Fermi

potential, Q_{ss} is the sheet charge at the interface, Q_B is the space charge and C_{ox} the gate

capacitance per area ($C_{ox} = \epsilon_o \epsilon_{r,ox}/d_{ox}$, d_{ox} is the oxide thickness, ϵ_o and $\epsilon_{r,ox}$ are the absolute and relative dielectric constants). ϕ_F and Q_B are given by:

$$\phi_F = \frac{kT}{q} \ln \frac{N_B}{n_i} \qquad (5.2)$$

$$Q_B = \sqrt{2q\, \epsilon_o \epsilon_{r,Si} N_B\, (2\phi_F + V_{BB})} \qquad (5.3)$$

$\epsilon_{r,Si}$ is the relative dielectric constant of silicon, N_B the background doping concentration and V_{BB} the reverse voltage of the substrate in respect to the source.

From Eq. (5.1), it is easily seen which method can be used for the adjustment of V_T. Changing the gate material will alter ϕ_{MS}, the doping concentration will alter Q_B, the orientation of the silicon will change Q_{ss}, and a different insulator will change ϵ_r.

The most simple way, however, is to introduce a charge at the interface by ion implantation. By this method, this quasi sheet charge, Q_{ss} can be continuously increased or decreased, depending on the type of carriers introduced, and thus, V_T can be adjusted continuously. The change in V_T is given approximately by

$$\Delta V_T = K\, q\, \frac{N_\Box}{C_{ox}} \qquad (5.4)$$

The constant K accounts for the number of implanted carriers influencing the threshold, e.g., for the portion of the implanted ions which are in the channel and are electrically active. In Fig. 5.3, different possibilities for realizing this threshold adjust are shown. Whereas a) is closest to the model of compensating Q_{ss}, one needs a lower dose for b), and in the case of c), the influence of variations in

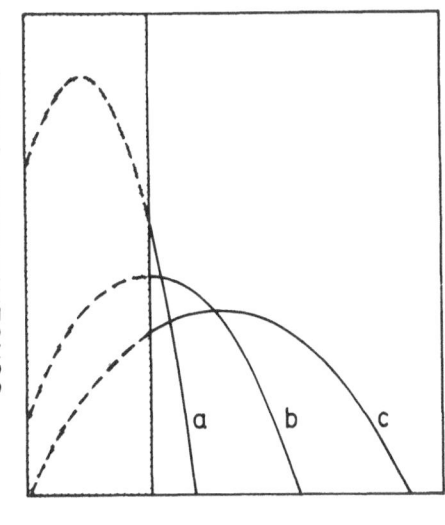

Fig. 5.3 - Schematic description of threshold adjust; (a) maximum of implanted distributions in oxide, (b) maximum at the interface, (c) maximum inside the semiconductor.

the oxide thickness is lowest. An expression more exact than Eq. (5.4) accounts for the profile shape of the implant. Assuming a rectangular profile with the doping concentration of $N_A = KN_\square/x_j$, it is true that

$$\Delta V_T = K q \frac{N_\square}{C_{ox}} + \frac{kT}{q} \ln \frac{N_B}{N_A} + q \frac{N_A}{2 \epsilon_o \epsilon_{r,Si}} (1 - \frac{Q_B}{KqN_\square})^2 \qquad (5.5)$$

x_j is the junction depth. Many publications deal with special problems of the adjustment of V_T, assuming realistic profiles.[5.2-5.5] For p-channel devices, a p-dopant (usually boron) is implanted to reduce the threshold voltage.

Of course, not only the reduction of the threshold voltage to lower power consumption is possible. By increasing the implanted dose, normally-on transistors can also be produced. These transistors are called depletion transistors, and are required as load transistors instead of standard enhancement transistors. This application results in a reduced power consumption and a switching speed enhanced by a factor 2 to 3. In Fig. 5.4, the dose de-

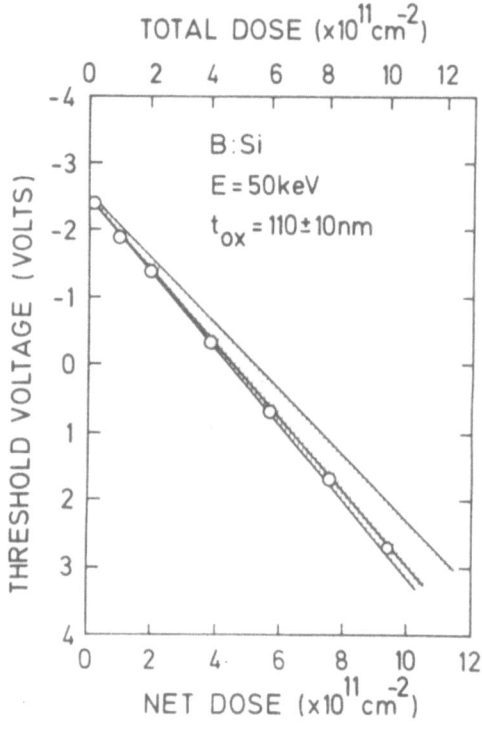

Fig. 5.4 - Threshold voltage as a function of boron dose for p-channel transistors.[5.6]

pendence of the threshold voltage is given for a 50 keV boron implant through 110 nm SiO_2. The shaded area indicates the scatter produced in V_T by a ± 10 nm variation in the oxide thickness. To reduce this sensitivity, the boron energy has to be selected according to the oxide thickness. This is shown in Fig. 5.5, where the sensitivity of ΔV_T to oxide variations is depicted schematically.

Depletion and enhancement transistors can be easily fabricated together with enhancement transistors by masking them with photoresist during a second implantation process. In Fig. 5.6, the typical fabrication procedure is shown. After opening windows in the SiO_2, the

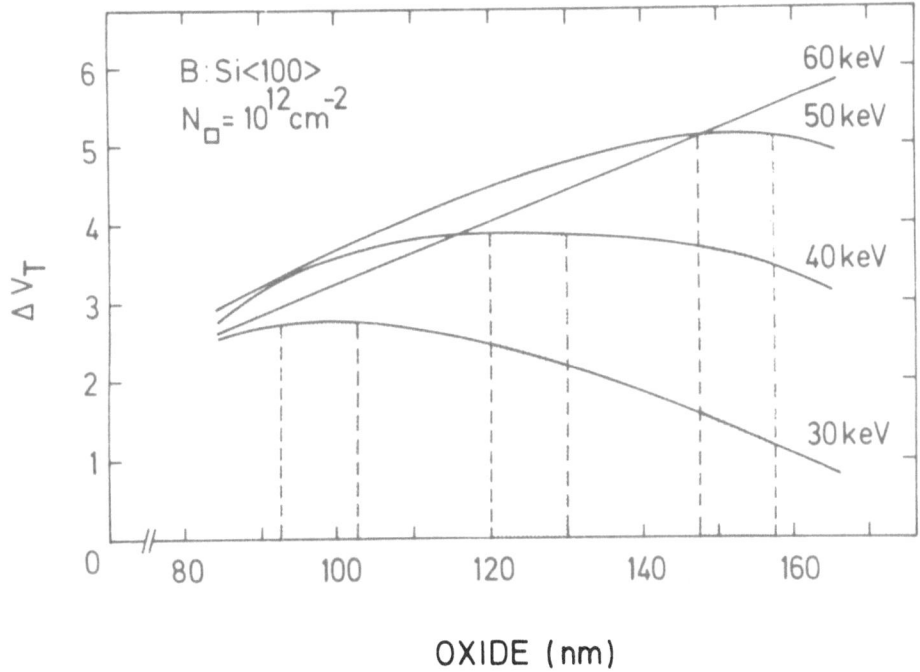

OXIDE (nm)

Fig. 5.5 - Dependence of ΔV_T on oxide thickness as a function of implantation energy.

source and drain areas are implanted and diffused in an oxidizing atmosphere. Subsequently, the gate areas are also opened in a second lithographic step and reoxidized to the desired gate oxide thickness (50-120 nm). Boron (in the case of p-channel devices) is implanted with a dose depending on the desired threshold voltage. The field oxide masks all other areas against the beam. After this implant, a photoresist layer is deposited and structured so that it masks the enhancement transistors. This is a non-critical step, since the alignment tol-

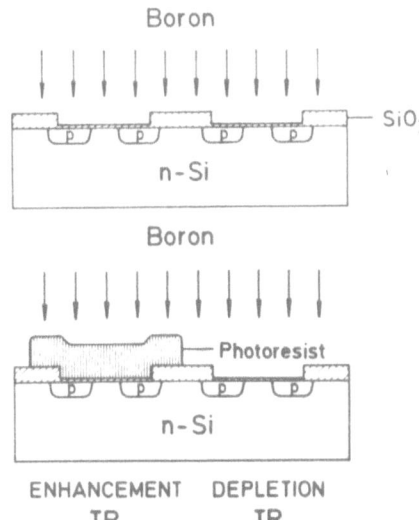

Fig. 5.6 - Fabrication procedure for enhancement and depletion transistors.

erances are very large, and a third implantation is performed to set the threshold of the

depletion transistors.

An industrial standard is currently represented by n-channel devices (NMOS). n-channel transistors are usually (depending on the
oxide thickness and the gate material) depletion devices. Boron is
implanted to increase the threshold
in order to obtain enhancement
transistors; phosphorus is implanted to adjust the depletion threshold. In Fig. 5.7, the dependence of
V_T on the substrate bias and on
the boron dose, for a boron implantation with 50 keV through
150 A of SiO_2, is shown.

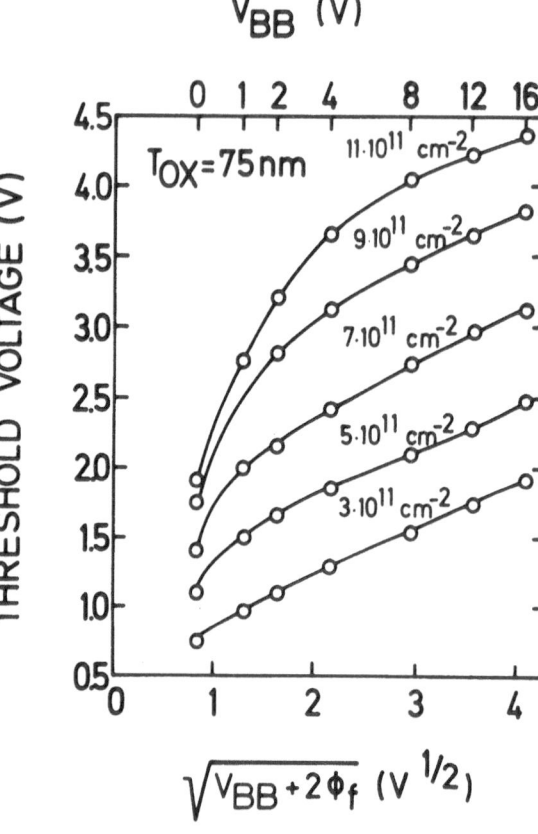

Fig. 5.7 - Dependence of enhancement threshold on boron dose and bias for n-channel
transistors.[5.7]

The dependence of the shift for zero substrate bias, depending on the boron and phos-

phorus dose, is shown in Fig. 5.8. The linear dependence of ΔV_T on the dose is clearly seen,

justifying the simple expression of Eq. (5.4). The sensitivity to oxide thickness variations is

similar to that described in Fig. 5.5; for phosphorus, however, the optimum energies are

approximately a factor of 3 higher for the same oxide thickness.

The calculated dependence of enhancement and depletion threshold voltage shifts,

respectively, for boron-implanted p- and n-channel transistors, is shown in Fig. 5.9. From this

figure, one can see the strong sensitivity of the threshold shift to the energy at low energy

levels (case (a) of Fig. 5.2) for this 100 nm oxide, again showing that oxide thickness and

energy have to be optimized.

MOS circuits can operate with greatly varying supply voltages. At high voltages, the silicon below the conductor lines can be inverted, and parasitic transistors and channels can thus be formed. To avoid this, one can increase the so-called field threshold voltage by ion implantation. For p-channel devices an n-dopant (phosphorus or arsenic), for n-channel devices a p-dopant (boron) is used. In Fig. 5.10, the process for adjusting the field threshold is depicted schematically. After growing a thin oxide

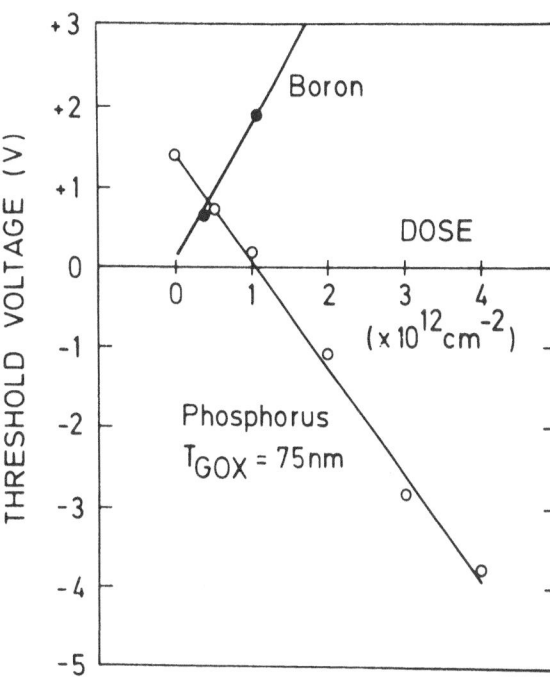

Fig. 5.8 - Dependence of depletion and enhancement threshold at zero bias on ion dose; according to Ref. 5.7.

(≈ 50 nm), Si_3N_4 is deposited and structured. Photoresist is used to mask against the boron or phosphorus implants. Typical doses are 10^{13} cm^{-2}. The dependence of the field threshold on dose and substrate bias for a 0.5μm thick oxide is shown in Fig. 5.11.

5.3 CMOS Devices

In CMOS devices, both n-channel and p-channel transistors are fabricated on the same chip. For most CMOS devices, all the above-mentioned techniques apply, and additionally, the doping of the p-well (or, sometimes, the n-well). Using a photoresist mask, boron (for p-wells) or phosphorus (for n-wells) is implanted with doses of 0.8 to 10×10^{12} cm^{-2}, and driven into the silicon to a depth of several microns at temperatures between 1150 and 1200°C. A typical diffusion time is 16h. After this step, complementary MOS transistors are fabricated, using standard techniques.

In Fig. 5.12, such a CMOS process is shown schematically. The threshold voltages of the transistors in the well depend on the implantation doses for the well itself, and also on those for the adjust implantation.

Examples for the dependence of the enhancement and depletion thresholds on the boron and phosphorus doses, in the case of n-well technology, are given in Fig. 5.13.[5.9] For example, doses of 1.5×10^{12} cm^{-2} phosphorus and 4×10^{11} cm^{-2} boron result in threshold voltages of $V_{Tn} = -V_{Tp} = 0.7$ V.

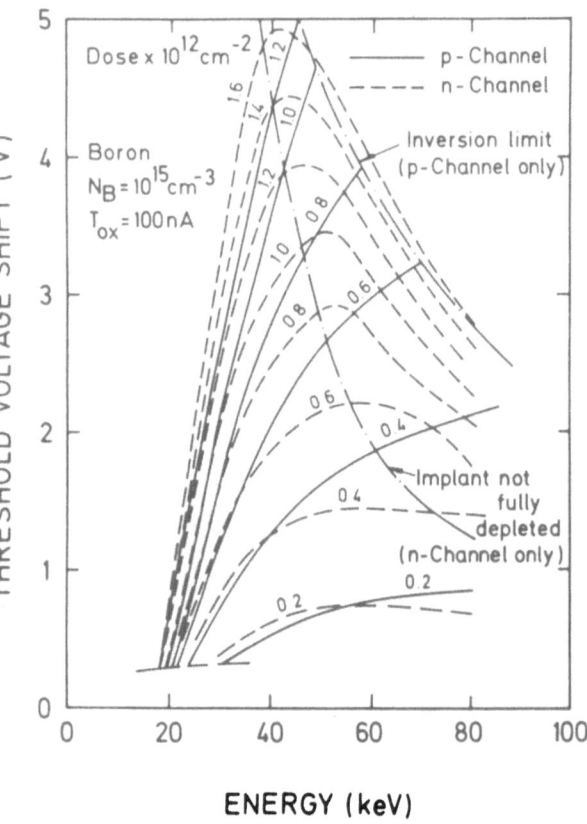

Fig. 5.9 - Threshold voltage shift vs. implant energy and dose for p- and n-channel devices.[5.8]

5.4 Charge-Coupled Devices

Charge-coupled devices (CCD's) are related to MOS devices. They use the same technology, but their operation depends on the storage of minority carriers in potential wells, instead of the conduction of majority carriers. The first application of implantation to CCD's was the manufacture of drain electrodes to avoid blooming.[5.10] CCD's are used for the detection of optical signals (line and area images),

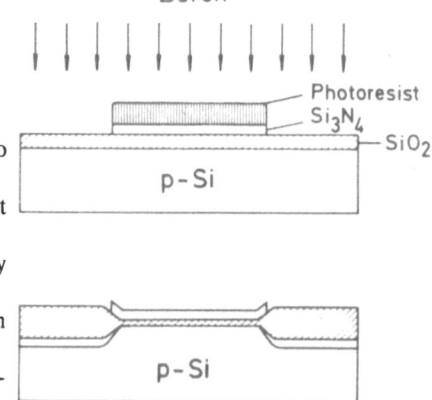

Fig. 5.10 - Schematic description of field threshold adjust.

for memories, or for analog signal processing.[5.11] The application of ion implantation is mainly the formation of buried channels to reduce the influence of the surface (trapping, etc.), by implanting low doses, e.g. 5×10^{12} cm^{-2}, of boron ions for p-channel devices[5.12] at fairly high energies. The charge transfer is enhanced due to fringing fields, but the area consumption is slightly larger than with surface channel CCD's. At the surface, a channel stop implant using phosphorus or arsenic

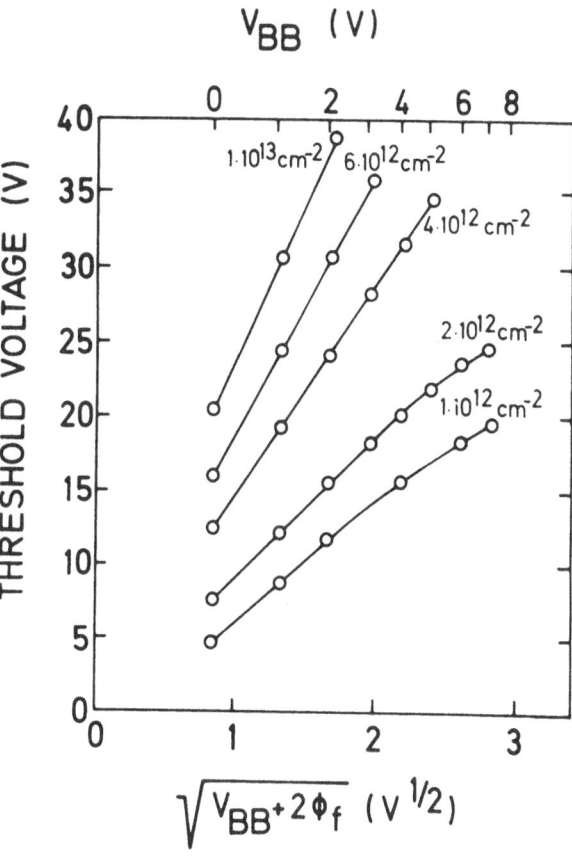

Fig. 5.11 - Dependence of field threshold on boron dose and bias.[5.7]

can also be performed. From the standpoint of density considerations, the two-phase CCD is the most important for storage applications. A cross section of a high-density two-phase coplanar electrode structure is given in Fig. 5.14.[5.13] This structure consists of two levels of polysilicon gates. A potential well is formed under one half of each phase electrode by selectively implanting arsenic ions in the vicinity of the SiO_2-Si interface through a photoresist mask. Polysilicon is deposited and etched in order to be offset from the implanted arsenic. After etching the gate oxide to self-align the arsenic implant to the first gate level, the gate oxide is regrown, and a similar process is used to define a potential well under the second polysilicon gate electrode. After this, a diffusion step is used to drive the arsenic into the

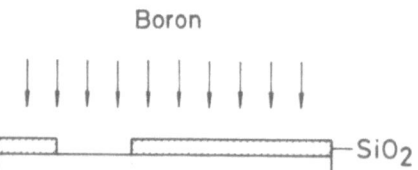

Boron

↓↓↓↓↓↓↓↓↓↓

— SiO₂

n-Si

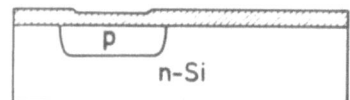

p

n-Si

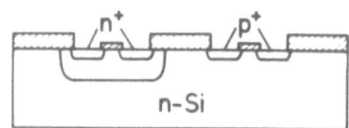

n⁺   p⁺

n-Si

Fig. 5.12 - Schematic description of CMOS process.

TABLE 5.2   One-micron MOS process; according to 5.14

1.	Starting substrate - p, 5Ωcm (100)
2.	Back-side I/I-B^{11}, 200 keV, 3×10^{15} cm^{-2}
3.	Back-side I/I-Ar^{40}, 350 keV, 5×10^{15} cm^{-2}
4.	Mask 1: defines alignment marks
5.	Mask 2: defines active area
6.	I/I field - B^{11}, 65 keV, 5×10^{12} cm^{-2}
7.	Growth of semi-recessed field oxide
8.	I/I enhancement channel - B^{11}, 20 keV, 9.3×10^{11} cm^{-2}
9.	Growth of gate oxide
10.	Mask 3: defines depletion channel I/I mask
11.	I/I depletion channel - As^{75}, 72 keV, 1.9×12 cm^{-2}
12.	Mask 4: defines polysilicon gate
13.	I/I source/drain - As^{75}, 100 keV, 1×16 cm^{-2}
14.	Mask 5: defines contact hole
15.	Mask 6: defines metallization

Nominal vertical dimensions in finished devices are:

Gate oxide - 25 nm

Field oxide under polysilicon - 305 nm

Polysilicon - 320 nm

Contact hole stack over source/drain and polysilicon - 320 nm

Field oxide under aluminum - 435 nm

Junction depth - 350 nm

Aluminum - 500 nm

silicon. To fabricate a buried-channel CCD, a phosphorus implant is performed prior to the formation of the CCD structure.

5.5 Typical MOS Processes

An advanced process for fabricating fully-implanted one-micron MOS circuits, using 6 implantation steps, is summarized in Table 5.2.[5.14] E-beam exposure is used throughout. Back-side ion implantation of boron and argon, the first step in wafer processing, is used for both gettering and ohmic back-side contacts. The third implant is the field implant, to avoid the formation of inversion channels. The threshold of enhancement transistors is adjusted using a boron implant. Source and drain are implanted using the gate oxide as a screen. An etch-back procedure of the thick oxide is used to avoid problems with low thick-oxide parasitic thresholds, believed to be caused by trapped holes in the arsenic-implanted field oxide. This has been done using the etch-rate ratio of 10 to 1 for implanted unannealed SiO_2 to thermal SiO_2, to minimize undercut of gate and field oxide beneath poly. This process involves no diffusion step, and is an example of a typical advanced VLSI technology for NMOS. In the future, the LOCOS process might be modified, using ion implantation of nitrogen to further increase di-

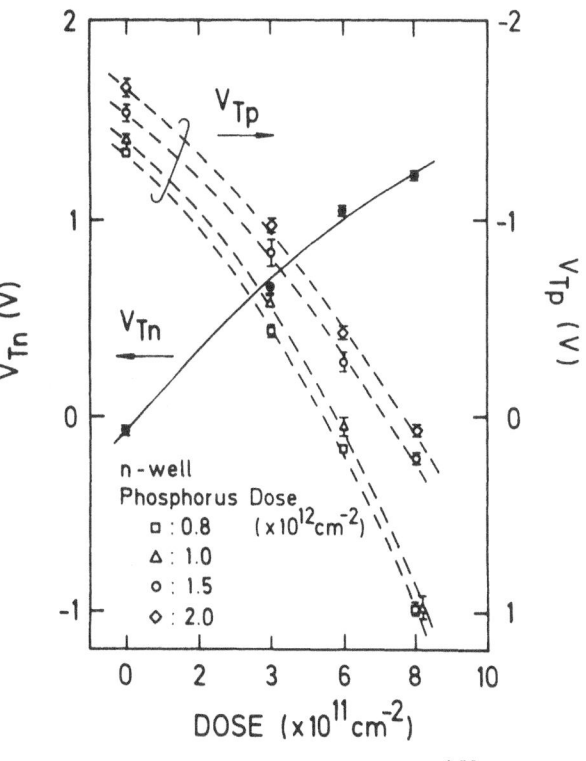

Fig. 5.13 - Threshold voltage T_{Tn} and V_{Tp} of n- and p-MOSFET's as a function of boron dose for threshold-voltage control.[5.9]

mensional control and thus decrease the dimensions.

Another very interesting process is a fully implanted CCD/CMOS technology developed by Motorola.[5.15] It consists of a n-channel CCD, which has a higher speed than a p-channel CCD, with double-polysilicon electrodes over two oxide thicknesses to permit a two-phase operation. In Fig. 5.15, a cross section of the structure is shown. The main process steps are listed in Table 5.3. At first, an argon getter implant on the back surface of the wafer is performed. For the substrate-adjust and p-well implant, a special sequence is used; see Fig. 5.16. A nitride layer on a thin oxide layer protects the p-well regions during the phosphorus implant in order to increase the minority carrier lifetime in these regions. After a subsequent oxidation, the boron p-well implant is performed. Both implants are then driven in at the same time. After stripping all oxide, a second argon getter implant is performed. In the next

TABLE 5.3   All-implanted CCD/CMOS process; according to 5.15

1.	Starting substrate: n, 8-12 Ωcm, (100)	20.	Mask 5: tap (if necessary)
2.	Back-side I/I - Ar, 100 keV, 3×10^{15} cm^{-2}	21.	I/I tap (if necessary) - As
3.	Oxide nitride layer 1	22.	Poly 1 deposition
4.	Mask 1: defines field oxide	23.	I/I poly 1 doping - P, 100 keV, 2×10^{15} cm^{-2}
5.	I/I substrate adjust - P, 150 keV, 1.8×10^{12} cm^{-2}	24.	Mask 6: poly 1 patterning
6.	Growth of thick oxide (5500Å), nitride strip	25.	Gate oxidation 2 (3200Å)
7.	I/I p-tub -B, 60 keV, 2×10^{18} .7.198 1cm^{-2}	26.	Poly 2 depositions
8.	Drive-in diffusion	27.	Mask 7: poly 2 patterning
9.	Oxide strip	28.	Mask 8: p-source/drain
10.	Back-side I/I - Ar, 100 keV, 3×10^{15} cm^{-2}	29.	I/I source/drain B
11.	Oxide-nitride-layer 2	30.	Mask 9: n-source/drain
12.	Mask 2: active area	31.	I/I source/drain - P, 105 keV, 7×10^{15} cm^{-2}
13.	Mask 3: p-field adjust	32.	Annealing
14.	I/I p-field adjust in p-tub - B	33.	TEOS, Phossil glass deposition
15.	Mask 4: n-field	34.	Mask 10: contact holes
16.	I/I n-field adjust - P	35.	Metal deposition
17.	Field oxidation	36.	Mask 11: metal patterning
18.	Strip of oxide-nitride layer 2	37.	Passivation glass
19.	Gate oxidation 1 (900Å)	38.	Mask 12: contact holes

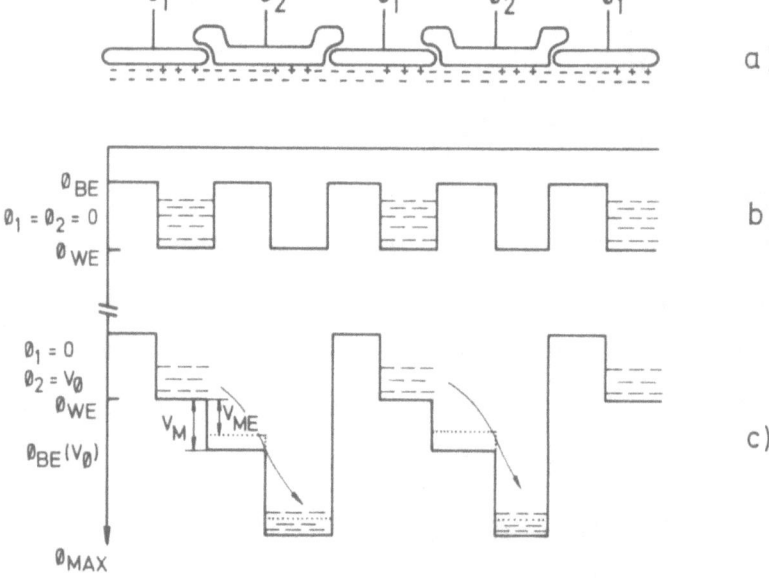

Fig. 5.14 - Two-phase coplanar CCD; (a) schematic cross section, (b) lateral potential distributions for the "store" condition (both ϕ_1 and ϕ_2 clocks are low), (c) lateral potential distribution for the "transfer" condition, to transfer charge from ϕ_1 to ϕ_2.[5.13]

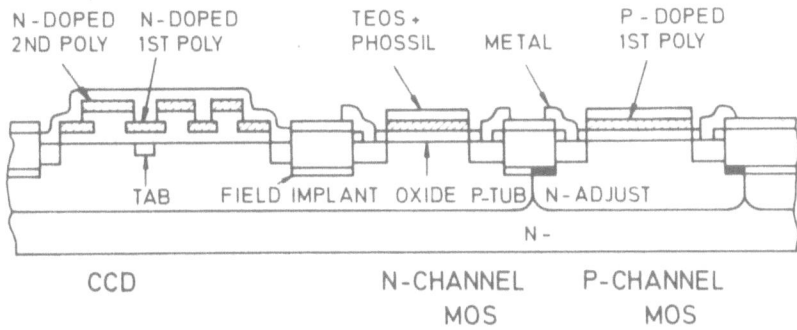

Fig. 5.15 - Cross section of CCD/CMOS structure.[5.15]

steps, the boron and phosphorus field implants are made in the p- and n-regions. An oxide-nitride pattern is used to protect the active areas during these implants as well as the subsequent field oxidation. After nitride stripping, the first gate oxide is grown. The arsenic-tap implant is then performed if needed. After the first poly-deposition, a phosphorus blanket implant is performed, giving a sheet resistance of 1 kΩ/$\square$. After patterning, as well as a second gate oxidation, and a second poly-deposition with patterning, the sheet resistance is reduced by the

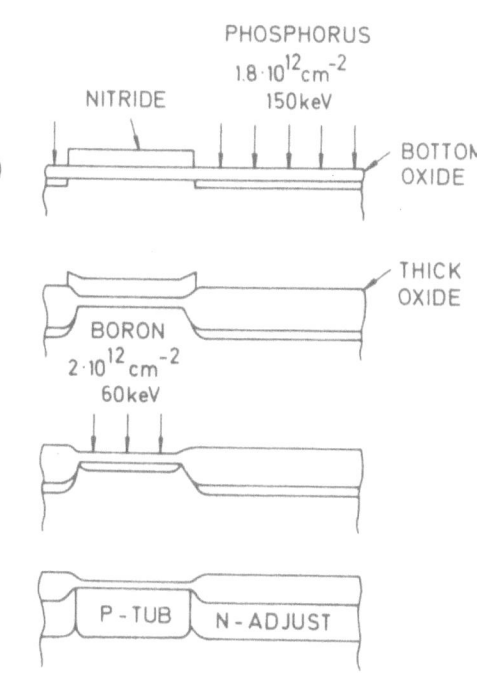

Fig. 5.16 - Self-aligned substrate adjust/p-tub sequence, (a) substrate adjust implant, (b) thick oxide growth, (c) nitride strip and p-tub implant, (d) tub drive-in.[5.15]

phosphorus drain/source implant to 35 Ω/$\square$ and, in the area of p-devices, it is reversed in conductivity by the boron drain/source implant to p-type with 107 Ω/$\square$. The end of the process is fairly standard with TEOS and PSG deposition, contact definition, metallization, passivation and mounting. This process involves the exceptional number of 10 implantation steps and, of course, no diffusion step.

The application of fluorine implantation to obtain polysilicon growth on SiO_2
as well as epitaxial growth on single-
crystalline silicon, for the manufacture of
high-density MOS circuits (BOMOS), is a
further example.[5.16] The process is shown
in Fig. 5.17. Several ion species such as
boron, BF_2, phosphorus, arsenic and fluorine were tried at various doses and energies. No significant differences were observed between species, which are equally
effective on the enhanced nucleation.
Among them, only the fluorine ion keeps
the deposited silicon film neutral in polarity after successive annealing, and fluorine

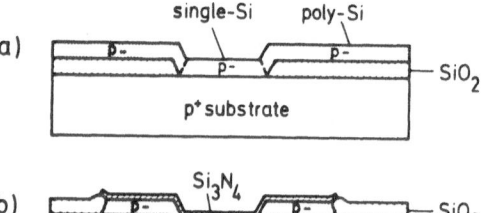

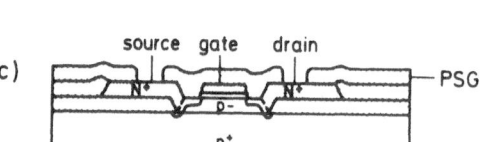

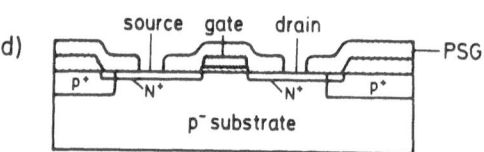

Fig. 5.17 - Schematic diagram of the production of BOMOS transistors (a,b,c) in comparison to a bulk MOS transistor (d).[5.16]

is fortunately available from boron trifluoride (BF_3), which is routinely used as a source gas

for BF_2 implantation in MOS production. The physical reason for the enhanced nucleation

effect is not yet well understood, but the ion-induced damage or internal stress near the oxide

surface seems to produce strong nucleation sites or traps for migrating Si atoms on the surface.

Fluorine implants of 5×10^{13} to 1×10^{14} cm^{-2} give the best result for a moderate taper-etch of

the oxide and the nucleation effect. The threshold is adjusted using boron implantation after

local oxidation and gate oxidation.

6. APPLICATION OF ION IMPLANTATION TO BIPOLAR DEVICES

The first experiments for the fabrication of bipolar transistors using implantation were not

very successful. Attempts were made to use boron for the base and phosphorus as the emitter

dopant. The breakthrough came with the use of arsenic as the emitter dopant.[6.1] Since that time, much work has been done on bipolar transistors, implanting boron for the base and arsenic for the emitter.

It is possible to implant either the emitter first or the base first. If the base is implanted first, then it diffuses during the emitter drive-in. Therefore, it is more advantageous to implant the emitter first, to drive in at a relatively high temperature (950 to 1000°C) and to subsequently implant the base, which has to be annealed at moderate temperatures (850 to 900°C). In this way, a good control of the process is maintained, and a coupled diffusion of boron and arsenic is avoided.

6.1 Current Gain

The current gain of an implanted transistor can be adjusted very easily by changing the base dose. The collector current of a transistor is given by:

$$I_C = \frac{q n_i^2 A_E D_n}{Q_B} \exp\left(\frac{q V_{BE}}{kT}\right) \tag{6.1}$$

with A_E the emitter area, V_{BE} the base-emitter voltage, D_n the diffusion coefficient of the electrons in the base and Q_B/D_n the base Gummel number. The base current is given by:

$$I_B = \frac{q n_i^2 A_E}{(Q_E/D_p)_{eff}} \exp\left(\frac{q V_{BE}}{kT}\right) \tag{6.2}$$

$(Q_E/D_p)_{eff}$ is the effective emitter Gummel number. Since the current gain h_{FE} is given by:

$$h_{FE} = \frac{I_C}{I_B} = \frac{(Q_E/D_p)_{eff}}{Q_B/D_n} \tag{6.3}$$

and

$$Q_B \propto N_\square \tag{6.4}$$

the current gain is inversely proportional to the dose of the base implantation (if the emitter has a constant technology):

$$h_{FE} \propto \frac{1}{N_\square} \tag{6.5}$$

Therefore, for a constant emitter efficiency, the current gain is adjustable by varying the base dose. An example for this relation is given in Fig. 6.1, where the current gain is plotted vs. the inverse base dose[6.2] for a 5 GHz transistor. The constant of proportionality in Eq. (6.4) is a function of the emitter Gummel number and the electron diffusion coefficient in the base, and is better found experimentally. A more detailed discussion of this problem is given by Archer.[6.3]

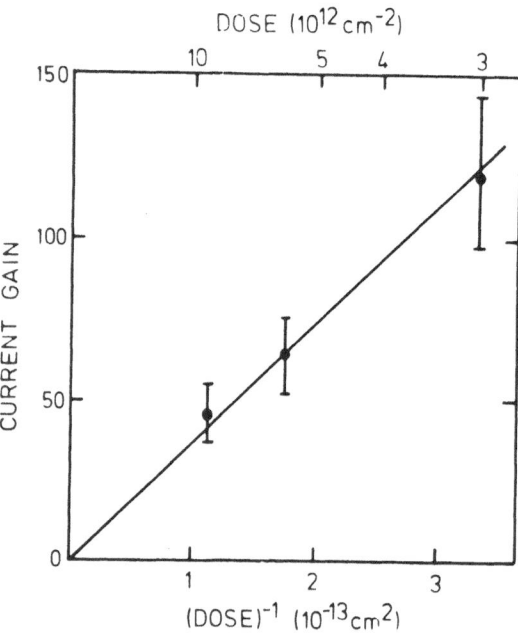

Fig. 6.1 - Current gain vs. the inverse of the deep-base implantation dose $(N_\square^{-1})$ for a 5-GHz all-implanted transistor.[6.2]

6.2 Cut-off Frequency

The cut-off frequency of a bipolar transistor is defined as the frequency at which the small-signal common-emitter short-circuit current gain is one. It is calculated as the inverse of the total emitter-to-collector delay time, i.e., $f_T = 1/2 \, \pi \tau_{ec}$. These delay times consist of the charging time of the emitter-base depletion layer, the transit time in the collector-base transition layer, the charging time of the collector capacitance and the base transit time. The main contribution is due to the base transit time, which is given by:

$$\tau_B = \frac{W_B^2}{\eta \, D_n} \tag{6.6}$$

for an n-p-n transistor where W_B is the base width and η is a measure of the average built-in electric field. Since, to the first order, the range of the ions is proportional to the energy, this offers a simple means of adjusting W_B and, thus, of adjusting the cut-off frequency by changing the implantation energy of the base. At the same

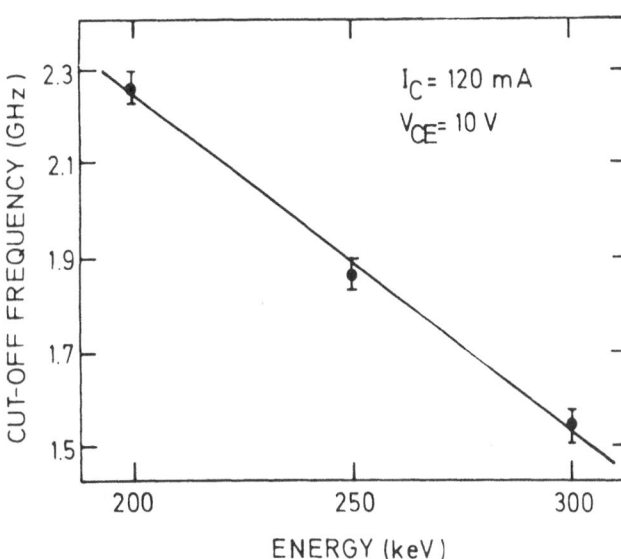

Fig. 6.2 - Cut-off frequency vs. the energy of the deep-base implantation for a 2-GHz all-implanted transistor.[6.2]

time, the current gain changes somewhat, since the number of ions in the base also changes a bit. A complete decoupling of cut-off frequency and current gain is therefore impossible.

An example for the variation of the cut-off frequency with varying implantation energy is given in Fig. 6.2.[6.2] For transistors operating at high frequencies (or clock-rates), two base implants are usually used. These are an active base implant as just discussed, in order to adjust the current gain, and an inactive implant to provide for a low resistance ohmic contact to the base. In order to avoid a compensation of the emitter, this implant is often performed outside the emitter region.

6.3 Subcollector

To reduce the series resistance of the collector, highly-doped subcollectors are used in IC technology. Instead of diffusing such subcollectors with antimony, it is possible to use the implantation of antimony or arsenic. The advantage is again a better reproducibility (in the case of antimony), or the compatibility with planar technology as well as lower sheet resistivities (in the case of arsenic, which can only be diffused in closed systems, but has a higher solubility than antimony). After implantation, an inert preannealing is required, followed by a drive-in step in an oxidizing atmosphere in order to avoid the formation of stacking faults during the deposition of the epi layer.

Using high-energy ion implantation, buried layers can be directly implanted, thus avoiding the epitaxial deposition of the collector region. An example of such a structure is given in the next chapter. A problem with this technique, however, is the considerable heating of the samples during implantation, if the implants are to be performed in a realistic time.

6.4 Typical Bipolar Processes

An example for an advanced $1\mu m$ I^2L technique was presented by Evans *et al.* at the 1979 IEDM Meeting.[6.4] They used 10 e-beam patterning steps, and made full use of ion implantation. The process sequence is given in Table 6.1. The first implant is arsenic for the collector; it is partially driven before resistor definition to minimize loss of arsenic during the plasma-oxide etching which defines the resistor. After the resistor implant and annealing, the

TABLE 6.1   One-micron bipolar VLSI process 6.4

PROCESS STEP MINIMUM GEOMETRY (μm)	CONDITIONS
(1) alignment marks; 10.0	$5\mu m$ Si etch (CF_4, O_2)
(2) isolation; 1.6	5500Å anisotropic plasma Si etch 11000Å oxidation at 950°C
(3) collector; 3.8	As; 80 keV, 3×10^{15} cm^{-2} (50Ω/□) anneal at 1000°C
(4) resistor; 2.0	As; 100 keV; 2.3×10^{13} cm^{-2} (500 Ω/□); anneal at 1000°C
(5) p$^+$ extrinsic base; 1.4	implant mask - 6300Å plasma oxide; B; 50,70 keV; 1.5×10^{15} cm^{-2} (57 Ω/□)
(6) p$^-$ intrinsic base; 8.1	implant mask - 16000Å resist; B; 300 keV; 2×10^{12} cm^{-2} (20 KΩ/□); anneals and passivation at 900°C
(7) contacts; 1.0	
(8) 1st level metal; 3.0	Pt (200Å)/TiW (1750Å)/Al+Cu (4000Å)
(9) vias; 1.1	6000Å plasma oxide dielectric
(10) 2nd level metal; 3.8	TiW (1750Å)/Al+Cu (6000Å)

base contact is made, using a double-energy implantation at 50 and 70 keV to obtain a flatter profile. The active base is made by implanting boron with 300 keV. For comparison, I^2L circuits with Schottky outputs were also made. In that case, no collector implantation was performed, but rather, a 150 keV phosphorus implant to increase down beta and a 60 keV arsenic implant to lower the PtSi barrier height.

High-voltage transistors can be produced on the same chip as I^2L circuits, if the I^2L emitter is doped using an additional phosphorus implant,[6.5] as shown in Fig. 6.3. A different approach to obtain high-speed ECL and I^2L circuits on a chip is given in Fig. 6.4.[6.6] The high

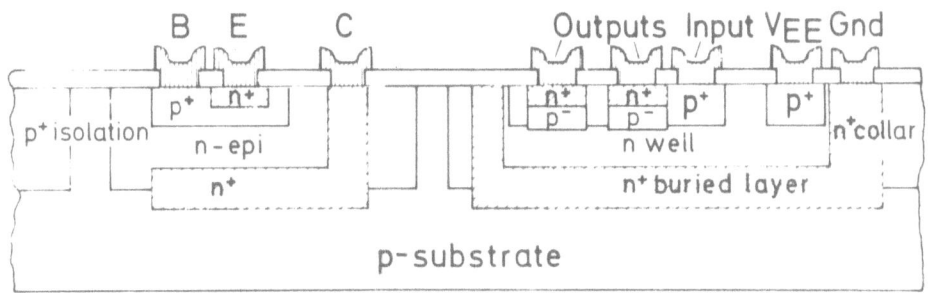

Fig. 6.3 - A schematic cross section of a high speed I^2L gate and a high-voltage linear npn transistor.[6.5]

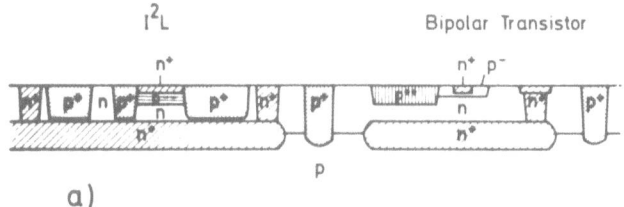

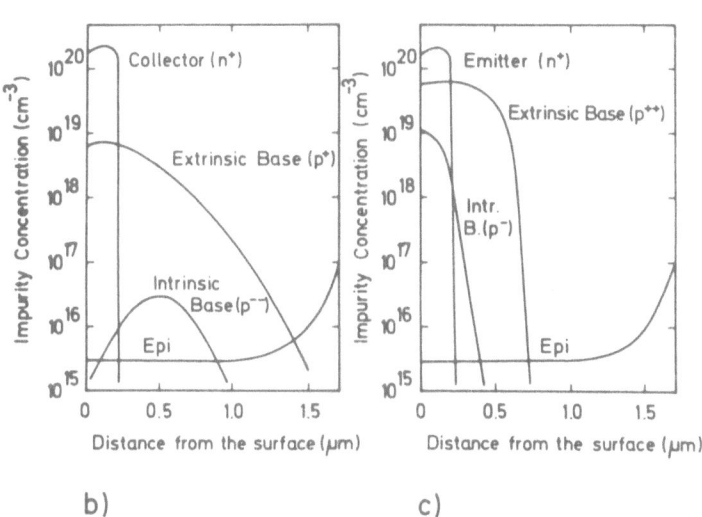

Fig. 6.4 - Cross section (a) and doping profiles of I^2L (b) and linear (c) transistors.[6.6]

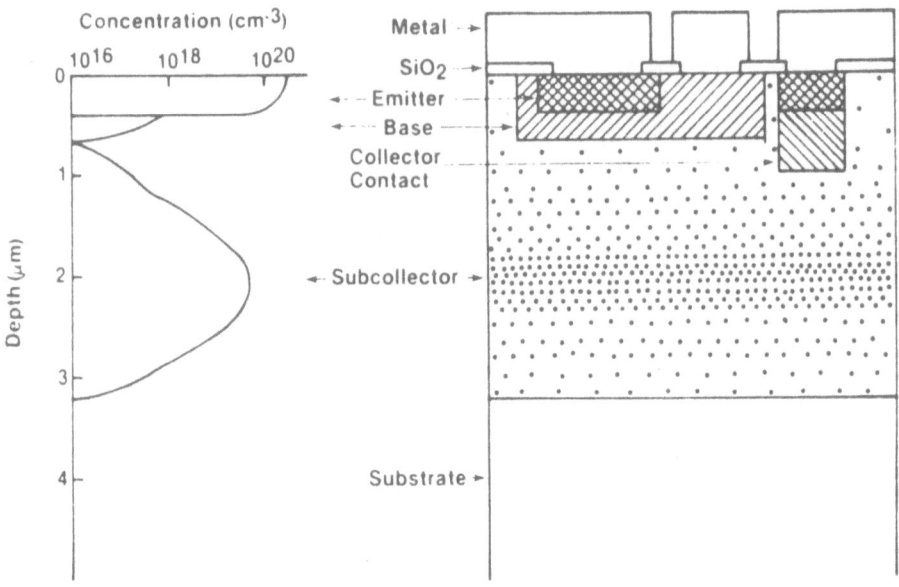

Fig. 6.5 - Schematic diagram of a non-epitaxial bipolar structure produced by high-energy ion implantation, together with the common emitter transfer characteristic of this structure.[6.7]

upward gain for I^2L inverters and moderate downward gain for linear transistors were obtained using four different implants for the active and inactive base regions of the two structures. The implants were 260 keV with 10^{12} cm^{-2} and 60 keV with 3.4×10^{14} cm^{-2} for I^2L, as well as 40 keV with 1.7×10^{14} cm^{-2} and 40 keV with 10^{15} cm^{-2} for ECL circuits, respectively.

A transistor completely fabricated using low and high energy implantation was developed at IBM.[6.7] A cross section of this transistor, together with a doping profile, is given in Fig. 6.5. Phosphorus with 2.9 MeV and a dose of 3.5×10^{15} cm^{-2} was implanted into p$^+$-silicon, forming the buried layer and the collector. Subsequently, boron and arsenic were implanted to form the base and emitter. For the collector, an additional phosphorus reach-through implant was used. A single high-temperature step at 1000°C was employed. The transistors showed a current gain of 120. The sheet resistivity of the subcollector was 22 $\Omega/\square$. Such buried layers can be produced selectively, and with higher yield than extremely thin epitaxial layers.

ACKNOWLEDGEMENTS

This work was partly assisted by the "Deutsche Forschungsgemeinschaft" and the "Bundesministerium für Forschung und Technologie" of West Germany. I am indebted to my co-workers J. Götzlich, K. Haberger, K. Hoffmann, H. Kranz, and G. Prinke for their contributions. The technical assistance of B. Schmiedt, J. Bosch, M. Bleier, E. Traumüller, and A. Sixt is very much appreciated.

7. APPENDIX: RANGE DATA

In the following tables, projected range and range straggling of antimony, argon, arsenic, boron, and phosphorus in silicon, SiO_2, Si_3N_4 and PMMA-resist are given, calculated according to the theory of Biersack (J. Biersack, Nucl., Inst. & Methods 182/183, 199 (1981)).

Table 7. 1   Range and range straggling of various ions vs in silicon (ρ = 2.33 g/cm^3)

Energy (keV)	Antimony (Sb) R_p (μm)	ΔR_p (μm)	Argon (Ar) R_p (μm)	ΔR_p (μm)	Arsenic (As) R_p (μm)	ΔR_p (μm)	Boron (B) R_p (μm)	ΔR_p (μm)	Krypton (K) R_p (μm)	ΔR_p (μm)	Phosphorus (P) R_p (μm)	ΔR_p (μm)
10							.0309	.0193				.0072
20							.0633	.0319				.0126
30							.0956	.0415				.0176
40							.1274	.0492				
50							.1574	.0524				
60							.1866	.0610				.0313
70							.2152	.0657				
80							.2426	.0690				.0395
90							.2696	.0735				
100							.2966	.0764				
110							.3209	.0798				
120							.3454	.0825				
130							.3697	.0849				
140							.3932	.0872				
150							.4163	.0893				
160							.4386	.0913				
170							.4610	.0931				
180							.4827	.0948				.0722
190							.5041	.0964				
200							.5251	.0979				
220							.5661	.1007				
240							.6059	.1032				
260							.6446	.1055				
280							.6823	.1076				
300							.7192	.1096				
320							.7572	.1113				
340							.7906	.1130				
360							.8252	.1146				
380							.8562	.1160				
400							.8927	.1174				
420							.9254	.1187				
440							.9580	.1199				
460							.9900	.1211				
480							1.0215	.1222				
500							1.0526	.1232				

Table 7.2 Range and range straggling of various ions vs in SiO_2 (ρ = 2.27 g/cm^3)

Energy (keV)	Antimony(Sb) R_p (μm)	ΔR_p (μm)	Argon (Ar) R_p (μm)	ΔR_p (μm)	Arsenic (As) R_p (μm)	ΔR_p (μm)	Boron (B) R_p (μm)	ΔR_p (μm)	Krypton (K) R_p (μm)	ΔR_p (μm)	Phosphorus (P) R_p (μm)	ΔR_p (μm)
10	.0087	.0025	0.0117	0.0048	.0094	.0031	.0322	.0192	0.0097	0.0029	.0125	.0059
20	.0140	.0038	0.0213	0.0084	.0157	.0050	.0602	.0335	0.0158	0.0047	.0236	.0105
30	.0187	.0050	0.0310	0.0118	.0216	.0068	.1076	.0454	0.0214	0.0063	.0350	.0149
40	.0231	.0061	0.0407	0.0152	.0273	.0085	.1457	.0551	0.0268	0.0079	.0467	.0191
50	.0273	.0072	0.0506	0.0185	.0329	.0102	.1935	.0632	0.0322	0.0094	.0584	.0232
60	.0315	.0082	0.0607	0.0217	.0385	.0118	.2603	.0703	0.0374	0.0109	.0707	.0272
70	.0355	.0092	0.0708	0.0248	.0441	.0135	.2563	.0764	0.0427	0.0124	.0830	.0310
80	.0395	.0102	0.0811	0.0279	.0497	.0151	.2915	.0818	0.0479	0.0138	.0955	.0347
90	.0435	.0112	0.0915	0.0310	.0553	.0167	.3285	.0866	0.0531	0.0152	.1080	.0383
100	.0474	.0122	0.1020	0.0339	.0609	.0183	.3644	.0910	0.0583	0.0167	.1207	.0418
110	.0513	.0132	0.1125	0.0368	.0665	.0198	.3914	.0940	0.0635	0.0181	.1334	.0451
120	.0552	.0141	0.1231	0.0397	.0721	.0214	.4222	.0965	0.0688	0.0195	.1461	.0484
130	.0591	.0151	0.1338	0.0424	.0778	.0229	.4545	.1015	0.0740	0.0209	.1589	.0515
140	.0630	.0160	0.1445	0.0451	.0834	.0245	.4549	.1045	0.0792	0.0223	.1717	.0545
150	.0668	.0170	0.1552	0.0478	.0891	.0260	.5140	.1074	0.0845	0.0237	.1845	.0574
160	.0706	.0179	0.1661	0.0504	.0948	.0275	.5441	.1104	0.0897	0.0250	.1973	.0603
170	.0745	.0188	0.1769	0.0529	.1005	.0291	.5729	.1127	0.0950	0.0264	.2101	.0630
180	.0783	.0198	0.1877	0.0554	.1063	.0306	.6013	.1149	0.1003	0.0278	.2229	.0656
190	.0822	.0207	0.1986	0.0578	.1120	.0321	.6291	.1171	0.1056	0.0291	.2356	.0682
200	.0860	.0216	0.2094	0.0602	.1178	.0336	.6545	.1191	0.1109	0.0305	.2484	.0707
220	.0936	.0234	0.2312	0.0647	.1294	.0365	.7104	.1224	0.1215	0.0331	.2737	.0754
240	.1013	.0252	0.2529	0.0691	.1410	.0394	.7627	.1262	0.1322	0.0358	.2988	.0799
260	.1090	.0271	0.2745	0.0733	.1527	.0423	.8130	.1243	0.1430	0.0384	.3238	.0840
280	.1166	.0288	0.2961	0.0773	.1645	.0452	.8625	.1321	0.1538	0.0410	.3466	.0880
300	.1243	.0306	0.3177	0.0811	.1763	.0480	.9109	.1347	0.1646	0.0436	.3731	.0917
320	.1320	.0324	0.3391	0.0848	.1882	.0507	.9589	.1371	0.1755	0.0462	.3974	.0952
340	.1397	.0342	0.3604	0.0883	.2001	.0535	1.0046	.1393	0.1864	0.0487	.4215	.0986
360	.1475	.0359	0.3817	0.0916	.2120	.0562	1.0504	.1414	0.1974	0.0512	.4453	.1018
380	.1552	.0377	0.4028	0.0949	.2240	.0589	1.0952	.1434	0.2084	0.0537	.4689	.1048
400	.1630	.0394	0.4238	0.0980	.2360	.0615	1.1392	.1452	0.2195	0.0561	.4923	.1077
420	.1707	.0411	0.4446	0.1009	.2481	.0641	1.1826	.1470	0.2305	0.0585	.5154	.1104
440	.1785	.0429	0.4653	0.1038	.2601	.0667	1.2254	.1488	0.2416	0.0609	.5383	.1130
460	.1863	.0446	0.4859	0.1066	.2722	.0692	1.2675	.1504	0.2527	0.0633	.5610	.1155
480	.1941	.0463	0.5064	0.1092	.2844	.0717	1.3091	.1516	0.2639	0.0656	.5835	.1179
500	.2020	.0480	0.5267	0.1118	.2965	.0741	1.3502	.1530	0.2750	0.0680	.6057	.1202

Table 7.3 Range and range straggling of various ions vs in Si_3N_4 (ρ = 2.9 g/cm³)

Energy (keV)	Antimony (Sb) Rp (μm)	ΔRp (μm)	Argon (Ar) Rp (μm)	ΔRp (μm)	Arsenic (As) Rp (μm)	ΔRp (μm)	Boron (B) Rp (μm)	ΔRp (μm)	Krypton (K) Rp (μm)	ΔRp (μm)	Phosphorus (P) Rp (μm)	ΔRp (μm)
10	.0069	.0020	0.0093	0.0038	.0074	.0024	0.0258	0.0149	0.0077	0.0023	.0098	.0047
20	.0110	.0030	0.0169	0.0066	.0122	.0040	0.0554	0.0261	0.0125	0.0037	.0186	.0083
30	.0147	.0040	0.0245	0.0094	.0170	.0054	0.0859	0.0351	0.0169	0.0050	.0275	.0118
40	.0182	.0048	0.0323	0.0120	.0215	.0068	0.1162	0.0425	0.0212	0.0063	.0367	.0152
50	.0215	.0057	0.0401	0.0146	.0259	.0081	0.1460	0.0486	0.0255	0.0075	.0460	.0184
60	.0249	.0065	0.0480	0.0171	.0303	.0094	0.1749	0.0539	0.0296	0.0086	.0556	.0216
70	.0280	.0073	0.0561	0.0196	.0347	.0107	0.2030	0.0584	0.0338	0.0098	.0652	.0246
80	.0311	.0081	0.0642	0.0221	.0391	.0120	0.2304	0.0624	0.0379	0.0110	.0750	.0276
90	.0342	.0089	0.0724	0.0245	.0435	.0132	0.2569	0.0659	0.0420	0.0121	.0849	.0304
100	.0373	.0097	0.0807	0.0268	.0479	.0145	0.2827	0.0690	0.0461	0.0132	.0948	.0332
110	.0404	.0105	0.0890	0.0291	.0522	.0157	0.3079	0.0719	0.0503	0.0143	.1048	.0359
120	.0435	.0112	0.0974	0.0313	.0566	.0170	0.3324	0.0744	0.0544	0.0155	.1148	.0384
130	.0465	.0120	0.1059	0.0335	.0612	.0182	0.3562	0.0768	0.0585	0.0166	.1249	.0409
140	.0496	.0127	0.1143	0.0357	.0657	.0194	0.3796	0.0789	0.0626	0.0177	.1349	.0433
150	.0527	.0135	0.1228	0.0378	.0701	.0206	0.4023	0.0809	0.0668	0.0188	.1450	.0457
160	.0558	.0142	0.1313	0.0398	.0746	.0218	0.4246	0.0827	0.0709	0.0199	.1551	.0479
170	.0588	.0150	0.1399	0.0418	.0791	.0231	0.4464	0.0844	0.0751	0.0210	.1651	.0501
180	.0617	.0157	0.1484	0.0438	.0834	.0243	0.4678	0.0860	0.0793	0.0220	.1752	.0522
190	.0647	.0164	0.1570	0.0457	.0881	.0254	0.4888	0.0874	0.0834	0.0231	.1852	.0542
200	.0677	.0172	0.1656	0.0476	.0927	.0266	0.5093	0.0888	0.0876	0.0242	.1952	.0562
220	.0737	.0186	0.1827	0.0512	.1018	.0290	0.5493	0.0913	0.0960	0.0263	.2151	.0600
240	.0797	.0201	0.1999	0.0546	.1109	.0313	0.5880	0.0936	0.1045	0.0284	.2349	.0635
260	.0855	.0215	0.2170	0.0579	.1201	.0334	0.6255	0.0956	0.1130	0.0305	.2546	.0669
280	.0918	.0229	0.2341	0.0611	.1294	.0355	0.6618	0.0974	0.1215	0.0326	.2741	.0700
300	.0978	.0243	0.2511	0.0641	.1387	.0375	0.6971	0.0991	0.1301	0.0346	.2934	.0730
320	.1039	.0257	0.2680	0.0670	.1480	.0403	0.7315	0.1006	0.1386	0.0366	.3125	.0758
340	.1100	.0271	0.2848	0.0698	.1574	.0424	0.7651	0.1020	0.1473	0.0386	.3315	.0785
360	.1171	.0285	0.3016	0.0724	.1668	.0445	0.7978	0.1033	0.1559	0.0406	.3502	.0810
380	.1201	.0299	0.3183	0.0750	.1762	.0457	0.8299	0.1046	0.1646	0.0426	.3688	.0834
400	.1262	.0313	0.3348	0.0774	.1857	.0479	0.8612	0.1057	0.1733	0.0445	.3872	.0857
420	.1343	.0327	0.3513	0.0798	.1951	.0499	0.8919	0.1068	0.1820	0.0464	.4055	.0879
440	.1402	.0341	0.3677	0.0821	.2046	.0520	0.9220	0.1077	0.1908	0.0483	.4235	.0900
460	.1463	.0354	0.3839	0.0843	.2141	.0545	0.9515	0.1087	0.1996	0.0502	.4414	.0920
480	.1427	.0368	0.4001	0.0864	.2237	.0556	0.9805	0.1096	0.2084	0.0521	.4591	.0940
500	.1530	.0381	0.4161	0.0884	.2334	.0555	1.0090	0.1104	0.2172	0.0539	.4766	.0958

Table 7.4 Range and range straggling of various ions vs in PMMA (ρ = 1.19 g/cm³)

Energy (keV)	Antimony(Sb) R_p (μm)	ΔR_p (μm)	Argon (Ar) R_p (μm)	ΔR_p (μm)	Arsenic (As) R_p (μm)	ΔR_p (μm)	Boron (B) R_p (μm)	ΔR_p (μm)	Krypton (K) R_p (μm)	ΔR_p (μm)	Phosphorus (P) R_p (μm)	ΔR_p (μm)
10	0.0161	0.0032	0.0192	0.0054	0.0158	0.0037	0.0540	0.0171	0.0158	0.0036	0.0213	0.0063
20	0.0248	0.0048	0.0349	0.0096	0.0261	0.0060	0.1113	0.0294	0.0257	0.0057	0.0397	0.0114
30	0.0325	0.0062	0.0504	0.0135	0.0358	0.0082	0.1678	0.0386	0.0348	0.0077	0.0583	0.0163
40	0.0397	0.0076	0.0659	0.0174	0.0451	0.0102	0.2222	0.0458	0.0436	0.0095	0.0772	0.0209
50	0.0467	0.0089	0.0817	0.0212	0.0542	0.0123	0.2743	0.0514	0.0522	0.0114	0.0963	0.0255
60	0.0535	0.0101	0.0975	0.0249	0.0633	0.0142	0.3241	0.0560	0.0607	0.0132	0.1156	0.0298
70	0.0601	0.0114	0.1136	0.0285	0.0724	0.0162	0.3718	0.0599	0.0691	0.0149	0.1351	0.0340
80	0.0667	0.0126	0.1297	0.0320	0.0814	0.0181	0.4177	0.0632	0.0775	0.0167	0.1547	0.0380
90	0.0732	0.0138	0.1460	0.0355	0.0904	0.0200	0.4619	0.0660	0.0858	0.0184	0.1744	0.0419
100	0.0796	0.0150	0.1623	0.0388	0.0994	0.0219	0.5045	0.0685	0.0942	0.0201	0.1941	0.0456
110	0.0860	0.0162	0.1787	0.0421	0.1084	0.0238	0.5457	0.0706	0.1026	0.0218	0.2138	0.0492
120	0.0923	0.0173	0.1952	0.0453	0.1175	0.0256	0.5857	0.0726	0.1109	0.0235	0.2335	0.0526
130	0.0987	0.0185	0.2117	0.0484	0.1265	0.0275	0.6246	0.0743	0.1193	0.0251	0.2532	0.0559
140	0.1049	0.0196	0.2283	0.0514	0.1356	0.0293	0.6623	0.0759	0.1276	0.0268	0.2729	0.0590
150	0.1112	0.0208	0.2449	0.0543	0.1447	0.0311	0.6992	0.0773	0.1360	0.0285	0.2924	0.0620
160	0.1175	0.0219	0.2614	0.0572	0.1538	0.0329	0.7351	0.0786	0.1444	0.0301	0.3120	0.0650
170	0.1237	0.0230	0.2780	0.0600	0.1629	0.0347	0.7702	0.0798	0.1528	0.0317	0.3314	0.0678
180	0.1300	0.0241	0.2946	0.0627	0.1721	0.0365	0.8045	0.0809	0.1612	0.0333	0.3508	0.0705
190	0.1362	0.0253	0.3112	0.0654	0.1812	0.0383	0.8382	0.0820	0.1696	0.0349	0.3700	0.0731
200	0.1424	0.0264	0.3278	0.0680	0.1904	0.0400	0.8712	0.0829	0.1781	0.0365	0.3892	0.0756
220	0.1548	0.0286	0.3608	0.0729	0.2089	0.0435	0.9353	0.0847	0.1950	0.0397	0.4273	0.0803
240	0.1673	0.0308	0.3937	0.0776	0.2274	0.0469	0.9974	0.0862	0.2120	0.0429	0.4649	0.0847
260	0.1797	0.0329	0.4264	0.0821	0.2460	0.0503	1.0575	0.0876	0.2290	0.0460	0.5020	0.0888
280	0.1921	0.0351	0.4590	0.0864	0.2647	0.0536	1.1159	0.0889	0.2462	0.0490	0.5386	0.0927
300	0.2046	0.0372	0.4914	0.0904	0.2834	0.0569	1.1728	0.0900	0.2633	0.0520	0.5750	0.0963
320	0.2170	0.0393	0.5235	0.0943	0.3022	0.0601	1.2284	0.0911	0.2806	0.0550	0.6109	0.0997
340	0.2295	0.0414	0.5554	0.0980	0.3210	0.0633	1.2828	0.0920	0.2978	0.0580	0.6462	0.1029
360	0.2420	0.0435	0.5871	0.1015	0.3399	0.0664	1.3360	0.0929	0.3152	0.0609	0.6812	0.1060
380	0.2545	0.0456	0.6186	0.1049	0.3588	0.0695	1.3882	0.0938	0.3325	0.0638	0.7157	0.1088
400	0.2670	0.0477	0.6498	0.1081	0.3778	0.0726	1.4374	0.0945	0.3499	0.0667	0.7498	0.1115
420	0.2795	0.0498	0.6807	0.1112	0.3968	0.0755	1.4899	0.0953	0.3674	0.0695	0.7835	0.1141
440	0.2921	0.0518	0.7115	0.1142	0.4158	0.0785	1.5395	0.0960	0.3848	0.0723	0.8168	0.1165
460	0.3047	0.0539	0.7419	0.1170	0.4348	0.0814	1.5883	0.0966	0.4023	0.0750	0.8498	0.1189
480	0.3173	0.0559	0.7722	0.1198	0.4538	0.0843	1.6365	0.0973	0.4199	0.0777	0.8823	0.1211
500	0.3299	0.0579	0.8022	0.1224	0.4728	0.0871	1.6841	0.0979	0.4374	0.0804	0.9145	0.1232

REFERENCES

1.1   W. Shockley, U.S. Patent No. 2787, 564 (1957).

1.2   T. Alväger and N. J. Hansen, Rev. Sci. Inst. $\underline{33}$, 367 (1962).

1.3   W. J. King, J. T. Burrel, S. Harrison, F. Martin and C. M. Kellett, Nucl. Inst. & Methods $\underline{38}$, 178 (1965).

1.4   J. Lindhard, M. Scharff, and H. E. Schiott, Kgl. Danske Videnskab.   Selskab., Mat.-Fys. Medd $\underline{33}$, 14 (1963).

1.5   R. W. Bower and H. G. Dill, IEEE Int. Electron Devices Meeting, Washington (1966).

1.6   K. G. Aubuchon, Int. Conf. on Prop. and Use of MIS Structures, Grenoble (1969).

1.7   R. S. Payne and R. J. Scavuzzo, IEEE Int. Electron Devices Meeting, Washington (1971).

2.1   J. Lindhard, M. Scharff and H. E. Schiott, Kgl. Danske Videnskab. Selskab., Mat. Phys. Medd. $\underline{33}$, 14 (1963).

2.2   J. P. Biersack, Nucl. Inst. & Methods $\underline{182/183}$, 199 (1981).

2.3   J. B. Sanders, Can. J. Phys. $\underline{46}$, 455 (1968).

2.4   J. F. Gibbons, W. S. Johnson and W. S. Mylroie, "Projected Range Statistics in Semiconductors", Dowden Hutchinson and Ross, Academic Press, Stroudsburg (1975).

2.5   B. Smith, "Ion Implantation Range Data for Silicon and Germanium Device Technologies", Learned Information (Europe) Ltd., Oxford (1977).

2.6   W. K. Hofker, Philips Res. Rep., Suppl. No. 8 (1975).

2.7   J. F. Gibbons and S. Mylroie, Appl. Phys. Lett. $\underline{22}$, 568 (1973).

2.8   N. L. Johnson and S. Kotz, "Continuous Univariate Distributions", Vol. 2, Wiley , New York (1970).

2.9   V. G. K. Reddi and A. Y. C. Yu, Solid State Tech. $\underline{15}$, 35 (1972).

2.10   J. F. Gibbons, in "Handbook on Semiconductors", Vol. 3 (Ed. T.S. Moss), North Holland Publishing Company, Amsterdam (1980).

2.11   F. Jahnel, H. Ryssel, J. Biersack, H. Haberger, K. Müller and Henckelmann, Nucl. Inst. Methods $\underline{182/183}$, 223 (1981).

2.12   H. Matsumura and S. Furukawa, Jpn. J. Appl. Phys. $\underline{14}$, 1983 (1976).

2.13   H. Runge, Phys. Stat. Sol. $\underline{39}$, 595 (1977).

2.14   L. Christel, Second Int. Conf. Ion Beam Modification of Materials, Albany, 1980.

2.15   H. Ishiwara, S. Furukawa, J. Yamada and M. Kawamura, in "Ion Implantation in Semiconductors", Ed. S. Namba, Plenum Press, New York (1975).

2.16   R. A. Moline and A. G. Cullis, Appl. Phys. Lett. 26, 551 (1975).

2.17   G. Fischer, G. Carter and R. Webb, Rad. Effects 38, 41 (1978).

2.18   P. Sigmund, J. Appl. Phys. 50, 726 (1979).

2.19   T. Hirao, K. Inoue and Takayanagi, J. Appl. Phys. 50, 193 (1979).

2.20   A. Goetzberger, IEDM, Washington (1975).

2.21   R. A. Moline, G. W. Reutlinger and J. C. North in "Atomic Collisions in Solids", Vol. 1 (Eds. J. Datz, B.R. Appleton, C.D. Moak), New York (1975).

2.22   W. K. Chu, M. J. Sullivan, S. M. Ku and M. Shatzkes, First Int. Conf. on Ion Beam Modification of Materials, Budapest (1978).

2.23   T. Ito, S. Hijya, H. Nishi, M. Shinoda and T. Furuya, Jpn. J. Appl. Phys. 17, 201 (1978).

2.24   J. M. Poate and T. C. Tisone, Appl. Phys. Lett. 24, 391 (1974).

2.25   H. Nishi, T. Sakurai, T. Akamatsu and T. Furuya, Appl. Phys. Lett. 26, 337 (1974).

2.26   Z. L. Liau and J. W. Mayer in "Treatise on Materials Science and Technology", Ed. J. K. Hirvonen, Academic Press, New York (1980).

2.27   B. Y. Tsaur, Z. L. Liau and J. W. Mayer, Appl. Phys. Lett. 34, 167 (1979).

2.28   J. Lindhard, Kgl. Danske Videnskab Selskab, Mat.-Fys. Medd. 34, 14 (1965).

2.29   G. Dearnaley, J. H. Freemann, G. A. Gard, M. A. Wilkins, Can. J. Phys. 46, 587 (1968).

2.30   R. A. Moline, G. W. Reutlinger in "Ion Implantation in Semiconductors" (Eds. I. Ruge, J. Graul), Berlin, Heidelberg, New York (1971), p. 58.

3.1   P. Sigmund and J. B. Sanders, Proc. Int. Conf. Appl. Ion Beams Semicond. Technol. Ed. P. Glotin, Grenoble, 1967, p. 215.

3.2   P. V. Pavlov, D. E. Tetel'baum, E. I. Zorin and V. I. Alekseev, Sov. Phys. Solid State 8, 2141 (1967).

3.3   D. K. Brice, Rad. Effects 11, 227 (1971).

3.4   D. K. Brice, "Ion Implantation Range and Energy Deposition Distributions, Vol. 1: High Incident Ion Energies," Plenum Press, New York, 1975.

3.5   K. B. Winterborn, "Ion Implantation Range and Energy Deposition Distributions, Vol. 2: Low Incident Ion Energies," Plenum Press, New York, 1975.

3.6   T. Tsurushima and H. Tanoue, J. Phys. Soc. Jpn. 31, 1965 (1971).

4.1   G. Soncini, these proceedings

4.2   I. Ruge, H. Müller and H. Ryssel in "Advances in Solid State Physics XII," Pergamon-Vieweg, Ed. O. Madelung, 1972.

4.3   L. Csperegi, E. F. Kennedy and J. W. Mayer, J. Appl. Phys. 49, 3906 (1978).

4.4   L. Csepregi, J. W. Mayer and T. W. Sigmon, Appl. Phys. Lett. 29, 92 (1976).

4.5   L. Csepregi, E. F. Kennedy, T. J. Gallagher and J. W. Mayer, J. Appl. Phys. 48, 4234 (1977).

4.6   E. F. Kennedy, L. Csepregi, J. W. Mayer and T. W. Sigmon, J. Appl. Phys. 48, 4241 (1977).

4.7   S. Prussin and A. M. Ferm, J. Electrochem. Soc. 122, 830 (1975).

4.8   S. Prussin in "Ion Implantation in Semiconductors," Ed. S. Namba, New York, 1975, p. 499.

4.9   J. J. Comer and S. A. Roosild, Rad. Effects 25, 275 (1975).

4.10   S. Hasegawa, K. E. Forward and H. Hartnagel, Electron Lett. 11, 53 (1975).

4.11   S. Prussin, J. Appl. Phys. 45, 1635 (1974).

4.12   T. E. Seidel, R. S. Payne, R. A. Moline, W. R. Costello, J. C. C. Tsai and K. R. Garnder, IEDM Digest, p. 581 (1975).

4.13   L. Csepregi, W. K. Chu, H. Müller and J. W. Mayer, Rad. Effects 28, 227 (1976).

4.14   D. A. Antoniadis, these proceedings.

4.15   G. Declerck, these proceedings.

4.16   G. J. van Gurp, J. W. Slotboom and F. J. B. Smolders, ESSDERC 1979.

4.17   K. Tsukamoto, Y. Akasaka and K. Kijima, Jpn. J. Appl. Phys. 19, 89 (1980).

4.18   H. Ryssel, K. Haberger, K. Hoffmann, G. Prinke, R. Dümcke and A. Sachs, IEEE Trans. Elec. Devices ED-27, 1484 (1980).

4.19   J. Gyulai, L. Csperegi, T. Tagy, J. W. Mayer and H. Müller, Le Vide 174, 416 (1974).

4.20   H. Ryssel, K. Müller, K. Haberger, R. Henkelmann and F. Jahnel, Appl. Phys. 22, 35 (1980).

4.21   T. Tamura, Appl. Phys. Lett. 23, 51 (1975).

4.22   S. Mader and A. Michel, J. Vac. Sci. Technol. 13, 391 (1976).

5.1   H. G. Dill, R. W. Bower and T. N. Toombs in "Ion Implantation", Eds. F. H. Eisen and C. S. Chadderton, p. 349, Gordon and Breach, London (1971).

5.2   T. W. Sigmon, Proc. IEEE <u>63</u>, 1619 (1975).

5.3   (R. J. Swanson and J. D. Meindl, IEEE J. Solid State Cir. <u>SC-7</u>, 146 (1972).

5.4   B. Höfflinger and L. Gabler in "Ion Implantation in Semiconductors," Ed. S. Namba, Plenum Press, New York (1975), p. 717.

5.5   T. Mashahara and J. Itoh, IEEE Trans. Elec. Devices <u>ED-21</u>, 79 (1974).

5.6   M. Komoshida and O. Kudoh, Appl. Phys. Lett. <u>24</u>, 501 (1974).

5.7   J. T. Clemens, R. H. Doklan and J. J. Nolen, IEDM, p. 299, (1975).

5.8   K. Board, Int. J. Elec. <u>43</u>, 151 (1977).

5.9   T. Okzone, H. Shimura, K. Tsugi and T. Hirao, IEEE Trans. Elec. Dev. <u>ED-27</u>, 1789 (1980).

5.10   C. H. Sequin, Bell Syst. Tech. J. <u>51</u>, 1923 (1972).

5.11   W. F. Kosonocky, Wescon 1974, Technical Session 2 (1974).

5.12   S. Shimizu, S. Iwamatsu and M. Ono, Appl. Phys. Lett. <u>22</u>, 286 (1973).

5.13   P. K. Chatterjee, C. W. Taylor and A. F. Tasch, Jr., IEEE Trans. Elec. Dev. <u>ED-26</u>, 871 (1979).

5.14   W. R. Hunter, L. M. Ephrath, W. Grobman, C. M. Osburn, B. L. Crowder, A. Cramer and H. E. Luhn, IEDM (1978).

5.15   De Witt Ong, IEEE Trans. Elec. Dev. <u>ED-28</u>, 6 (1981).

5.16   J. Sakurai, IEEE J. Solid State Cir. <u>SC-13</u>, 468 (1978).

6.1   R. S. Payne and R. J. Scavuzzo, IEDM, Washington (1971)

6.2   R. S. Payne, R. J. Scavuzzo, K. H. Olson, J. M. Nacci and R. A. Moline, IEEE Trans. Elec. Dev. <u>Ed-21</u>, 273 (1974).

6.3   J. A. Archer, Solid State Electron. <u>17</u>, 387 (1974).

6.4   S. A. Evans, S. A. Morris and J. Englade, IEDM, page 196, Washington (1979).

6.5   O. Ozawa, S. Kameyama, Y. Sasaki, Y. Tokumaru, M. Nakai and Teruo Tanji, IEDM, page 188, Washington (1979).

6.6   K. Kanzaki, M. Taguchi, G. Sasaki, A. Furukawa and K. Acki, IEDM, page 328, Washington (1979).

6.7   A. E. Michel and W. H. Dexter, unpublished.

SILICON EPITAXY

D. A. Antoniadis

Massachusetts Institute of Technology
Cambridge, Massachusetts   02139

1. INTRODUCTION

Epitaxy or homo-epitaxy of silicon is the process of growing crystalline silicon layers on a single crystal silicon substrate. The merits of the process for device technology are discussed in other sections of this lecture series. Here we will concentrate on epitaxy by chemical vapor deposition (CVD), which is to be contrasted with physical vapor deposition where atoms to be deposited are emitted by a heated source and are incident on the substrate without any chemical reaction taking place. An example of the latter process is "molecular beam epitaxy" or MBE. For a detailed review of silicon CVD readers are referred to a recent article by Bloem and Giling.[1]

Although there are various reactor configurations that are used for silicon epitaxy by CVD, here we will concentrate on the horizontal reactor type. The basic principles are equally valid for other reactor types. Figure 1 is a schematic of a typical horizontal reactor. After being inserted at the right, the gases travel along the reaction chamber and reach the heated susceptor on

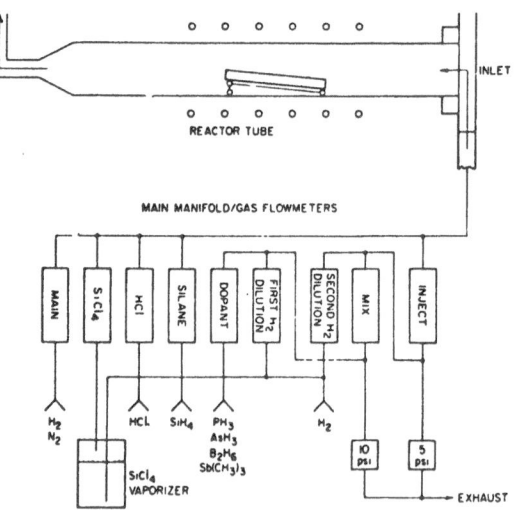

Fig. 1 - Schematic of the horizontal epitaxial reactor.

which the silicon wafers rest. In many reactors, and in the one in Figure 1, the wafers are heated by inductively coupling radio frequency (RF) energy to the susceptor, which in turn heats the wafers. The walls of the reactor are kept cold.

The energy imparted to the gas by the heated wafer and susceptor causes the gases to react at the surface of the substrate or slightly above it so that atoms or molecules of the reaction product are formed on the substrate. Table 1 describes the various source gases and their functions, and Table 2 summarizes the key features of several silicon gaseous sources. Table 3 presents a typical reactor sequence used during epitaxial silicon growth. The following sections will elaborate on the fundamental models that describe the kinetics of silicon

Table 1

GASES USED IN EPITAXIAL DEPOSITION

Gas	Function	Comments
N_2	main flow	Purges out explosive/poisonous gases prior to opening the reactor tube to air
H_2	main flow	Most common ambient for growth or epitaxial layers
$SiCl_4$	Si source	Common liquid Si source Vaporized in an H_2 bubbler Corrosive vapor
SiH_4	Si source	Common gaseous Si source Pyrophoric gas
HCL	Si etchant	Most common Si etchant used for substrate preparation Corrosive poison gas
PH_3	Si dopant	Most common phosphorous source for doping epitaxial silicon Flammable poison gas
AsH_3	Si dopant	As PH_3
$Sb(CH_3)_3$	Si dopant	A liquid antimony source used as a vapor at a concentration of a few hundred ppm in H_2 Used because SbH_3 is unstable Poisonous vapor
B_2H_6	Si dopant	As PH_3

Table 3

TYPICAL EPITAXIAL GROWTH CYCLE

Step	Time	Temperature	Gas Concentration	Comment
N_2 purge	2'	R.T.	---	Purge out O_2
H_2 purge	2'	R.T.	---	Charge to H_2 ambient
Heat	2'	1200°C	---	In H_2 ambient
HCL etch	2'	1200°C	1%	Etch 0.26 μ of silicon
H_2 purge	2'	to 1050°C	---	To lower temperature and remove HCL
Growth	8'20"	1050°C	0.05% SiH_4 0.3 ppb PH_3	Growth of 5 μ, 1 Ω-cm P-doped Si layer
H_2 purge	1'	1050°C	---	Purge reactants prior to cooling
Cool	4'	R.T.	---	In H_2 ambient
N_2 purge	2'	R.T.	---	Before opening to air

Table 2

SOURCES OF SILICON IN EPITAXY

Source	Typical Conditions		Comments
	Temperature (°C)	Rate (μ/min)	
SiH_4	1000 to 1050	0.2 to 1.0	Pure gaseous source Pyrophoric gas Low-temperature deposition Low autodoping Moderate growth rates Surface quality sensitive to O_2
$SiCl_4$	1150 to 1200	0.5 to 1.5	Corrosive liquid source High-temperature deposition Moderate autodoping, outdiffusion Moderate-high growth rates Most common source for linear bipolar integrated circuits Easy to obtain good crystal quality on thick layers
$SiHCl_3$	1150 to 1200	1.0 to 10	Corrosive liquid source High-temperature deposition Moderate autodoping Very high growth rates Most common source of poly-Si dielectric isolation Very high purity epitaxial layers used in high-voltage devices
SiH_2Cl_2	1050 to 1100	≥ 1.0	Gaseous source at 7 psi Properties: intermediate $SiCl_4$, SiH_4
$SiBr_4$	---	---	Rarely used source
$Si(CH_3)_4$	1150	0.4	Rarely used source

growth, and dopant inclusion.

2. FILM GROWTH

Figure 2 shows a cross section of the reactor and identifies three regions: 1) the main gas stream, 2) the boundary layer, and 3) the adsorbed layer. The main gas stream consists of the carrier silicon- and dopant-containing gases flowing by forced convection. It can be thus assumed that at any point along the deposition zone, the temperature and composition of gas are uniform in a direction perpendicular to the primary gas flow. It can also be assumed that

no chemical reactions occur in this region. It has been shown both experimentally and

theoretically[2(a),(b)]

that for this flow con-
figuration a
"boundary layer" of
relatively static gas
exists adjacent to the
growing surface. The

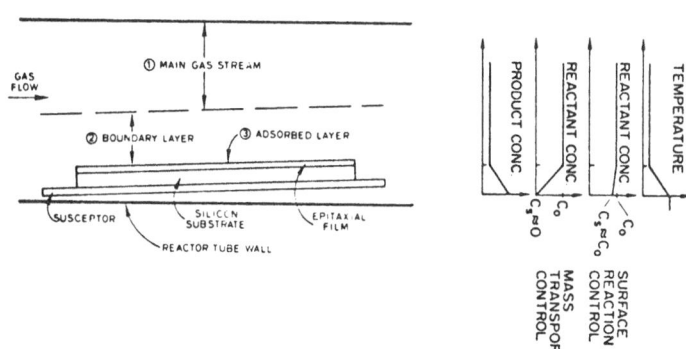

thickness of this layer
is about 0.5 mm.

Fig. 2 - Cross section of horizontal reactor with thermal and chemical profiles during growth.

The last region, i.e., the adsorbed layer consists of a population of hydrogen, silicon-, and

dopant-containing species that occupy adsorption sites and are capable of moving on the

silicon surface.

Silicon epitaxial growth is in general terms the result of the reaction of the silicon

containing gas at the substrate surface. In this discussion we will consider only silane as the

source gas. Then, the relevant reaction is thermal decomposition of silane as

$$SiH_4 \xrightarrow{\text{High T}} Si + 2H_2 \tag{1}$$

Epitaxial growth proceeds by the following steps:

(1)   mass transfer of the reactant molecules (such as SiH_4) by diffusion from the

        turbulent-layer reservoir across the boundary layer to the silicon surface.

(2)   adsorption of reactant atoms on the surface.

(3)   one or more chemical reactions at the surface.

(4)   desorption of product molecules (such as H_2).

(5)   mass transfer of the product molecules by diffusion through the boundary

        layer, back to the turbulent layer.

(6)   lattice arrangement of the adsorbed silicon atoms (may occur as part of 3).

Concentration gradients across the boundary layer are such that there is a diffusive flux of product molecules away from the surface. The turbulent layer is depicted as having no concentration or temperature gradient.

The overall deposition rate is determined by the slowest of processes 1 to 6 above. Figure 2 also shows idealized reactant concentration profiles for the case of surface reaction and mass transport limited growth. The expected temperature dependence of the growth rate varies markedly for different controlling mechanisms. If diffusion of either reactant or product across the stagnant layer (so-called mass-transport control) is the slowest part of the reaction, the growth rate will not depend on the temperature to first order. On the other hand, if a surface chemical reaction is the slowest process (surface-reaction rate control), one would expect the deposition rate to have the same temperature dependence as the chemical reaction, which is an exponential function of inverse temperature. Furthermore, because the various silicon sources undergo different chemical reactions with different activation energies, one would expect them to have diverse growth-rate temperature dependences under surface-reaction rate control.

The curve shown in Fig. 3 displays a low-temperature reaction regime in which surface-reaction rates dominate the growth rate and a high-temperature non-activated reaction regime in which mass transport dominates the growth rate. One striking feature is that similar plots for $SiCl_4$, SiH_4, SiH_2Cl_2, and $SiHCl_3$ have the same activation energy; (roughly 39 kcal/mole or 1.6 eV/molecule). This common activation energy in the temperature-dependent region implies that the dominant surface "reaction" is step 6; (the lattice arrangement of the silicon atoms). Indeed, the act-

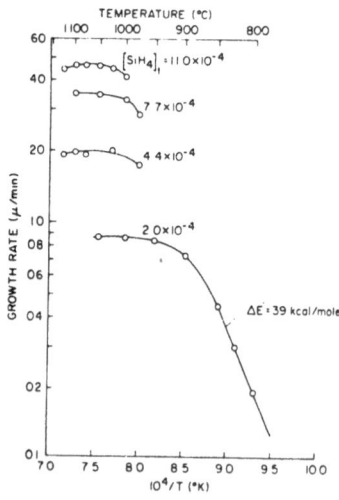

Fig. 3 - Arrhenius plot of epitaxial growth rate in SiH_4.

ivation energy for surface diffusion of adsorbed silicon atoms on silicon has been measured to be 36 ± 6 kcal/mole.[3]

At higher temperatures, the growth reaction is non-activated, suggesting that the slowest process is mass transport of either products or reactants. The rate-limiting step is probably reactant transfer. This hypothesis is supported by two considerations. First, in the boundary layer under reactant mass-transport limited conditions, the growth rate is proportional to the diffusion coefficient of the reactant species in hydrogen. This coefficient, in turn, is proportional to the inverse square root of the molecular weight of the reactant molecule. Second, the large temperature gradient in the boundary layer tends to retard the gaseous diffusion of reactants toward the susceptor and to increase the diffusion of products away from the susceptor.

In summary, boundary-layer theory and empirical observations produce the following qualitative results for the epitaxial growth rate of common silicon gaseous sources.

(1) The growth rate indicates two distinct regions, a low-temperature region in which the growth rate fits an Arrhenius plot and a high-temperature region in which the growth rate does not depend on temperature.

(2) In the low-temperature region: a) activation energy is approximately 39 kcal/mole, a number consistent with the activation energy for surface diffusion or adsorbed silicon atoms; b) activation energy is independent of source type; c) for low source concentrations, the growth rate is directly proportional to the input silicon source concentration; d) the linear growth rate/source concentration relationship applies at high growth rates only for SiH_4. The chlorine-containing sources do not obey this relationship because a reverse reaction, HCl etching, occurs.[4]

(3) In the high-temperature region: a) the temperature insensitive growth rate is controlled by mass transport of the silicon source from the turbulent layer to the silicon interface by diffusion; b) for low source concentrations, the growth

rate is directly proportional to the input silicon source concentration; c) the growth rates for chlorine-containing sources are not linear functions of source concentration at high concentrations because of HCl etching.

The changeover between the surface-controlled and mass-transport controlled regimes of SiH_4 deposition can be described by the balance of molecular fluxes because no reverse reactions complicate the SiH_4 problem. The growth rate, g, is directly proportional to the SiH_4 concentration at the silicon surface, $[SiH_4]_I$, as given by the following equation:

$$g = K_I [SiH_4]_I \qquad (2)$$

The surface-reaction rate constant K_I has the characteristic activation energy of 39 kcal/mole and the units of the growth rate (cm/sec) because $[SiH_4]_I$ is a unitless ratio of gas flows. The flux of SiH_4 atoms across the boundary layer F_I is proportional to the SiH_4 concentration gradient across the layer, which is the difference between SiH_4 concentration in the turbulent layer $[SiH_4]_T$ and that at the interface $[SiH_4]_I$,

$$F_I = h(SiH_4) [[SiH_4]_T - [SiH_4]_I] \qquad (3)$$

where $h(SiH_4)$ is the mass-transport coefficient for SiH_4 and is a temperature-insensitive constant that relates molecular flux to the concentration gradient; it is proportional to the SiH_4 diffusion coefficient in H_2 and is inversely proportional to the boundary-layer thickness. The flux of silicon atoms being incorporated into the lattice is

$$F_2 = N_{Si}g \qquad (4)$$

in which N_{Si} is the atomic density of silicon, $N_{Si} = 5.0 \times 10^{22}$ cm^{-3}, and g is the epitaxial growth rate. A simultaneous solution of Eqs. (2) through (4) obtains the interface silane concentration,

$$[SiH_4]_I = [SiH_4]_T \left[1 + \frac{N_{Si}}{h(SiH_4)} K_I \right]^{-1} \qquad (5)$$

There are two limiting cases:

Case 1: Surface-reaction control.

At low temperatures, K_I is small enough so that $[SiH_4]_I \sim [SiH_4]_T$ (see Fig. 2). In this case, from Eq. (2),

$$g = K_I [SiH_4]_T \qquad (6)$$

Case 2: Mass-transport control.

At high temperatures, K_I becomes large, and $[SiH_4]_I \sim 0$; then from Eqs. (3) and (4),

$$g = \frac{h(SiH_4)}{N_{Si}} [SiH_4]_T = K_M [SiH_4]_T \qquad (7)$$

The rate constant for mass-transport control is K_M (cm/sec).

Although Eqs. (6) and (7) predict a first-order growth-rate dependence on silane concentration, the surface-reaction rate constant K_I is exponentially dependent on temperature and the mass-transport rate constant K_M has only a linear dependence on temperature (essentially no dependence over the narrow 1000° to 1100°C range commonly used for SiH_4 epitaxy).

Experimental SiH_4 growth rate is plotted vs. temperature in Fig. 3. The surface-reaction region has an activation energy of $\Delta E = 39$ kcal/mole. Transfer to the mass-transport region is nearly complete at 1000°C for all growth rates from 0.1 to 0.5 μm/min. From 1000°C, the maximum variation observed is only 10%.

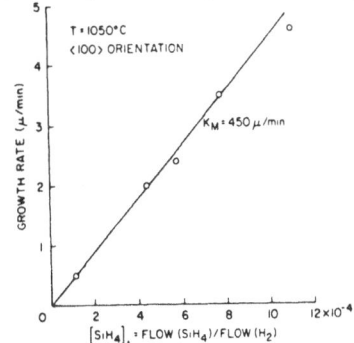

Fig. 4 - SiH_4 growth rate vs. SiH_4 concentration.

Figure 4 is a graph of growth rate vs. silane concentration for $T = 1050°C$. The slope of the line is the mass-transport rate constant value $K_M = 7.5 \times 10^{-4}$ cm/sec = 450 μm/min. These values for activation energy[3] and the rate constant[2(b)] are consistent with the literature.

3. DOPANT INCORPORATION IN EPITAXIAL LAYERS

The primary reason for deposition of one layer of single-crystal silicon on another by CVD is to obtain layers with different dopant concentrations. In general, a well-controlled dopant concentration can be introduced into a depositing epitaxial film by addition of suitable gases containing the desired dopant atoms during the deposition. The dopant molecules react simultaneously with the silicon-containing molecules, and the resulting dopant atoms are incorporated into the depositing film. However, this process is not straightforward and has only recently been addressed with satisfactory results by Reif *et al.*[5,6,7] The discussion in this section follows closely that of Ref. 6.

During the initial stages of growth, the dopant incorporation process goes through a transient period, and 2-3 minutes are required[7] before the steady-state epitaxial doping level is established. As a result, a transition layer corresponding to this initial transient develops, and the epitaxial dopant concentration within this layer is either higher or lower than expected, depending on the initial conditions prior to the deposition cycle (e.g., substrate surface concentration, prebake time, prebake temperature, etc.). A special case of this initial transient problem is the commonly known "autodoping" phenomenon, which occurs when lightly-doped films are deposited on heavily-doped substrates. The extent of this transition layer imposes severe limitations on the fabrication of submicron epitaxial films, which will be required with the development of VLSI technology. In order to derive the epitaxial doping profile resulting from a given time-varying gas-phase composition during growth, the epitaxial doping model must account for thermal redistribution of dopant atoms in silicon during epitaxial deposition. Because the redistribution of impurities within the solid is controlled by diffusion, the starting mathematical framework is the impurity continuity equation throughout the silicon, from a plane very deep inside the substrate up to the silicon surface. Assuming no loss or generation of dopant atoms in the solid, the continuity equation for dopant atoms between two points y_1 and y_2 can be written as follows:

$$\frac{d}{dt} \int_{y_1}^{y_2} C(y)dy = -[F(y_2) - F(y_1)] \tag{8}$$

where F is the dopant flux, C is the dopant concentration in the solid and y and t are the spatial and time variables, respectively. As shown in Fig. 5 the y-direction is perpendicular to the silicon surface, and y_f is defined as the location of the moving gas-solid interface. For more details on the continuity equation see the "Computer Simulation of Complete IC Fabrication Process" chapter in this series.

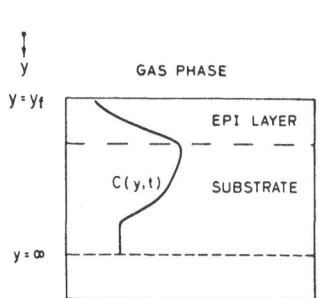

Fig. 5 - Schematic cross section of a silicon wafer for the purpose of solving the dopant continuity equation.

To solve the continuity equation the form of F as well as the initial and boundary conditions must be specified. With the exception of the two boundaries the dopant flux is diffusive and is given by

$$F(y) = \frac{d}{dy} (D(y) \cdot C(y)) \quad \text{for } y_f > y > \infty \tag{9a}$$

where D is the dopant diffusivity. The initial condition is given by

$$C(y,0) = f_1(y) \tag{9b}$$

where $f_1(y)$ represents the distribution of impurities in the substrate just before epitaxial deposition.

The boundary conditions are

$$F(y = \infty) = 0 \tag{10}$$

$$F(y_f) = f_2(t) \tag{11}$$

Equation (10) indicates that the impurity diffusive flux at a plane very deep in the silicon substrate is zero. This is a reasonable boundary condition because silicon wafer thicknesses are much larger than diffusion lengths in bulk silicon.

Equation (11) indicates that the impurity diffusive flux in the solid at the gas-solid interface is, during epitaxial growth, a function of time. As will be shown later, an expression for $f_2(t)$ can be derived from a basic understanding of the mechanisms controlling the incorporation of impurities in the epitaxial silicon during growth. The time-dependence of Eq. (11) is related to: 1) the transients associated with the establishment of a steady-state deposition process, and 2) the time-variation (if any) of the gas-phase composition in the reactor. Notice that the autodoping problem falls in the first category. The right-hand side of Eq. (11) is a function of dopant partial pressure, epitaxial growth rate, deposition temperature, and to a lesser degree: reactor geometry, hydrogen velocity, etc.

In order to derive an expression for $f_2(t)$ in Eq. (11) for the epitaxial doping model, the behavior of dopant species in each important region of an epitaxial reactor is studied in detail, both in terms of gas-flow dynamics and chemical kinetics. For simplicity, we will assume that the silicon deposition is limited by transport of the silicon-containing molecules through the boundary layer, since most epitaxial deposition processes are designed to operate in this mass-transport-limited regime for good control of the deposition rate. It is important to note that, although the silicon deposition is mass-transport-limited, the doping process occurring simultaneously need not be so, and we must consider processes occurring at the surface as well as in the gas phase.

Main Gas Stream

In most epitaxial reactors, gas-phase depletion of dopant species in the main gas stream is almost negligible. Therefore, the partial pressure of dopant species in the main gas stream region can be assumed to be independent of position, and nearly equal to the partial pressure of dopant species at the reactor input. Moreover, the time constant associated with the transport of dopant species in the main gas stream by forced convection is of the order of one second,[5] while the time constant of the overall doping in a horizontal reactor is of the order of forty seconds.[7] Therefore, it can be assumed that any time-variation of the dopant partial pressure at the reactor input is transmitted "instantaneously" throughout the main gas stream. Consequently, the following expression can be used to describe the behavior of dopant species in the main gas stream region

$$P_D(t) \cong \overset{\circ}{P}_D(t) \tag{12}$$

in which P_D is the partial pressure of dopant species in the main gas stream region, and $\overset{\circ}{P}_D$ is the input dopant partial pressure. A more detailed discussion of the derivation of Eq. (12) is given in Ref. 6.

Boundary Layer

The time constant associated with the transport of dopant species through the boundary layer by diffusion is of the order of 0.1 seconds,[7] which is much shorter than that of the overall doping process. Therefore, it is reasonable to assume that the flux of dopant species leaving the boundary layer by adsorbing on the silicon surface, F_a, follows closely any time-variation of the dopant flux entering the boundary layer from the main gas stream, $F_y(\bar{y})$, i.e.,

$$F_a(t) \cong F_y(\bar{y}, t) \tag{13}$$

Again, a more detailed discussion of the derivation of Eq. (13) is present in Ref. 6.

The two fluxes in Eq. (13) can be expressed in terms of deposition parameters. The flux of dopant species leaving the main gas stream toward the wafer surface, $F_y(\bar{y})$, can be approximated by

$$F_y(\bar{y}) = k_m \left[\overset{\circ}{P}_D - \overset{*}{P}_D \right] \tag{14}$$

in which k_m is the boundary layer mass transport coefficient of dopant species in hydrogen, and P_D^* is the dopant partial pressure just above the gas-solid interface. k_m is a function of reactor geometry, hydrogen velocity, gas-phase temperature, etc. An expression for $F_a(t)$ is obtained in the next section.

Adsorbed Layer

In order to determine F_a, the sequence of steps taking part in the doping process and occurring at the silicon surface must be considered in detail. We will focus now the discussion to the case of arsenic dopant atoms from an arsine dopant gas. The dopant incorporation steps are shown in Fig. 6. When an arsine molecule in the gas phase is close to the silicon surface it undergoes a process of adsorption (step (1)). The arsine molecule, once adsorbed, decomposes chemically yielding elemental arsenic (step (2)) which then diffuses on the surface until it finds an incorporation

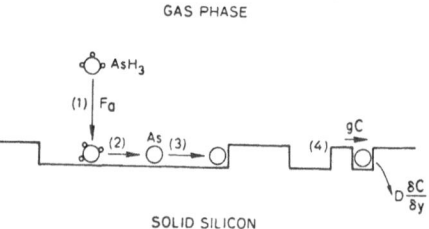

Fig. 6 - Steps taking part in the doping process, and occurring at the silicon surface: 1) arsine adsorption; 2) arsine chemical decomposition; 3) arsenic surface diffusion and site incorporation; and 4) covering of incorporated arsenic by subsequently arriving silicon atoms. [After Ref. 6]

site and attaches to it (step (3)). The arsenic atom, now incorporated in the silicon lattice, is quickly covered by subsequently arriving silicon atoms (step (4)), and diffuses into (or out of) the bulk silicon. For the treatment presented here, it is sufficient to select one of the above steps as the rate limiting step, leaving all other steps in the sequence near thermodynamic equilibrium. If step (1) above is assumed to be the slowest in the sequence, the surface steps can be summarized in the following two reactions:

i) Arsine adsorption

$$AsH_3(g) \; + \; s \; \rightleftarrows \; AsH_3-s \qquad K_1 \; = \; \frac{(\theta_D N_s)}{P_D^*(\theta N_s)} \; = \; \frac{k_F}{k_R} \qquad (15)$$

ii) Chemical decomposition, site incorporation, and covering of arsenic by silicon

$$AsH_3-s \; \underset{\longrightarrow}{\longleftarrow} \; As(ss) \; + \; s \; + \; \frac{3}{2} \, H_2(g) \quad K_2 \; = \; \frac{(k_H N) \, (\theta N_s) P_{H_2}^{3/2}}{(\theta_D N_s)} \qquad (16)$$

where s represents a vacant adsorption site on the surface, AsH_3-s represents an arsine molecule occupying an adsorption site, K_1 and K_2 are equilibrium constants, θ_D is the fraction of adsorption sites occupied by arsine, N_s is the surface density of adsorption sites per unit area, k_F and k_R are the forward and reverse reaction rate constants for arsine adsorption, θ is the fraction of adsorption sites which are vacant, k_H is Henry's law constant, and P_{H_2} is the hydrogen pressure ($\sim$1 atm). An expression for F_a can now be obtained by using Eqs. (15) and (16)

$$F_a \; = \; k_f \left[P_D^* - C/K_P \right] \qquad (17)$$

where $k_f \equiv k_F \, (\theta N_s)$ is a kinetic constant associated with arsine adsorption, and $K_P \equiv K_1 K_2 / k_H$ is a thermodynamic constant relating dopant species in the gas phase and the solid silicon. Equations (14) and (17) can now be used to obtain the following expression relating F_a and the input dopant partial pressure P_D^o

$$F_a \; = \; k_{mf} \left[P_D^o - C/K_P \right] \qquad (18)$$

in which $k_{mf} \equiv \left[1/k_m + 1/k_f \right]^{-1}$.

As discussed earlier, the times associated with gas-phase mass transport of dopant species are negligible as compared to the time constants measured for the overall doping process. Therefore, the only mechanisms that could be responsible for these long time constants ought to be associated with the adsorbed layer. This point will become evident in the following discussion of an expression describing the behavior of dopant-species in the adsorbed layer.

By considering mass balance of dopant species in the adsorbed layer the following equation is obtained

$$F_a - gC(y_f) + F(y_f) = \frac{d(\theta_D N_s)}{dt} \tag{19}$$

where g is the epitaxial growth rate. In Eq. (19), F_a represents the rate at which the adsorbed layer increases its population of dopant species. The second term in Eq. (19) represents the rate at which the adsorbed layer decreases its population of dopant species due to the silicon covering step (see Fig. 6), and the third term represents diffusive exchange between dopant atoms in the adsorbed layer and the bulk silicon. The right hand side of Eq. (19) represents the rate of change of the density of dopant species per unit area in the adsorbed layer, and becomes zero when the overall doping process reaches steady state. By substituting Eq. (18) into Eq. (19), the following expression is obtained

$$k_{mf} \left[P_D^o - C(y_f)/K_P \right] - gC(y_f) + F(y_f) = K_A \frac{dC(y_f)}{dt} \tag{20}$$

in which Eq. (16) was used with $K_A \equiv k_H \, (\theta N_s) \, P_{H_2}^{2/3}/K_2$. Equation (20) relates the epitaxial dopant concentration at the silicon surface $C(y_f)$ and the input dopant partial pressure P_D^o. It is clear from Eqs. (19) and (20) that any abrupt variation of P_D^o with time during epitaxial growth is not transmitted "instantaneously" to $C(y_f)$ because some time is required before the population of dopant species in the adsorbed layer $\theta_D N_s$ accommodates to the new steady state condition. This storage-like behavior of the adsorbed layer is responsible for the relatively long time constants measured experimentally for the overall doping process. Equation (20) can be rearranged to give the boundary condition $f_2(t)$ needed in Eq. (11), i.e.,

$$F(y_f) = f_2(t) = - k_{mf} \left[P_D^o - C(y_f)/K_P \right] + gC(y_f) + K_A \frac{dC(y_f)}{dt} \tag{21}$$

The continuity Eq. (8) can now be solved subject to the initial and boundary conditions given by Eqs. (9b) , (10) and (21), respectively. This can be carried out numerically as described by Reif and Dutton.[6] However, values for the parameters in the doping model need to be determined. In the next section the technique used to determine the values of k_{mf}, K_p and K_A is discussed.

Epitaxial Doping Model Parameter Determination: k_{mf}, K_p and K_A

Numerical values for k_{mf}, K_p and K_A can be determined experimentally. In this section, the technique employed to obtain values of these parameters is described. It consists of comparing the predictions of the epitaxial doping model described in the previous section to experimental results. The values of k_{mf} and K_p are extracted by properly fitting theory to experiments under "steady state". The epitaxial doping model is contained mathematically in Eq. (20), which relates $C(y_f)$ to $\overset{o}{P}_D(t)$. This equation can be simplified, however, when only steady-state deposition conditions are considered. The term "steady state" is used here to describe deposition conditions for which the gas-phase composition in the reactor has remained unchanged for a time longer than several time constants of the epitaxial system. Under these conditions, the epitaxial dopant concentration is uniform and Eq. (2) can be simplified to

$$0 = k_{mf} \left[\overset{o}{P}_D - C/K_p \right] - gC \qquad (22)$$

in which C is the uniform epitaxial doping level, and $\overset{o}{P}_D$ and g are time-independent. Equation (22) can be used to determine the uniform epitaxial dopant concentration that results from deposition conditions in which the total deposition time is much longer than the time constant of the system, and the gas flows entering the reactor are kept constant throughout the whole deposition cycle. Equation (22) can be rearranged as

$$\overset{o}{P}_D/C = 1/K_p + g/k_{mf} \qquad (23)$$

which indicates that a plot of $\overset{o}{P}_D/C$ vs. g generated experimentally under steady-state deposition conditions should yield a straight line from which values of k_{mf} and K_p can be extracted.

Figure 7 shows a plot of $\overset{o}{P_D}/C$ vs. g with the circles showing experimental results carried out under steady-state conditions.[6] In these experiments the hydrogen, silane, and arsine flows were unchanged throughout the depos-ition cycle. The thicknesses of the epitaxial

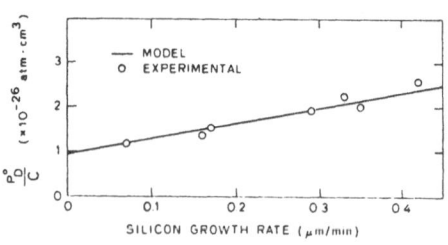

Fig. 7 - Plot of $\overset{o}{P_D}/C$ vs. silicon growth rate used to determine k_{mf} and K_p. [After Ref. 6]

layers were between 7 and 9 μm so as to neglect the influence of the initial transient period. The arsine partial pressure was varied to produce epitaxial doping levels in the 10^{15} - 10^{17} cm^{-3} range, and the silane partial pressure was varied to produce growth rates between 0.07 and 0.42 μm/min. Higher growth rates were accompanied by a loss of growth rate and doping level uniformity along the length of the susceptor, with downstream wafers having lower growth rates and higher resistivities than upstream wafers. These nonuniformities were apparently caused by a parasitic homogeneous decomposition of silane in the gas phase. The substrates were (100)-oriented silicon wafers, boron-doped, with resistivities ranging from 1 to 4 Ω-cm. The epitaxial layer thicknesses were measured by a groove and stain technique, and the resistivity of the layers was determined by the four point probe technique. By fitting Eq. (23) to the experimental data in Fig. 7, the following values were obtained: $K_p = 1.05 \times 10^{26}$ cm^{-3} atm and $k_{mf} = 4.85 \times 10^{19}$ cm^{-2} sec atm, for the experimental conditions described above.[6]

In order to determine the value of K_A a different technique is required. In this case, the continuity equation (Eq. (8)) is solved to determine the epitaxial doping profile that results when $\overset{o}{P_D}(t)$ is varied with time during epitaxial growth. At this point, the profile depends on the value of K_A (see Eq. (20)), and a family of profiles can be obtained by varying K_A. Next, the deposition conditions simulated above are carried out experimentally, and the resulting epitaxial doping profile is measured. The value of K_A can now be determined by selecting, from the family of calculated profiles, which gives the best match to the measured profile. For details, see Ref. 6.

4. EXAMPLES OF USE OF THE EPITAXIAL DOPING MODEL

The epitaxial doping model presented above has been incorporated into the process simulation program SUPREM,[8] by Reif and Dutton.[6] It was thus used to simulate experiments carried out by these authors on a Unipak VI horizontal reactor. The model parameters for this reactor were determined as: $K_p = 1.05 \times 10^{26}$ cm^{-3} amt^{-1}, $k_{mf} = 4.85 \times 10^{19}$ cm^{-2} sec^{-1} atm^{-1} and $K_A = 5.7 \times 10^{-5}$ cm. In fact, the last parameter, K_A, could be determined only after incorporation of the model to a numerical simulator, because it does not yield to closed form solution of the model equations.

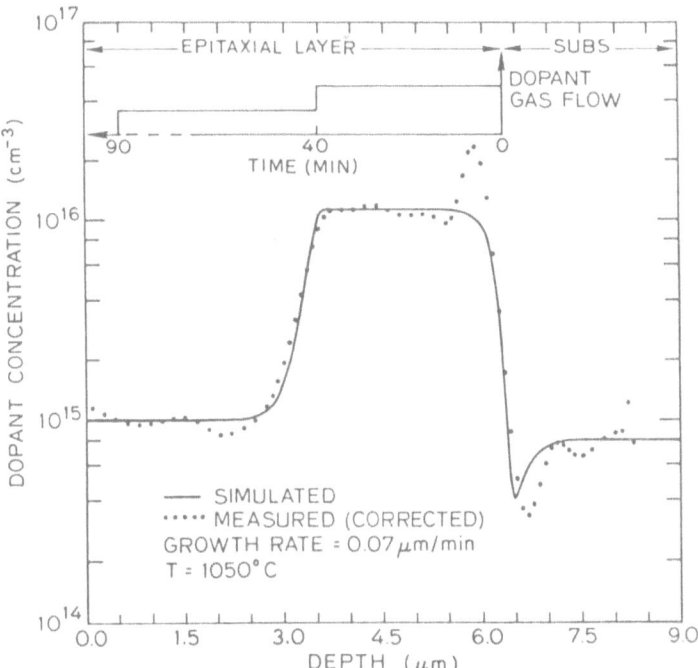

Fig. 8 - Measured and simulated doping profiles resulting from the decreasing step change in arsine flow indicated in the inset. [After Ref. 6]

Figure 8 compares measured by spreading resistance (dotted line), and simulated (solid line) doping profiles resulting from the decreasing step change of arsine flow indicated in the inset. The agreement is very good, particularly taking into consideration that the peak of measured data at about 6 μm is an artifact of the algorithm that converts spreading resistance profiles to doping profiles.[6]

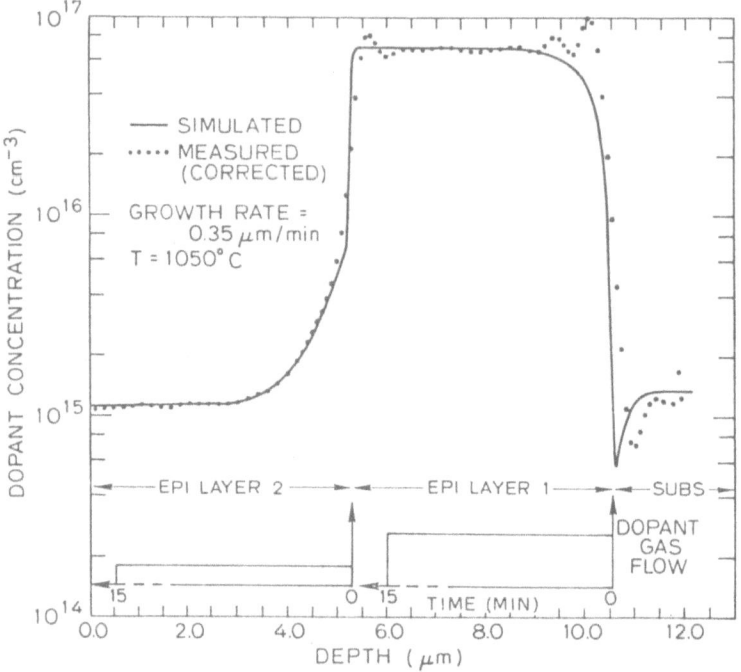

Fig. 9 - Measured and simulated doping profiles corresponding to two consecutive epitaxial depositions. [After Ref. 6]

Figure 9 shows another comparison between the doping profile simulated by SUPREM (solid line) and that measured by the spreading resistance technique (dotted line). In this experiment,[6] two independent and consecutive arsenic-doped epitaxial films were deposited, as indicated in the inset. The arsine flows corresponding to the first and second layers were adjusted to produce epitaxial doping levels of approximately 10^{17} cm^{-3} and 10^{15} cm^{-3}, respectively. The reactor was purged with hydrogen for eight minutes at 1050°C between the end of the first deposition cycle and the beginning of the second. The transition between the high doping level of layer 1 and the low doping level of layer 2 is typical of that encountered when a lightly doped epitaxial film is deposited on a heavily doped substrate or buried layer. It consisted of a very abrupt transition followed by a more gradual transition until the required impurity concentration is reached. The gradual transition is caused by the "autodoping" phenomenon. Autodoping describes the unwanted introduction of dopant atoms into the growing film from the gas phase and should be distinguished from the solid-state outdiffusion of dopant atoms from the underlying substrate into the growing epitaxial film.

This agreement between the simulated and measured profiles in Fig. 9 is excellent. However, in order to simulate this autodoping profile, another parameter must be defined, and its value must also be determined. This new parameter is the concentration of dopant species present in the adsorbed layer prior to the epitaxial deposition step, $(\theta_D N_s)^o$. It determines the initial value of $C(y_f)$ in Eq. (21), i.e.,

$$C(y_f, t = 0) = K_A(\theta_D N_s)^o$$

As mentioned earlier, the initial condition of the epitaxial system is given by the distribution of impurities in the solid silicon just before epitaxial deposition. However, in cases like that shown in Fig. 9 in which the substrate doping level is much higher than the required epitaxial doping level, $(\theta_D N_s)^o$ is not negligible and must also be taken into account. This is because the dopant species present in the adsorbed layer prior to epitaxial deposition are responsible for the autodoping phenomena.[9]

The technique used to determine $(\theta_D N_s)^o$ is similar to that used to determine K_A. Several SUPREM simulations, each with a different value of $(\theta_D N_s)^o$, were carried out until a good fit was obtained to the measured profile.[6] The simulated doping profile in Fig. 9 was obtained with $(\theta_D N_s)^o = 4 \times 10^{11} \text{ cm}^{-2}$.

The cause of the autodoping tail in Fig. 9 is that the dopant concentration in the epitaxial film reaches the required steady-state doping level slowly, because some time is required to remove the excess dopant species from the adsorbed layer. This excess species are produced by the evaporation of dopant atoms from the substrate prior to epitaxial growth. Part of the excess dopant species in the adsorbed layer is incorporated in the growing film and produce the gradual transition shown in Fig. 9, and the rest is released to the gas-phase by desorption.

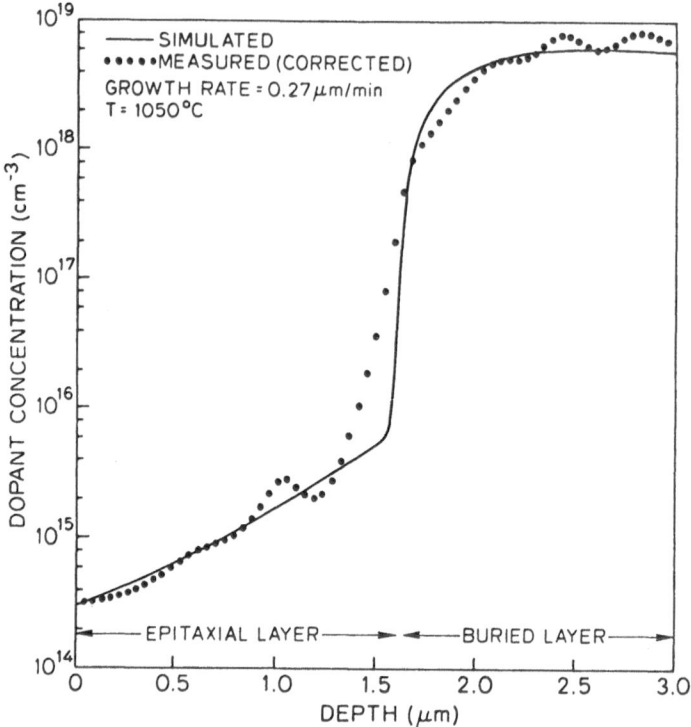

Fig. 10 - Measured and simulated profiles corresponding to a
typical autodoping situation. [After Ref. 6]

Finally, Fig. 10 shows the measured (dotted line) and simulated (solid line) profile of a

more typical autodoping case. In this experiment arsenic was implanted $(3 \times 10^{15}$ cm^{-2} 100

keV) into a boron-doped, 10 Ω-cm, (100) silicon wafer, and then redistributed for 2 hours at

1250°C. The substrate was vapor etched with HCl (0.5% by volume, 2 min., 1200°C) and

then baked in hydrogen (32 min, 1200°C) before the epitaxial deposition step. The epitaxial

layer deposited in this experiment was intended to be intrinsic, i.e., no arsine entered the

reactor. The growth rate was approximately 0.27 μm/min and the deposition time was 6 min.

The simulated profile was obtained with the above $(\theta_D N_s)^{\circ}$. It was determined that the

experimental results and thus $(\theta_D N_s)^{\circ}$ were not sensitive to prebake times and temperatures.[6]

Although the discussed treatment of autodoping is often sufficient, it is by no means

complete. There are other sources of autodoping atoms that cannot be treated by the

presented one-dimensional model. For instance, atoms previously adsorbed on the walls of the reaction chamber may desorb during a subsequent deposition, enter the gas stream and eventually be incorporated into the growing epitaxial films in an unpredictable manner. This source of dopant atoms is especially important if the walls of the system are hot during deposition and is of small importance in a conventional, cold-wall RF heated reactor. A more serious source of unwanted dopant atoms may be the heated susceptor. To avoid this, the susceptor is generally sealed with an impervious coating of silicon carbide. Careful inspection of this coating is necessary when a source of unexpected dopant atoms is sought. An additional measure of precaution is the frequent etching of silicon films deposited on the susceptor, that may harbor unwanted dopant impurities. Finally, we should note that dopant atoms may travel laterally during the autodoping process so that they are incorporated into regions of the film which are not deposited directly over heavily doped regions of the substrate. This displacement is called "lateral autodoping". In the same class also belongs the so-called backside autodoping which results when the back of the wafer is heavily doped.

Besides the precautions already mentioned, other means for reducing autodoping include 1) operating at lower temperature so that less dopant evaporates, 2) avoiding HCl as a reaction byproduct since HCl can lead to significant dopant transport, and 3) using a dopant species with low vapor pressure to dope the substrate. For example, arsenic with vapor pressure of 1 atm at 630°C produces more serious autodoping than antimony which has the same vapor pressure at 1600°C.

REFERENCES

1.   J. Bloem and L. I. Giling, "Mechanisms of the Chemical Vapor Deposition of Silicon," in Current Topics in Materials Science, Vol. 1, Ed. E. Kaldis, North-Holland Publishing Co. (1978).

2.   (a) S. Berkman, "An Analysis of the Gas Flow Dynamics in a Horizontal Reactor," Heteroepitaxial Semiconductors for Electronic Devices, Chapter VI, Eds. G. W. Cullen and C. C. Wanof, Springer, NY (1975) and (b) F. C. Eversteyn, *et al.*, J. Electrochem. Soc. 117, 925 (1970).

3.   R. F. C. Farrow, J. Electrochem. Soc. 121, 899 (1974).

4.   H. C. Theurer, J. Electrochem. Soc. 108, 649 (1961).

5.   R. Reif, T. I. Kamins and K. C. Saraswat, J. Electrochem. Soc. 126, 653 (1979).

6.   R. Reif and R. W. Dutton, J. Electrochem. Soc. 128, 909 (1981).

7.   R. Reif, T.I. Kamins and K. C. Saraswat, J. Electrochem. Soc. 125, 1860 (1978).

8.   D. A. Antoniadis, S. E. Hansen and R. W. Dutton, Stanford Electronics Labs., Stanford University, Tech. Rep. SEL 78-020, June 1978.

9.   M. Tabe and H. Nakamura, J. Electrochem. Soc. 126, 822 (1979).

COMPUTER SIMULATION OF COMPLETE IC FABRICATION PROCESS

D. A. Antoniadis

Massachusetts Institute of Technology
Cambridge, MA 02139

1. INTRODUCTION

This article summarizes the current state-of-the-art in complete process modeling. Important aspects of both technology modeling and computer simulation which make it possible to numerically simulate multiple diffused species -- arsenic-boron and phosphorus-boron -- as well as redistribution effects associated with moving boundaries in oxidation and epitaxy will be discussed.

This article is organized as follows. An overall computer program structure which makes it possible to simulate a complete sequence of fabrication steps is described. Next, the features of process models for ion implantation, impurity migration, oxidation and segregation phenomena and epitaxy are discussed with emphasis on physical effects such as extrinsic diffusion and enhanced diffusion and oxidation rates. These effects are crucial for realistically modeling current and future technologies. Last, numerical aspects of the modeling are discussed.

2. PROCESS SIMULATION

Figure 1 illustrates the schematic structure of the process simulation program SUPREM II.[1] The program is designed so that steps can be simulated either individually or sequentially, just as they would occur during the actual fabrication of an IC. The output of the program, available at the end of each step, consists of the one-dimensional profiles of all the dopants present in the silicon and silicon-dioxide materials. These profiles may be displayed in various formats including line-printer output, line-printer plots, and high-resolution plot. It is understood that, in sequential step simulation, the output of a processing step constitutes the initial conditions for the subsequent step. The junction depths and sheet resistances of all n or

p layers formed during the process are also calculated.

The fabrication step simulation is based on several process models. Typical models

implemented in a simulator include:

   (a)   ion implantation

   (b)   chemical predeposition through the surface (gaseous or solid)

   (c)   oxidation

   (d)   epitaxial growth

   (e)   etching

   (f)   oxide deposition

Diffusive migration of
impurities must be fully
accounted for in all of
the above models involv-
ing high temperature (b,
c, d). Physical parame-
ters for the models are
stored in the program for
user convenience and
constitute the default
values used by the mod-
els. However, these val-
ues may be overrriden by
user-specified parameters.

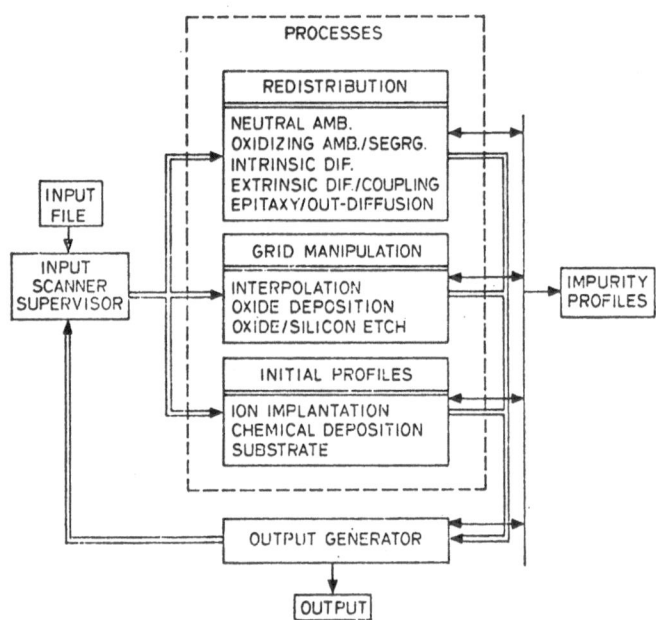

Fig. 1 - General block diagram of the SUPREM process simulator.

The various process models are implemented as modular subprograms, each consisting of

a number of subroutines and special functions. Figure 1 is a schematic of this arrangement.

Input parameter specifications are designed to resemble actual process runsheet data and documentation. A run may consist of a series of process steps. The sequence of steps and the specification of the correct model parameters are controlled by a supervisor program that evokes the appropriate step subprogram. All communication between the various subprograms is directed through the common variable area of the computer memory.

One essential part of the common area contains the impurity concentration arrays. For example, a capability for handling up to three different impurity species has been found to be suitable.[1] The impurity concentration is stored in terms of discrete profile with some maximum number of points (400 is typical[1]). Each concentration value corresponds to a point in the discrete space (spatial grid) defined along a vertical axis, with its origin at the surface of the material -- silicon (Si) or silicon dioxide (SiO_2). Very often, during processing, the physical dimensions of the simulated discrete space may change, as happens for example during oxidation, etch, deposition and epitaxy. The distance between spatial grid points may not be uniform during any of the above processes. A cubic spline interpolation, in the Grid Manipulation block of Fig. 1, is used at the end of each of the processing steps to restore uniformity of the spatial grid.

3. PROCESS MODELS

This section describes a series of process models. The overall thrust is to bring forward current thinking about models for technology steps such as ion implantation, diffusion, oxidation and epitaxy as they relate to practical process simulation.

Ion Implantation

Two areas are considered here for implanted impurities in silicon: first order double half-gaussian approximations for arsenic and phosphorus, and a modified Pearson IV distribution for boron.

The simplest description of an implanted impurity profile in silicon or silicon dioxide is a symmetrical gaussian curve with first two moments, the projected range, R_p, and the standard deviation, σ_p, calculated from the LSS theory.[2] However, experimental distributions of many ions, such as boron or arsenic, are found to be asymmetrical. The simple gaussian approxima-

tion of those implanted profiles is often inadequate so that higher-

order moments must be used to construct range distributions.

Gibbons and Mylroie[3] have shown that the third central moment

is enough to provide sufficient information to construct accurate

distributions when the asymmetry is not excessive (less than the

standard deviation). In these cases, the distribution can be repre-

sented by two half-gaussian profiles, each with a different stand-

ard deviation, σ_1 and σ_2, joined together at a modal range R_m as

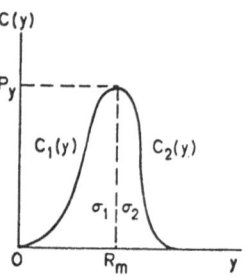

Fig. 2 - The joint half gaussian representation of as-implanted impurity profiles.

shown in Fig. 2. This method can be used for profiles such as arsenic and phosphorus.

However, for boron a modified Pearson IV distribution[1] is found to be more realistic. For the

joint half-gaussian distribution, the two sides are given by

$$
\begin{aligned}
C_1(y) &= P_x \, \exp \left[- (y - R_m)^2 / 2\sigma_1^2 \right] && 0 < y < R_m \\
C_2(y) &= P_x \, \exp \left[- (y - R_m)^2 / 2\sigma_2^2 \right] && R_m < y < \infty
\end{aligned}
\tag{1}
$$

The parameters are determined from table look-up and interpolation.[3,4]

Hofker et al.[5] have shown that the implanted boron profiles in amorphous silicon, before

annealing, may be described by a Pearson type IV distribtuion. Later Ryssel et al.[6] (see also

Ryssel's article on "Ion Implantation" in this series), found that this is true for other elements

and targets as well. Although this distribution describes well the profile of boron from the

surface to a distance somewhat beyond the peak of the distribution, Fig. 3 shows that the

approximation fails to model the observed exponential tail that is due to a small residual

random scattering of boron ions along channeling directions even when silicon wafer targets

are properly tilted to avoid channeling. Based on experimental results[7-9] an empirically

modified Pearson IV distribution can be used by adding an exponential tail with a fixed

characteristic length (0.045μm), independent of dose, energy and crystalline surface

orientation.[1] The tail is attached to the shoulder of the standard Pearson IV distribution where the concentration drops to 50% of the peak value. Of course, after the addition of the tail, renormalization of the distribution to the implanted ion dose is necessary. Typical resulting profiles from this modification are shown in Fig. 3.

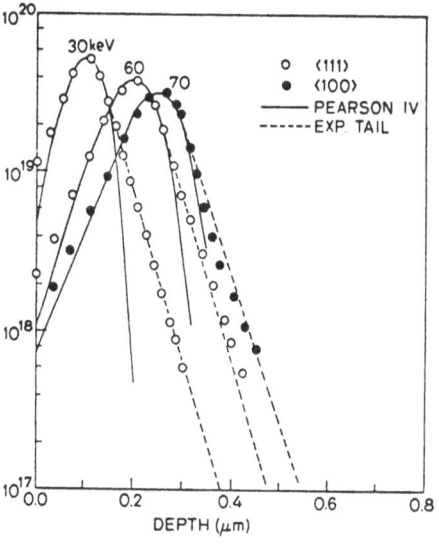

Fig. 3 - Boron as-implanted profiles in <111> and <100> silicon in a random direction. Pearson IV and modified Pearson IV distributions for representation of these profiles. Dose is 10^{13} cm^{-2}.

Impurity Migration During Thermal Processing

The redistribution of impurities in the space of the silicon-silicon dioxide system, during thermal processing, is governed by the general continuity equation which can be written as

$$\frac{d}{dt} \int_{V(t)} C dV = \int_{V(t)} (g - \ell) dV - \int_{S(t)} \vec{F} \cdot \vec{n} dS \qquad (2)$$

where C = impurity concentration in atoms per unit volume, S(t) = closed surface (function of time, t), V(t) = volume enclosed by S(t), $\vec{F}$ = impurity flux vector, $\vec{n}$ = outward unit normal to S(t), g = impurity generation rate per unit volume, and ℓ = impurity loss rate per unit volume.

The left-hand side of Eq. (2) represents the time rate of change of the impurity content in a volume, V(t), and it is equated to the net impurity generation rate in the same volume minus the net impurity outflow through the surface S(t) enclosing V(t). The reason for using the continuity equation in integral form is that it simplifies treatment of volume changes which occur during silicon oxidation and epitaxy processes. The generation and loss mechanisms have been included to account for the exchange of impurity atoms between different states in the silicon lattice as in the case of arsenic where atoms may coexist in substitutional and clustered states.

For one-dimensional flow along the y-axis perpendicular to the silicon surface pointing inward, Eq. (2) can be written

$$\frac{d}{dt} Q(y_1, y_2) = U(y_1, y_2) - [F(y_2) - F(y_1)] \tag{3}$$

where

$$Q(y_1, y_2) = \int_{y_1}^{y_2} C(y) dy \tag{4}$$

is the impurity atom content between y_1 and y_2,

$$U(y_1, y_2) = \int_{y_1}^{y_2} (g - \ell) dy \tag{5}$$

is the net generation between y_1 and y_2 and the flux $F(y)$ is positive in the y-direction. Physically, the impurity flux may arise from solid-state diffusion, from interface phenomena such as evaporation or segregation, and from the motion of interfaces, as in the case of silicon oxidation and epitaxy.

In the sections that follow we discuss the models that describe the physical processes in Eq. (3) as well as the numerical implementation of this equation.

Solid state diffusion is the physical mechanism responsible for impurity migration within the silicon body during high-temperature processing steps. At any point, y, the diffusive flux $F_D(y)$, of impurities is related to their concentration and diffusivity gradient by the modified Fick's first law.[10] For one-dimensional flow the relation is

$$F_D(y) = -\frac{d}{dy} (D(y) \cdot C(y)) \tag{6}$$

where $D(y)$ is the diffusion coefficient of the impurity. Under the assumption of single-state migration there is no generation or loss within the material. Under the further assumption of uniform diffusivity, the well known Fick's second law can be derived from Eq. (2), i.e.,

$$\frac{\partial C}{\partial t} = D \frac{\partial^2 C}{\partial y^2}. \tag{7}$$

Fick's second law is generally adequate for the calculation of diffusive impurity migration under low impurity concentration conditions. However, this approximation fails as the impurity concentration increases to or above the intrinsic carrier concentration, $n_i(T)$, in the semiconductor at the process temperature. Figure 4 is a plot of n_i vs. temperature according to Morin and Maita,[11] as used in SUPREM II. In addition Eq. (7) may fail even at low concentration conditions if another impurity species is present in the silicon at high concentration. We refer to silicon in which all impurities exist at concentrations lower than $n_i(T)$ at the process temperature, as <u>intrinsic</u> silicon. If the opposite is true then we refer to it as <u>extrinsic</u>.

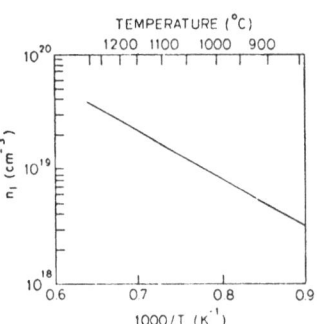

Fig. 4 - Intrinsic carrier concentration in silicon vs temperature.

One of the first attempts to explain diffusive flux under extrinsic conditions was to include the "electric field effect" of the introduced free carriers on the impurity ion migration,[12] in a way similar to ambipolar diffusion in plasmas. Thus, the effective diffusion coefficient in Eq. (6) becomes a function of impurity concentration given by

$$D = D_i f_e = D_i \left\{ 1 + \left[1 + 4 \left(\frac{n_i}{C} \right)^2 \right]^{-1/2} \right\} \tag{8}$$

where D_i is the intrinsic diffusion coefficient. As can be seen, the maximum value of f_e is 2, for $C \gg n_i$ and this is clearly inadequate to explain diffusivity enhancements of the order of 10 to 20 often observed with most of the common impurities at high concentrations.

At present, it is generally accepted that extrinsic diffusion phenomena are the result of impurity migration by interaction with charged point defects in silicon.[13] All common impurities diffuse in silicon by means of interaction with the lattice point defects such as silicon atom vacancies and interstitials. Thus, the diffusion coefficient is proportional to the concentration of such point defects. Although the concentration of neutral defects at any given temperature is independent on the impurity concentration (so long as it does not approach that of silicon atoms), the concentration of defects at various charge states (which have been identified within the silicon bandgap), depends on the Fermi level position in the bandgap and thus is a function of impurity concentration. Recently, Fair[14] summarized some of the latest work on diffusion in silicon and suggested that the effect diffusion coefficient should be the sum of several diffusivities, each accounting for impurity interactions with different charge states of lattice point defects, presumed but not conclusively shown to be vacancies. The effective diffusion coefficient can thus be expressed as

$$D_e = f_e\{D^x + D^-[V^-] + D^=[V^=] + D^+[V^+]\} \tag{9}$$

where D^c is the diffusivity due to each identified charge state, c, of vacancies, (c: =, -, x, +, i.e., doubly negative, singly negative, neutral and positive), and $[V^c]$ is the concentration of vacancies in each charge state, normalized to the intrinsic concentration of that state. Using the Boltzmann approximation it can be easily shown that these normalized concentrations may be given by

$$[V^-] = \frac{n}{n_i}, [V^=] = \left(\frac{n}{n_i}\right)^2 \text{ and } [V^+] = \frac{n_i}{n} \tag{10}$$

where n is the free electron concentration. Thus, under intrinsic conditions ($n = n_i$), Eq. (9) becomes

$$D_i = D^x + D^- + D^= + D^+ \tag{11}$$

i.e., the intrinsic diffusivity is the sum of the diffusivities resulting from the various vacancy charge states.

The above model is the basis of the appropriate process simulation model for diffusion. However, with the exception of phosphorus, it has been assumed that only the neutral and one charged defect state is responsible for the diffusivity of impurity atoms. The specific form used for computer implementation is:

$$D = D_i(1 + \beta f_v)/(1 + \beta) \tag{12}$$

where D_i is the measured intrinsic diffusivity and $f_v = n/n_i$ for donors and n_i/n for acceptors. Thus, under intrinsic conditions $f_v = 1$ and $D = D_i$. On the other hand for extrinsic conditions the physical meaning of parameter β can be derived by combining Eqs. (9), (10) and (12), to obtain

$$D^x = D_i \frac{1}{1 + \beta} \text{ and } D^v = D_i \frac{\beta}{1 + \beta} \tag{13}$$

Thus $\beta = D^v/D^x$, is an index of the effectiveness of charged vacancies relative to neutral ones in impurity diffusion. Figure 5 is a plot of normalized diffusivity vs. f_v. Although it might be expected that β for any impurity element is a function of temperature, no definite characterizations exist at present. Typical values[1] are $\beta = 3$ for boron and $\beta = 100$ for arsenic while for phosphorus a completely different model is used as described later.

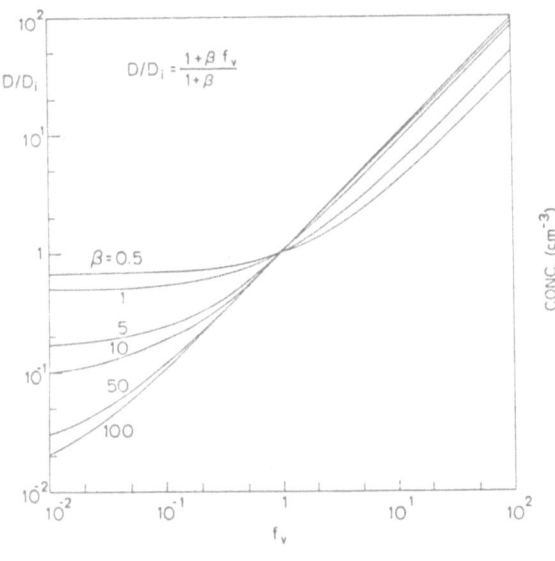

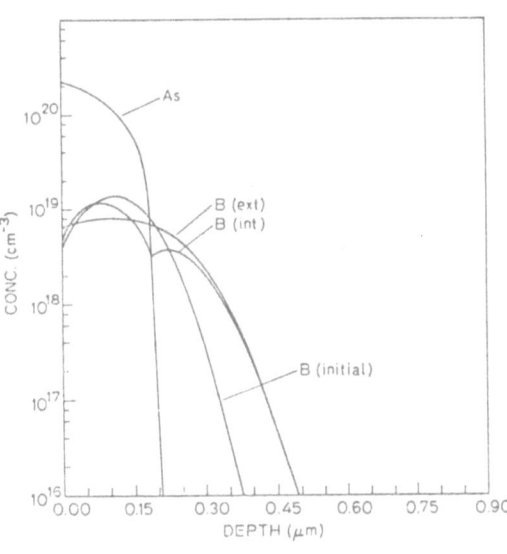

Fig. 5 - Normalized diffusivity vs normalized carrier concentration for different values of β.

Fig. 6 - Illustration of effects of heavy arsenic doping on the redistribution of boron.

It is well known that when different impurity atoms are present in silicon there is direct interaction among them.[15] Figure 6 illustrates one such simulated case where high concentration arsenic affects the distribution of boron. The interaction is directly modeled through the use of Eq. (6) for the calculation of impurity flux. For the example shown in Fig. 6 the diffusive flux of boron in the region which is heavily n-doped is reduced because $f_v(\text{boron}) = n_i/n \to 0$, while at the edge of the arsenic profile the dip is produced by the rapid spatial change of f_v and thus of diffusivity for boron.

In SUPREM II, the diffusive flux for arsenic and boron is modeled by Eq. (6). However, for phosphorus a slightly different form is used to simplify the application of the phosphorus model as described next. This form is

$$F_D(y) = - D_e(y)\frac{d}{dy}C(y) \tag{14}$$

It has been shown that for β relatively large, i.e. greater than 5, the above equation is essentially equivalent to Eq. (6).[10]

Phosphorus

A model for the diffusive migration of phosphorus has been presented by Fair and Tsai.[16] The model predicts with reasonable accuracy the phosphorus kink formation as well as the base push effect, commonly observed during heavy emitter diffusions in bipolar technology. According to this model the physical explanation of these "anomalous" effects lies in the enhancement of vacancy concentration in the silicon caused by the dissociation of the phosphorus-doubly ionized vacancy pairs that flow from the surface into the silicon bulk. A typical high concentration phosphorus profile is composed of

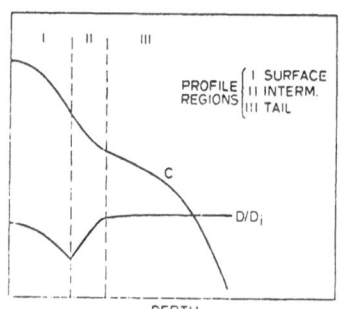

Fig. 7 - A typical phosphorus doping profile with a demarkation of the three regions considered by the Fair and Tsai model. Also shown is the local diffusivity profile derived by the model.

three regions as shown schematically in Fig. 7. In **SUPREM II**, each region is identified according to the criteria outline below. For more detailed discussion the reader is referred to Ref. 16, 17, and 18 and to the article on "Diffusion" by Declerck in this series. Briefly, the diffusivity in the three regions is given as:

I) Surface Region: Region between surface and position where electron concentration, n(y), becomes equal to n_e, given by

$$n_e = 4.65 \times 10^{21} \ \exp\left(-\frac{0.39 \ eV}{kT}\right) cm^{-3} \tag{15}$$

Phosphorus diffusivity given by

$$D_{eI} = f_e\left\{D^x + D^=\left(\frac{n}{n_{ie}}\right)^2\right\} \tag{16}$$

where n, the electron concentration is given by

$$C = n + 2.04 \times 10^{-41} n^3 cm^{-3} \tag{17}$$

and n_{ie}, the effective intrinsic carrier concentration accounting the heavy doping stress-induced bandgap narrowing is given by

$$n_{ie} = n_i \ \exp\left[\frac{1.5 \times 10^{22}(C_{TS} - 3\times10^{20} \ cm^{-3})eV}{kT}\right] \tag{18}$$

for C_{TS}, total phosphorus surface concentration greater than $3 \times 10^{20} \ cm^{-3}$.

II. Intermediate Region: Region starts where $n = n_e$. Diffusivity here is given by

$$D_{eII} = f_e\left\{D^x + D^=\left(\frac{n_e}{n_{ie}}\right)^2\right\}\left(\frac{n_{ie}}{n}\right)^2 \tag{19}$$

Region ends where $D_{eII} = D_{eIII}$, where D_{eIII} is given below.

III) Tail Region: Region begins where $D_{eII} = D_{eIII}$, where

$$D_{eIII} = f_e \left\{ D^x + D^- \frac{n_s^3}{n_e^2 n_{ie}} \left[1 + \exp \left(\frac{0.3 \text{ eV}}{kT} \right) \right] \right\} \tag{20}$$

where n_s is the surface electron concentration. This region is assumed to extend to infinity. The diffusiivity of all impurities in this region is multiplied by an enhancement factor f_{enh} given by

$$f_{enh} = \frac{D_{eIII}}{f_e D^x} \tag{21}$$

Arsenic

Almost all dopants may exist in silicon in more than one state particularly when the concentrations are high. Typically one of these states is substitutional and therefore mobile while the others, if present, may be some form of precipitate or cluster which is immobile. Exchange between these states gives rise to the generation and loss terms in the continuity Eq. (20) of the mobile species. Clustering of arsenic has been discussed by many authors, [e.g. 19, 20, 21, 22]. It is now discussed as an essential model for arsenic diffusion. The assumed chemical reaction is SUPREM II (version 05) is according to Tsai et al.[21]

$$3As^+ + e^- \underset{k_d}{\overset{k_c}{\rightleftarrows}} As_3^{+2} \rightarrow 25°C \ As_3 \tag{22}$$

where k_c and k_d are the clustering and declustering rate coefficients. Defining the concentration of clustered atoms, C_c, as

$$C_c = C_T - C \tag{23}$$

where C_T is the total concentration and C the substitutional concentration, the conservation equation for C_c may be written as

$$\frac{\partial C_c}{\partial t} = k_c n C^3 - \frac{1}{3} k_d C_c = \ell - g \tag{24}$$

where $n \sim C + 2/3 \, C_c$ and ℓ and g refer to the loss and generation terms in Eq. (2). From simple mass action law, the equilibrium clustering coefficient, k_c, is given by

$$k_c = \frac{k_c}{k_d} = \frac{C_c}{3nC^3} \tag{25}$$

Equation (24) together with Eq. (2) can be used to describe the thermal migration of arsenic atoms in silicon and their partition into clustered and substitutional populations. Actually, SUPREM II does not use this dynamic scheme for arsenic diffusion calculation. Instead it assumes that the clustered atoms are always in equilibrium with unclustered ones. This is a good approximation for temperatures above 900°C.

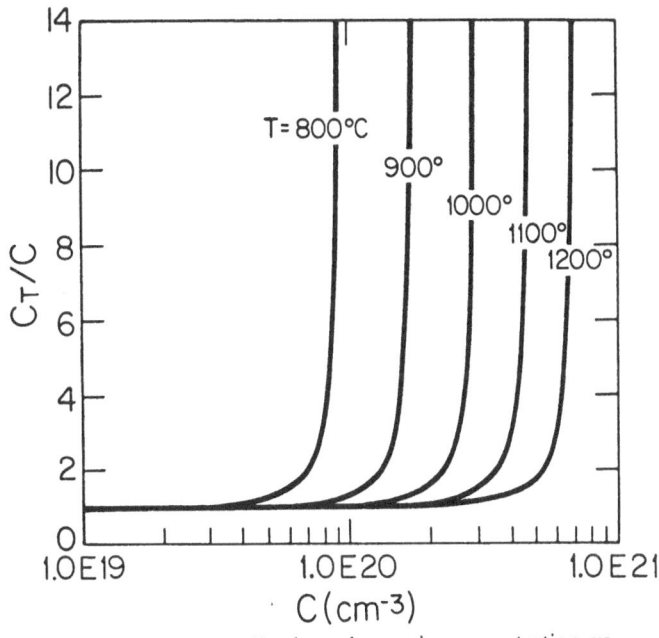

Fig. 8 - Normalized total arsenic concentration vs the substitutional arsenic concentration, resulting from clustering model, as a function of process temperature, under thermal equilibrium.

Then using Eqs. (23) and (24) the following relationship is derived

$$C_T = \frac{C + k_e C^4}{1 - 2k_e C^3} \tag{26}$$

Using this equation, plots of C_T/C for different temperatures are shown in Fig. 8. The k_e

used is the same as that in Tsui *et al.*[21] Equation (6) can now be rewritten as

$$F_D = -D\frac{dC}{dC_T}\frac{dC_T}{dy} - C\frac{dD}{dy} \tag{27}$$

Assuming that for concentrations where clustering is important ($C_T > 5 \times 10^{19}$ cm^{-3}), $D = D_i\frac{n}{n_i}$ (a very good approximation for arsenic), Eq. (27) yields

$$F_D = -D\left[\frac{dC}{dC_T} + \frac{C}{n}\frac{dn}{dC_T}\right]\frac{dC_T}{dy} \tag{28}$$

Thus, the problem of duffusion of arsenic is simplified into a single population problem (total arsenic) diffusing with effective diffusivity given by

$$D_e = D_i\frac{n}{n_i}\left[\frac{dC}{dC_T} + \frac{C}{n}\frac{dn}{dC_T}\right] \tag{29}$$

Equation (28), is then used in Eq. (3), with U=0, to model the migration of arsenic.

Although the clustering model predicts formation and disolution of clusters during predeposition and subsequent drive-in, an assumption must be made about the initial cluster concentration when arsenic is implanted. In this case, it is arbitrarily assumed that the initial clustered arsenic profile is simply that determined by thermal equilibrium at the annealing temperature, given the known implanted profile of total arsenic atom concentration.

Oxidation Enhanced Diffusion

It has been observed by several authors [e.g. 23, 24] that the diffusivity of boron and phosphorus is enhanced when the silicon surface is oxidized. Recently the same effect has been observed for arsenic.[25] This phenomenon of oxidation enhanced diffusion (OED) is generally attributed to enhancement of silicon point defects due to the oxidation. Perhaps the most plausible model has been proposed by Hu[26] and subsequently quantified by Antoniadis *et al.*[24] and more recently by Antoniadis.[27] This model relates OED with oxidation stacking fault (OSF) growth, by invoking a dual diffusion mechanism for impurities in silicon whereby both vacancy and interstitialcy effects are responsible for diffusion, and postulating an enhancement, due to oxidation, of the concentration of silicon self-interstitials. Thus, according to the model, during oxidation the interstitialcy component of impurity diffusivity is enhanced leading to OED while increased interstitial precipitation leads to OSF. For boron, in the present version of SUPREM II, a temperature dependent but time independent OED is assumed for any ambient causing silicon oxidation. For phosphorus a fixed enhancement of a factor 1.8 is assumed for dry O_2 and 3.3 for wet O_2. Recent work[28] has shown that for both boron and phosphorus OED is related to the rate of oxidation. An improvement to SUPREM II has been suggested by Taniguchi *et al.*,[29] where the diffusivity of phosphorus and boron at low concentrations under oxidizing conditions is given by

$$D = D_i + E \left(\frac{dZ_{ox}}{dt} \right)^{0.3} \exp \left(-\frac{y}{25\mu m} \right) \exp \left(-\frac{2.08 eV}{kT} \right) \tag{30}$$

where $E = 1.7 \times 10^{-5}$ cm^2/sec for <100> Si and 6.1×10^{-6} cm^2/sec for <111>.

Thermal Oxidation

The rate of SiO_2 growth on silicon is described by the well-known formula,[30]

$$Z_{ox}^2 + AZ_{ox} = B(t + \tau) \tag{31}$$

where Z_{ox} is the oxide thickness, t is time and A and B are related to the linear and parabolic growth coefficients K_L and K_P and the normalized partial pressure, P_{O_2}, of O_2 by

$$A = P_{O_2} K_P / K_L$$

$$B = P_{O_2} K_P$$

The parameter τ is related to the initial oxide thickness by

$$\tau = \frac{Z_{ox}^2(t=0) + AZ_{ox}(t=0)}{B} \tag{32}$$

Under relatively low dopant concentrations, K_P and K_L depend only on silicon crystal orientation and on the oxidizing ambient, and they are singly activated functions of temperature.

It is, however, well known that under high surface concentration conditions, such as in MOST source and drain or bipolar emitter region, the oxidation rate of silicon is enhanced. A detailed description of the phenomenon has been given by Ho et al.[31] and by Declerck in his article on "Oxidation" in this series.

According to the model, the linear rate coefficient can be written as

$$K_L = K_L^i[1 + \gamma(C^T - 1)] \tag{33}$$

where K_L^i is the intrinsic (i.e., low concentration) coefficient, γ is an experimentally determined parameter given by

$$\gamma = 2.62 \times 10^3 \exp\left[-\frac{1.10eV}{kT}\right], \tag{34}$$

and C^T the normalized total vacancy concentration given by

$$C^T = \frac{1 + C^+\left(\frac{n_i}{n}\right) + C^-\left(\frac{n}{n_i}\right) + C^=\left(\frac{n}{n_i}\right)^2}{1 + C^+ + C^- + C^=} \tag{35}$$

with

$$C^+ = \exp\left[(E^+ - E_i)/kT\right] \quad ; \quad E^+ = .35 \text{ eV}$$

$$C^- = \exp\left[(E_i - E^-)/kT\right] \quad ; \quad E^- = E_g - .57 \text{ eV}$$

$$C^= = \exp\left[(2E_i - E^=)/kT\right] \quad ; \quad E^= = E_g + E^- - .11 \text{ eV}$$

The above expressions should be recognized as the normalized intrinsic concentrations of vacancies in the three charge states with their corresponding state energies in the silicon bandgap. Finally the silicon energy bandgap, E_g, and intrinsic level, E_i, are given as functions of temperature by

$$E_g(T) = 1.17 - 4.73 \times 10^{-4}[T^2/(T + 636)] \text{ eV}$$

$$E_i(T) = E_g/2 - kT/4$$

For n-type dopants the enhancement of the parabolic oxidation rate has also been obtained by Ho and Plummer.[32] K_P is given as

$$K_P = K_P^i(1 + \delta C_T^{0.22}) \tag{36}$$

where

$$\delta = 9.63 \times 10^{-16} \exp\left[-\frac{2.83\text{eV}}{kT}\right] \tag{37}$$

K_P^i is the intrinsic parabolic rate and C_T is the total n-type dopant atom concentration.

Since during oxidation the impurity surface concentration changes due to diffusion and segregation distrubtion, the calculated enhanced values of K_L and K_P may generally be time dependent. Thus, for simulation purposes, using the classical oxide growth Eq. (31) an incremental form of the same equation is used, namely

$$\Delta Z_{ox} = \frac{1}{2}\left[-(2Z_{ox} + A) + \sqrt{(2Z_{ox} + A)^2 + 4B\Delta t}\right] \tag{38}$$

Thus, as the simulation time proceeds in small time increments, Δt, and the impurities redistribute in the silicon, the coefficients A and B are obtained from the surface impurity concentrations in each time increment, and from those values the corresponding increment of oxide thickness, ΔZ_{ox}, is calculated.

Impurity Transfer Across Interfaces

As has been suggested earlier, impurity segregation plays a major role during oxidation, in determining the impurity profiles in silicon and the resulting device properties -- sheet resistance and inversion threshold. In addition to the SiO_2 - Si surface, segregation-type phenomena can occur at gas-solid interfaces including the cases of inert as well as silicon chemical vapor deposition environments. Although the actual details of the chemistry taking place at such interfaces may not be known, generally the dopant atom flux across interfaces may be phenomenologically described by means of a first order kinetic model as

$$F_s = h(C_1 - C_2/m_{eq,1-2}) \tag{39}$$

where F_s is the dopant flux defined positive from region 1 to 2, C_1 is the dopant concentration of the interface in region 1 and C_2 the same in region 2. The factor $m_{eq,1-2}$ is the well-known equilibrium segregation coefficient for the specific impurity species in the system of regions 1-2 and is defined as

$$m_{eq,1-2} = \frac{C_2}{C_1} \tag{40}$$

Finally, h which has units of velocity, is the surface mass-transfer coefficient.

The form given in Eq. (29) can serve for simulating a broad range of process steps including the following:

(a)   Evaporation. The mass-transport coefficient becomes the impurity evaporation of coefficient which is a function of temperature. The same evaporation coefficient is used for SiO_2 and for silicon.[1] Also, C_1 is assumed to be zero and $m_{eq} = 1$.

(b)   Chemical deposition. Chemical deposition is simply modeled by assuming (arbitrarily) that $h \rightarrow \infty$, (actually, in SUPREM, h is set equal to 1 μm/sec), $m_{eq} = 1$ and C_1 equal to either the dopant solid solubility or to any other specified concentration. Thus, the simulated surface concentration of silicon becomes very rapidly equal to C_1. It is recognized that the model may be overly simplistic but there exist too many different processes by which dopants may be deposited in silicon, making it impractical to attempt to accurately model each of them specifically.

(c) SiO_2 - Si interfacial flux. Under non-oxidation conditions the SiO_2 - Si interface is stationary and Eq. (39) is sufficient to model the impurity flux exchanged between the two regions. Unfortunately, there exists practically no characterization of the flux in this stationary system with the exception of the phosphorus doped SiO_2 - Si system (used for silicon doping), which has been carefully explored by Ghoshtagore.[33] In this case h was derived as a singly activated function of temperature, while m_{eq} was assumed infinite. Actually, the exact value of m_{eq} would not alter the observed result as long as it is kept large (say > 50).

In the case of a moving interface as in silicon oxidation there also exists a motion-induced interfacial flux resulting from the different dopant concentration across the interface. This flux denoted by F_b, is given by

$$F_b = -v_{ox}(C_1 - \alpha C_2) \tag{41}$$

where $v_{ox} = dZ_{ox}/dt$, is the oxide growth rate and α is the ratio of oxidized silicon to resulting oxide thickness (equal to 0.44). Generally this flux competes with the flux F_s. If $h \gg v_{ox}$, then $C_2 \rightarrow m_{eq}C_1$, while if $h \ll v_{ox}$, then $C_1 \rightarrow \alpha C_2$. In all characterizations of the moving interface system to date, the first condition has been implicitly assumed, i.e., that the equilibrium segregation condition prevails. Thus, in the absence of any meaningful values h has been arbitrarily assumed equal to $0.1 \mu m/min$ and thus the condition $h \gg v_{ox}$ is always satisfied.[1]

(d)   Silicon epitaxy. Impurity redistribution during epitaxial growth includes both a diffusive component as well as an interfacial flux, F_s. A detailed derivation of the form of F_s has been recently given by Reif and Dutton,[34] and is included in the article on "Silicon Epitaxy" in this series. Instead of F_s, the symbol $F(y_f)$ is used there. Briefly, the interfacial flux is given by

$$F_s = K_{mf}[P^\circ_D - C_I/K'_P] - gC_i - K_A \frac{dC_I}{dt} \qquad (42)$$

where C_I is the impurity concentration at the solid surface, g is the epitaxial growth rate, P°_D is the input dopant partial pressure, and K_{mf}, K'_P and K_A are parameters related to the specific epitaxial reactor system and must thus be determined for each such system. A fourth parameter must be specified for proper simulation of autodoping, namely, the initial value of C_I. Procedures for the determination of these parameters are given in the above cited references.

4. NUMERICAL SIMULATION OF IMPURITY MIGRATION

The previous section discussed the models that may be used to describe impurity fluxes in a SiO_2 - Si system. In the present section, the numerical implementation of these process models is discussed with particular emphasis on the discrete formulation of the impurity continuity equation under the moving boundary conditions encountered during silicon oxidation and epitaxy. However, in order to establish a basis for the discussion of this issue, the numerical solution of the continuity equation under stationary conditions is outlined first.

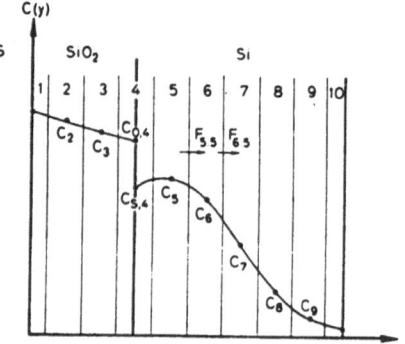

Fig. 9 - Illustration of the partitioning of the simulation space into discrete cells with impurity fluxes across cell boundaries.

Stationary Boundaries

The space over which the impurity continuity Eq. (3) is to be solved is partitioned into discrete cells. The impurity concentration, C(y), is evaluated at points (or nodes) lying in the middle of each of the discrete cells. Figure 9 illustrates this space discretization. For instance, taking cell i = 6 as an example, the discrete continuity equation (neglecting generation-loss terms) is written as

$$\frac{d}{dt}Q_6 = -[F_D(y_{6.5}) - F_D(y_{5.5})] \tag{43}$$

where $F_D(y_{6.5})$ and $F_D(y_{5.5})$ are the impurity diffusive fluxes at the right- and left-hand boundaries of cell 6, and Q_6 is the impurity content of cell 6. Thus, for the general ith cell not lying at any of the space boundaries, the continuity equation discretized in space becomes

$$\frac{d}{dt}Q_i = -[F_D(y_{i+1/2}) - F_D(y_{i-1/2})] \tag{44}$$

Where F_D is evaluated at the cell boundary from Eq. (6) or (14) and Q_i is given by

$$Q_i = \int_{y_{i-1/2}}^{y_{i+1/2}} C_i \, dy \tag{45}$$

For simplicity we have ignored generation-loss terms in (44). Both the spatial derivative in Eq. (6) and (14) and the integration in (45) are carried out using numerical approximations. Specifically, the flux F_D is evaluated by replacing (6) or (14) by a difference equation while Q_i is evaluated using midpoint integration. The set of relevant discrete equations is given in the Appendix.

The cells at the two extreme boundaries, as well as at the SiO_2 - Si boundary (when it exists), deserve attention at this point.

Top Boundary

The first discrete cell is actually a half-cell with its node at the physical boundary. Generally, at this boundary an evaporation or incorporation flux, $F_s(0)$, governed by Eq. (39) as discussed above, is assumed. At the inner boundary of this cell there exists a diffusive flux that may be evaluated as for all other cells, from Eq. (6) or (14). The continuity equation for this cell becomes

$$\frac{d}{dt} Q_1 = -[F_D(y_{1+1/2}) - F_s(0)] \tag{46}$$

Deep Boundary

This boundary usually lies inside the silicon substrate at the point where the simulated space terminates. The last cell in this end is also a half-cell similar to the first one. Typically it is convenient to assume a reflecting boundary (implied by setting $h = 0$ in Eq. (39)) at this point. Because the depth of simulation is often specified rather arbitrarily, care must be taken to ensure that the presence of a reflecting boundary at that point does not affect the simulation results. On the other hand, this reflecting boundary acquires actual physical significance in simulations of Si on sapphire or of poly-Si on SiO_2, where it may be desirable to study the effect of reflecting deep boundary on impurity redistribution. The continuity equation for this last cell becomes

$$\frac{d}{dt} Q_n = F_D(y_{n-1/2}) \tag{47}$$

SiO_2 - Si Interface

Since this interface is a boundary point where impurity concentrations must be evaluated it must always lie on a node ($i = I$). This node is shared by two half-cells, the one in SiO_2 and the other in Si. Since the impurity concentration is generally discontinuous across the interface (because of thermodynamic segregation), the interface node contains two different concentrations, one for the oxide half-cell and one for the silicon half-cell. An interfacial flux, F_s, described by Eq. (39), is assumed to flow between these two cells, while diffusive flux calculated from Eq. (6) or (14) with the appropriate diffusivities for the two materials, is flowing across the other two boundaries. The continuity equation for the two interfacial half-cells becomes

$$\frac{d}{dt}Q_{1,ox} = -[F_s - F_D(y_{1-1/2})] \tag{48}$$

$$\frac{d}{dt}Q_{1,Si} = -[F_D(y_{1+1/2}) - F_s] \tag{49}$$

The established boundary cell Eqs. (46), (47), (48) and (49) together with the set of Eq. (44) for all other cells constitute a system of equations that describes the temporal evolution of the impurity distribution. Numerically, this evolution is obtained by replacing the time derivative in this system of equations by a discrete approximation. All the above equations are of the type

$$H_i(t) = \frac{d}{dt}Q_i(t) \tag{50}$$

where $H_i(t)$ denotes in abbreviated form the cell boundary flux difference due to either F_s or F_D's. Assuming that at time t_o the concentration distribution is known, the distribution at a future time t_1 may be derived by solving the equation

$$\int_{t_o}^{t_1} H_i(t)\, dt = Q_i(t_1) - Q_i(t_o) \tag{51}$$

Various methods exist for performing this integration numerically. A suitable method is the second-order implicit method, which assumes that during the time interval $(t_1 - t_o)$, fluxes are constant, i.e., $H_i(t) = [H_i(t_o) + H_i(t_1)]/2$. Then Eq. (51) becomes

$$[H_i(t_o) + H_i(t_1)]/2 = [Q_i(t_1) - Q_i(t_o)]/(t_1 - t_o) \tag{52}$$

There are as many equations of the form given above, as the number of discrete space cells. Also, since for each cell the flux function H_i, is evaluated at t_1 (i.e., at the future time), it involves not only the unknown cell concentration $C_i(t_1)$ but also the two neighboring cell concentrations $C_{i \pm 1}(t_1)$. Thus, the resulting equations are mutually coupled and form a system with a tridiagonal matrix. This system of equations is solved for the unknown C, successively in small (simulated) time increments by means of Gaussian elimination. Also, since the system may be nonlinear due to the dependence of diffusivity on impurity concentration, Newton-Raphson interations of the solution must be performed until convergence of the results (concentrations) is achieved.

Moving Boundary: SiO_2/Si

When silicon is oxidized the interface between SiO_2 and Si into the silicon and the oxide layer expands with a velocity $v_{ox}(t)$. Under these conditions the moving boundary induced impurity flux, F_b, given by Eq. (41) also flows across the interface.

The presence of the moving boundary complicates the numerical formulation of the continuity equation in two ways. First, because of the nonunity volumetric ratio of Si to SiO_2 ($\alpha = 0.44$), there is expansion of the discrete volume cells as they become part of the SiO_2.

Figure 10 illustrates this effect; for the cells around the interface, volume is a function of time and the integration boundaries of Eq. (45) change between times t_0 and t_1. Second, the existence of the two interfacial fluxes F_s and F_b creates a jump discontinuity in the impurity flux that propagates through space during oxidation. In the discrete space-time domain, as illustrated in Fig. 11, the interface moves by small steps in short intervals of time within which the system may be considered linear. It is therefore possible to consider the effect of interfacial fluxes as a superposition of two distinct processes

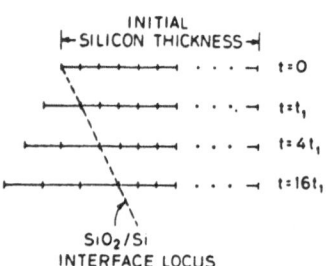

Fig. 10 - Relationship of SiO_2 and Si volumes when silicon is oxidized at four time increments. Here oxidation rate is assumed purely parabolic. The tick marks identify nodes.

each containing only one of the two flux terms. The first process consists of instantaneous motion of the interface at time t_0^+ to its new node position, I. Thus, the interfacial flux F_s is considered as flowing across the boundary I for the entire time interval (t_0, t_1). The same is true for the diffusive flux $F_D(y_{I-1/2})$, flowing across the cell boundary between the two SiO_2

cells near the interface. On the other hand, the second process consists of the redistribution of the impurity contents of these two SiO_2 cells, due to the moving interface flux, F_b. As shown in Fig. 11 the moving interface crosses the cell boundary at time t in the interval (t_o, t_1). Thus, in the interval (t_o, t_1), F_b is only internal to the cell $(I - 1/2)$, and only in (t, t_1) does it give rise to impurity flux between the two cells. Superposition of the two processes discussed above requires that for moving boundary conditions, Eq. (51) be modified only for the cells I and $(I - 1/2)$ by the addition of F_b to H_i for these two cells. Carrying out the integration as before and remembering that F_b is nonzero at the boundary only in (t, t_1) yields

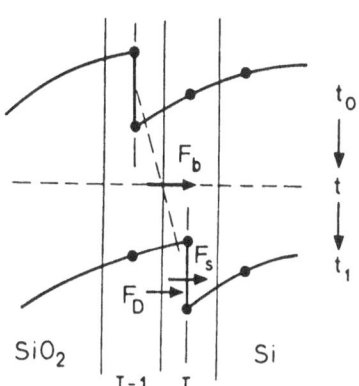

Fig. 11 - Interfacial cells and moving bundary flux at two increments of time. For simplicity the volumetric ratio, α, is assumed unity in this figure.

$$[H_i(t_o) + H_i(t_1)]/2 \pm \frac{t_1 - t}{t_1 - t_o} F_b(t_1) = [Q_i(t_1) = Q_i(t_o)]/(t_1 - t_o) \qquad (53)$$

Referring to Fig. 11, i in the above equation may be equal to I or to $(I - 1)$, where I identifies the interface node. The + sign is used when the equation represents the continuity in the interfacial cell I in SiO_2 and the - sign when the continuity in cell $(I - 1)$, also in SiO_2, is represented.

To complete this discussion, it remains now to outline the interrelation of time and space discretization that arises from the consideration of the moving boundary. Given the constraints that the interface must always lie on a node and that the number of oxide nodes may increase by at most one node in any time step (while at the same time the silicon node number is reduced by one), two distinct possibilities for advancement of the interface exist, depending on the time step and on the oxidation rate which is typically a quadratic function of time. These two possibilities are shown in Fig. 12 (a and b). This figure illustrates the discete impurity distribution for times t_o and t_1. The solid vertical lines indicate the cell boundaries and thus trace the evolution of the discrete system space.

The arrows show the motion of the SiO_2/Si interface. In

Fig. 12(a) the number of oxide nodes increases by one, while

in Fig. 12(b) it remains fixed. In the first case, (a), the

interface crosses the left-hand boundary of the cell that con-

tains the interface at time t_1, while in the second case, (b), it

does not cross any boundary. When case (a) arises, Eq. (53)

must be used in the two cells that share the crossed boundary

as already discussed, while in case (b) it should not be used

because the cell boundary in SiO_2 never sees F_b; and thus,

Eq. (52) is valid just as in all other cells. Referring back to

Fig. 12, it can be seen that due to the volumetric difference

between SiO_2 and Si, the volume of three cells changes as Si

gets converted to SiO_2. In both (a) and (b) the impurity

content of these cells is crosshatched, and the hatching polar-

ization is used to indicate the corresponding volumes at the

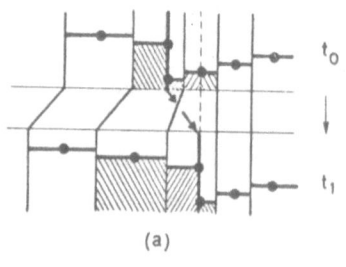

(a)

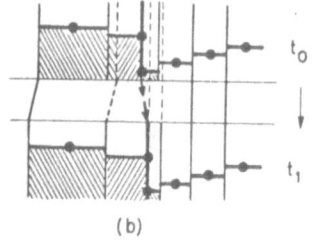

(b)

Fig. 12 - Examples of interface motion (a) when the number of oxide nodes increases by one and (b) when the number does not increase. In case (a) the interface crosses a cell boundary while in (b) it does not.

two times. Since volume expands from t_0 to t_1, the cell boundaries and thus the limits of

integrations of Eq. (45) must be traced from time t_1 back to time t_0. Given that at t_1 the cell

boundaries must lie half-way between nodes, in case (b) the cell boundaries around the

interface do not trace back ontojthe boundaries at t_0. The same is true also in case (a) where

the two interfacial half-cells at time t_0 become a single SiO_2 cell at t_1 while the full Si cell at

t_0 becomes two half-cells at t_1. Using always midpoint integration to establish $Q(t_1)$, the

hatching polarization and broken lines serve to indicate the impurity volume content at t_0,

$Q(t_0)$, that is used in the calculation of the content difference in Eq. (52) and/or (53). Of

course other integration rules may also be applied, but midpoint has been found satisfactory.

Epitaxy

Figure 13 shows the discretization of the simulation space for epitaxy as implemented in SUPREM.[34] In Fig. 13a the solid line separates the gas phase and the solid phase. P_D is the dopant partial pressure in the gas phase which can be approximated by P_D°.[34] The solid silicon is shown partitioned into discrete cells with broken lines delineating the cell boundaries. The dopant concentration within each cell (i.e. C_i, C_{i+1}, C_{i+2}, etc.) is considered uniform. The coupling between the surface boundary condition (Eq. (42)) and the main body of SUPREM is carried out in two steps, and is described below.

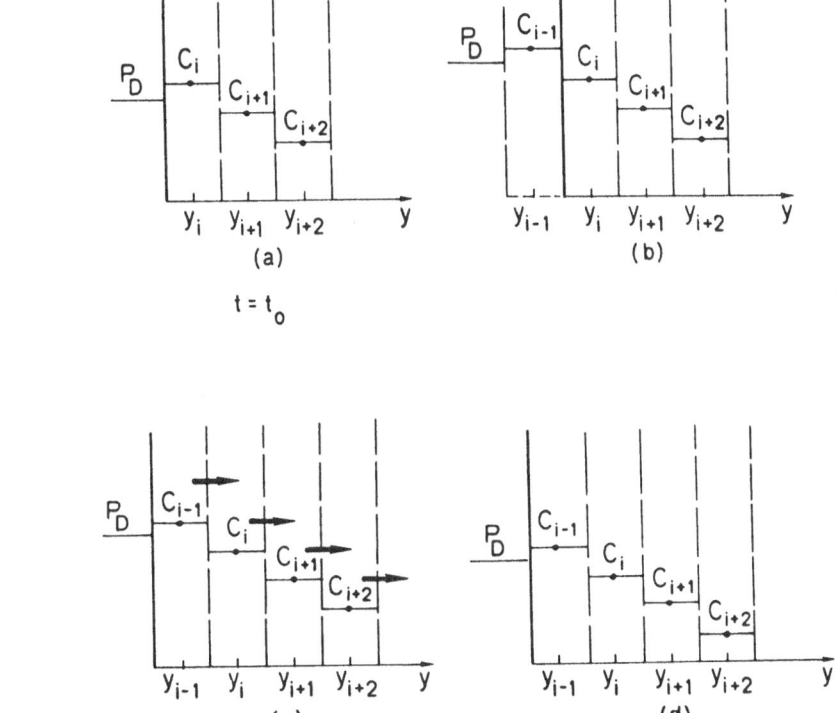

Fig. 13 - Implementation of the numerical technique used to solve the continuity equation with the surface boundary condition dictated by the epitaxial process. For explanation see text.

Step 1:

At time $t = t_0$ (Fig. 13a) the doping profile in the solid silicon is known. This profile is either the initial condition or the result of the simulation up to the time t_0. The cycle now starts by adding a new cell y_{i-1} (Fig. 13b). In order to calculate the dopant concentration C_{i-1} of this new cell a numerical routine is used to solve Eq. (42) with the left-hand side of the equation set to zero, i.e.

$$0 = k_{mf}\left(P^\circ_D - \frac{C_{i-1}}{K'_P} \right) - gC_{i-1} - K_A \frac{dC_{i-1}}{dt} \tag{54}$$

This is equivalent to accounting only for dopant introduction into the newly added cell without computing the simultaneous impurity redistribution in the solid silicon. This step is illustrated in Fig. 13b, with the arrow representing the net flux of dopant atoms entering the new cell.

Step 2:

The thermal redistribution of impurities that occurs during the growth of cell y_{i-1} is now computed. This is done by entering the impurity profile shown in Fig. 13b into SUPREM, which then computes the thermal redistribution of impurities during the interval Δt under consideration. This is illustrated in Fig. 13c. The arrows in the figure represent diffusive fluxes crossing cell boundaries. Notice that no flux is shown crossing the gas-solid interface. This is because the net introduction of impurities during this interval, Δt, was already considered in connection with Fig. 13b and Eq. (54) when C_{i-1} was first determined. After the dopant concentration in each cell is rearranged due to these diffusive fluxes, one time increment Δt has been advanced (Fig. 13d), and the cycle of operations just described is repeated.

APPENDIX

The second-order implicit difference equation for the diffusive flux resulting from Eq.

(14) is

$$F_D(y_{i+1/2}) = -1/2 \left[f_E \ (C_{i+1/2}^{m-1}) \ D \ (C_{i+1/2}^{m-1}) \ \frac{C_{i+1/2}^{m-1} - C_i^{m-1}}{y_{i+1} - y_i} \right. $$

$$\left. + \ f_E \ (C_{i+1/2}^{m}) \ D \ (C_{i+1/2}^{m}) \ \frac{C_{i+1}^{m} - C_i^{m}}{y_{i+1} - y_i} \right] \qquad (A-1)$$

where i = 1, 2, ..., (n-1), m is the discrete time step index, and $C_{i+1/2} = (C_{i+1} + C_i)/2$.

The impurity content of the general ith cell using midpoint integration is given by

$$Q_i \ = \ C_i(y_{i+1} - y_{i-1})/2 \qquad (A-2)$$

where $i \neq 1, I, n$. For the various boundary cells the impurity content equations are

$$Q_1 \ = \ C_1(y_2 - y_1)/2 \qquad (A-5)$$

$$Q_{I-1/2} \ = \ C_{I,ox}(y_I - y_{I-1})/2 \qquad (A-4)$$

$$Q_{I+1/2} \ = \ C_{I,Si}(y_{I+1} - y_I)/2 \qquad (A-5)$$

$$Q_n \ = \ C_n(y_n - y_{n-1})/2 \qquad (A-6)$$

REFERENCES

1. D. A. Antoniadis, S. E. Hansen, and R. W. Dutton, "SUPREM II -- A Program for IC Process and Simulation," SEL 78-020, Stanford Electronics Labs, Stanford University, June 1978.

2. J. Lindhard, M. Scharff and M. Schiot, Mat. Fys. Medd. Dan. Vid. Sclsk. 33, 1 (1963).

3. J. Gibbons and S. Mylroie, Appl. Phys. Letts. 22, 568, June 1973.

4. Mayer, Erihasen and Davies, Ion Implantation in Semiconductors, Academic Press, New York (1974).

5.   W. K. Hofker, D. P. Oosthoek, N. J. Koelman and H. A. M. De Grefte, Rad. Effects 24, 223 (1975).

6.   H. Ryssel, K. Haberger, K. Hoffmann, G. Prinke, R. Dumcke and A. Sachs, IEEE, ED-27 8, 1484 (1980).

7.   W. K. Hofker, H. W. Werner, D. P. Oosthoek and H. A. M. De Grefte, Appl. Phys. 2, Springer-Verlag, 265 (1973).

8.   H. Ryssel, H. Kranz, K. Muller, R. A. Henkelmann and J. Biersack, Appl. Phys. Lett. 30, 399, April 1977.

9.   H. Ryssel, private communcation.

10.   F. Morehead, Extended Abstracts, Electrochem. Soc. Meeting, Abstract No. 139, p. 366, Spring 1980.

11.   F. J. Morin and J. P. Maita, Phys. Rev. 96, 28 (1954).

12.   K. Lehovec and A. Slobodskoy, Solid-State Electron 3, 45 (1961).

13.   J. S. Makris and B. J. Masters, J. Appl. Phys. 42, 3750 (1971).

14.   R. B. Fair, Proc. of Third International Symp. on Silicon Materials Science and Technology (77-2), The Electrochem. Soc., p. 968, May 1977.

15.   A. F. W. Willoughby, J. Phys. D: Appl. Phys. 10, 455 (1977).

16.   R. B. Fair and J. C. C. Tsai, J. Electrochem. Soc. 124, 1107 (1977).

17.   R. B. Fair, J. Appl. Phys. 50, 860 (1979).

18.   D. A. Antoniadis and R. W. Dutton, IEEE, ED-26, 490 (1979).

19.   R. O. Schwenker, E. S. Pan and R. F. Lever, J. Appl. Phys. 42, 3195 (1971).

20.   R. B. Fair and G. R. Weber, J. Appl. Phys. 44, 280 (1973).

21.   M. Y. Tsai, . F. Morehead, J. E. E. Baglin and A. E. Michel, J. Appl. Phys. 51, 3230 (1980).

22.   R. B. Fair, Proceedings of the Fourth International Symposium on Silicon Materials Science and Technology, H. R. Huff, R. J. Kriegler and Y. Takeishi Editors, the Electrochemical Society, p. 963 (1981).

23.   G. Masetti, S. Solmi and G. Soncini, Solid State Electron 16, 1419 (1973).

24.   D. A. Antoniadis, A. G. Gonzalez and R. W. Dutton, J. Electrochem. Soc. 125, 813 (1978).

25.   D. A. Antoniadis, A. M. Lin and R. W. Dutton, Appl. Phys. Lett. 33, 1030 (1978).

26.   S. M. Hu, J. Appl. Phys. 45, 1567 (1974).

27.   D. A. Antoniadis, Proceedings of the Fourth International Symposium on Silicon Materials Science and Technology, H. R. Huff, R. J. Kriegler and Y. Takeishi Editors, The Electrochemical Society, p. 947 (1981).

28.   A. M. Lin, D. A. Antoniadis and R. W. Dutton, J. Electrochem. Soc. $\underline{128}$, 1131 (1981).

29.   K. Taniguchi, K. Kurosawa, and M. Kashiwagi, J. Electrochem. Soc $\underline{127}$, 2243 (1980).

30.   B. E. Deal and A. S. Grove, J. Appl. Phys. $\underline{36}$, 3770 (1965).

31.   C. P. Ho, J. D. Plummer, J. D. Meindl and B. E. Deal, J. Electrochem. Soc. $\underline{125}$, 813 (1978).

32.   C. P. Ho, J. D. Plummer, J. Electrochem. Soc. $\underline{126}$, 1523 (1979).

33.   R. N. Ghoshtagore, Solid State Electron $\underline{17}$, 1065 (1974).

34.   R. Reif and R. W. Dutton, J Electrochem. Soc. $\underline{128}$, 909 (1981).

BEAM PROCESSING TECHNIQUES APPLIED TO CRYSTAL SILICON SUBSTRATES

Giovanni Soncini

CNR - Instituto LAMEL
Via Castagnoli 1
40126 Bologna, Italy

1. INTRODUCTION

Conventional isothermal heat treatments in open or closed tube furnaces are extensively used in Integrated Circuits (IC) processing. Most of the high temperature fabrication steps are rate determined by a solid-state diffusion process (e.g. oxidation, dopant diffusion and redistribution, radiation damage annealing, gettering,...) whose diffusion constant is exponentially dependent on temperature with activation energies of the order of a few eV. Reliable production of IC devices in large numbers thus requires a tight control of the main process parameters, i.e. temperature and time. Modern semiconductor furnaces operate in a wide temperature range, which extends up to the silicon melting point, with a temperature control better than $\pm\ 1\,°C$; and for processing times $t > 10^2$ sec, this lower limit being related to the time required for insertion and extraction of the furnace boat which carries the batch of silicon slices to be heat-treated.

During the last few years, growing interest has been raised among semiconductor technologists by beam processing techniques, where the heat treatment is induced by energetic beams of either photons or electrons or, more recently, ions. Beam processing allows isothermal heat treatments at temperatures up to the melting point with processing times of the order of one sec or less; it also allows non-isothermal heat treatments at temperatures well above the silicon melting point, i.e. with transient liquid-phase formation, and with processing time down to 10^{-8} sec or less. Spatial and temporal control of the temperature profile induced within the semiconductor by the absorption of the beam energy allows a great selectivity of the heating effects, both laterally and in depth, down to distances of the order of 1μm, i.e. with a resolution on the scale dimension of the smallest device structures currently produced.

Beam processing can be performed with a great variety of sources. These include pulsed and CW laser or electron guns, light flashes and, more recently, ion accelerators; it offers new experimental possibilities, both for basic as well as applied solid-state research, and the subject has become one of increasing international activity.[1-4] There is little doubt that the main driving force behind much of this growth has been the realization that beam processing may have a significative impact on future IC fabrication.

This course is primarily devoted to silicon IC, and this review, which is by no means exhaustive, will examine beam processing in the light of its potential usefulness for silicon device technology. To this purpose, Fig. 1 summarizes the dominant IC planar fabrication process, and indicates the main areas where beam processing techniques might have an impact on conventional fabrication methods. The great num- ber and variety of potential ap-

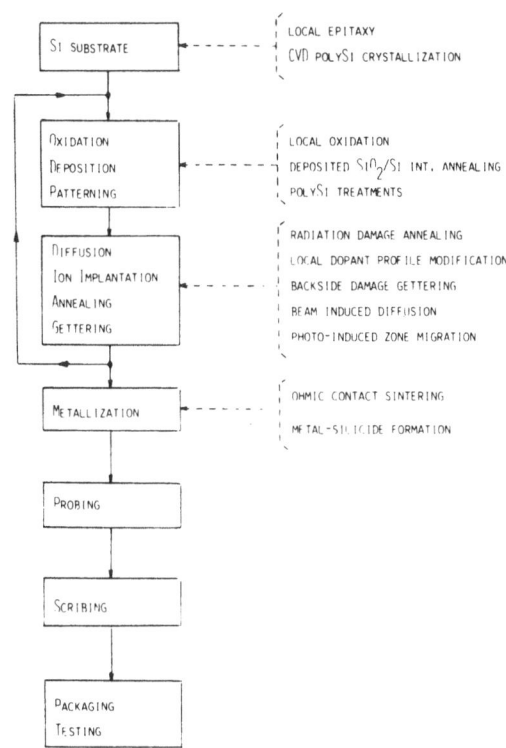

Fig. 1 - Flowchart of the main IC fabrication steps. Potential areas of beam processing implementation are also indicated.

plications testifies of the many new possibilities opened by the research activity carried out in the last few years in this field, and prevents us from attempting a discussion of all the cited topics. We'll consider here only beam processing of single crystal silicon substrates,* and will

---

* Beam processing of silicon layers deposited on amorphous substrated will be reviewed in G. Rozgonyi's lecture of this course.

focus mainly on electrical activation and radiation damage annealing of ion implanted layers. This subject has been most deeply investigated and, due to the ever increasing role of Ion Implantation in IC technology, should be considered first as a potential useful technique for being implemented in existing fabrication procedures; and in fact most of the published work dealing with applications refers to beam processed, ion implanted devices.

2. THERMAL EFFECTS INDUCED BY BEAM PROCESSING

In order to understand beam processing experimental results as well as to get quantitative forecast it is of primary importance to know the sample temperature distribution as a function of the main processing parameters. Being these beam induced heating cycles very fast, experimental measurements of the specimen temperature distribution are generally impossible, so we must rely on numerical simulation of thermal effects.

Three different main heating cycles can be induced within the target by beam processing, and these will be referred to as quasi-adiabatic, thermal flux and isothermal.[5] Their main features and the resulting temperature profiles are summarized schematically in Fig. 2. Although the interaction mechanisms with the sample are very different depending on whether light or electrons or ion beams are used,[6] the induced heating effects are similar so that many of the processes to be considered, being thermally activated,

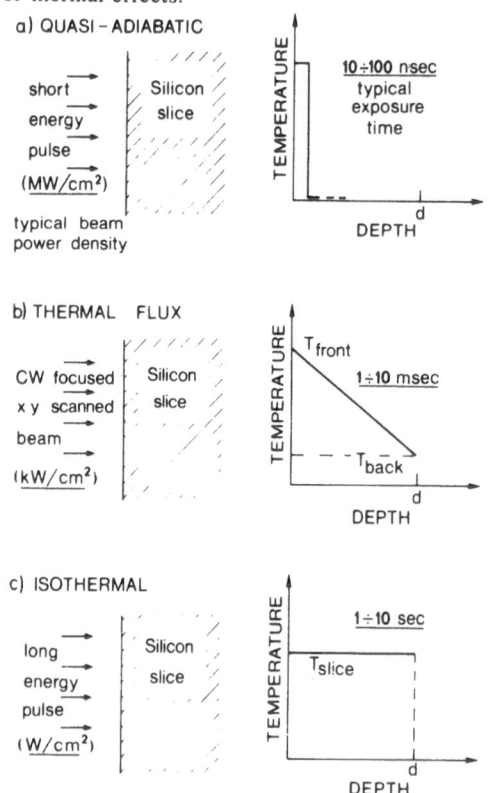

Fig. 2 - Schematic representation of the main thermal effects induced by beam processing. Typical beam power densities and exposure times are also indicated. From C. Hill, Ref. 5.

can be carried out by using, in the appropriate way, different beam sources.

Beam induced heating effects will be described first with reference to pulsed and CW

lasers, being these the most commonly used processing sources. Then we'll briefly discuss the

use of pulsed or CW electron beams, since these sources are gaining increasing importance in

semiconductor processing. Flash light, ions and other particle sources, being in their early

stage of development, will not be considered.

2.1 Laser Processing

i) Quasi-adiabatic processing: In quasi-adiabatic processing very short (less than 100

nsec) pulse energy beams of high intensity $(1 - 2 \text{ j/cm}^2)$ are used. To avoid "hot spot"

problems related to local inhomogeneities of the multimode non-uniform laser beam cross

section, a beam homogenizer is usually employed, which consists of a curved quartz rod with a

ground input face to introduce scatter.[7] The beam is randomized by the multiple internal

reflections and at the output a uniformity better than 10% over the all spot area can be

achieved. Large area substrates can be processed by partially overlapped laser spots.

The absorbed beam pulse energy is nearly instantaneously converted in local heat, which

diffuses into the sample by conventional thermal conduction. The temperature transient T(zt)

can be calculated by solving, subject to appropriate boundary conditions, the heat conduction

equation:

$$\rho C \frac{\partial T(zt)}{\partial t} = \frac{\partial}{\partial z} K \left[\frac{\partial T(zt)}{\partial z} \right] + E(zt) \tag{1}$$

where ρ = density (gr/cm^3); K = thermal conductivity (Watt/cm °K); C = specific heat (j/gr

°K).

The unidimensional model is usually applicable, being the beam diameter much larger than

the heat penetration depth on the time scale of interest.

The volume heat source E(zt) results from the energy loss exponential decay of the light beam within the sample, thus becoming

$$E(zt) = -\frac{\partial I(zt)}{\partial z} = (1 - R) I_0(t) \alpha e^{-\alpha z} \quad (Watt/cm^3)$$ (2)

being α and R the absorption and reflectivity coefficients of the silicon substrate at the laser

wavelength, and $I_0(t)$ the laser intensity.

The heat Eq. (1) has to be solved numerically, since structural (i.e. single crystal or

amorphous) dependence of the absorption and reflection coefficients, and temperature

dependence of thermal conductivity and specific heat must be taken into account. Possible

changes of phase, i.e. melting, should also be considered.[8]

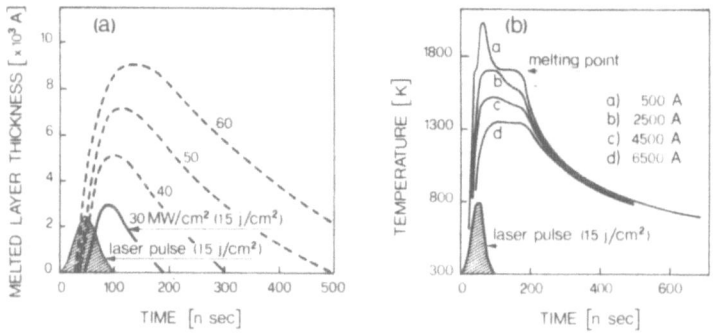

Fig. 3 - Time evolution of the solid-liquid interface for increasing laser
beam power densities (a) and of the temperature profile at different depth
(b) inside the amorphous silicon surface layers. Calculations have been
carried out with reference to a 50 nsec Q-switched ruby laser pulse.
From P. Baeri *et al.*, Ref. 8.

Typical results of these calculations carried out for a silicon substrate amorphized by

heavy implant and beam processed by a single 50 nsec, 1.8 j/cm^2 Q-switched ruby laser pulse

is shown in Fig. 3. Energy is absorbed most strongly at the silicon surface, and the surface

melt first, followed by successively deeper melting to a maximum just over 2500Å. Heat

diffusing into the underlying material subsequently cools this melted layer and leads to

resolidification. Thermal transient affects a substrate thickness of the order of 1μm, i.e.

comparable with the absorption depth of the laser light, thus evidentiating the negligible role

of the heat diffusion during the time scale of interest. Temperature of the remaining substrate is almost unaffected. This avoids lifetime degradation, usually observed when silicon is heated up to temperatures of the order of 1000°C, i.e. typical of conventional furnace processing.[9]

By using focused laser beam pulses, localized heating both in depth and laterally is also feasible. Beam spot diameters of the order of a few μm are easily achieved; in this case, however, the heat diffusion length

$$L_{TH} = \sqrt{D_{TH}t} \quad (cm) \tag{3}$$

where $D_{TH} = K/\rho C$ (cm^2/sec) is the specimen thermal diffusivity, becomes comparable to the spot size, thus influencing the total heat-treated volume. Being $D_{TH} \simeq 0.1$ (cm^2/sec) for silicon at melting temperture, heat diffusion lengths of the order of 1μm are in fact to be expected for exposure times of 10 - 100 nsec.

The results of thermal transient analysis of quasi-adiabatic processing of silicon slices evidences the presence of two contiguous phases: the underlying solid crystal and the liquid layer which extends from the surface to about 2500Å and whose permanence exceeds the pulse laser duration. The solid-liquid interface movement within the semiconductor as a function of pulse energy density is also reported in Fig. 3. The maximum thickness of the melted region and the permanence of the surface in the liquid phase are mainly related to the total energy absorbed by the target and increase with the beam pulse energy density. This is because little heat is lost by diffusion during the very short pulse interaction time, and therefore almost all the energy absorbed by the target is utilized in raising the absorption layer temperature up and above the melting point, thus justifying the name given to this kind of beam processing technique.

The very rapid refreezing of the melted layer, which resolidifies at velocities of the order of 1 - 2 m/sec must also be underlined. This corresponds to a quench-rate of the order of 10^{10}°C/sec, i.e. exceeding by order of magnitudes the values achievable by conventional rapid-quench techniques.

Results reported in Fig. 3 apply for Q-switched Ruby laser pulses; however they are fully
representative of thermal transient and related phenomena induced by quasi-adiabatic processing of semiconductor targets.*

Experimental evidence of the overall picture here summarized has been given by Auston
et al.,[13] who measured the transient optical reflectivity of the silicon surface treated by a
Q-switched, 30 nsec Nd-YAG laser. Their results are reported in Fig. 4.

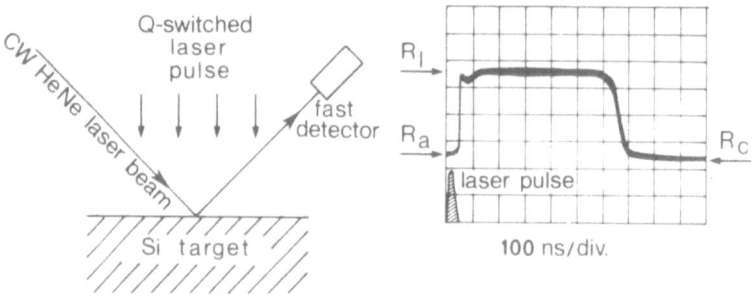

Fig. 4 - Time-resolved reflectivity experiment. R_a, R_l and R_c refer to
amorphous, liquid and crystalline silicon reflectivity. From D. H. Auston *et
al.*, Ref. 13.

The reflectivity raises abruptly as the surface melts and remains at the constant level R_l
typical of the liquid silicon phase until the surface refreezes, Since the CW He-Ne monitoring
laser light penetrates only $\simeq$ 100Å in molten silicon, the flat top of the reflectivity curve
indicates the total duration of the melting phase while the fall time mesures the time required
for the liquid/solid interface to move through the optical skin depth and allows an estimate of
the silicon resolidification velocity. Finally a slight reflectivity decrease is observed when the
silicon crystal surface cools down to room temperature.

Phase changes are accompanied by material density changes as well, which must be
considered since they may play an important role on the residual damage within the resolidi-

---

*   The melting model here outlined is not universally accepted. A non-thermal process
      involving the formation of a persistent hole-electron plasma has been suggested to
      alternatively explain the observed beam induced effects.[10] At this time, however, the
      simple melting model appears to gain more general consensus, and recent experiments
      specifically designed to discriminate between the two hypothesis give further evidence in
      favor of a thermally activated melting mechanism.[11,12]

fied melted layer, and will be a major problem to be solved for applying adiabatic beam processing to multilayer, selectively engraved strucutres like I.C. The molten silicon density is in fact about 10% larger than that of crystalline silicon, which is a similar amount denser than the amorphous phase. Beam processing induced phase changes thus create internal stresses which, in the liquid-phase regime, can induce various hydrodynamic effects.[14]

ii) Thermal Flux Processing: In thermal flux processing scanned CW laser beams with a spot diameter d of the order of $50 \div 100 \mu m$ and scanning velocity v of a few cm/sec are used to heat up the semiconductor surface. These experimental conditions lead to a spot permanence or dwell or exposure time $t \simeq v/d$ of the order of msec. The laser beam is passed through a lens and deflected by x (horizontal) and y (vertical) mirrors onto the silicon sample, which is mounted in the lens focal plane. The x mirror is driven by a triangular waveform, while the y mirror is driven by a staircase waveform. The beam is scanned across the target in the x direction, stepped vertically by a controlled y increment and then scanned back across the target in the reverse x direction. Scanning velocity and scan-lines overlap are easily controlled by adjusting the triangular and staircase waveform parameters.[15,16]

In thermal flux processing, heat is continuously supplied to the slice front surface by the scanned focused beam and continuously removed from the slice back surface by the heat sink. Dwell times of the order of msec are much longer than the thermal time constant of the silicon volume radiated by the beam, so that the semiconductor surface has adequate time to reach thermal equilibrium with the moving heat source. This makes it possible to calculate, by solving the heat equation in a steady state regime, the temperature that will be reached at the center of the beam spot. Calculations of this type have been carried out and the results are shown in Fig. 5.

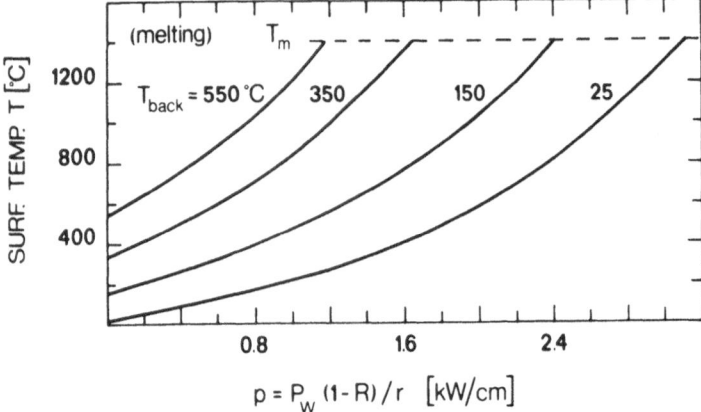

Fig. 5 - Silicon front surface temperatures vs normalized beam power for increasing back surface temperature. P_W, R and r represent the incident beam power, silicon reflectivity and spot radius. From Y. I. Nissim *et al.*, Ref. 18.

Let's notice that the front surface temperature is a function of the ratio of the absorbed beam power to spot radius,[*] and consequently this is the parameter which needs to be controlled in thermal flux annealing experiments. Let's further observe the noticeable influence of the thermal sink temperature, thus suggesting that an accurate control of the slice back surface temperature is needed to obtain reproducible results. This also suggests the possibility of reducing the laser beam power by applying thermal bias to the back surface. In this way, temperature gradient within the sample is also reduced.

Thermal flux processing of silicon substrates is usually carried out in the solid phase regime; however curves of the Fig. 5 suggest that by increasing the power to radius ratio of the beam, liquid phase regime can be achieved.

iii) Isothermal processing: In isothermal processing large diameter CW laser beams and long exposure times, up to a few sec, are used to process thermally isolated targets. Due to the high value of silicon thermal diffusivity and to the low value of beam power density here considered (tens of W/cm^2) the heat generated by beam absorption at the front surface is

---

[*] This is due to the essentially hemispherical simmetry of the heat flow when, as usual, the irradiated sample is thicker than the spot diameter. As a result, the surface temperature becomes a function of the beam power divided by the spot radius rather than power per unit area.[15]

rapidly distributed through the slice, whose temperature may be considered uniform at any time. Isothermal beam processing can thus be identified as a "fast furnace" heat treatment.

Being the slice thermally isolated, heat loss is only possible by radiation, and its temperature is determined by the balance between input beam power density P_W and output radiated power according to equation:

$$\rho \, C \, d \, \frac{dT}{dt} \; = \; P_W - 2\epsilon \, \sigma \, (T^4 - T_o^4) \tag{4}$$

where d is the slice thickness, ϵ and σ are the effective emissivity of the sample's surface and the Stefan-Boltzmann constant respectively, and T_o is the temperature of the external sample surroundings. Solution of Eq. (4) allows the slice temperature as a function of time to be deduced for different beam power densities and exposure times, as shown in Fig. 6. When beam processing ends, $P_W = 0$ and the specimen cools down loosing energy by the radiative process.[19] Temperature raise transient times can be effectively shortened by controlling the incident beam power density during exposure, thus allowing heat treatments well below one sec to be reproducibly carried out.

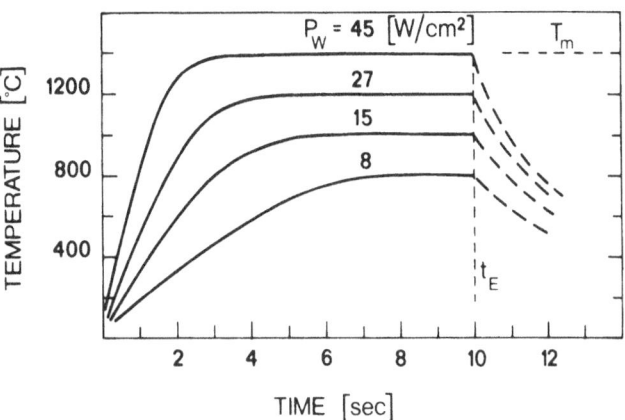

Fig. 6 - Silicon slice temperature vs exposure time for increasing beam power density in isothermal processing

2.2 Electron-Beam Processing

As stated previously, electron sources are also to be considered for heat-treating semiconductor samples.

Electron beam processing of semiconductors is attracting growing interest mainly because, as compared to the previously described technique based on laser beams, has the following advantages:

i)   the energy absorption by the specimen is more predictable, being less depend-
        ent on variables such as surface structure and reflectivity, than energy absorp-
        tion from light beams;

ii)   uniform processing of large area multilayer substrates is feasible at low cost
        and high throughput;

iii)   the beam main parameters, i.e. electron energy E and current density J are
        easily controllable over a wide range of values.

Thermal effects induced within the silicon sample by electron beam processing can be
modelled by solving the unidimensional heat Eq. (1). In this case, however, the volume heat
source term E(zt) deserves particular attention, since it has to be estimated by a Monte Carlo
simulation of the electron-solid interaction,[20] being related to the electron beam energy loss
profile within the sample. This can be obtained by memorizing the amount of energy lost by
each penetrating electron along its trajectory and by finally displaying the histogram of the
energy loss distribution vs. depth. Examples of energy loss profiles in silicon processed by 10
and 30 KeV electron beams are shown in Fig. 7. In computing these profiles electron
back-scattering has also been considered. Once the heat source term is known, heating effects
are readily calculated.[21] Results which refer to thermally isolated silicon targets whose front
surface is uniformly irradiated by either the 10 KeV or 30 keV electrons at various beam
current densities are also reported in Fig. 7. Temperature distributions refer to the moment t_m
immediately prior to the beginning of melting.

At 10 KeV, the electron beam energy is absorbed within $1\mu m$ thick layer, while at 30
KeV a layer of about $7\mu m$ is needed.

For very high beam current densities (10^3 Amp/cm^2, typical for pulsed generators) a
"quasi-adiabatic" temperature increase takes place, and its distribution reflects the trend of
the energy loss profile, the heat diffusion within the sample playing a negligible role. Melting
is reached in times of the order of 50 nsec in the under-surface layer which coincides with the
energy loss peak; then the melted region spreads inwards and outwards to the front surface.

Outwards movement of the liquid/solid interface is more rapid, since surface layers receive heat by conduction also from the already liquified layers beneath. This phenomenon becomes particularly evident in the higher energy case (E = 30 KeV), where at first melting takes place at about 2.5 μm below the front surface of the silicon sample. Front surface melting by high density electron beams thus requires the formation of a molten layer whose minimum thickness is related to the energy loss peak depth Δ_{PS} and, consequently, to the electron beam energy

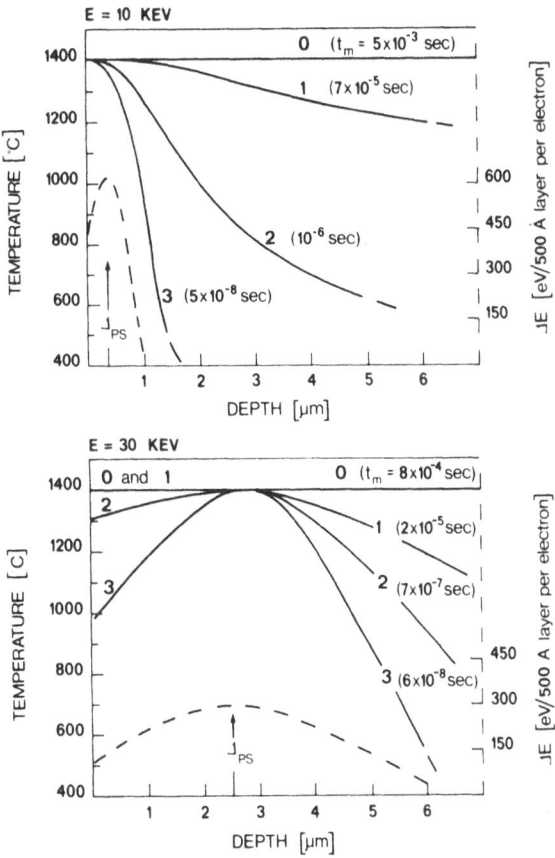

Fig. 7 - Temperature profile within the silicon sample at different beam current density for 10 KeV and 30 KeV electron energies at the moment t_m immediately prior to the beginning of melting. Curves indicated by 3, 2, 1 and 0 refer to current densities of 10^3, 10^2, 10 and 1 Amp/cm^2 respectively. Dotted curves indicate the energy loss profile ΔE within the silicon target. ΔE is reported in eV/500Å layer per electron. From P. G. Merli, Ref. 21.

E. Total melted layer thickness can be reduced by reducing the energy of the beam (but 10 KeV represent a practical minimum for field effect emission pulsed electron sources[21]) or, more conveniently, by changing the beam incidence angle.[22] This could be achieved in practice by the use of a variable magnetic field.

For lower beam current densities, heat diffusion plays an ever increasing role, and temperature profiles become more and more smooth until, as shown in Fig. 7, at the lowest beam current density here considered (1 Amp/cm^2), isothermal processing takes place. The sample temperature increases uniformly, and its steady-state value depends on the power density impinging on the front surface, while it becomes independent of the beam energy loss profile within the sample. Finally, the use of a focused, electromagnetically scanned electron beam on a target whose back surface is kept at a fixed temperature by a heat sink, allows thermal flux processing. Also in this case the target surface temperature, when the unidimensional model applies, becomes a function of the ratio of the beam power to spot radius.

Results of the heat equation solution here outlined emphasize the great versatility of the electron beam processing technique; since it can operate, by changing the beam parameters and exposure time, in quasi-adiabatic and in isothermal or thermal-flux mode. Liquid or solid phase processing can also be considered. Furthermore, by operating in the quasi-adiabatic mode, it offers the possibility of changing the thickness of the melted layer by varying electron beam energy and incidence angle, while by operating in the solid-phase regime it offers the possibility of an accurate control of the sample surface temperature. This applies also to multilayer, selectively engraved structures like IC, since local thermal properties, differently to the optical ones, do not vary appreciably.

Adiabatic beam-processing of silicon slices has been demonstrated by using a specially designed electron gun capable of delivering high intensity, large diameter 100 nsec electron pulses of 12 KeV mean energy,[23] while thermal flux and isothermal processing has been carried out by electromagnetically scanned CW electron beams. Electron welders and modified scanning electron microscopes have been used in these preliminary experiments, with results that are very similar to those obtained by the conventional CW scanned laser source. Recently, a purposely designed more versatile electron beam apparatus capable of high speed multi-scan annealing has been used.[24] A new machine capable of delivering large area, $10^{-3} \div 10$ sec duration, high intensity electron pulses has been designed and is under construction in our laboratory.[25]

3. BEAM PROCESSING OF ION IMPLANTED LAYERS

Doping of semiconductors by ion implantation produces heavy crystallographic radiation damage, and the as implanted layer can be regarded as nearly amorphous. Thermal annealing is needed to restore the layer crystallinity and to electrically activate the implanted dose. This is conventionally carried out in open or closed tube furnaces at temperatures of 900 - 1000°C for times of the order of tens of min.[26]

Recovery of the radiation damage and electrical activation of the implanted dose can also be achieved by beam annealing, as demonstrated by the early works carried out by Russians[27] and Italians[28] which led to the present intensive research effort in the semiconductor beam processing area.

Beam annealing of ion implanted layers can be carried out both in the solid-phase regime (i.e. by thermal flux or isothermal heating) or in the liquid-phase regime (i.e. by quasi-adiabatic heating). This lecture will briefly outline the general behaviour of the recovery of radiation damage and electrical activation of the implanted layers by beam processing. In our examples, we'll refer to phosphorus, however all the other main silicon dopants, i.e., arsenic and boron, behave similarly. For further details and for a deeper analysis of the many new interesting phenomena now under investigation the reader should refer to the specialized literature.[1-4]

3.1 Solid-Phase Regime

i) Regrowth mechanism: Recovery of the radiation damage is obtained, as outlined in Fig. 8, by thermally activated solid-phase epitaxial regrowth, the underlying perfect crystal acting as a recrystallization seed. Experiments on time-resolved Rutherford Back-Scattering of damage annealing in implanted layers demonstrate that recrystallization velocity V in (100) oriented silicon is related to the substrate temperature T by the simple Arrhenius-type relationship:

$$V = V_0 e^{-E_a/kT} \tag{5}$$

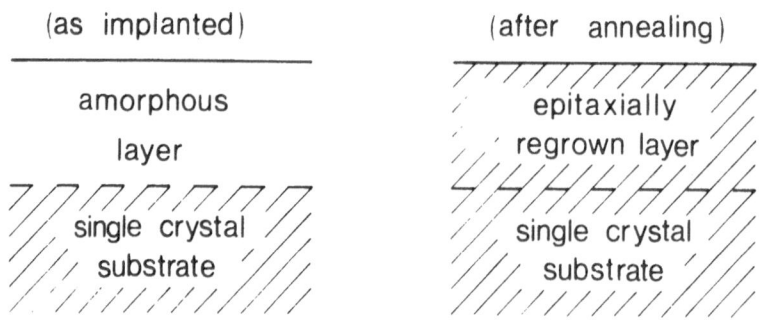

Fig. 8 - Implanted layer regrowth by thermally activated solid-phase epitaxy.

For (100) oriented silicon we'll assume, as suggested by the most recently published data:[29] $V_o = 2.9 \times 10^{19}$ cm/sec and $E_a = 2.7$ eV.

Regrowth velocity vs. substrate temperature is shown in Fig. 9, where, for convenience, the time regrowth t of a $0.1 \mu m$ thick amorphous layer is also reported. This value should be considered as the minimum regrowth time or, equivalently, the minimum beam processing exposure time needed to allow complete recrystallization of the amorphous layer up to the front surface. Clearly, if an exposure time shorter than t is applied, the partially regrown layer is overlaid by a remaining amorphous surface layer. This phenomenon is easily observed by changing the scanning velocity (i.e. the dwell or exposure time) in thermal flux processing of ion implanted layers.

The plots of Fig. 9 are of common use in practice, since the minimum exposure time is related to the temperature and, by using the computer results of Fig. 5 for thermal flux or Fig. 6 for isothermal processing, to the beam power needed to carry on a complete annealing of the implanted layer. These estimates are usually consistent with experimental results.

i.i) Residual Radiation Damage: Solid-phase beam processing, being analogous to the conventional furnace annealing with scaled-down times may leave residual disorder within the recrystallized implanted layer. This is usually reduced with respect to conventional annealing due to the greatly reduced time available for nucleation and growth of extended defects.[30]

Residual radiation damage in beam annealed implanted layers is strongly influenced, as in the conventional furnace processing, by substrate, implant and annealing parameters. Particu-

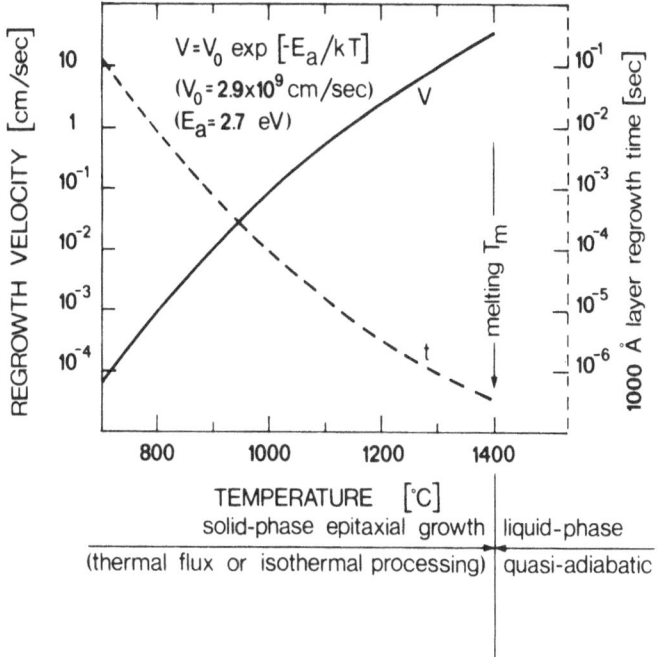

Fig. 9 - Solid-phase regrowth velocity vs. temperature. Dotted line indicates the time required for the regrowth of 1000Å thick layer. From D. H. Auston, Ref. 29.

lar care is needed when thermal-flux heating induced by scanned CW laser or electron beams is applied, since in this case both exposure time and partial overlap between adjacent scanning lines become critical. Furthermore, thermal bias of the back silicon slice surface is usually advantageous to reduce beam power and thermal gradient across the substrate. In this way the residual disorder within the recrystallized implanted layer is also reduced.[31]

An example of the residual crystallographic disorder in a phosphorus implanted silicon layer annealed by a multi-scan electron-beam induced thermal flux regime (MEBA) is shown in Fig. 10. By increasing the beam current, i.e. the surface layer temperature, the very small point defect clusters, 50Å in diameter, turn out into larger clusters and final dislocation loops. A further increase in beam current density tends to increase the dislocation loops diameter which eventually disappear when front surface temperatures approaching the silicon melting point are achieved.[32] Electrically active point defects are also detectable by Deep Level Transient Spectroscopy (DLTS) measurements, especially when high beam power to radius

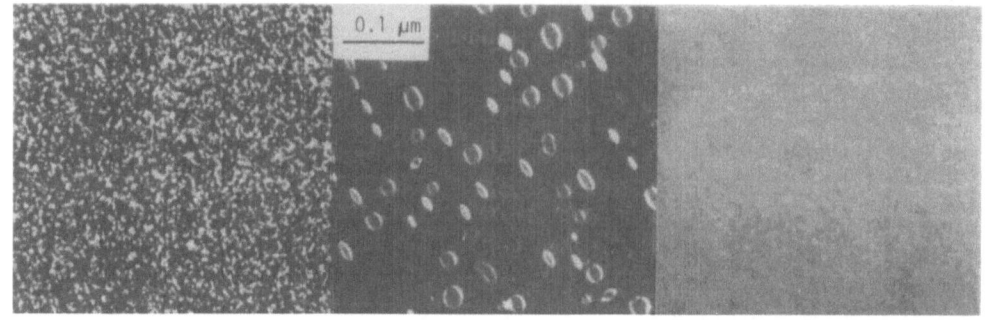

MEBA at increased beam current density

Fig. 10 - Residual crystallographic disorder in a phosphorus implanted silicon layer after Multi-scan Electron Beam Annealing (MEBA) at increasing beam current density. From M. Servidori *et al.*, Ref. 32.

ratios, i.e. annealing temperatures near the melting point are used.[33,34] These defect, whose origin has not yet been completely identified, are usually found below the implanted junction, thus acting as effective generation-recombination centers.

Besides residual crystallographic disorder, attention should be paid to substrate carrier lifetime which is also affected by the annealing procedure. Conventional single step furnace annealing is known to induce, by a still not completely known mechanism, substrate lifetime degradation;[9] and multi-steps time consuming annealing sequences are to be applied to avoid this phenomenon.[35] Beam processing, being carried out at higher temperature but with greatly reduced annealing times, appears to avoid lifetime degradation, and bulk lifetime enhancement has also been reported.[36] This peculiarity is of considerable interest in processing lifetime sensitive devices like ion implanted solar cells.

i.i.i) Doping Profile Redistribution: The reduced time-scale of beam processing allows recrystallization of the implanted layer and electrical activation of the implanted dose without inducing appreciable modification in the as-implanted doping profile. An example of this remarkable result is shown in Fig. 11, which refers to a phosphrous implanted, MEBA annealed layer.[36] The doping profile resulting from conventional 30 min, 900°C furnace

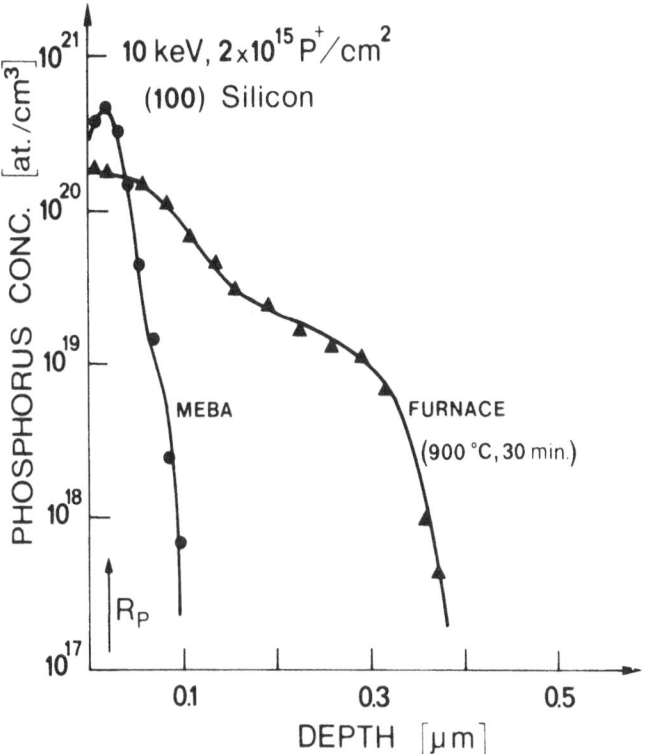

Fig. 11 - Phosphorus implanted doping profiles after MEBA and conventional furnace annealing. From G. G. Bentini *et al.*, Ref. 36.

annealing is also shown for comparison. The profile broadening due to solid state diffusion, usually enhanced by the presence of crystallographic defects, is fully evident.

3.2 Liquid-Phase Regime

i) Regrowth Mechanism: Recovery of the radiation damage by quasi-adiabatic beam processing is obtained, as outlined in Fig. 12, by liquid-phase epitaxial regrowth, the underlying perfect crystal acting as a recrystallization seed for the

Fig. 12 - Implanted layer regrowth by liquid-phase epitaxy.

melted layer. The threshold-like behaviour becomes evident, since a melting exceeding the amorphous layer thickness has to be achieved for epitaxial regrowth during subsequent solidification. Below threshold the substrate crystal seed is not reached and a polycrystalline layer results from solidification.

The pulse energy threshold value for ion implanted layer annealing depends on implant parameters. An example is shown in Fig. 13 where Rutherford Back-Scattering yield for a phosphorus implanted substrate treated by Q-switched ruby laser pulses of increasing energy density is shown.[37] In this case a pulse energy exceeding 1.5 j/cm^2 has to be used for annealing the implanted layer. Pulse energy higher than 2.2 j/cm^2 induce surface damage, while below threshold a polycrystalline layer is obtained.

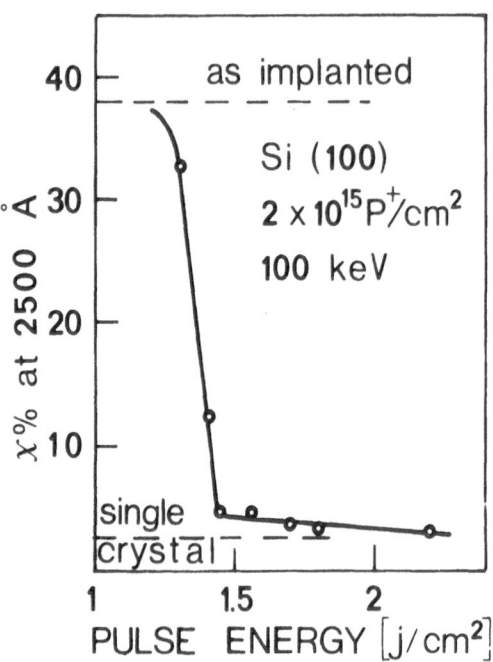

Fig. 13 - Rutherford Back-Scattering dechanneling fraction X measured at the damaged/undamaged interface of a phosphrous implanted silicon crystal. Annealing was performed by a Q-switched laser pulse at increasing energy densities. 1.5 j/cm^2 : implanted layer annealing threshold; 2.2 j/cm^2: surface damage threshold.

i.i) Residual Radiation Damage: A distinct advantage of annealing in liquid-phase regime is the avoidance of any extended residual crystallographic defect within the implanted layer, and the 100% electrical activation of the implanted dose. These remarkable results apply to both (100) and (111) substrate crystallographic orientation, for high and low implant doses and for all the main silicon dopant impurities here considered.[38] No bulk lifetime degradation is induced since, as stated previosuly, substrate remains essentially unaffected by the quasi-adiabatic heating.

The extremely high crystal regrowth and quench rate however, typical of the Q-switched laser induced melting regime, while effectively preventing extended defects to be nucleated, causes a high density of electrically active point defects to remain within the regrown layer

and in the channeling tail region. Heat-treatments at high temperature ($\gtrsim 700°C$) are usually

needed to remove these point defects. A remarkable temperature reduction can be, however, achieved by carrying out the point defects passivation in an atomic hydrogen atmosphere, while molecular hydrogen does not seem to have any significant effect.[39]

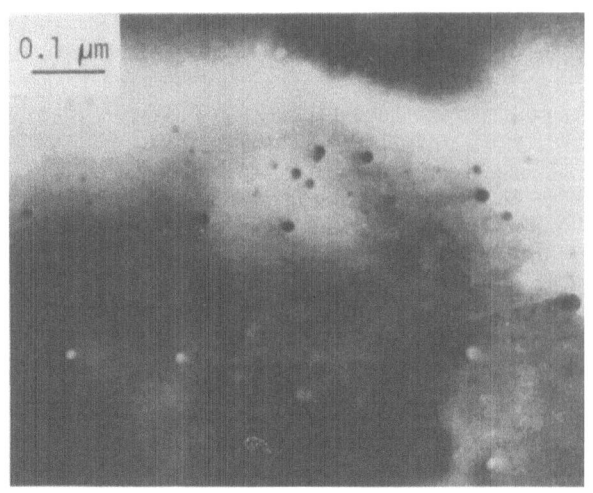

Fig. 14 - TEM micrograph of microvoids formed by the rapid resolidification of the melted layer. From G. G. Bentini *et al.*, Ref. 36.

Microvoids, related to the rapid refreezing of the melted region, may also be observed both in pulsed laser and electron beam quasi-adiabatically annealed samples.[36] An example of these defects is shown in the TEM micrograph of Fig. 14.

i.i.i) Doping Profile Redistribution: A significative redistribution of the dopant atoms is to be expected in the metling regime annealing, due to the high diffusion coefficients in the liquid silicon,

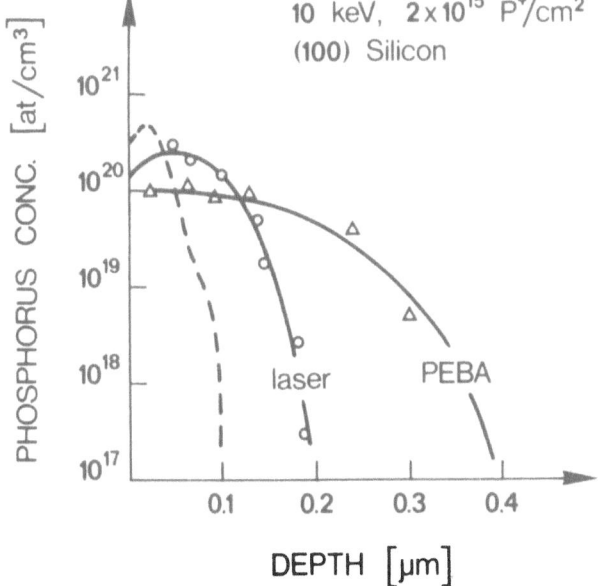

Fig. 15 - Doping profile redistribution of phosphorus within the implanted layer annealed by a Q-switched ruby laser pulse or a 12 KeV, 100 nsec electron pulse (PEBA) As implanted profile shown in dotted line.

which exceed by orders of magnitude the corresponding values in the solid silicon at melting temperature.

Figure 15 shows this effect for 10 KeV phosphorus implanted silicon annealed by either a 1.6 j/cm^2, 15 nsec laser pulse or a 12 KeV mean energy, 100 nsec, 0.84 j/cm^2 electron pulse. Final profiles are related to the melted layer thickness, and are explainable in terms of conventional thermal diffusion in the molten silicon, if duffusion coefficients of $10^{-4} \div 10^{-5}$ cm^2/sec are assumed. In both cases, a complete electrical activation of the implanted dose is achieved, and the regrown crystal appears free of extends defects. From DLTS measurements, now in progress in our laboratory, it seems that a reduced number of electrically active point defects are observed in the electron beam annealed samples, most probably related to the deeper melting and lower crystallization velocity achieved by this technique.

Surface dopant concentrations greatly exceeding equilibrium solubility limits can also be obtained by quasi-adiabatic beam processing, as evidentiated in Fig. 16 where a very high phosphorus concentration peak of 5×10^{21} at/cm^3 (corresponding to 10% of Si atoms) has been obtained by laser annealing of the most heavily implanted layer. Also in this case a complete electrical activation and absence of extended

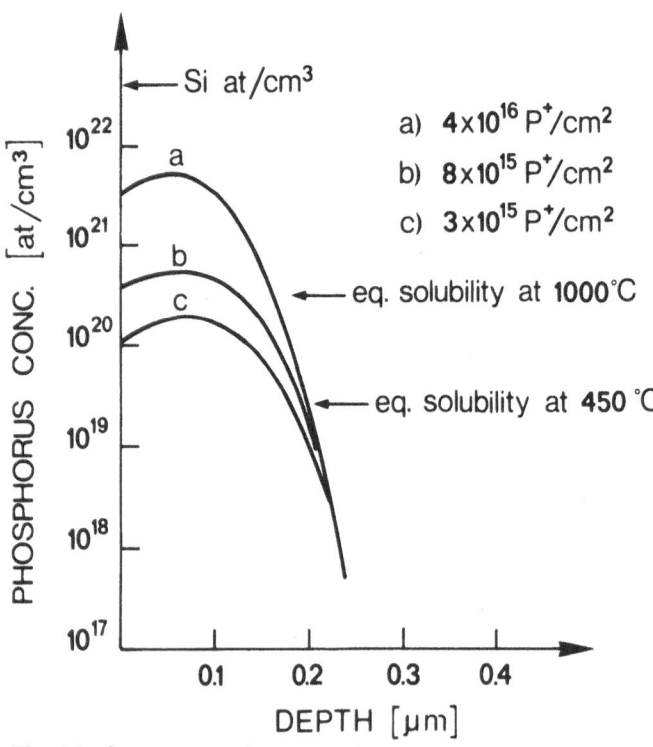

Fig. 16 - Supersaturated profiles obtained by pulse laser annealing of heavily doped phosphorus implanted layer. From M. Finetti *et al.*, Ref. 40.

defects was observed. A junction depth X_j lower than $0.3\mu m$ and a sheet resistance $R_\square = 8$ ohm/$\square$ was measured, corresponding to a "quality factor" q.f. $= (R_\square X_j)^{-1} \simeq 0.45 \mu m^{-1}$ ohm^{-1}, i.e. about four times higher than the one obtainable by conventional furnace processing.

These highly supersaturated solutions are far from equilibrium and their thermal stability, a necessary condition for practical application, depends on the kinetics of precipitation of the silicon phosphide phase. The occurence of precipitation was extensively studied and the main results published elsewhere.[40]

4. BEAM PROCESSED DEVICES: PROBLEMS AND PERSPECTIVES

From the previous discussion, it appears that beam processing offers a variety of new options to the device technologist. In order to really exploit these possibilities, however, the basic problem of selective control beam processing induced effects in multilayer, selectively engraved structures, must be solved. So far only beam annealed ion implanted MESA diodes[41] and solar cells[42] have been fabricated successfully, while, to the author's knowledge, beam processing of more complex devices, like bipolar and MOS transistors, are still in the early experimental stage.

To focus on the main problems one has to face, if beam processing has to be applied to a multilayer, multimaterial structure, let's briefly consider the most simple and common situation, i.e. the monochromatic laser light absorption by an oxidized silicon surface. The oxide layer acts as an antireflection coating, thus inducing non-uniform transmission at various oxide thicknesses, as shown in Fig. 17, which refers to a ruby laser light ($\lambda = 0.69\mu m$) at normal incidence angle. Quasi-adiabatic laser beam processing in liquid-phase regime of selectively photoengraved IC structures thus induces severe beam power absorption non-uniformities at the boundary between regions of different oxide thickness, where differences in the melted layer thickness are also observed. Hydrodynamic forces are here induced which generate waves within the molten layer on which the hot oxide is floating. The following rapid resolidification freezes-in the surface waves, thus giving rise to an oxide-rippling effect which

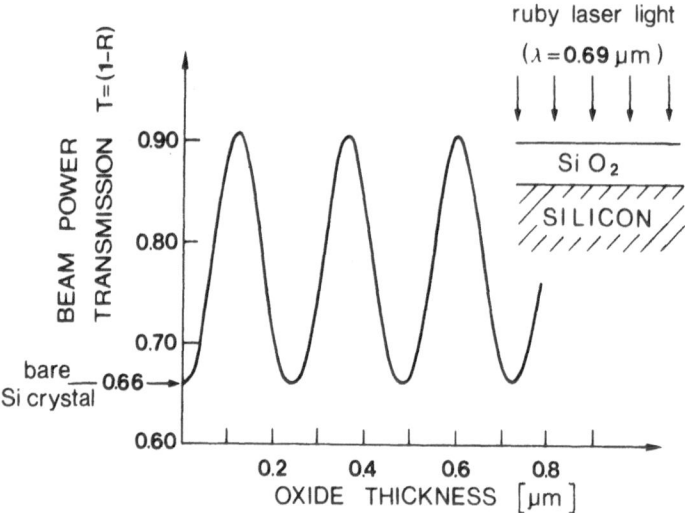

Fig. 17 - Ruby laser light transmission in an oxidized silicon surface at
normal incidence angle.

is particularly evident in the thinner oxide region, as schematically shown in Fig. 18. Similar

effects are also observed in pulse electron beam annealed multilayer structures.[14]

To apply quasi-adiabatic beam processing to IC fabrication it appears therefore unavoida-

ble to protect the oxide and the silicon underneath where beam processing induced transfor-

mations are not wanted. This implies the presence of some kind of autoregistered protective

layer to prevent damage; and consequently the IC fabrication process which is to take

advantage of this new technique has to be adequately redesigned. An interesting example of

this "strategy" is offered by the short channel MOS transistor developed at Hitachi Laborator-

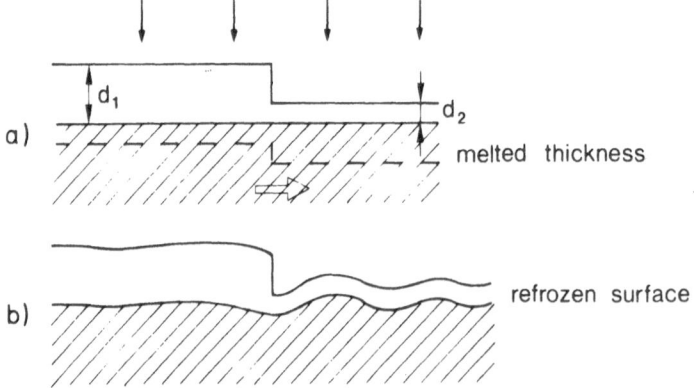

Fig. 18 - Rippling effect in quasi-adiabatic beam processing of an
oxidized silicon surface.

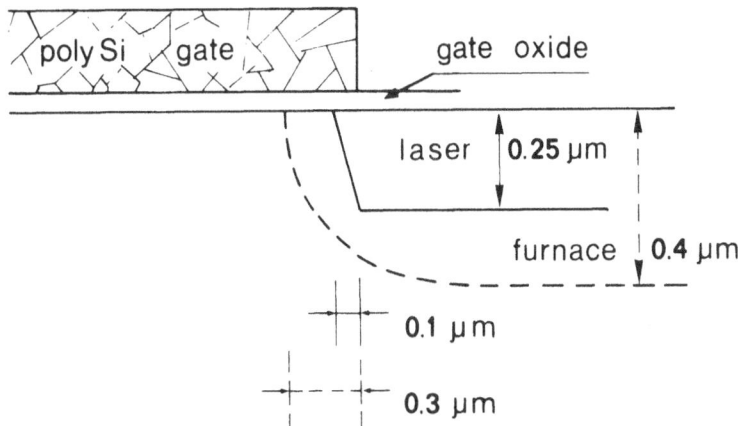

Fig. 19 - Short-channel, laser pulse annealed MOS transistor. Doping profile of furnace annealed source and drainregion is also shown in dotted line.

ies, where quasi-adiabatic laser pulse processing has been used to anneal the arsenic implanted source and drain regions, the polysilicon layer acting in this case as a protective mask for the gate oxide.[43] By this self-aligned annealing procedure both source and drain junction depth and capacitive overlaps with the polysilicon gate is significantly reduced, as shown schematically in Fig. 19, thus allowing a better high frequency performance and a reduced threshold voltage dependence on channel length. Pulse laser beam annealing of emitter and base regions of bipolar transistors has also been reported.[44] The extremely short annealing time and the liquid phase regime prevents any interaction between simultaneous diffusing impurities, thus avoiding emitter push-out effects. This technique may be of interest for reliable production of high speed, shallow junctions bipolar transistors. Similar results are achievable by quasi-adiabatic pulse electron beam processing.

Thermal flux or isothermal processing in solid phase regime avoids hydrodynamics and consequent rippling effects, thus allowing an easier implementation in existing IC fabrication procedures. Also in this case the presence of oxide overlayers of different thicknesses imply different transmitted fractions of the incident beam power, and unwanted damage or degradation of the oxide/silicon interface may arise. Besides these difficulties however, more and more examples of successful beam processed devices in solid-phase regime appear in the literature. Gate threshold[45] and source and drain[46] implants of MOS transistors are recent samples.

5. CONCLUSIONS

The basic characteristics of beam processing technique applicable to silicon IC fabrication have been outlined, with emphasis to potential advantages in annealing ion implanted devices. The continuing trend toward reduced dimensions will offer increasing opportunities to beam processing as a cost effective new technique for higher performance IC fabrication. It might also play a fundamental role in switching from today batch-processing to future fully automated serial IC processing.

ACKNOWLEDGEMENTS

The author is indebted to the many colleagues who are or have been involved in the beam processed, ion implanted solar cells research program carried out at LAMEL, who contributed to many of the results presented in this work. In particular, he wishes to thank Dr. P. G. Merli for the computer simulation of electron beam induced thermal effects and Dr. M. Servidori for his informations on residual radiation damage in ion implanted layers.

REFERENCES

1.   S. D. Ferris, H. J. Leamy and J. M. Poate, "Laser-Solid Interactions and Laser Processing," Academic Press (1979).

2.   C. W. White and P. S. Peercy, "Laser and Electron Beam Processing of Materials," Academic Press (1980).

3.   C. L. Anderson, G. K. Celler and G. A. Rozgonyi, "Laser and Electron Beam Processing of Electronics Materials," Electrochemical Society (1980).

4.   J. F. Gibbons, L. C. Hess and T. W. Sigmon, "Laser and Electron Beam Solid Interactions and Materials Processing," North-Holland (1981).

5.   C. Hill, Ref. 4, p. 361.

6.   M. Von Allmen, Ref. 2, p. 6.

7.   A. G. Cullis *et al.*, J. Phys. E.: Scient. Instrum. <u>32</u>, 688 (1979).

8.   P. Baeri *et al.*, J. Appl. Phys. <u>50</u>, 788 (1979).

9.   M. Finetti *et al.*, Revue de Phys. Applique <u>13</u>, 809 (1978).

10.   J. A. Van Vechten, Ref. 2, p. 53.

11.   M. J. Nathan *et al.*, Appl. Phys. Lett. <u>36</u>, 512 (1980).

12.   G. G. Bentini *et al.*, Phys. Rev. Lett. <u>46</u>, 156 (1981).

13.   D. H. Huston *et al.*, Ref. 1, p. 11.

14.   C. Hill, Ref. 3, p. 26.

15.   J. F. Gibbons, Ref. 3, p. 1.

16.   A. Gat and J. F. Gibbons, Appl. Phys. Lett. <u>32</u>, 142 (1978).

17.   H. E. Cline and T. R. Anthony, J. Appl. Phys. <u>48</u>, 3895 (1977).

18.   Y. J. Nissin *et al.*, J. Appl. Phys. <u>51</u>, 274 (1980).

19.   M. Bruel *et al.*, Rad. Effects, <u>44</u>, 173 (1979).

20.   A. De Salvo and R. Rosa, Material Chemistry <u>4</u>, 495 (1979).

21.   P. G. Merli, Optik <u>56</u>, 205 (1980).

22.   P. G. Merli and R. Rosa, Optik <u>58</u>, 201 (1981).

23.   A. R. Kirkpatrick, Ref. 3, p. 108.

24.   R. A. McMahon and H. Ahmed, Ref. 3, p. 123.

25.   P. G. Merli, to be published.

26.   H. Ryssel; this course.

27.   I. B. Khaibullin *et al.*, Rad. Effects <u>36</u>, 225 (1978).

28.   G. Foti *et al.*, J. Appl. Phys. <u>16</u>, 189 (1977).

29.   D. H. Auston *et al.*, Ref. 1, p. 11.

30.   A. Gat *et al.*, Appl. Phys. Lett. <u>32</u>, 276 (1978).

31.   G. A. Rozgonyi, *et al.*, Ref. 1, p. 457.

32.   M. Servidori, to be published.

33.   N. M. Johnson *et al.*, Appl. Phys. Lett. <u>34</u>, 704 (1979).

34.   M. Mizuta *et al.*, Appl. Phys. Lett. <u>37</u>, 154 (1980).

35.   E. C. Douglas and R. V. D'Aiello, IEEE Trans. ED-27, 792 (1980).

36.   G. G. Bentini *et al.*, Ref. 2, p. 272.

37.   F. Zignani, *et al.*, Proc. of the 2th European Solar Energy Photovoltaic Conference, Berlin 1979, Reidel Publishing Co.

38.   J. S. Williams, Ref. 3, p. 249.

39.   L. C. Kimerling and J. L. Benton, Ref. 2, p. 385.

40.   M. Finetti *et al.*, J. Electrochem. Soc., to be published.

41.   K. L. Wang *et al.*, Ref. 1, p. 569.

42.   G. Soncini, Proc. of the 3rd European Solar Energy Photovoltaic Conference, Cannes 1980, Reidel Publishing Co.

43.   M. Miyao *et al.*, Jap. J. Appl. Phys. 19, 129 (1980).

44.   N. Natsuaki *et al.*, Ref. 4, p. 375.

45.   G. Zimmer, Electronics Lett. 15, 184 (1979).

46.   J.D. Speight *et al.*, Ref. 4, p. 383.

LITHOGRAPHY SYSTEMS FOR VLSI

A.N. Broers

IBM Thomas J. Watson Research Center
Yorktown Heights, NY   10598

1. INTRODUCTION

This article describes the different lithographic approaches being used to fabricate microcircuits. All approaches are directed towards reducing dimensions and packing components closer together. The accuracy with which structures can be placed is as important as how small the structures can be.

Resolution and Overlay

In microcircuit manufacturing, minimum linewidth is set by the ability to control linewidth and not simply by the ability to reproduce a given linewidth. In order to establish the minimum linewidth, patterns containing a range of linewidths are exposed on many wafers and hundreds of measurements made across the wafers. The distribution of widths for each nominal linewidth is fitted to a Gaussian distribution and the standard deviation (σ_ℓ) determined. The device designer then chooses the minimum linewidth that gives him an acceptable ratio of standard deviation to linewidth. The linewidth is usually ten to fifteen times the standard deviation. Devices typically have to be made tolerant to $\pm 3\sigma_\ell$ variations in linewidth. Uniform deviations from the nominal linewidth can generally be compensated for by altering the linewidth in the mask, or by altering the written linewidth with direct electron beam exposure.

Errors in the position of one pattern with respect to another are known as overlay errors, and are also measured at many points across a wafer and over many wafers. Any systematic offset error (δ) due to faulty adjustment of the alignment system is determined and subtracted to leave the random errors. The random errors are again fitted to a Gaussian distribution and the standard deviation (σ_a) calculated. In order to get a high yield of good devices, devices are designed to be tolerant to alignment errors of up to $\pm(\delta + 3\sigma_a)$. When laying out a microcircuit, the distance that must be left between structures in order for them not to overlap is determined by both linewidth control and overlay accuracy.

The end result of this type of analysis is that devices must be designed to be tolerant to errors two to three times greater than might be measured in a few laboratory experiments. For example, early laboratory experiments with electron beams produced both linewidth and overlay of $\sim 0.1\mu$ but the best systems today have difficulty producing microcircuits with dimensions below about 0.5μ and with 3σ overlay errors of less than $\pm 0.3\mu$. The reader should bear in mind the difference between initial experimental results and the eventual performance obtained in manufacturing environments when interpreting data published on high resolution lithography.

2. CONTACT/PROXIMITY PRINTING

Contact printing remains a widely used wafer exposure technique. Today masks are more often "hard-surface," typically chromium on glass, than photographic emulsion, but the basic ultra-violet (UV) contact printing process has remained the same for many years. In some instances a small gap is maintained between mask and wafer in order to reduce mask damage and to eliminate the picking up of resist from the wafer onto the mask. When a gap is left, the term proximity printing is often used. Resolution in a shadow image is set by diffraction between the mask and the bottom of the resist layer. Thick resists, or gaps between mask and resist, degrade resolution. In practice the minimum useable linewidth W(m) can be approximated from

$$W = 15 \sqrt{\frac{\lambda_S}{200}} \tag{1}$$

where λ(m) is the wavelength of the radiation used to expose the resist (it is assumed that the wavelength is the same in the resist and in the gap), and S(m) is the distance between the mask and the bottom of the resist. This is the condition where the intensity at the center of an isolated line matches the background intensity. This criterion is derived from the degradation in resolution due to Fresnel diffraction.[1] More rigorous computations of exposure profiles for proximity printing have been made by Lin.[2]

Linewidth versus gap for deep UV radiation and soft x-rays is plotted in Fig. 1. Good agreement has been established with experiment, at least in the region of 0.5μ to 2μ. For example, 0.5μ linewidth has been produced in 1μ of PMMA resist by employing 2000Å-2600Å radiation and maintaining intimate contact between mask and wafer.[2] It is

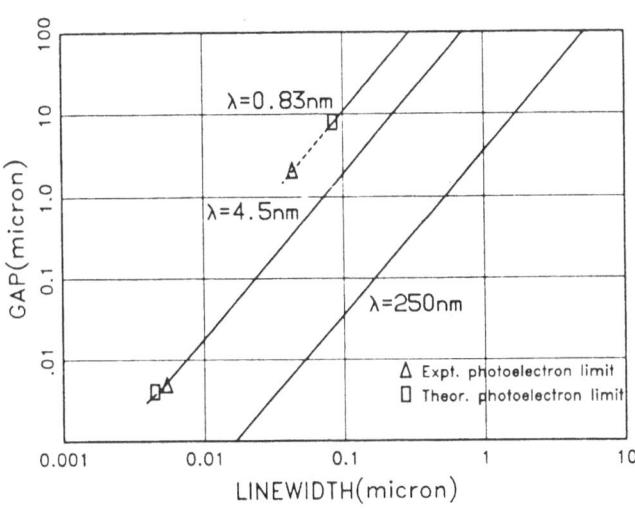

Fig. 1 - Linewidth versus gap for deep UV and x-ray proximity printing. Theoretical points correspond to the Gruen range for the maximum energy photoelectrons. Experimental points were measured by Feder and Spiller.[10]

obviously difficult to maintain perfect contact over large areas because of contaminating dust particles, and a few square centimeters is probably a practical limit even for experimental devices. Ultimately, the size of the structure is limited only by the thickness of the imaging layer. For example, it should be possible to produce 250nm dimensions in 100nm resist over a small fraction of a square millimeter.

In principle, positioning of a mask with respect to a sample can be made extremely precise ($\geq 0.1\mu$) by using optical detection methods. Accuracy beyond the Rayleigh criterion is possible by threshold detection because, in general, the S/N ratio in the detected signal is favorable. Methods that employ diffracting components on mask and wafer have also proven effective in the laboratory,[3,4] although in some instances errors arise if the mask to wafer spacing varies.[5] Overlay accuracy over the whole sample will, of course, depend on temperature control and on the ability to control mask and wafer distortion. No fundamental limits can be identified in this instance. The major drawbacks of contact or proximity printing are the low yield it produces due to mask/wafer damage, the need for an expensive mask with a short life, and limited resolution. They have led to the replacement of most contact printers with optical projection cameras and although deep UV illumination allows larger gaps between mask and wafer, it is unlikely that there will be a return to contact printing.

X-Ray Lithography

X-ray lithography is proximity printing using soft x-rays with wavelengths between 0.4nm and 5nm.[6,7] The wavelength is so short that diffraction effects are negligible, at least down to linewidths of about 0.1μ, if it is assumed that the mask to wafer spacing is 50μ. The mask can no longer be made on a quartz plate because such a plate would absorb the soft x-rays. Instead it is formed on a thin ($< 10\mu$) membrane of materials such as silicon, mylar, polyimide, or a combination of two materials. When two materials are used, one chosen for dimensional stability (silicon, silicon carbide, boron nitride, etc.) and the other for ruggedness (e.g. polyimide).

X-ray lithography offers one outstanding advantage over other methods. Image fidelity is preserved in very thick resist layers. This allows high aspect ratio resist patterns to be produced (see Fig. 2) and means that there are no proximity effects as there are with electron exposure. On the other hand, the mask is fragile and may not be dimensionally stable, and the large divergence of conventional x-ray sources gives rise to image distortion if the mask or wafer is not flat. Despite

Fig. 2 - High apsect ration 1μ linewidth pattern in PMMA resist obtained in x-ray lithography (Spiller *et al.*[10]).

these difficulties, Lepselter *et al.*[8] have recently built a conventional source, full wafer (75mm

diameter), X-ray system capable of exposing 75 wafers per hour. The system has produced

FET's with final gate lengths of 1μ and below. Overlay accuracy has not been published in

the normal form used for microcircuit manufacturing, but appears to be $\pm$ 0.25μ (1σ). Mask

and wafer are aligned with respect to each other using a dual wavelength optical microscope.

The mask is in focus at one wavelength and the wafer at the other. A more recent method

employs two zone-plate lenses. The mask consists of a gold pattern on a composite membrane

of boron nitride and polymide.

For overlay accuracy better

than $\pm$ 0.5μ (3σ), a step and re-

peat approach will probably have

to be used, particularly for larger

(125mm) wafers. With step and

repeat, the mask can be limited in

size to the maximum stable area as

set by mask and/or wafer instabil-

ities. The smaller mask will also

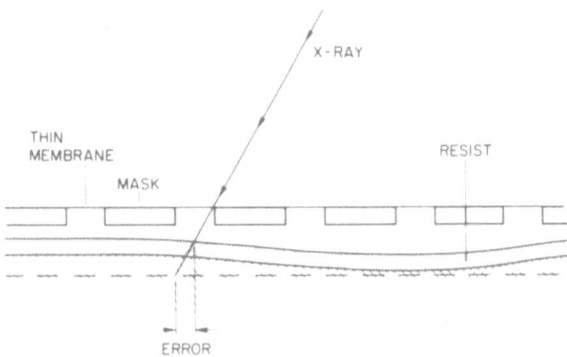

Fig. 3 - With a conventional x-ray source, the position of the shadow image varies with the mask to wafer spacing. Wafers frequently become buckled in hot processing.

greatly alleviate the run-out error that arises with a divergent source (see Fig. 3). Unfortunate-

ly, exposure times with presently available resists and conventional sources is too long for a

small area step and repeat approach to be economic. Exposure time would have to be reduced

to a few seconds for a source to wafer distance of about 50 cm. Today, exposure times are

about a minute. The source to sample distance determines the run-out error and the penum-

bral blurring. Penumbral blurring depends on the source size, and the ratio of the mask/wafer

spacing, to the source to mask distance. Mask to wafer spacing has to be greater than about

20μ if the mask is to be safe from contact damage with the wafer.

One source that produces an x-ray beam with all the properties needed for a step and

repeat system is the electron storage ring. For several years physicists have been using the electron storage ring as the "brightest" short wavelength light source available to man. A storage ring consists of a circular stainless-steel vacuum tube in which electrons circulate at speeds very close to the speed of light. The electrons are injected into the ring from an accelerator, and their energy is increased and maintained by passing them through a highly excited microwave cavity every time they pass around the ring. They are held in the circular orbit by magnets. Every time the electrons are deflected (accelerated) by one of these magnets, they emit radiation with a spectrum of wavelengths that extends from visible light to that of hard x-rays. By choosing the appropriate electron energy, copious soft x-rays are emitted of the wavelength needed for x-ray lithography.

At first it would seem difficult to understand how such an expensive device could be economical for manufacturing microcircuits. A storage ring suitable for lithography is predicted to cost $5-10M. The answer lies in the vast total x-ray flux that emanates from the ring. Enough radiation is emitted at each of the typically sixteen bending magnets to expose 40 125mm diameters wafers per hour, a production rate similar to that of one of today's optical projection cameras.[9] Therefore, fully utilized, a storage ring could produce the equivalent of about sixteen projection cameras. A projection camera costs $500K to $1M, so provided one uses all sixteen ports of the storage ring, it becomes competitive with optics and, of course, can produce higher resolution.

The radiation from a storage ring is emitted in a broad sweep with a very narrow vertical spread. At a distance of about ten meters from the ring, the beam is typically 1cm high and thirty centimeters wide. An oscillating, grazing incidence mirror will be used to broaden this beam out in the vertical direction. Because the angular divergence of the beam is very much smaller than that of a conventional x-ray source, the problem of misalignment with non-flat wafers and masks is avoided. The very intense radiation should also allow exposure times to be so short that it will be possible to expose relatively small areas at one time. In this way, the size of the exposed area can be made so small that distortions of the mask and wafer are not important.

X-ray lithography with a storage ring offers short exposure time and insensitivity to distortions, but there remains difficulty of finding a means of aligning the mask to the sample with adequate accuracy. Many laboratories are working on high resolution optical sensors that in theory provide accuracies of a tenth of a micron and it appears likely that a solution will be found. Provided it is, x-ray lithography should provide a means of producing dimensions below a micron at a cost comparable to that provided above 1μ by contact or projection printing.

Two factors set the resolution for x-ray proximity printing: (1) diffraction between mask and wafer, and (2) the range of the photoelectrons formed when the x-ray photon energy is absorbed in the resist. The same diffraction criterion used above for deep UV shadow printing can be applied to x-rays to give the relationships shown in Fig. 1. Photoelectron range has been measured experimentally[10] and estimated from the Gruen range.[11] In all cases, ultimate resolution is about 0.02μ. For microcircuit applications, the significant factor to be observed from Fig. 1 is that a gap in excess of 100μ can be tolerated for 1μ linewidth. In most semiconductor applications, penumbral blurring due to finite source size is more likely to limit resolution.

X-ray lithography is perhaps the most promising method for large volume production of devices with dimensions below 1μ. However, it may also be possible to produce dimensions below 1μ with optical projection cameras, and it is impossible to tell at what point a change-over will take place.

4. OPTICAL PROJECTION

Two forms of optical projection have been used successfully for semiconductor device fabrication; 1:1 scanning projection, and step and repeat reduction projection. Their main advantage over contact or proximity printing is that the mask and wafer are completely out of contact and the potential for damage is eliminated. To date, the scanning cameras have used reflecting lenses and the step and repeat cameras have used refracting lenses. There is no particular reason for this, however, and in particular it may be advantageous to use a large numerical aperture mirror lens in a step and repeat camera. Modern refracting lenses offer highest resolution although reflecting systems have proven more economic in manufacturing environments and ultimately should produce highest resolution because they can be used at shorter wavelengths. In practice, 60% modulation has been considered necessary for satisfactory resist exposure. For an incoherently illuminated system, 60% contrast is obtained for a linewidth of $(\lambda/1.28 \times (N.A.))$, where N.A. is the numerical aperture of the lens. Figures 4 and 5 show the modulation transfer functions (M.T.F.) for various optical systems.

$$M.T.F. = \frac{2}{\pi}(\phi - \cos\phi\,\sin\phi)$$

$$(2)$$

$$\phi = \cos^{-1}\frac{\lambda}{4\ell\,(N.A.)}$$

where ℓ(nm) is the linewidth. Higher contrast can be obtained for relatively large linewidths by using a partially coherent illumination system, that is by arranging that the image of the source

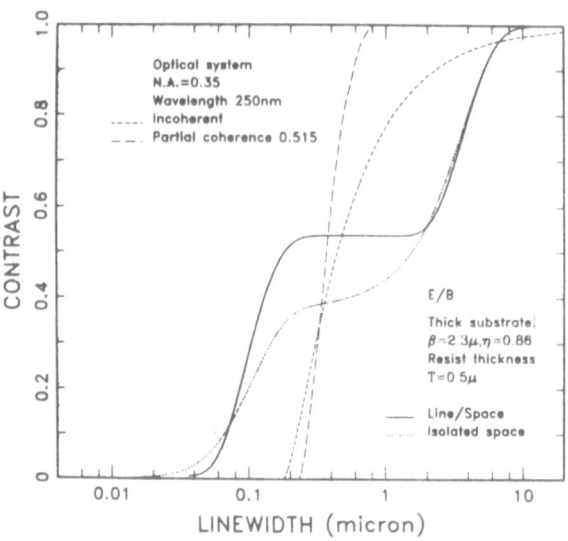

Fig. 4 - Contrast versus linewidth for ultimate resolution optical projection and electron beam (25kv) lithography. Electron backscattering coefficients for silicon substrate were obtained from Ref. 55.

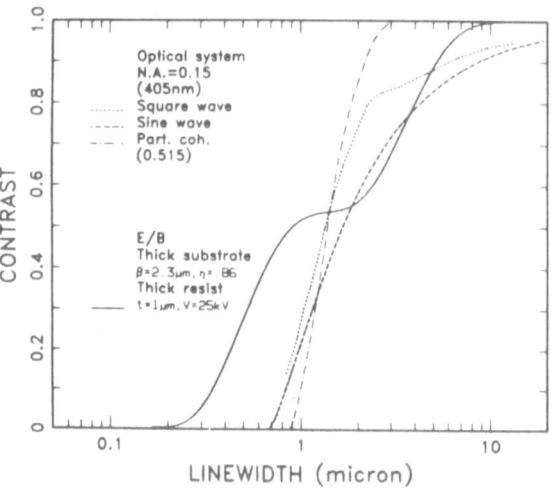

Fig. 5 - Contrast versus linewidth for conventional optical projection and electron beam in use today. Minimum linewidth for routine operation is best estimated by doubling the linewidth at which 30% contrast is obtained. This avoids being overly pessimistic in determining the performance of electron beam systems, or being overly optimistic for optical systems using partially coherent illumination.

only partially fills the pupil of the projection lens. Partial coherence produces higher contrast for linewidths above that at which 35% contrast is obtained in the incoherent case. 30%-70% filling of the pupil has proven optimum and for example, can increase contrast from 60% for the incoherent case to more than 80% (see Figs. 4 and 5). Contrast for partially coherent illumination is obtained from Offner.[12] Too high a degree of coherence can give rise to undesirable interference effects between lines, and increases exposure time. A major advantage of partial coherence is that it effectively increases the depth of field.[13]

Depth of field for the incoherent case is given approximately by $\pm \lambda/2$ (N.A.)^2 but depends on the substrate reflectivity, the degree of coherence of the illumination and the minimum feature size.[13] Two layer resist processes in which the image is formed in a thin, flat resist layer on top of a much thicker planarizing layer, alleviate the need for a large depth of field and make it easier to form high resolution high aspect ratio resist patterns.[14-16] These resist systems make it possible to obtain satisfactory results at contrast levels considerably lower than 60%.

REDUCTION PROJECTION PRINTER

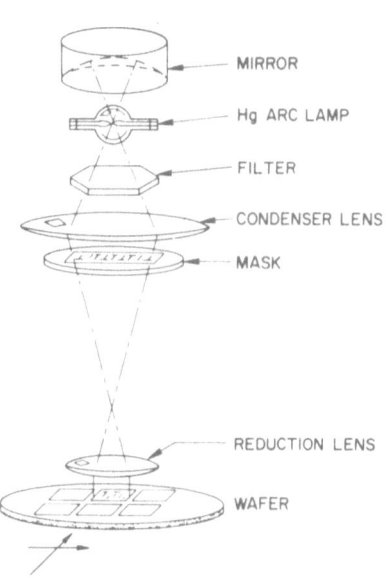

- MIRROR
- Hg ARC LAMP
- FILTER
- CONDENSER LENS
- MASK
- REDUCTION LENS
- WAFER

Refracting Lenses

Figure 6 shows the general layout for a step and repeat camera with a refractive lens. Several commercial step and repeat cameras are available.[17]

Fig. 6 - Concept for typical step-and-repeat optical projection camera. The mask is generally 5 to 10 times larger than the image.

The highest performance refracting lenses built to date for microcircuit cameras have numerical apertures of about 0.35 and can cover fields of about 1cm × 1cm. Higher numerical aperture lenses have been designed but are yet to be evaluated for microfabrication applications. For example, Wilczynski and Tibbetts have designed a 0.41 N.A. lens that covers 6.3mm × 6.3mm and is corrected for operation at 405 nm (see Fig. 7).[18] Minimum linewidth for this lens with coherent illumination should be about 0.8μ, and the depth of field ± 1.2μ.

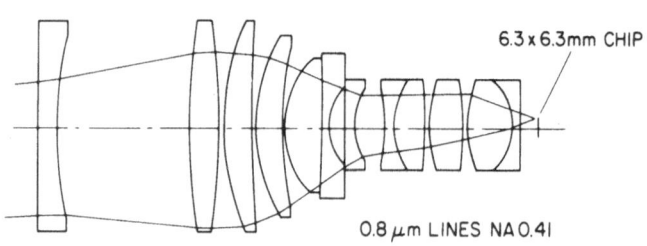

Fig. 7 - Microcircuit lens designed by Wilczynski and Tibbetts.[18] Lens is capable of imaging 0.8μ features over a 6.3mm x 6.3mm chip.

The image for most high performance microcircuit lenses is typically distorted from an absolute grid by 0.25μ to 0.5μ, and this distortion limits overlay accuracy when different cameras are used for the different layers of a single sample. Because the field size for all high performance refracting lenses is very much smaller than a silicon wafer, they are always used in cameras that operate in the step and repeat mode. The sample position is generally tracked by a laser interferometer although systems that align at every chip site are soon to become available.[17,19] Alignment at every chip will avoid errors due to drifts between the wafer and the interferometric reference point. A persistent problem with all optical alignment systems is that at certain resist thickness light incident on the sample is totally reflected and the alignment marks "disappear". To avoid this it is necessary to carefully control resist thickness or to remove the resist from the alignment marks. Both solutions complicate the process.

In the limit, microscope objectives with 0.95 N.A. can be used for microfabrication and, provided very small fields (200μ x 200μ) are adequate, linewidths < 0.4μ should be achievable under carefully controlled laboratory conditions, and in very thin resist layers. Depth of field will be reduced to about ± 0.2μ. Deep UV (λ = 200nm - 260nm) refracting lenses will be difficult to build because of the lack of materials that are transparent at these wavelengths and yet have relatively high refractive indices. Elements made from low refractive index glass must have smaller radii of curvature and this increases higher order aberrations.

Reflecting Lenses

The most commonly used projection aligner today uses a mirror lens system with a numerical aperture of 0.16.[12] A crescent on the mask is illuminated from a curved capillary arc lamp and this portion of the mask is imaged onto the wafer (see Fig. 8). Full wafer exposure is accomplished by scanning mask and wafer

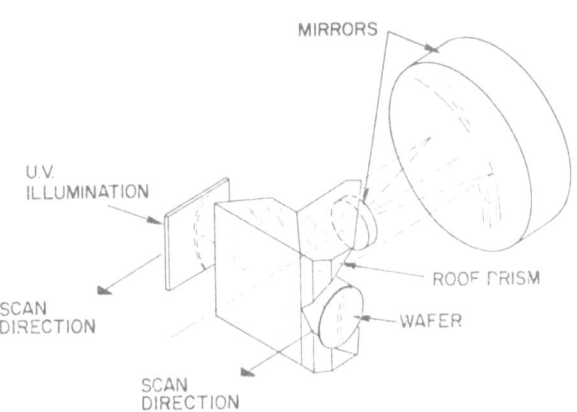

Fig. 8 - 1:1 optical scanning projection camera built by Perkin Elmer.[20] This camera which is widely used today can reproduce linewidths down to about 2μ.

passed the optical system. The newest version of this camera is capable of exposing more than 100 125mm diameter wafers in an hour, and distortions in the image have been reduced to less than 0.25μ over the 125mm field.[21] This high degree of correction has been accomplished by using phase sensitive interferometry to measure the perfection of both the individual component mirrors, and the integrated optical system. It is proposed in the future to compensate for the inevitable microscopic changes in wafer size that arise during device processing by changing the magnification of the lens system. Image size in the direction of the slit will be changed by small motions of a pair of weak refractive elements in the optical path, and correction in the other direction will be made by micro-scanning the mask during exposure. Micro-scanning will slightly increase or decrease the scan length for the mask compared to that for the wafer. Provided distortions of the sample are isotropic, this approach should provide comparable overlay accuracy to the step and repeat systems.

Resolution with mirror lenses can be improved further by merely operating at shorter wavelengths, the only difficulty being that of finding resist/lamp combinations that give satisfactory exposure times. Several experimental resists have already proven adequate for operation at 300nm and there are no fundamental reasons why it should not soon be economi-cal to operate down to 250nm.

The present 1:1 full wafer scanning mirror is limited to a numerical aperture of 0.16 by the need to cover a 125mm wafer in a single scan, and at the same time avoid having the mask or wafer vignette the optical system. If the full wafer capability is sacrificed and a system designed to only expose a portion of the wafer in a single scan, or a partial scan, then it should be possible to increase the numerical aperture to 0.35 (Wilczynski and Tibbetts have already patented a scanning mirror camera with N.A. = 0.35[22]) or higher. Ultimately it may be possible to build a system with a numerical aperture of 0.35 which, at a wavelength of 250nm, would give the contrast shown in Fig. 4, and be able to produce 0.5μ linewidths. Depth of field will be very small ($\pm 0.8\mu$ for the incoherent case), but methods such as capacitive sensing are available to monitor the wafer surface to $< \pm 0.1\mu$ and provide a means for continuously correcting focus.

The possibility of optical cameras reaching sub-micron resolution presents a formidable challenge to competitive lithographic methods using electrons, ions, or x-rays. In general, optical systems deliver very much more exposure flux to the sample and provide much shorter exposure times per unit area, or per resolution element. All the high resolution optical systems, however, need a precision mask and unity magnification systems need an electron beam fabricated mask.

As with UV shadow printing, positioning of a projected image with respect to marks on a sample should be achievable to an accuracy of $\pm 0.1\mu$ through a variety of methods, and fields can be joined together with similar accuracy by laser interferometry. Overlay accuracy over the whole sample will depend on temperature control, distortion in the imaging optics (this will be particularly important for large field refractive lenses when all devices levels cannot be exposed on a single camera), and by distortions of the sample.

Scanning Electron Beam Systems

In scanning electron beam systems, the pattern is written with a small electron beam which is controlled (deflected and turned on and off) by a computer. It is the only method available today for generating complex VLSI patterns with dimensions beyond the capabilities of existing optical pattern generators ($\leq 1\mu$). Electron beams can be used to write directly on wafers, or to make masks. In the first instance, writing speed is of paramount importance and for mass production of identical devices the technique remains too expensive. In cases where many different parts are to be made, the high cost can be offset by the elimination of masks. Personalization of logic gate arrays is such an application, and IBM has been successfully using electron beams for this application for several years.[23] For mask making, scanning electron beam is superior in every way to optical systems and electron beam mask makers are being used extensively despite their expense. Initially, because of accuracy problems, many electron beam mask makers are being used only for making 10X reticles. Standard photo-repeaters then produce the 1X mask.

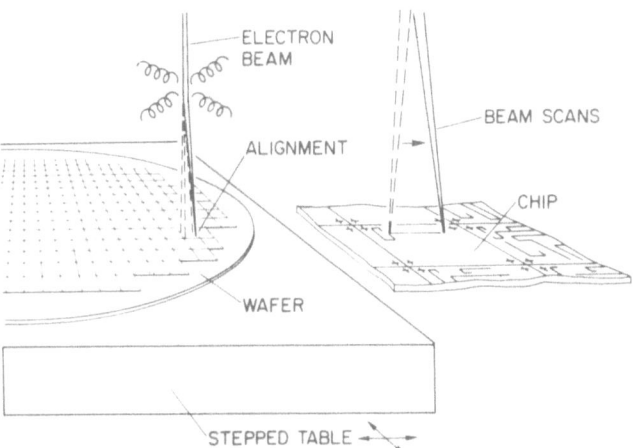

Fig. 9 - Vectorscan writing with electron beam. The beam is electronically scanned to write each chip. The sample is then mechanically moved to the next chip size.

Figure 9 shows the basic operating mode of the most common type of electron beam system.[23,24] The electron beam cannot be scanned over the whole wafer because of deflection aberrations and deflection system noise, so electronic scanning has to be combined with mechanical movement to cover the sample. A method must therefore be found to accurately maintain the position of the beam with respect to the sample. In Fig. 9 a direct beam-to-sample reference measurement is made before each complete section (chip) of the pattern is exposed. The measurement is made by collecting electrons back-scattered from four marks at the corner of each field. It is used to check for beam position, scan rotation and orthogonality, and to calibrate scan amplitude.

In cases where the chip is written by pure electronic scanning, two basic methods have been employed; raster and vector. With raster scanning, the beam scans the entire pattern area and is turned on where required. In vector writing, the beam is directed only to points where exposure is required. Vector scanning is more efficient because no time is wasted scanning areas not requiring exposure, although it places more stringent requirements upon the deflection system in terms of dynamic accuracy and in particular eddy current errors. Raster scanning makes it easier to apply corrections for deflection aberrations and pattern distortion. Vector scanning allows more efficient data compaction and is more suitable when proximity corrections have to be applied.

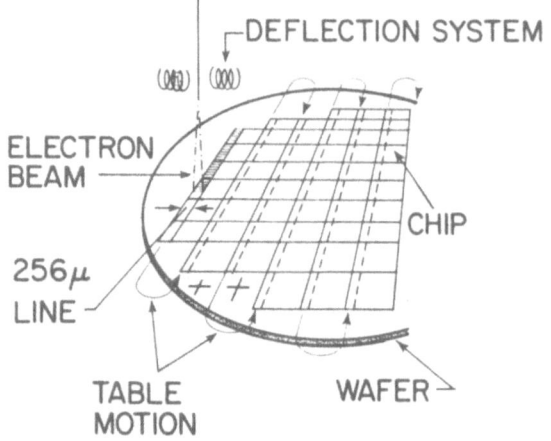

- I - DIMENSIONAL LINE SCAN - (RASTER)
- CONTINUOUSLY MOVING TABLE
- FULL WAFER ALIGN
- LASER INTERFEROMETER FEEDS
  TABLE POSITION BACK TO E-BEAM

Fig. 10 - Operating node for the EBES electron beam mask-maker.[25] The beam is electronically scanned over a line 256μ long and the sample moved in a serpentine manner under the beam to complete wafer exposure. A laser interferometer tracks the exact position of the sample.

Figure 10 shows the EBES electron beam mask maker in which the beam is electronically scanned in one direction only, and continuous mechanical movement is used for the other direction.[25] Chips are written strip by strip, the same strip on every chip being written before proceeding to the next strip. The position of the beam on the sample is initially checked with a direct beam to sample measurement, but after this a laser interferometer keeps track of the sample stage. Errors in position are corrected by feeding signals to the electron beam deflection coils. EBES type systems are available commercially and are extensively used for producing masks with dimensions down to about 1μ. They can produce 1 to 2 masks for 100mm wafers in an hour and deliver about $\sim 10^{-6}$ C/cm^2 to the resist.

The throughput of scanning electron beam systems was originally limited by electron optical performance; that is by the beam current/resist sensitivity criterion. However, this situation has largely been alleviated by the discovery of resists with improved sensitivity, and by electron optical advances in the following three areas.

(i) Improved electron gun design and new cathode materials, in particular lanthanum hexaboride, have increased brightness from about 2×10^5 A/cm^2 steradian (20kV) to 10^6 A/cm^2 steradian.[26] Thermal field emission cathodes yield a further improvement to 10^7 A/cm^2 steradian but so far have only been used in the Hewlett-Packard electron beam system.[27]

(ii) Computer-aided design of lens/deflection system has led to an increase in field size of about five times, and to a three-fold increase in beam aperture.[28,29] The latter yields an increase in beam current of a factor of 9. It is now possible to deflect an electron beam, without unacceptable beam growth, over a larger field (50,000 times the beam diameter) than the noise/bandwidth characteristics of the best deflection drivers allow. For a 10 MHz beam incrementing rate, the maximum field size that can be accurately addressed at present is about 20,000 times the beam diameter.

(iii)   Shaped-beam systems have been developed which increase the number of
          pattern elements exposed by the beam at each beam position. Such systems
          increase the beam current for a given edge sharpness by more than an order of
          magnitude over typical round-beam systems. However, because of electron-
          electron interactions, which result from the large beam currents, lower electron
          brightnesses have had to be employed for shaped spots above about 1μm in
          size and this reduces the beam current advantage to about ten times.

The early shaped beam systems used a fixed beam shape and this meant that linewidths
that were uneven multiples of the beam size could only be written by introducing overlaps.
With many resist processes, overlaps produce undesirable bulges in the pattern. This problem
was overcome through the use of a column which alters the shape of a rectilinear beam at high
speed. This advance also allows larger areas of the pattern to be exposed at a given beam
position (see Fig. 11), and further enhances another advantage that shaped beams have over
round beams; that the number of beam addresses in a given pattern is reduced. The variable-
shaped-beam illumination system used by Pfeiffer[56] is shown in Fig. 12. The variably

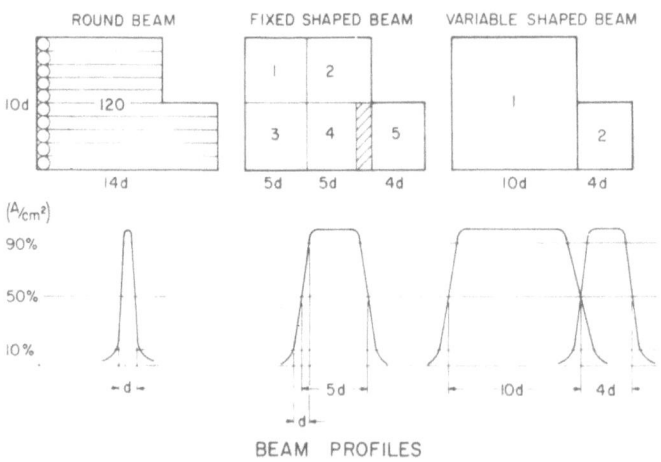

Fig. 11 - Methods in which pattern elements are filled with
round beams, fixed square beams and variably shaped beams.
Note the reduction of beam addresses with the shaped beams.

illuminated aperture is used as the source for the same type of column that produces the fixed shape beam.

A problem with shaped beams that remains even with the new variable-shape concept is that it is not possible to produce, efficiently, angle lines and other non-rectangular shapes. This is mainly of importance in magnetic-bubble devices. Most silicon devices can be built entirely with rectilinear shapes.

Overall throughput for scanning electron beam systems is complex, and in addition to electron optical performance and resist sensitivity, de-

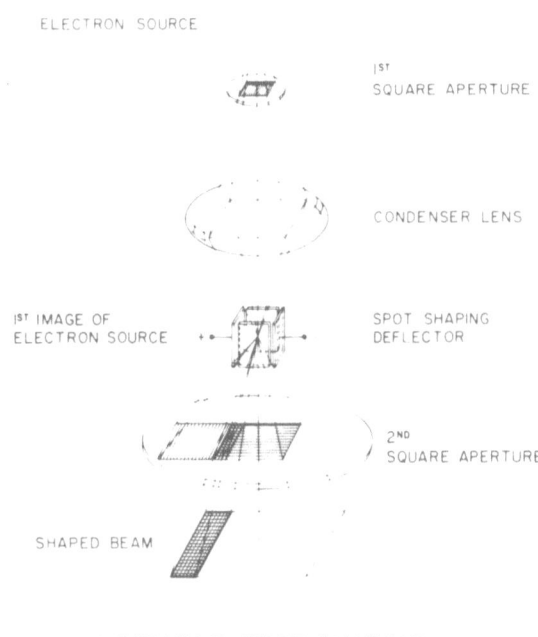

VARIABLE SPOT SHAPING

Fig. 12 - Concept used in variable-shape electron beam columns. The image of the first aperture is moved across the second square aperture to form a variable aspect ratio rectangular source.

pends on the noise bandwidth characteristics of the deflection system, digital to analog conversion rates, data transfer rates, switching speed for digital pattern generation hardware, alignment time, mechanical movement of the sample, and the time taken to load the samples into the vacuum. The relative importance of these different factors depends on the overall system configuration. The assessment of the system capital and operating costs is similarly complex. It is therefore not possible from a cost-performance point of view to compare individual features, for example raster and vector, round beam and shaped beam, laser reference and direct reference, in an absolute manner. Instead, it is necessary to consider the overall system. An approximate assessment for today's systems can be made by considering data presented for three types of high-throughput system. System cost is not discussed because costs have not been published. Unfortunately this makes comparisons of limited use because cost may be as large an influence as throughput.

The first is the conventional vectroscan-type system shown in Fig. 9.[23] Here the pattern is written in a vector manner with a round beam, the diameter of which is typically one quarter the minimum linewidth. This means that the beam must be incremented to sixteen positions to write a minimum image square. Chip writing time can be estimated if one knows the beam incrementing rate and the average fraction of the chip area that needs to be exposed. For example with a 1μ minimum image, a chip of 5mm×5mm contains a total 25×10^6 minimum images or $16\times25\times10^6$ beam positions. If, on average, 25% of the chip must be written, then 10^8 beam positions must be written per chip. At 10 MHz, which is the maximum beam incrementing rate used in I.C. fabrication to date, this will take 10 seconds. Time taken for table stepping, registration, wafer loading, and data transfer must be added to the writing time to come up with the overall wafer writing time. In the most advanced systems, these overhead times can be kept below 1 second per chip. If the wafer is 75mm in diameter and contains 160 chips, the total time per wafer is about 30 minutes.

The second system is the high throughput EL3 system recently built at IBM and described by R. Moore *et al.*[30] The key specifications for EL3 are given in Table 1. EL3 combines a high performance variable shaped beam column with a dual deflection system.

TABLE I: EL-3 Key Specifications

Minimum Image	1μ	2μ
Field Size	5mm	10mm
Beam Edge Resolution	0.25	0.5
Overlay (Mean + 3σ)	$0.4\mu m$	$0.7\mu m$
Wafer/Mask Capabilities	(57 →	165mm)
Current Density	$(50A/cm^2)$	
Throughput W/Hr. at 10 $\mu C/cm^2$ 3" Wafers	10-20	20-45

Beam size can be varied over a range of 4 to 1. A highly accurate but relatively slow magnetic deflection coil deflects the beam in a raster-like sequence to the center of an array of overlapping sub-fields located on 75 micron centers. 10mm × 10mm can be covered with a resolution of $0.02\mu m$. After arriving at the center of each sub-field, the variable shaped beam is vector-addressed inside each sub-field with electrostatic deflection plates. The sub-fields overlap each other so that there are no discontinuities, and the sub-fields are small enough to dispense with the need for a super-precision deflection. Shapes can be positioned anywhere in the written field on a 0.1μ grid. Errors in both magnetic and electrostatic deflection are "learned" by scanning a reference grid, storing corrections in the control computer, and feeding the corrections back during wafer writing.

Because in EL3 the current density is very high (50 A/cm^2), the edge sharpness of the larger spot sizes is degraded by electron-electron interactions in the beam. To avoid this problem, the edges of shapes are generally written with narrow rectangular beams. These smaller beams "frame" the shape and ensure that the edges are sharply defined.

The third system is that developed at Hewlett Packard.[27] While the EL3 is the latest in a succession of evolved systems, the HP system is more radical. It uses a sub-field electrostatic deflection system with a relatively small beam aperture and gains back beam current by using a high brightness thermal field emission cathode. Even so the total beam current is less than a tenth the maximum current of EL3 (600 nA vs. 6μA for EL3). Because the beam angle is very small, beam size is dominated by electron-electron interactions (Coulomb repulsion) rather than gun brightness or lens/deflector aberrations.

The pattern is written by means of two electrostatic deflectors. The first, an octopole, deflects the beam to the center of an array of 75μ by 75μ sub-fields, and provides astigmatism correction. The octopole is calibrated by scanning 25 fiducial marks distributed over the 5mm×5mm deflection field. Calibration includes corrections for astigmatism, focus, and distortion. The second deflector is a quadrupole which provides a high speed raster deflection to fill in the sub-fields. The calibration procedure provides magnification and trapezoidal corrections for the sub-fields. The beam in incremented over the sub-field at 300 MHz.

It is difficult to estimate the throughput of the HP system in the manner described above for vectorscan systems because the address increment is so large (0.5μ). This means that shapes can only be placed on a 0.5μ grid and individual images smaller than about 2μ must have rounded corners. If one assumes that this difficulty can be overcome then one can estimate, using the published system design specifications, that it should be possible to expose approximately 20 75mm diameter wafers per hour for 1μ minimum linewidth. This assumes that the beam diameter must be 250nm, that is four times smaller than the minimum linewidth, and that 25% of the deflection field is exposed. It is also assumed that there ar 170, 5mm×5mm fields per wafer, field to field stepping takes 250ms, registration takes 50ms, and the beam current is 245nA. A resist with a sensitivity of 1.3×10^{-6} C/cm^2 is required for this writing rate.

So far the HP system has been used to fabricate surface wave transducers with minimum linewidth of 0.3μ. Surface wave transducers are single layer devices that do not require overlay capability.

Resolution for Electron Beam Lithography

Ultimately, resolution in electron beam lithography is set by the range over which the primary electrons interact with the resist. That is by the range over which low energy secondary electrons can be created (the resist is exposed mainly by secondaries) and by the subsequent straggling of the secondaries into the resist. For thin resists and thin substrates (thin compared with the primary electron penetration distance) this resolution limit falls at about 12nm. For thin resists on thick substrates, backscattered electrons from the substrate decrease contrast and the minimum dimension increases to about 40nm. For thick resists and samples thick compared to the primary electron penetration range, electron scattering in the resist (forward-scattering) and backscattering of electrons from the substrate becomes more important than the electron interaction range. In these cases, in order to produce an ideal exposure distribution exposure dosage must be altered according to the local pattern density to compensate for variations in the backscattered electron dose (proximity effect).[31] Compensation for this effect is valuable even at dimensions of 1μ-2μ, particularly for thick (> 500nm) resist layers.

Figure 13 shows the contrast for electron beam exposure versus linewidth (assuming an infinite array of lines and spaces) for different resist thicknesses. It can be seen that with a thick substrate and a thin resist, the contrast is the same for 0.05μ lines as it is for 1μ lines, provided the resist is thin ($< 0.1\mu$). This is because the backscattered distribution is much wider than either the secondary distribution or the forward scattered distribution. For the thicker resist cases, it is assumed that the lateral scattering in the resist can be approximated by a Gaussian distribution.[32]

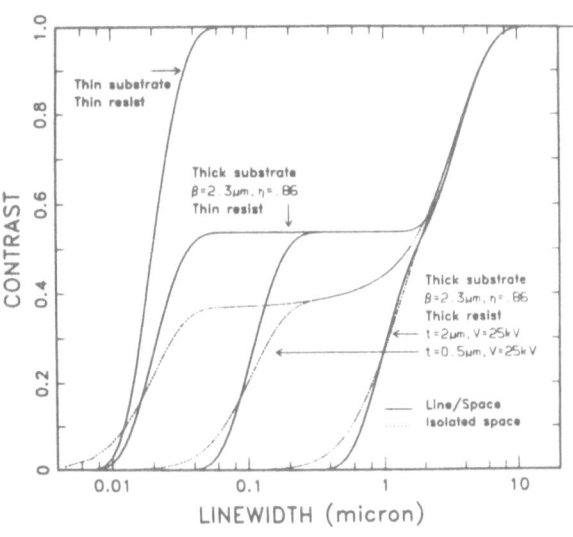

Fig. 13 - Contrast for line/space patterns and for isolated spaces written with electron beam. This substrate effectively eliminates back-scattered electrons and provides highest contrast for smalles linewidths.

It should be noted that the fraction of the exposure due to backscattered electrons is reduced for multilayer resist processes where the image is formed in a thin top layer that is backed by a thicker underlying layer. The underlying resist is of lower atomic weight than silicon and consequently backscatters fewer electrons. The contrast for such a sample would be higher than that in Figs. 4, 5 and 13. All this discussion relates to the contrast obtained for an infinite array of lines and spaces. When the total pattern area is small compared to the area from which the backscattered electrons emerge, resolution on bulk substrates should be as good as that on thin membranes. For example, the backscattered contribution to an isolated 25nm line on a silicon substrate is only 0.004 times the incident exposure.[32]

Although it is possible to build very small devices on thick substrates, (see, for example Ref. 33) most sub-tenth micron operational devices built to data have used thin ($< 0.1\mu$) Si_3N_4 membranes as the substrates.[34] Vapor deposited monolayers of silicone oil have also been used as resist. Isolated metal lines 8nm wide have been formed[35] and devices such as the niobium nanobridge SQUID's (Superconducting QUantum Interference Devices) shown in Fig.

14 have been made and tested.[36] The most significant advantage of thin substrates is that they allow the samples to be examined in a transmission electron microscope. It is almost impossible to examine such samples in a conventional secondary electron SEM because the same straggling of secondary electrons that limits resolution in the resist pattern also limits the resolution of the SEM.

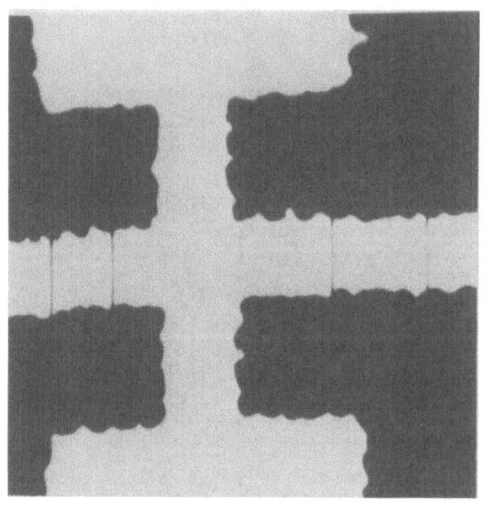

Fig. 14 - Transmission electron micrograph of two nanobridge SQUID's. A SQUID consists of a superconducting ring containing two "weak links". In this instance, the weak links are niobium wires 40nm wide fabricated by electron beam.

Contrast for Optical and Electron Beam Exposure

The relationship between the contrast function and the minimum reproducible linewidth is difficult to establish for optical or electron beam exposure because it depends on variations in resist contrast, substrate reflectivity or scattering power, resist thickness variations, and many other variables. Several workers have modelled the exposure and development process for specific cases taking many of these variables into account,[37-39] but few have made direct comparisons between optical and electron beam exposure. Figures 4 and 5 make such comparisons, but many factors must be considered in interpreting these data. In particular, it is not sufficient to consider just the contrast at the minimum required linewidth. The contrast for smaller linewidths must also be considered. Contrast at these higher spatial frequencies determines the edge slope on the resist and the sharpness of corners, etc. In practice, it appears that at least 30% contrast has to be maintained at a linewidth half that of the nominal minimum linewidth. There is considerable difference between electron beam and optical exposure in this respect. For electron beam, contrast is constant over a broad range of linewidths. This is because there is a large difference between the widths of the secondary electron and backscattered electron distributions. In cases where the forward scattering spread function is wider than the secondary distribution, the forward scattering distribution is still generally much narrower than the backscattered distribution. In the region of constant contrast, the backscattered electrons in effect "fog" the image, but the image remains relatively sharp. A definition of 60% is unnecessarily pessimistic because the resist process can be adjusted to be insensitive to the background fog.

Figure 4 shows that a 0.35 N.A. (λ = 250nm) optical system has higher contrast for linewidths above 0.5μ that a 25kV electron beam exposing 0.5μ thick resist on a thick substrate. At 1.25μm linewidth, the contrast of the optical system is 85% compared to 54% for the electron beam. For the reasons just discussed, however, the exposure transition at the boundary between a line and a space is sharper for the electron beam case than it is for the optical system even though the nominal contrast is lower. Contrast for the electron beam does not fall to 30% until a linewidth of 0.15μ is reached which suggests that the smallest useful linewidth is 0.3μ. Many electron resists have high enough contrast to produce satisfactory patterns at an exposure contrast of 50%. For the optical system linewidth at 30% contrast is 0.3μ suggesting a minimum useful linewidth of 0.6μ.

Parallel Imaging Electron Beam Systems

Three types of parallel imaging electron beam exposure system have been built and evaluated; the reduction projection system, the 1X photocathode projector, and the electron beam proximity printer. The first, the reduction projection system,[40] is the electron optical analogue of the optical step and repeat camera. It reduces the image 4 to 10 times and uses a stencil mask because no practical substrate is thin enough to avoid deleterious electron scattering. Unfortunately, difficulties with fabricating stencil masks, distortions in the projected image, and the absence of a satisfactory method way to compensate for proximity effects have precluded its successful competition with other approaches.

1:1 photocathode electron projection systems of the type shown in Fig. 15 have been under development for many years. They employ a photocathode which is masked with a thin metal pattern. The photocathode is illuminated with ultraviolet light and the photoelectrons are accelerated

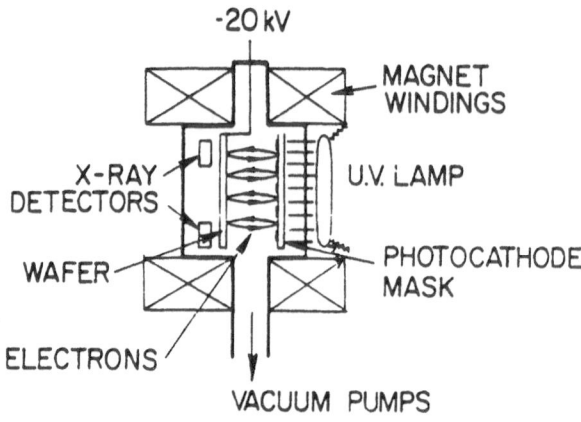

Fig. 15 - Photocathode electron beam projection system originally described by O'Keefe *et al.*.[57]

from the cathode to the anode with an accelerating potential of about 20kV. A uniform magnetic field focuses the electrons.

This approach offers the resolution and depth of field of electron optics together with the parallelism and simplicity of proximity printing. Problems with the lifetime of the patterned photocathode delayed progress for many years but the introduction of cesium iodide as the photoemitter largely removed this problem.[41] Successful methods for aligning the projected image to the sample have also been developed. The problems that remain are pattern distortion, reduction of image contrast because of backscattered electrons being returned to the sample surface, and correction for proximity effects.

The sample is part of the electron optical system, and distortion is introduced if it is not absolutely flat. Calculations show that a 30μ wafer bow will introduce a 1μ error in pattern position.[41] Electrostatic wafer chucking is being applied to solve this problem. So far 2μ linewidth LSI devices have been successfully made with a 1:1 electron beam projector, and it is predicted that 60 10cm diameter wafers can be exposed in one hour.[30] For dimensions below 1μ, the dimensional stability required to satisfy overlay requirements will become extremely severe. Demonstration of a method for correcting proximity effects by adjusting the shapes in the mask has been reported.[30]

In the electron beam proximity printer, a flood beam of electrons passes through a stencil mask to expose the resist. The mask is placed far enough away from the wafer surface to prevent damage, but close enough to allow the beam to have acceptable divergence. The current density in the beam is proportional to the square of the divergence angle. As with the reduction projection printer, the mask cannot have a substrate because there is no substrate that would not scatter electrons. The stencil problem of unsupported areas falling out is overcome by making a double exposure with two complementary masks. Because of the fragility and potential dimensional instability of the mask, exposure is made in a step and repeat manner. The mask is 2cm-5cm in size.

Bohlen *et al.*[55] have developed a stencil mask technology that is capable of linewidths well below 1μ. The pattern is first formed in a gold layer on a silicon wafer that has previously been boron doped to a depth of about 3 microns. The silicon is then reactively ion etched to a depth in excess of 3 microns. The reactive ion etching is anisotropic and produces slots in the silicon that have vertical walls. The wafer is then etched away from the back until the boron doped region is reached. This leaves a silicon membrane about 3 microns thick with holes wherever exposure is required.

The advantages of electron beam proximity printing over X-ray lithography are that the beam can be tilted to provide fine alignment adjustments and scanned to produce an alignment signal. It is also relatively wasy to achieve short exposure times. The disadvantages are the need for a double exposure and the inability to vary the exposure dose in order to overcome proximity effects.

Ion Beams

Focused ion beams can be used to "write" diffusion patterns in semiconductor substrates or to cut patterns in thin films through sputter etching. These techniques could simplify semiconductor device production and perhaps reduce cost. High energy ions have also been used to expose resist materials (> 100 keV ions are needed to penetrate typical resist layers) with the promise of less image blurring due to lateral scattering (proximity effect) than electrons. In the first two cases, ion charge densities are needed that are at least ten times greater than the electron charge densities required to expose electron resist. In the case of resist exposure, however, experiments indicate that 200 keV ions are almost two orders or magnitude more efficient than 20kV electrons.[42] This is because the energy of the ions is more completely absorbed in the resist layer.

Until recently, direct writing with ions did not look attractive because of the low brightness and large chromatic spread of the best conventional ion source, the duoplasmatron.[43] This source has a maximum brightness of 200 A/cm^2 steradian (12 keV argon) and an energy spread of 12 eV. Thermal electron sources produce brightness of 5×10^4 - 2×10^6 A/cm^{2}ster (12kV) and an energy spread of 1eV - 3eV. Conventional ion microprobes using a duoplasmatron produce beam diameters of 0.25μm - 0.5μ with beam current of 10^{-11} - 10^{-10} ampere.[44] Electron probes focus 10^{-7} - 10^{-6} ampere into similar beam diameters. The overall speed of a conventional ion microprobe for sputter machining, or standard semiconductor ion implantation, would be 10^4 - 10^5 times slower than an electron beam resist exposure system. Electron beam systems are only just fast enough to be economical so only specialized applications could be addressed economically with ions.[45]

Two new types of ion source, the field ion source, and the liquid metal source (two versions of this source have been built: the needle type[46] and the capillary type[47]) have produced brightness 10^4 - 10^6 times greater than the duoplasmatron. This makes it possible, in principle, to produce ion currents similar to those available in electron beams. Chromatic spread is also reduced for the field ion source, making it potentially feasible to produce beam sizes of 10nm with adequate current (10^{-11} amp) for laboratory microfabrication experiments. The liquid metal sources should provide adequate current ($\sim$ 1 nanoamp) into larger beam diameters ($\sim$ 0.25μ) to make them interesting for semiconductor device processing applications. Experiments have already been carried out in which 40nm wide slots have been machined in thin gold films using an ion beam with a diameter of approximately 100nm.[48]

Significant problems remain with both the new types of ion source, however. The most important are that the field ion source only produces a total emitted ion current about 10^{-8} ampere limiting final beam currents to the order of 10^{-10} ampere, and the liquid metal sources only operate with liquid metal ions and have a relatively large chromatic spread (5eV - 14eV). Recent results with a low temperature field-ion source promise higher total ion currents,[49] though, and liquid sources using eutectic alloys have produced boron, arsenic, and silicon ion beams, but at reduced brightness.[50] An alloy source has been used to produce GaAs FET's using direct ion implantation.[51] The energy spread of the liquid metal sources limits the beam size to several hundred angstrom even with short focal length, low aberration, electrostatic lenses but this should not be a problem for microcircuit applications. The stability of the emitted current from both sources needs to be improved for routine lithography applications.

Focusing and deflection systems for ions are in a very early stage of development compared to those for electrons. It is, therefore, premature to assess their ultimate capability in terms of field size and accuracy. With electrons, large deflection field and large beam aperture are obtained simultaneously by overlapping a distributed deflection field with the focusing field produced by an axially symmetric lens. Many aberrations are cancelled in this way. Overlapping of the deflection yoke and the lens is possible because magnetic fields are used. The fields produced by these components, however, are not large enough to focus and deflect ions. Electrostatic lenses and deflection plates are generally used with ion beams, but they cannot be overlapped. Multi-pole lenses (quadrupoles and octopoles) are used for focusing high energy ion beams, and some combination of these may offer a solution to this difficulty, but large aperture designs are yet to be completed. The easiest way to cover a large field will be to move the sample mechanically and join together smaller fields. Laser interfero-metry can be used to track the sample position and feed back errors in table position to the ion beam deflection system in the same way used in electron beam mask-makers. This approach, in principle, allows field size to be unlimited.

Ions can be used for shadow printing just as workers have used UV light, x-rays and electrons. In this case, a bright source is not needed, only a source that produces adequate total current. If necessary, the beam can be scanned across the mask and sample to complete the exposure. Masks for ion printing have been made from silicon membranes, and it is possible in principle to make freely suspended stencil masks similar to those used for electron beam proximity printing.[55] Ions are laterally scattered in passing through silicon membranes and this scattering increases linewidth by about 0.1μ limit for a mask to resist separation of 15μ - 25μ.[52] This separation would be marginal from the point of view of mask to wafer damage if step and repeat exposure was needed to obtain adequate overlay accuracy. Mask heating will be greater than for ultra-violet or x-ray lithography, although calculations have shown that it should be possible to keep thermal distortions below 0.1μ.[52] Ultimately in the absence of these practical difficulties, resolution and overlay will be the same for shadow printing as for direct writing. Ion beam proximity printing has recently been used to fabricate operating transistors.[53]

Ultimately, resolution in a sputtering process is limited by ion penetration into the substrate and/or by the range over which momentum can be transferred effectively enough to remove atoms from the sample surface. The diameter from which atoms can be sputtered has been reported to be about 10nm for incident ion energi$s up to 12kv.[54] This still means that it should be possible to fabricate structures as small as the ion beam (i.e., potentially 10nm). As already mentioned, Seliger[48] has produced 38nm wide slots in a gold films by sputtering. For resist exposure, the resolution limit will be set by the range over which the ions interact with the sample. As with electron beam exposure, ions will create secondary electrons up to several nanometers away from the beam, and these secondary electrons can travel further before their energy is absorbed. Ultimate resolution will probably be about 10nm - 20nm, as it is for electron beam exposure. At present this limit is academic because the smallest ion beams are too large to allow this resolution limit to be reached.

REFERENCES

1. P. Tischer, Electronics in Microelectronics, Ed. W. A. Kaiser and W. E. Proebster, North Holland, 1980.

2. B. J. Lin, J. Vac. Sci. Technol. 12, 1317 (1975); J. Vac. Sci. Technol. 15, 1012 (1978).

3. B. Fay, *et al.*, J. Vac. Sci. Technol. 16, 1954 (1979).

4. D. C. Flanders, *et al.*, Appl. Phys. Lett. 31, 426 (1977).

5. D. Kern and D. Nelson, Proc. 9th Int. Conf. on Electron and Ion Beam Technology, Ed. R. Bakish, Electrochem. Soc., Inc., Pennington, NJ, p. 491, (1980).

6. D. L. Spears and H. I. Smith, Electron. Lett. 8, 102 (1972).

7. E. Spiller and R. Feder, Sci. Am., Nov. (1978).

8. M. P. Lepselter, IEDM Digest, IEEE, NY, p. 42 (1980).

9. W. D. Grobman, IEDM Digest, IEEE, NY, p. 415 (1980).

10. E. Spiller and R. Feder, "X-Ray Lithography in H. J. Queisser X-Ray Optics", Springerverlag, Berlin, 1977.

11. A. E. Gruen, Z. Naturforsch 12a, 89-95 (1957).

12. A. Offner, Photogr. Sci. Eng. 23, 374 (1979).

13. M. C. King, "Principles of Optical Lithography", Chapter in VLSI Electronics Microstructure Science, Vol. 1 & 2, Ed. Einsbruch, Academic Press, 1981.

14. B. J. Lin, SPIE Proc., 174 (1979).

15. K. L. Tai, *et al.*, J. Vac. Sci. Technol. 16, 1977 (1979).

16. M. Hatzakis, *et al.*, IBM J. Res. Dev. 14, 452 (1980).

17. See papers by S. Wittekock, J. Lauria, *et al.*, G. Dubreoueq, *et al.*, H. E. Mayer, *et al.*, and J. Dey, *et al.*, Microcircuit Engineering 80, Ed. R. P. Kramer, Delft University Press, Hollant, p. 155-210 (1980).

18. R. E. Tibbetts and J. S. Wilczynski, Proc. 1980 International Lens Design Conference, SPIE Vol. 237, Ed. R. E. Fischer, Soc. of Photo-Optical Instrumentation Engineers, Bellingham, Washington, DC, p. 321 (1980).

19. J. S. Wilczynski, J. Vac. Sci. Technol. 16, 1929 (1979).

20. D. A. Markle, Solid. St. Tech. 17, 50 (1974).

21. D. A. Markle, private communication, 1981.

22. J. S. Wilczynski, *et al.*, U.S. Patent No. 4,171,871.

23. H. S. Yourke and E. V. Weber, Proc. International Electron Devices Conf. Washington, D.C., p. 431 (1976); T. H. P. Chang, *et al.*, Proc. 7th Int. Symp. on Electron and Ion Beam Science and Tech., Ed. R. Baksih, Electrochem. Soc., Inc. Pennington, NJ p. 97, (1974).

24. G. L. Varnell *et al.*, Proc. 6th Int. (IEEE) Conf. on Electron Ion and Laser Beam Tech., Ed. R. Bakish, Electrochem. Soc. Inc., Princeton, NJ p. 126 (1980).

25. D. R. Herriott, *et al.*, IEEE Trans. Electron Devices, ED-22, 385 (1975).

26. A. N. Broers, SEM 1974, ITT Research Inst., Chicago, IL, p. 9, (1975).

27. J.C. Eidson, *et al.*, Hewlett-Packard Journal, May 1981, p. 3.

28. E. Munro, J. Vac. Sci. Technol. 12, 1146 (1975).

29. D. P. Kern, J. Vac. Sci. Technol. 16, 1686 (1979).

30. R. D. Moore, to be published in J. Vac. Sci. Technol. (1981).

31. T. H. P. Chang, J. Vac. Sci. Technol. 12, 127 (1975).

32. A. N. Broers, J. Electrochem. Soc. 128, 166 (1981) and Proc. 9th Int. Conf. on Electron and Ion Beam Science and Tech., Ed. R. Bakish, Electrochem. Soc., Inc., Princeton, NJ, p. 396 (1980).

33. R. E. Howard, *et al.*, Appl. Phys. Lett. 36, 592 (1980).

34. R. Laibowitz, *et al.*, Appl. Phys. Lett. 35, 891 (1979).

35. A. N. Broers, *et al.*, Appl. Phys. Lett. 29, 596 (1976).

36. R. Voss, *et al.*, Appl. Phys. Lett. 37, 656 (1980).

37. F. H. Dill, IEEE Trans. Electron Devices ED-22, 440 (1975).

38. J. S. Greeneich, Chapter in Electron Beam Technology in Microelectronic Fabrication, Ed. G. R. Brewer, Academic Press, NY, 59 (1980).

39. A. R. Neureuther, IEDM Digest, IEEE, NY p. 214 (1980).

40. M. B. Heritage, Proc. 13th Symp. on Electron, Ion and Photon Beam Technol., Ed. R. Bakish, Electrochem. Soc., Princeton, NJ, p. 1135 (1975).

41. J. P. Scott, J. Vac. Sci. Technol. 15, 1016 (1978).

42. T. M. Hall, *et al.*, J. Vac. Sci. Technol. 16, 1889 (1979).

43. H. Leibl, J. Phys. E. 8, 797 (1975).

44. A. N. Broers, Proc. 5th Int. Conf. on Electron and Ion Beam Science and Tech., Ed. R. Bakish, Electrochem. Soc. Inc., Princeton, NJ, p.3, 1972.

45. R. L. Seliger and J. W. Ward, J. Vac. Sci. Technol. 12, 1378 (1973).

46. R. Clampitt, *et al.*, J. Vac. Sci. Technol. 12, 1208 (1975).

47. V. E. Krohn and G. R. Ringo, Appl. Phys. Lett. 27, 479 (1975).

48. R. L. Seliger, *et al.*, J. Vac. Sci. Technol. 16, 1610 (1979).

49. Siegel

50. K. Gamo, *et al.*, Microcircuit Engineering 80, Ed. R. P. Kramer, Delft University Press, Holland, p. 283 (1980).

51. R. L. Kubena, *et al.*, Electron Device Letters, June 1981.

52. D. B. Rensch, *et al.*, J. Vac. Sci. Technol. 16, 1897 (1979).

53. J. L. Bartett, *et al.*, to be published in J. Vac. Sci. Technol. 1981.

54. J. A. McHugh, "Secondary Ion Spectrometry", Chapter in Methods and Phenomena; Methods of Surface Analysis, Ed. S. P. Wolsky and A. W. Czanderna, Elserier Publishing Co., Amsterdam, Holland 1975.

55. W. D. Grobman and A. J. Speth, Proc. 8th Int. Conf. on Electron and Ion Beam Tech., Ed. R. Bakish, Electrochem. Soc., Princeton, NJ, p. 276 (1978).

56. H. C. Pfeiffer, J. Vac. Sci. Technol. 15, 887 (1978).

57. T. W. O'Keefe, J. Vine and R. M. Handy, Solid State Electron. 12, 841 (1969).

CHAPTER IV

BIPOLAR DEVICES

BIPOLAR LINEAR INTEGRATED CIRCUITS

Franco Bertotti and Bruno Murari

SGS-ATES Componenti Elettronici Spa
20019 Settimo Milanese
Milano, Italy

1. INTRODUCTION

Today, almost all linear electronic functions have been integrated using most of the technology and circuits available. Up to now the semiconductor manufacturer has always been willing to respond to customer demands and realize any given function with specific characteristics. This has resulted in the large number of integrated circuits presently available.

In the future, to maintain and improve an economic edge for both customer and manufacturer, the electronic system as a whole must be examined for integration. This trend implies a new and closer working relationship between user and supplier.

To achieve this relationship, the former must have a better understanding of the parameters and limits of semiconductor technologies, while the latter must improve his knowledge concerning the circuits and systems requested by the market.

Once a dialogue has been established, it will be possible to utilize to the maximum tomorrow's semiconductor technology in the realization of reliable and cost effective systems.

2. THE GENERAL SYSTEM

Any small electronic system, whether for a television, a radio, a telephone, a typewriter or an automobile, can be divided into 4 basic functional blocks as shown in Fig. 1. Each of the blocks shown can be described very precisely by its technology requirements and performance.

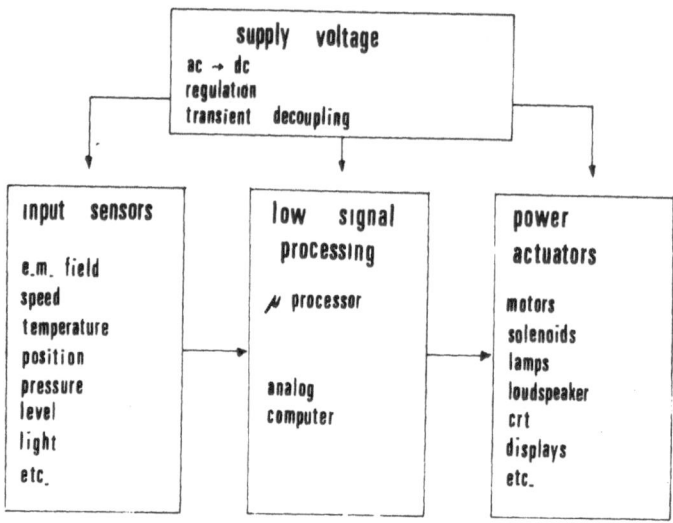

Fig. 1 - Typical system partitioning

2.1 Power Supply

The function of this block is to convert energy from the outside world (transformer, battery, etc.) in the most effective way to provide the supply for the rest of the system. This block can include rectifiers, filters and regulators together with circuits to suppress transient coming from the power source. The circuits in this block must have high efficiency power characteristics otherwise heat dissipation induced reliability problems will be encountered or, alternatively, it will be necessary to incur the high cost and inconvenience of using large heat sinks.

A high voltage and high current rugged power technology is necessary for integration of this block. Requirements are power transistors with optimized geometries and packages with built-in copper heat sinks.

2.2 Input Sensors

Sensors are those elements that measure a particular physical parameter and convert it into an electrical signal, either voltage or current. In the main these sensors are analog since the world in which they operate is an analog world.

Typically sensors can be used for detecting or measuring temperature, pressure, magnetic flux, light, etc. In general, they convert the physical dimension to an electrical signal with good linearity but very poor efficiency. It follows, therefore, that the entire system performance depends on three factors: the signal to noise ratio, linearity and drift of the elements included in this block.

The technologies used are different for each different type of sensor but nonetheless they are all typified by their low voltage operation.

2.3 Low Signal Processing

This block carries out all the control and calculation functions for the entire system. The techniques used can be digital, after suitable A/D conversion of the input signals provided by the sensors, or, alternatively, completely analog, whichever is more cost effective. The basic characteristic of the circuits in this block is that they can handle a high volume of data at small signal levels. It is in this block that microprocessors find their place. Analog functions are used within this block often only to provide for interface to the microprocessor although small systems may be realized with purely analog circuits. A Large Scale Integration technique is the preferred technology for this block since it permits the integration of a large number of functions in a small space.

Up to now the enormous progress made in the field of microlithography has been used almost exclusively for LSI MOS technologies, which are essentially surface structures. The same is not directly possible for bipolar volume structures as the reduction in the scale of devices is matched by a corresponding reduction in the operating voltage. In subsequent sections, the reason for these limitation will be given along with some proposed solutions to overcome them.

2.4 Power Actuators

Power actuator circuits can be defined as those power circuits that drive motors, coils, loudspeakers or displays after amplifying an input signal that can be digital, for which they act as a simple switch, or analog as in the case of an audio amplifier driving a loudspeaker. It is important to note that these circuits are supplied by input signals of the order of 2 to 3V that can be easily produced by the low signal processing circuits described in 2.3 operating at around 5V.

The circuits in this block, in common with those of the power supply, are characterized by their efficient utilization of the power available. Again the preferred technology is high voltage, high current power technology. In order to obtain the maximum possible efficiency in this block, it is better not to include any voltage regulation but to have, at the most, only some form of transient protection (such as Zener clamping) to hold transients to safe levels. This protection could alternatively be included in the main supply block.

As with the power supply block, all the features to ensure reliability apply, that is optimized semiconductor technology, circuit complexity and packages with built-in copper heatsinks. To further enhance reliability it is wise to build in a certain amount of excess power capacity and to include circuits that protect the power stages against overloads caused by short circuits, overvoltages or overtemperature conditions.

3. THE PLANAR PROCESS

The planar process, invented by Fairchild engineers at the beginning of the 60's, exploits the property of silicon dioxide that it will protect an underlying silicon body from penetration by dopants at high temperature. This property had been discovered sometime earlier by researchers at Western Electric.

The planar structure is shown below:

As can be seen, isolation between components is obtained by means of a reverse biased junction.

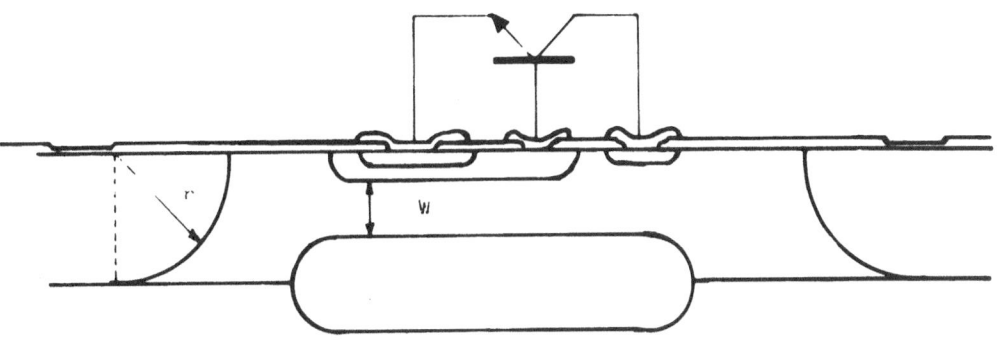

Fig. 2 - Standard NPN structure

This classic structure has dominated the bipolar world until the present day with slight variations introduced only in the last few years.

3.1 Different Methods of Isolation

To eliminate some of the problems of the planar process (parasitic capacitance, outdiffusion of the buried layer during the isolation cycle, parasitic currents to the substrate) various isolation techniques have been developed. The most important of these techniques are described below:

a) Polycrystalline Isolation Process

The flow-chart of the process is shown in Fig. 3. The outdiffusion of the buried layer is limited by the reduction in the isolation and collector diffusion cycles permitted by the high diffusion coefficient of the boron and phosphorus in the polycrystalline zones.

We can thus obtain a BV_{CEO} of around 100 V with h_{FE} from 100 to 250 using epitaxial thicknesses of around 18μ. There should be no great difficulty in increasing the BV_{CEO} to 150V with epitaxial thicknesses of around 25μ.

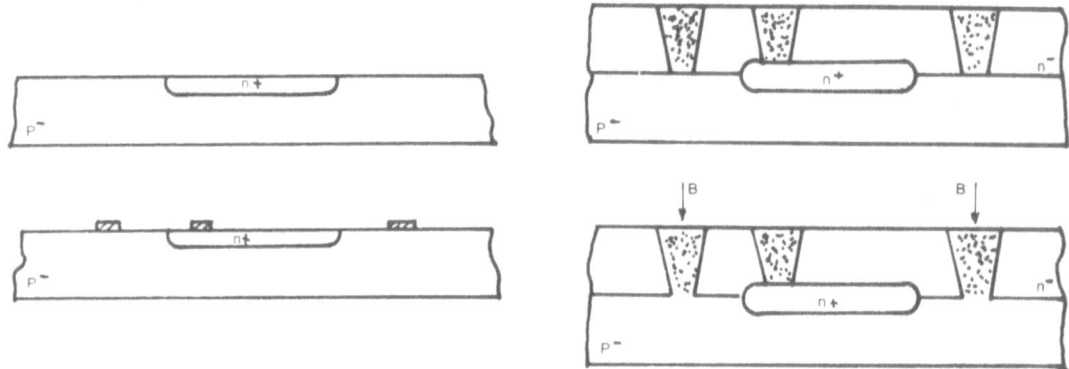

Fig. 3 - Poly process isolation

b) **Ion Implantation Process**

Fig. 4 shows the flow chart of the process. The insulation and sinker produced from top and bottom make it possible to reduce considerably the diffusion times and therefore the out-diffusion of the buried layer.

This process gives the same results as the polycrystal process with equivalent epitaxial thicknesses.

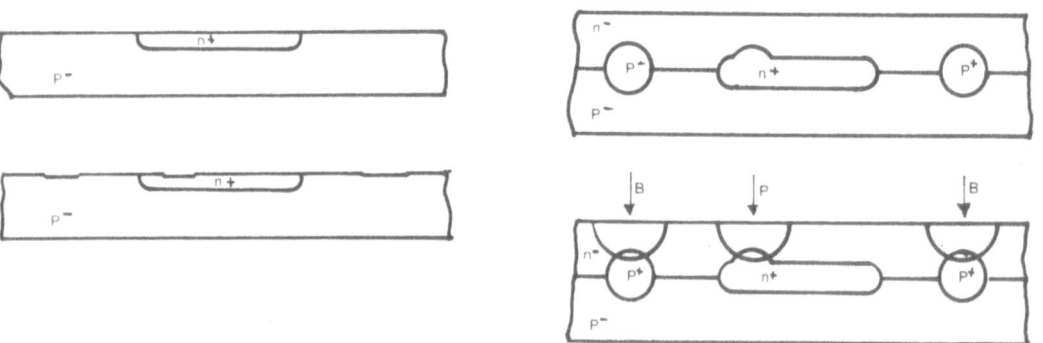

Fig. 4 - Ion implanted isolation

The implanted isolation structure with diffusion from above and below allows a significant reduction in size of individual components thanks to two important factors:

1) lateral diffusion is reduced as a result of the reduced isolation time; and 2) surface isolation is obtained with the same mask, i.e., it is a self-aligning structure.

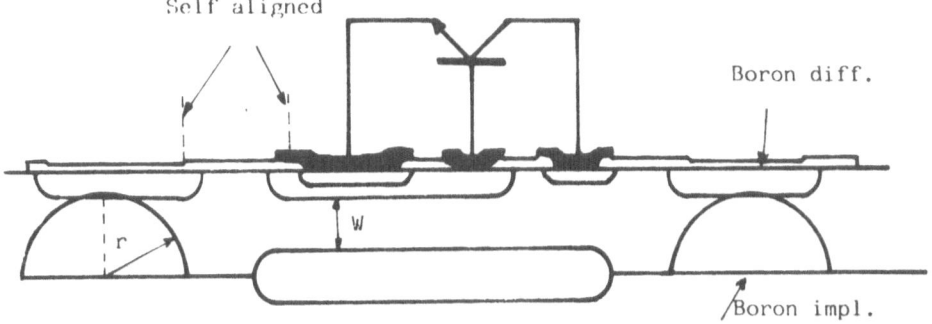

Fig. 5 - Ion implanted self-aligned NPN structure

A further benefit of this structure is that it is possible to realize transistors that operate at high voltages as well as low voltage elements on the same die, with a consequent saving in chip area simply by reducing surface tolerances.

Note that the field plates on the base and isolation regions avoid breakdown caused by curvature of the electric field lines and allows a substantial increase in VCBO.

This process has been used to realize an automative alternator regulator capable of withstanding overvoltages of ± 120 V with notable circuit complexity since it is capable of signalling numerous supply system failures. This circuit contains 150 transistors, 200 resistors and a 500 mA transistor capable of withstanding 120 V/1A for 50 ms.

It should be noted that this process is conceptually relatively old but it can only be realized practically now that ion implanters are available to prevent autodoping phemonena during epitaxial growth.

322

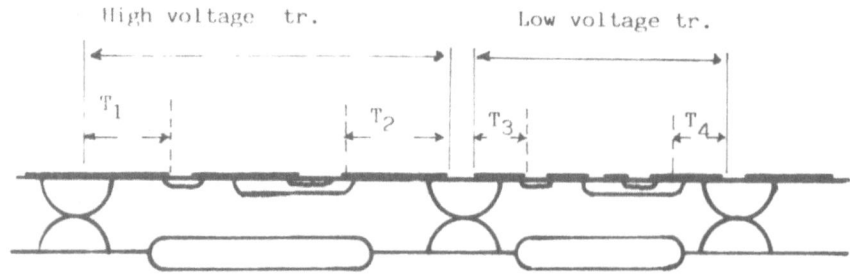

Fig. 6 - The tolerances are different for low and high voltage NPN transistors

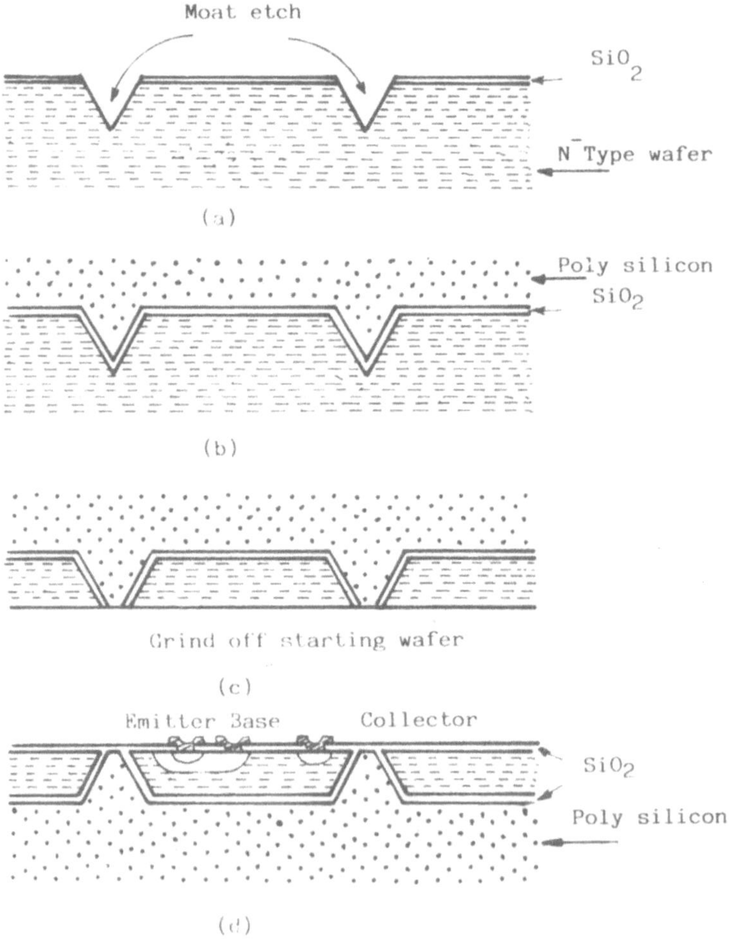

Moat etch

SiO$_2$

N$^-$Type wafer

(a)

Poly silicon
SiO$_2$

(b)

Grind off starting wafer

(c)

Emitter Base   Collector

SiO$_2$

Poly silicon

(d)

Fig. 7 - Dielectric isolation process

c) **Dielectric Isolation**

With dielectric isolation, components are isolated by an oxide layer. This approach greatly reduces the collector-substrate capacitance of the transistors. The process flow-chart is shown in Fig. 7.

Starting with an n-type substrate, grooves are formed on the back by preferential etching that permits a good definition of depth.

Oxide is grown on the surface of the grooves and a layer of polycrystal silicon of $\sim 200\mu$ is deposited. Then the original wafer is lapped until isolation regions are formed.

After oxidation the process becomes similar to the standard

d) **The Isoplanar Process**

With this type of process the lateral capacitance of components is eliminated and the area is drastically reduced.

This process is used essentially for low voltages since it is difficult to grow thick oxide layers.

Starting with a p-type substrate an n^+ buried layer is diffused and an epitaxial layer of about $1.5 - 2\mu$ thickness is grown.

This is followed by an oxidation, deposition of a nitride layers and an oxide deposition to define the geometry of the nitride (a).

By means of masking and etching the structure shown in (b) is obtained.

At this point the silicon is etched to a depth equal to 55% of the oxide that will be grown (c). Taking advantage of the different speeds of oxidation of silicon and nitride (25:1) the structure of Fig. (d) is obtained. Components can therefore be realized on the isolated islands using standard processes.

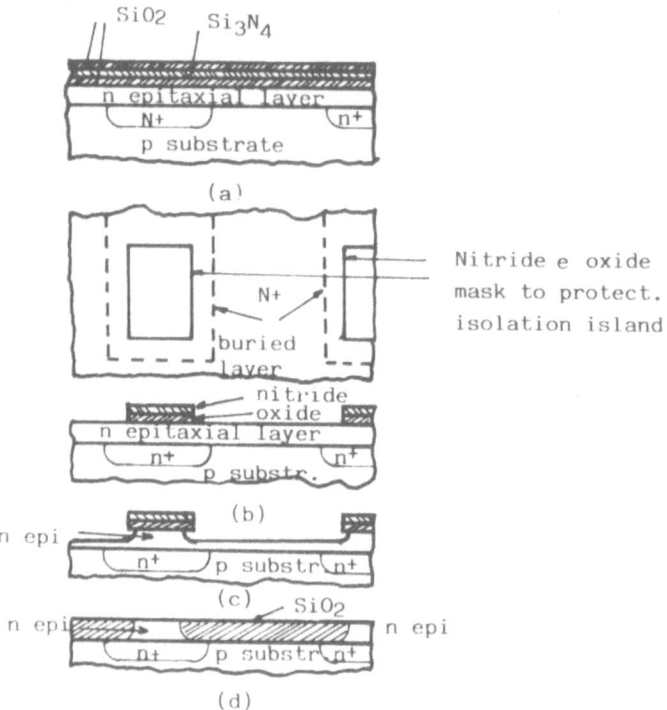

Fig. 8 - Isoplanar process

3.2 Integrated Components

Resistors

In integrated circuits, resistors can be realized using various structures or diffused regions or special techniques such as ion implantation.

The possibilities are:

- base diffusion resistors   ($\pm$ 25% spread, 100 to 200 $\Omega/\square$)

- emitter diffusion resistors ($\pm$ 25% spread, 2 to 4 $\Omega/\square$)

- epitaxial layer resistors   ($\pm$ 50% spread, 1 to 5 K$\Omega/\square$)

- pinch resistors using base emitter diffusion

$$(- 50\% + 100\% \text{ spread, } 3,5 \text{ to } 15 \text{ K}\Omega)$$

- high $\rho\,\square$ (1 to 2 K$\Omega/\square$) implanted resistors ($\pm$ 15% spread)

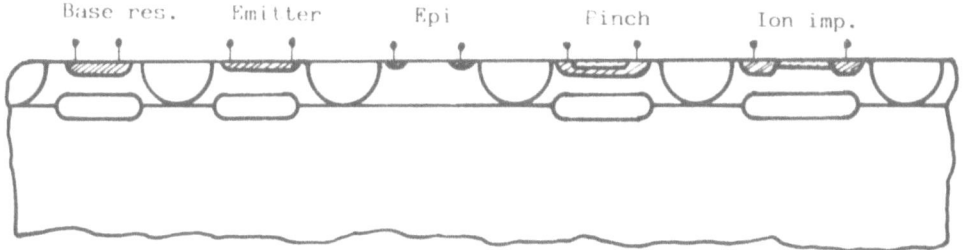

Fig. 9 - Different structures resistor

NPN Transistor

High performance double diffused NPN transistors can be realized using the planar process.

Working on the structure NPN transistors for currents from several mA up to several amps can be obtained.

Typical geometries for small signals are:

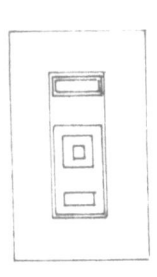
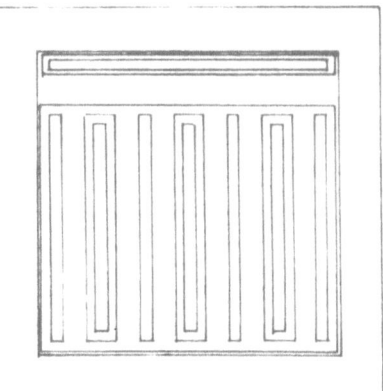

Fig. 10 - Minimum size and interdigited NPN

For high currents it is necessary to increase the emitter area and introduce an additional diffusion (sinker) to reduce the saturation resistance. To improve the robustness of the device, emitter ballast resistors are also introduced.

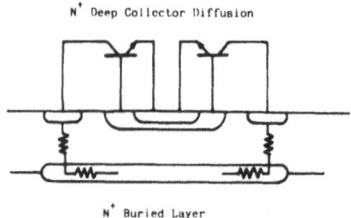

Fig. 11 - Power integrated- circuit structure with deep-collector N^+ diffusion

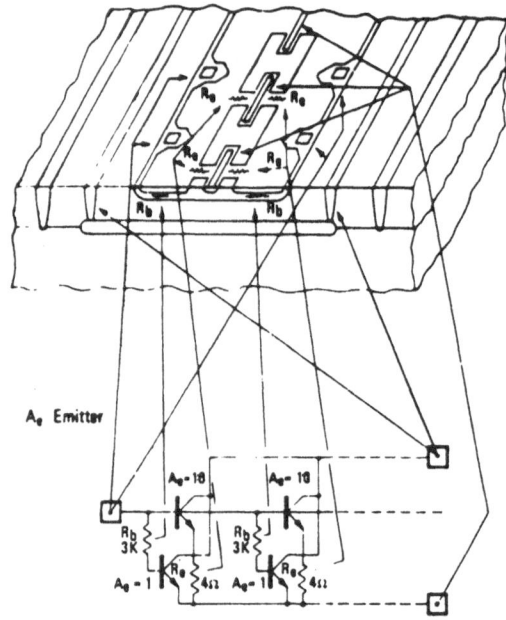

Fig. 12 - Section showing power device with emitter ballast resistors

A further NPN type with high F_E (1000-5000) and V_{CEO} = 3 to 10V can be realized with a second, deep emitter diffusion.

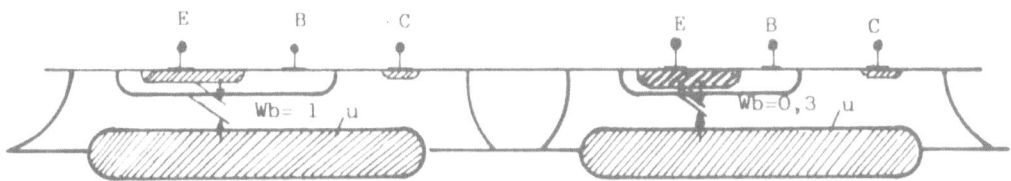

Fig. 13 - Standard and super β transistor

PNP Transistors

Two types of PNP transistor can be realized with the planar process: vertical and lateral.

Their greatest limitation is the poor frequency response due mainly to the width of the

base ($f_t \simeq 3$ to 10 MHz)

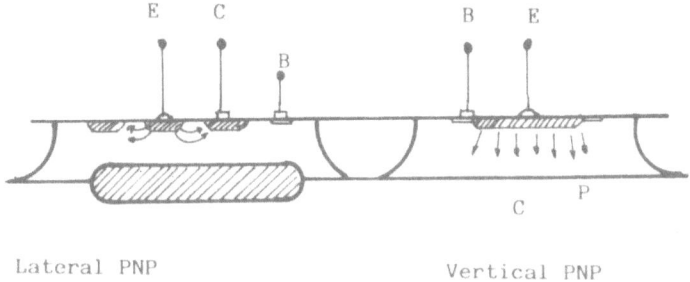

Lateral PNP   Vertical PNP

Fig. 14 - Lateral and vertical PNP

The emitter and collector of the lateral PNP transistor are realized with the NPN base

diffusion hence the efficiency of the resulting emitter is rather low because the emitter charge

and the thickness are low too. Naturally this can be improved by an additional P^+ diffusion.

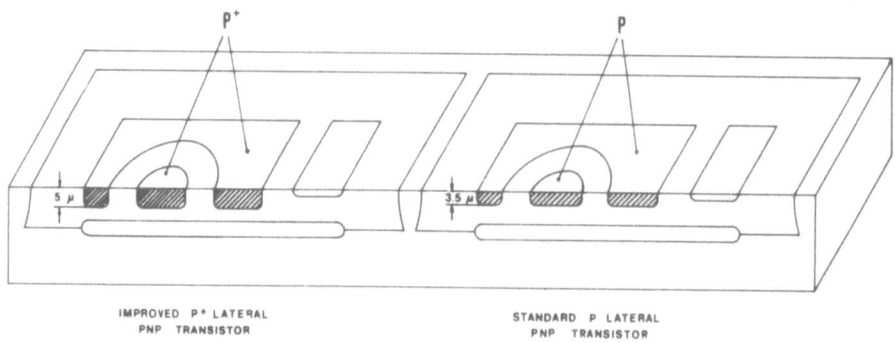

Fig. 15 - P$^+$ and P standard lateral p-n-p transistor structure

The improved performance of a P$^+$ lateral PNP is shown in Fig. 16

For currents higher than 0,5:1 mA, multidisk structures derived from the one shown in Fig. 15 are used. It is thus possible to obtain PNP devices for operating currents up to 100 mA with $h_{FE} \geq 10$ (PNP with P$^+$).

Zener

The most common type of Zener is obtained with the base and emitter diffusions of the NPN transistors (6.5 - 9V). The disadvantage is that the breakdown zone is very close to the surface thus the long term stability is not very good. Better results are obtained with bulk Zeners realized with P$^+$ ion implantation.

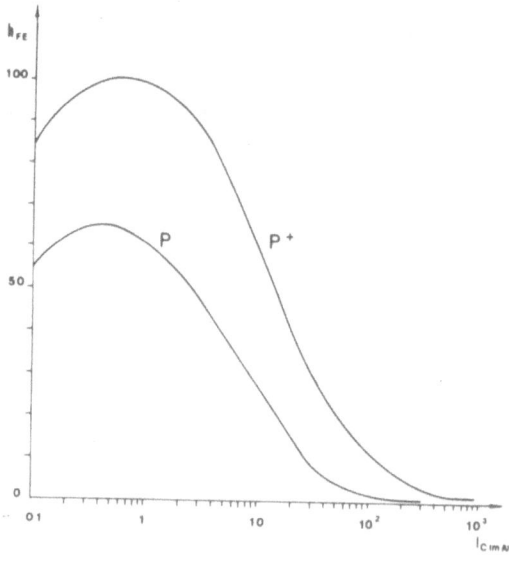

Fig. 16 - h_{FE} comparison between standard and P$^+$ 20-disk lateral p-n-p transistors

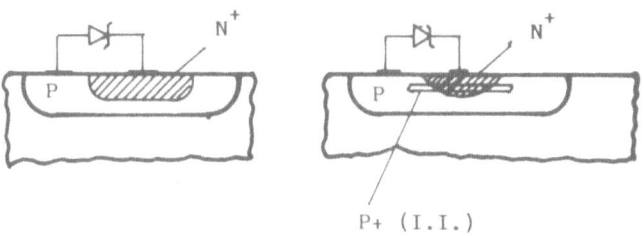

Fig. 17 - Standard base emitter Zener and I.I. bulk type

Ion Implanted J-FET

The use of ion implantation enables integrated J-Fets with very precise pinch-off voltages of around 1V and fairly high drain-gate breakdown voltages (50 to 60V) to be obtained. The structure of these devices is shown in Fig. 18. Base diffusion of NPN transistors is used to produce the source and drain while the channel and upper gate are realized with P implantation and N implantation, respectively.

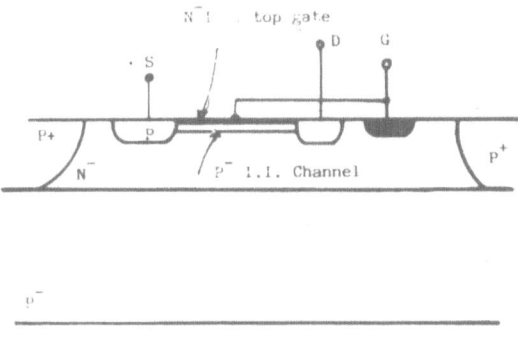

Fig. 18 - Ion implanted J-FET

4. BIPOLAR TECHNOLOGIES

It is difficult to find an orderly way of grouping together the different bipolar technologies, particularly those used in linear devices. We have found that the most logical grouping is by breakdown voltage of the integrated NPN transistors. In any integrated circuit there is more than one breakdown voltage and different circuits can withstand different voltages according to the structure used. The lowest limit in a transistor (whether integrated or not) is the collector-emitter breakdown voltage with open base, VCEO (Fig. 19).

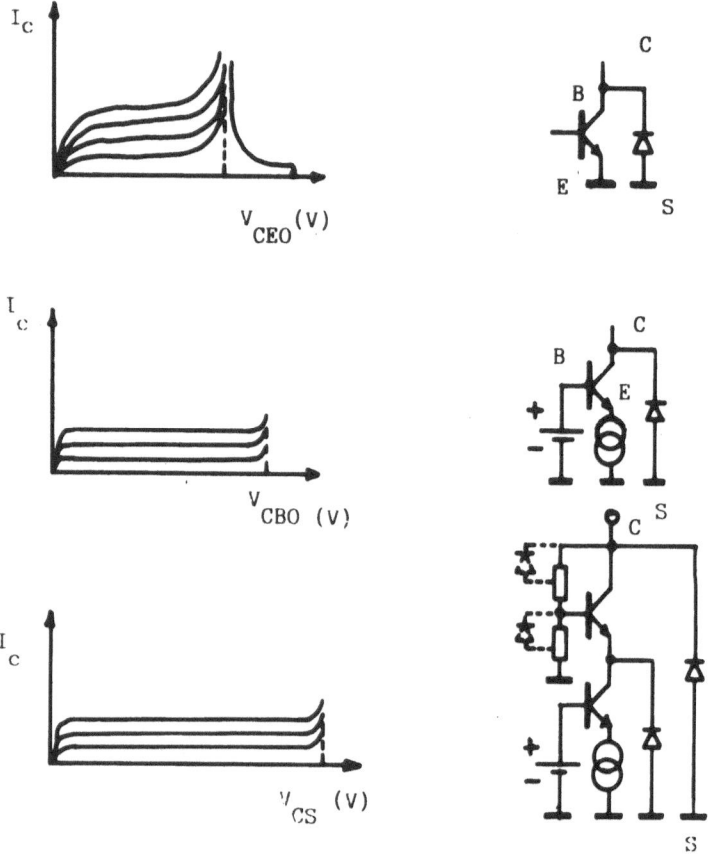

Fig. 19 - Breakdown voltage of integrated NPN transistors

A second limit, higher than VCE, is the collector-base breakdown voltage with open emitter, VCBO, for a single transistor when it is operated in grounded base configuration. This is the same as the breakdown voltage between a diffused resistance and the epitaxial layer in which it is built. The value of VCBO is greater than VCEO by a ratio of 1,5 to 3 and has a ratio of:

$$\text{VCEO} = \text{VCBO}/h_{\hat{\text{FE}}}\,(1/n)$$

where n is a function of the technology used.

It is possible to achieve higher breakdown values although this involves the use of "circuit tricks." For example, by putting transistors in cascade, it is possible to increase V_{cc} at the expense of saturation voltage. Figure 19 shows a stage which can withstand voltages higher than VCBO because the two transistors T1 and T2 split the supply voltage into equal points. A breakdown voltage higher than VCEO and VCBO is the collector-substrate breakdown voltage VCD (Fig. 20).

To subdivide bipolar technologies for integrated circuits we consider the smallest of the three breakdown voltages, VCEO, which is the same for either a discrete or integrated transistor, listing the VCEO value that is met by 95% of production devices.

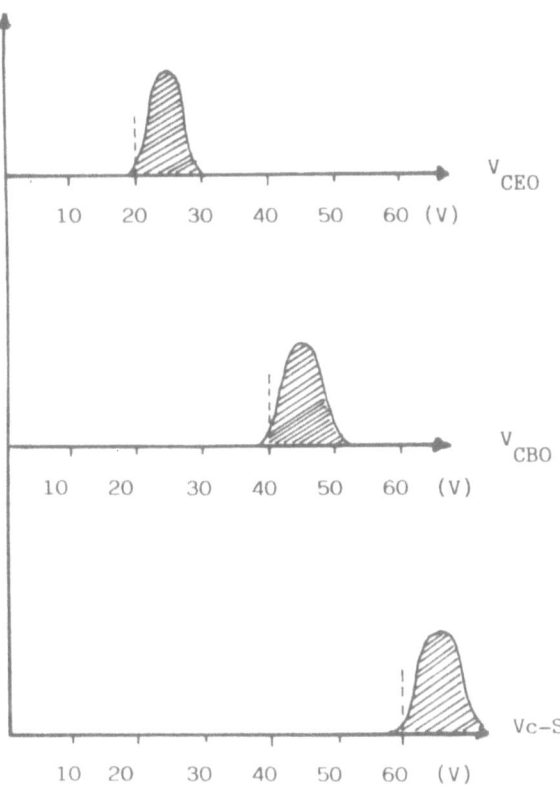

Fig. 20 - Gaussian distribution of breakdown voltages
for a 20V technology NPN transistor

Table 1 shows the different technologies available in production:

TABLE 1

VCEO(V)	VCBO(V)	VC–S(V)
10	30	50
20	40	60
30	50	70
40	80	90
60	90	120
120	240	300

The limits to scale reductions of a bipolar structure are better understood with the aid of the diagram shown in Fig. 21.

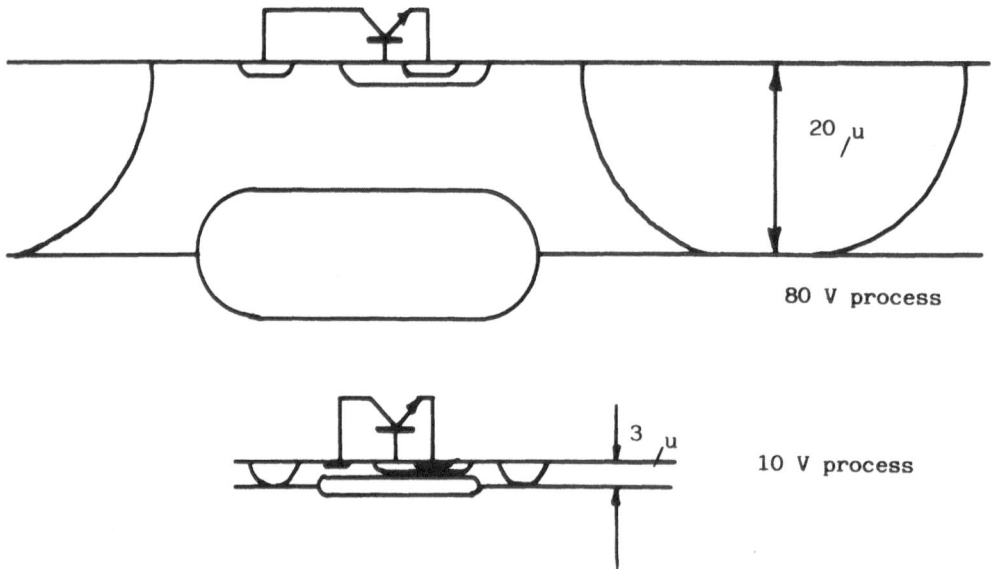

Fig. 21 - Bipolar transistor cross-section

The breakdown voltage VCB is determined mainly by the depth of the epitaxial layer between base and collector and by its doping level.

It is obvious from the cross sections in Fig. 21 that circuits which must support high voltages have inherently larger surface areas. Figure 22 shows a graph of minimum transistor surface area as a function of VCEO for different technologies and Fig. 23 shows the number of transistors per mmsq. of silicon surface area as a function of VCEO.

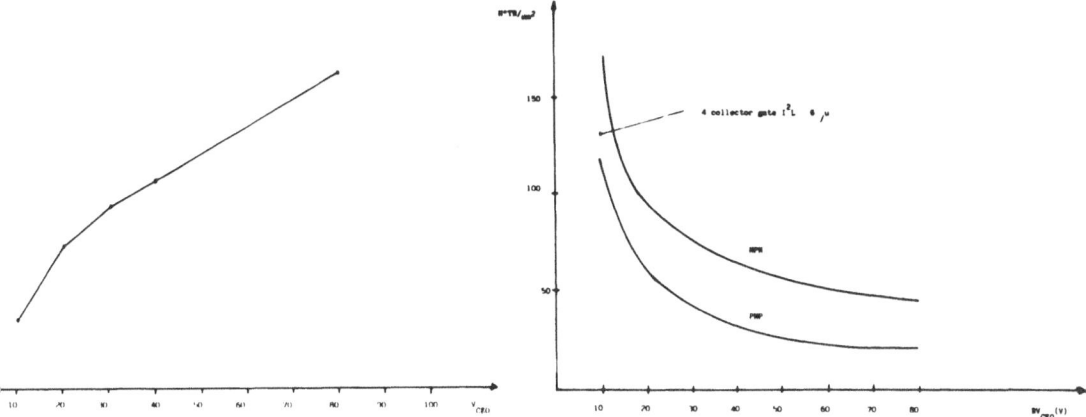

Fig. 22 - Minimum transistor area vs. V_{CEO}

Fig. 23 - Small geometry NPN devices vs. BV_{CEO} for a mm^2 active area

It is evident that to produce a bipolar LSI device taking advantage of the latest production equipment, which can produce windows in the oxide less than 4 micron, it is essential to accept a reduced device operating voltage. We are thus left with two conflicting aims: we need large dimensions for power circuits while to produce LSI we need low voltages. It now becomes clear that to make the most efficient use of the silicon it is necessary to split the functions and hence the decision to partition the functions as described earlier.

By operating at a low voltage level (e.g. 5V) within the sensor 5V level brings with it a notable reduction in the power dissipated per function integrated.

In fact, the power dissipation limit to circuit complexity is improved by a factor of around 4. In addition, the possibility of using a mixed analog/digital circuit design is of fundamental importance. An I^2L digital structure can be inserted in a 5V (10V breakdown) linear technology with just one extra mask and implantation.

So it is possible to obtain compatible linear and digital circuits with low cost, high yield, high reliability and wide operating temperature range (see Fig. 24).

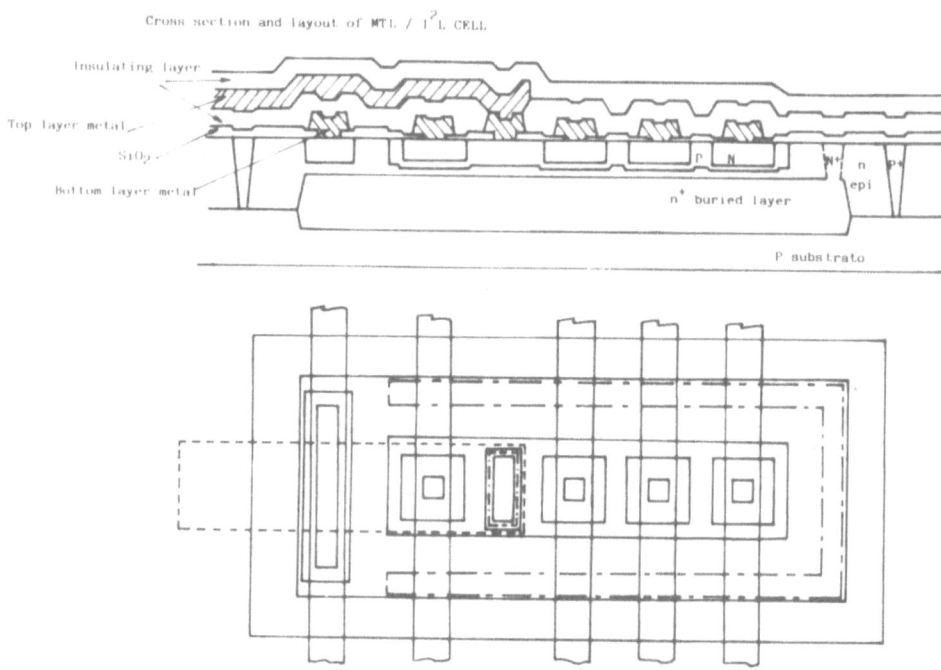

Fig. 24 - Cross section and layout of MTL/I^2L cell

5. PARASITICS IN I.C.

5.1 Surface Effects and Reliability

The reliability of high voltage devices is closely linked to careful design of the layout to prevent undesired surface effects.

In fact, the threshold voltage of the epitaxial layer in the zones with thicker oxide varies from 20 to 40 V in the high-voltage processes. It is therefore necessary to use the following techniques if devices are to be used at higher voltages: 1) lateral PNP channel protection obtained with the emitter metal; 2) use of the emitter N^+ channel stopper in zones where the epitaxial layer is at high voltage and is crossed by metal at low voltages; 3) use of a field-plate obtained by extending the metal connected to the P-diffusion to avoid inversion channels towards the insulation.

It should be remembered that to be really effective these protection expedients must take into consideration the movement of the gate in time due to the presence of mobile charges on the surface of the oxide accelerated by the temperature (see Fig. 25). For operating voltages greater than 100V it is also necessary to protect the insulation junction (P^+) epi (N^-) from the effect of the metals crossing it at high voltage. These can change the field lines in the corner zone and cause localized breakdown (Fig. 26).

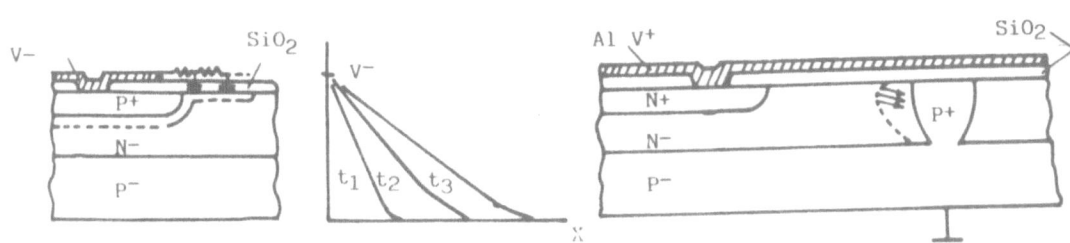

Fig. 25 - Electrical equivalent circuit and voltage diagram of ion surface migration

Fig. 26 - Collector isolation diffusion breakdown on standard structure

The phenomenon is eliminated by protecting the epi-insulation junction using a polycrystalline field plate anchored to the insulation (Fig. 27). Using the protection described, it is possible to obtain integrated devices with operating voltages greater than 200V (BV_{CEO}) and long term stability.

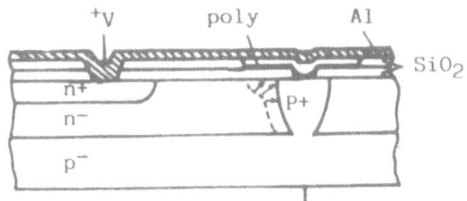

Fig. 27 - Improved breakdown voltage with poly field plate structure

5.2 Lateral NPN Transistors

When an IC island is more negative than the (grounded) substrate a "lateral NPN" can be produced of the form shown above. The collector of the NPN transistor can be formed by an adjacent island or even islands further away (a function of the electron lifetime in the substrate and epi layer). This can cause circuit malfunctions. A layout technique used to avoid the parasitic PNP is shown in Fig. 29.

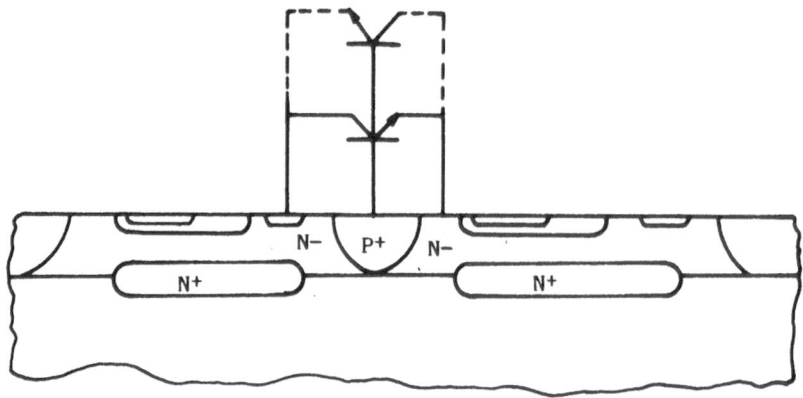

Fig. 28 - Parasitic lateral NPN

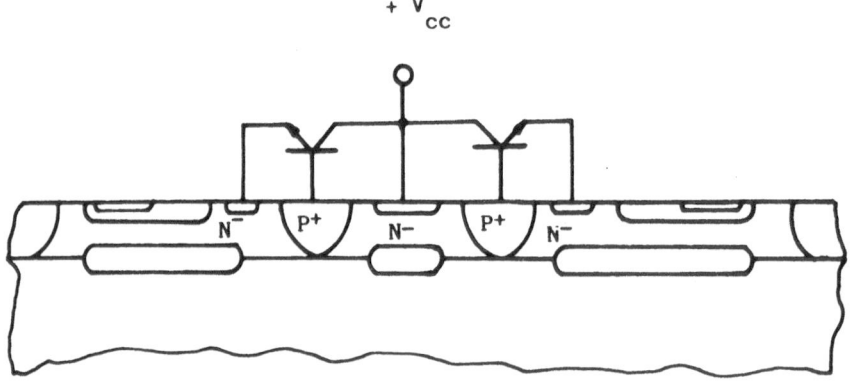

Fig. 29 - A method to avoid the parasitic lateral NPN

The island that could go negative is surrounded by a barrier island connected if possible to the most positive point.

5.3 <u>SCR</u>

The structure of an SCR is shown in Fig. 30.

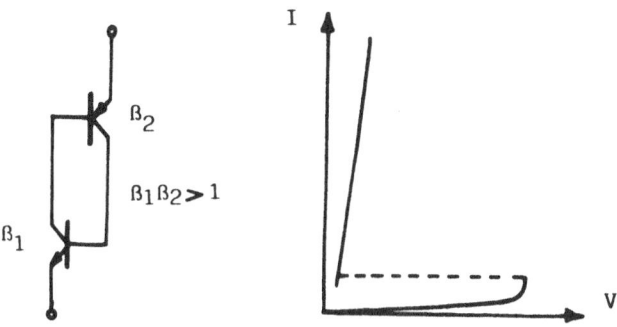

Fig. 30 - Parasitic SCR

This structure can be realized easily in integrated circuits by placing an NPN and a PNP transistor in the same island.

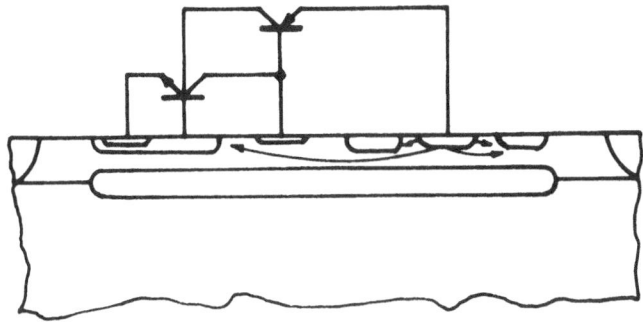

Fig. 31 - NPN and PNP in the same island realize a parasitic SCR

In this case $\beta_1 = 100 - 200$ and $\beta_2 = 0.01$. Triggering of the SCR can be avoided simply by drastically reducing β_2 using a sinker diffusion between the NPN and the PNP.

6. LIMITS TO PRECISION

6.1 Absolute Value and Matching of Components

The most important parameters of components realized in analog and analog/digital integrated circuits are: voltage references, op-amp offsets, comparator offsets and resistor matching.

6.2 **Voltage References**

The best known and simplest voltage reference is the Zener realized by a reverse-biased base-emitter junction. The disadvantage of this structure is the ±5% spread of the absolute values, noise (particularly high at low frequencies) and the random noise variation with current. Noise levels of the order of several millivolts peak-to-peak are typical in this type of structure.

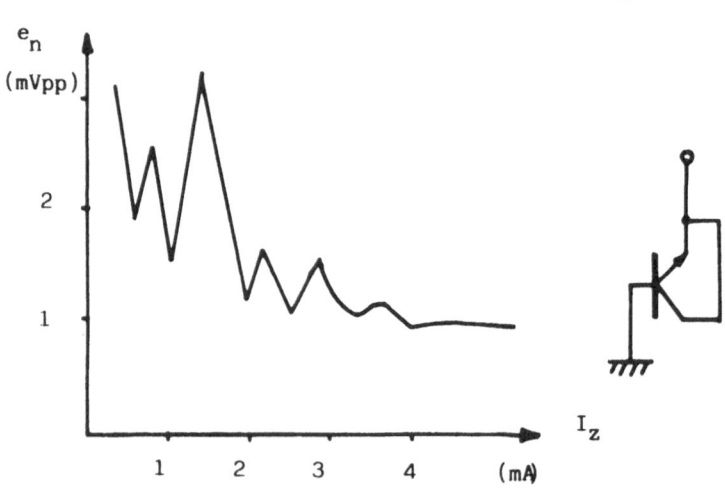

Fig. 32 - Voltage noise on base-emitter zener

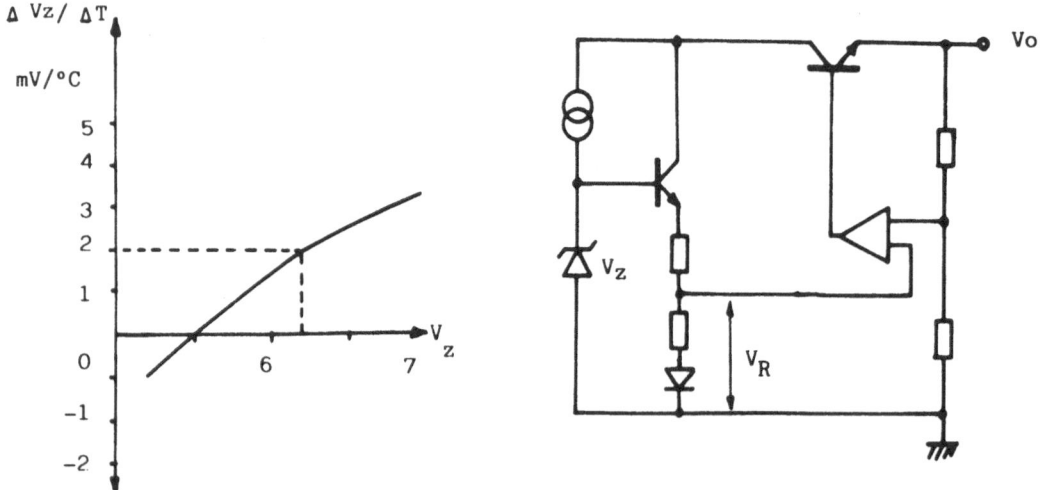

Fig. 33 - Voltage regulator with compensated zener (μA 723)

Its thermal drift, a function of its absolute value, can be compensated easily with the use of forward biased junctions as in the μA 723. A variant of this version is the bulk Zener which, having buried junctions, suffers less from surface phenomena and is thus more stable with time. Examples of thermally compensated bulk Zeners indicate values of thermal drift of about 30 ppm and stability of 10 μV/year.

6.3 Band Gap

Another reference element is the band gap which exploits a normal forward biased VBE which has a negative temperature coefficient, compensated by a special circuit that sums the resistive voltage drop $R_2 I_2$ with a positive coefficient (Fig. 35). It is known that a forward biased VBE behaves as shown in Fig. 34.

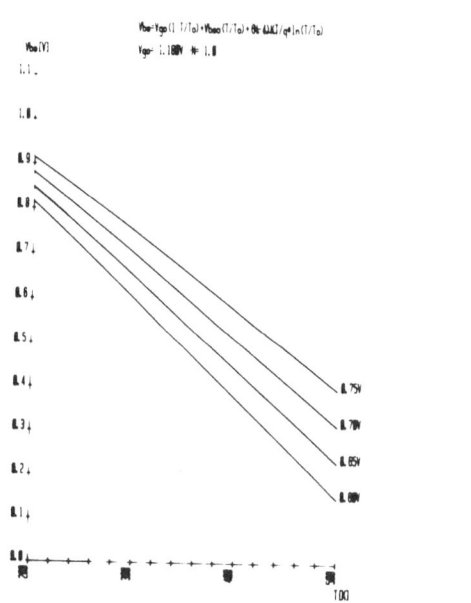

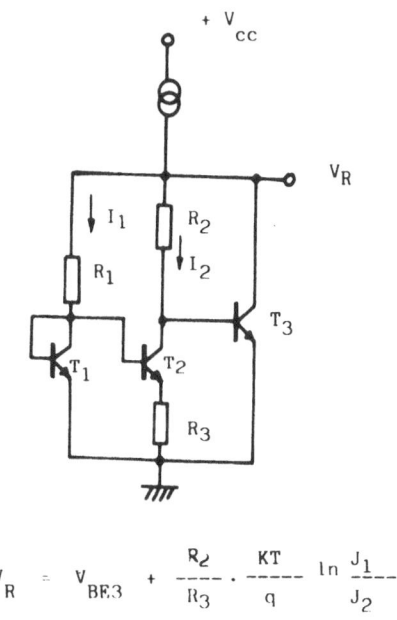

$$V_R = V_{BE3} + \frac{R_2}{R_3} \cdot \frac{KT}{q} \ln \frac{J_1}{J_2}$$

Fig. 34 - V_{BE} and ΔV_{BE} vs. temperature

Fig. 35 - Band gap voltage reference

A positive coefficient voltage drop is summed with the VBE to compensate as far as possible the negative coefficient of the VBE itself. The positive coefficient of the current I_2 is obtained by operating the transistors T_1 and T_2 at different current densities.

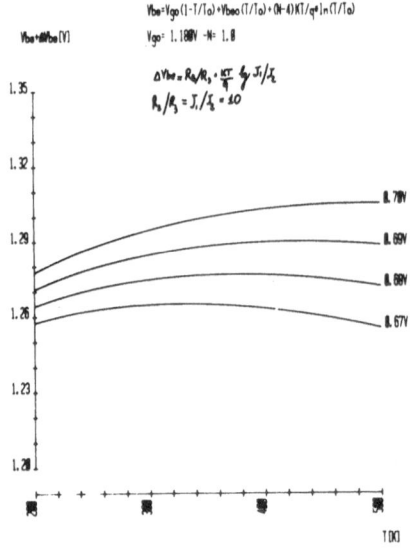

Fig. 36 - The T.C. is related to absolute value of V_{BE}

Excellent examples of the band gap reference exist on the market (LM-108 μA 7800).

6.4 **Op-Amp** **and** **Comparator** **Offsets**

Other precision elements are op-amps and comparators. These devices utilize differential structures that exploit the inherent matching of similar components located close together on the die.

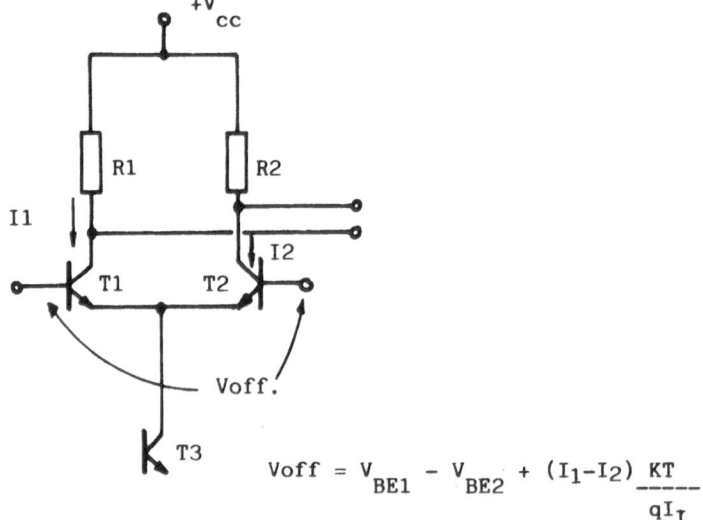

$$Voff = V_{BE1} - V_{BE2} + (I_1 - I_2)\frac{KT}{qI_I}$$

Fig. 37 - Differential amplifier

The thermal drift of two transistors that operate at equal current levels is a function of the static offset in accordance with the curve given in Fig. 38.

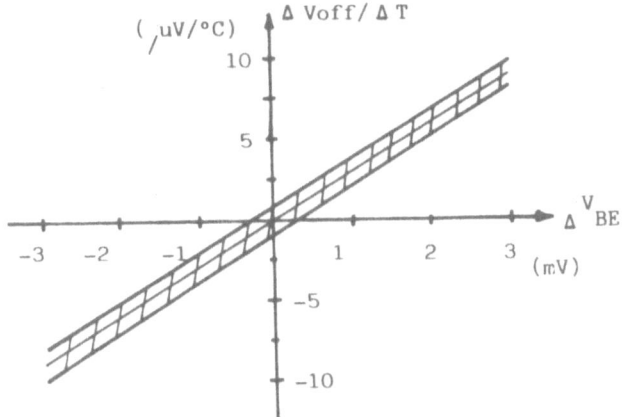

Fig. 38 - Offset and drift correlation

A good op-amp is characterized by the drift of the first stage only. Therefore it is possible to optimize the drift by optimizing the offset.

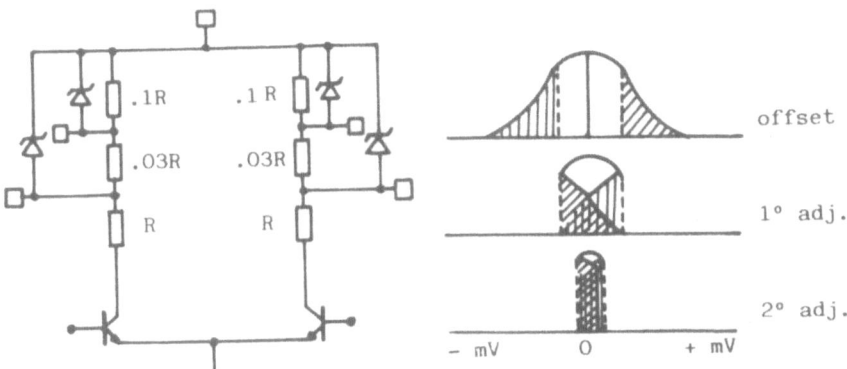

Fig. 39 - Zener zap offset adj.

6.5 Laser Trimming

Today there are wafer test machines that carry out the automatic trimming of the op-amp offset using a laser to break metal connections between resistances.

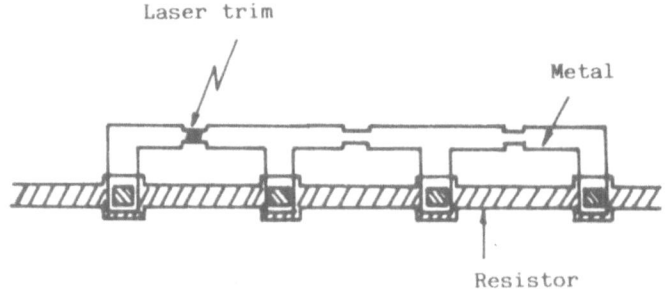

Fig. 40 - Laser trimming

6.6 Zener Zap

A cheaper and more reliable trimming system has been realized using elements that form a short circuit between metal areas. This method consists of placing a Zener, realized with a reverse biased base-emitter junction, between the two points to be short-circuited.

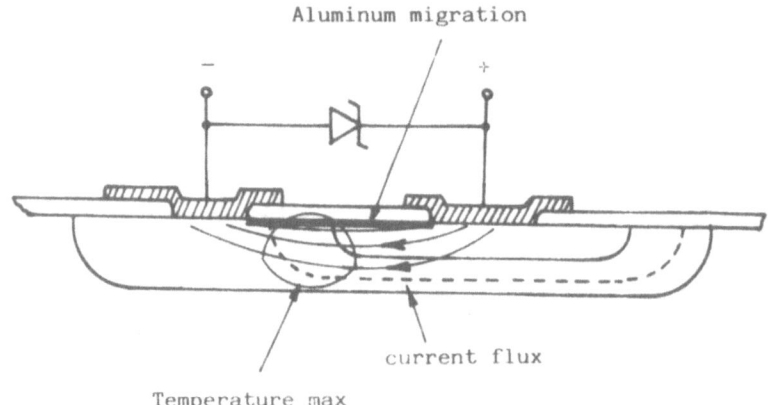

Fig. 41 - Zener zap

Forcing current in the Zener up to 100-200 mA creates a violent local heating effect which causes the overlying aluminium to melt and migrate under the oxide thus forming a stable short circuit between the anode and the cathode of the Zener with a resistance of 1-2Ω. This method is very reliable because there are no projections of material in the vicinity and the connection remains covered by the surface oxide guaranteeing long term stability.

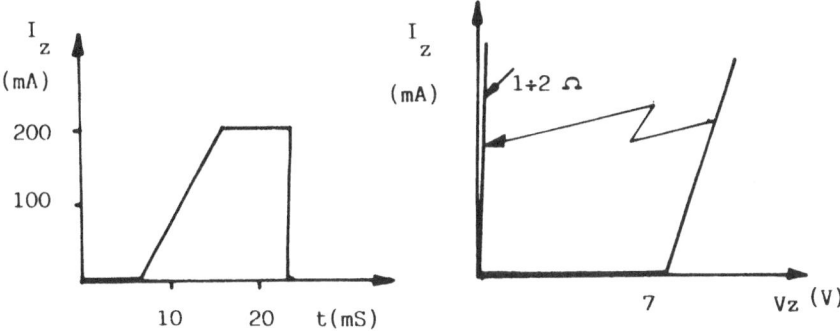

Fig. 42 - Firing current and zener characteristics

6.7 **Resistor Matching**

The structure of integrated resistors has been described in Sec. 3.2. The absolute value of diffused resistors has a tolerance of the order of $\pm 25\%$, falling to $\pm 15\%$ for implanted resistors. Thermal coefficients range from 2 to $3\%/^\circ$C.

The precision of the absolute value is determined by variations in resistivity from wafer to wafer and between different points on the same wafer, masking tolerances, contact resistance and variations in mobility due to mechanical stress.

Using large dimension resistors (20 μ), divided into interconnected modules to minimize resistance gradient effects, matching of 0.3% or better can be obtained. Thin resistors (8 μ) allow 1 to 2% matching.

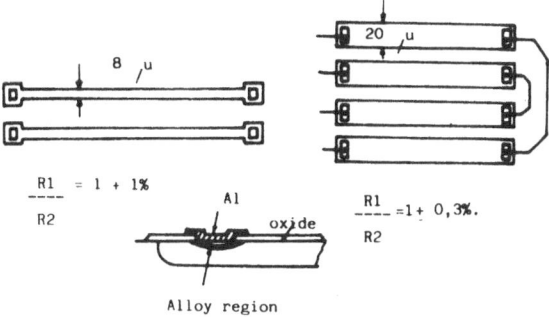

Fig. 43 - Matching of resistors

Another phenomenon that adversely affects the precision of resistors is mechanical stress. Further on, we will see how mechanical stress affects charge mobility, and hence resistor values, and examine methods to reduce this effect.

Contact resistance can also influence component values, particularly when the number of contacts is limited. A typical contact resistance value is 500 $\Omega/\mu m^2$.

6.8 **Thermal Feedback**

Very often in an integrated circuit the matching of electrical parameters in devices close together is used to obtain thermal compensation (differential amplifiers, band gap, etc.). This is valid if the temperature between the two elements is effectively equal but in practice the temperature on the die may be non-uniform, particularly when there are elements that dissipate power.

We will consider an integrated circuit die soldered to a copper support by a thin layer of solder.

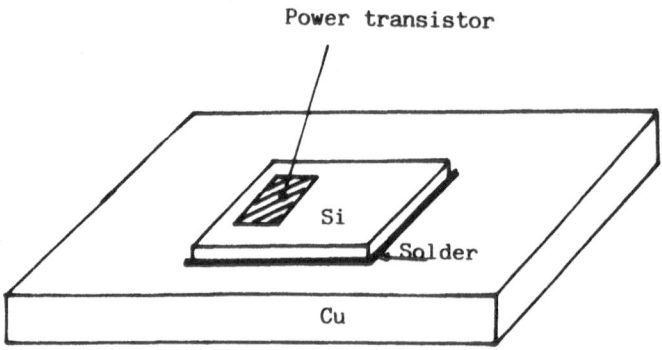

Fig. 44 - Power I.C. soldered to copper slug

In one part of the circuit, a power transistor for example, power is dissipated as a step function

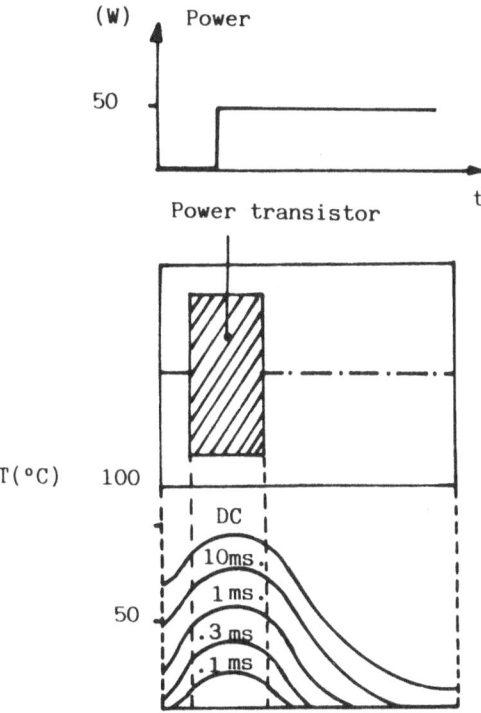

Fig. 45 - Thermal profile of power I.C.

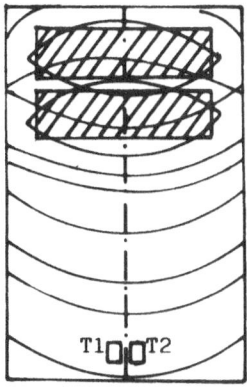

Fig. 46 - Thermal map
on power op-amp

Analysing the variation of temperature in time over a section of the chip, clearly defined

trends can be noted depending on position and time.

Familiarity with thermal phenomena in silicon makes it possible to optimise the arrange-

ment of sensitive elements and minimise the effect by exploiting thermal symmetry.

With a good layout, offsets in the order of 100 μV per watt dissipated by the power stage can

be obtained.

6.9 Mechanical Stress

It is known that charge mobility in silicon is a function of the mechanical stress to which

the device is subjected. Sensitivity to this stress is not uniform in all directions but depends

on the crystal orientation.

Mechanical stress can be introduced in various phases of manufacturing. Rapid removal

from the furnace can cause curvature of the wafer due to the different thermal expansion of

silicon dioxide and nitride with respect to silicon.

Fig. 47 - Mechanical stress caused by different thermal
coefficient from silicon and dioxide

Hard soldering (eutectic-gold-silicon) on material with a different coefficient of linear

expansion can cause tension on the surface with a consequent change in charge mobility.

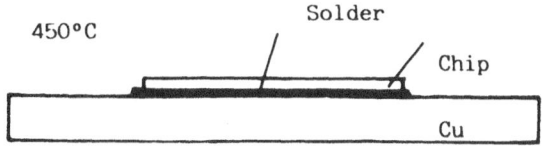

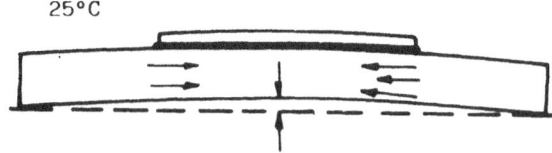

Fig. 48 - Mechanical stress after soldering

Another mechanical stress is that caused by the contraction of the resin when it polymerises.

All of these causes of mechanical pressure demand a careful study the stress behaviour and

sensitivity of integrated components. The effects of mechanical stress can be minimized by a

45° orientiation of resistors in a "100" substrate but in "111" substrates there is no non-

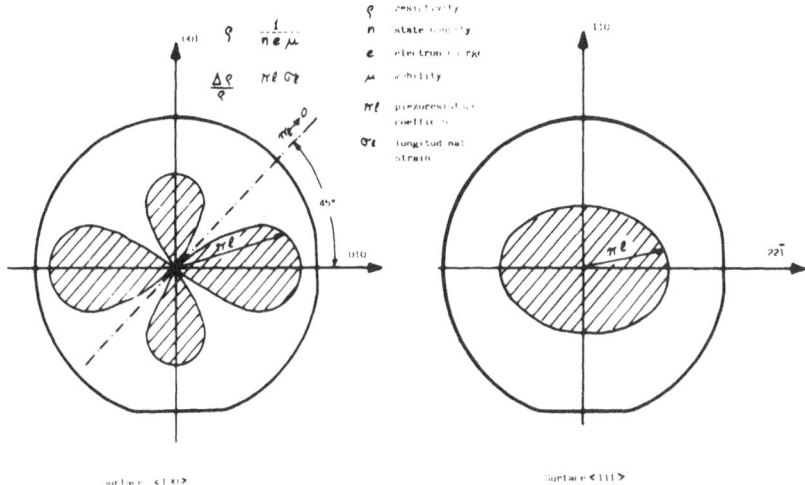

Fig. 49 - Piezo resistive sensitivity for different types of silicon wafers

7. CONCLUSIONS

The race to reduce the dimensions of bipolar devices differs for four types of circuits: purely digital circuits, purely linear circuits, mixed digital/linear circuits and power circuits. For purely digital circuits the trend towards infinitely smaller dimensions will continue with improvements in lithography and with the realization of multiple interconnection layers.

As far as the purely linear circuit is concerned the road towards greatly reduced dimensions is limited by certain other factors. The first of these is the supply voltage. Systems designers have to be convinced to change from today's standard $\pm 15V$ or 12V circuits to 5V circuits and to realize a multitude of analog circuits capable of operating correctly in these conditions.

The second limiting factor is the deterioration in the electrical performance of individual active components if dimensions are reduced beyond a certain level.

The third and last factor is the still limited knowledge of the mechanical and thermal stress phenomona inside the chip which affect precision.

In mixed digital analog circuits the possibility of reducing the dimensions will be linked to the precision which must be reached for the analog parts as we saw previously.

Finally, in power devices the physical limit which must be overcome is the possibility of extracting heat from a single silicon crystal. In this sector, the emphasis is on improving efficiency and reducing the thermal resistance caused by the silicon itself and its package container as opposed to reducing dimensions.

To conclude, we can say that in the 80's the bipolar sector, in particular the analog or analog digital part of it, will be characterized by the realization of high complexity custom circuits, i.e., systems oriented towards satisfying customers' requirements and realizing standard functions which make full use of the possibilities offered by existing technologies and by those still under development.

REFERENCES

1. B. Murari, "Power Integrated Circuit: Problems, Trade-offs and Solutions", IEEE J. Solid State Circuits **SC-13**, 3 (1978).

2. P. Antognetti, G. R. Bisio, F. Curatelli and S. Palara, "Three Dimensional Transient Thermal Simulation: Application to Delayed Short Circuit Protection in Power IC's", IEEE J. Solid State Circuits **SC-15**, 3 (1980).

3. R. C. Dobkin, "5A Regulator with Thermal Gradient Controlled Current Limit", IEEE Inter. Solid State Circuits Conf. (1979).

4. V. Prestileo, "High Voltage Technology for Display Drivers", in Electronic Inform. Display 2nd Course, Inter. School of Physic for Industry, presented at Erice, Italy, June (1977).

5. P. Selini and G. Vignola, "Reliability Improvement Through Design as Applied to a Family of Monolithic Lin. Operat. Amplif." , Congresso Varna Bulgaria, Sept. (1977).

6. F. Oettinger, D. Blackburn and S. Rubin, "Thermal Characterization of Power Transistors", IEEE Trans. Electron Devices **ED-23**, 8 (1976).

7. D. Bowler and F. Lindholm, "High Current in Transistor Collector Regions", IEEE Trans. Electron Devices **ED-20**, 3 (1973).

8. M. Felici, "Rottura Secondaria (Second Breakdown) nei Transistori di Potenza" AEI Milano Giugno (1980).

BIPOLAR DIGITAL CIRCUITS

W. Holt

Plessey Research (Caswell) Ltd.
Towcester
United Kingdom

I. INTRODUCTION

A typical bipolar integrated circuit process provides n-p-n transistors as the primary active device, various diode and resistor structures, and lateral p-n-p transistors of rather poor performance. This range of components gives the bipolar process the flexibility to make a wide variety of digital and analogue circuits on the same process, or even the same chip.

This inherent flexibility contrasts with the MOS case in which circuit type and process are more closely linked by the active device type or types involved, i.e., PMOS, NMOS and CMOS. Given the highly sophisticated and competitive nature of the integrated circuit industry process variants have been optimised to meet the requirements of specific types of logic circuit, but strictly speaking there is no such thing as an "ECL process" or a "TTL process". Later on in this lecture I will be describing a process developed for subnanosecond ECL, which has also been used extensively for high frequency analogue circuits and also some I^2L and mixed function circuits.

Before returning to the technology aspects of bipolar digital circuits, I will briefly trace the development of the main bipolar logic circuit families.

2. DEVELOPMENT OF BIPOLAR LOGIC

2.1 Resistor Transistor Logic

Resistor Transistor Logic (RTL) was one of the first integrated circuit logic families to be produced. RTL was also one of the first circuit configurations optimised for integration rather than merely an integrated form of an existing discrete component circuit.

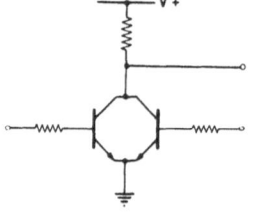

Fig. 1 - RTL gate

The circuit diagram of basic 2 input RTL gate is shown in Fig. 1. This gate performs the NOR function for positive logic. RTL gates use common collector transistors, thus saving on the area needed to isolate transistors. Only two components, plus a common load, are needed per gate input. Using the process technology current in the mid-1960's a basic RTL gate operating off a 3 volt supply gave a typical delay of 30nsecs for a power consumption of 2mW.

The layout of more complex RTL circuits is greatly facilitated using the built in crossover available at each input by the simple expedient of running a metallisation line over the input resistor. This feature was exploited by Plessey in a range of custom designed circuits for digital frequency synthesis. The most complex of these circuits used 80 RTL gates to give a 20MHz variable decade divider on a chip size of 1.75 x 2mm in 1968, using the relatively coarse geometry process capability of that period.

The disadvantages of RTL as a general purpose logic family lie in the poor interface properties. The circuit requires close tolerance components for operation over a wide temperature range, and with a logic swing of 0.8 volts noise margins are low. In addition, basic gate fan-outs are limited to 4 with the standard 3 volt supply rail.

2.2 Diode Transistor Logic

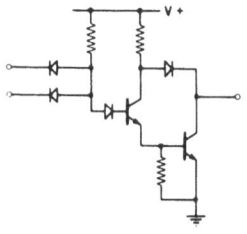

The Diode Transistor Logic (DTL) family was the next integrated circuit logic type to appear after RTL, and the first circuits were very much based on discrete component designs. The basic DTL gate shown in Fig. 2 performs the NAND function for posit-

Fig. 2 - DTL gate

ive logic. The DTL gate is made up of a diode AND gate cluster followed by a transistor inverter. The voltage translation or level shifting between the diode gate and inverter in the discrete version was normally carried out by a resistor with a capacitor across it to help in the removal of stored charge in the transistor inverter when turning it off. Some of the first integrated DTL circuits achieved the required voltage translation by means of two series diodes. To achieve the same effects as the capacitor, it was then necessary to make the diodes store enough charge to remove that in the base of the transistor. However, stored charge was an undesirable feature in the gate diodes. Before the development of Schottky diodes, this required selective gold doping, a very difficult process to control. The version of DTL shown here was used to produce gate delays of 20 - 30 nS, with noise margins of around 1 volt and high fan out capability (greater than 20). These features soon established DTL as the first general purpose logic family. However the packing density of DTL is inferior to RTL which still finds some applications today, usually in L.S.I. circuits with TTL interfaces.

2.3 Transistor-Transistor Logic

Transistor-Transistor Logic (TTL or T^2L) is very much the natural extension of DTL into a form more suitable for integrated circuit realization. In the basic gate circuit shown in Fig. 3, the input diode cluster of DTL is replaced by a multi-emitter transistor,

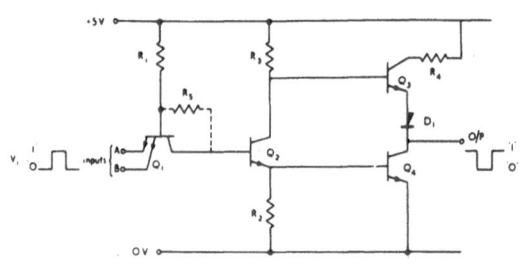

Fig. 3 - 2 input TTL NAND gate

each emitter being a gate input. The output stage gives a low impedance output capable of driving high capacitance loads when the gate changes from either state.

The operation of the circuit is as follows: With either input at 0 volts, Q_1 and Q_3 are on, Q_2 and Q_4 are off and the output voltage is high; when both input voltages increase to one diode drop, Q_2 starts to conduct and the output voltage starts to droop as shown in the region a to b of the transfer characteristic in Fig. 4; when both input voltages exceed two diode drops, Q_4 starts to turn on; finally, the emitter base junctions of Q_1 become reversed biased, Q_2 and Q_4 are on and Q_3 is off. The diode D_1 ensures that under steady state conditions the output transistors Q_3 and Q_4 cannot conduct simul-

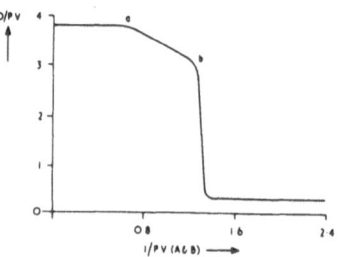

Fig. 4 - TTL voltage transfer characteristic

taneously. However, this does not prevent a large current surge, limited only by R_4, flowing

during the switching transient. This self-generated noise is a distinct disadvantage of TTL,

requiring careful layout and liberal use of decoupling capacitors.

RTL, DTL and TTL are all forms of saturating logic, which rely on the use of gold

doping to reduce minority carrier lifetime and, hence, storage times. With the development of

a reliable Schottky barrier diode technolgy, an implored form of TTL, Schottky TTL was

introduced.

Charge is stored by minority carriers in the base and collector regions of a transistor in

saturation. By adding a Schottky barrier clamping diode, connected between the base and

collector, saturation can be prevented because the forward drop of the Schottky diode is lower

than that of the collector base junction. Hence, excess base current is diverted from the

transistor through the Schottky diode, which being a majority carrier device, has negligible

minority carrier storage. These composite Schottky clamped transistors can be made with

storage times of 1-2ns, compared to 5-10ns for conventional gold-doped transistors. Also the

higher lifetime of non-gold doped structures gives higher gain.

The circuit of a 2 input Schottky TTL NAND gate is given in Fig. 5. Diodes D_1 and D_2 protect Q_1 from reverse voltage spikes on the input. The undesirable droop in the transfer characteristic of conventional TTL has been eliminated by the inclusion of Q_3 which ensures that Q_2 does not conduct before the output transistor Q_4 (see Fig. 6). The Darlington connection Q_5, Q_6 gives greater drive capability for capacitive loads. Q_6 is only the unclamped transistor and is prevented from entering the saturation region by the action of Q_5. With a logic zero input Q_2 and

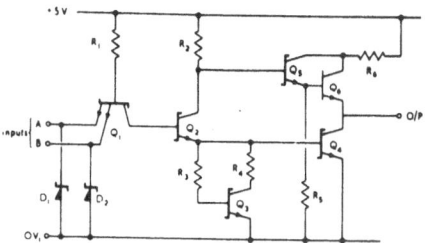

Fig. 5 - 2 input Schottky TTL NAND gate

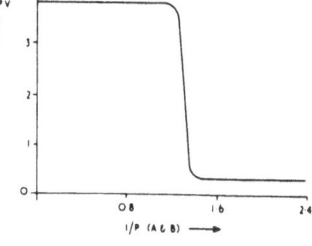

Fig. 6 - Schottky TTL voltage transfer characteristic

Q_4 will be off and Q_5, Q_6 on. With both inputs at logic 1, Q_2, Q_3, and Q_4 are on and Q_5, Q_6 off. The overall effect in the first generation of Schottky TTL was to improve the gate delay from 10 to 3 nsecs at some increase in power consumption, compared with the standard TTL.

2.4 Emitter Coupled Logic

Emitter Coupled Logic (ECL) is a non-saturating current steering circuit form which was used before the advent of integrated circuit technology for high speed applications. ECL depends on having well matched transistor parameters which the integrated process readily provides. A typical ECL NOR/OR gate is shown in Fig. 7. The gate is made up of a current mode switch, voltage reference and emitter follower outputs. The logic swing is from $-1V_{BE}$ to -2_{BE} or approximately ±400mV on either side of the reference voltage of

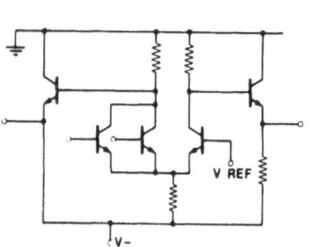

Fig. 7 - ECL gate

-1.2 volts. When all of the inputs are low the tail current flows from the Q_4 and a low appears at the OR output emitter follower Q_5. If any of the inputs goes high the tail current is diverted from Q_4 to that transistor and the OR output goes high. The second emitter follower coupled to the common collector of the input transistors produces the NOR output. The transfer characteristic of an ECL gate is shown in Fig. 8.

In addition to the availability of complementary outputs, ECL provides the wired-OR function (emitter dotting) with less than 10% increase in gate delay. The wired-AND function (collector dotting) can also be produced.

The non-saturating nature and low voltage swing of ECL make this one of the fastest forms of bipolar logic, capable of producing gate delays well below 1ns. Although the noise margin is small, around 200mV, the current drain is nearly constant, resulting in very small internal noise generation.

Reference 1 gives a detailed description of the earlier forms of bipolar logic.

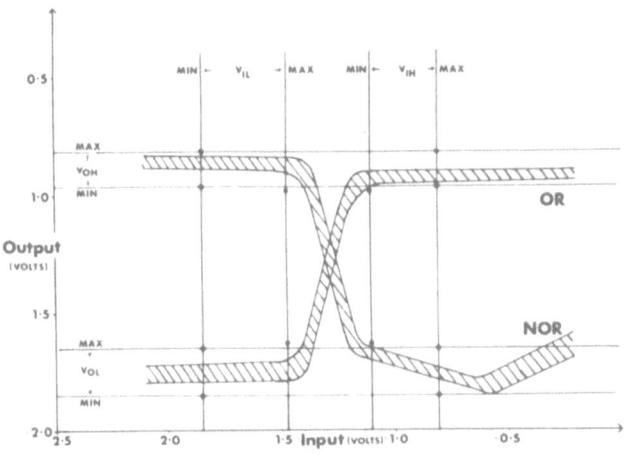

Fig. 8 - ECL transfer curves

2.5 **Integrated Injection Logic (I^2L)**

The logic families described so far compare unfavourably with respect to MOS in terms of packing density. The simultaneous invention by Philips[2] and IBM[3] of I^2L, termed "Merged Transistor Logic" by the latter, at last provided a bipolar logic form capable of matching MOS packing density.

The basic circuit consists of a lateral p-n-p current source, the injector, and multi-collector p-n-p transistor operating in the inverse mode, i.e., with the normal emitter acting as collector, and vice versa. In the physical realisation of this structure the p-n-p base and n-p-n emitter are merged into one region, and also the p-n-p collector and the n-p-n base. The structure and equivalent circuit of a basic 4 collector gate are shown in Figs. 9 and 10. The gate operates as follows: When T_1 is on, due to a logic high at its base, the collector of T_1 is at $V_{CE}(sat) \simeq 0$ volts), thus sinking the base current being supplied to T_2 by the p-n-p current source. When T_1 is turned off by applying a logic low ($\simeq 0$ volts) its collector rises, the $V_{BE}(on)$ of T_2 is reached and T_2 is driven into saturation. Thus the gates can be directly

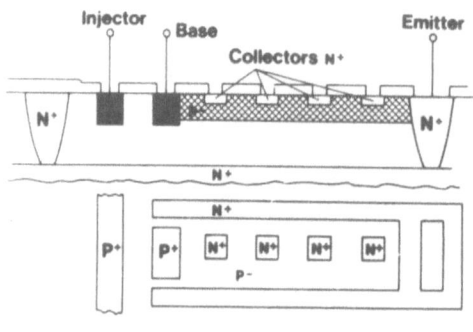

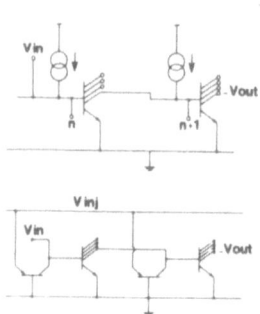

Fig. 10 - I²L logic circuit

Fig. 9 - I²L gate structure

connected without the need for any level shifting circuitry. The basic I²L gate is a multi-output inverter with a wired AND capability at each output.

One problem with I²L is the provision of sufficient current gain through the structure. The p-n-p injector has a typical current gain in region 0.5 to 1 and, because the n-p-n

transistor is operated in the inverse node, its current gain is also low at around 5 to 10. The gain of an I²L gate as a function of current is shown Fig. 11. The difference between the near and far collector gains is caused by the voltage drop due to lateral current flow debiasing the far collector.

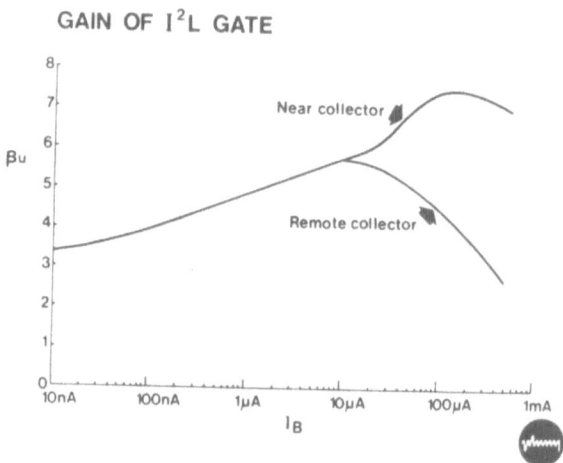

Fig. 11 - Gain of I²L gate

The main virtues of I²L are high packing density, around 200 gates/mm², and low power delay product, around 1 pJ. The main drawback is the fact that both p-n-p and n-p-n transistors saturate so that charge is stored in all regions of the device. Stored charge in the epitaxy region beneath the remote collector has to travel through the base resistance to the contact so that discharge of this collector will be slow. This gives rise to a differential delay, depending on which collector used, a significant problem for circuit designers. This effect can be reduced at some expense in area by providing a low resistance 'ladder' diffusion around the collectors.

2.6   Progress in Speed and Complexity

The progress in the speed performance of bipolar logic from the first integrated circuits to modern ECL is summarized in Fig. 12. Starting with the early Fairchild RTL and 50nS gate delay, this shows the progression through DTL, TTL and then a series of ECL developments extending down to 250 pS. The fastest Schottky TTL at 1.5nS lies somewhat above the line, being introduced around 1978. Note the competition from gallium arsenide circuits, which at the moment is potential rather than actual as this technology has not yet reached production status.

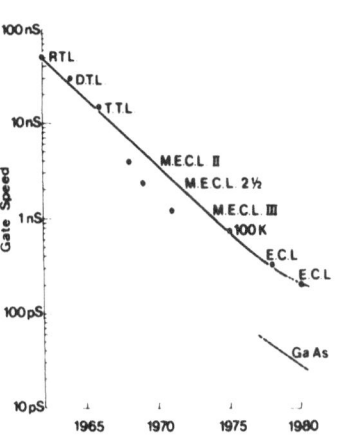

Fig. 12 - Logic performance

On the same time scale complexity has increased from 6 or 8 gates per chip to several thousand gates per chip. In the race for even higher scales of integration, bipolar technology has been overtaken by MOS with 20K and more gates per chip.

3.   AN ION IMPLANTED BIPOLAR PROCESS

The ion implanted bipolar process described in this section was developed primarily for making subnanosecond ECL, although it has also been used for high performance linear circuits for use up to 1GHz, data conversion products and I^2L. The development was based on an earlier all diffused process with an npn transistor f_T of 2.5 GHz producing 1nS ECL gates. In its initial form the new process used the same 4 micron minimum feature size as the all diffused process.

In the new process ion implantation is used as a controlled low-temperature deposition process followed by a conventional drive in. In other words the final profile is not controlled by ion implantation, but is a combination of ion implantation substantially modified by subsequent heat treatments.

The main changes from the earlier process are:-

1.   Ion implanted arsenic replaces diffused phosphorus for the emitter.   The concentration dependent diffusion coefficient of arsenic results in steeply graded profile, in marked contrast with the shallow phosphorus profile.

2.   The base and p^+ regions are ion implanted to improve control.

3.   The epitaxial layer thickness is reduced from 3.5 to 2.6 microns.

4.   Photoresist masking is used to define implanted regions.

The process sequence, shown in outline in Figs. 13-20, is as follows:

1.   First oxidation, 5000 Å on 5 ohm cm P type - (100) substrate
2.   Buried N^+ photoengraving
3.   Buried N^+ deposition - arsenic from a spin on source
4.   Buried N^+ drive in - 10 ohms per square
5.   Epitaxial layer growth - 2.6 microns of 1.2 ohm cm n-type
6.   Second oxidation - 5000Å
7.   Isolation photoengraving
8.   Isolation deposition - implanted boron - 50 ohms per square
9.   Isolation drive in - regrow oxide
10.   Collector sink photoengraving
11.   Collector sink deposition and drive - phosphorus 2 ohms per square
12.   P window photoengraving
13.   P clearance photoengraving - print resist only
14. p^+ boron implant - 65 ohms per square
15.   Resist removal
16.   Boron Base implant - 550 ohms per square
17.   Silox deposition - 4000Å
18.   Contact photoengraving
19.   N clearance photoengraving - print resist and etch aluminum implant mask
20.   Arsenic Emitter implant - 20 ohms per square
21.   Resist removal
22.   Drive in emitter, base and p^+.
23.   Metallisation magnetron sputtered 94.5% Al 1.5% Si 4% Cu - 1 micron thick

24.   Passivation

25.   Photoengrave bonding pads

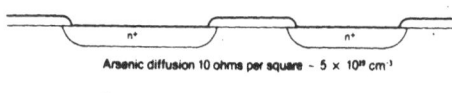

Fig. 13 - Ion implanted bipolar process
(wv) - buried n$^+$

Fig. 14 - Ion implanted bipolar process
(wv) - epitaxy

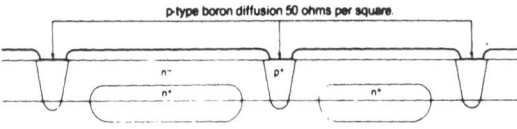

Fig. 15 - Ion implanted bipolar process
(wv) - isolation diffusion

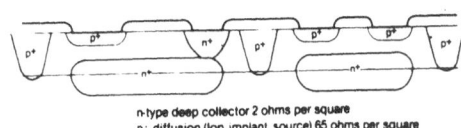

Fig. 16 - Ion implanted bipolar process
(wv) - deep collector, p$^+$ diffusions

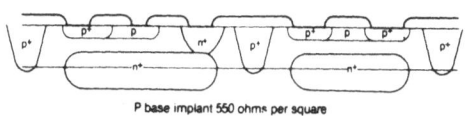

Fig. 17 - Ion implanted bipolar process
(wv) - base implant, deposited oxide,
contacts photoengrave

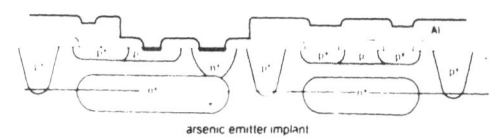

Fig. 18 - Ion implanted bipolar process
(wv) - select n contacts - aluminum,
emitter implant

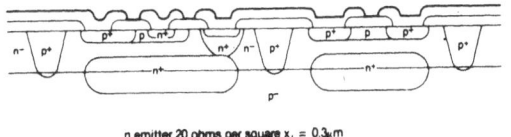

Fig. 19 - Ion implanted bipolar process
(wv) - drive-in and metallise

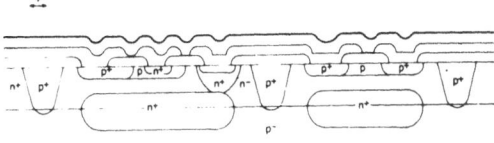

Fig. 20 - Ion implanted bipolar process
(wv) - passivate

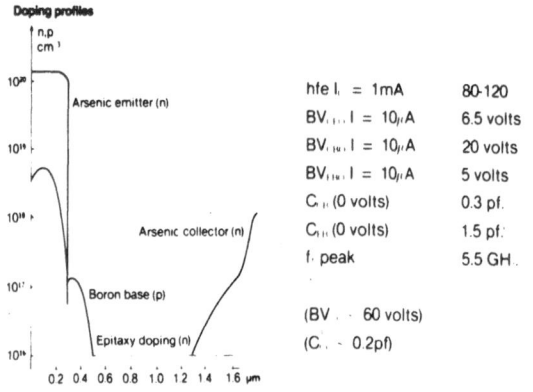

hfe I. = 1mA	80-120
BV. ,.. I = 10,A	6.5 volts
BV. ,,. I = 10,A	20 volts
BV. ,,,. I = 10,A	5 volts
C. ,, (0 volts)	0.3 pf.
C. ,, (0 volts)	1.5 pf.
f. peak	5.5 GH.
(BV . . 60 volts)	
(C. . 0.2pf)	

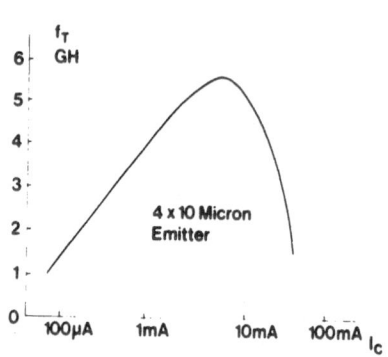

Fig. 21 - Ion implanted bipolar process (wv)

Fig. 22 - Transistor characteristics (typical values)

Fig. 23 - Ion implanted bipolar process (wv) - f_T vs. I_C

The concentration profile through the n-p-n transistor is shown in Fig. 21. The extremely steep arsenic concentration gradient is the main factor in doubling the f_T compared with the all diffused process.

The main characteristics of the n-p-n transistor are given in Fig. 22. The variation of f_T and H_{FE} with collector current for an n-p-n transistor with a 4 x 10 micron emitter are given in Figs. 23 nd 24.

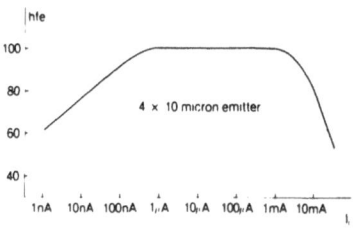

Fig. 24 - Ion implanted bipolar process (wv) - hfe vs I_c

4.   CURRENT STATUS BIPOLAR LOGIC

4.1   Market Share

In order to put the current status of bipolar logic into perspective, it is useful to survey the total digital market. The market share estimates shown in Fig. 25 are intended as a rough indication only, and are based on various published[4] and unpublished sources. Accurate data on the major captive producers (e.g., IBM) is not available, and categorization is not always easy or consistent.

MOS	68 %
TTL	26 %
ECL	3 %
Other Bipolar	3 %

Major Growth Areas –
MOS Memory and Microprocessors

Fig. 25 - 1981 digital IC market (includes logic, memory and microprocessors)

The main feature is the predominance of MOS, with 68% of the market; MOS continues to show high growth rate in both absolute and market share terms. In particular, new mass markets for MOS memory and microprocessors have been developed over the last decade and these are certain to be major growth areas for the next decade.

In the 32% share held by bipolar circuits, TTL with 26% is the dominant logic family. This figure could be somewhat misleading in that many LSI circuits, e.g., random access memories, are classified as TTL when TTL interfaces are provided, even if the internal circuitry is different (e.g., ECL).

4.2   TTL - The Dominant Standard Logic Family

The development of Schottky TTL has helped to expand the range of TTL applications. With a catalogue running in many hundreds of parts, ranging from simple gates to bit slice microprocessors, 4 bit arithmetic logic units (ALUS) and high speed memories, plus a variety of speed/power options, TTL is **the** general purpose standard logic family. The combined effects of many years of volume production experience, and competition from the multiplicity of vendors, produces a "buyers' market." In addition, the 15 years of accumulated user experience provide a built-in demand.

The TTL logic families now available are summarized in Fig. 26. In addition to the standard and low power Shcottky ranges introduced some years ago, several new high performance TTL families have been developed recently. Smaller device geometries, ion implantation and improved circuit design features making more use of Schottky have been com-

Family	Power per gate. MW	Gate delay. NS	Power delay product. pJ
Standard. TI 54/74	10	10	100
Low power. TI 54/74L	1	35	35
Schottky. TI 54/74S	20	3	60
L P Schottky. TI 54/74LS	2	10	20
Fast. Fairchild	4	2	8
LS². National	2	5	10
Advanced Schottky. TI 54/74AS	22	1 5	33
Advanced L P. Schottky. TI 54/74ALS	1	4	4

Fig. 26 - TTL logic families

bined to produce gate delays down to 1.5nS (Advanced Schottky - TI 54/74 AS) and power delay products down to 4 pJ (Advanced L.P. Schottky - TI 54/74 ALS). Whilst these new ranges will undoubtedly help extend the life of the TTL family, the increasing competition from other approaches must be recognised.

In terms of packing density and, hence, cost per function, I^2L, NMOS and CMOS have a substantial advantage. Modern CMOS processes in particular are starting to match most of the features of TTL with higher packing density, lower speed power product and the unique CMOS feature of automatic powering down in the stand by mode. If sheer speed is the primary requirement then the ECL still outperforms TTL by about a factor 4. The advantages and limitations of TTL are summarized in Figs. 27 and 28.

Over 350 functions available
Basic gates ———→ 4 bit A.L.U.
Range of speed / power options
High volume, multi-sourced production
~ 15 years of accumulated user experience

Low packing density (cf I^2L, MOS)
x 4 slower than E C L
Power - delay product ~ x 10 greater than I^2L
No ultra low power stand by mode as provided by CMOS

Fig. 27 - TTL logic - advantages

Fig. 28 - TTL logic - limitations

4.3 ECL and the Gate Array Trend

Any standard logic family suffers from the problem that once a reasonable range of small and medium scale integration (S.S.I. and M.S.I.) parts has been developed, plus a few obvious large scale integration (L.S.I.) parts like random access memories, there is semi-infinite potential variety of application specific or custom designed L.S.I. functions. This first became apparent in the main frame computer market as manufacturers strove to increase operating speed and to maintain and develop unique system architectures. As basic gate speeds increased, the finite time delay introduced by interchip connections became more significant. In addition, the higher logic swings required to maintain system noise immunity and the greater capacitance of interchip wiring required higher power dissipation. This in turn aggravated the already difficult thermal management problem.

The obvious solution is the integration of more functions on a single chip, using lower on chip logic swings and partitioning the system to maximize the ratio of logic functions to pins. Unfortunately this generated a large number of unique custom designs, each required in low production volumes. In fact this variety of types and characteristic low volume exactly mirrors the printed circuit boards these custom integrated circuits are intended to replace. The cost and long timescales of the full custom design approach, plus the shortage of experience I.C. designers prevent this being the general answer. This has led to a variety of "semi-custom" design methods. Historically, one of the first approaches was the TI "discretionary wiring" concept. Partly to overcome yield problems, circuit elements were pre-tested and then a unique second layer interconnection pattern to produce the required circuit was devised by computer. This concept never progressed beyond the R & D stage.

Field programmable read only memories (PROMS) can be considered as an alternative to custom logic in certain cases. The same programming techniques, fusing metal or polysilicon links, are now finding more applications in field programmable logic arrays.

The main contender for the semi-custom logic market at this time is the gate array (see Ref. 5). In mainframe and large minicomputers, where the highest possible speed is required, this means ECL or similar current steering logic with two or more layers of metallisation. Multilayer metallisation is essential to achieve high-density interconnections without the speed penalties of alternative forms of crossover. In addition, the extra layer(s) of metallisation ease the circuit layout problem and hence reduce design turnaround time.

A reliable high yield multi-layer metallisation process has been developed as an add-on feature to ion implanted bipolar process described in Sec. 3. The process used the same magnetron sputtered first layer

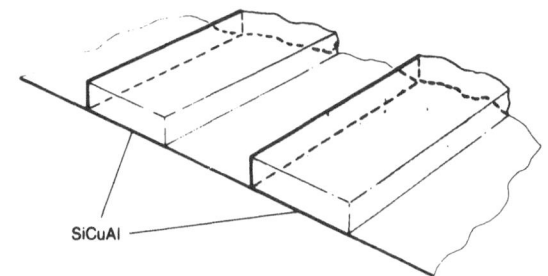

SiCuAl

Fig. 29 - Multi-level metallisation - 1st level

metal alloy of aluminum, silicon and copper as the standard process (Fig. 29). The inclusion of copper suppresses the formation of hillocks which can rupture the interlayer dielectric, as well as improving resistance to electromigration. The inter-layer dielectric is polyimide, an organic material which is now available in a highly pure semi-conductor grade and is stable up to 450°C (Fig. 30). The poly-imide provides as smooth surface for the second, identical metal layer (Fig. 31). The polyimide-metal sequence can, in principle, be repeated to give any required number of interconnection lev-els. In practice, the process has been demonstrated up to four levels and most circuits use only two. A stereoscan picture of a via connection in a two-level cir-cuit is shown in Fig. 32.

During the development of the process, a series of tests on the electromigration resistance of various systems was carried out with the results shown in Fig.

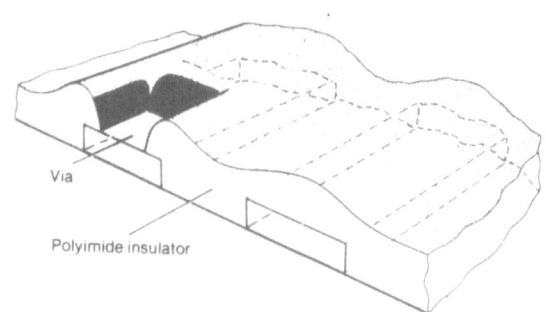

Fig. 30 - Multi-level metallisation - interlevel isolation

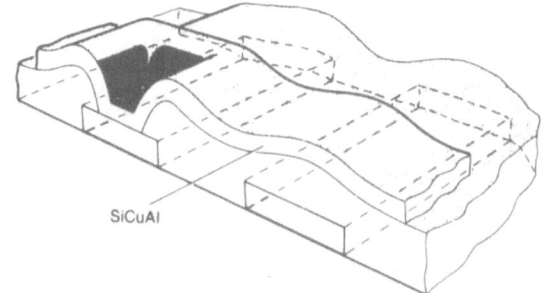

Fig. 31 - Multi-level metallisation - 2nd level

Fig. 32 - Multi-level metallisation - SEM picture

33. The test tracks run over steps, and for a given alloy the step resistance R_S is a good indicator of the electromigration properties of the system. The mean time to failure of the

preferred system is 7700 hours at 180°C and at 15 ma micron^{-2} this corresponds to a mean life-time of 200K hours at 180°C under normal operation condi-tions of 3 ma micron^{-2}.

Using this multilayer proc-ess a family of ECL gate arrays has been developed. These are based on a minor cell containing

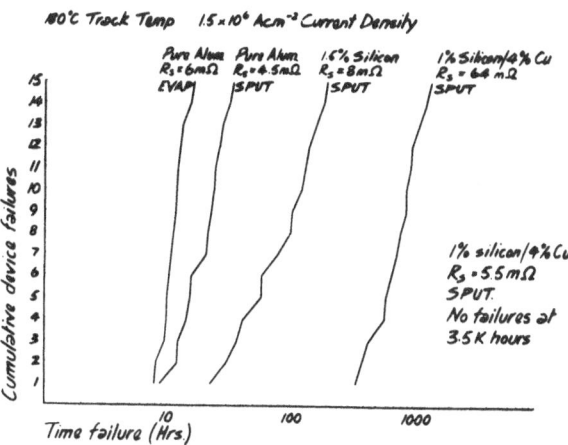

Fig. 33 - Electromigration vs. composition

3 dual emitter transistors, 3 normal transistors and 8 resistors arranged for, but not connected as, an ECL gate. Four of these cells form a major cell and in the simplest member of the range, there are 36 cells (see Fig. 34). The logic power of this array is equivalent to about 100 simple gates. The array has 28 bonding pads, each having a buffer transistor available capable of driving a 50 ohm line at ECL logic levels. Gate delays are around 500 picosecs. for internal gates. This is increased by 80 picosecs. for the emitter dot function and by 200 picosecs. for collector dot function.

A larger version using 144 of the same cells in a 64 pin package has also been produced. This array, which is equivalent to 400 simple logic gates, is available in three speed versions, offering gate delays 0.5nS, 1nS and 2nS at power levels of 3 watts, 1.2 watts and 750 mW, respectively. For higher levels of complexity a multilevel gate array is being used to provide 700 gates.

As part of the development of a reduced geometry process the 100 gate array has been reduced in size by 25%, i.e., from 4 to 3 minimum features (see Fig. 35). The performance of these two versions is compared in Fig. 36. The speed improvements are roughly propor-tional to the square of the shrink factor.

These ECL examples emphasize extreme speed, but the gate concept is now being exploited over a wide range of speeds, circuit types and technologies, including MOS.[5]

4.4 I^2L – Bipolar VLSI

The invention of I^2L was hailed by many as the bipolar answer to high density MOS logic. The initial promise has not bee fulfilled for a variety of reasons. Although I^2L can be made on almost any bipolar processes, to achieve only moderate worst case gate delays of around 25nS demands carefully optimised processing. The gain and speed variations with collector position referred to in

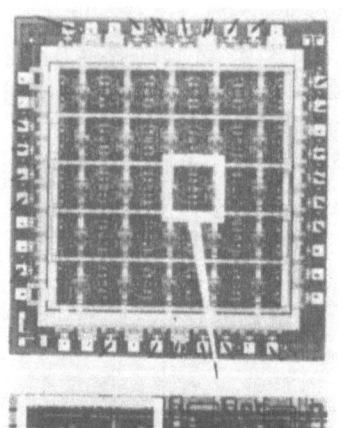

Silicon chip layout
0·150 inch square

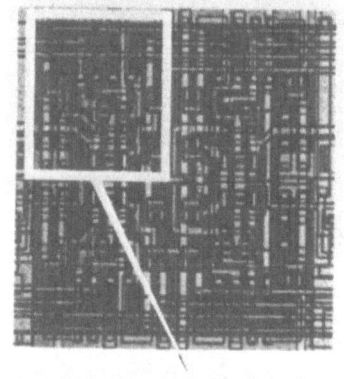

Major cell

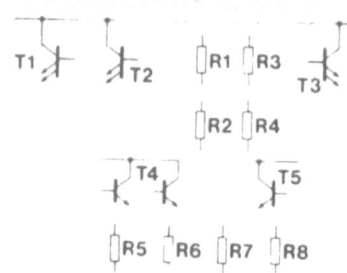

Minor cell components

Fig. 34 - ECL gate array

section 2.4 can be minimised, but at the expense of increased process complexity and reduced packing density. Another problem is the adverse effect of scaling on I^2L performance. The p-n-p transistor gain is partly a function of the collector to emitter area ratio, which is less than unity in the inverter structure. In shrinking device size edge effects to tend to make this ratio reduce still further. Furthermore, as I^2L is a form of saturating logic, storage effects limit the speed potential.

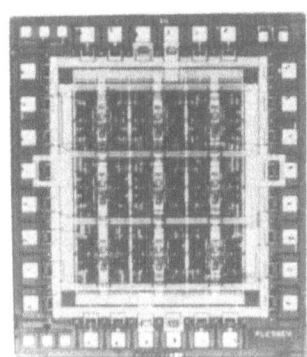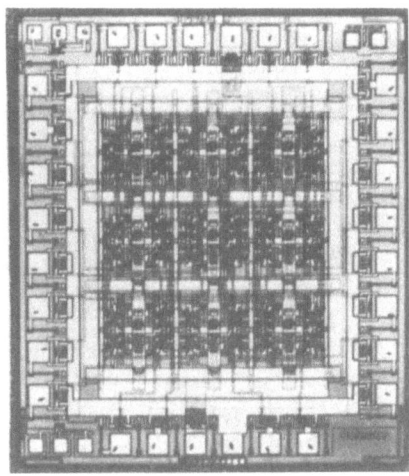

Fig. 35 - 3/4 size and full size gate array chips

Despite these limitations some L.S.I. devices have been produced in I^2L. Typical examples are the T.I. SBP9900 16 bit microprocessor and the Fairchild 16 bit 9440 microprocessor. Compared with the NMOS technology used in the majority of microprocessors, I^2L

	4μ Geometry	3μ Geometry
Intrinsic gate delay	500 p secs	375 p secs
Wire 'OR' delay	80 p secs	40 p secs
Wire 'AND' delay	200 p secs	100 p secs
Maximum Flip-Flop clock rate	500 MHz	950 MHz

Fig. 36 - Performance of full size and 3/4 size gate arrays

has the advantage, being a bipolar technology of wider temperature range and better resistance to radiation, both important in military applications. The radiation hardness and general properties of commercial NMOS and I^2L processes are compared in the table at 4 - 5 micron geometries.

	I^{2}L	NMOS
Packing Density Gates/mm^2	200	150
Minimum Gate Delay	~ 20 nS	20 nS
Power Delay Product	.7-2 pJ	3pJ
Neutrons n/cm^2	~ 3 x 10^{13}	10^{15}
Total Ionizing Dose Rads (Si)	3 x 10^6	10^4-10^5
Transient Dose Rate Upset Rads (Si)/S	> 10^9	10^5
Transient Dose Rate Survival Rads (Si)/S	~10^{10}	10^5

Various attempts have been made to improve on the basic I^{2}L concept. For example, Plessey developed "Substrate Fed Logic" (SFL) in which a double epitaxial layer process was used to form a vertical p-n-p.[6] Schottky diodes were added to the input to give a multi-input multi-output gate. Although working LSI circuits were made, the process complexity limited possibilities for scaling and speed limitations common to other forms of I^{2}L resulted in the abandonment of this process.

Other forms of Schottky I^{2}L have been described in the literature, but one of the most promising new bipolar LSI structures, Integrated Schottky Logic (ISL) used a different principle.[7] In ISL the n-p-n transistors are used in the normal mode, the Schottky diodes to reduce the voltage swing. This gives higher speeds, but lower packing density than I^{2}L.

On current evidence there does not seem to be a bipolar logic form to seriously challenge MOS in the VLSI stakes. However, I^2L is very useful in mixed function circuits as high density, low performance logic, compatible with ultra fast ECL and analogue circuits.

4.5 Data Conversion

Many of the inputs and outputs of total electronic systems are analogue rather than digital. The increasing sophistication and lower costs of digital processing is leading to the replacement of many analogue functions by digital equivalents, e.g., digital audio recording and digital T.V. studio equipment. The analogue to digital and digital to analogue converter (ADC and DAC) functions are becoming increasingly important. As with the logic area, MOS devices predominate at low speeds, with bipolar in high speed applications.

The three basic approaches to high speed data conversion, all parallel, parallel series and successive approximation, are summarized in Fig. 37. A four-bit all-parallel circuit, A-D converter capable of operating at over 100MHz in 8 bit systems is shown in Fig. 38.[8] The device, the SP9754, is made on the ionimplanted bipolar process described in Sec. 3.

The same process has been used to make 8[9] and 10 bit DAC's. The principle of operation, shown in Fig. 39, is the use of equal current sources. All devices in the array are matched and switch equal currents at equal current densities, and therefore achieve very closely matched speeds (Fig. 40). This is not generally possible in R-2R

ALL PARALLEL:	
ADVANTAGES	DISADVANTAGES
High speed	Large no. of comparators
Self sampling	

PARALLEL SERIES	
ADVANTAGES	DISADVANTAGES
Uses fewer comparators	Lower speed
Can be optimised for speed/hardware	Analogue design more difficult

SUCCESSIVE APPROXIMATION	
ADVANTAGES	DISADVANTAGES
Simpler design	Lower speed
Fewer comparator C.F. parallel	

Fig. 37 - Data conversion methods for high speed

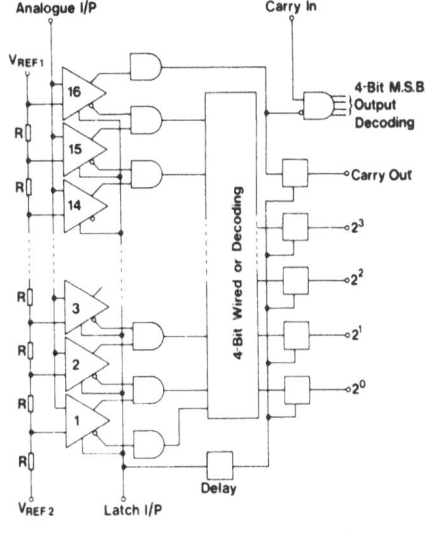

Fig. 38 - SP9754 4-bit expandable ADC

ladder networks, where each stage operates at half the current of its predecessor. The 8 bit DAC, SP9758, has a settling time of 5nS to $\pm 1/2$ LSB into a 50 ohm load.

The 8 bit DAC is also used at basis for a 15 MHz 8 bit ADC[10] as shown in Fig. 41.

One important high volume application of data conversion is at the interfaces between analogue telephones and digital switching and transmission systems (Fig. 42). By using the same ion implanted bipolar process, a multichannel CODEC has been developed.[11] The original version used a separate NMOS logic control circuit, but by using I^2L the logic and analogue function have been integrated into a single chip 8 channel CODEC. The block diagram of the circuit is shown in Fig. 43 and the performance is summarized in Fig. 44.

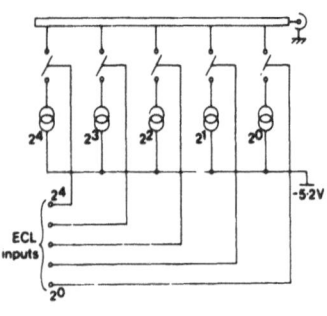

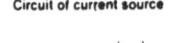

Circuit of current source

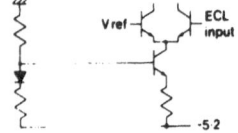

Fig. 39 - Schematic diagram of 5-bit DAC

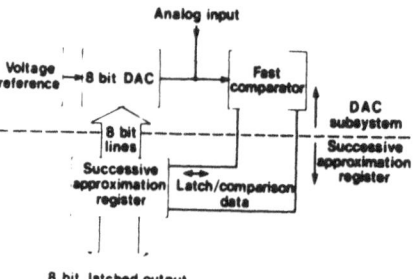

Fig. 40 - Successive approximation analogue to digital conversion

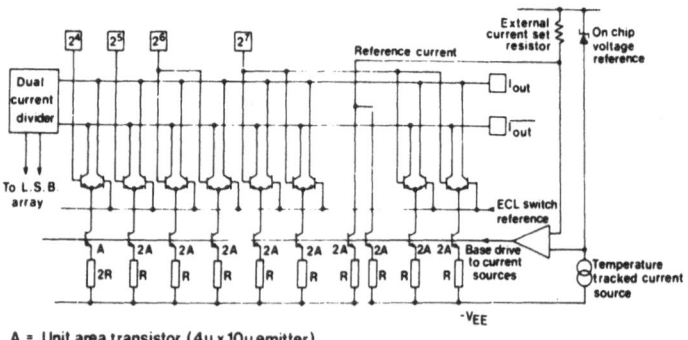

A = Unit area transistor ($4\mu \times 10\mu$ emitter)
R = Unit value resistor (400Ω)

Fig. 41 - Current source array

FUNCTION   Converts analog speech signals into a series of 8 bit digital codes

PURPOSE   Interfaces the subscribers analog signal to the digital format used within the exchange

CONSTRAINTS   1 Demanding specification on noise, linearity and stability

          2 Low power consumption per subscriber

          3 Low cost per subscriber

Fig. 42 - CODEC - voiceband coder/decoder

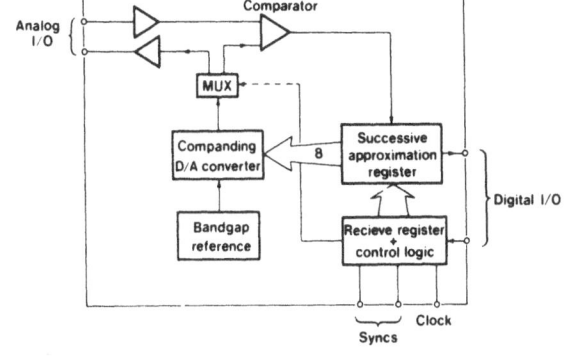

Fig. 43 - Single chip multi channel CODEC block diagram

Technology	High speed bipolar/I^2L
No. of channels	Encodes/decodes up to 8 channels
Supply	± 5 V
Power	200 mW (25mW/channel)
I/O Data rate	2.048 Mb/s
Linearity & distortion	Exceeds CCITT G711/712 specifications
Chip size	100 x 130 thou. (1625 sq. thou/channel)

Fig. 44 - Single chip multi channel CODEC performance

5. CONCLUSIONS - THE FUTURE ROLE FOR BIPOLAR

1. For Low speed logic, below about 1MHz, MOS will continue to expand, eventually eliminating bipolar circuits. Many systems built in the past from TTL or SSI/MSI CMOS have already been replaced by software programmed MOS microprocessors. The availability of lower cost single chip microcomputers which are easy to program, will accelerate this trend.

2. The Medium speed, general purpose standard logic market currently dominated by TTL will also fall to MOS. Improved MOS speeds will be combined with gate array and other custom and semi-custom design techniques to gradually replace standard TTL. Improved CAD has already brought the cost of a semi-custom LSI design down too the same level as that of the PCB it replaces, thus removing one obstacle to on chip integration.

3. Bipolar logic in the form of ECL gate arrays will continue to provide the highest speed logic for demanding applications in computing, signal processing, data transmission and instrumentation.

4. Small ultra fast bipolar memories will continue to be used, but for large (4K) and above memories MOS will maintain market dominance. Indeed, as MOS processes are sealed down in size, large fast bipolar memories will be displaced.

5. In the important areas of data conversion and other mixed function products bipolar will continue to be used. High density bipolar logic like I^2L will find its main application here, rather than as a direct competitor to MOS.

REFERENCES

1. "Analysis and Design of Integrated Circuits" Meyer, Lynn and Hamilton (McGraw-Hill).

2. K. Hurt and A. Slob, IEEE J. Solid State Circuits **SC-7**, 346, (1972).

3. H. H. Berger and S. K. Wiedmann, IEEE J. Solid State Circuits **SC-7**, 340, (1972).

4. "World Market Forecast," Electronics, 13 January 1981, pages 121-142.

5. "Gate Array - A Special Report," J. G. Posa, Electronics, 25 September 1980, pages 145-158.

6. V. Blatt, P. S. Walsh and L. W. Kennedy, IEEE J. Solid State Circuits **SC-10**, 336, (1975).

7. J. Lohstroh, "Performance comparison of ISL and I^2L", IEEE International Solid-State Circuits Conference 1979.

8. P.H. Saul, A. Fairgrieve and A. Fryers, "Monolithic Components for 100 MHz Data Conversion", IEEE J. Solid State Circuits **SC-15**, 3, June (1980).

9. P. H. Saul, P. J. Ward and A. J. Fryers, "An 8-Bit, 5ns Monolithic D/A Converter Subsystem", IEEE J. Solid State Circuits **SC-15**, 6, (1980).

10. P.H. Saul, "Successive Approximation Analog-to-Digital Conversion", IEEE J. Solid State Circuits **SC-16**, 3, (1981).

11. P. Schwarz, V. Blatt and C.C.A. Priest, "A Multi-channel CODEC and Filter Subsystem", ISSCC 79, February 14, 1979, pp. 30-31. in 0

ADVANCED BIPOLAR DEVICES AND RELATED PROBLEMS

K. Kimura and T. Takahashi

Nippon Electric Company, Ltd.
Kawasaki, Japan

1. INTRODUCTION

There exists an increasing demand for new data processing systems which feature high performance, smaller system size, low cost and high reliability.

To fulfill these requirements, highly integrated bipolar LSI with subnanosecond gate delay is indispensable especially in the field of medium and large scale computer mainframe application. The large-computer manufacturers generally prefer custom designed devices. The middle approach uses the bit-slice to make up the major block of the system with a relatively large gate-array to create uniqueness.

Table 1 shows the recently announced bipolar LSI's in production and laboratory level.[4-10] Bipolar LSI's with up to 1,000 gates integration level are already available on volume production basis. Furthermore, more advanced bipolar LSI's with up to 12,000 gates integration level are recently announced at the international conference during these two

Table 1 - Recently announced bipolar LSIs.

Production Level

Category	Code	Mfr.	Function
Logic	AM2903	AMD	4-bit CPU
	TDC1010J	TRW	16 X 16 bit Multiplier
	MAC-1	MOTOROLA	Gate Array
	8A2000	SIGNETICS	Gate Array
	TAT008	TI	Gate Array
	B450D	NEC	9kbit FPLA
Memory	10470	FC	4kbit RAM
		HITACHI	4kbit RAM
		FUJITSU	4kbit RAM
	MB7141	FUJITSU	32kbit PROM
	MM76320	HARRIS	32kbit PROM
	82S321	SIGNETICS	32kbit PROM
Linear	AD1140	AD	16bit A/D Converter
	TDA1540	PHILIPS	14bit D/A Converter

Laboratory Level

Function	Mfr.	Performance
370/138 MPU	IBM	5,000 Gate
Convolber	TRW	10,000 Gate
32bit CPU	NTT	12,000 Gate
16kbit RAM	HITACHI	25nS, ECL
16kbit RAM	IBM	45nS, TTL
18kbit DRAM	IBM	75nS, TTL
64kbit DRAM	IBM	50nS, TTL

years. They are still on a trial production basis now, but will sure to be used practically in the near future. It seems that the device technology is running ahead of the system designer's capability to realize it. In high speed computer systems, it has generally been accepted that the overall gate delay within a complete system is made up of two constituent parts, the intrinsic gate delay and the interconnections delay. For example, in the conventional systems using ECL-10K devices and multi-layer printed circuit board technology, the total system gate delay is typically 4 nsec. which is made up of 2 nsec. due to the gate itself and 2 nsec. due to interconnections. Although ECL-100K is available offering 0.7 nsec. gate delay at the same level of integration, it is difficult for the system manufacturer to improve his overall perform-ance without improving his interconnection technology.

It is clear that the best solution for these problems is to increase the level of integration per chip. However, this approach brings the following problems when conventional ECL technology is applied.

First, the power dissipation per chip increases significantly since the ECL gate relatively dissipates a lot of power. This high power requires an expensive special cooling system. Second, chip size is also increased since the conventional planar transistor structure requires a relatively large area for its location. This large chip size decreases the chip yield significantly resulting in increased chip costs.

To overcome these problems and realize improved bipolar LSI's for the next generation, the power dissipation per gate should be decreased without increasing the gate delay time. In other words, the power delay-product of a gate must be decreased. A smaller sized transistor with higher performance is required for this improvement. Such a sophisticated bipolar transistor can not be realized without improved process technology. Consequently, some advanced bipolar device technologies are trying to be developed.

The following section will first discuss the limiting device parameters for a CML gate delay, and the advanced bipolar device technologies proposed recently. Then a more detailed study will be done citing Isoplanar and PSA processes.

Finally, the related problems are cleared and the performance of the bipolar logic in the future will be estimated assuming that the device scaling technique is applied.

2. LIMITING FACTORS FOR SPEED IN CML

Recently, the performance of the MOS device has improved rapidly by applying device scaling technique. They have achieved a performance that was once considered possible only with the bipolar process. The bipolar TTL device which has a few nanoseconds gate delay may be replaced by the advanced MOS device in the future. Therefore, the subnanosecond delay performance for bipolar devices should be kept even if the level of integration is improved.

Although some modified DTL type logic such as ISL and STL are proposed recently with advanced level of integration, the gate delay time is still in a few nanosecond range. In this point of view, in this section, the gate delay time is estimated by considering the CML as the circuit type.

It is important to know the limiting factors for delay time in the device when performance is tried to be improved. Propagation delay time for a basic CML can be estimated using the expression shown in Fig. 1. The first term represents the base time constant composed by a base resistance and the equivalent input capacitance. This input capacitance consists of junction capacitances including the Miller-Effect and diffusion capacitance which evaluates the transit excess carrier in the base region. The second term represents the collector time constant composed by a collector load resistance and load capacitances which consist of isolation and wiring stray capacitances and equivalent input capacitance of the next stage. The first term is almost independent of the gate current but depends on the inherent device parameters. The second term depends on the gate current level since the value of a collector load resistor is inversely proportional to the gate current. These expressions indicate that decreasing the stray and input capacitance is important to improve the delay performance when the level of integration is increased. The next section will introduce the effective device technologies for this improvement.

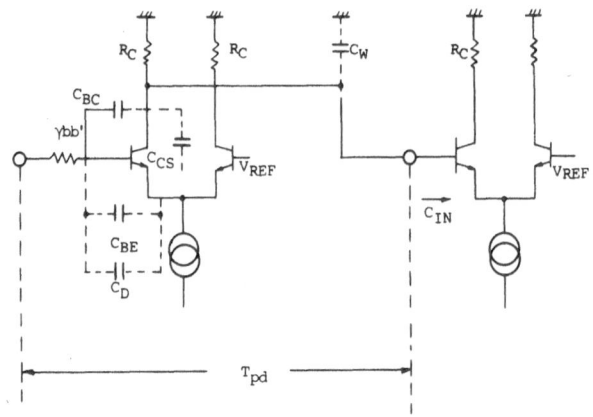

$$T_{pd} \propto \tau_B + \tau_C$$

$$\tau_B \propto \gamma_{bb'} (C_{BE} + G \cdot C_{BC} + C_D)$$

$$= 0.7 \cdot \gamma_{bb'} \left(\frac{C_{BE}}{2} + 2 \cdot C_{BC} + \frac{1}{2\pi f_T R_C} \right)$$

$$\tau_C \propto R_C (C_{CS} + C_W) + R_C \cdot C_{IN}$$

$$= 0.7 R_C (C_{CS} + C_W)$$

$$+ 0.7 R_C \left(\frac{C_{BE}}{2} + 2 C_{BC} + \frac{1}{2 f_T R_C} \right)$$

$$\therefore \ T_{pd} = 0.7 \gamma_{bb'} \left(\frac{C_{BE}}{2} + 2 C_{BC} + \frac{1}{2 f_T R_C} \right)$$

$$+ 0.7 R_C \left(C_{CS} + C_W + \frac{C_{BE}}{2} + 2 C_{BC} + \frac{1}{2 f_T R_C} \right)$$

Fig. 1 - Typical CML gate schematic and expressions for tpd estimation.

3. ADVANCED DEVICE TECHNOLOGIES FOR IMPROVEMENT

Particular processing techniques and overall process flows have been developed to improve the performance limitations discussed above as well as the other fundamental issues of yield and reliability.[3] Major approaches are listed below.

1.   Oxide isolation

2.   Shallow junction technology

3.   Fine-line lithography

4.   Self-aligning technology

Oxide isolation is a new isolation approach, in which the active p-type diffusions that isolate conventional bipolar devices are replaced by passive insulator-oxide regions; there is no need to separate the isolation region from the transistor base. Hence the oxidized isolated device achieves a considerable size reduction over its diode-isolated counterpart.

As a result, collector substrate stray capacitance due to junction isolation can be significantly reduced. Technology for a thin epitaxial layer is indispensable when oxide isolation is formed actually. Various approaches such as Isoplanar, CDI, VIP and V-ATE are presented. Details of the oxide isolation will be introduced citing the Isoplanar process in the following section.

Shallow junction technology is effective for the following two improvements. First, the base emitter and base collector junction capacitances are decreased due to a reduced side junction area. Second, improved f_T values can be achieved by the following advanced technologies. Ion implantation is an essential tool for realizing high-performance vertical impurity profiles. This process enables the exact control of junction depth, and contributes to the shallow junction formation. Both electron beam annealing of ion-implanted regions and low-temperature high-pressure oxidation hold the promise of enabling the fabrication of a very thin base region with a low density of emitter-collector defects. It is well known that the f_T value also depends on the amount of free carrier storage in the emitter space charge layer. This storage decreases when the doping gradient in the junction is increased. Higher values of f_T can be obtained by the use of arsenic, instead of phosphorus, as an emitter diffusion impurity. This improvement is attributed to the absence of a push-out effect and the increased doping gradient of the arsenic emitter.

Fine-line lithography is another important approach for improving the level of integration. Fine-line lithographic systems, such as deep UV printing, electron-beam writing and X-ray will make smaller lateral geometries possible with very low defect densities. In addition, advanced processing techniques and equipment for plasma etching and ion-beam milling will allow the fully automatic dry processing.

As a result, junction capacitances can be significantly reduced due to smaller lateral geometries of a transistor. Fine-line lithography will also decrease metal width and spacing.

This reduction results in lower wiring capacitances and increased density by allowing device structures to be packed more closely. Consequently, on chip interconnection delay can be decreased.

Self-aligning technology is also a powerful approach for increasing the performance of a basic bipolar transistor by reducing the intrinsic collector base capacitance and shrinking the lateral geometries. When this method is applied, masks for oxidation of a silicon layer also fix the other basic patterns of all IC elements such as electrodes, transistors and interconnections. Therefore, the other masks do not require exact alignment. A finer pattern and a smaller transistor can be realized easily using conventional rules which require no critical processing technologies resulting in higher yield. Practical application of the self-aligning technology is introduced with details citing the PSA process in the next section.

4. ISOPLANAR AND PSA PROCESS

In order to realize the bipolar device with improved performance, various unique process sequences have been presented by applying the advanced processing technologies as shown in the previous section. Two of the powerful, and at the same time different, approaches are the Isoplanar process, presented by Fairchild, and the PSA process, presented by NEC. The following will describe in detail these two technologies.

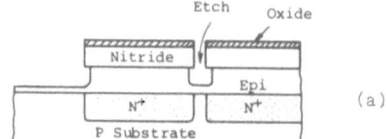

Cross section of Isoplanar II devices during the isolation process (isolation etch).

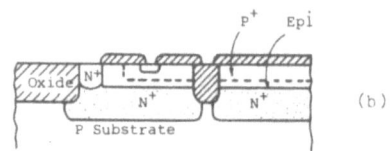

Cross section of Isoplanar II devices just after the emitter diffusion process.

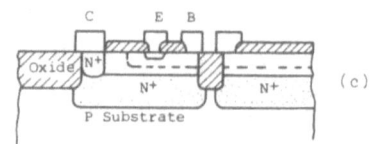

Cross section of completed Isoplanar II devices after definition of metal iterconnection.

1. Isoplanar

Fig. 2 - Cross section of Isoplanar-II transistor.

Isoplanar is basically a kind of oxidized isolation technology. The original version called Isoplanar-I was developed first. Then the more sophisticated process called Isoplanar-II[1] was developed by applying a walled emitter structure which is effective in reducing transistor size. A brief fabrication procedure of an Isoplanar-II transistor is shown in Fig. 2.

As in the Isoplanar-II process, n^+ buried collected diffusion has been made into a p-substrate. Then a thin n-type epitaxial layer is grown over the wafer surface. Next, a layer of silicon nitride and thin oxide is deposited. These layers are masked to define the isolation, and in turn provide a mask for etching away part of the epitaxial silicon layer. The silicon nitride does change into silicon dioxide, only very slowly, leaving plenty of time for oxide to be grown. Areas now stripped of nitride are etched still more deeply, right through to the buried layer, in order to reduce the surface step heights of the thick isolation oxide.[a] Next follows a long low-temperature oxidation, which fills the deeply etched areas with isolation oxide, but leaves the nitride-covered areas unoxidized. A deep n^+ diffusion contacts the buried collector and provides low collector resistance.[b] The base diffusion, indicated by the dashed line, can also be used to make p-type resistors. Oversized masks are used for easier mask alignment. Finally, emitter diffusion and all the electrodes are formed.[c] Because the emitter size is defined by the coincident area of the mask and the silicon island in Isoplanar-II technology, the minimum emitter size that can be easily produced is smaller than with conventional technologies. Comparison of later geometries and cross section of a transistor using conventional planar diffused isolation, Isoplanar-I and Isoplanar-II technology is shown in Fig. 3. The dashed line indicates the center of the isolation. Planar transistors achieve isolation between adjacent devices by a reversed biased p-n diode as previously described. The minimum distance between devices is determined by masking tolerances, two depletion layer widths and the width of the isolation p^+ region. The spacing between the base and the p^+ isolation diffusion, shown in the cross section, must be large enough so that the two depletion layers do not meet.

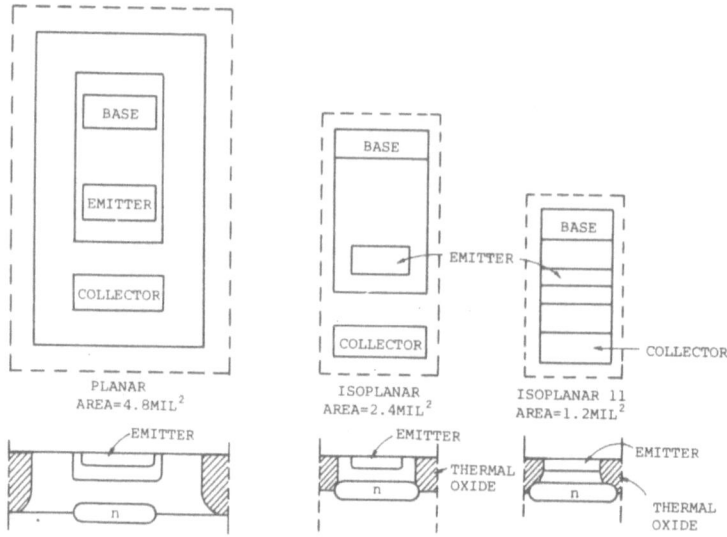

Fig. 3 - Comparison of Isoplanar and planar transistor.

In the Isoplanar process, a thick oxide is selectively grown between devides in place of the p^+ isolation region of the planar process. This insulating SiO_2 isolation needs no separation from base and collector regions, resulting in a substantial reduction in device and chip size. Both the base region and the base contact are designed to terminate at the isolation wall. However, the emitter is separated from the edge of the base.

In the Isoplanar-II process, the emitter ends terminate in the oxide wall of the isolation. This "Walled Emitter" structure allows reduction of the silicon area of an IC transistor of more than 70 percent compared to a conventional planar transistor and over 40 percent compared to an Isoplanar-I transistor. For a given emitter size suitable for internal circuits on a chip, the collector base area is reduced more than 60 percent when compared with both Planar and Isoplanar-I technologies. This reduced collector base junction area resulting in corresponding reduction in collector base capacitance has a significant impact on the high speed capabilities of bipolar transistors.

Since the active transistor is only the area under the emitter, all capacitance values of area outside the emitter, and resistance value of base, emitter and collector regions outside this area, are parasitic elements to be reduced. The reduction of collector base capacitance that can be achieved with the walled emitter structure is the result of elimination of inactive transistor area outside the emitter. This walled emitter transistor structure is the first significant improvement in collector capacitance since the invention of the planar transistor.

In the transistor design for the subnanosecond circuit, parasitics were further reduced by taking advantage of masking alignment latitude resulting from the self-aligning nature of the structure at almost every mask level. The advantage offered by reduction in capacitance was used to improve speed or reduce power dissipation of integrated circuits. While collector base capacitance reduction is the most significant aspect of this structure, the reduced collector substrate capacitance resulting from small device size is also important in high speed circuits. Consequently, gate propagation delay of 750 picoseconds can be attained with 3.3 pJ the power delay product by applying these advanced process technologies.

2. PSA Process

A new polysilicon process, called the polysilicon self-aligned (PSA) process described in this section, features a unique application of polysilicon to a self-aligned process.[2] This self-alignment process is realized by applying selective thermal oxidation technology for single crystalline silicon to a newly introduced polysilicon process. Small components and fine interconnections are easily obtained. The polysilicon layer also provides decreased-parasitic-capacitance resistors which replace standard single crystalline silicon resistors. This feature satisfies the requirement for high-value resistors for a low power individual gate without increasing the gate area and the parasitic capacitance. As a result, a gate with low power dissipation is attained without increasing delay time. In short, the PSA process has successfully achieved high packing density, high speed and low power.

PSA Concept and Fabrication Steps

The major difference between the PSA method and the conventional method lies in the processing sequence. In the conventional process, electrodes and interconnections are

constructed with aluminum after the formation of all p-n junctions for transistors and resistors
(Fig. 4(a)). However, with the PSA method, they are constructed in a newly introduced
polysilicon layer by selective thermal oxidation technology using Si_3N_4 as a mask before
emitter base junctions are formed (Fig. 4(b)).

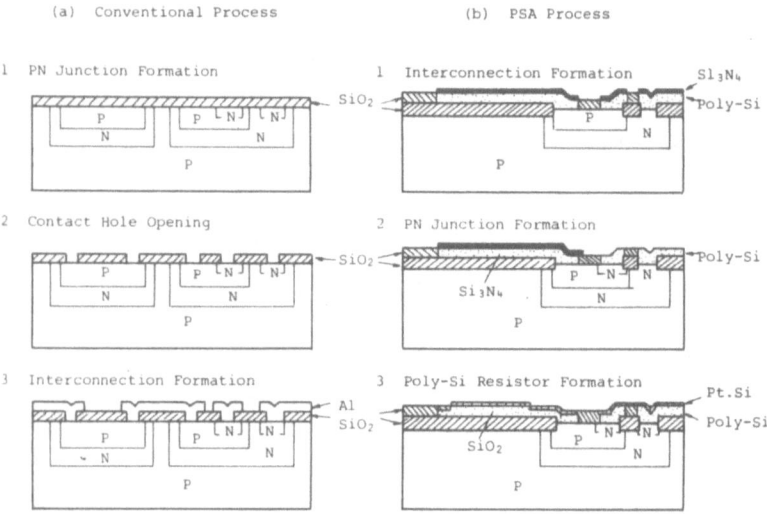

Fig. 4 - Comparison of PSA process and planar process.

Figures 5(a) to (e) show cross-sectional views of these new fabrication steps. The
starting material is a p-type substrate with 250 $\Omega/\square$ ion-implanted collector regions followed
by diffusion or with n^+ buried layer on which a 2 μm, 0.2 Ω/cm n-type epitaxial layer is
deposited. After p^+ channel stopper diffusion, the surface of the wafer is selectively oxidized
to 1 μm thickness using the nitride oxide sandwich structure as a mask to oxidation as shown
in Fig. 5(a).

Then as shown in Fig. 5(b), boron is ion-implanted into the base region with an acceleration energy of 100 keV and a dose of 2×10^{14} cm^{-2} using photoresist as a mask. After Si_3N_4
films are removed, the surface of the wafer is covered with non-doped CVD polysilicon to a
thickness of 5000Å. The selective thermal oxidation technology used for single crystalline
silicon, as shown in Fig. 5(a), is also applied to this polysilicon layer, as shown in Fig. 5(c).

Then, as shown in Fig. 5(d), Si_3N_4 films used as a selective thermal oxidation mask for polysilicon are selectively removed, and phosphorus (n^+) is diffused through the polysilicon layer to get emitter base junctions and n^+ collector contacts. As a result, an electrode built-in transistor with a walled emitter structure is formed. If a transistor with a low internal base resistance (R_{BB}) is required, boron (p^+) diffusion is newly added before the phosphorus (n^+) diffusion. After the removal of the remaining Si_3N_4 films, boron (p^-) with 50 keV acceleration energy and 3.5×10^{13} cm^{-2} dose is ion-implanted into the polysilicon layer for polysilicon high-value resistor formation without using a mask, resulting in a 2 kΩ/□ sheet resistivity. As shown in Fig. 5(e), the thin SiO_2 film formed in Fig. 5(c) and (d) steps is etched off, except over the polysilicon resistors. Then the exposed polysilicon surface is alloyed with platinum into Pt-Si in a self-aligned fashion. This Pt-Si formation aims at the distinc-

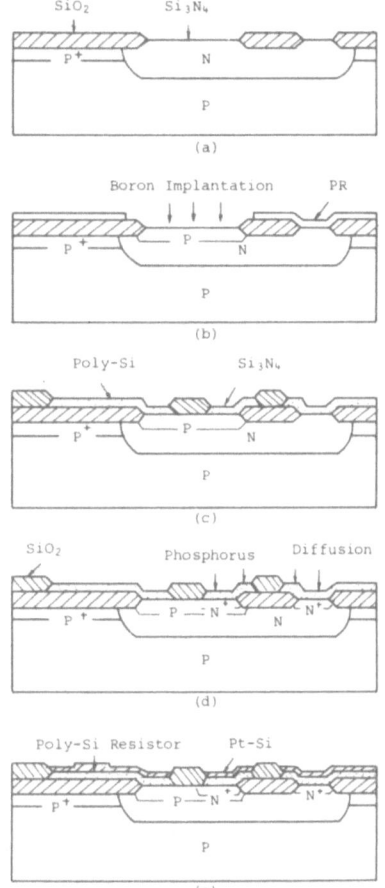

Fig. 5 - Fabrication steps for PSA transistor and polysilicon resistor.

tion of polysilicon resistors from polysilicon electrodes and/or interconnections and reduction in sheet resistivity for the latter elements.

Figure 5(e) shows the final schematic cross section of a typical PSA transistor and a polysilicon resistor. The emitter region is constructed by diffusion through the polysilicon electrode. The number of masking steps to obtain this PSA transistor and polysilicon resistor is the same as with the conventional process (i.e., six or seven steps). Because of the increased mask alignment ease, the PSA process is superior to the conventional process.

A top view of the PSA transistor is shown
in Fig. 6. In this figure, dotted lines show the
effective patterns of single crystalline silicon
and polysilicon after selective thermal oxidation. Shaded areas show the effective area for
emitter, base contact and collector contact. As
a result, a very small transistor with 2×12 μm
base regions can be easily obtained, even if the
conventional 4 μm design rule is applied. This
is because the original photo mask pattern is
uniformly shrunk by selective thermal oxidation treatment.

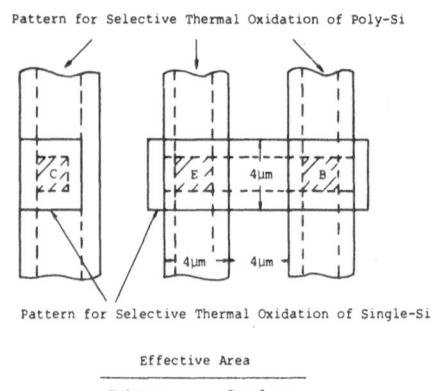

Pattern for Selective Thermal Oxidation of Poly-Si

Pattern for Selective Thermal Oxidation of Single-Si

Effective Area	
Emitter	2 x 2 μm
Base	2 x 12 μm

Fig. 6 - Top layout view of a PSA transistor.

5. RELATED PROBLEMS IN ADVANCED BIPOLAR LSI

Increasing level of integration with advanced process technology also brings some
problems which are not previously seen. Major problems are described in the following.

1. Increased Number of Pins

The number of pins for a LSI chip tends to increase in proportion to the level of integration, especially in the case of random logic LSI's for general purpose such as a Gate-Array.
Generally, the number of pins can be roughly estimated by the following expression.

$$p = \alpha \cdot G^{\beta}$$

where P is the number of pins, G is the number of gates and α, β is the constant. Up to five
hundred signal pins are estimated to be required in a chip if possible. The conventional wire
bonding approach can not be employed any more because of its poor productivity for such a
large chip. One of the solutions for this problem is a batch bonding technology using flexible
tape as a chip carrier. In the application, called TAB technology, lead frames are made by
photoresist etching of copper foil, which

is laminated onto a 35mm width film carrier, and by gold tin plating. The chips are assembled on the film carrier by thermal compression bonding or eutectic bonding between the gold pad on the chip and lead frames. A photomicrograph of the film carrier assembly with LSI chips is shown in Fig. 7. The chips with lead frames are punched off from the film carrier and assembled onto an ultimate ceramic substrate or a chip carrier. Each chip occupies about 10

Fig. 7 - Film carrier bonded with LSIs using TAB technology.

mm square on the substrate, which results in higher packing density compared to conventional dual in line package.

Another advanced bonding technology is called Controlled Collapse Bonding (CCB) technology presented by IBM. By this process, solder on the chip and a tinned electrode on the substrate are placed in juxtaposition and the system is reflowed, resulting in a single solder joint, or pad. Surface tension insures good alignment and provides positive support for this chip. Spreading is limited on the chip by a specially designed contact, the "ball-limiting metallization," and typically on the substrate by a glass dam. This CCB technology also enables assembly with high packing density.

2. Thermal Radiation Problem

When the level of integration is increased, a chip dissipates larger power which should be removed. Effective heat sinking is obviously of great importance for achieving a high packing density without exceeding the tolerable device temperature. The common situation is that of a large IC chip with a planar array of many individual heat sources. In this case, device is conventionally cooled by transfer of heat from semiconductor to air. However, air cooling is limited to a power density of about $1W/cm^2$. Although liquid cooling system can improve this value up to 10-20 W/cm^2, it requires relatively higher costs. New cooling technology which break through the $1W/cm^2$ limit with lower costs is desired for recent advanced bipolar devices with high packing density. One of our practical approach is multi-chip hybrid assembly technologies as shown in Fig. 8. In this case, LSIs are bonded onto an ultimate ceramic substrate using TAB technology. This package has finned stads and about $1W/cm^2$ the power density is achieved with forced air cooling. As a result, packing density for assembling is significantly increased using conventional cooling system.

Fig. 8 - Multi-chip assembly using TAB technology.

3. Circuit Testability

The most positive attribute of LSI - significantly greater circuit densities and functionality on chips - has created the most outstanding problem of testing these large clusters of circuits through chip inputs and outputs; the problem is particularly severe for sequential networks. One solution for this problem is the scan-path-technique. Although this technique needs three extra pins for control and some additional logic is required for shift register operation, testability is significantly improved. In this technique, all the flip-flops in a chip are connected serially and form a shift register. Each flip-flop has an input data selector, which is controlled by shift mode control input. This structure allows the flip-flops to be used as ordinary flip-flops in the normal mode and shift registers in the chip test mode. In the normal mode, normal data is loaded to the flip-flop. In the shift mode, output from the previous flip-flop in the shift register chain is loaded.

This shift register approach enables the sequential logic circuit to be changed to a combinational one. This is because all the states of the circuit are completely determined by the shift input data and other terminal data. A typical application of the scan-path-technique is shown in Fig. 9. The Level Sensitive Scan Design (LSSD) proposed by IBM also employs the similar approach which can improve the testability successfully. Required numbers of binary patterns for testing can also be decreased considerably by applying these approaches.

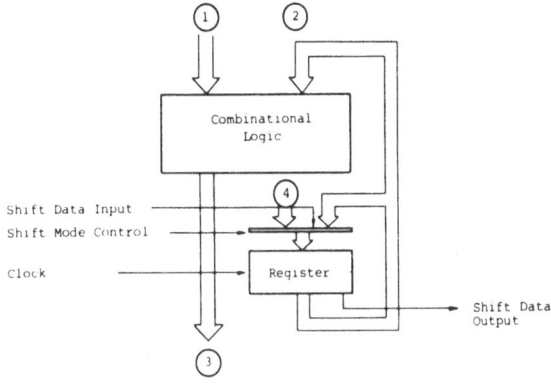

Fig. 9 - Application example of Scan-Path-Technique.

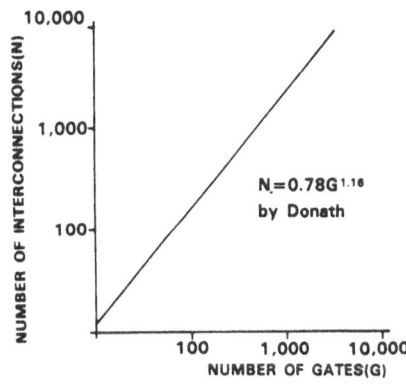

$N = 0.78G^{1.16}$
by Donath

Fig. 10 - Number of gates vs. number of interconnections.

4. Problem for Interconnections

The increased level of integration causes rapid increase in interconnections on chips. The typical relationship between the number of gates and the number of required interconnections is shown in Fig. 10. This increased wiring requires larger areas for wiring resulting in increased chip size which decreases yield. The best solution for this problem is the multi-layer metallization. Although this approach requires a few additional processing steps for metallization, the chip size can be reduced significantly resulting in higher yield.

Using the polysilicon layer for the first level interconnection, applied in the PSA process, is an effective approach to realize multi-level interconnections. In our application, three-layer metallization is applied for the recent random gate LSI's using aluminum and gold for the second and the third layer, respectively. Furthermore, fine-line metallization will bring new problems. First, since the metal lines must carry a higher current with respect to cross-sectional area, electron migration becomes a major factor with today's aluminum-copper-silicon metallization. As a result, a new metallization scheme must be employed to accommo-date the higher current density in the future, especially as metal width approaches 1 μm. In addition, new methods for depositing metal are being investigated to reduce defect levels. Second, wet etching also becomes useless for such fine metal definition. Consequently, this process is being replaced by dry plasma etching, which will provide extremely fine metalliza-tion patterns.

Another problem is stray capacitance or wiring. Average line length on a chip tends to increase in proportion to the level of integration. On the contrary, power dissipation per gate decrease resulting in poor drivability for wiring capacitances. Therefore, average gate delay is determined by the interconnections rather than the gate itself even on a chip. Consequently, stray capacitances due to interconnections should be reduced. In this point of view, oxide isolation approach is effective for this purpose since interconnections run on the thick oxide layer resulting smaller stray capacitances.

5. Designability

Manpower and terms required for designing recent highly integrated custom LSIs are on a rapid increase. Manual design becomes almost impossible due to its lower efficiency and big possibility of mistakes. Circuit complexity brings significant increases in cost and turn-around-time.

Solution for these problems is automatic design approach using CAD system combined with gate-array design approach. The CAD system is indispensable for designing such LSI's

manual designing. The ideal CAD system is required to support all the activities from logic design to chip testing
including logic simulation, automatic
layout and routing, dynamic simulation,
test pattern generation and fault simulation. Usually, gate-array contains
basic cells, areas for automatic wiring,
peripheral I/O circuits and pads. Required functions can be customized on
the gate-array using the CAD system.
Although this approach requires some
redundancy for the level of integration,
designability and turn-around-time can
be significantly improved. A flow
chart for a typical LSI design applying
the CAD system is shown in Fig. 11.

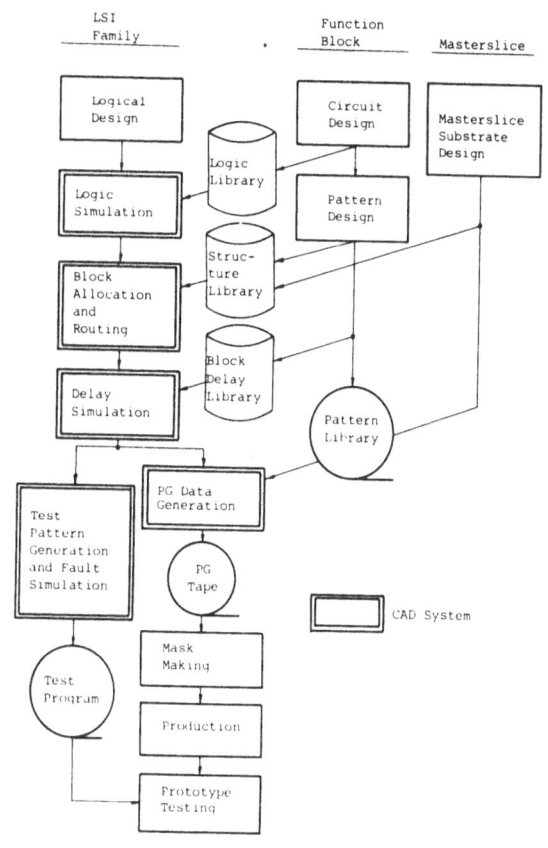

Fig. 11 - Design flow of a typical masterslice LSI using CAD system.

A microphotograph of an example masterslice LSI deisgned by the CAD system is shown in Fig. 12.

6. ESD Protection

Electrostatic discharge (ESD) damage had been considered a problem peculiar to the MOS device. However, this kind of attention should be paid for the recent advanced bipolar devices because of its very small sized shallow junctions. Ever since the problem of ESD damage was identified for MOS structure, several protection schemes have been proposed. The most popular approach today is the combination of resistors and diodes to limit the excess

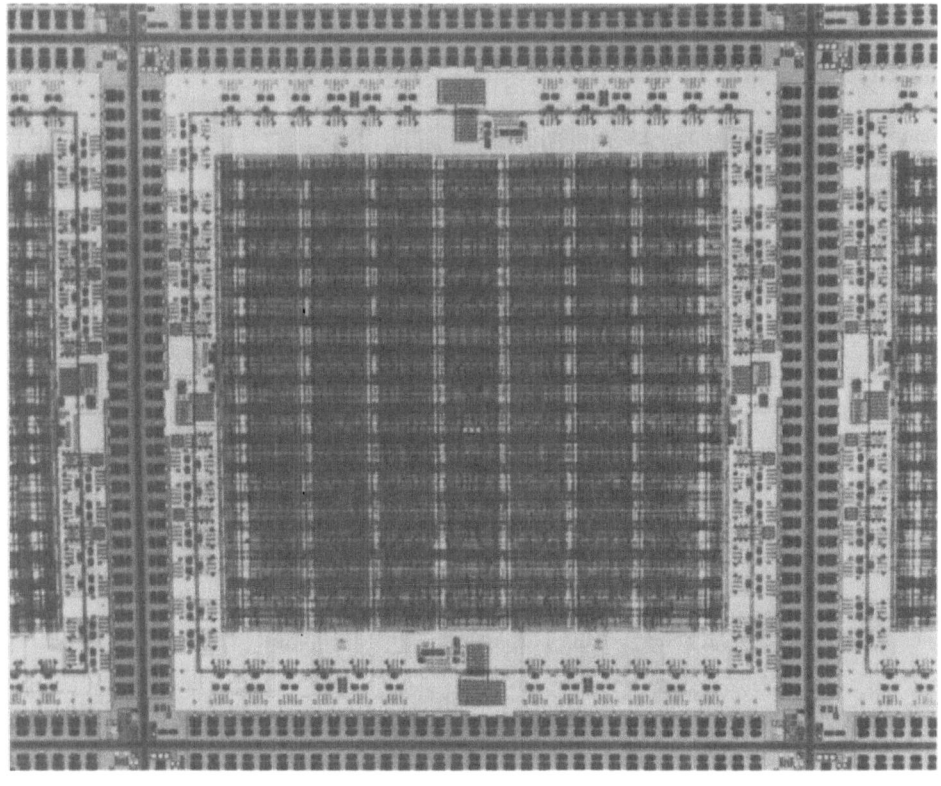

Fig. 12 - A micrograph of a masterslice LSI designed by CAD system.

voltage and shunt the current away from the weak MOS gate structure during an ESD pulse at

the IC pins. Since these elements provide a compatibility with the other types of complex

integrated circuit structures, extra process steps are not required. This approach is also

effective for the recent advanced bipolar LSI's resulting in enough resistivity for ESD damage.

The only problem here is the undesirable extra circuit delay caused by the additional elements

to prevent ESD damage. Therefore, additional circuits for ESD protection must be designed

carefully so as not to degrade the inherent high-speed performance of bipolar logic.

6. BIPOLAR DEVICE SCALING

Recent developments in photolithography, notably direct-step-on-wafer (DSW) techni-

ques, are the key to scaling down geometries. The new DSW equipment permits geometries of

1 μm, whereas projection alignment has a practial limit of about 2 μm. A scaling principle has

been used to design miniaturized MOS devices by co-ordinated changes in dimension, voltages,

and doping concentrations. However, bipolar devices are more complex, and it is difficult to

obtain simple scaling rules to design bipolar devices with smaller physical dimensions.

The horizontal and vertical designs of small geometric bipolar transistors are governed by

various factors, such as voltage and current operating levels, base collector and emitter base

capacitances, diffusion capacitance and base resistance. In addition, in bipolar scaling,

voltages can not be reduced much, since they are already close to their lower limit. We

estimated the performance limit in bipolar devices by assuming utilization of device scaling

technology. The following shows the assumed scaling rules and estimated results in perform-

ance.

A CML gate with a logic swing of 0.4 volts is assumed a circuit type. Assumed transistor

structure which employs oxide isolation and double base structure is shown in Fig. 13. Highly

doped p$^+$ diffusion is added for the base contacts to decrease the extrinsic base resistance. It

is assumed that all the lateral geo-

metries and vertical dimensions

are scaled properly in proportion

to the emitter strip width W_E

which changes from 2 μm to 0.25

μm. The numbers in Fig. 13 mean

the dimensional multiple factor of

emitter strip width.   Junction

depth is reduced in proportion to

the emitter width. In accordance

with the scaling level, all the dop-

ing level for impurity are also in-

creased to keep the normal junc-

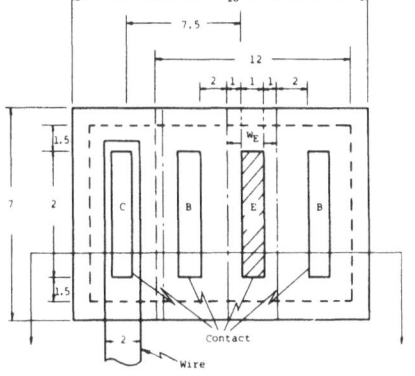

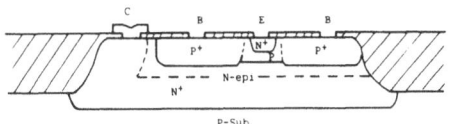

Fig. 13 - Assumed transistor structure for device scaling.

tion profiles. Assumed or calculated device parameters are listed in Table 2.

Junction capacitances and transit time, f_T are significantly improved as one scales down. Base resistance is also slightly decreased but is rapidly increased at scaling limit. This results from the rapid decrease of base width which cancels the decreased lateral geometries of the base area. On the

Table 2 - Calculated and estimated parameters for a scaled transistor.

PARAMETERS	UNIT	ESTIMATED VALUE				
Emitter Width	μm	2.0	1.5	1.0	0.5	0.25
Epi thickness	μm	1.5	0.6	0.4	0.3	0.1
Xje	μm	0.14	0.11	0.07	0.03	0.015
Xjc	μm	0.29	0.22	0.15	0.07	0.04
ρ_{epi}	Ω-cm	1.0	0.6	0.3	0.1	0.04
Wiring Pitch	μm	6.0	4.5	3.0	1.5	0.75
f_T	GHz	4.2	8.8	16	32	58
γbb'	Ω	590	520	480	440	920
Cje	fF	37	29	20	9	6.5
Cjc	fF	54	41	25	13	6
Ccs	fF	56	41	27	12	5.4
Cw	fF	12	13	15	17	17

contrary, the wiring capacitance in not changed and even slightly increased in spite of its decreased metal width. This would result from the reduced dielectric thickness due to scaled dimensions. It indicates that the stray capacitances due to wiring will become critical parameters when the scaling technology is practically applied in the future.

Figure 14 shows the simulation result obtained by using these device parameters. In order to estimate the average gate delay in practical use, CML gate with F/I = F/O = 2 an average output wire length of 600 μm is assumed. The result shows that the power delay product could be significantly reduced by the device scaling. The result also predicts that the average system gate delay could be limited to around 80 picoseconds when bipolar technology is applied. The integration level shown in Fig. 14 is a calculated value under the following assumptions.

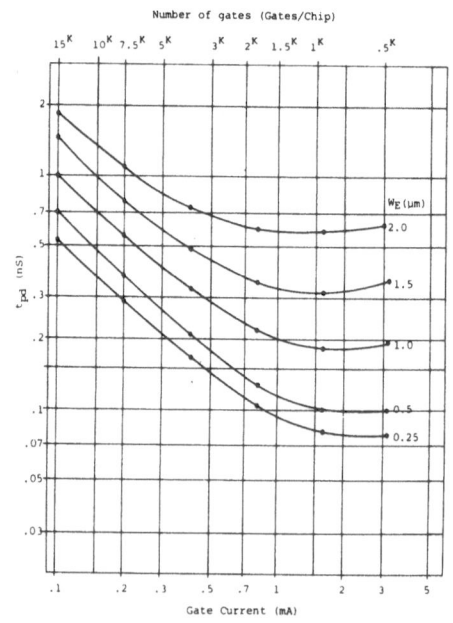

Fig. 14 - CML gate delay estimation with device scaling (F/I = F/O = 2, L_{wire} = 600 μm).

That is, a two level series gated CML is employed for a basic gate with source voltage of -3.3 volts and the maximum available power dissipation for internal cells is limited at 2 watts. Figure 15 shows the simulation result for a basic CML gate delay under the condition F/I = F/O = 1, and output wire length of 100 μm. In this case, limited gate delay times of 57 picoseconds are estimated.

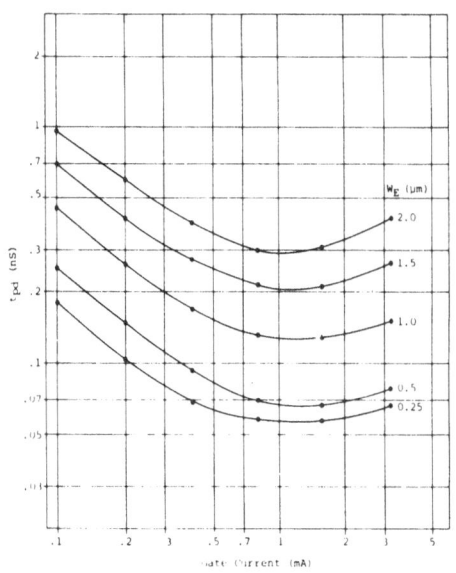

Fig. 15 - CML gate delay estimation with device scaling (F/I = F/O = 1, L_{wire} = 100 μm).

7. CONCLUSION

High-speed LSI logic is obviously a very important subject to most designers, especially in the computer industry. Recent progress in high-speed custom LSI's, bit-slice microprocessors and gate-arrays makes the broad impact on computer logic. SSI or MSI devices can not improve the total system delay any more due to the large interconnection delay required for connecting chips. The higher speed performance is still required while the level of integration is significantly increased in LSI and VLSI. Once subnanosecond speed is required, CML is the only way to get it. Therefore, advanced device technology is indispensable in order to improve the basic gate delays when the increased level of integration requires smaller power dissipations for a gate.

The circuit simulation for a basic CML gate shows that all the device parameters, especially junction and stray capacitance, must be decreased to improve the power delay product performance. The major effective process approach for this improvement is oxide isolation, shallow junction technology, fine-line lithography and self-aligning technology. The summary of each approach was discussed and some of these in detail were studied citing practical process sequence of Isoplanar and PSA. These advanced technologies contribute to the realization of high-speed LSI's. The rapid movement into LSI and VLSI with advanced bipolar process also brings some related problems which should be solved. First, the increased number of pins requires advanced bonding technology such as the batch bonding process. TAB technology and CCB technology were introduced as a solution for this problem.

The thermal radiation problem is also important when the level of integration is increased both in chip and system level. An effective heat radiation method is required without usage of expensive liquid cooling system. Multi-chip assembly on an ultimate multi-layer substrate with fins is certainly an effective solution for this problem.

In addition, device testability is also a serious problem. The logic for a new LSI should be designed considering testability it requires some additional logic for this purpose. The scan-path-technique was introduced for solving this problem.

Tremendous increase of interconnection in LSI's require advanced metallization for wiring. Multi-layer interconnections and finer patterning are indispensable for improvement. The wiring stray capacitances, which become a dominant factor of gate delay as power decreases, should also be reduced.

Recent LSI's, which contain complex logic function with many devices, can not be realized without applying CAD systems. If not, required man-power and terms for development become increased rapidly resulting in cost explosion.

For wider usage of LSI and VLSI in actual systems, these above-mentioned problems should be solved and the intrinsic gate performance should be improved.

Our estimated results in device scaling technique for bipolar devices utilizing advanced fine-line lithography system will promise great improvement for the total device performance. It seems that about 60 picoseconds is the theoretical speed limit when bipolar device technology is applied. However, actual problems caused by the physical and production limit will have to be solved before the scaling technology is widely used in practice.

It seems that there is still much room left for improvement in bipolar device performance. However, the progress in device technology and advanced equipment promise further improvement in bipolar devices.

8. ACKNOWLEDGEMENT

The authors wish to express their thanks to Mr. Matsumura, Mr. Goto and Mr. Sasaki of Nippon Electric Company Ltd. who have made various suggestions for this work.

REFERENCES

1.   V. A. Dhaka, *et al.*, "Subnanosecond Emitter-Coupled Logic Gate Circuit Using Isoplanar-II", IEEE J. Solid State Circuit, SC-8, 5 368-372 (1973).

2.   K. Okada, *et al.*, "A New Polysilicon Process for a Bipolar Device - PSA Technology", IEEE J. Solid State Circuit SC-14, 2, 307-311 (1979)

3.   C. P. Snapp, "Bipolar Quietly Dominate," MSN, p. 45-67, Nov. (1979).

4.   C. Davis, *et al.*, "IBM System/370 Bipolar Gate Array Micro Processor Chip," Proc. of ICCC, 2, 669-673 (1980).

5.   A. H. Dansky, "Bipolar Circuit Design for VLSI Gate Arrays," ibid, 674-677.

6.   M. Fener, *et al.*, "The Layout and Wiring of a VLSI Microprocessor," ibid, 678-679.

7.   R. F. Penoyer, *et al.*, "An 18K Bipolar Dynamic Random Access Memory Chip," Digest of ESSCIRC, 164-165, Sept. (1980).

8.   K. Steven, *et al.*, "A One-Micron Bipolar VLSI Convolver," Digest of ISSCC, 226-227 Feb. (1981).

9.   Y. Horiba, *et al.*, "A Bipolar 2500-Gate Subnanosecond Masterslice LSI," ibid, 228-229.

10.   J. E. Selleck, *et al.*, "64K Dynamic 1/N Fractional Device Bipolar Memory," ibid, 220-221 Feb. (1980).

CHAPTER V

MOS DEVICES

MOS TECHNOLOGIES AND DEVICES FOR LSI

Bernd Hoefflinger

University of Dortmund
Lehrstuhl Bauelemente der Elektrotechnik
Postfach 500 500
D-4600 Dortmund 50

ABSTRACT: MOS technologies and devices have matured to a leading status in very large scale integration (VLSI). Silicon gate and ion implantation technology, as well as a host of other recent processing techniques, have led to MOS technologies, of which new CMOS processes are outstanding examples. Devices are operated both in weak- and strong-inversion or barrier- and drift-conrol modes. These are described by advanced models, which help to optimize devices and to stimulate circuits with high performance on one side and limitations in gain and noise margin on the other side. Nevertheless, an overall powerful technology is presented in the face of the 1 μm barrier.

1. MORPHOLOGY OF MOS DEVICES

Metal-oxide-semiconductor transistors are operated by controlling a potential barrier between source and drain. In a cross section this barrier in its basic properties is shown in Fig. 1 for four different conditions: flat-band, accumulation, depletion and inversions. The most important modifications are introduced by ion implantation into the semiconductor. If this results in counter-doping, the band diagrams shown in Fig. 2 may result. Threshold voltages are shifted and a transition from normally-off to normally-on occurs as the implanted dose is increased. Another important role is played by work function differences between various gates and various substrates as shown in Fig. 3. A degree of freedom or a restriction may lie in the use of Al, p^+- or n^+-poly-Si and p- or n-substrates. The choice becomes particularly important in the design of a complementary MOS (CMOS) technology.

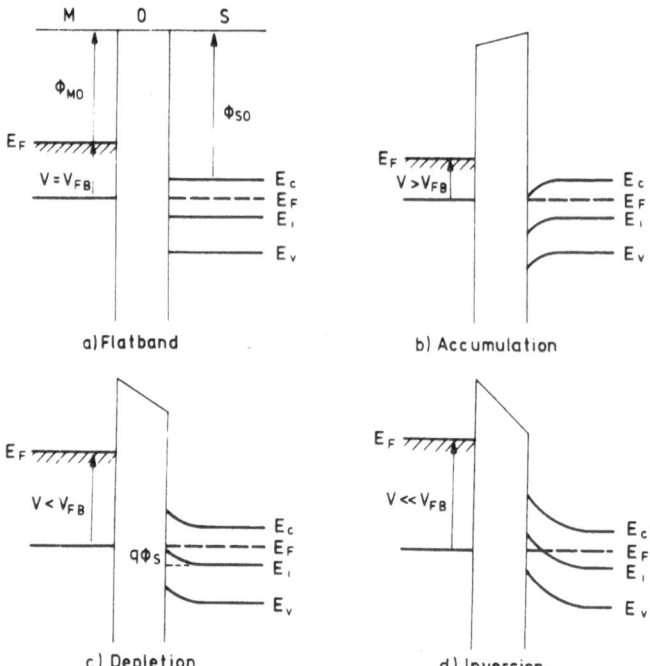

Fig. 1 - M̲etal O̲xide S̲emiconductor band diagram.

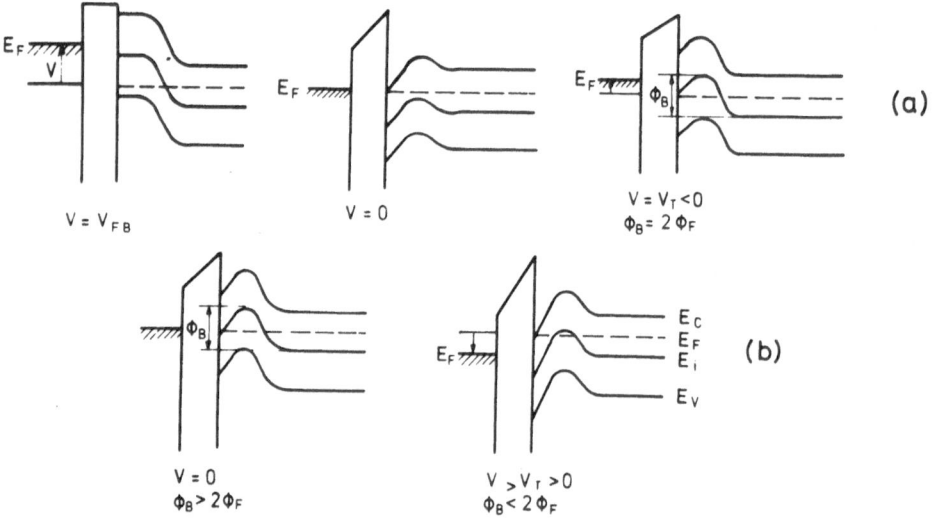

Fig. 2 - MOS band diagram for the flatband condition in an n-doped substrate with a p-doped surface layer; (a) implantation dose small enough for normally-off (enhancement) type; (b) larger dose for normally-on (depletion) type.

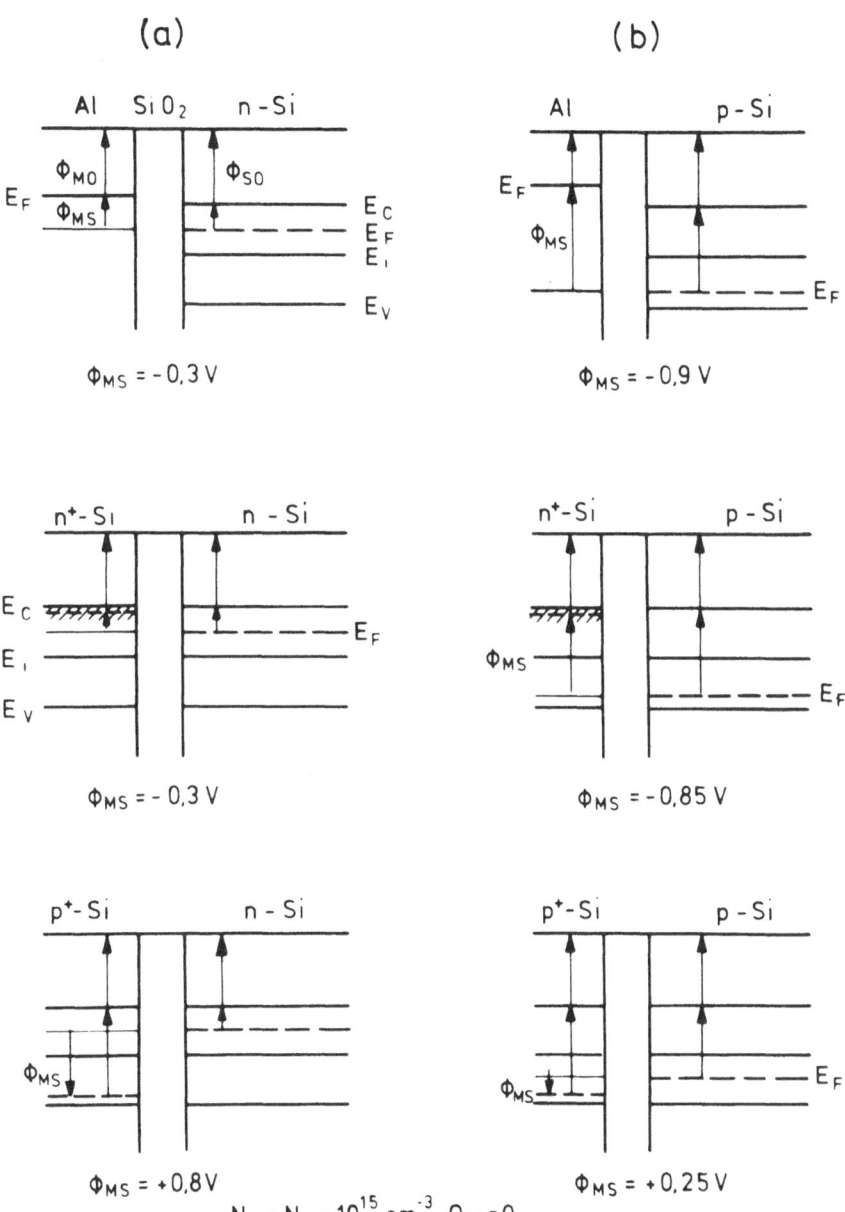

Fig. 3 - Metal-semiconductor work function difference in various MOS structures.

2. MOS TECHNOLOGIES

2.1 A CMOS Process as a Case Study

The use of complementary MOS transistors has appeared to be attractive for circuit designers already in the early period of MOS technology, when PMOS technique was developed for the first LSI technology. Before the application of ion implantation to solid state devices, the technology arsenal was restricted to conventional doping techniques like high-temperature predeposition and diffusion processes and thermal oxidation techniques with the capability of low oxide charges for good threshold voltage control.

Threshold voltage requirements for active and parasitic CMOS transistors forced the process engineers to choose appropriate doping concentrations for the substrate and the diffused wells. Thus tradeoffs between other transistor parameters could not be avoided. Well surface concentration has to be about one order of magnitude higher than substrate doping density to achieve narrow threshold voltage tolerances. The necessity for NMOS-enhancement transistors with $V_T \approx 1$ V imposed putting them into a p-well with a doping concentration in the range of 10^{16} cm^{-3} diffused into a lower doped n-silicon substrate with (100)-orientation.

Relatively low parasitic thresholds were eliminated by diffused channel stops allowing only limited device density. This initial process concept established the main direction in bulk CMOS technology. Drawbacks and tradeoffs have been overcome by more recent technological feasibilities like ion implantation, local oxidation, LPCVD and dry etching.

Besides the progress in CMOS processing, the advances in NMOS enhancement-depletion technology enabled an ever increasing functional capacity of very large integrated circuits.

Meanwhile it is desirable to provide various technologies in a compatible process sequence, which may be selected for an optimum implementation of a given circuit function. This means compatible integration of NMOS, CMOS and bipolar transistors. An attractive solution for this problem of a compatible NMOS, CMOS, bipolar technology was demonstrated[18,31] by the extension of NMOS technology placing the PMOS transistors into an

n-well.

The process steps are listed in Table 1 and the process is illustrated in Fig. 4 by the cross

section.   The process sequence mainly differs from standard metal gate CMOS by using an

n-well for the p-channel transistors in a p-substrate and by the high-energy boron implantation

through the field oxide.

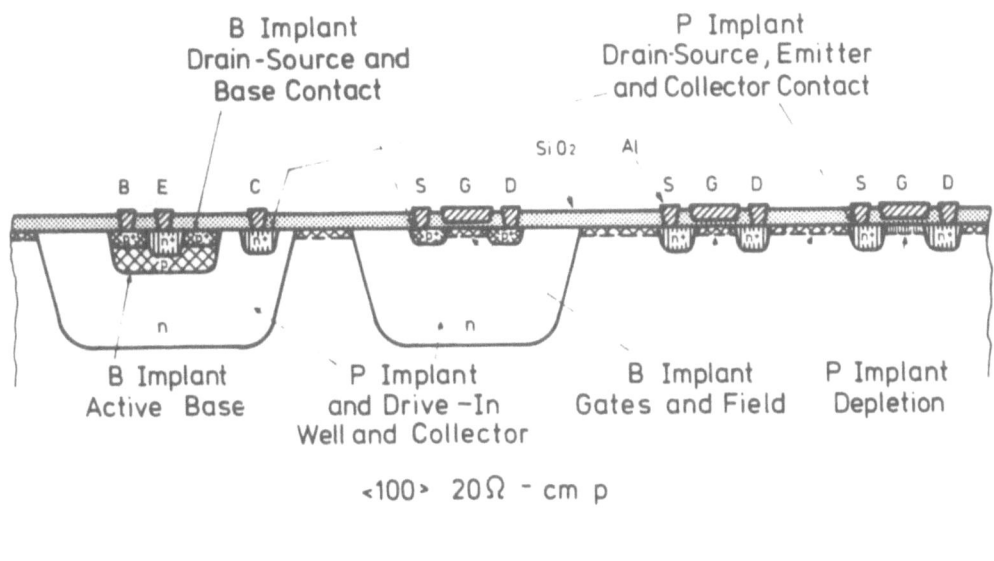

CMOS – NMOS E / D – Bipolar Process

Fig. 4 - Cross section of aluminum gate n-well CMOS process with 4-implant NPN
bipolar transistor.

The threshold voltages of p- and n-channel enhancement transistors and n-channel field

transistors are adjusted by this common implantation step.   This is possible because these

three thresholds must be shifted to a more positive value.   As shown in Fig. 4, the field region

of the n-well is protected by photoresist masking during the implantation, because the

threshold voltage of the field region is originally high enough.   Thus in an n-well-CMOS-

process only a single field implant is necessary, whereas in p-well-processes two field implan-

tation steps are required, particularly for circuits with supply voltages $V_{DD} \geq 12$ V.

Table 1

	Process Step	Oxide Thickness(μm)	Energy (keV)	Dose (cm^{-2})	Final Junction Depth(μm)	Conditions
1	initial oxide	0.1				
2	P implant (n-well)		150	$7 \cdot 10^{12}$		
3	well drive-in				7.5	19h, 1200C, N$_2$
4	first masking oxide	0.3				60', 1024C, O$_2$ wet
5	B implant		40	$4 \cdot 10^{13}$		
6	base anneal, and drive-in				2.3	80', 1200C, N$_2$
7	B implant		40	$2 \cdot 10^{15}$		85', 1024C, N$_2$
8	anneal and 2nd masking oxide	0.3			1.2	60', 1024C, O$_2$ wet
9	P implant		80	$5 \cdot 10^{15}$		
10	field oxide	0.85			1.0	150', 960C, H$_2$/O$_2$
11	B implant (field, p- and n-channel enhancement)		350 B$^+$ (175 B^{++})	$1.7 \cdot 10^{12}$		
12	gate oxide	0.07				60', 960C, H$_2$/O$_2$
13	P implant (depletion)		80	$1.5 \cdot 10^{12}$		
14	annealing					15', 960C, N$_2$
15	metallization					

Table 2

		unit	min.	typ.	max.	standard deviation
1.	NMOS					
	V_{TN}	V	1.07	1.13	1.19	0.03
	V_{TDN}	V	-2.65	-2.85	-3.0	0.08
	V_{TFN}	V		20		
	B_{ON}	μAV^{-2}		30		
	$\rho'_{S,D}$	ohm/□		20		
	K_{1N}	$\sqrt{V}$		0.4		
2.	PMOS					
	V_{TP}	V	-1.31	-1.15	-1.0	0.08
	V_{TFP}	V		-18		
	B_{OP}	μAV^{-2}		12.5		
	$\rho'_{S,D}$	ohm/□		70		
	K_{1P}	$\sqrt{V}$		1.0		
3.	NPN Bipolar					
	β_F		55	85	130	16
	BV_{BEO}	V	6	6.5	8	
	BV_{CEO}	V	29	31	33	
	BV_{CBO}	V	50			
	V_A	V		107		
	$V_A \times \beta_F$	V		9000		
	ρ'_b	ohm/□		70		
	r'_c	ohm		15		2

The reason is the negative threshold shift by the oxide charge Q_{OXF}. This is illustrated in Fig. 5. Assumptions are $Q_{OXF} = 10^{11}$ cm^{-2} at the field oxide-substrate interface and a minimum field threshold of $|16\,V|$ for both field regions with 1 μm thick field oxide. For the low doped n-substrate as well as the high doped p-well the field threshold is too low and must be enhanced by a phosphorus and a boron implantation, respectively. In case of a high resistivity p-substrate only a boron implantation serves for adjustment of the field threshold. This is necessary in NMOS technology too. Thus full compatibility is achieved between NMOS and the n-well CMOS-process.

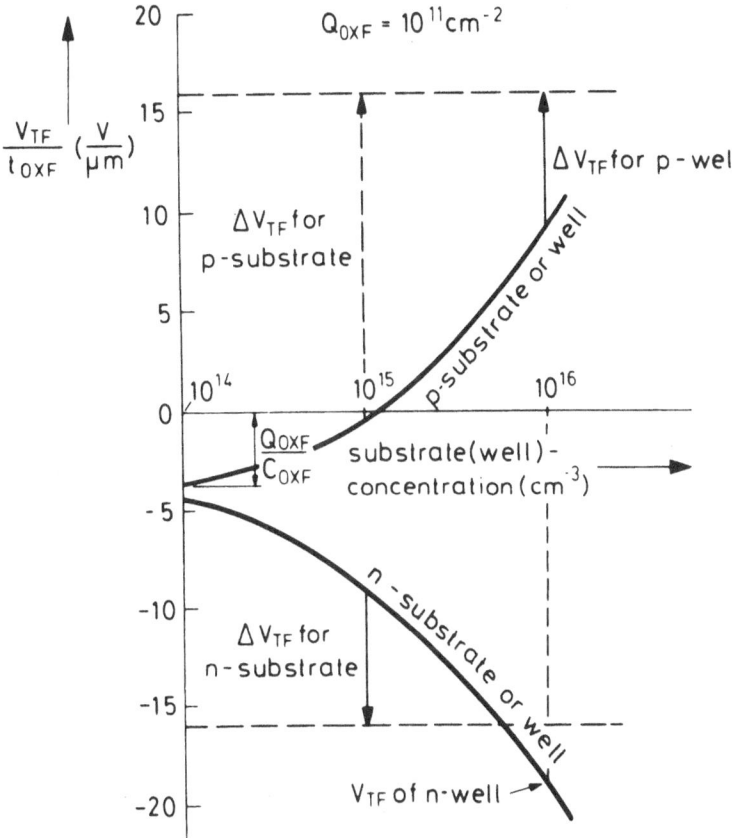

Fig. 5 - Normalized field-threshold voltages V_{TF} as functions of substrate or well concentration and shifts ΔV_{TF} due to ion implants.

For symmetrical inverter characteristics, one should have the same absolute value of p- and n-channel threshold voltages. Therefore, the doping concentration of the n-well has to be set by phosphorus implantation and drive-in so that the difference between the p- and n-channel threshold equals two CMOS-transistor threshold voltages. The following boron implantation shifts the threshold voltages to the symmetrical value.

This process design scheme is realized in Fig. 6. The symmetrical transistor threshold voltages are plotted for three gate oxide thicknesses against the well doping concentration and the resulting boron dose. The comparatively high values of V_{TFN}, V_{TFP} and the drain-source breakdown voltage $BV_{DS,P}$, enable a wide range of supply voltage, e.g., from 2.4 V to 24 V with $N_W = 10^{17}$ cm^{-3} for a high voltage process or from 0.4 V to 8 V with $N_W = 10^{15}$ cm^{-3}

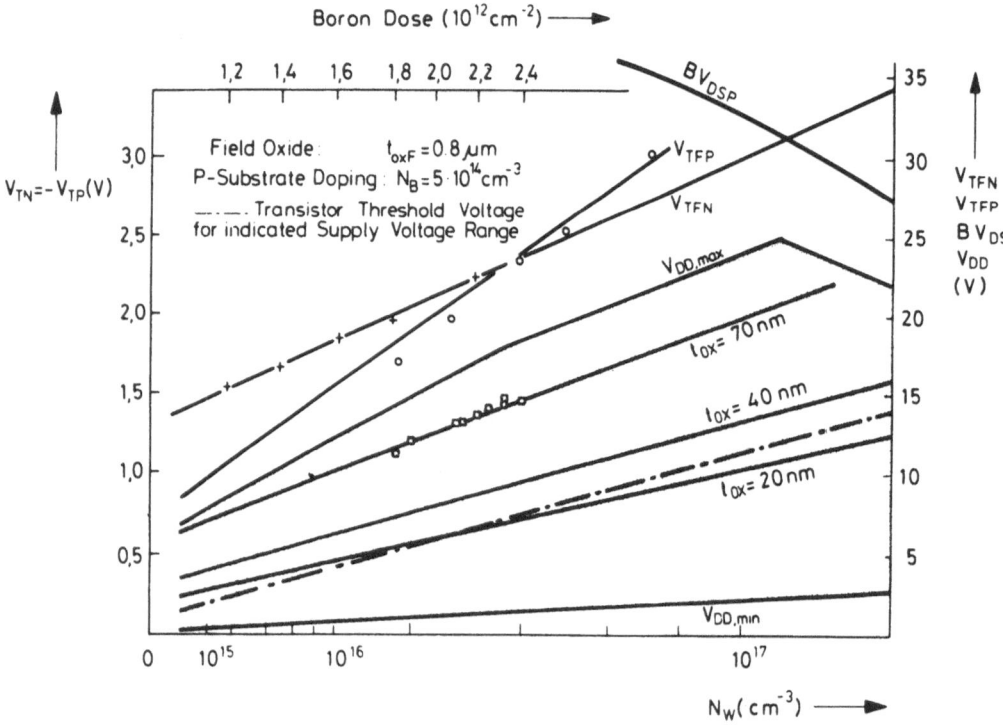

Fig. 6 - Threshold voltages, breakdown voltages and maximum supply voltages.

for a low voltage process. The latter one yields high performance p-channel transistors with respect to transconductance and body effect. The dotted line indicates the threshold voltage for the marked supply voltage range.

The boron implantation converts the PMOS transistors from surface- to buried-channel mode.[12] Therefore, the mobility in the channel is increased by 30% to an average value of 250 cm^2/Vs for an n-well concentration of 10^{16} cm^{-3}. These data are demonstrated by the measured gain constant histograms of Fig. 7.

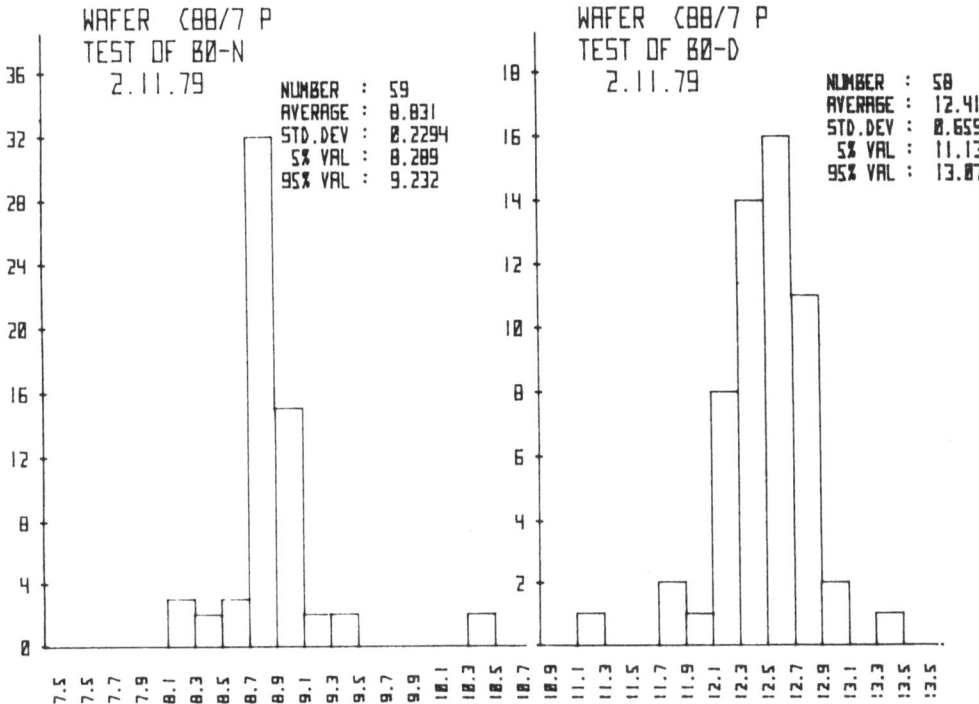

Fig. 7 - Histograms of PMOS gain constants $\mu_0 \epsilon_{ox}/t_{ox}$ in $\mu A/V^2$ for surface (left) and buried-channel mode, respectively.

In the n-well process, the source-to-substrate bias effect is considerably lower for the NMOS-transistors due to the lower doped substrate. This is beneficial for the speed in

NAND-gate logic, because PMOS-substrate bias is zero and the NMOS-transistors can get a source-to-substrate bias up to V_{DD}.

2.2 Silicon Gate Process

Further improvement of the CMOS/bipolar process is achieved by making use of the benefits of silicon gate technology. The process sequence cross section is given in Fig. 8. The sequence shows the formation of the NMOS on the left, PMOS in the middle and the bipolar transistor on the right side. The implantations for the NMOS enhancement- and depletion-transistors are drawn into the same transistor cross section on the left part of the figure. The process steps corresponding to the sequence in Fig. 8 are listed in Table 3 and the mask count in Table 4. The process is again based on the planar standard NMOS silicon gate enhancement/depletion technology containing the process steps (2)-(4), (6)-(8), (10), (11), (14) and the masks M2-M4, M6, M7, M10, M11.

Table 3

(1) Well implantation and drive-in

(2) Field oxidation

(3) Boron implantation for V_{TNE}, V_{TP} and V_{TFN}

(4) Gate oxidation

(5) Implantation of active boron base for NPN-transistor

(6) Phosphorus implantation for V_{TND}

(7) Polysilicon deposition

(8) Arsenic NMOS-drain/source-implantation

(9) Boron PMOS-drain/source-implantation

(10) Phosphorus glass deposition

(11) Deep n^+-contact formation

(12) Formation of capacitor oxide (Fig. 3.12)

(13) Deep p^+-contact formation

(14) Aluminum evaporation

Additional steps (1), (5), (9), (12), (13) provide n-wells, PMOS-drain-source regions, active base and Al-Poly capacitors for analog functions. Patterning of these additional device elements occurs by the additional masks M1, M5, M8, M9.

Table 4

M1	n-Well
M2	V_{TNE}, V_{TP} and V_{TF} threshold implant
M3	Active areas
M$\bar{2}$	V_{TND} threshold implant
M4	Poly/n^+-contacts
M5	Active base boron implant
M6	Polysilicon
M7	n^+-drain/source-implant
M8	p^+-drain/source-implant
M9	n^+-contact holes and Al-Poly-capacitors
M10	contact holes
M11	metallization

The benefits of this CMOS process are high device density due to poly/n^+-contacts, good device performance of MOS transistors with respect to punch-through, subthreshold current and geometry-dependence of threshold voltage due to shallow implanted junctions ($x_j = 0.3$ μm) and the deep boron profile in the channel region.

In Fig. 9, the two-dimensional profile simulation[48,55] of the implanted and not implanted P-channel transistor is plotted. Figure 10 shows the electric field of the transistor (a) without and (b) with the boron channel implant. It can be seen that the electric field of the implanted transistor becomes zero inside the silicon substrate leading to a buried channel with essentially increased mobility. The field at the drain end of the channel is lower than for the not implanted transistor raising the drain-source voltage capability.

Figure 11 is a plot of the NMOS transistor doping profiles. The channel implantation is deeper than the drain source junctions preventing punch-through in scaled-down short-channel devices.

This CMOS process concept can be based on the planar Si gate NMOS-process as described with a single implant for the field and transistor threshold adjustment as outlined in Fig. 12(a). It is also possible to use the Si gate isoplanar process as the basic technology.

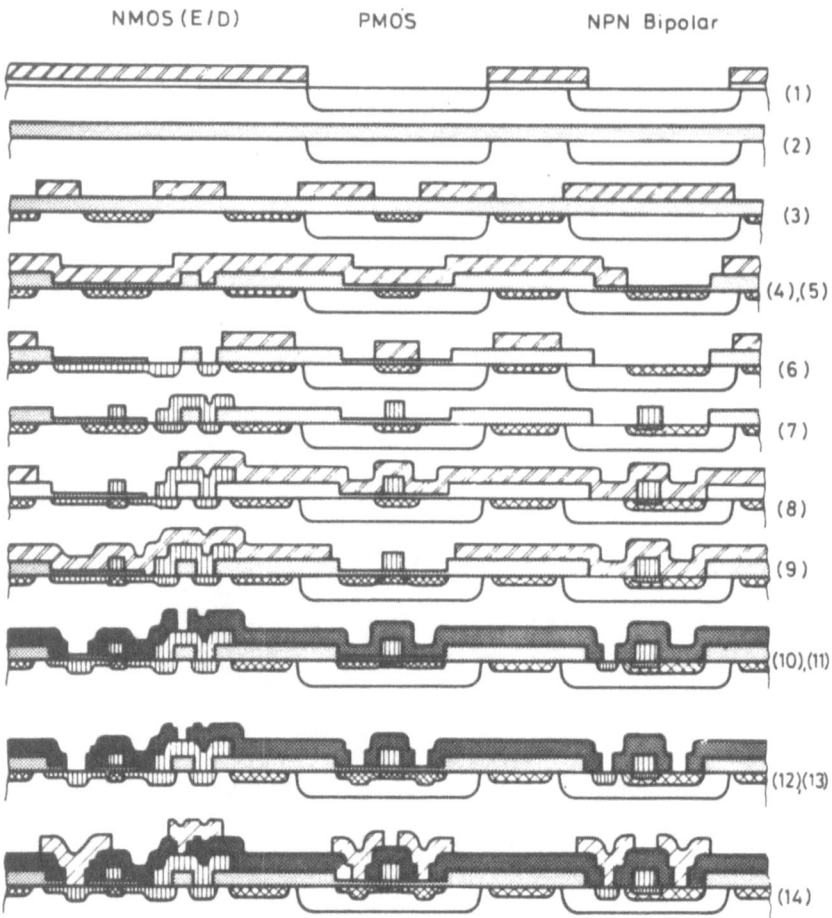

NMOS (E/D)   PMOS   NPN Bipolar

Fig. 8 - Cross sections for a N-well silicon gate CMOS/bipolar process
sequence.

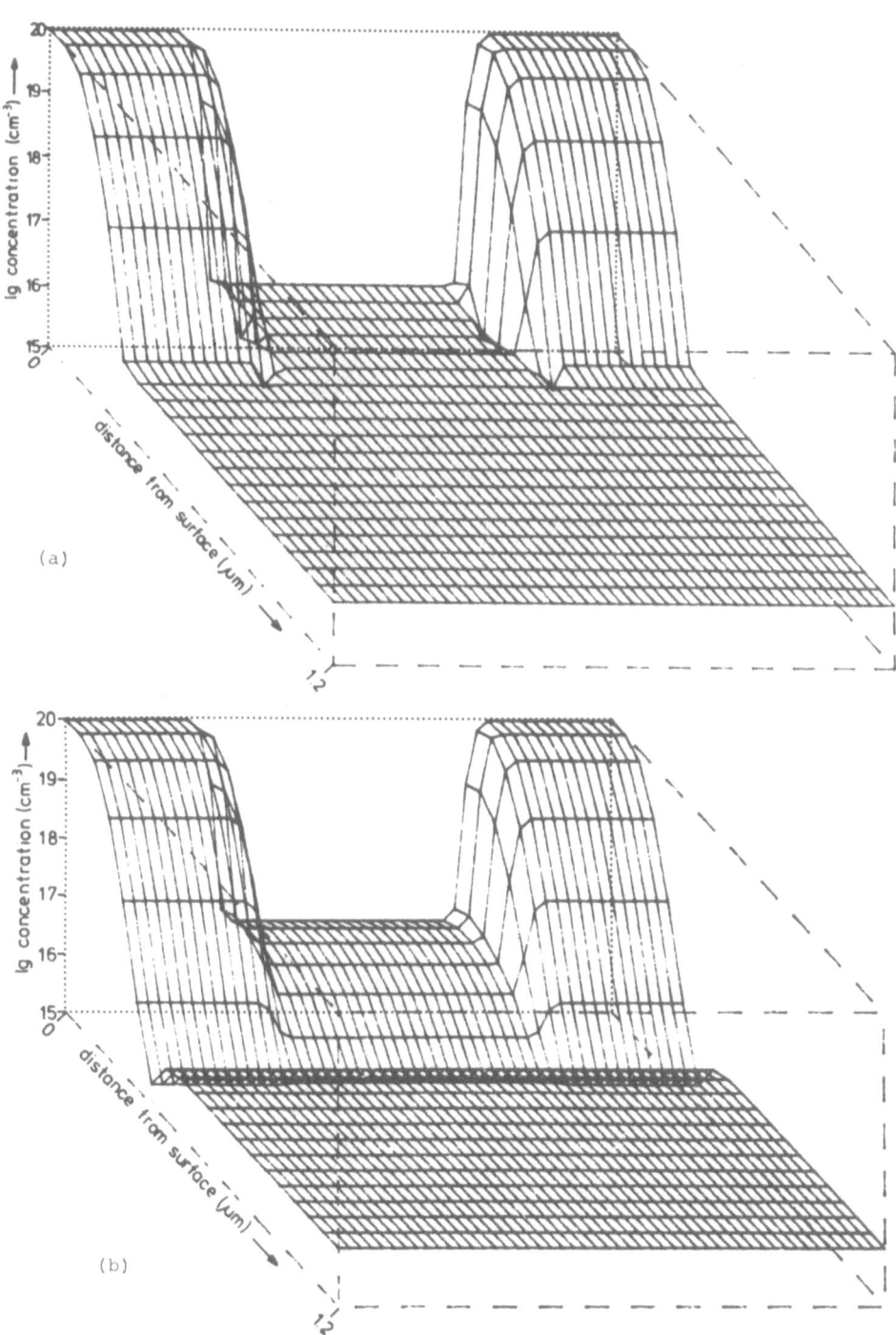

Fig. 9 - Two-dimensional profile simulation of not-implanted (a) and implanted (b) PMOS transistor.

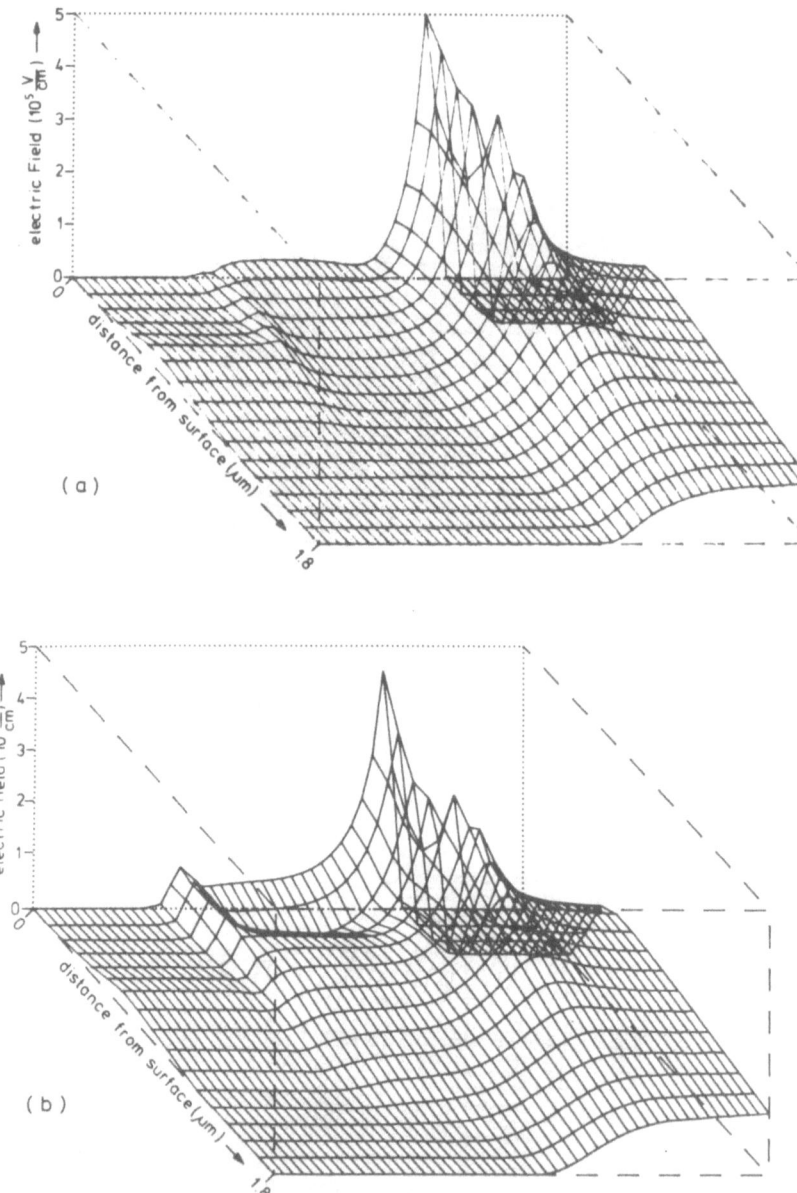

Fig. 10 - Two-dimensional magnitude of the electric field in not-implanted (a) and implanted (b) PMOS transistor.

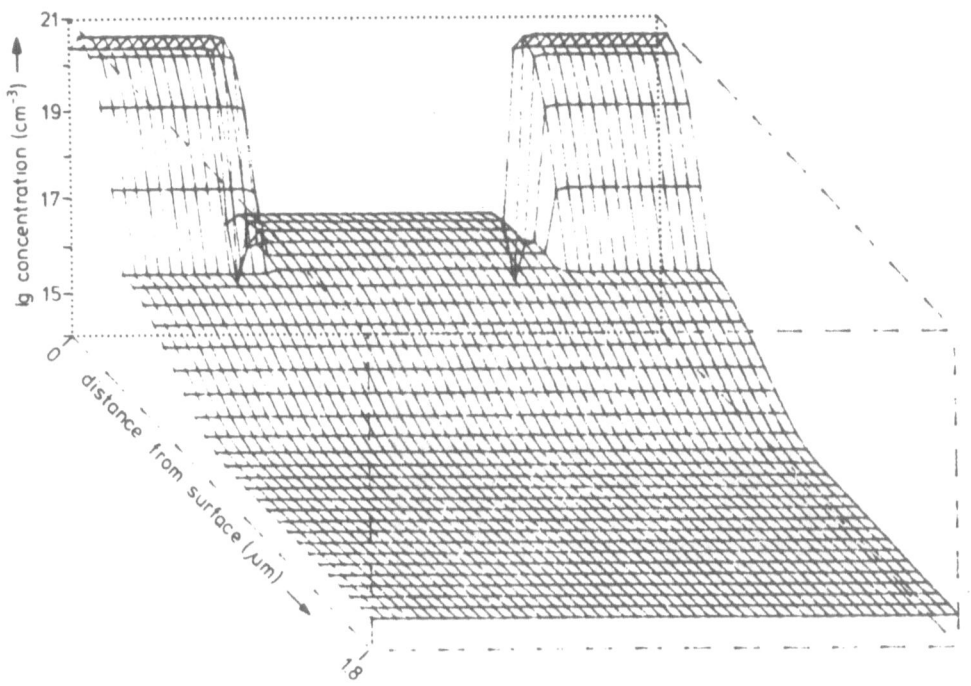

Fig. 11 - Two-dimensional profile simulation of implanted NMOS transistor.

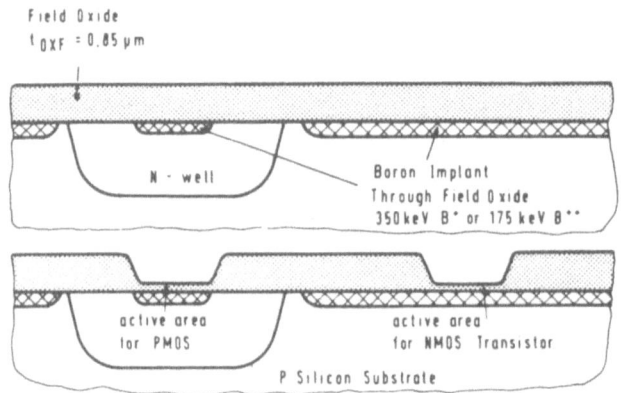

Field Oxide
$t_{OXF} = 0.85\,\mu m$

N - well

Boron Implant
Through Field Oxide
350 keV B⁺ or 175 keV B⁺⁺

active area
for PMOS

active area
for NMOS Transistor

P Silicon Substrate

(a) Field and Transistor   Threshold-Implant for Planar Process

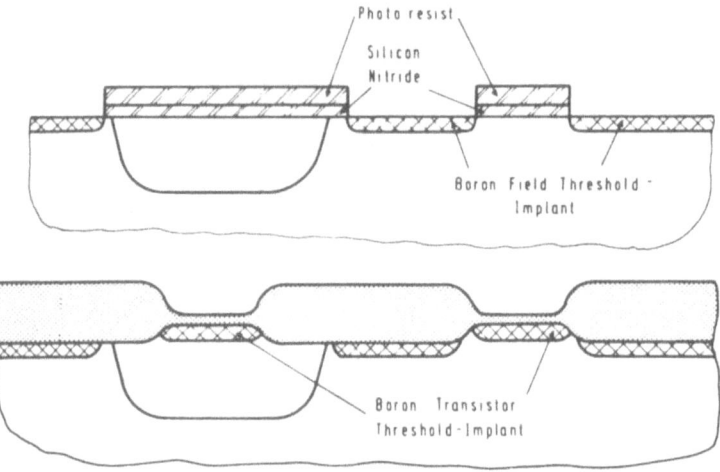

Photo resist

Silicon
Nitride

Boron Field Threshold -
Implant

Boron Transistor
Threshold-Implant

(b) Field and Transistor   Threshold-Implants for Isoplanar Process

Fig. 12 - Cross sections showing threshold implants for planar
(a) and iso-planar (LOCOS) (b) processes.

Then the field and transistor threshold adjustments are carried out independently by two boron implantations (Fig. 12(b)). The second implantation through the gate oxide provides the adjustment of the n-channel transistors for symmetric threshold voltages.

2.3 New MOS Technologies

Increasing speed of integrated circuits can be obtained by decreasing parasitic capacitances. This is intended in SOS technology by working on a 1 μm thick monocrystalline silicon layer deposited on an insulating substrate. However, technological and cost problems prevented the breakthrough of CMOS-SOS to high volume production. Nevertheless, the benefits of an isolated substrate are attractive further on. As an alternative, CMOS technology on buried oxide[63] could become important in the future.

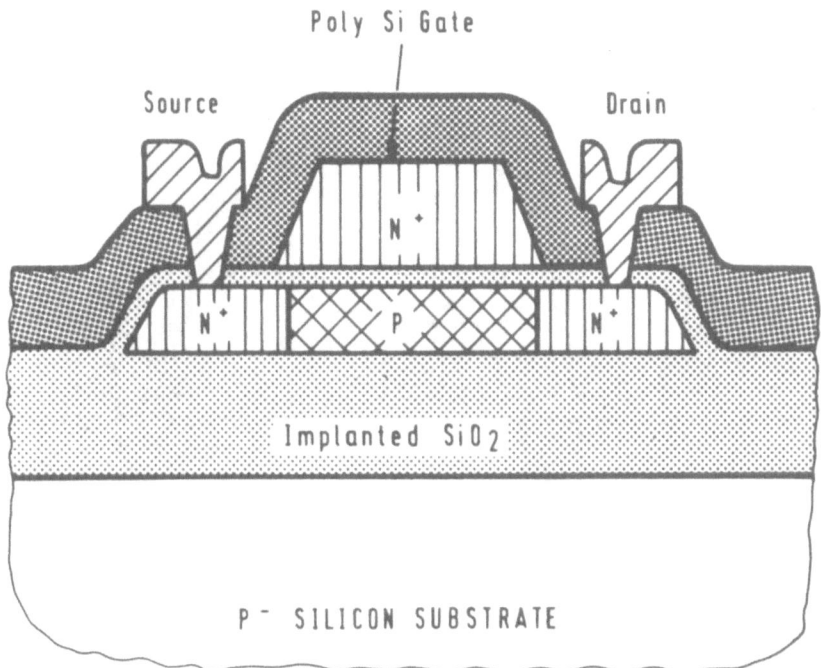

Fig. 13 - NMOS transistor on buried, implanted SiO_2.

A buried oxide layer is formed in a bulk silicon substrate by a high-dose oxygen implantation and high-temperature annealing. An epitaxial layer is grown on the thin monocrystalline silicon layer, which remains on the buried oxide after annealing. Si gate CMOS transistors are realized in the epitaxial layer similar to SOS technology (Fig. 13). Izumi *et al.*[63] have reported data for propagation delay and power dissipation of a CMOS ring oscillator, which are competitive to the existing technologies. One drawback is the high dose of $2 \cdot 10^{18}$ cm^{-2} for the oxygen implantation, which overstrains even existing high current implanters with respect to economical production facility.

Another new approach in CMOS technology is the attempt by Gibbons and Lee[44] to stack a second active device layer upon the monocrystalline transistor structure. Thereby a vertical CMOS inverter strucure is originated (Fig. 14). The NMOS transistor on the top is fabricated in laser-recrystallized polysilicon and is separated from the underlying PMOS transistor by the joint gate.

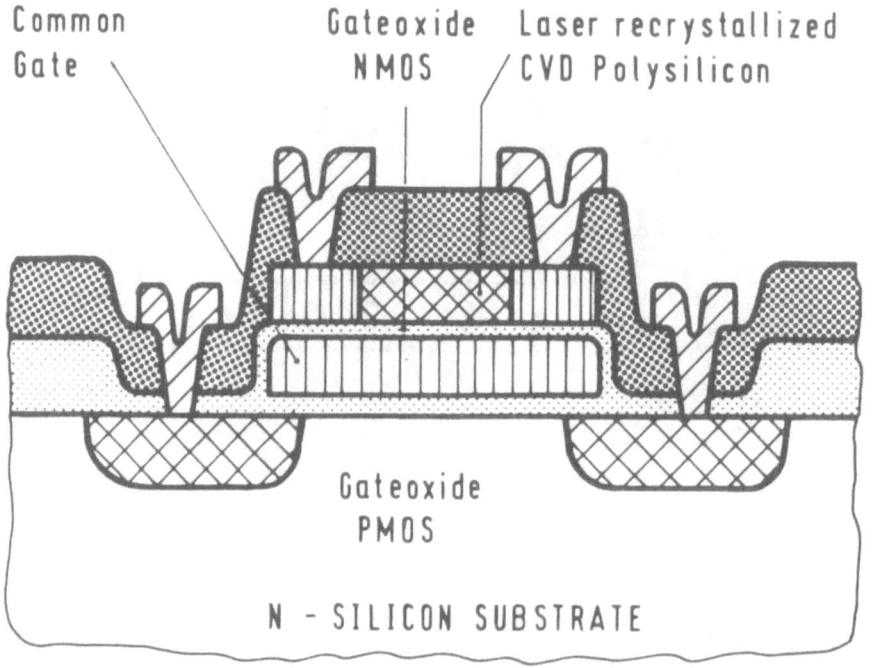

Fig. 14 - Vertical CMOS inverter structure.

3. MOS DEVICE MODELS

3.1 <u>Strong-Inversion</u> or <u>Drift-Control</u> <u>Mode</u>[27,46,55]

Figure 15 shows the schematic cross section of an NMOS transistor and the schematic distribution of several quantities along its channel. They are the surface potential $\Phi(O,y)$, the charge density $Q'(y)$, the longitudinal electric field along the interface $E_y(O,y)$, the transverse electric field along the interface $E_x(O,y)$ and along the bottom of the channel $E_x(x_C,y)$, the longitudinal and transverse field gradients along the interface and the longitudinal electron drift velocity v_y. The selected operating condition is that of "pinch-off," which is difficult to define and probably the most essential limiting condition in small MOS transistors. In order to organize the approach, it is assumed that the longitudinal electric field is E_G, at the drain end

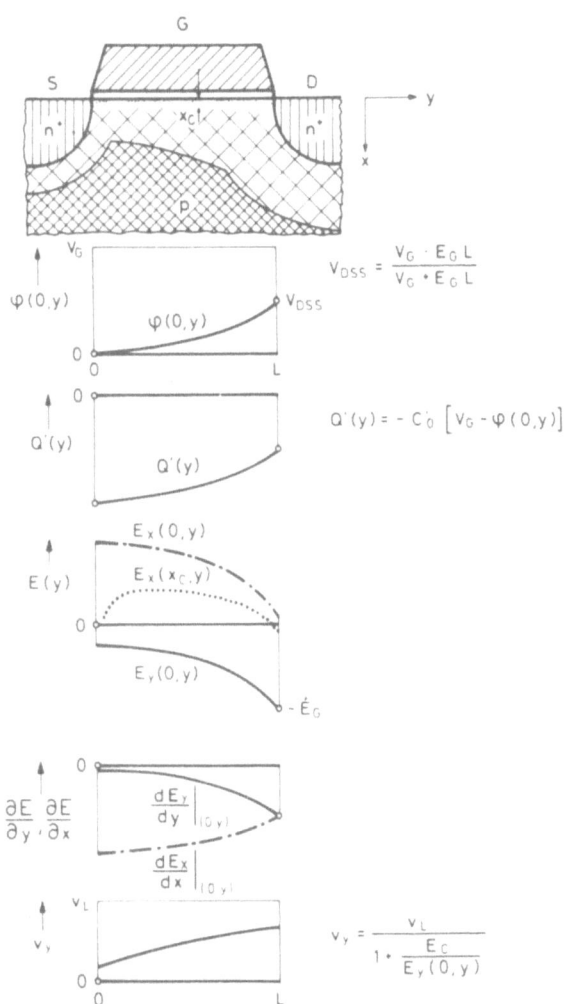

Fig. 15 - Schematic cross section and schematic distributions of potential, charge density, electric field, field divergences and longitudinal velocity in our NMOS transistor.

$y = L$ of the channel and that the drain-source voltage drop in this case is $V_{DS} = V_{DSS}$. In

Fig. 15, $V_{DSS} < V_G$. This suggests that the current available will be less than the famous

$$I_{DO} = \frac{\beta_0}{2} V_G^2, \quad \beta_0 = \frac{\varepsilon_I \mu_0}{t_I \dfrac{W}{L}}$$

This reduction is very significant in short-channel transistors, making a quantitative interpretation rather essential.

In order to see the velocity effect, neglect all other effects like threshold voltage dependence on V_{DS} and V_{SB} and the mobility reduction due to transverse surface fields. In the triode region, i.e. for $V_{DS} < V_{DSS}$,

$$I_D = \frac{\beta_0}{1 + V_{DS}/(E_C L)} \left(V_G - \frac{V_{DS}}{2} \right) V_{DS} \tag{1}$$

Observing the channel at $y = L$, the current can also be written as

$$\begin{aligned} I_D &= Q'(L) \, W \, v_y \, (L) \\ &= \frac{\varepsilon_I}{t_I} \, (V_G - V_{DS}) \, W \, \frac{\mu_0 \, E_y \, (L)}{1 + E_y \, (L)/E_C} \\ &= \frac{\beta_0}{1 + E_y \, (L)/E_C} (V_G - V_{DS}) \, E_y \, (L) \, L \end{aligned} \tag{2}$$

The combination of Eq. (1) and (2) yields a relationship between V_G, V_{DS} and $E_y(L)$:

$$V_G = V_{DS} \, \frac{E_y \, (L) \, L + \dfrac{1}{2} \, V_{DS}[E_y \, (L)/E_C - 1]}{E_y \, (L) \, L - V_{DS}}$$

At pinch-off, $V_{DS} \rightarrow V_{DSS}$ and $E_y \, (L) \rightarrow E_G$:

$$V_G = V_{DSS} \, \frac{E_G L + \dfrac{1}{2} V_{DSS} \, (\gamma - 1)}{E_G L - V_{DSS}} \tag{3}$$

$$\gamma = E_G/E_C. \tag{4}$$

The current can also be written as

$$I_D = \frac{\beta_0}{2} V_{DS}^2 \frac{1}{1 - V_{DS}/[E_y(L)\, L]} \tag{5}$$

The transconductance is, following Eq. (1),

$$g_m = \frac{\partial I_D}{\partial V_G} = \frac{\beta_0 V_{DS}}{1 + V_{DS}/(E_C L)}$$

and together with Eq. (5),

$$g_m = \frac{2 I_D}{V_{DS}} \frac{1 - \dfrac{V_{DS}}{E_y(L)\, L}}{1 + \dfrac{V_{DS}}{E_C L}}$$

The output conductance is, following Eq. (2),

$$g_D = \frac{\partial I_D}{\partial V_{DS}} = \frac{I_D}{E_y(L)\, L\, [1 + V_{DS}/(E_C L)]}$$

The available voltage gain is

$$\frac{g_m}{g_D} = 2 \left[\frac{E_y(L)\, L}{V_{DS}} - 1 \right]$$

At pinch-off, $V_{DS} \rightarrow V_{DSS}$, $E_y(L) \rightarrow E_G$, $g_m \rightarrow g_{mS}$, $g_D \rightarrow g_{DS}$ so that

$$g_{mS} = \frac{2I_{DS}}{V_{DSS}} \; \frac{1 - \dfrac{V_{DSS}}{E_G L}}{1 + \dfrac{V_{DSS}}{E_C L}} \quad, \tag{6}$$

$$g_{DS} = \frac{I_{DS}}{E_G L \left[1 + V_{DSS}/(E_C L) \right]} \quad, \tag{7}$$

and the available voltage gain

$$A_v \equiv \frac{g_{mS}}{g_{DS}} = 2 \left(\frac{E_G L}{V_{DSS}} - 1 \right) . \tag{8}$$

This model offers expressions for such important short-channel characteristics as reduced current, closed-form pinch-off voltage as a function of effective gate voltage, and a finite output resistance as well as a finite available voltage gain. The major control parameter is the pinch-off field E_G. The extreme $E_G \rightarrow \infty$ yields the ultimate maxima of saturated current, $I_{DS\infty}$ and pinch-off voltage, $V_{DS\infty}$:

$$\lim_{E_G \to \infty} V_{DSS} \equiv V_{DS\infty} = \sqrt{E_C L(2V_G + E_C L)} - E_C L \tag{9}$$

associated with a maximum current

$$\lim_{E_G \to \infty} I_{DS} \equiv I_{DS\infty} = \frac{\beta_\theta}{2} V_{DS\infty}^2$$

$$= k v_L \frac{\epsilon_1}{t_1} W V_G \tag{10}$$

$$k \equiv 1 - \frac{E_C L}{V_G} \left(\sqrt{1 + 2 \frac{V_G}{E_C L}} - 1 \right)$$

In some recent publications,[38,53] velocity saturation effects were modeled as limiting the

current to

$$I_{D\infty} = \frac{\varepsilon_I}{t_I} WV_G v_L.$$

This estimate is too optimistic as clearly shown by Eq. (10) and its graphical representation in Fig. 16. This figure shows k as a function of the effective gate voltage normalized to the product of hot-carrier critical field E_C and channel length L. Large errors are incurred by using $I_{D\infty}$, when current drive capability of small MOS transistors is estimated. It also shows that the discrepancy improves only slowly with increasing gate voltage: At $V_G/(E_C L) = 1$, k ≈ 0.3 and a ten times increase to $V_G/(E_C L) = 10$ makes k ≈ 0.6.

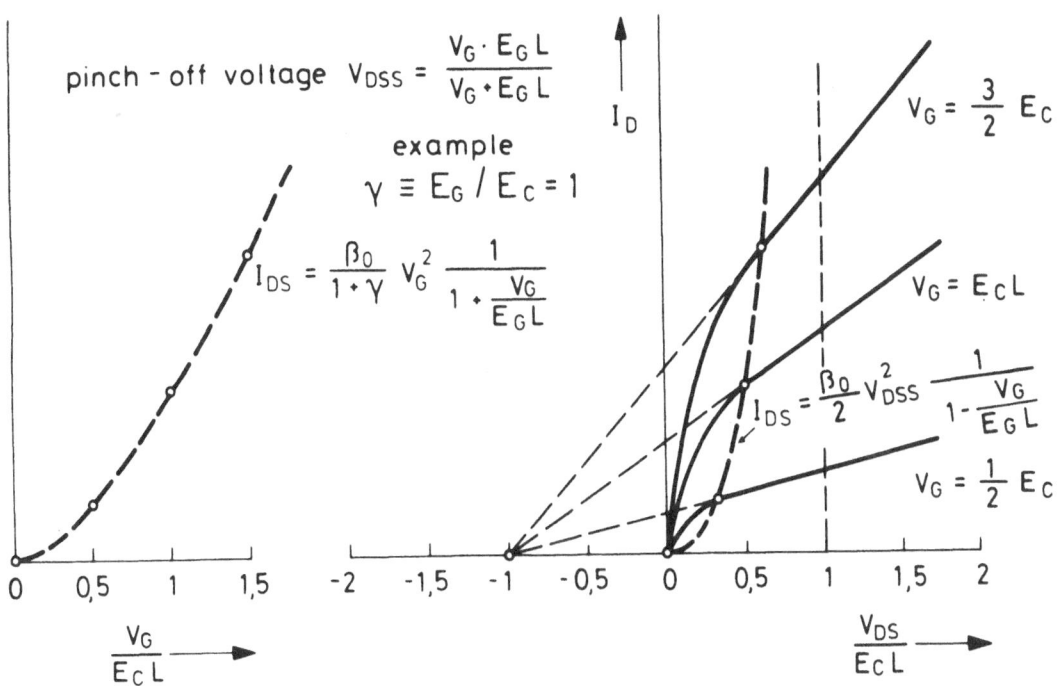

Fig. 17 - Model characteristics for a short-channel transistor, if the pinch-off field E_G is chosen to be the hot-carrier critical field E_C.

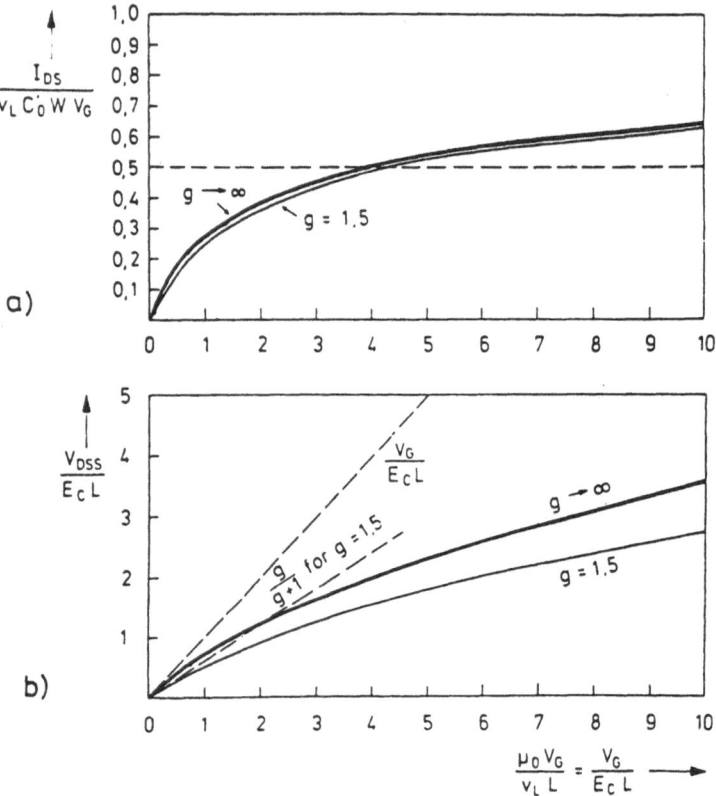

Fig. 16 - Normalized pinch-off current I_{DS} (a) and pinch-off
voltage V_{DSS} (b) for two gain constants g.

Frequently, the pinch-off field E_G has been identified with the hot-carrier critical field E_C. There is no physical reason for having this particular constant pinch-off field, but it yields very simple characteristics as plotted in Fig. 17. If an Early voltage V_A is defined by writing,

$$g_{DS} = \frac{I_{DS}}{V_{DSS} + V_A}$$

this would have the constant value $V_A = E_C L$, if $E_C = E_G$. The available voltage gain would be

$$A_v = 2 \frac{E_C L}{V_G}.$$

A variety of suggestions to identify pinch-off have appeared in the literature. With reference to Fig. 15, they can be described as follows:

A) $E_x(o,y) = -E_y(o,y) = -E_G$ [66]

B) $E_x(o,y) = 0$ [67]

C) $E_x(s_c,y) = 0$ [68]

D) $\dfrac{dE_y}{dy}_{(o,y)} = \dfrac{1}{f}\dfrac{dE_x}{dx}_{(o,y)}$ [60,70,17,27]

(11)

All of these conditions are likely to occur at pinch-off within a small interval near $y = L$, and for $V_{DS} > V_{DSS}$, these loci will move towards the source and cause channel-length modulation. Assumption D) can be interpreted as follows: $dE_x/dx \gg dE_y/ky$ means gate control, $dE_y/dy \gg dE_x/dx$ means drain control, and the transition between these cases marks pinch-off. This equation can also be written in fairly simple terms so that it was selected before[27] and is exploited further in the following. From,[27]

$$\frac{dE_y}{dy} \approx \frac{E_C}{2L}\left(\frac{E_G L}{V_{DSS}} - 1\right),$$

(12)

which is a fair approximation for $V_G < 12E_CL$. If free carriers are neglected,

$$\frac{dE_x}{dx} = \frac{qN_B}{\varepsilon_s}$$

so that

$$E_GL = (g_o + 1) V_{DSS} \tag{13}$$

with

$$g_o = \frac{2qN_BL}{f\varepsilon_sE_C}. \tag{14}$$

An optimization of short-channel current and available voltage gain requires a large g_o and Eq. (14) indicates some directions in design. However, there are tradeoffs: a large N_B decreases μ_o and, correspondingly, increases E_C. Corresponding transistor characteristics, Fig. 18, are quite realistic, and a good correspondence with experimental results was possible.[17,18] The available voltage gain is

$$A_v = 2g_o.$$

As mentioned, some experimental data shows that insulator thickness as well as channel thickness play a role. This will be pursued in the following section.

In the g model, the pinch-off voltage is, if $E_GL = (g + 1) V_{DSS}$,

$$V_{DSS} = \frac{g + \frac{1}{2}}{g + 1} E_CL \left\{ \sqrt{1 + \frac{2g(g+1)}{(g+\frac{1}{2})^2} \frac{V_G}{E_CL}} - 1 \right\}. \tag{15}$$

The short-channel transistor model is completed by adding well-known effects in field-effect

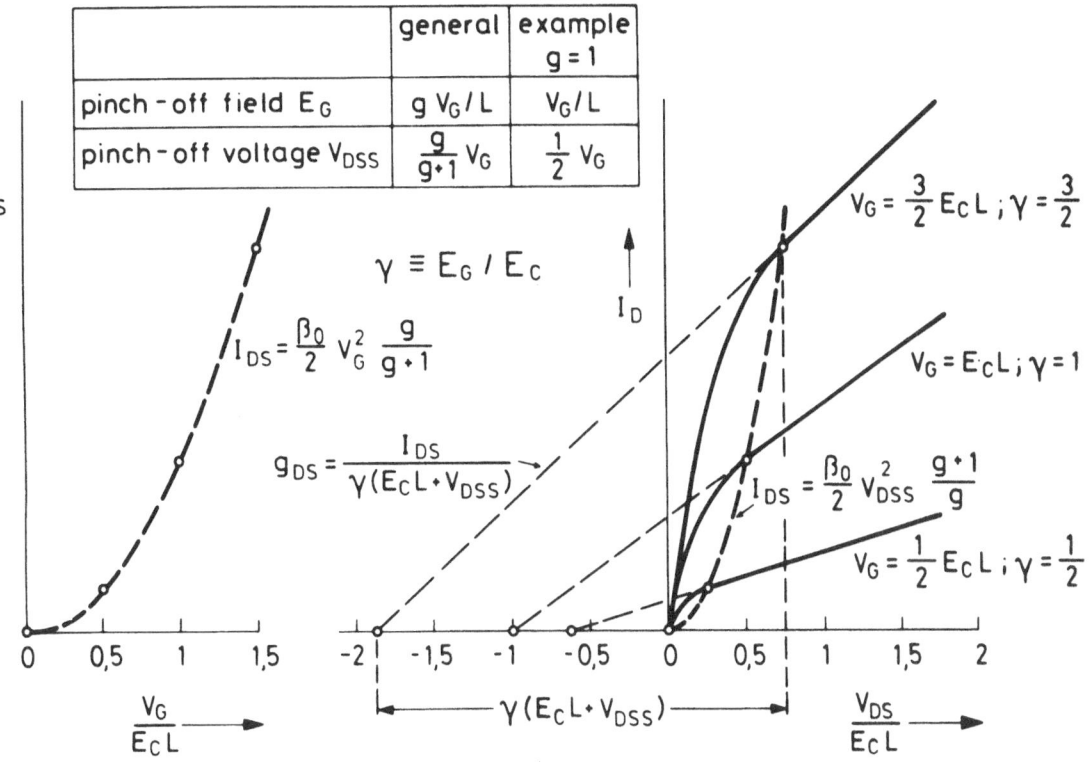

Fig. 18 - Model characteristics for a short-channel transistor, if the pinch-off field E_G is proportional to the gate voltage.

transistors. They are the mobility reduction due to transverse electric fields modeled as

$$\mu_0 = \frac{\mu_{00}}{1 + \Theta\,(V_{GS} - V_{FB})} \tag{16}$$

and the dependence of the threshold voltage on the drain voltage, modeled as

$$V_T = V_{T_0} - K_3\,(\sqrt{V_{DS} + \Phi_D} - \sqrt{\Phi_D})\,, \tag{17}$$

where V_{FB} is the flat-band voltage and Φ_D the junction built-in voltage.

With Eq. 16, 17, the drain current I_D is

$$I_D = \frac{\mu_0}{1 + \frac{\mu_o V_{DS}}{v_L L}} \frac{\varepsilon_I}{t_I} \frac{W}{L} \left[(V_{GS} - V_T) V_{DS} - \frac{1}{2} V_{DS}^2 \right] \qquad (18)$$

for $V_{DS} \leq V_{DSS}$.

The pinch-off voltage is preferably modeled in the form of Eq. (15) and the pinch-off current is

$$I_{DS} = \frac{\beta_0}{2} V_{DSS}^2 \frac{g+1}{g} . \qquad (19)$$

For $V_{DS} \geq V_{DSS}$, the simplest model is

$$I_D = I_{DS} + g_{DS} (V_{DS} - V_{DSS}),$$

with $g_{DS}^{-1} = r_{DS}$ from Eq. (7) so that

$$I_D = I_{DS} \left[1 + \frac{V_{DS} - V_{DSS}}{(g + 1) V_{DSS} \left(1 + \frac{V_{DSS}}{E_C L} \right)} \right] . \qquad (20)$$

A more complex version for $V_{DS} \geq V_{DSS}$ is based on actually modelling channel-length modulation.[27]

The total parameter set now includes

μ_{00} :   flat-band mobility

v_L :   scattering-limited velocity

Θ :   mobility reduction factor for transverse fields

V_{To} :   threshold voltage for zero drain bias

K_3 :   drain-effect constant

g :   gain parameter

The result is a model with 6...7 parameters supplemented by material and geometry data like ε_s, ε_I, t_I, W,L.

Its application will be described in section 3.3.

3.2 <u>Weak-Inversion</u> or <u>Barrier-Control</u> <u>Mode</u>

In a potential or band diagram description, the MOS transistor channel region presents a barrier between source and drain. The height Φ_B of this barrier is controlled by gate, drain and bulk potentials. Whenever the barrier height is reduced to zero, carriers move predominantly under drift control, the transistor is in its familiar ohmic or strong-inversion regime, which was described so far in this chapter. However, there are increasingly important modes of current flow across any sufficiently lowered barriers between source and drain. They have been called sub-threshold, weak-inversion, punch-through and static-induction modes.[11,22,48,56] Their characteristics are exponential current-voltage relations:

$$I_D \sim e^{\Phi_B/V_t}$$

In long-channel transistors,

$$\Phi_B \sim \frac{1}{n} V_{GS}, \quad n \approx \frac{C_o + C_d}{C_o} , \tag{21}$$

where C_o is the gate capacitance and C_d the capacitance of the surface depletion-layer. The transconductance in this case would be

$$g_m = \frac{I_D}{nV_t}. \tag{22}$$

For a full small-signal characterization, the output conductance is needed, too:

$$g_{DS} \approx \frac{K_2 I_D}{2LF\sqrt{V_{DS} + \Phi_D}}. \tag{23}$$

Distinct features occur in the control of the barrier, if the transistor channels become short. Figure 19 shows a schematic representation of a short-channel configuration. Here, a) is a plot of equipotential and field lines, b) shows sections of field and potential along the barrier in the x direction and c) shows a section of the potential along the interface. It is important to note that field lines running across the barrier start at the drain, whereas in long-channel devices, these field lines start at the gate and terminate on substrate impurities. The result is drain-induced barrier lowering (DIBL),[56] which has also been called the second-gate effect or static induction.[11] Its simplest model is

$$\Phi_B = \Phi_0 + AV_{GS} + BV_{DS}. \qquad (24)$$

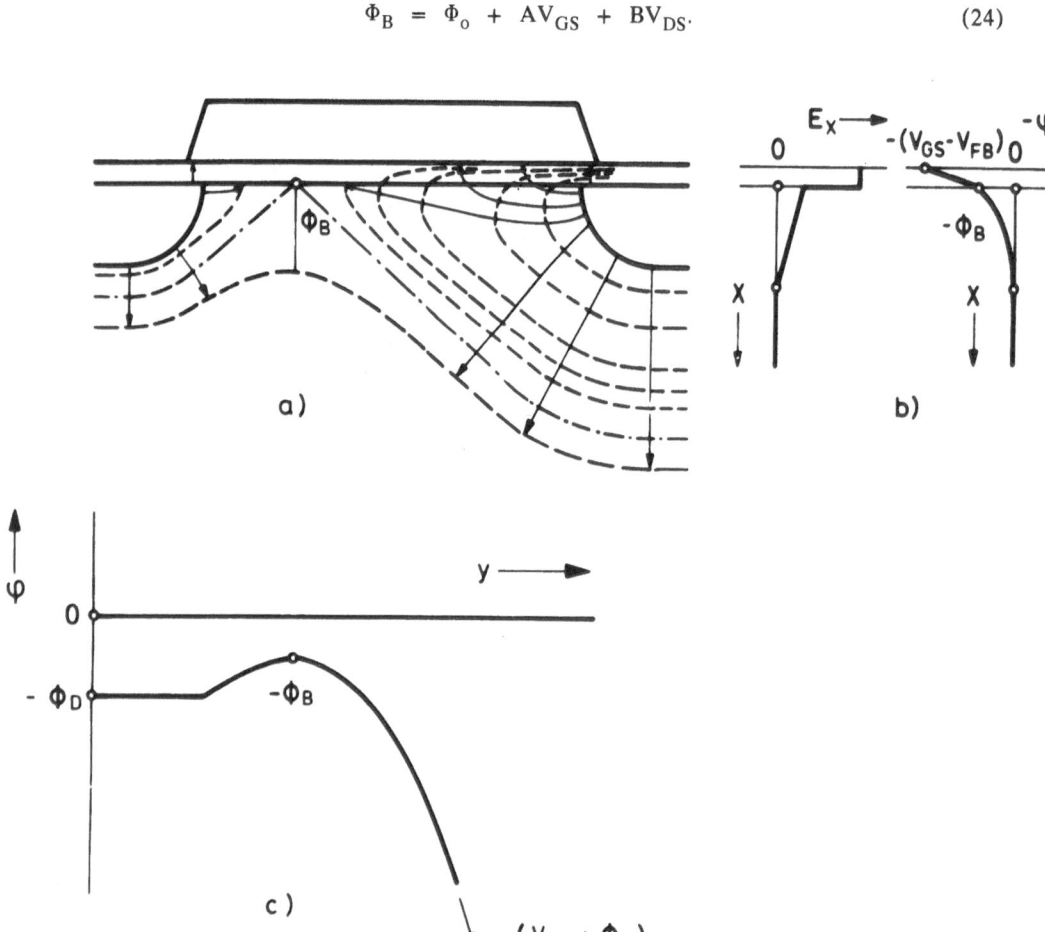

Fig. 19 - Schematics of barrier control in a fairly short transistor.

V_{GS} is the gate voltage relative to the flat-band voltage, Φ_o is a reference potential, $A < 1$ and usually $B << A$. An interpretation of A and B was given in,[10] based on drain field lines following a quarter-circle with radius L to the barrier. Then,

$$A = 1 - \xi$$

$$B = \xi \qquad .$$

$$\xi = \frac{2 \varepsilon_{Si} \, t_l}{\pi \varepsilon_l L}$$

More advanced modeling[45] relates the parameters to technology data and it shows useful parameter correlations:

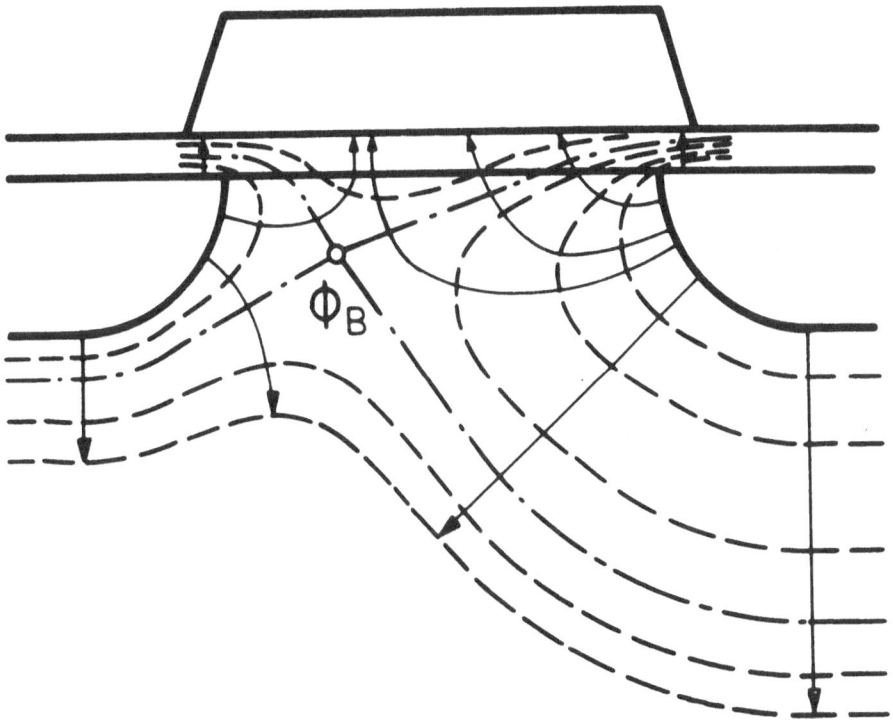

Fig. 20 - Schematics of barrier control in a short transistor and for potentials that cause a buried ridge of height ϕ_B.

$$A = 1 - 2\eta$$

$$B = \eta \tag{25}$$

$$\eta = \frac{\varepsilon_{Si}}{\varepsilon_I} \frac{t_I}{\ell\,(V_{DS}, V_{SB})} \sqrt{\frac{N_1}{N_2}}$$

ℓ is related to and somewhat larger than L. N_1 and N_2 are average shallow- and deep-implant concentrations in the channel, respectively.

With progressively shorter channels (or larger drain voltages), a transition occurs to a situation, shown schematically in Fig. 20. All field lines terminate on the gate. The barrier minimum Φ_B occurs at some depth under the interface as a saddle point in the potential contour diagram. A three-dimensional plot of quantitative two-dimensional analysis[48] of a 0.8 μm multiple-implant transistor is shown in Fig. 21. This mode has been called the punch-through mode, which will eventually have a transition into the space-charge-limited (SCL) mode,[5] if the mobile carrier densities become larger than the impurity densities. As long as barrier control is dominant, both the transfer and output characteristics will be exponential functions as shown in Fig. 22 in a semilog plot. These characteristics have also been called triode-like. A simple model was given in:[48]

$$\Phi_B = \Phi_o + AV_{GS} + BV_{DS} - CV_{GS}V_{DS}. \tag{26}$$

This extension of Eq. (24) is illustrated by the thin lines in Fig. 8. In this barrier-control mode, the transconductance is

$$g_m = \frac{I_D}{(A - CV_{DS})V_t} \quad , \tag{27}$$

and the output conductance is

$$g_{DS} = \frac{I_D}{(B - CV_{GS})V_t} \quad , \tag{28}$$

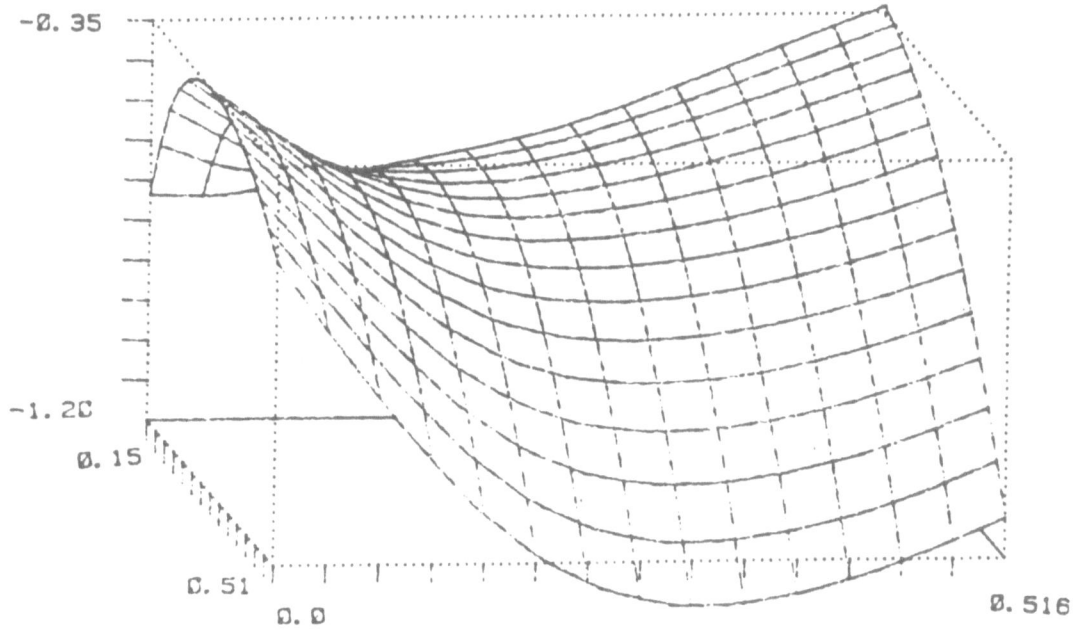

Fig. 21 - Close-up of a two-dimensional buried potential ridge.

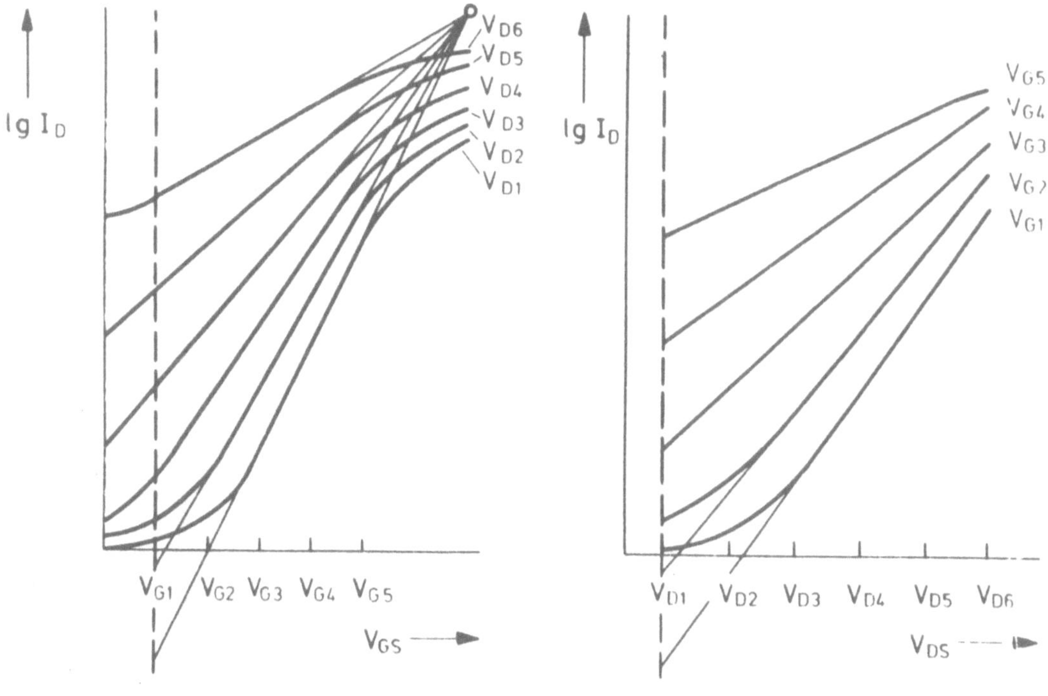

Fig. 22 - Semilog plots of transfer and output characteristics of a 0.8μm MOS transistor in the barrier-control regime.

so that the available voltage gain

$$A_{V,\ max} = \frac{A - CV_{DS}}{B - CV_{GS}}. \tag{29}$$

In the simplified model of Eq. (29),

$$A_{V,\ max} = \frac{1}{\eta} - 2. \tag{30}$$

In the short-channel case, it is then required to make L/t_I as large as possible and to implement the benefits of double channel implants. It should be noted that a small parameter η, Eq. (25), also means minimum short-channel and drain-bias effects on the transistor threshold voltage.

The current in the barrier-control mode can be modeled as

$$I_D = \frac{qD_n x_c W}{L_B}\ n_S\ , \tag{31}$$

where

$$n_S = n_o e^{\Phi_B}$$

and the channel thickness

$$x_c = \frac{V_t}{E_X(o)}\ . \tag{32}$$

The barrier or base width L_B is only a fraction of the source-drain length L. Actually, L_B and

x_C are functions of V_{GS}, V_{DS} and V_{SB}. Approximate capacitances can be derived as

$$C_{GS} = \frac{\partial Q_n}{\partial V_{GS}} \approx g_m \frac{L_B^2}{2D_n} ,$$

$$C_{GD} = \frac{\partial Q_n}{\partial V_{GD}} \approx g_{DS} \frac{L_B^2}{2D_n} .$$

$$(33)$$

Two-dimensional analyses are required for recent short-channel transistors. It should be noted that $L_B \ll L$ in devices with $L \leq 2\mu m$, so that such devices may have large gain-bandwidth products.

A simple field-line model is supplemented in order to explain short-channel barrier control. Figure 23 shows the situation for a short-channel surface barrier of height Φ_s schematically. An NMOS transistor is assumed with shallow and deep impurity concentrations N_1 and N_2, respectively. Field lines running across the barrier Φ_s emerge at the gate and eventually hit an E=0 point at some depth y_c. This is also the end point of field lines emerging at the source and drain, respectively. Figure 23(b) and (c) shows the electric field and the potential along the paths AB and AC. With reference to this figure,

$$V_{DS} = \frac{qN_2}{2\epsilon_{Si}} (x_D^2 - x_S^2),$$

$$L = x_D + x_S,$$

$$x_D = \frac{L}{2} + \frac{\epsilon_{Si} V_{DS}}{qN_2 L},$$

$$(34)$$

$$V \equiv V_{DS} + \Phi_{BI} - V_{GS} = \frac{qN_2}{2\epsilon_{Si}} x_D^2 - \frac{qN_1}{2\epsilon_{Si}} y_C^2 - \frac{qN_1 y_c}{C_o'} ,$$

$$(35)$$

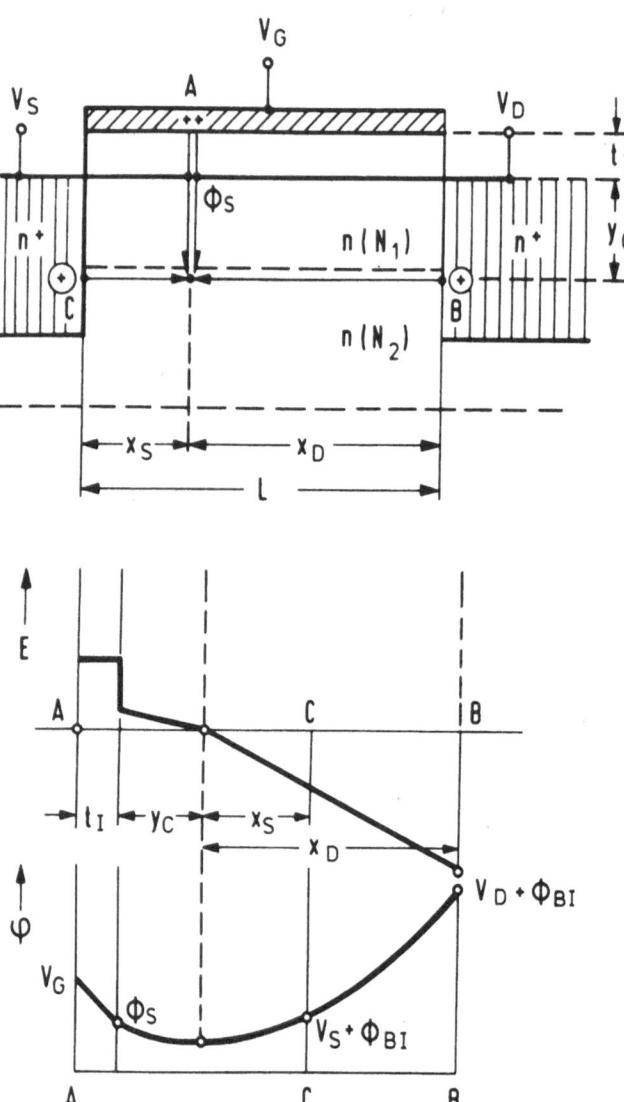

Fig. 23 - Simplified field line model for drain-induced barrier lowering.

$$\Phi_S = V_{GS} - \frac{qN_1 y_c}{C_0'} \quad . \tag{36}$$

From Eq. (35),

$$y_c = \sqrt{\frac{N_2}{N_1} x_D^2 - \frac{2\varepsilon_{Si} V}{qN_1} + \frac{\varepsilon_{Si}^2}{C_0'^2}} - \frac{\varepsilon_{Si}}{C_0'}. \tag{37}$$

With $x_D^2 \approx \dfrac{L^2}{4} + \dfrac{\varepsilon_{Si} V_{DS}}{qN_2}$ and eta identical $< \varepsilon$ sub Si $>$ over $< C$ sub o prime L $>$ sqrt $< <$ N sub 1 $>$ over $<$ N sub 2 $> >$

inserted in Eq. (37) and the latter inserted in eq. (36),

$$\Phi_S = \frac{qN_1 \varepsilon_{Si}}{C_0^{2\prime}} + V_{GS} - q \frac{\sqrt{N_1 N_2} L}{2C_0'} \sqrt{1 - \frac{4\varepsilon_{Si}}{qN_2 L^2}(V_{DS} + 2\Phi_{BI} - 2V_{GS}) + 4n^2} \tag{38}$$

If the square-root term is expanded,

$$\Phi_S \approx \frac{qN_1 \varepsilon_{Si}}{C_0'} (1-\eta) + 2\eta\Phi_{BI} - q \frac{\sqrt{N_1 N_2} L}{2C_0'} + (1-2\eta)V_{GS} + \eta V_{DS}. \tag{39}$$

This expression for the barrier has been used in Eq. (24), (25). If Eq. (38) is expanded further,

$$C = \frac{4\varepsilon_{Si}}{qN_2 L^2} \eta$$

is obtained. This may serve as an interpretation of Eq. (26).

3.3 Parameter Correlation and Device Characterization

Some measured transistor characteristics are shown in Figures 24-27. Figure 24 shows the output characteristics of a 1 μm NMOS transistor in strong inversion, where the dashed curves represent the model. The model parameters were extracted from the output conductance g_{DS} in Fig. 25, the maximum available voltage gain in Fig. 26 and the weak-inversion characteristics in Fig. 27. The latter yields the barrier-control parameters I_0, A, B, C.

In Fig. 26,

$$A_{V,O} \approx \frac{A}{B - C(V_{GS} - V_{FB})}. \tag{40}$$

Furthermore, Fig. 27 allows an extraction of the threshold voltage V_T. At a constant reference current,

$$A(V_T - V_{FB}) + BV_{DS} - C(V_T - V_{FB}) V_{DS} = A(V_{T0} - V_{FB}).$$

This hyperbolic relation between V_T and V_{DS} can be approximated by Eq. (17). Next, mobility data is extracted from Fig. 25. The output conductance at $V_{DS}=0$ yields μ_{00} and Θ. Another strong-inversion quantity, the gain constant g, is available from Fig. 26. The strong decrease of $A_v = 2g$ with increasing gate voltage suggests that the mobility term E_C in Eq. (14) may be responsible. With a constant v_L and $v_L = \mu E_C$,

$$A_V = 2g = \frac{4qN_B L\mu_{00}}{f\varepsilon_{Si}v_L[1 + \Theta(V_{GS} - V_{FB})]}. \tag{41}$$

Figure 26 suggests

$$A_V = \frac{A_{V,0}}{1 + \Theta (V_{GS} - V_{T0})} \tag{42}$$

with $A_{v,o}$ from Eq. (4) being the maximum available gain in the barrier-control regime. If $C=0$, $A_{v,o}=1/\eta$ with η from Eq. (25). With $A_v=2g$ from eq. (42), the most significant improvement of the strong-inversion model resulted in the contents of Fig. 24. Theory and experiments indicate the following most basic parameter set for both the barrier- and drift-control models (with dependent parameters in parenthesis).

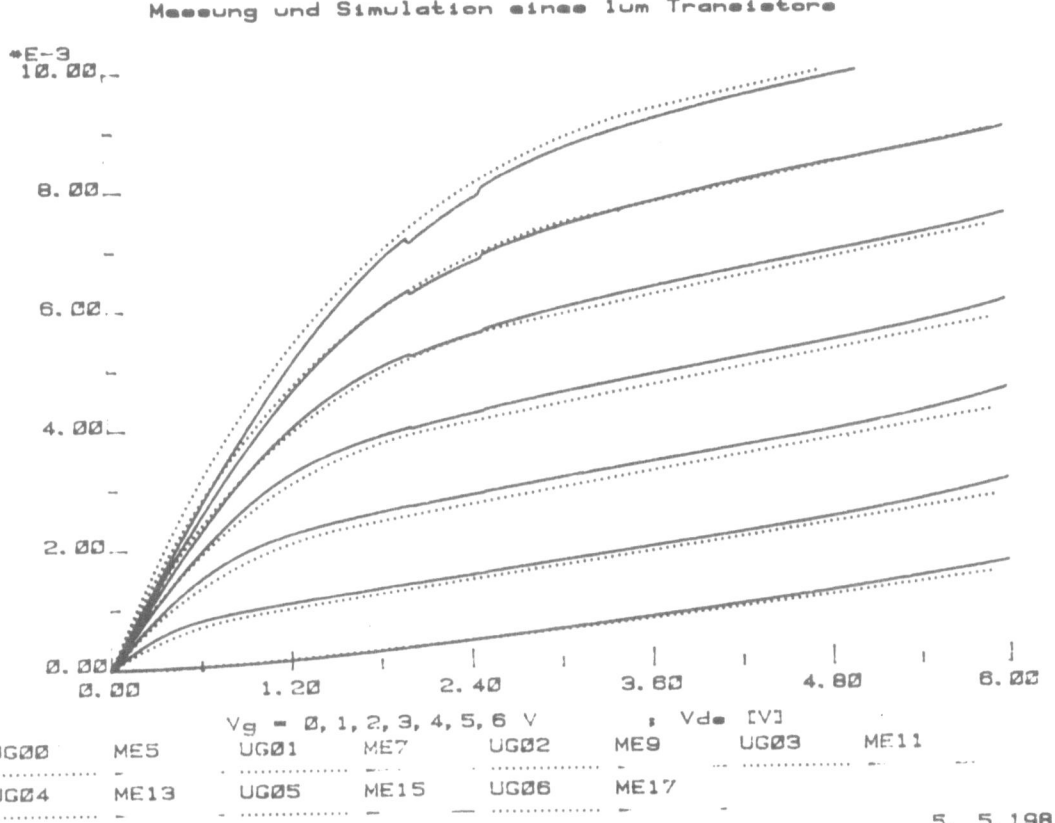

Fig. 24 - Output characteristics of a 1μm silicon gate NMOS transistor. Full lines: measurement. Dotted lines: simulation.

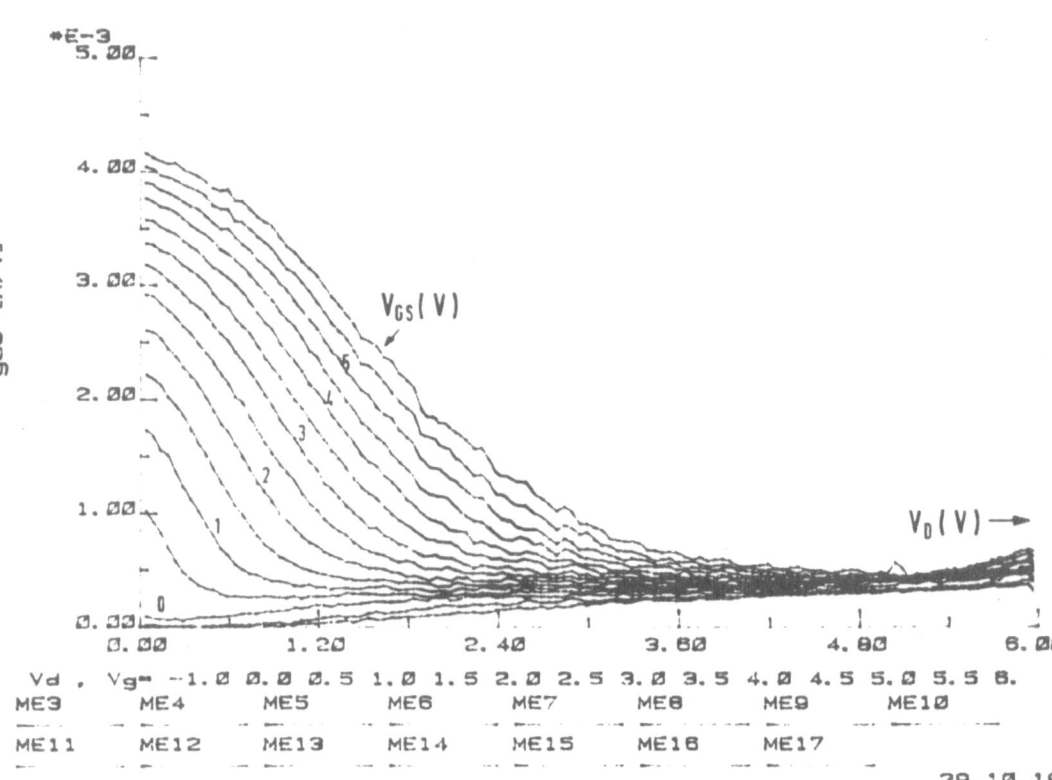

did / dVde ve vde of an 1u Si-gate transistor

Fig. 25 - Measured output conductance of the transistor shown in Fig. 24.

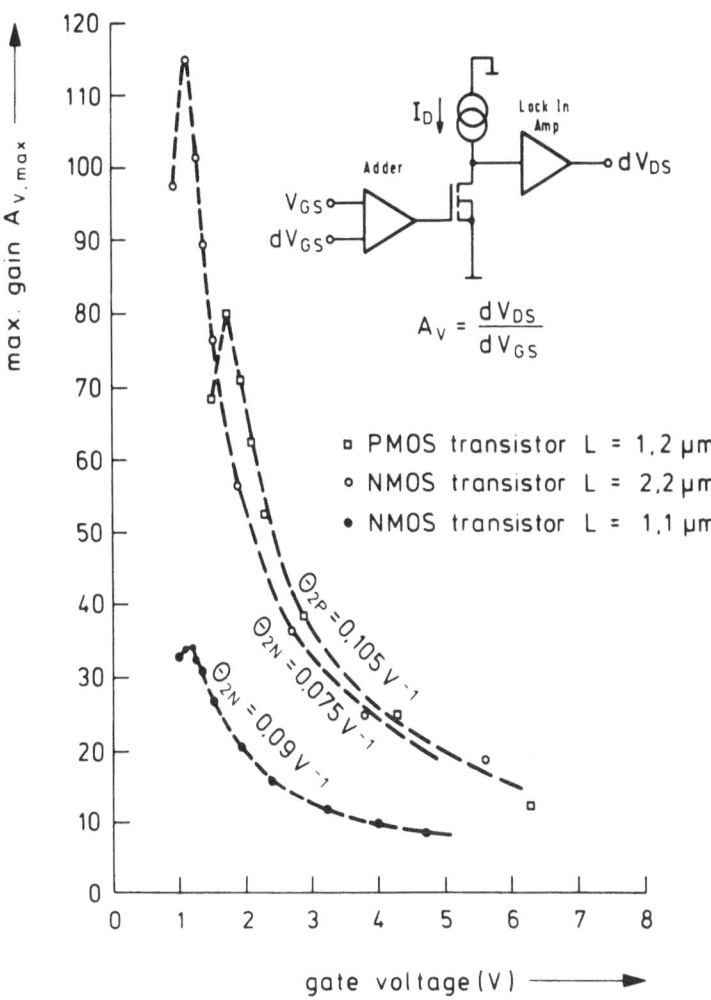

Fig. 26 - Maximum available voltage gains of various short-channel MOS transistors.

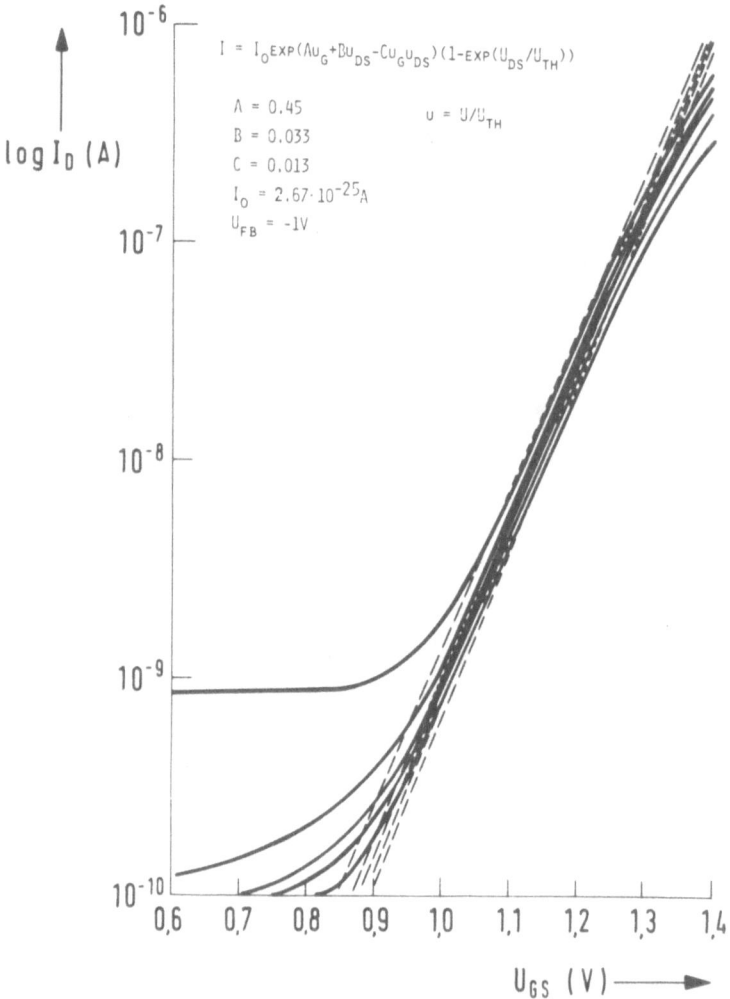

Fig. 27 - Weak-inversion characteristics of a short-channel NMOS transistor.

$$I_o$$

$$\eta(A,B,C,A_{V,O},K_3)$$

$$V_{To}$$

$$\mu_{00}$$

$$\Theta$$

Further work will have to be done to clarify these dependencies, but a physical basis is apparent, which may lead to the important issue of parameter correlation and from there to the statistical simulation and design of MOS devices and circuits.

4. MOS DEVICE PERFORMANCE

4.1 Gain-Bandwidth Products

Gain bandwidth products can be summarized as follows:[55] drift control, long-channel

$$\frac{g_m}{2\pi C_o} \approx \frac{\mu_0 V_G}{2\pi L^2} \sim \frac{\mu_0}{2\pi L}$$

drift control, short-channel

$$\frac{g_m}{2\pi C_O} \approx \frac{\mu V_G}{2\pi L^2} \frac{g}{(g+1)(1+\frac{V_{DSS}}{E_c L})} < \frac{v_L}{2\pi L}$$

barrier control, long-channel

$$\frac{g_m}{2\pi C} \approx \frac{D_n}{\pi(L-L_D)^2} = \frac{\mu_0 V_t}{\pi(L-L_D)^2}$$

barrier control, short-channel

$$\frac{g_m}{2\pi C} \approx \frac{D_n}{\pi L_B^2} \quad ; L_B < L$$

4.2 MOS Inverters

VLSI-type, small MOS transistors, as described in the previous chapter, were found to provide marginal available voltage gains, as expressed in equations (14) and (30). This will cause problems in inverters with respect to the necessary gain and noise margin as discussed in general terms by Keyes.[13] These problems will be treated in the following sections.

A drift-controlled short-channel driver transistor and a resistive load are assumed as a first model inverter. Inverter transfer, driver transfer and driver output characteristics are plotted in Fig. 28. The model transistor of Fig. 18 has been inserted with g = const. The inverter voltage gain A_v is equal to the available voltage gain $2g$ of the driver. Moreover,

$$v_{OH} = \frac{g}{g + 1} (v_{IH} - v_{TE})$$

$$\approx \frac{g}{g + 1} \frac{1}{A_v} = \frac{1}{2(g + 1)}$$

Voltages are normalized to V_{DD}. In order to obtain the noise margin of this inverter, its transfer characteristic has been redrawn in Fig. 29 together with its mirror image. The noise margin NM has been inserted following the definition of Hill.[57] In normalized form,

$$nm = \frac{2g-1}{2g+1} v_{TE} + \frac{g-1}{2g \, (2g + 1) \, (g + 1)}$$

(43)

$$\approx \frac{A_v}{A_v+1} v_{TE}$$

A 25 percent noise margin can be obtained with $v_{TE}=0.375$ and $A_v = 5$.

A driver in the barrier-control mode is assumed in the inverter of Fig. 30. The supply voltage is 0.7V. The mirror characteristic and the noise margin has been added in a). Due to the triode-like output characteristics of the driver with $A_{v,max} < 3$, the noise margin is < 20 percent.

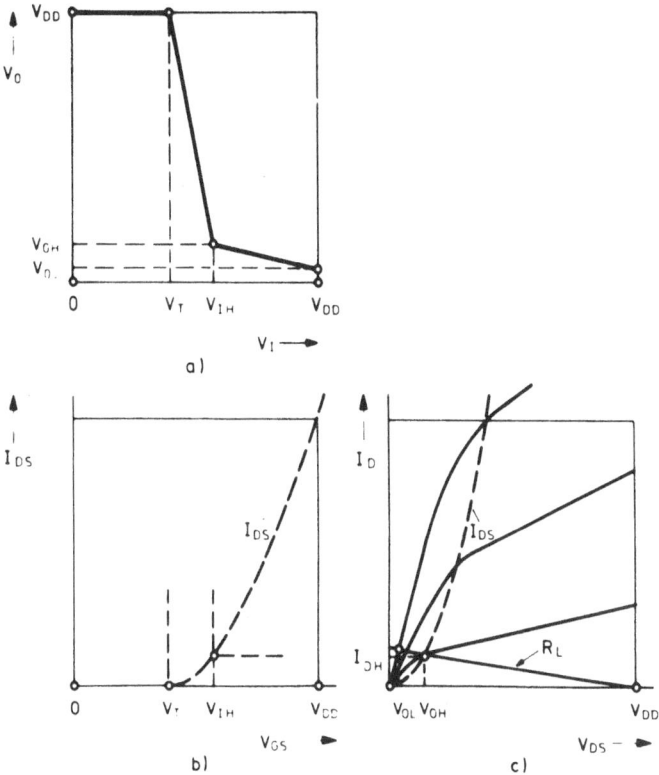

Fig. 28 - Schematics of an NMOS inverter with resistive load and short-channel driver.

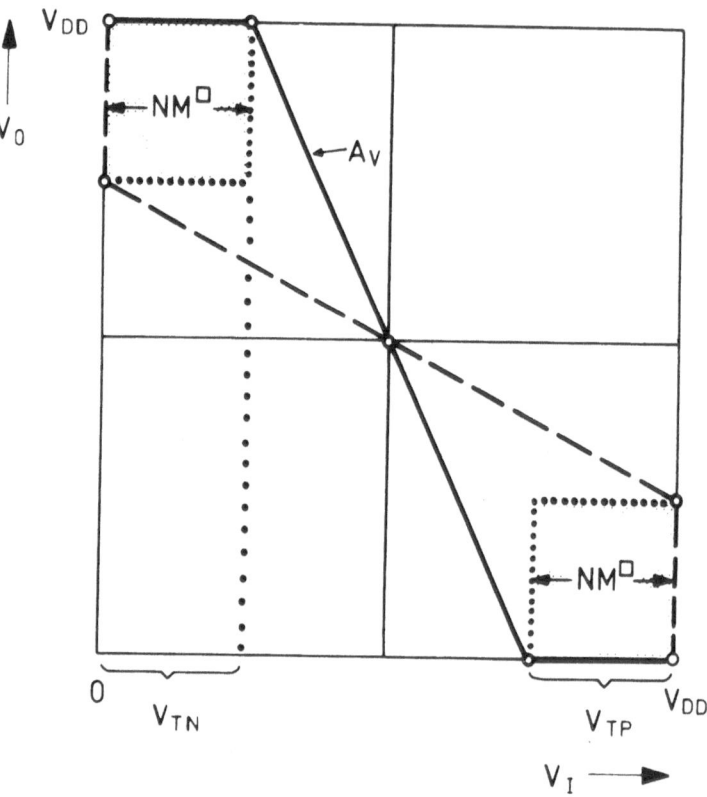

Fig. 29 - Schematics of a CMOS inverter transfer characteristic with noise margin NM.

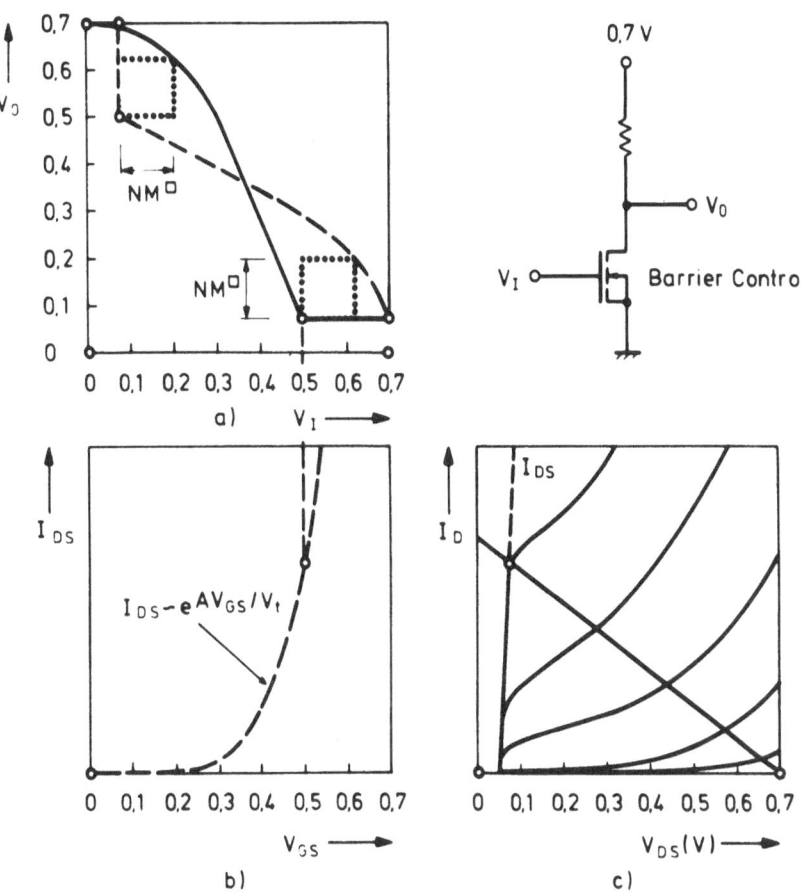

Fig. 30 - Schematics of an NMOS inverter with resistive load and short-channel driver in barrier-control mode.

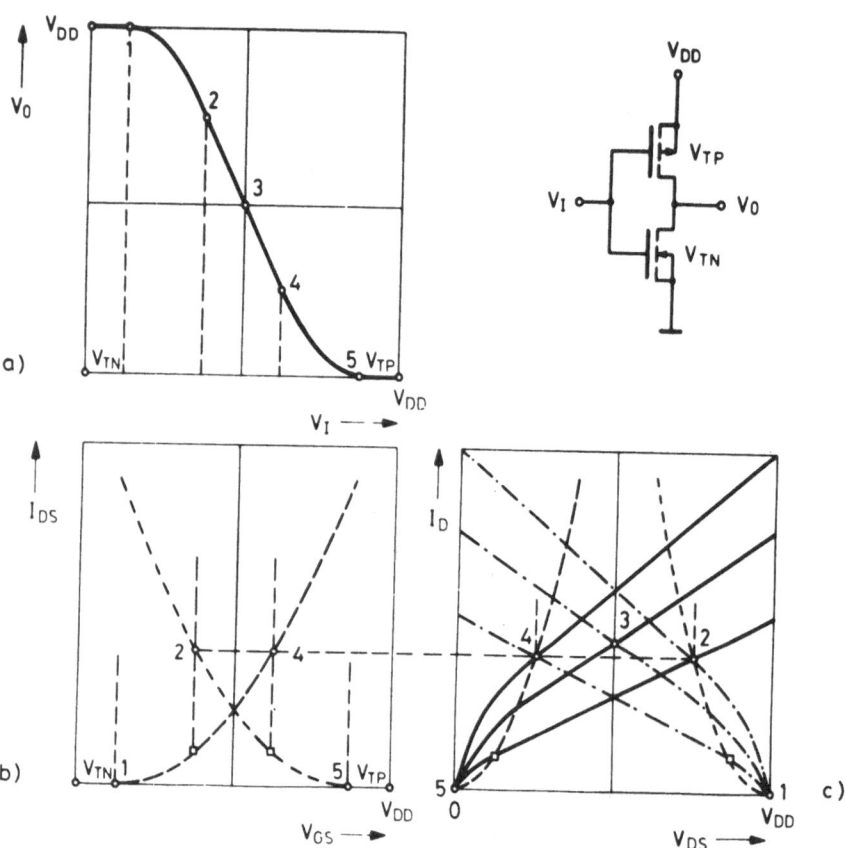

Fig. 31 - Schematics of a short-channel CMOS inverter.

The more severe non-idealities of these inverters are voltage swings smaller than the supply voltage and the unsymmetry as well as temperature dependence of the inverter transfer characteristics.

Adopting the short-channel model of Fig. 18 for both PMOS and NMOS short-channel transistors, the CMOS inverter model of Fig. 31 is obtained, assuming symmetrical transistors for clarity. The available transistor voltage gain is $A_{v,max} = 2$ in the diagrams. It shows up as the slope of the transfer characteristic between points 2 and 4 in Fig. 31.

Despite the low available voltage gain, the characteristics are still usable. Quantitatively,

$$V_{DD, min} = \frac{2V_T}{1 - 1/A_v} \tag{44}$$

in the symmetrical case $V_{TN} = V_{TP} = V_T$. The noise margin $NM = V_T$ in the $V_{DD,min}$ case, and its relative value is

$$\frac{NM}{V_{DD, min}} = \frac{1 - 1/A_v}{2}, \tag{45}$$

which means that a 25 percent margin is obtainable with an internal transistor voltage gain as low as $A_{v,max} = 2$. This is a considerably lower requirement than in the single-channel inverters, where $A_{v,max} = 5$ is needed to obtain the same noise margin. Assuming the same process, the CMOS channel length can be scaled down to $1/3$ of the length required for a single-channel technology. The CMOS inverter allows the relatively lowest supply voltage. Taking $A_v = 2$, $V_{DD,min} = 4 \, V_T$.

For strong-inversion operation and minimized currents in the high- or low-state of the inverter, $V_{T,min} \approx 0.2$ V and $V_{DD,min} \approx 0.8$ V, assuming that about 100 mV of gate voltage are required to raise or lower the current by one decade in the sub-threshold region. Because this region cannot be avoided or excluded in low-voltage design, it might as well be considered for beneficial use. A short-channel situation has to be considered, and the barrier-control model of the previous section may serve to analyze it. Again for the sake of simplicity, symmetrical transistors are assumed. With the available transistor voltage gain of Eq. (29), the inverter voltage gain is

$$A_v = \frac{2A - CV_{DD}}{2B - C\,(V_{DD} - V_{FBN} - V_{FBP})}.$$ (46)

To maintain a 25 percent noise margin, $A_v \geq 2$ and $V_{FBN} = V_{FBP} = V_{DD}/4$. With these figures, Eq. (46) can be satisfied for $A \geq 2B$ or

$$\eta = \frac{\varepsilon_{Si} t_I}{\varepsilon_I L} \sqrt{\frac{N_1}{N_2}} \leq \frac{1}{4},$$ (47)

an important and simple design for this potentially attractive low-voltage mode of operation. In the short-channel domain, $L < 2$ μm, barrier-control modes mean effective base widths L_B smaller or much smaller than L, allowing high-speed operation with very low supply voltage and yet a sufficient noise margin.

This section and the previous one have shown that the available voltage gain of VLSI MOS transistors is an important practical limitation in the evolution of low-power, high-speed digital circuits. Drift-control and barrier-control modes of operation have to be considered, and the present status of modelling can be summarized in the following examples:

1.   Drift control, Eq. (14)

$$A_{v,\,max} = \frac{2qN_B L}{f \varepsilon_{Si} E_c}$$

$\approx 5L$ (μm.), if $N_B = 5 \times 10^{15}$ cm^{-3}, $E_c = 10^4$ Vcm^{-1}, $t_I < 40$ nm, $f \approx 1$

2.   Barrier control, Eq. (12, 30)

$$A_{v,\,max} = \frac{\varepsilon_I L}{\varepsilon_{Si} t_I} \sqrt{\frac{N_1}{N_2}} - 2$$

$\approx 25L$ (μm), if $N_1/N_2 = 10$ and $t_I = 40$ nm.

Actually, two-dimensional process and device simulation are necessary to design technologies at the 1 μ m scale. However, directions are apparent from the simple models discussed so far.

4.3 Delay Times

Inverter delay times t_D depend on the supply voltage V_{DD}, the load capacitance C and the current available for charging and discharging:

$$t_D \sim \frac{V_{DD} C}{I_{DH}} \, .$$

Here, $I_{DH} = I_{DS} (V_{IH})$ has been introduced as a reference current. In the resistive load inverter, the longest time is the rise time t_R, during which the load gate is charged through the equivalent load resistor,

$$R_L = \frac{V_{DD} - V_{OH}}{I_{DH}}$$

of the test gate:

$$t_R \approx 3 \, R_L C = 3 \, \frac{V_{DD} C}{I_{DH}} \frac{A_v + 1}{A_v + 2} \, .$$

The fall time t_F can be approximated by assuming an average discharge current $I_{DM}/2$:

$$t_F \approx 2 \, \frac{V_{DD} C}{I_{DM}} \, .$$

Thus, the delay time will be

$$t_D \approx \frac{t_F + t_R}{2} \approx \frac{V_{DD} C}{2} \left[\frac{3}{I_{DH}} + \frac{2}{I_{DM}} \right] \, .$$

The assumption of an average current is made in CMOS inverters as well[2]:

$$t_R = t_P = 2 \frac{V_{DD}C}{I_{DP}}$$

$$t_F = t_N = 2 \frac{V_{DD}C}{I_{DN}} .$$

The most optimistic current assignment is Eq. (10). This would offer for the resistive-load inverter:

$$I_{DH} = k_H v_L C'_o W \frac{V_{DD}}{A_v} .$$

If $C = F_o C'_o WL$, where F_o is the fan-out and $I_{DM} = k_M v_L C'_o WV_{DD}(1 - v_T)$:

$$t_D \approx \frac{F_o}{2} \frac{L}{v_L} \left[\frac{3A_v}{k_H} + \frac{2}{k_M (1 - v_T)} \right]$$

With the minimum transit-time $\tau_{min} = L/v_L$, $A_v = 5$, $k_H = k_M = 1/3$, and $v_T = 1/3$ this yields

$$t_D \approx 25F_o \tau_{min} .$$

In the CMOS inverter,

$$I_{DP} = k_P v_L C'_o W_P V_{DD} (1 - v_{TP})$$

$$I_{DN} = k_N v_L C'_o W_N V_{DD} (1 - v_{TN})$$

and with $L_P = L_N = L$:

$$\tau_D = \frac{L}{v_L} \left[\frac{C}{C'_o W_P L k_P (1 - v_{TP})} + \frac{C}{L C'_o W_N k_N (1 - v_{TN})} \right].$$

In the case of no parasitics and equal transistor areas, the delay time would be

$$\tau_D \approx 2 F_o \frac{L}{v_L} \left[\frac{1}{k_P (1 - v_{TP})} + \frac{1}{k_N (1 - v_{TN})} \right].$$

If $k_P \approx k_N = 1/3$, $v_{TP} \approx v_{TN} = 1/3$,

$$\tau_D \approx 12 F_o \, \tau_{min}.$$

With actual parasitics, the advantage of the CMOS inverter will not be as clear. However, it should be kept in mind that the CMOS inverter can offer the same noise margin as the single-channel inverter with channel lengths of $1/3$ of the single-channel case.

It has been pointed out[38] that, if C is dominated by parasitic capacitance $C_P >> F_o C'_o WL$,

$$\tau_D \sim \frac{C_P}{C'_o W k v_L}$$

so that decreasing the channel length L would offer no advantage anymore.

So far, drift-control currents have been assumed in the estimates of delay times. In the case of barrier-control currents, minimum transit-times are

$$\tau_{min} = \frac{L_B}{2 D_n} > \frac{L_B}{v_t} \approx \frac{L_B}{4 v_L}.$$

These time constants are larger than the scattering limit L/v_L, until the outer limit of a Brownian motion with the thermal velocity v_t across a very short barrier L_B is reached. In silicon, this limit occurs for barrier lengths $L_B < 0.1 \, \mu m$. Delay times as defined above are shorter than the actual time τ_I for information transfer through a gate.[14] For MOS circuits, τ_I $\approx 2\ \tau_D$. Thus, for a $1\mu m$ NMOS transistor, $\tau_I \approx 200...400$ ps in the drift-control mode, i.e. for $V_{DD} > 1$ V. In a barrier control mode, the response is slower. However, the power-delay product

$$Pt_D \approx CV_{DD}^2$$

may be improved due to a small V_{DD}.

REFERENCES

1.   J.D. Meindl and R. N. Swanson: Potential Improvements in Power-Speed Perform-ance of Digital Circuits, IEEE Proceedings, pp. 815-816. May 1971.

2.   A. K. Rapp: Applications of Field-effect Transistors in Digital Circuits, in Field-effect Transistors (Wallmark, J. T. Johnson, H., ed.) Prentice-Hall, pp. 312-358, 1966

3.   A. J. Strachan and K. Wagner: LOCOS CMOS. Technology/Design System for LSI in CMOS, IEEE Int. Solid-State Circ. Conf., Philadelphia, pp. 60-61, February 1974.

4.   R. N. Swanson and J. D. Meindl: Ion-implanted Complementary MOS Transistors in Low-voltage Circuits, IEEE J. Solid-State Circuits, **SC-7**, pp. 146-153, 1972

5.   R. Zuleeg: A Silicon Space-Charge-Limited Triode and Analog Transistor, Solid-State Electronics, **10**, pp. 449-460, 1966.

6.   T. I. Kamins and R. S. Muller: Statistical Considerations in MOSFET Calculations, Solid-State Electronics, **10**, pp. 423-431, 1967.

7.   B. Hoeneisen and C. Mead: Fundamental Limitations in Micro-electronics I. MOS Transistors, Solid-State Electronics, **15**, pp. 819-829, August 1972.

8.   J. T. Wallmark: Fundamental Physical Limitations in Integrated Electronic Circuits, in Solid-State Devices, The Institute of Physics, London, pp. 133-167, 1975.

9.   J. Borel: Advanced MOSFET Technologies: A Review, Solid-State Circuits, Editions du Journal de Physique, 1976, pp. 69-87.

10.   R. N. Swanson and J. D. Meindl: Fundamental Performance Limits of MOS Integrat-ed Circuits, IEEE Int. Solid-State Circuits Conf., Philadelphia, Dig. Techn. Papers, pp. 110-111, February 1975.

11.   J. Nishizawa, T. Terasaki and J. Shibata: Field-effect Transistor Versus Analog Transistor (static induction transistor), IEEE Trans. Electron Dev., **ED-22**, pp. 185-197, April 1975.

12. W. Schemmert, L. Gabler and B. Hoefflinger: Conductance of Ion-Implanted Buried-Channel MOS Transistors, IEEE Trans. Electron Dev., **ED-23**, pp. 1313-1319, December 1976.

13. R. W. Keyes: Physical Limits in Digital Electronics, Proc. IEEE, **63**, pp. 740-767, May 1975.

14. R. Müller, H. Pfleiderer and K. U. Stein: Energy per Logic Operation in Integrated Circuits: Definitions and Determination, IEEE J. Solid-State Circ., **SC-11**, pp. 657-661, October, 1976.

15. O. G. Folberth and J. H. Bleher: Grenzen der Digitalen Halbleitertechnik, Nachrichtentechn. Zeitschrift, NTZ, **30**, pp. 307-314, 1977.

16. Stein, K.U.: Noise-Induced Error Rate as Limiting Factor for Energy per Operation in Digital IC's, IEEE J. Solid-State Circ., **SC-12**, pp. 527-530, October 1977.

17. B. Hoefflinger, H. Sibbert, G. Zimmer, E. Kubalek, and W. Menzel: Model and Performance of Hot-Electron MOS Transistors for High-Speed, Low Power LIST, IEEE Int. Electron Devices Meeting, Washington, Techn. Digest, pp. 463-467, December 1978

18. J. Schneider, B. Hoefflinger, G. Zimmer: A Compatible NMOS, CMOS Metal Gate Process, IEEE Trans. Electron Dev., **ED-25**, pp. 832-836, July 1978.

19. K. Wagner and J. G. M. Klomp: LOCMOS-CAD, ein Wirtschaftliches Und Produktionsgewichtetes System Für den Entwurf von Digitalen LSI-Schaltungen, in Grossingtegration - Technologie, Entwurf, Systeme (Hoefflinger, B., ed.), Oldenbourg, 275-334, 1978.

20. H. Nihira, M. Konaka, H. Iwai and J. Nishi: Anomalous Drain Current in N-MOSFET's and its Suppression by Deep Ion Implantation, IEEE Int. Electron Dev. Meeting, Washington, pp. 487-491, December 1978.

21. T. Nakamura, M. Yamamoto, H. Ishikawa and M. Shinoda: Submicron Channel MOSFET's Logic Under Punch-Through, IEEE J. Solid-State Circ., **SC-13**, pp. 572-577, October 1978.

22. --: Punch-Through MOSFET for High-Speed Logic, IEEE Int. Solid-State Circuits, Conf., Dig. Tech. Papers, pp. 22-23, 1978.

23. J. Nishiuchi, H. Oka, T. Nakamura, H. Ishikawa and M. Shinoda: A Normally-Off Type Buried Channel MOSFET for VLSI Circuits, IEEE Int. Electron Dev. Meet., Washington, pp. 26-29, December 1978.

24. R. H. Dennard, F. H. Gaensslen, E. J. Walker and P. W. Cook: MOSFET Designs and Characteristics for High Performance Logic at Micron Dimensions, IEEE Int. Electron Dev. Meet., Washington, pp. 26-29, December 1978.

25. B. Hoefflinger: Systemintegration, in Grossintegration - Technologie, Entwurf, Systeme, Oldenbourg, pp. 359-377, 1978.

26. W. T. Alexander, S. J. Boardman, P. Krebs, and A. T. P. MacArthur: High Density Uncommitted Arrays Using an Advanced CMOS Technology, ESSCIRC, Southhampton, pp. 76-78, 1979.

27. B. Hoefflinger, H. Sibbert and G. Zimmer: Model and Performance of Hot-Electron MOS Transistors for VLSI, IEEE Trans. on Electron Devices, **ED-26**, pp. 513-520, or Journal on Solid-State Circuits, **SC-14**, pp. 435-442.

28. B. Hoefflinger, K. Schumacher and H. Sibbert: Some Design Aspects of MOS LSI Operational Amplifiers, IEEE J. Solid-State and Electron Devices, **SSBD-3**, pp. 31-40.

29. P. Shah, D. Laks and A. Wilson: High Performance, High Density MOS Process Using Polyimide Interlevel Insulation, IEEE Int. Electron Devices Meeting, Washington, pp. 465-468, December, 1979.

30. D. N. Wollesen: CMOS LSI: Comparing Second-Generation Approaches, Electronics, Sept. 13, pp. 116-123, 1979.

31. G. Zimmer, B. Hoefflinger and J. Schneider: A Fully Implanted NMOS, CMOS, Bipolar Technology for VLSI of Analog-Digital Systems, IEEE, J. Solid-State Circuits, **SC-14**, pp. 312-318, or Trans. Electron Dev., **ED-26**, 390-396, April 1979.

32. F. H. Gaensslen: Geometry Effects of Small MOSFET Devices, IBM J. Res. Dev., **23**, pp. 682-688, November 1979.

33. W. Fichtner and H. W. Pötzl: MOS Modelling by Analytical Approximations I., Int. J. Electronics, **46**, pp. 33-35, January 1979.

34. P. W. Cook, S. E. Schuster, D. R. Freedman, J. T. Parrish and V. Di Lonardo: One Micron MOSFET PLA's, IEEE Int. Solid-State Circ. Conf., Philadelphia, Dig. Tech. Papers, pp. 62-63, 278, February 1979.

35. T. W. Houston, C. L. Everett, H. M. Darley and G. W. Taylor: Silicon MESFET Circuit Performance for VLSI, ibid., pp. 80-81.

36. T. Yamaguchi, M. L. Lust, S. Ragsdare and S. Sato: Submicron Channel MOS IC Technology, ibid., pp. 82-83, 283.

37. J. T. Wallmark: A Statistical Model for Determining the Minimum Size in Integrated Circuits, IEEE Trans. Electron Dev., **ED-26**, pp. 135-142, February 1979.

38. J. L. Moll: Outer Limits of VLSI, IEEE Semicond. Interface Specialists Conf., New Orleans, December 1979.

39. A. E. Ruehli: Survey of Computer-Aided Electrical Analysis of Integrated Circuit Interconnections, IBM J. Res. Dev., **23**, pp. 629-639, November 1979.

40. P. W. Cook, S. E. Schuster, J. T. Parrish, V. Di Lonardo and D. R. Freedman: 1 μm MOSFET VLSI Technology: Part III - Logic Circuit Design Methodology and Applications, IEEE Trans. Electron Dev., **ED-26**, pp. 333-346, April 1979.

41. H. Sibbert, B. Hoefflinger and G. Zimmer: Analytisches Modell, Leistungsfähigkeit und Skalierung kleinster MOS-Transistoren für höchstintegrierte Digitalschaltungen, NTG-Fachberichte, **68**, pp. 128-134, March 1979.

42. W. Schemmert: Ladungstransport in Ionenimplantierten Buried-Channel MOS-Transistoren, Archiv Electronik u. Übertrag. Techn., **AEÜ-33**, pp. 23-31, January 1979.

43. N. Ohwada, T. Kimura and M. Doken: LSI's for Digital Signal Processing, IEEE J. Solid-State Circuit, **SC-14**, 214-220, April 1979.

44. J. T. Gibbons and K. F. Lee: One-Gate-Wide MOS Inverter on Laser-Recrystallized Polysilicon, IEEE Electron Devices Letters, **EDL-1**, pp. 117-118, June 1980.

45. B. Hoefflinger: A Barrier-Control Model for Submicron MOS Transistors, to be published.

46. B. Hoefflinger: Output Characteristics of Short Channel Field-Effect Transistors, IEEE Transactions on Electron Devices **ED-28**, pp. 971-976, August 1981.

47. J. H. Kroeger and O. N. Tozun: CAD Pits Semicustom Chips Against Standard Slices, Electronics, July 3, pp. 119-123, 1980.

48. S. Liu, B. Hoefflinger, and D. O. Pederson: Interactive Two Dimensional Design of Barrier Controlled MOS Transistors, IEEE Trans. Electron Devices, **ED-27**, pp. 1550-1558 August 1980.

49. A. D. Lopez and H. F. Law: A Dense Gate Matrix Layout Style for MOS LSI, IEEE Int. Solid-State Circ. Conf., San Francisco, Dig. Tech. Papers, pp. 212-213, February 1980.

50. Y. Sakai, T. Masuhara, O. Minato and N. Hashimoto: MOS Buried Load Logic, IEEE Int. Solid State Circuit Conf., San Francisco, Dig. Tech. Papers, pp. 56-57, February 1980.

51. K. N. Ratnakumar, J. D. Meindl and D. J. Bartelink: Performance Limits of E/D NMOS VLSI, ibid., pp. 72-73, 260.

52. T. Ito, T. Nozaki, H. Ishikawa and Y. Fukukawa: Thermal Nitride Gate FET Technology for VLSI Devices, ibid., pp. 74-75.

53. K. Lehovec and R. Zuleeg: Analysis of GaAs FET's for Integrated Logic, IEEE Trans. Electron Dev., **ED-27**, pp. 1074-1091, June 1980.

54. C. Mead: Physics of Computational Systems, in Introduction to VLSI Systems (Mead, C., Conway, L., ed.), Addison-Wesley, pp. 333-371, 1980.

55. B. Hoefflinger and G. Zimmer: New CMOS Technologies, in Solid-State Devices (Carroll, J., ed.), The Institute of Physics, London, pp. 85-139, 1981.

56. R. Troutman: VLSI Limitations from Drain-Induced Barrier Lowering, IEEE Trans. Electron Dev., **ED-26**, pp. 461-469, April 1979.

57. C. F. Hill: Noise Margin and Noise Immunity in Logic Circuits, Microelectronics, **1**, pp. 16-21, April 1968.

58.   R. Hoshikawa, H. Kikuchi, S. Baba, S. Sato, K. Kawato, N. Inui and O. Wada: A 10.000 Gate CMOS LSI Processor, IEEE Int. Solid-State Circuit Conf., Dig. Tech. Papers, pp. 106-107, February 1980.

59.   T. Ohzone, T. Hirao, K. Tsuji, S. Horiuchi and S. Takayanagi: A 2 k x 8 bit Static MOS RAM with a New Memory Cell Structure, IEEE J. Solid Cir., **SC-15**, pp. 201-205, April 1980.

60.   T. Iizuka, H. Nazawa, Y. Mizutani, H. Kaneko and S. Kohyama: Variable-Resistance Polysilicon for High Density CMOS RAM, IEEE Int. Electron Dev. Meet., Washington, Tech. Dig., pp. 370-373, December 1979.

61.   T. Masuhara, O. Minato, T. Sasaki, H. Nakamura, Y. Sakai, T. Yasiu and K. Uchibori: 2 K x 8 b HCMOS Static RAM's, ibid., pp. 224-225, 277, February 1980.

62.   J. G. Posa: Gate Arrays, Electronics, 53, No. 21, pp. 145-158, September 25, 1980.

63.   K. Izumi, M. Doken and H. Ariyoshi: CMOS Devices Fabricated on Buried SiO_2 Layers Formed by Oxygen Implantation into Silicon, Electronics Lett., **14**, 593-594, August 1978.

64.   C. Mulder, C. Niessen and R. M. G. Wijnhoven: Layout and Test Design of Synchronous LSI Circuits, IEEE Int. Solid-State Circuit. Conf., Dig. Tech. Papers, pp. 248-249, February 1979.

65.   S. Das Gupta, E. B. Eichelberger and T. W. Williams: LSI Chip Design for Testability, ibid., pp. 216-217, February 1978.

66.   B. Hoefflinger: MOS Circuits, Lectures and Tutorials: Digital Technology, Status and Trends, Edited by H. Painke, Oldenbourg Verlag, Munich pp. 75-113, 1981.

67.   G. Zimmer, H. Fiedler, B. Hoefflinger, E. Neubert and H. Vogt, Performance of a Scaled Si Gate n-Well CMOS Technology, Electronics Letters, September 1981.

SILICON ON INSULATOR INTEGRATED CIRCUITS

Michel Montier

THOMSON - EFCIS
Grenoble, France

1. INTRODUCTION

In the past, high complexity SOS IC chips were used in specialized applications, requiring high speeds and high levels of radiation hardness, which cannot be attained with standard silicon microcircuits.

Two other capabilities of SOS ICs are the high packing density (500-900 gates per square millimeter) and the low speed-power product (Power-delay product near 0.1 picojoules).

For these major reasons several government programs are being carried out using advanced SOS IC technologies and very recently very high complexity microprocessors have appeared mostly in the U.S. and Japan. Of course, in SOS there is a need for high performance technology processing as in NMOS. One to two micron geometries and all new improvements for doping and deposition are being used in research and development laboratories.

In my presentation today I will address mostly silicon on sapphire but, due to its recent appearance, I shall also briefly speak about the more general problem of silicon on Insulator.

The major part of my presentation will be technology-oriented. Modelling and circuits applications will be approached later.

1.1 Advantages of CMOS/SOS

The major speed limitation of bulk MOS IC's are related to the parasitic capacitances.

An evaluation of these capacitances as functions of the oxide thicknesses, the design rules and the geometrical arrangements, gives the relative contributions of the different elements.

You can see the evaluation on Fig. 1.1 where PMOS structure is taken as example. References (1), (2), (3), (4), (5) of the transistor cross section are indicated on the table and the drawing.

Nearly one-half of the contribution can be eliminated by an insulating substrate as: only lateral junctions remain; the second electrode of interconnection capacitances is eliminated.

ORIGIN	RELATIVE CONTRIBUTION	SOLUTION
INTERCONNECTIONS ON SiO₂ (1)	1	INSULATING SUBSTRATE
DIFFUSED INTERCONNECTIONS (2)	2	INSULATING SUBSTRATE
GATE DRAIN (3)	2	SELF ALIGNMENT
DRAIN DIFFUSION (4)	1	INSULATING SUBSTRATE
CROSS-OVER	1/6	INTERCONNECTIONS WIDTH
INTRINSIC GATE (5)	1	SHORT CHANNELS

From the LSI and VLSI point of view, calculations have some interest to learn if changes appear with reduction in geometries. The purpose here is to find out if CMOS/SOS keeps its superiority in low parasitic capacitance compared to bulk CMOS, when the gate length decreases (Fig. 1.2).

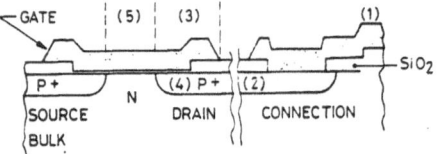

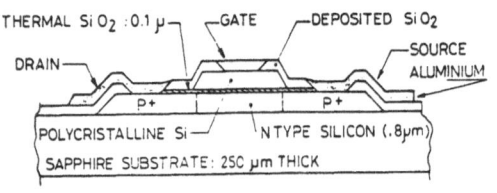

Fig. 1.1 - Parasitic capacitances on bulk silicon

-   All parasitic capacitances are taken into consideration: Those shown before and, in addition, capacitances between interconnection lines.

-   Realistic design rules are also used W/L = 2 (width to length ratio). Interconnection length = 10 L; width and space = 1,5 L. Junction width = 2 L.

-   Realistic decrease in field oxide thickness is also considered (dashed lines) and this case will normally occur for micro-pattern lithography requirements.

The results suggest that the SOS technology keeps its superiority even in the submicron range.

If we consider the density of logic circuits in bulk and SOS CMOS, we can see (Fig. 1.3) that bulk structures suffer from several constraints.

Thus, in layout design rules we have to consider:

- The sideways diffusion of P-well. (6 to 7 μm junctions depths are normally used.) This is imposed by source and drain junction depth and their space charge extension.

- Spacing between P-well and P^+ diffusions is difficult to improve, because of the necessity of guard bands implantation. This space can be used for metal interconnections in many circuits elements but, as a consequence, the designer does not have the same freedom. (For example, in 5-6 μm CMOS design rules, the distance between PMOS and N.MOS may reach 19 μm).

The advantage of SOS/CMOS lies mainly in natural islands isolation and selective doping possibilities for active regions.

For many practical applications we can conclude that surface saving with SOS lies near 30% with the same design rules.

Due to the photocurrents created in the silicon substrate the reduced volume of active regions offer a lower radiation sensitivity in SOS devices. From the transient radiation tolerance point of view Fig. 1.4 indicates the higher hardness of SOS. We observe that aluminum gate CMOS-SOS offers a one and one-half decade advantage. A more sophisticated technology-like silicon gate deep depletion CMOS/SOS, leads to one decade more in maximum dose permitted.

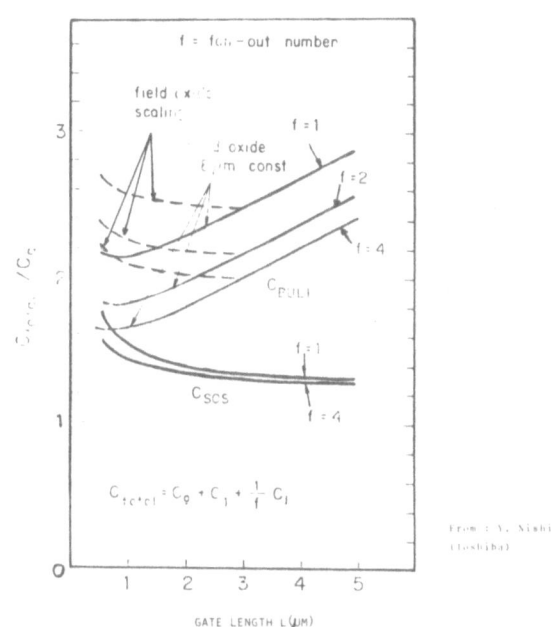

Fig. 1.2 - Parasitic capacitances between bulk and SOS

It was found that, in some cases, bulk CMOS devices could be latched. This destructive effect was attributed to the parasitic bipolar transistors present in CMOS structure, and it was demonstrated that the breakdown was initiated by noise pulses or radiation induced currents.

One condition to induce latch-up current is the following β_{PNP} x $\beta_{NPN} > 1$. Recently, various techniques have been employed to prevent the latch-up problem. They include gold doping to reduce current gains or epitaxial structure to shunt the emitter base junctions.

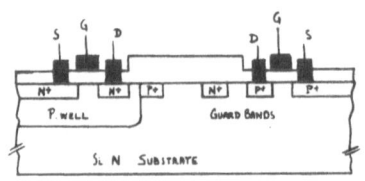

SELF ALIGNED GATE   CMOS ON BULK SILICON

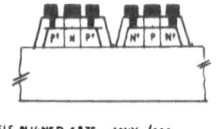

SELF ALIGNED GATE   CMOS /SOS

Fig. 1.3 - CMOS/SOS vs. bulk MOS density

It is evident that this effect cannot exist in SOS/CMOS because the MOS transistors are isolated.

1.2 Limitations of CMOS/SOS

CMOS/SOS suffer several limitations coming from the poor silicon layer crystal quality. Three major reasons explain this situation:

-       Interatomic crystal lattice difference between Si and sapphire.

-       Compressive stresses in silicon, due to thermal expansion difference with sapphire, which induce large defect density (stacking faults and twins).

-       Aluminum auto doping effect induced during the early stage of silicon epitaxy.

You can see in Fig. 1.5 two cross sections of a SOS transistor. It shows two very important charge effects which strongly influence electrical characteristics of MOS devices. The first one is SIDE CHANNEL INVERSION

- During chemical silicon etching <111> crystalline plans appear (because of selective etching rate).

- Higher charge density in <111> plans induce different inversion threshold between the edges and the top transistors.

The second phenomenon is seen in the second cross section which indicates fixed positive charges in the sapphire.

- The charge number depend on Si growing conditions, sapphire and subsequent thermal sequences.

- High charge concentrations can induce inversion layer in silicon and leakage current between source and drain.

The charges which have been described before are responsible for some anomalies in electrical current characteristics.

If we observe a cross section in the channel zone of a MOS transistor it is possible to detect 4 separate contributions for the source to drain current (Fig. 1.6).

The first zone is related to the normal MOS current flow near the silicon surface.

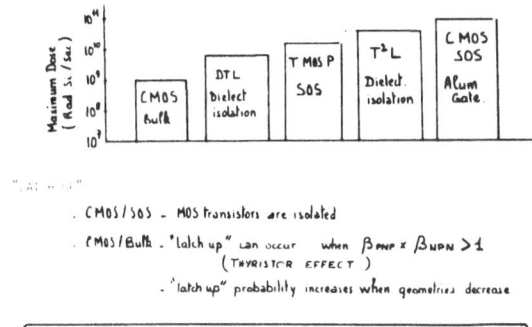

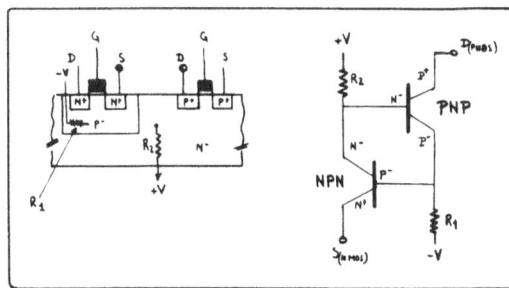

Fig. 1.4 - Radiation hardness and isolation advantages on SOS

The second zone is the bulk contribution (minority carriers generation-recombination).

The third zone (or back-channel zone) is induced by the interface charge between the silicon epitaxial layer and the sapphire substrate.

In most cases sapphire is positively charged and induces negative charges in silicon near the interface. But we mentioned that the interface charge density is inherent with the manufacturing process.

In particular it has been established that epitaxy conditions, oxidation, and annealing steps can highly change the interface charge density.

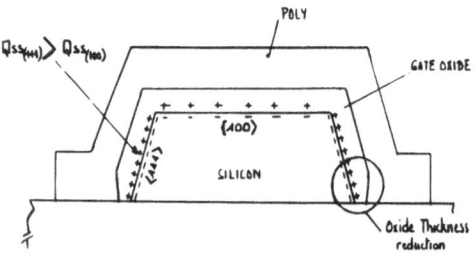

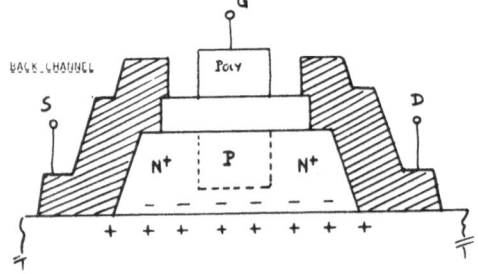

Fig. 1.5 - Interface charges

The fourth zone represents the contribution of the lateral MOS transistor. We note that this conductance is not very important (small width) but it will be sufficient to modify the total drain current in the weak inversion mode.

The lower part of the picture shows a schematic example of gate transfer characteristics for a complementary pair of MOS transistors (at constant V_D).

Experimental cases are represented on Fig. 1.7 and three observations can be made:

-    V_{TN} is shifting to 0 when the substrate doping increases and its control becomes difficult.

For low doping concentration it is clear that

1) The presence of a parasitic transistor increases drastically the I_{DS} current for 0 gate voltage (wafer number 4).

2) The charge density near the silicon-sapphire interface strongly changes from wafer to wafer, and induces back channel leakage currents (increasing effect with lower impurity concentrations).

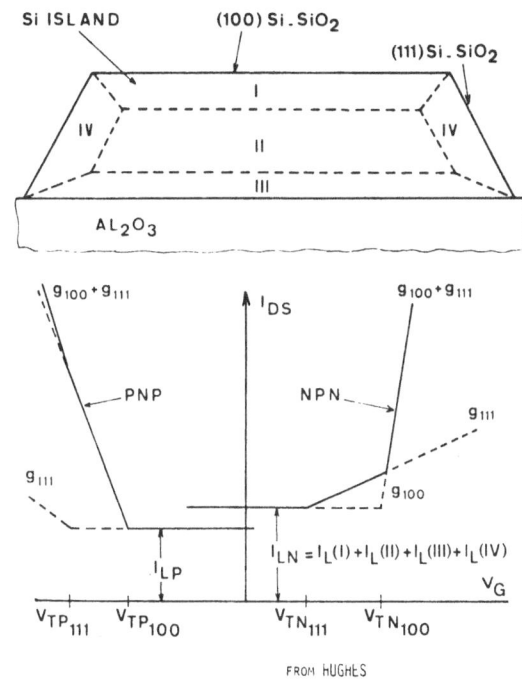

Fig. 1.6 - Ideal SOS MOST characteristic

Channel length is also an important parameter as far as leakage current is concerned.

Figure 1.8 shows that the back channel effect strongly changes (see before) but can also drastically increase when the gate length decreases.

In optimized technologies, changes are not so important because of doping adjustments on the edges and at the silicon-sapphire interface, and consequently good control of back-channel and diode leakage currents are achieved.

2. TECHNOLOGY

2.1 SOS Substrate Preparation

One of the present limitations of SOS technologies is the cost of sapphire. To lower the substrate cost several solutions has been investigated: Edge Fine Growth

(EFG) - Flame Fusion growth (verneuil) - Schmid Viechnicki growth. But, at the moment, the Czochralski method is mostly used because of its higher crystal quality, its right crystal orientation after growing, its crystal diameter capability.

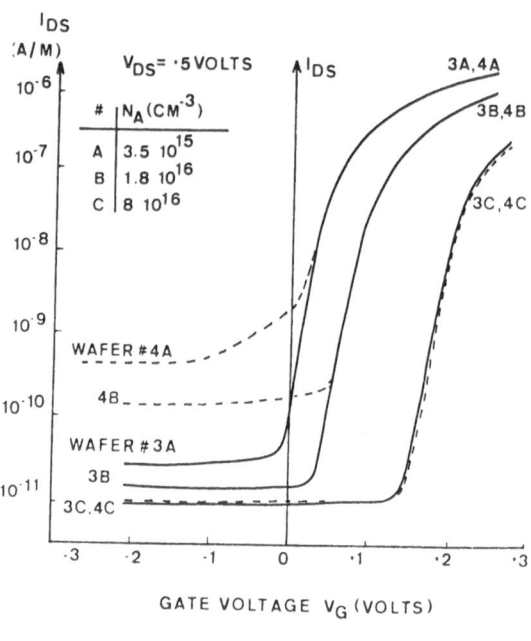

Fig. 1.7 - Typical n-channel I_{DS}-V_G characteristics

2.2 Silicon Layer Epitaxy

Silicon epitaxy on sapphire is done by conventional atmospheric pressure CVD methods, using silane as the gas source and vertical or barrel reactors. But the silicon layer growth is quite different from silicon on silicon epitaxy.

In fact, the early stage of growth shows a discontinuous silicon layer. This is shown by scanning electron microscopy (Fig. 2.1). In these cases, deposition was stopped after a short time ($\approx$ 2 seconds) and various examinations show isolated islands of silicon. Two observations can be made:

- diffraction pattern and electron micrographs indicate 3 island orientations

  - $\{100\}$ is the most important

  - Two kinds of $\{110\}$ orientations are always present and these misorientations lead to twins and stacking faults at the island boundaries.

- Impurity analysis indicates a high aluminum concentration near the silicon sapphire interface. It was earlier demonstrated that chemical reduction of alumina by silane in hydrogen atmosphere is responsible for that, the reaction stops when a continuous layer exists and at this time aluminium contamination decreases rapidly.

In another way Fig. 2.2 shows that silicon crystal quality increases strongly when carbon contamination decreases at the sapphire surface. And it will be noted that sapphire is difficult to polish and that very few reactive agents can etch this material.

2.3 X-Ray Silicon Characterization

The experimental device used is a symmetrical arrangement of two crystals (Fig. 2.3). The slits F_1 and F_2 are used to separate the $K\alpha_1$ and $K\alpha_2$ radiations on the monochromator, the X-ray beam area can be adjusted with F_3.

The detector measures the diffracted intensity. The "Rocking Curve" is the graph of the reflected and incident intensity ratio when the angle Θ varies near the theoretical Bragg angle Θ_B. It was demonstrated by several authors that the full width at half maximum for a given thickness can be used as a good measurement of the crystal quality (Fig. 2.4). Furthermore, the integrated reflecting power has been correlated to the silicon thickness.

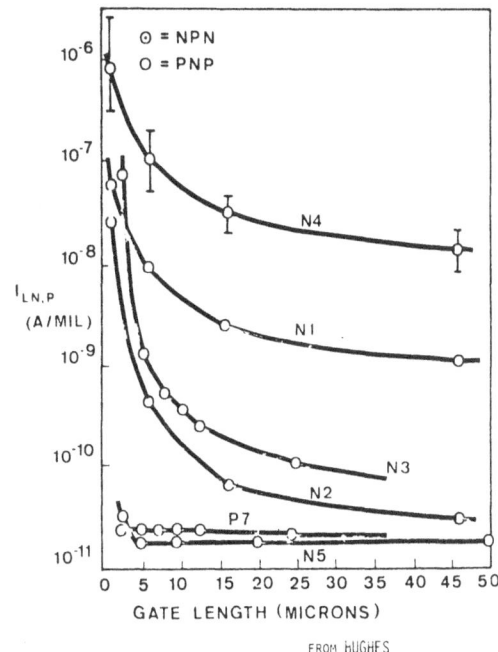

Fig. 1.8 - Leakage current as a function of channel length

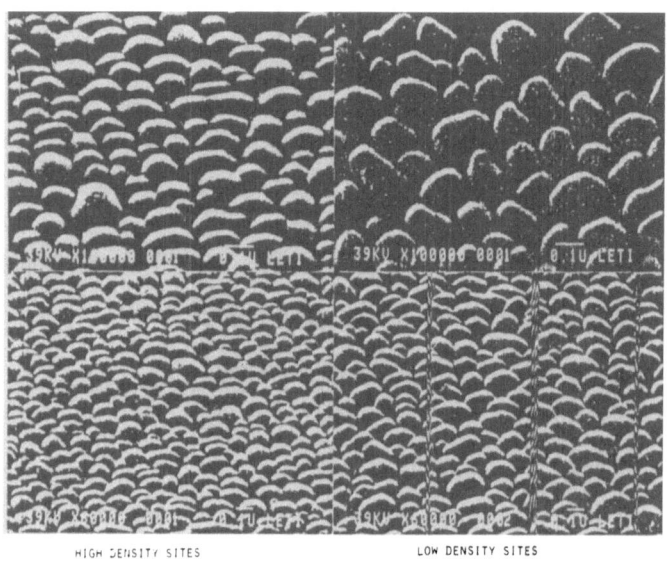

HIGH DENSITY SITES   LOW DENSITY SITES

Fig. 2.1 - Surface nucleation

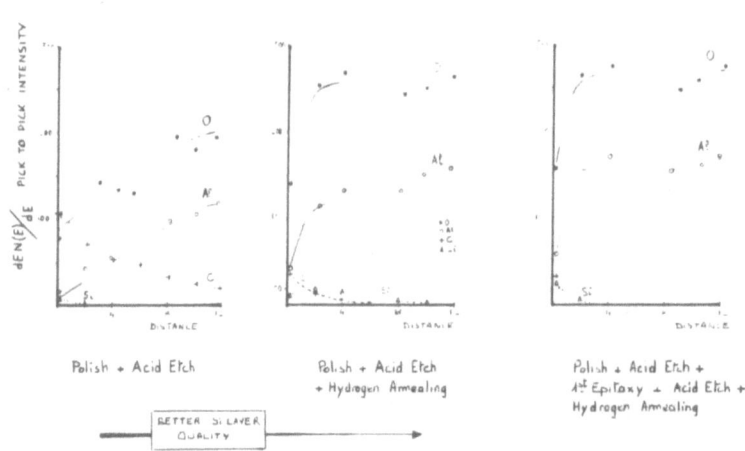

Fig. 2.2 - Surface preparation before Si layer growing (Auger analysis)

For its powerful capabilities, we have already proposed the R.C. measurement as an everyday characterization of SOS wafers and we have defined an experimental "rocking curve full width at half maximum" as acceptance criterion.

2.4 Other Silicon on Insulator Substrates

When SOS technology was first developed, many people predicted it to be a large step toward standard devices. Results to date, indicate that SOS devices did not reach the target.

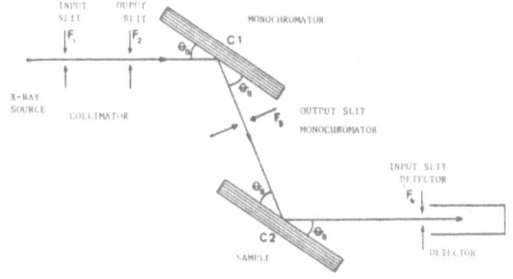

Fig. 2.3 - Principle of the experimental device with two crystals.

We can expose two reasons to explain this evolution:

- First, production yields have not reached expected values, and it has appeared that chip areas are limited;

- secondly, the reduction in substrate cost has not followed the predicted values.

At the same time, new improvements in laser capabilities have occurred and fabrication of MOS devices in laser annealed polysilicon on silica have been demonstrated. All silicon devices seem very attractive if one considers the high electrical performances of silicon and its silicon dioxide interface. Many problems are to be solved in this field, but excellent performances have been recently obtained on SOI substrates.

A method for manufacturing silicon on insulator devices consists of using a buried insulating layer formed by ion implantation. A practical solution developed by NTT in Japan is described on Fig. 2.5. Electrical results in these experiments show that diodes and MOS transistors are comparable to those of SOS technology.

Another solution can be the Graphoepitaxy technique developed in the MIT laboratory. This technique consists of using an artificial surface relief structure to induce the desired orientation in the polysilicon layer.

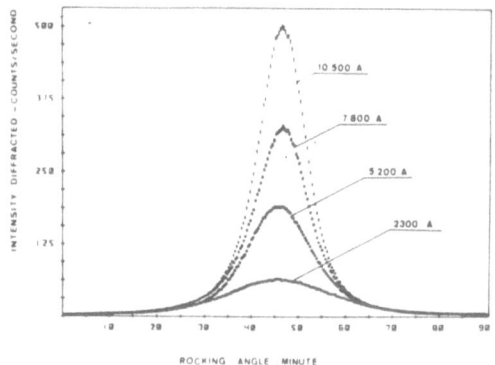

Fig. 2.4 - Rocking angle vs. intensity diffraction

The theoretical model assumes that:

-     One particular plane parallel to the wafer surface exists in the crystal grains.
-     A grating structure is capable of reducing the random orientation of grains which normally exist on a flat surface.

Experimentally it has been demonstrated that achieving a (100) oriented silicon layer was realistic, and the pictures describe the process used.

Actual results indicate crystalline defects (roughness microcracks and large spread in crystallographic orientations.)

In the same way Gibbons et al have demonstrated promising results with CW argon laser crystallization of polysilicon islands on Si_3N_4. Smooth surfaces and maximum 2 x 20 μm^2 single crystal patterns have been obtained.

The last new technique is the silicon bridging epitaxy developed by M. Tamura et al. In this case, silicon dioxide patterns are used and lateral recrystallization is obtained from bare silicon areas. At that time, maximum lateral growth distance on SiO_2 is ~ 1 μm and device characteristics are not mentioned.

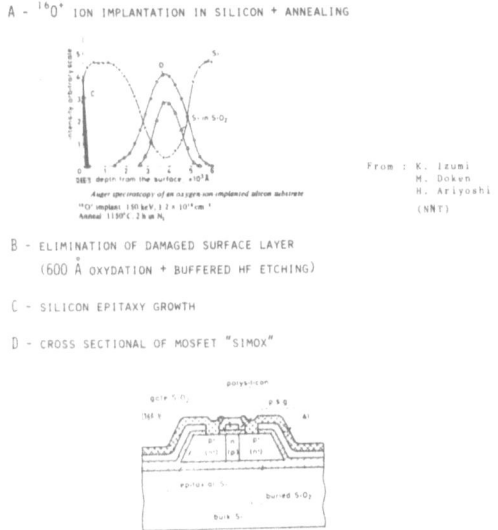

A - $^{16}O^+$ ION IMPLANTATION IN SILICON + ANNEALING

From : K. Izumi
M. Doken
H. Ariyoshi
(NNT)

B - ELIMINATION OF DAMAGED SURFACE LAYER
(600 Å OXYDATION + BUFFERED HF ETCHING)

C - SILICON EPITAXY GROWTH

D - CROSS SECTIONAL OF MOSFET "SIMOX"

Fig. 2.5 - Isolation utilizing buried SiO_2 layers formed by ion implantation in silicon (separation by implanted oxygen: "SIMOX")

2.5 CMOS-SOS Technologies

2.5.1 - The final thickness of the silicon layer in SOS structures (o,6 μm) allows "deep depletion" and enhancement mode transistors with a uniform substrate doping.

A CMOS inverter is shown in Fig. 2.6 in the case of N⁻ as grown silicon layer. For negative potentials the $N^+/N^-/N^+$ transistor is completely depleted and in the "off" state. This kind of device never uses the classical surface inversion conduction, but the $P^+/P^-/P^+$ is a normal enhancement mode transistor.

This CMOS "deep depletion" technology has several advantages coming from its process simplicity (only two impurity implants), the natural symmetry, and low values of P and N channel threshold voltages, and of course all capabilities of SOS. So for 3-4 μm IC features good yields and high performances are obtained.

2.5.2 - If we consider now the capability for "deep depletion" CMOS-SOS technology to stay in VLSI competition, reduction in gate length and other dimensions must be studied.

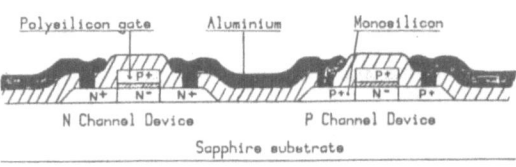

Unfortunately, results indicate that leakage currents tend to increase and yields to decrease. We have observed that back-channel and edge currents were very sensitive to the channel length and this is attributed to the low doping level of the silicon layer.

Monosilicon thickness	4700 Å	
Gate oxide thickness	650 Å	
P+ polysilicon thickness	6000 Å	
	N channel	P channel
Threshold voltages	+ 0.8v ± 0.2v	- 0.6v ± 0.2v
Mobilities	450cm²/v.s ± 10%	220cm²/v.s ± 10%
Leakage currents (channel length=4µm VG=0v, VD=5v)	< 100 pA/µm	< 50 pA/µm

Essential Electrical Parameters

To eliminate these failures an accurate doping control near silicon-sapphire interface must be

Fig. 2.6 - The basic deep depletion CMOS inverter

used. It needs extra doping steps to implant selectively N and P wells and afterwards enhancement MOS transistors are easier to control.

Figure 2.7 shows how "back-channel" compensation (Boron 200 KeV) and threshold adjustment (Boron 60 KeV) are performed on the N channel transistor. Figure 2.8 indicates the deep phosphorus implant to prevent the punch-through effect and the low depth boron implant to control the threshold in the P channel device.

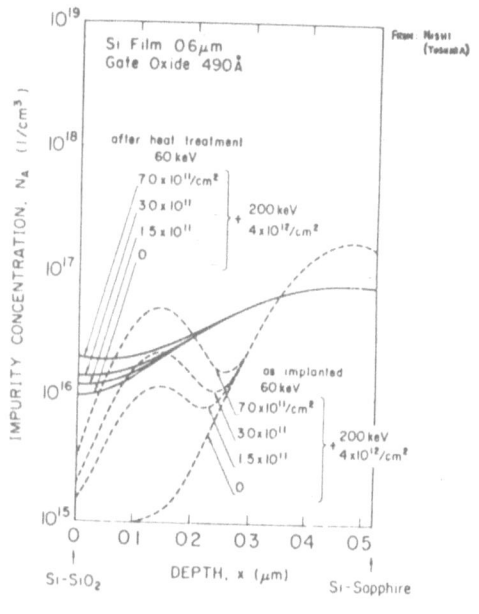

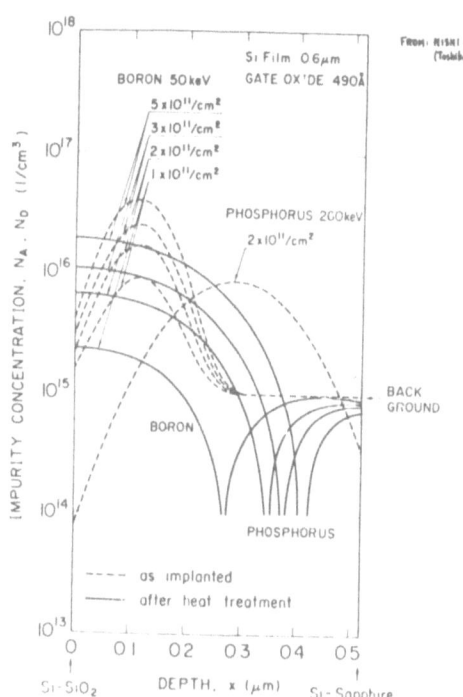

ig. 2.7 - Back channel compensation in E/E   Fig. 2.8 - Breakdown voltage adjustment in E/
MOS/SOS   CMOS/SOS

With N^+ poly gates and right doping concentrations, it has been shown that 1 to 2 μm

channel length were practical and that E/E CMOS/SOS technology can be a VLSI competi-

tor.

3. DEVICE CHARACTERIZATION AND TESTING

3.1 Device Currents in CMOS-SOS Si Gate

Drain current versus gate voltage measurement is a powerful method to characterize MOS

devices, especially for SOS applications.

To obtain the experimental graphs shown on Fig. 3.1 we have used an automatic data

acquisition system driven by a HP 9830 calculator with a 32 Kbits memory. Commercially available instrumentation allows the low level current measurements. One can observe for

"deep depletion" CMOS:

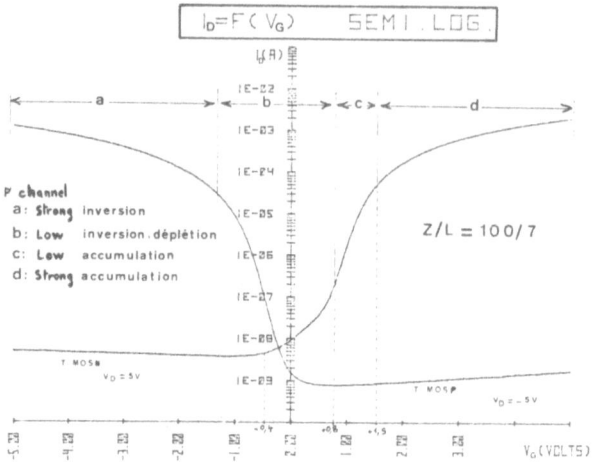

Fig. 3.1 - Typical p and n-channel I_{DS}-V_{GS} characteristics - deep depletion poly p$^+$ gate

-     Drain current for 0 gate voltage

-     Edge transistor contribution in the (b) region

-     Residual leakage currents

  -     in strong accumulation region for P channel transistor (I_{FP} = Diode

    leakage current).

  -     in depletion region for N channel (I_{FN} = Diode leakage + "back

    channel" currents)

-     Using drain voltage (V_D) as a parameter it is possible to obtain also the

  punch-through effect visualisation.

For VLSI applications, switching characteristics will be studied in the 1÷2 μm channel

range. In Fig. 3.2, the limitation of "deep depletion" CMOS is indicated, and one can see that

the drain current (for 0 gate voltage) reaches a prohibited value (100 pA/μm) for 2,5 μm

channel length.

3.2 Threshold Voltage in CMOS/SOS Devices

Threshold voltage becomes another important parameter to control when the channel

length decreases. Figure 3.3 shows that in CMOS/SOS the short channel effect is less

important than in bulk technologies, but an increase in threshold voltage is associated with

silicon thickness variations. This last effect is well explained by poor silicon quality and

consequently SiO_2 quality in thin layers.

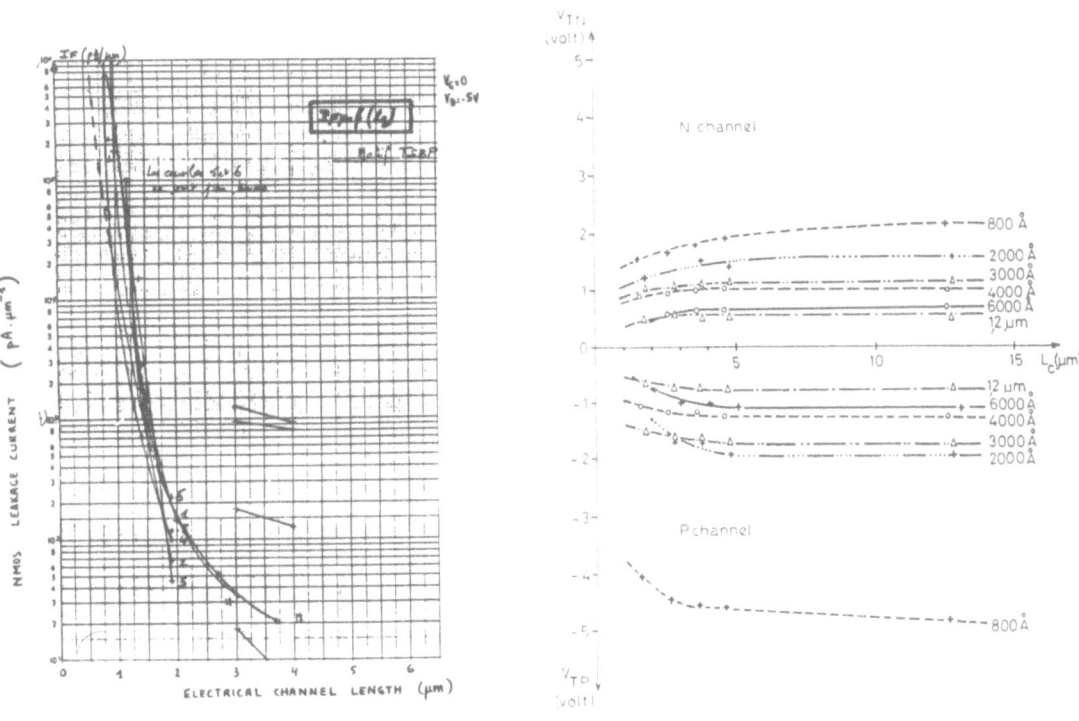

g. 3.2 - Channel length reduction effects in D/D
MOS - $I_{FN} = F(L_E)$

Fig. 3.3 - Threshold voltage vs. channel length

We report now a very specific behaviour of "deep depletion" MOS structures. Figure 3.4

shows a strong variation in N channel threshold voltage after annealing of source and drain

implants.

- for lower temperatures, annealing is not sufficient to activate ion implanted

  impurities and to eliminate interface states.

-       for Zone II progressive activation of doping centers take place. This variation indicates a nonequilibrium state in the silicon layer and presumably at the silicon-sapphire interface.

-       for temperatures higher than the epitaxy one, a large degradation appears and boron can diffuse from polysilicon through the gate oxide.

It is assumed that the large amount of crystal defects and contaminations in SOS substrates may explain these phenomenas, but we have demonstrated that threshold voltages are very stable at room temperature and under electrical and temperature stresses.

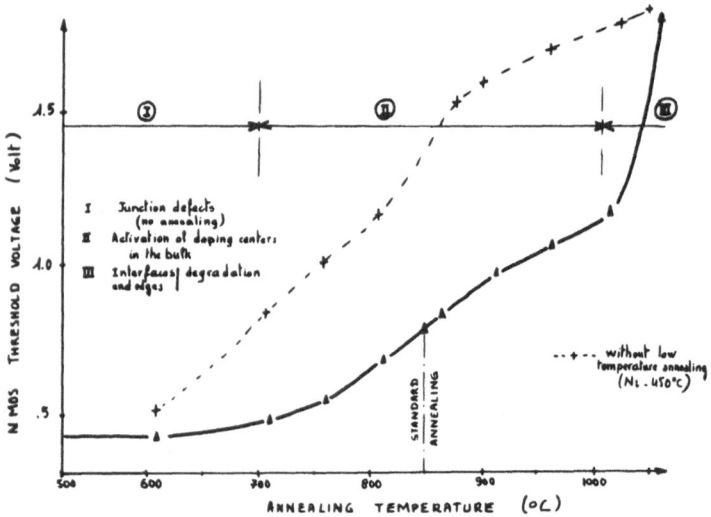

Fig. 3.4 - Threshold voltage vs. annealing temperature in D/D CMOS/SOS

3.3 CMOS/SOS Reliability

In the actual state of the art the CMOS/SOS reliability level seems comparable with that of bulk devices.

In order to compare different SOS technologies we illustrate in Fig. 3.5 some typical results under electrical stress.

After high positive stress is applied to the devices (see conditions on the graph) a significant shift can appear in the current characteristics. It is shown here that, for each technology, an anomalous conduction appears for different critical voltages, and that lower annealing and oxidation temperatures induce a higher electrical stability.

As shown before, such a parasitic conduction appears preferentially in parasitic edge transistors and the defect in this case can be explained by electron injection into the oxide layer.

The low temperature "deep depletion" process seems to offer more than a factor of 2 in electrical stability and appear as a competitive technology in SOS business.

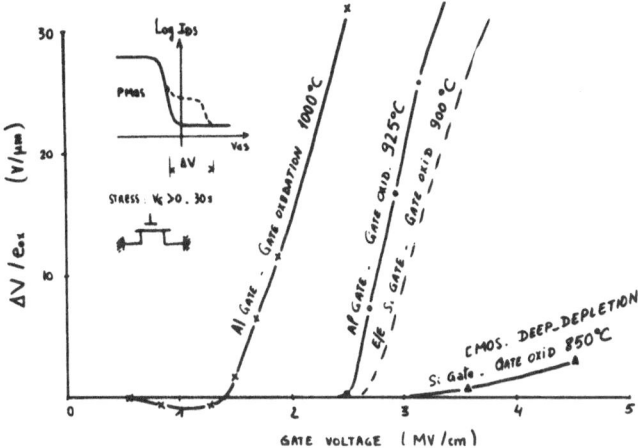

Fig. 3.5 - Deep depletion reliability (electrical stress)

4. DEVICE MODELLING

The aim in SOS devices is to obtain:

- A CAD static model of MOS transistors in the strong inversion region, with channel lengths in the range of 2 μm.

476   *Montier*

Parameters easily correlated with technology and topology measured on a single device with bulk not externally available.

Enhancement CMOS/SOS SiN$^+$ Gate is taken as an example. Electrical channel lengths vary from 1 to 6 μm. In SOS, the substrate is electrically floating and its potential is determined by the flow of leakage current through B-S and B-D diodes (Fig. 4.1). As soon as we reach the saturation region, some multiplication current appears and adds itself to this leakage current which biases the substrate. With this bias both the threshold voltage and transistor current change.

As the multiplication current I_M itself depends on I_{DS}, it becomes evident that we can reach the value of each point on the characteristic through an iterative method only, described at the bottom of the Fig. 4.1.

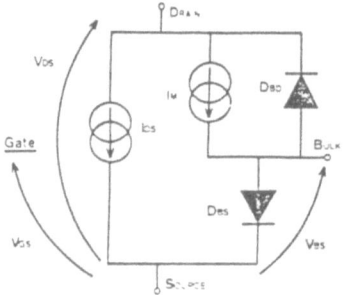

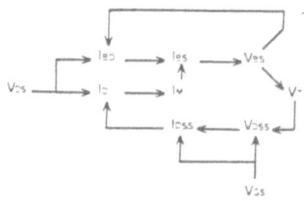

Fig. 4.1 - Electrical model of SOS/MOST -- Calculation mode of the theoretical characteristics

So:

- the multiplication current must be accounted for, not just when it becomes as high as the MOS current itself, but also as soon as it starts growing (i.e., just after saturation occurs).

- before describing any parameter of the intrinsic transistor it is necessary to characterize the B.S. and B.D diodes.

Three currents are differentiated in Fig. 4.2:

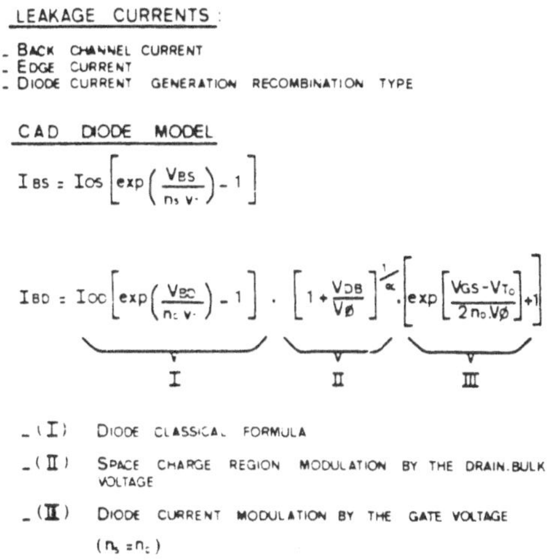

LEAKAGE CURRENTS :
- BACK CHANNEL CURRENT
- EDGE CURRENT
- DIODE CURRENT GENERATION RECOMBINATION TYPE

CAD DIODE MODEL

$$I_{BS} = I_{OS}\left[\exp\left(\frac{V_{BS}}{n_1 \, v}\right) - 1\right]$$

$$I_{BD} = I_{OC}\left[\exp\left(\frac{V_{BC}}{n_2 \, v}\right) - 1\right] \cdot \underbrace{\left[1 + \frac{V_{DB}}{V_{\beta}}\right]^{\frac{1}{\alpha}}}_{} \cdot \underbrace{\left[\exp\left[\frac{V_{GS} - V_{T_0}}{2\,n_0\,V_{\phi}}\right] + 1\right]}_{}$$

$$\underbrace{}_{I} \quad \underbrace{}_{II} \quad \underbrace{}_{III}$$

- (I) DIODE CLASSICAL FORMULA
- (II) SPACE CHARGE REGION MODULATION BY THE DRAIN.BULK VOLTAGE
- (III) DIODE CURRENT MODULATION BY THE GATE VOLTAGE ($n_1 = n_2$)

Fig. 4.2 - Bulk-source and bulk-drain diodes model

- the back channel current flowing along the interface of silicon and sapphire

- the edge current due to parasitic MOS transistor on the edge of the silicon islands.

- the diode current which might be of generation-recombination type

We made the assumption that the substrate biasing is only due to the diode current and then used the classical diode current equations (part I) modified by:

- the space charge region modulation by the drain bulk voltage (part II)

- the modulation by the gate voltage (empirical approach). (part III).

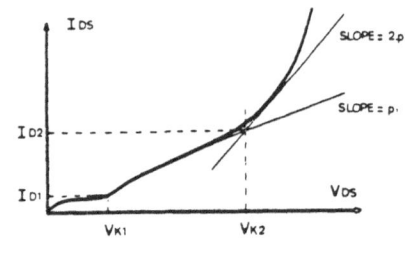

$$I_M = I_{DS} \cdot K_M \cdot (V_{DS} - V_{DSS})^3 \cdot \exp\left[\left(-\frac{V_M}{V_{DS} - V_{DSS}}\right)^3\right]$$

PARAMETER ACQUISITION IS BASED ON FIRST KINK AND SECOND KINK DETECTION

FIRST KINK : $I_M = I_{BD}/10$

SECOND KINK : $I_M = I_{DS}/10$

Fig. 4.3 - Multiplication current model and parameter acquisition

The multiplication current can be modelled over a wide range with this empirical formula, including 2 parameters: KM and VM.

The acquisition is based on the first kink and the second kink detection (Fig. 4.3). At the saturation point, the multiplication current begins to grow. When $V_{DS} = V_{K1}$ it becomes larger than the diode current, and when $V_{DS} = V_{K2}$ it becomes larger than the intrinsic transistor current. With these assumptions, it's easy to get the value of the parameters. But, be careful, every variable is accessible only through an iterative calculation.

To be sure that the model was credible we have measured the bulk voltage. (V_B) on a special device with a bulk contact.

So we have compared the experimental and theoretical behaviour of the floating substrate bias (Fig. 4.4).

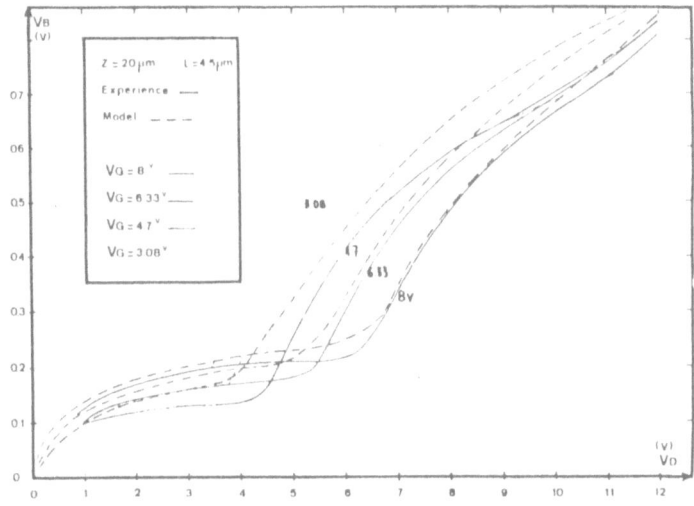

Fig. 4.4 - Floating substrate bias vs. drain and gate voltages

Measurements show good agreement, and the deviations observed for low and high drain voltages may be attributed to measurement limitations. (Measurements have been made with probes directly on wafers and bulk contact can affect the behaviour of the device).

Figure 4.5 is an example of diode and transistor parameters acquisition, N and P type. One can see a good agreement between the theory and experiment. The low current levels related here can explain the difficulty to eliminate noise problems with wafer measurement.

The average error is less than 5% even for large drain voltages and with very important kink effect.

5. PERFORMANCE COMPARISON WITH OTHER TECHNOLOGIES

5.1 Speed and Power Consumption

As shown in Figure 5.1, the propagation delay time per gate of MOS devices looks even faster than that of the actual bipolar devices. In addition, CMOS/SOS devices seem to be competitive with scaled NMOS, and keep a net advantage in power consumption. Our results and simulations are included in this graph. From the production point of view, however, it is true that NMOS on bulk silicon keeps a certain advantage. But if we consider the total power dissipation in VLSI chips CMOS is, no doubt, superior to NMOS. So, a good selection in MOS technologies is quite difficult for future applications and each one offers particular advantages.

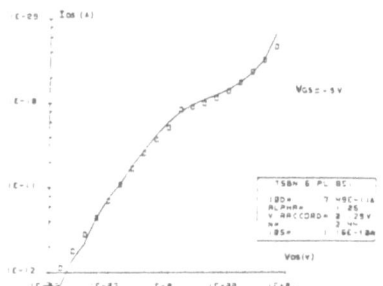

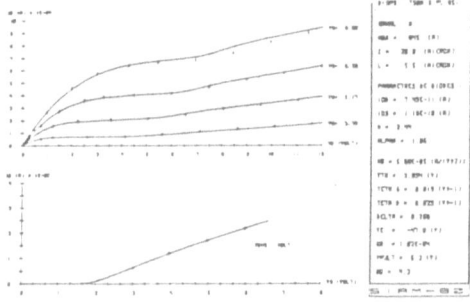

Fig. 4.5 - Comparison model - experience

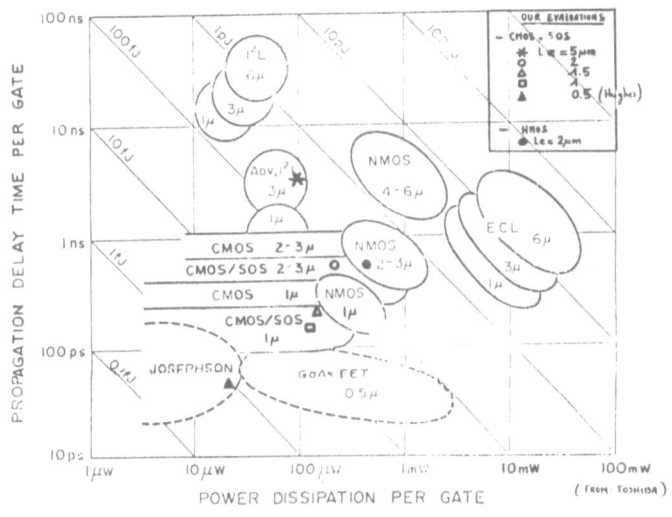

Fig. 5.1 - Speed and power consumption

5.2 Density Comparison

Figure 5.2 shows that CMOS/SOS is competitive with NMOS technology and has higher

density than bulk CMOS for same design rules. These evaluations take into account published

values and our practical experience.

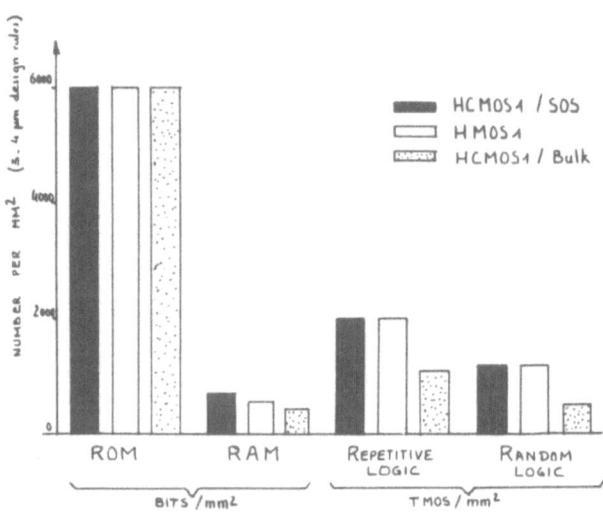

Fig. 5.2 - Density comparison

5.3 Economical Comparison

In Table 5.3 one can see that the SOS process maturity is not so high as that of bulk NMOS and CMOS technologies, but process complexity (the deep depletion CMOS would be the less complex) and speed power performances give some advantages to SOS.

In Table 5.4 it is shown also that, for high complexity and low consumption, CMOS/SOS IC's keep a comfortable advantage over AsGa but the maximum speed is lower by a factor of two.

| P R O P E R T Y | CURRENT TECHNOLOGY (1981) | | | FUTURE (85-90) |
| | N M O S | C M O S | | C M O S S O S |
		BULK	SOS	
RELATIVE PROCESS MATURITY (1 - 10)	9	8	4	2
PROCESS COMPLEXITY (NUMBER PROCESSING STEPS)	9-15	14-17	11-13	14-17
SPEED-POWER PRODUCT (PJ)	5-50	2-40	0.5-30	0.1-0.2

Fig. 5.3 - Summary IC properties

	SPEED	CONSUMPTION	DENSITY	COMPLEXITY MAX	NOISE (DB)
AsGa	2	1 A 10	1	1	1
CMOS/SSI	1	1	2 A 10	100	1.2 A 1.5 (SILICIUM)

Fig. 5.4 - Comparison AsGa/CMOS-SSI (relative values)

6.   CIRCUITS APPLICATIONS

In high speed applications CMOS/SOS is now credible and the Toshiba RAM memory with 18 ns access time is an evident example.   In Fig. 6.1 the circuit demonstrates the capability of SOS in small dimensions (L = 2 μ) and thin oxide layers (500 Å) which are comparable with the up-to-date HMOS technology.

Another example is the 4 K RAM memory (EFCIS design) with 3-4 μm features and a low active consumption (60 mW) - (Fig. 6.2).

Very recently high complexity (16 bits) microprocessors have appeared on the market and Table 6.3 shows their performances compared to the well-known MOTOROLA 68000 HMOS microprocessor.

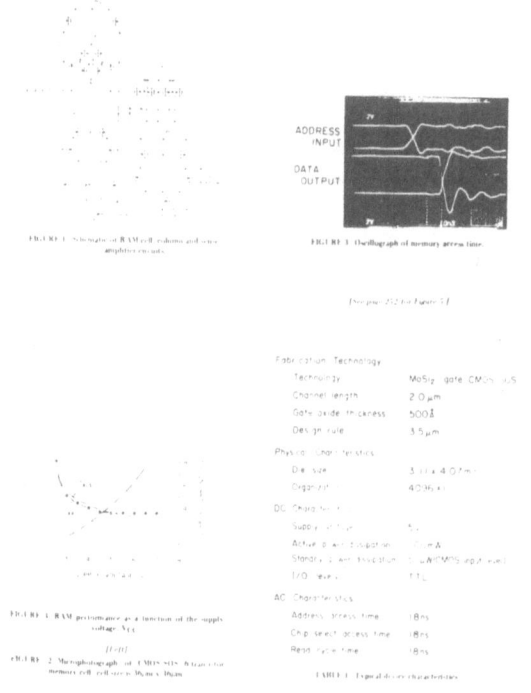

Fig. 6.1 - High speed circuits RAM - 18 MS - Toshiba

STATIC CHARACTERISTICS

SUPPLY VOLTAGE   :   4,5 TO 11 VOLTS

ACTIVE CONSUMPTION   :   60 MW

STANDBY CONSUMPTION   :   150 μM

IN / OUT LEVELS   :   TTL - LS

DYNAMIC CHARACTERISTICS

ACCES TIME   MAX   :   300 NS (v_{DD} = 4.5 V)   T = 125°C
              TYP   :   140 NS (v_{DD} = 5 V)   T = 25°C

READ CYCLE TIME   :   300 NS

WRITE CYCLE TIME   :   300 NS

WRITE PULSE TIME   :   150 NS

TECHNOLOGY

DESIGN RULES   :   3 - 4 μM

GATE OXIDE THICKNESS :   65 NM

TOPOLOGY

DIE SIZE   :   3,95 x 4,84 MM2   (19,1 MM2)

DENSITY   :   1238 TMOS/MM^2

Fig. 6.2 - 4K RAM (EFCIS) characteristics (CMOS/DD technology)

CHARACTERISTICS	ROCKWELL	TOSHIBA	MOTOROLA 68000
TECHNOLOGY	CMOS/SOS	CMOS/SOS-NMOS	HMOS 1
GATE LENGTH	2 μM	3,5 μM	3,5 μM
GATE MATERIAL	MoSi	Si	Si
OXIDE THICKNESS	65 NM	70 NM	65 NM
COMPLEXITY	67000 DEVICES	12000 GATES	70000 DEVICES
CHIP SIZE	5,37 x 6,25 MM	6,7 x 7,5 MM	44 MM
POWER CONSUMPTION	—	0,6 WATT	1,2 WATT
V_{DD}	5 VOLTS	5 VOLTS	5 VOLTS
F MAX			8 MHz

Fig. 6.3 - Comparison of 16 bit microprocessors

ADVANCED SILICON MOS DEVICES AND RELATED PROBLEMS

Robert H. Dennard

IBM Thomas J. Watson Research Center
Yorktown Heights, NY   10598

I. INTRODUCTION

The key to the tremendous progress in integrated circuits during the last decade has been the ability to reduce the size of individual devices and circuits through improvements in the photolithographic processing steps used to fabricate the structures. The projected era of Very Large Scale Integration (VLSI) in the near future relies on the continuation of this trend through further improvements in optical techniques or the application of newly developing electron-beam or X-ray patterning techniques with much higher resolution capability.

In this section the design approaches and problems in realizing miniaturized devices will be presented, starting with a review of the scaling principles. Experimental results obtained with a 1 μm silicon FET technology will be described, including discussions of performance capability for both logic and memory applications of this technology. Potential directions and problems for a 1 μm complementary MOS (CMOS) technology will be discussed.

Small geometry devices are particularly susceptible to transient disturbance due to ionizing radiation such as alpha or cosmic particles. This situation will be reviewed, emphasizing the results of newly developed models for the collection of excess carriers and discussing structural changes within the silicon substrate which can help in minimizing the amount of collected carriers.

Considerable activity in advanced MOS devices relates to processing innovations which can improve the device capability. These areas of work will be discussed, followed by a review of newer forms of FET's which operate on somewhat different principles.

II. REVIEW OF SCALING PRINCIPLES

A. Device Scaling

The concept of scaling gives a concise understanding of how to design miniaturized FET's, and directly indicates some of the fabrication challenges.[1,2] The basic idea is shown in Fig. 1. The small device on the right can be designed by scaling down all dimensions of the larger device, by reducing the applied voltages, and by increasing the substrate doping. All the changes are by a proportional amount as illustrated by the scaling factor α. The first equation shows that the depletion regions surrounding the source and drain junction are also reduced by a factor α, due to the reduced voltage and increased substrate doping. This is exactly true if the substrate bias is changed by more than a factor of α so that, even though the built-in junction potential remains nearly constant, the total potential across the junction can be scaled exactly. For this reason, the substrate bias is generally reduced disproportionately in miniaturized devices, and can be expected to go to zero for submicron devices.

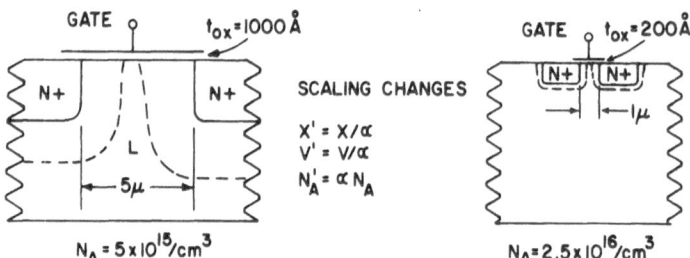

NEW DEPLETION THICKNESS =

$$X'_D = \sqrt{\frac{2\epsilon_{si}(V/\alpha + \Psi)}{q\,(\alpha N_A)}} \cong \frac{X_D}{\alpha}$$

NEW THRESHOLD VOLTAGE =

$$V'_t = \frac{1}{\epsilon_{ox}}\left(\frac{t_{ox}}{\alpha}\right)\left[\left(-Q_{eff} + \sqrt{2\epsilon_{si}\,q(\alpha N_A)\left(\frac{V_{s-sub}}{\alpha} + \Psi_s'\right)}\right] + (\Delta W_f + \Psi_s)\cong \frac{V_t}{\alpha}$$

NEW CURRENT =

$$I'_D = \frac{\mu\epsilon_{ox}}{t_{ox}/\alpha}\left(\frac{W/\alpha}{L/\alpha}\right)\left(\frac{V_g - V_t - V_d/2}{\alpha}\right)\left(\frac{V_d}{\alpha}\right) = \frac{I}{\alpha}$$

Fig. 1. - Principles of scaling.

As shown in the second equation of Fig. 1, the threshold voltage scales down directly with the oxide insulator thickness, t_{ox}, with the bulk charge in the silicon remaining constant. This is also true for channel-implanted devices if the depth of the implant region is reduced by the scaling factor. For aluminum or n^+ polysilicon gates the $\Delta W_f + \psi_s$ term is small in n-channel devices. The reduction of threshold voltage allows circuits to work at the reduced voltage levels. Moreover, control of the threshold is improved, in absolute terms, since any variation in interface charge or bulk charge has a diminished effect due to the thinner gate insulator.

The current in the scaled-down device, as shown in the last equation, is reduced by α times. Since all the electric fields are being held constant by scaling, the inversion layer density reaches the same value, and these carriers travel with the same velocity in the same lateral electric field; thus the current is reduced only because of the reduced device width, W. The reduction of current along with the reduced voltage gives a power reduction by a factor of α^2, a result which is crucial for VLSI circuits. This maintains constant power density if the circuit areas are also scaled by α^2 times by reducing interconnection line dimensions, so that many more circuits can be integrated on a chip without changing the cooling requirements. The speed of these circuits increases by the scaling factor α due to the reduction of all the device and wiring capacitances by that amount.

One device parameter that does not scale as desired is the subthreshold current, or the voltage swing required to turn off the device current to a given leakage level.[1,2] This current is dominated by diffusion of carriers from the source over the potential barrier in the channel region (at the surface) according to the Boltzman relationship. The barrier height required for a given leakage level can be scaled down only by lowering the operating temperature.[3] Design of dynamic RAM's requiring low leakage currents becomes difficult when the required barrier height is a significant fraction of the total applied supply voltage. RAM's using submicron devices are affected as the supply voltage is scaled to a few volts at room temperature. Logic circuits do not require such dramatic swings in channel current, and thus can work at lower voltages and smaller dimensions.

B. Interconnection Scaling

Scaling of interconnection lines for integrated circuits is illustrated in Fig. 2. If all the line dimensions including the spacing away from the ground plane (silicon substrate) are reduced by a factor α, the charac-

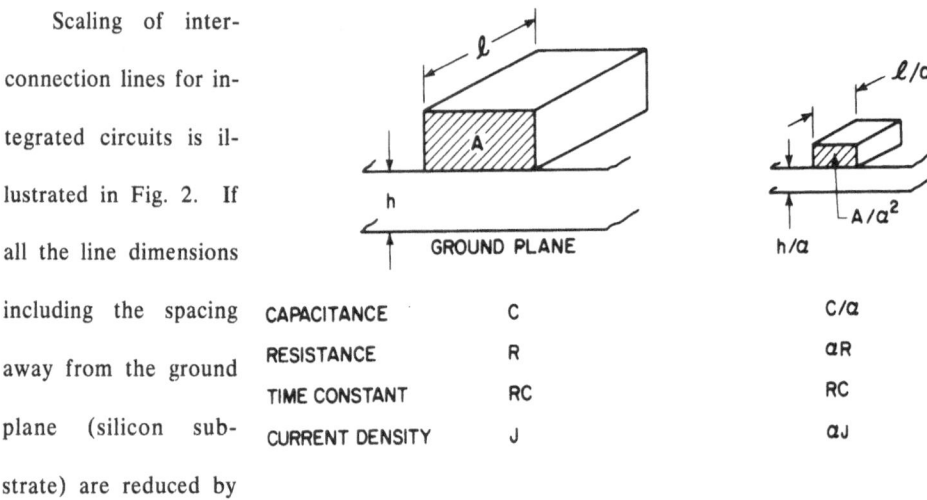

CAPACITANCE	C	C/α
RESISTANCE	R	αR
TIME CONSTANT	RC	RC
CURRENT DENSITY	J	αJ

Fig. 2. - Scaling of interconnection lines.

teristic impedance remains constant to first order. Thus the capacitance per unit length, from a line to the ground plane or between lines, remain constant. The total line capacitance is reduced because its length is reduced.

There are difficulties due to the fact that the cross-sectional area of a line goes down by α^2 times if the thickness of the film is reduced along with the width. The total resistance of the shorter line goes up by a factor of α. If the device currents flowing through the line are all scaled down by α, according to the previous section, then the voltage drop across the line is still the same. Thus the voltage drops in such lines become more significant in comparison to the scaled-down applied voltages. Also, the line is relatively more lossy, since the RC time constant is unchanged while the desired switching speed is increased by α. Further, the potential for electromigration is enhanced by the higher current density in the scaled-down conductor. Fortunately, FET's at 1μm dimensions are not severely affected by these problems with some care and with some evolution of gate materials as will be discussed. It is a fundamental problem, however, which will be felt more and more at submicron dimensions, particularly for high performance circuits. Low temperature must be mentioned again for its benefits in improving conductivity of pure metals and greatly reducing electromigration.[3]

III. 1μm NMOS TECHNOLOGY

This section reviews the results of comprehensive feasibility studies which have been done to demonstrate a 1μm MOSFET technology for high performance VLSI logic applications,[4-11] incorporating key results of more recent studies to give a complete summary of the status of such work. The processing of these advanced devices and prototype circuits used direct electron-beam lithography with double-layer resist systems, making extensive use of reactive-ion etching (RIE) to obtain high resolution structures.[8,9] This section will not consider the processing in depth but will discuss the device design considerations, key technology advances required for realizing usable VLSI structures, and experimental results which have been obtained.

Although the bulk of this work is aimed at normal room temperature digital systems, an interesting aspect is that designs for operation at liquid-nitrogen temperatures were also carried out and tested. In addition to the advantages pointed out in the previous section, liquid-nitrogen operation gives 2-3 times greater speed, as will be shown, due to the increased carrier mobility. It also makes dynamic FET circuits virtually static because of dramatic reductions in leakage current. Thus far, these advantages have not been sufficient to merit the large development effort that would be required to implement actual low-temperature systems, because other alternatives to high performance technology (bipolar, GaAs, Josephson) seem even more promising. However, it is an interesting avenue for future applications work, particularly when even smaller devices come into widespread use.

A. Device Design Considerations

1. Enhancement and Depletion Device Structures

A cross section of device structures designed for a 1μm NMOS technology is shown in Fig. 3. Generally this technology, which relies heavily on ion implantation, uses the same

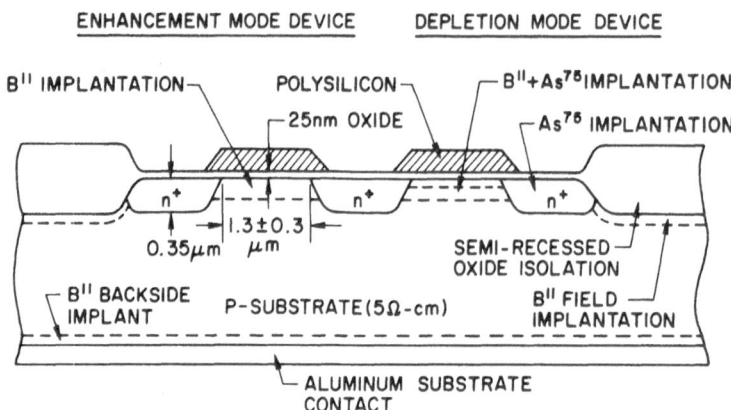

Fig. 3. - Cross-section of devices in 1μm NMOS technology.

process sequences which have also been widely adopted for most advanced commercial silicon

gate LSI products. However, all the processing steps are scaled down to give a much shallow-

er vertical structure by lowering the ion implantation energy, the oxidation temperatures and

times, and the film thicknesses.

To achieve the desired channel length in the 1μm range, the principles of scaling were

generally used - that is, the relatively thin insulator (25 nm of SiO_2), the junction depth (0.35

μm) and the substrate doping (corresponding to 5 ohm-cm) are all in the direction indicated

by scaling. The power supply voltage for logic circuits was chosen to be 2.5 volts, half the

value generally used today for devices down to 2 μm in length. Two-dimensional simulation

programs were used to do sensitivity analyses of these parameters to arrive at a final workable

design. The results are generally in agreement with other published work,[12] showing about

equal sensitivity to changes in the indicated three key parameters. The final design is weight-

ed toward maintaining good conductivity in the n^+ lines, which requires the indicated junction

depth. Another consideration is the depth of the boron channel doping in the enhancement

device. This is designed so that the depleted region under the gate, when the device is turned

on, extends through the channel implant into the lighter substrate doping for the minimum

applied source-to-substrate bias condition. This allows the device to be used in circuit

applications, such as transfer devices, where the source rises above ground during various circuit functions. The threshold gate voltage for turning on the device, V_t, does not increase substantially for this type of design, since the depletion region extending into the lighter doped substrate uncovers only a little additional bulk charge. Within this constraint, making the boron channel doping as deep as possible into the substrate helps somewhat to control the "short-channel effects" caused by spreading of the depletion regions from the source and drain into the channel region controlled by the gate. When the junctions are reasonably shallow, as in this case, the boron doping profile extends down to a similar depth and there is no advantage to introducing a second, lighter-doped channel implant below the first (as is commonly used in some commercial designs with very light substrate doping).

Design of a depletion device with good characteristics for a load device in logic applications has also been studied. This requires that the implanted n-type region, to shift the threshold negative, be made as shallow as possible. Figure 4 shows the results of two-

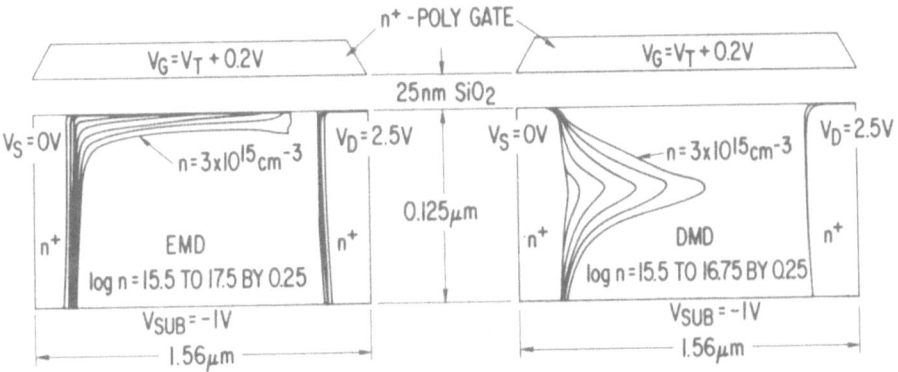

Fig. 4. - Electron concentration contours for enhancement (EMD) and depletion (DMD) devices operating in saturation. Contours step by 1/4 order of magnitude.

dimensional simulations for a depletion device in comparison to an enhancement device.[13,14] The contours of electron density show that the conduction in a depletion device with a large drain voltage occurs deep in the implanted n-region away from the surface. Since the carriers are further away from the modulating gate, the resulting increase in the length of electric field lines between gate and channel causes both a decrease in transconductance and a more negative shift of the threshold voltage. Figure 5 shows that the magnitude of the threshold shift for the saturation or pinched-off condition (2.5v on drain) is severe if the depth of the donor implant is increased. It is also more severe as the channel length is decreased. This type of behavior adversely affects the load-device characteristic, producing a deviation from the desired ideal "constant-current". However, the characteristics are still quite reasonable in this 1μm technology, particularly when somewhat longer channel lengths are chosen for the depletion device.

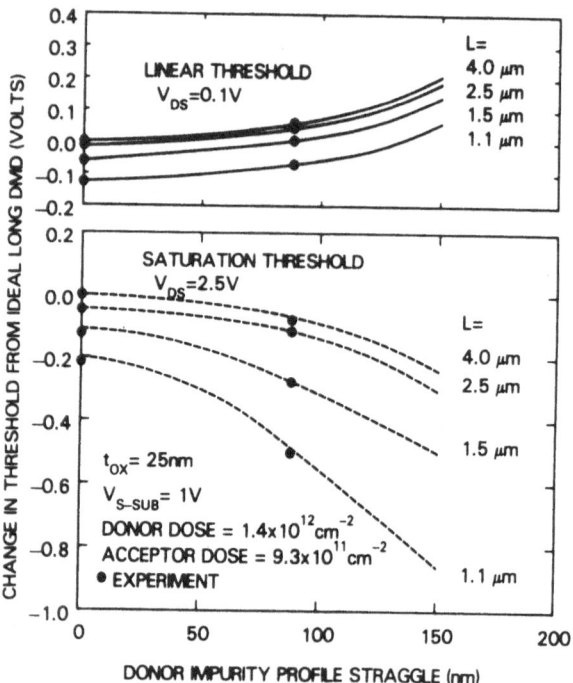

Fig. 5. - Threshold changes in a depletion device as a function of the standard deviation of a Gaussian donor doping profile peaked at the surface.

2. Low Temperature Design Considerations

The principal change in characteristics of an enhancement device cooled to liquid-nitrogen temperature is a large increase in transconductance due to the enhanced mobility. If the aim is to get higher speed, this increased conductance will be used simply to scale up the current levels in all circuits to proportionally reduce the switching delays caused by charging and discharging of capacitive loads. There is also a positive shift of threshold voltage in the order of 0.25 volts since the Fermi level in the p-substrate moves much closer to the valence band edge and also since the conduction band must be bent even closer to the Fermi level to remove the barrier to conduction at low temperature. To achieve the same threshold value used at room temperature, about 0.6 volts, the boron channel implant can be reduced in devices designed for liquid-nitrogen temperature.[5] Moreover, the substrate bias can be reduced to compensate for the appreciable change in built-in junction potential. In the referenced designs the substrate bias is changed from -1v to zero for the low temperature designs.

The enhancement device experiences freezeout of a large part of the p-carriers in the substrate; i.e., they are not ionized at liquid-nitrogen temperature. However, the p-carriers within the depletion region under the gate, which determine the threshold, are swept out by the electric field and therefore completely ionized. Moreover, the substrate remains reasonable conductive due to the high mobility of the carriers which do become ionized.

The depletion device, on the other hand, experiences a large positive threshold shift at liquid-nitrogen temperature due to freezeout of the donor doping impurities which were introduced to shift the threshold negative.[15] This happens because there is no electric field in a large portion of this n-type region, and any ionized carriers have a high probability of being recaptured at the impurity sites.

Since no satisfactory way was found to compensate for this large shift in threshold voltage, which greatly reduces the current in the depletion device, the circuit designs discussed in reference 5 used a positively-biased enhancement device which gives workable load-device characteristics.

B. Key Technology Problems

Fabrication of the structures shown in Fig. 3 and their practical use for VLSI circuits raise several areas of concern. An obvious potential hazard is the thin gate insulator. Indeed, serious consideration of scaled-down gate insulators was viewed with considerable pessimism prior to beginning this program. However, the experience to date has shown that high values of breakdown-field strength and low defect densities can be achieved with some careful process development. The reliability of thin gate insulators has been established by accelerated wear-out studies of test capacitors.[16] The results show that positively-biased capacitors with thin oxides (using polysilicon gates) are as reliable as thicker samples if the applied voltage is scaled down to maintain the same electric field across the gate insulator. A useful summary of insulator work is given in Ref. 17.

Contacts between the interconnection metal and the shallow n^+ regions are another serious concern. The junctions are sufficiently shallow that penetration of the metal through the shallow junctions due to localized alloying must be considered. Fairly high annealing temperatures of 400-500°C are required to achieve good interface charge density, particularly if radiation damage from electron-beam processing is present. Thus, simple metal systems, such as direct contact of the aluminum lines, are not adequate for scaled-down devices. Moreover, contact resistance in the very small contact areas is also a concern. Silicon doped aluminum is not sufficient in this respect. Metallurgy systems using barrier layers between the aluminum lines and the n^+ junctions have been developed which are reliable and also give very good contact resistance, in the order of 10 ohms for a 1 μm^2 area.[18]

As indicated in the section on scaling, the resistance of interconnection lines becomes acute when the film thickness and line widths are reduced. One way to address this problem is to add additional metal layers, using a thicker top layer for power supply distribution and for any heavily-loaded signal lines which require lower resistance. Some circuits require long lines interconnecting numerous gates, as in RAM arrays, which can be done with good layout density only by using a continuous line of the gate material. This requires much better conductivity than polysilicon conductors presently have, particularly if used in thinner films of 300nm or less as required for planarity prior to metal deposition. A successful solution to this problem has been to coat a thin polysilicon gate material with a metal-silicide layer with much higher conductivity.[10] Development of satisfactory processes steps, including capability for dry etching of small-dimension lines, have been successfully demonstrated.[19] Sheet resistance of 2-3 ohms/square have been obtained. Use of refractory-metal gate materials is being pursued in some areas for the same basic purpose.[20] These require considerably different processing since they cannot be oxidized. Passivation against corrosion and stability of the interface with the gate oxide are concerns in this case.

Another technology-related problem is how to deal with the radiation damage introduced by electron-beam pattern definition. This applies only to the final metal layer, where subsequent annealing is limited to a range of 400-500°C with present metallurgy systems. The problem has been shown to involve not only excess interface charge, but also includes the presence of radiation-induced trapping sites in the gate oxide.[11] These electrically neutral traps can be charged up during long-term operation due to capture of some of the electrons injected into the gate oxide (see next section regarding hot-electron injection). Figure 6 shows the results of recent annealing studies.[21] It has been found that annealing in pure hydrogen is effective at 400°C in reducing both positive interface charge and neutral trap densities. Use of higher annealing temperatures is even more effective, but introduces a hazard of increased junction leakage due to reactions of the junction contact metallurgy as discussed previously.

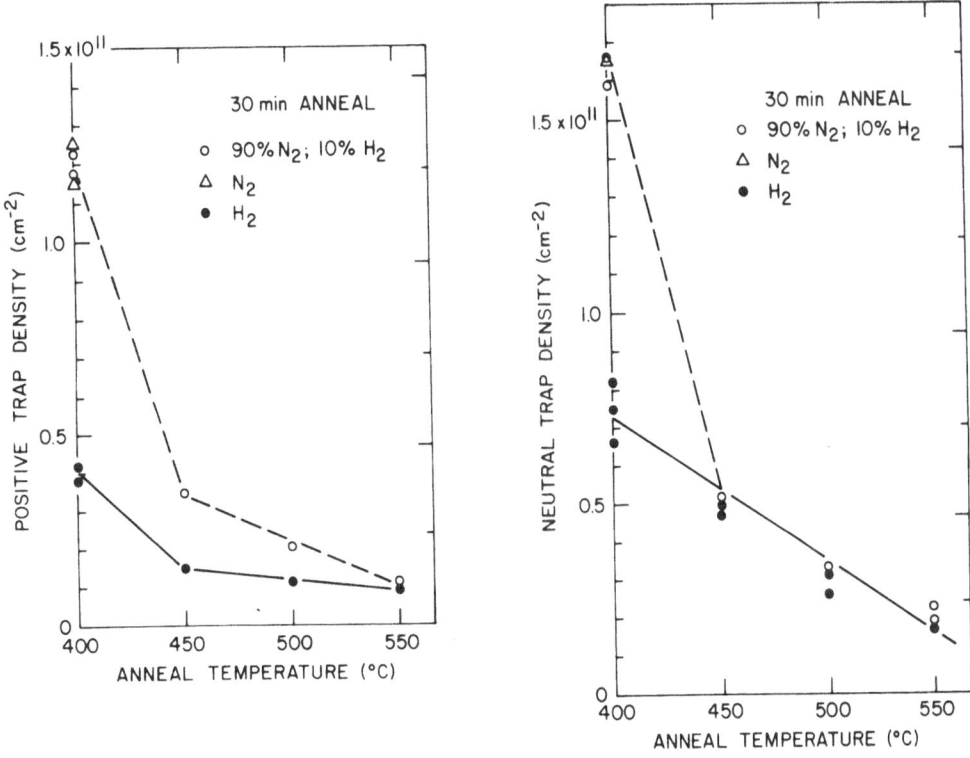

Fig. 6. - Annealing behavior of silicon-gate devices following electron-beam irradiation.

C. Experimental Device Results

For the devices discussed above, experimental results demonstrated successful operation. The device characteristics were indeed found to be substantially free of "short-channel effects". This is shown in Fig. 7, where the threshold voltage is plotted as a function of channel length. Even with the drain voltage parameter at its largest (2.5V)

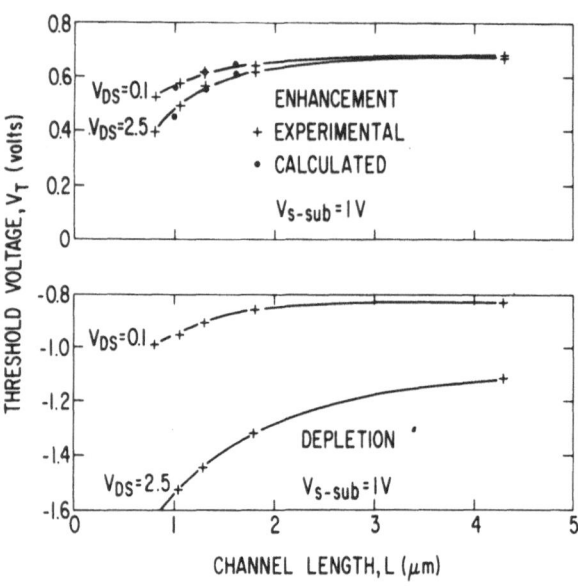

Fig. 7. - Short-channel threshold characteristics of enhancement and depletion devices in 1 μm NMOS technology.

value, the reduction in threshold voltage of the enhancement device from its nominal 0.6V. value is reasonable for channel lengths down to 0.8 μm. The depletion device has somewhat greater shift with channel length, in addition to the large shift at high drain voltage noted previously even for long channel lengths.

Narrow-channel behavior was also found to be reasonably well controlled by the same scaling changes required for controlling short-channel behavior. There seems to be a widespread belief that three-dimensional simulation programs are needed for small devices. This is not necessarily true if the devices are designed to be as free of dimensional effects as their larger counterparts.

The drain characteristics of the enhancement device also showed well-controlled behavior. The output current in the saturated condition over the intended operating range was shown to be reasonably constant, giving a high voltage gain as required to achieve sharp switching characteristics and good noise margins in NOR logic circuits with depletion loads.

The breakdown characteristics of these devices have also been found to exceed the requirements. Figure 8 shows the minimum "sustaining voltage" following breakdown which is defined in the inset figure, as a function of channel length. Characteristics of devices with larger dimensions, representative of a 2 μm technology with similar processing steps, are shown for comparison with the 1 μm results. The results clearly show that scaling is conservative with respect to breakdown characteristics, since the breakdown voltage decreases less than proportional to channel length. Though the electric field is supposedly held constant by scaling, breakdown voltages are actually improved relative to the scaled-down operating voltage. This is because the device breakdown mechanism requires impact ionization of hole current to forward bias the source junction. The energy required for impact ionization is greater than 1 volt, which becomes more significant in the scaled down devices.

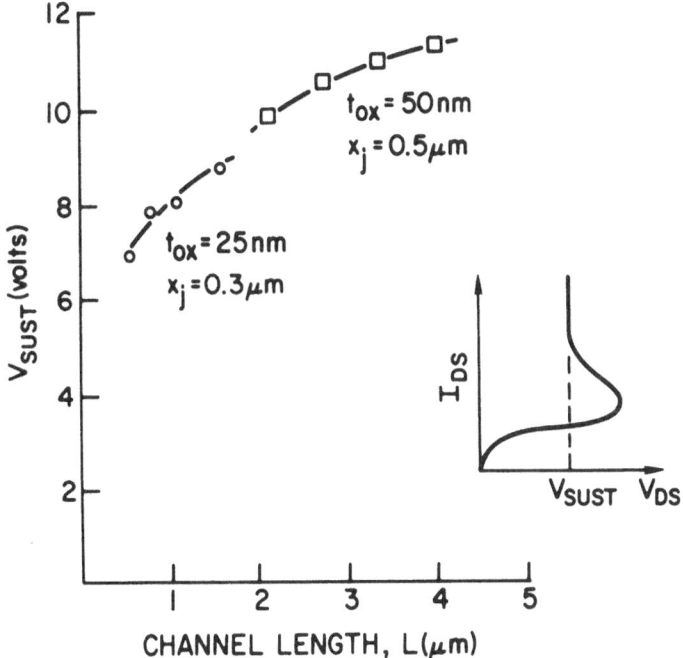

Fig. 8. - Drain-to-source voltage required to sustain breakdown as a function of channel length and technology level.

Degradation effects due to the injection of hot electrons from the device channel into the gate oxide were also investigated extensively for this 1 μm technology.[7] These effects were also found to be less severe in small devices as long as the operating voltage is scaled down. At room temperature, an operating range for avoiding hot-electron injection was defined which allows drain voltages up to about 6 volts.

Characterization of the devices designed for liquid nitrogen temperature also showed successful attainment of well-controlled characteristics.[5] The expected increase in transconductance was achieved, with the current higher by about a factor of four at low drain voltages and by about a factor of two at higher drain voltages where saturation velocity effects become important. The problem of hot-electron injection is aggravated at low temperature due to the higher channel currents and much reduced scattering effects. Still, a safe range of operation with drain voltages up to 4 volts was determined, which is adequate for most projected applications.[7]

D. Logic Circuit Performance

Test circuits using these devices were fabricated to establish circuit feasibility and measure the performance. Ring oscillators with relatively wide devices and minimum capacitance loading were built to test the fastest possible switching speed, while more representative circuits with fan-in and fan-out of 3 and a nominal wiring-capacitance load were also characterized. The results are given in Table I, which shows average switching delay and power consumption per stage of the ring oscillator test circuits. The best results were obtained for a channel length of 1.05 μm. The overall test results on devices and circuits indicate that this is a reasonable nominal value, although the circuits were originally designed with a more conservative 1.3 μm minimum length. It should be noted that circuit performance is very dependent on the tolerance allowed for channel length variations in the fabrication process and on noise margins built into the design to protect against coupling from other circuits, voltage drops in the power distribution network, etc. The results given here are thought to be realistic for large VLSI logic chips with a fabrication capability which can assure less than

TABLE I

RING OSCILLATOR RESULTS, AVERAGE DELAY AND POWER PER STAGE

	UNLOADED F.I., F.O.=1	LOADED F.I., F.O.=3 with $C_{wire}=50fF$
Room Temperature, Depl. Load		
$L=1.3\ \mu m,\ V_d=2.5\ V$	0.35 ns	1.9 ns
	0.48 mw	0.13 mw
$L=1.05\ \mu m,\ V_d=2.5\ V$	0.23 ns	1.1 ns
	0.63 mw	0.17 mw
Liquid Nitrogen Temp., Enh. Load		
$L=1.05\ \mu m,\ V_d=2.0\ V$	0.10 ns	0.46 ns
	1.3 mw	0.37 mw

± 0.3 μm variation in channel length. Much more aggressive performance levels have been reported recently for devices of similar vertical geometry, but much shorter channel lengths, based on more optimistic expectations for channel length control.[22,23]

The loaded circuit with fan-in and fan-out of 3, as shown in Table I has an average delay of about 1.1 ns. The power level of 0.17 mw is thought to be adequate for a VLSI logic chip with several watts of power dissipation. In such a chip, many circuits will not require the same performance level, and can be powered down by using smaller devices. Circuits with larger wiring capacitance can use push-pull output stages which consume no more power. Custom designed logic chips such as microprocessors make extensive use of structural elements such as programmed logic arrays (PLA's). It has been shown that high performance PLA's can be designed with lower power consumption than would be required for the equivalent number of NOR circuits.[6] PLA's with over a hundred product terms have been successfully fabricated in this technology with switching delays as low as 13 ns.

As expected, the circuits operated at liquid-nitrogen temperatures had smaller delays due to the increased device current. This also increases the power consumption. This may be reasonable since immersion in liquid nitrogen provides large power dissipation capability.

IV. MEMORY APPLICATIONS OF 1μm NMOS TECHNOLOGY

Silicon FET technology is strongly oriented toward high-density memory applications using dynamic one-device memory cells. Present day approaches to 16K bit or 64K bit chips typically use a double polysilicon gate process in which the first polysilicon layer forms the top electrode for storage capacitors and the second polysilicon layer forms the gates of the FET switches in the memory cell as well as the support circuits. One way to use 1μm technology for memory is to scale down the double polysilicon process. This has been done for an exploratory 256K bit RAM, achieving a cell area of about 70 μm^2. Since resistance becomes more important in the scaled-down lines, the second polysilicon layer was replaced by molybdenum to give much lower resistance for the long word lines which interconnect many FET gates.[20]

Another approach has been proposed to achieve high-density RAM's using only a single layer of gate material.[24] A cross-section of a memory cell is shown in Fig. 9. The material in this case is also a higher-

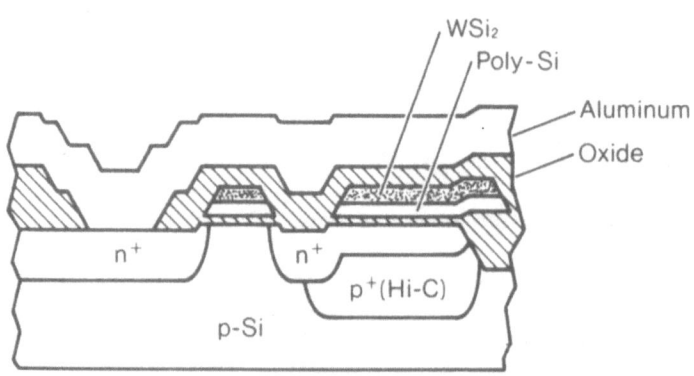

Fig. 9. - Cross-section of a dynamic RAM memory cell using a single gate-electrode level (composite polysilicon/tungsten-silicide layer).

conductivity layer using tungsten silicide as described earlier.[10] On the left is an FET switch with one n^+ electrode connected to an aluminum bit line for taking information to and from the cell. The gate of the device is part of a continuous word line running perpendicular to the bit line. Connected to the other n^+ electrode is a storage capacitor. The top electrode of this capacitor is another tungsten-silicide word line. In this arrangement the capacitor in each cell uses the adjacent wordline as a capacitor electrode. Along each wordline FET gates and capacitor electrodes are connected alternately. Since the capacitor electrode is connected to a wordline which is usually at ground potential, an inversion layer cannot be formed underneath it (as is normally accomplished in a double polysilicon layout using a positive bias on the first polysilicon storage electrode). Therefore an n-type ion implantation step is performed through a mask to dope the surface underneath all storage capacitors electrodes. At the same time, it is convenient to perform a lighter p-type implant enhancing the p doping underneath the storage junction to increase the junction capacitance which is in parallel with the main storage

504   *Dennard*

V. POTENTIAL 1 μm COMPLEMENTARY MOS (CMOS) TECHNOLOGY

There is considerable effort in bulk CMOS technology at the present time using 3-4 μm layout dimensions.[26] The potential for scaling down CMOS devices to 1 μm dimensions is very interesting and timely to consider. The general form of such a CMOS structure is illustrated in Fig. 10. The normal n-channel devices are supplemented with p-channel devices built in an n-type isolation pocket. This pocket, normally fabricated with ion implantation followed by

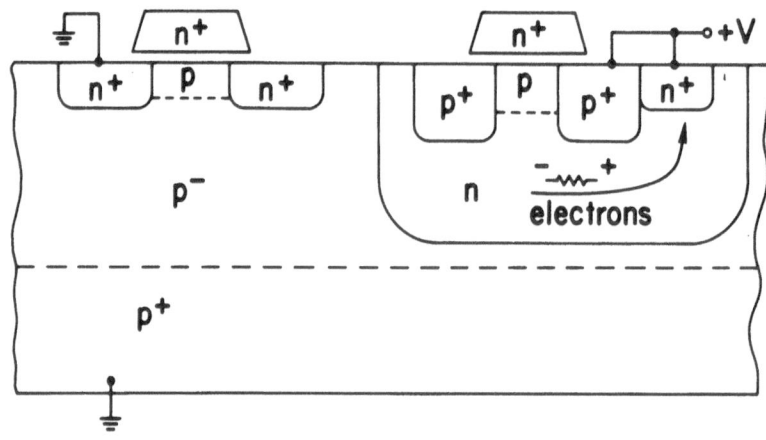

Fig. 10. - Illustration of possible scaled CMOS technology.

high temperature drive-in, would be reduced in depth and increased in doping concentration according to the normal scaling. The sheet resistance would be approximately maintained the same. The latchup problem in CMOS is related to the flow of collected excess electrons through this region toward the positive power supply. The resulting voltage drop along the n-type pocket, as shown in Fig. 10, can create a forward bias at the left-hand edge of the p^+ region connected to the same power supply. Latchup would occur if holes emitted from that p^+ region flow to the grounded n^+ source and forward bias that junction, causing emission of excess electrons flowing to the n pocket to sustain the process. It has been noted that since the potential drop required to forward bias these junctions do not scale, the latchup problem tends to become more containable with scaled-down supply voltages.[27] Moreover, while the pocket sheet resistance can be constant, as noted, the length of the resistive path under the critical p^+ region is reduced by scaling. According to current practice, an epitaxial layer would also be used as shown to provide lower substrate resistance shunting the n^+ "emitter".[26] This could also be more heavily doped and reduced in thickness. These are favorable factors in a problem which remains quite complicated to analyze.

In addition to the difficulties of fabricating a shallow epitaxial region, and the resultant yield hazard, some other technology and design problems are evident from consideration of Fig. 10. It has been pointed out that the sheet resistance of p^+ source/drain regions increases more than linearly as the depth is reduced.[28] Achieving shallow junctions with good conductivity will be difficult. Another problem relates to the design difficulty of the p-channel device if n^+ gate doping is used. To achieve a sufficiently low threshold magnitude, $|V_t|$, will require considerable counterdoping of the channel region with boron as shown. This buried-channel device has characteristics similar to the depletion device discussed with respect to Fig. 5. A shallow boron region - not easy to achieve - would be required to maintain control of the threshold and transconductance at high drain voltages. Using a p^+ gate for this device would require extra process development.

Thus, the development of workable CMOS structures at 1 μm dimensions is a challenging problem with several remaining unknowns. It is clear that success of such an endeavor would give much lower power dissipation than the NMOS depletion-load NOR circuits that were described. This could be important for many types of applications.

VI. RADIATION EFFECTS ON SMALL-GEOMETRY DEVICES

One key problem still facing 1 μm MOSFET technology is how to deal with various forms of radiation. Present day dynamic RAM's are subject to transient errors due to emission of alpha particles from radioactive impurities in the materials used to manufacture or package the devices.[29] Some problems remain even after efforts to purify the materials and shield the chips from the package. Scaled-down circuits are even more susceptible since the reduced voltage and capacitance both decrease the charge stored in a given node, while the charge collected from ionized carriers following an alpha-particle disturbance is not decreased nearly as much by the smaller collection area.[30] Even if the radioactive impurities can be eliminated, cosmic

radiation remains as an irreducible source of ionization bursts.[31]

The magnitude of the col-
lected charge from an alpha
particle striking a biased n^+ region on a 5Ω-cm silicon substrate is shown in Fig. 11, as a
function of the alpha-particle
energy and the width of the
long n^+ line. Measured results
are shown as well as the predictions of a Monte Carlo
model which has been developed to track the random motion (diffusion) of the ionized
electrons.[30] This model also

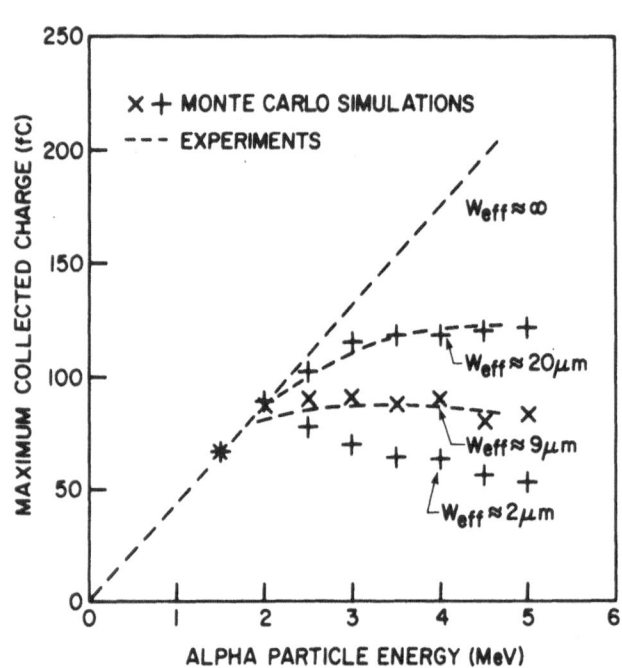

Fig. 11. - Peak charge collected in positively-biased n-regions with a periodicity W from alpha-particle hits of varying energy.

takes into account the results of finite element simulations for the early part of the collection

process, which show that the presence of the high excess-carrier density distorts the electric

field around the junction.[32] A resultant drift field is produced rather deep into the substrate

which causes quick collection of all the excess carriers down to a depth of 5 to 10 μm

depending on the substrate concentration. The projections of the model show that the

collected charge is not reduced greatly in smaller lines, as shown in the figure, and is still as

great as 80 fC for a 2 μm line struck by a 2 MeV alpha particle. The collected charge falls off

for higher energies because the most efficient ionization occurs deep in the substrate as the

particle slows down, and these excess carriers diffuse away to be collected by neighboring

lines.

Coping with these radiation problems is causing consideration of structural changes. The modeling results show that the collection of ionized carriers in the silicon is inversely related to the substrate doping concentration,[32] which may explain why epitaxial layers on heavily doped substrates are being reported in various areas. Also, CMOS devices built in isolation pockets are known to be less susceptible, since only ionized charge within the depth of the pocket is collected. Design approaches to provide redundancy and allow error correction can be effective because the error rates are actually quite low compared to operating speeds. Such approaches are not always easy to implement, particularly in logic circuits, but will undoubtedly become more popular with the advent of very small VLSI devices.

The classical radiation problem in a military enviornment is actually less severe in miniaturized MOSFET's. Scaling down of insulator thickness is known to help prevent device threshold changes and field inversion, since the ionized charge released in the insulator (and subsequently trapped) is reduced by the thinner layers. Reduction of the trapping sites at the oxide-semiconductor interface can be managed to some degree by controlling process conditions.[33]

Whereas the classical radiation problem considers a uniform dose across the device surface, there is recent interest in localized failures. For example, a model has been presented by which a single alpha-particle event could cause a significant threshold change in a small-dimension device.[34] Localized damage in the silicon from a high-energy nuclear interaction could also greatly affect the conduction properties of a small device.[35] Such possibilities remain to be investigated more fully.

VII. DIRECTIONS FOR PROCESS AND STRUCTURAL INNOVATION

Several recent exploratory efforts indicate directions in processing which may be important in future device applications. Generally, small devices require reduced processing temperatures to achieve shallower vertical dimensions. It has been shown that the temperature required for growing SiO_2 layers can be greatly reduced by using a plasma to enhance the growth rate. An application of this is to reduce the boron depth under a field region, which is driven deep during a normal field oxidation. Figure 12 shows the threshold voltage for inversion under a polysilicon line crossing a field oxide, as a function of the potential on an adjacent junction acting as a source.[36] It is shown that the boron doping under a field oxide processed at low temperature in an oxygen plasma is much shallower since the

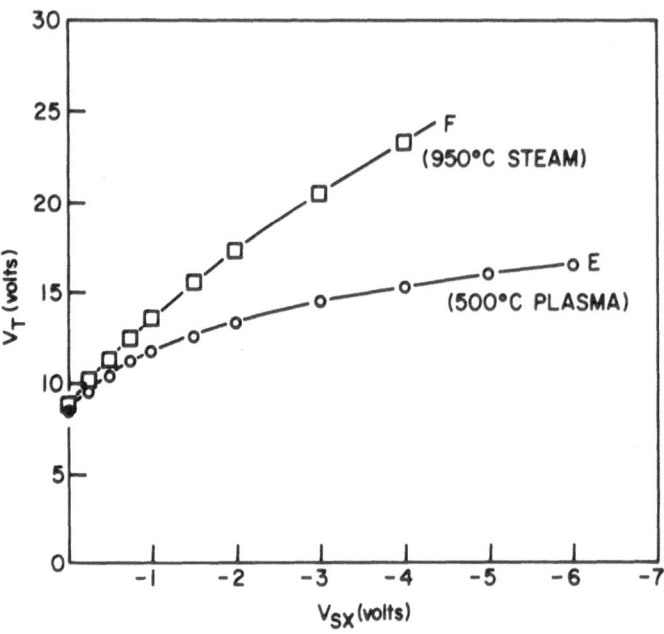

Fig. 12. - Threshold voltage required for inversion of isolation regions by an polysilicon conductor crossing the ~0.35μm thick field oxide, as a function of source-substrate voltage and processing condition. Field implant dose is adjusted to get similar minimum threshold value.

threshold rises slowly with the source-substrate voltage, indicating that the depletion region is extending into the lightly doped substrate. A much smaller boron implant dose achieves the same surface concentration. The "bird's beak" effect is also reduced with plasma field oxidation, because no oxide pad is required under the mask to prevent damage during oxidation. Generation of oxidation stacking faults should be greatly reduced with such processes.

Other important processing directions which achieve a physical change in a localized area are laser and electron beam annealing. These subjects are fully covered in other chapters of this publication.

Improved gate insulator processes may be needed in the future to achieve the thin insulators which will be required. Enhanced dielectric constants are very desirable to increase the charge storage in small-dimension dynamic RAM's, particularly because of the alpha-particle problem. Thermally-grown silicon-nitride layers have been shown to have potential for very thin insulators, and offer a somewhat larger dielectric constant.[37] Substantial increases in storage capacitance per unit area have been demonstrated using a tantalum-oxide insulating layer.[38]

Structural innovations are also being made using combinations of existing processes, using selectivity, or using directionality of processes (particularly RIE). An example is the self-aligned contact hole to a polysilicon gate, which is formed by using a silicon-nitride layer over the gate to allow selective oxidation of other areas.[39] Such a process has been used recently to obtain layout density improvements in a 32-bit processor chip.[40] Self-alignment of contact holes to junctions relative to a gate edge have also been accomplished.[41] Several schemes have been used to achieve shallow junctions near the gate edge, with deeper junctions spaced away from the gate edge.[41,42,43] The shallow junction can achieve a reduction in short-channel effects, and making it lightly doped increases the device breakdown voltages and reduces hot-electron effects due to reduced fields near the drain. The series resistance of the lightly-doped regions offsets these advantages to a degree.

VIII. NEWER DEVICE FORMS

Some new forms of FET's are emerging with various degrees of novelty. There are numerous recent papers on n-channel enhancement devices which use p-type gate electrodes.[44,45] The basic idea behind these devices is to reduce the electric field perpendicular to the surface and therefore increase the mobility of the carriers. In the extreme, the hope is to keep the carriers away from the surface and achieve bulk mobility. The operation of these devices is very similar to a depletion-type device as analyzed in Figures 4 and 5, except that there is a smaller n-type implanted dose and the gate work function shifts the threshold to a positive value. The p-channel device of the CMOS family (Fig. 10) is a p-channel version of the same idea. Although the mobility may indeed be higher, these devices will tend to have threshold lowering at high drain voltages and reduced carrier density due to the depth of the carriers from the interface. The design which brings the conduction closer to the surface should be better controlled.

Another device type receiving considerable attention is the vertical junction-gated FET, also called the static induction transistor.[46] This device is a short-channel vertical device operating in a punchthrough mode, that is with triode-like characteristics where the gate has control. Recently, it has been applied as a load device where the substrate drain connection also serves as the power supply distribution system for the chip.[47] Any punchthrough device is inherently sensitive to channel length variations, but the channel length can be controlled reasonable well in vertical structures. On the other hand, various analyses have also shown these devices to be quite sensitive to the spacing between the gate junctions.

One of the most novel approaches to dynamic RAM's in recent years has been the development of a buried n-channel device in which the storage mechanism is a potential well for holes at the surface.[48] In this structure, the surface well has to be isolated from the substrate by being enclosed by the buried n-type well, which can be achieved with recessed-oxide isolation between devices. Readout is nondestructive since the presence or absence of holes (in the surface well) modulates the conduction of electrons between source and drain in the buried layer. Resetting and writing the cell is more complicated and requires additional support circuits with more than one voltage level. The cell itself is fairly dense and has protection against disturbance from ionizing radiation.

IX. CONCLUSION

The area of advanced MOS devices is a fast moving field. Miniaturization of devices and circuits is the principal thrust, and faces numerous practical problems. The payoff in density and performance appears to be well worth the effort. Although many of the new ideas are only evolutionary, it seems likely that this evolution over a period of time will lead to considerable change and progress.

REFERENCES

1.   R. H. Dennard, F. H. Gaensslen, L. Kuhn, and H. N. Yu, "Design of Micron MOS Switching Devices," presented at IEEE International Electron Device Meeting, Washington, D.C., Dec. 1972.

2.   R. H. Dennard, F. H. Gaensslen, H. N. Yu, V. L. Rideout, E. Bassous, and A. R. LeBlanc, IEEE J. Solid State Circuits SC-9, 256 (1974).

3.   F. H. Gaensslen, V. L. Rideout, E. J. Walker and J. J. Walker, IEEE Tran. on Elec. Dev. ED-24, 218 (1977).

4.   H. N. Yu, A. Reisman, C. M. Osburn, and D. C. Critchlow, IEEE Tran. on Elec. Dev. ED-26, 318 (1979). (Also in IEEE J. Solid State Circuits SC-14, 1979)

5.   R. H. Dennard, F. H. Gaensslen, E. J. Walker, and P. W. Cook, ibid., p. 325.

6.   P. W. Cook, S. E. Schuster, J. T. Parrish, V. DiLonardo, and D. R. Freedman, ibid., p. 333.

7.   T. H. Ning, P. W. Cook, R. H. Dennard, C. M. Osburn, S. E. Schuster, and H. N. Yu, ibid., p. 346.

8.   W. R. Hunter, L. M. Ephrath, W. D. Grobman, C. M. Osburn, B. L. Crowder, A. Cramer, and H. E. Luhn, ibid., p 353.

9.   W. D. Grobman, H. E. Luhn, T. P. Donohue, A. J. Speth, A. Wilson, M. Hatzakis, and T. H. P. Chang, ibid, p. 360.

10.   B. L. Crowder and S. Zirinsky, ibid., p. 369.

11.   J. M. Aitken, ibid., p. 372.

12.   J. R. Brews, W. Fichtner, E. H. Nicollian, and S. M. Sze, IEEE Elec. Dev. Letters EDL-1, 1 (1980).

13.   M. R. Wordeman, Tech. Digest of the Int. Elec. Dev. Meeting, p. 26 Dec. (1979).

14.   M. R. Wordeman and R. H. Dennard, IEEE Tran. on Elec. Dev. ED-28, to be published (1981).

15.   F. H. Gaensslen, R. C. Jaeger, and J. J. Walker, Solid State Electron. 22, 423 (1979).

16.   C. M. Osburn and E. Bassous, J. Electrochem. Soc. 122, 89 (1975).

17.   P. Solomon, J. Vac. Sci. Technol. 14, 1122 (1977).

18.   C. Y. Ting and B. L. Crowder, ECS Extended Abstracts, ECS Spring Meeting, Minneapolis, Minn. (1981).

19.   M. Y. Tsai, H. H. Chao, L. M. Ephrath, B. L. Crowder, A. Cramer, R. S. Bennett, C. J. Lucchese, and M. R. Wordeman, Proceedings of the Fourth Int. Symp. on Silicon Materials Sci. and Technol., Semiconductor Silicon 1981, p. 573.

20.   T. Mano, K. Takeya, T. Watanabe, K. Kiuchi, T. Ogawa, and K. Hirata, IEEE ISSCC
       Dig. of Tech. Papers, p. 234 (Feb., 1980).

21.   D. J. DiMaria, J. M. Aitken, and L. M. Ephrath, ECS Extended Abstracts 80-2, 1381
       (1980).

22.   M. P. Lepselter, Tech. Digest of Int. Elec. Dev. Meeting, p. 42 (Dec., 1980).

23.   M. P. Lepselter, IEEE Spectrum 18, no. 5, 26 (1981).

24.   H. H. Chao, R. H. Dennard, M. Y. Tsai, M. R. Wordeman, and A. Cramer, IEEE
       ISSCC Dig. Tech. Papers, p. 152 (Feb., 1981).

25.   T.Ohzone, S. Kondo, K. Tsuji, T. Shiragasawa, T. Ishihara, and S. Horiuchi, IEEE
       ISSCC Dig. Tech. Papers, p. 236 (Feb., 1980).

26.   L. C. Parrillo, R. S. Payne, R. E. Davis, G. W. Reutlinger, and R. L. Field, Tech.
       Digest of Int. Elec. Dev. Meeting, p. 752 (Dec., 1980).

27.   D. Estreich, Ph.D Thesis, Stanford University (1980).

28.   D. B. Scott, 39th Device Research Conf. paper VI B (June, 1981).

29.   T. C. May and M. H. Woods, IEEE Tran. on Elec. Dev. ED-26, 1 (1979).

30.   G. A. Sai-Halasz and M. R. Wordeman, IEEE Elec. Dev. Letters EDL-1, 211 (1980).

31.   J. F. Ziegler and W. A. Lanford, Science 206, 776 (1979).

32.   C. M. Hsieh, P. C. Murley, and R. R. O'Brien, IEEE Elec. Dev. Letters EDL-2, 103
       (1981).

33.   G. F. Derbenwick and B. L. Gregory, IEEE Tran. on Nuclear Science NS-22, 2151
       (1975).

34.   T. R. Oldham and J. M. McGarrity, Late Paper at ECS Spring Meeting, Hollywood,
       FL. (1980).

35.   G. P. Mueller and C. S. Guenser, IEEE Tran. on Nuclear Science NS-27, 1474
       (1980).

36.   A. Ray and A. Reisman, J. Electrochem. Soc. 128, to be published (1981).

37.   K. Ohta, K. Yamada, M. Saitoh, H. Shiraki, A. Nakamura, K. Shimizu and Y. Tarui,
       IEEE ISSCC Dig. Tech. Papers, p. 66 (Feb., 1980).

38.   T. Ito, T. Nozaki, H. Ishikawa and Y. Fukukawa, ibid., p. 74.

39.   J. M. Mikkelson, L. A. Hall, A. K. Malhotra, S. D. Seccombe, and M. S. Wilson, IEEE
       ISSCC Dig. of Tech. Paper, p. 106 (Feb., 1981).

40.   K. Ohta, K. Yamada, M. Saitoh, K. Shimizu, and Y. Tarui, IEEE Tran. on Elec. Dev.,
       ED-27, 1352 (1980).

41. S. Ogura, P. J. Tsang, W. W. Walker, D. L. Critchlow, and J. F. Shepard, ibid., p. 1359.

42. S. Hsia, R. Fatemi, T. C. Teng, S. C. Sun, and C. Skinner, 39th Device Research Conf., paper III B (1981).

43. K. Nishiuchi, H. Ota, T. Nakamura, H. Ishikawa, and M. Shinoda, Tech. Dig. of Int. Elec. Dev. Meeting, p. 26 (1978).

44. E. Sun, B. Hoefflinger, J. Moll, C. Sodini, and G. Zimmer, Tech. Dig. of Int. Elec. Dev. Meeting, p. 791 (1980).

45. J. Nishizawa, T. Terasaki, and J. Shibata, IEEE Tran. on Elec. Dev., ED-22, 185 (1975).

46. Y. Sakai, T. Masuhara, O. Minato, and N. Hashimoto, IEEE ISSCC Dig. of Tech. Papers, p. 56 (1980).

47. P. K. Chatterjee, G. W. Taylor and M. Malwah, IEEE ISSCC Dig. of Tech. Papers, p. 22 (1979).

CHAPTER VI

RELIABILITY

SEMICONDUCTOR COMPONENT ACCELERATED TESTING AND DATA ANALYSIS

F. H. Reynolds

British Telecom Research Laboratories
Martlesham Heath
IPSWICH
IP5 7RE, England

1. INTRODUCTION: The field reliability of an electronic component - the probability of its survival over a stipulated period of service - is determined by the standards applied to its design, the control exercised over its manufacture, and the efficiency of any screening and batch-acceptance procedures which may be applied prior to its deployment. These topics have received intensive study in modern times yielding, for components of broadly comparable functional performance, (both active and passive) steadily improving reliability. Whilst this trend can be expected to continue, it can confidently be asserted that there will always be a need for reliability measurements. The component designer may wish to assess the elements of his component, a metallisation track, a transistor say, as a prior activity. The equipment maker or the equipment operator may decide to assess the reliability of a desired component type-code before its commitment, and in the production phase, they may need to identify burn-in treatments and the details of sampling tests. In the field, component reliability information is needed for such purposes as equipment design evaluation and spares and maintenance provisioning, and sometimes just for publication as a straightforward historical record.

The above objectives imply that the underlying purpose of reliability data collection is often one of prediction and although different routes can be followed, the common starting point is usually the observation of a group of components (or component elements) which are regarded as a random sample withdrawn from a population. In the classical manner of inferential statistics, the sample data is then used to draw conclusions about the component population so that when other members of the same population are deployed, their reliability behaviour can be inferred.

In all component reliability studies, the key component property is its **lifetime** which, essentially, is just the interval between its (real) **birth time**, when the component is operative, and a later **failure time** when it has become defective. In practice, this simple definition is not always easy to interpret; it may, for example, be desirable to discount part of the elapsed time due to intermittent operation, and some judgement might be needed to determine the birth time and to define defectiveness. For the present, it will be assumed that these, basically engineering, matters have been resolved and that a large population of components has been identified within which the lifetimes (which must not be confused with real times) of every member are known. Attention can then be focused on the way in which the lifetimes are distributed across the population, but before discussing specific lifetime distributions, some of the general properties of distributions characterized by a single continuous variable will be reviewed. Although much of the text is applicable to any type of electronic or electrical component, the examples given relate to semiconductor parts or elements of them.

The confusion surrounding the use of the word "test" in reliability literature should be recognized. The word will now be used only for a "life test" or "stress test," essentially having duration, with "measurement" used for a "snapshot" assessment of the state of a component. Care is also needed in respect of "parameter" which will always be a statistical quantity. The engineer's use of the term for an electrical characteristic such as a voltage will be avoided.

2. DISTRIBUTIONS[1-11]

If a property of the members of a population is represented by x, then the distribution of this variable, or **variate** (to use the proper term) is expressed by the **frequency density function** or **fdf** and is symbolized by $f(x)$. The fdf has the property that $f(x)\,dx$ is the fraction of the population having x in the range x to x+dx, and $f(x)$ must be zero at $\pm\infty$ as shown by the **distribution profile** of Fig. 1(a). The fraction of the population having an x-value less than that of a selected ordinate is

$$F(x) = \int_{-\infty}^{x} f(x)\, dx \qquad (1)$$

where $F(x)$ is called the **cumulative distribution function** or **cdf** shown on Fig. 1(b). By the definition of $f(x)$, $F(\infty) = 1$ and from Eq. (1),

$$f(x) = \frac{dF(x)}{dx} \qquad (2)$$

Sometimes the subscript p is added to the variate x, as in Fig. 1(b) to show that, at x_p, the cdf is equal to p, whence x_p is the **p-fractile** of the distribution or the $(p \times 100)^{th}$ **percentile**. The 0.5 fractile, or 50th percentile, is the **median** value of x. Very often, more than one distribution is discussed at the same time so that subscripts must be added to the fdf and cdf as $f_x(x)$ and $F_x(x)$ if confusion could arise.

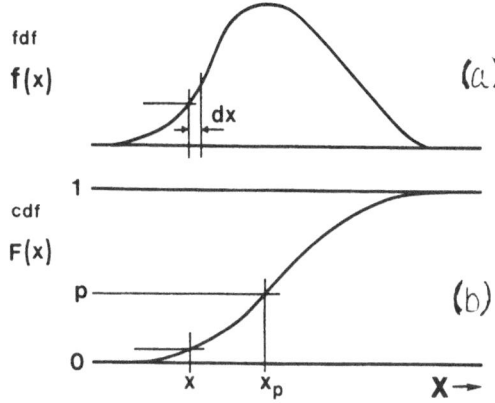

Fig. 1 - (a) The frequency density function (fdf): the distribution profile. (b) The cumulative distribution function (cdf). - with definition of x_p

There are two other useful population descriptors; one is the **mean**

$$m[x] = \int_{-\infty}^{\infty} x f(x)\, dx, \qquad (3)$$

and the **variance**

$$Var[x] = \int_{-\infty}^{\infty} (x - m[x])^2 f(x)\, dx, \qquad (4)$$

the square root of the variance being called the **standard deviation**.

If the members of a population are characterised by x of fdf $f_x(x)$, then they can also be described by another variate y of fdf $f_Y(y)$ where y is a function of x, or

$$y = g(x) \qquad (5)$$

If there is a simple one-to-one relationship between x and y, (g(x) could be, say, 2x but not x^2), then the fraction of the population members having x in the range x to x+dx also has y in the range y to y+dy whence

$$f_Y(y) = f_x(x) \frac{dx}{dy}. \qquad (6)$$

For the cdf, the fraction of the population having y in the range $-\infty$ to y, namely $F_Y(y)$ also has x in the range $-\infty$ to x, which is $F_x(x)$ so

$$F_Y(y) = F_x(x) \qquad (7)$$

which means that the cdf of the original distribution is equal to the cdf of the translated distribution at the corresponding value of the variate.

3. LIFETIME FUNCTIONS

Lifetime, denoted by t (as distinct from t which will be used later for real time) is a continuous variable so the expressions of Sec. 2 are immediately applicable. The frequency density function becomes f(t), and is sometimes alternatively called the **failure density function**, having the property that f(t) dt is the fraction of the component population having lifetimes in the range t to $t+dt$. The cumulative distribution function, also known as the **cumulative failure function** is then

$$Q(t) = \int_0^t ft \; dt \qquad (8)$$

where Q(t) represents the fraction of the population having lifetimes up to t. The lower limit of the integral is now, of course, zero. Clearly,

$$f(t) = \frac{dQ(t)}{dt}.$$ (9)

The fraction of the population having lifetimes greater than t is

$$R(t) = \int_t^\infty f(t)\, dt$$ (10)

where $R(t)$ is called the survival function or **reliability function** and so

$$R(t) + Q(t) = 1$$ (11)

Another lifetime function closely related to the fdf is the **hazard function**, $\lambda(t)$, which is such that $\lambda(t)\, dt$ expresses the number of lifetimes in the range t to $t+dt$ but now as a fraction of the number of components having lifetimes greater than t, which is $R(t)$. Hence,

$$\lambda(t) = \frac{f(t)}{R(t)} = \frac{f(t)}{1 - Q(t)}$$ (12)

Like the fdf, the hazard function can be integrated to give

$$H(t) = \int_0^t \lambda(t)\, dt$$ (13)

where $H(t)$ is the **cummulative hazard function**. Substituting Eqs. (9) and (12) into Eq. (13) gives

$$H(t) = \ell n \left[1 / (1 - Q(t)) \right]$$ (14)

showing that as the cumulative failure function approaches unity, the cumulative hazard function goes to infinity.

In summary, the lifetime functions $f(t)$, $Q(t)$, $R(t)$, $\lambda(t)$ and $H(t)$ have been defined, all being obtainable given any one of them. All the above symbols may be subscripted when any confusion can arise.

There are many mathematical expressions which can be fdfs. The formulae usually include independent coefficients or **parameters** which sometimes have names such as **location,** **scale** and **shape** in order to reflect their influence upon the distribution profile. For component lifetimes however, there are only three expressions which have been found to be widely applicable and then subject to an important proviso. These expressions describe the **lognormal,** **Weibull** and **exponential** distributions,[3-5] the lognormal and Weibull each having, in general, three parameters and the exponential, which is only a special case of the Weibull, having one. The proviso is that the populations may actually comprise several sub-populations each having a different set of parameter values, the population then being described as **multimodal**. For the moment, simple monomodal populations will be considered in which every component belongs to a distribution with one set of parameters.

3.1 **Lognormal Distribution**

The frequency density function for the lognormal lifetime distribution is given by

$$
\begin{aligned}
f_L(t) &= \frac{1}{\sigma(t-\gamma)\sqrt{2\pi}} \ \exp\left[-\frac{1}{2}\left(\frac{\ell n\left(\frac{t-\gamma}{\psi}\right)}{\sigma} \right)^2 \right]_{t \,>\, \gamma \,\geq\, 0} \\
&= 0 \qquad\qquad\qquad\qquad\qquad\qquad\qquad\qquad 0 \leq t \leq \gamma
\end{aligned}
\tag{15}
$$

where γ and ψ are the parameters of location and scale having the dimensions of time, whilst σ, the shape parameter, alternatively called the **dispersion**, is dimensionless. The restrictions shown above on γ imply that there are no lifetimes in the range 0 to γ. The specimen profiles of Fig. 2, drawn for $\gamma = 0$, reveal that the fdf is zero at $t = 0$ and $t = \infty$ so that a maximum always exists. In Fig. 2(a), it is seen that increasing the scale factor spreads the distribution out to higher values of lifetime whilst in Fig. 2(b), increasing the shape factor also extends

the distribution but the value
of σ has a profound effect on
the position of the peak. It
can be shown that the peak of
minimum magnitude occurs at
$\sigma = 1$. Any curve of Fig. 2 redrawn for a non-zero location
parameter would be translated
to the right by that amount.
The fdf is not directly integrable (as by Eq. (1)) for the cdf.
Suppose however, that the lifetime variate is converted to z
of fdf $f_z(z)$ where

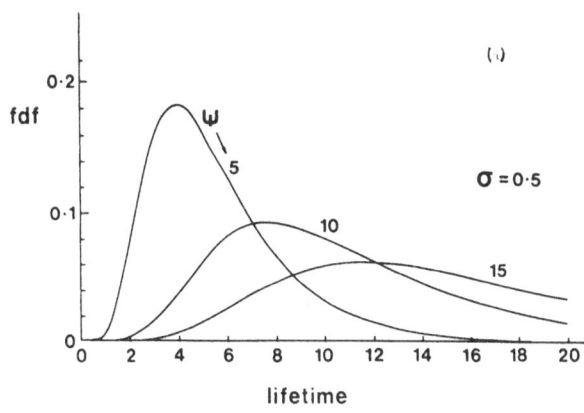

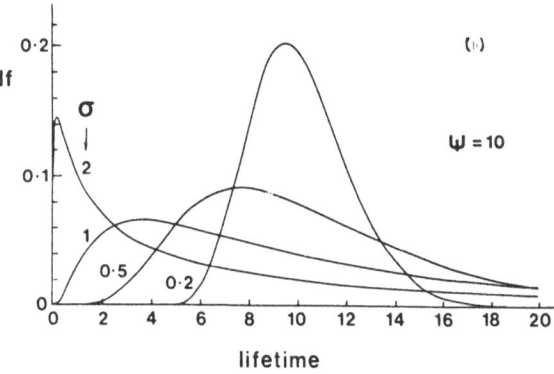

Fig. 2 - The frequency density function for the lognormal distribution: (a) effect of changing the scale parameter ψ; (b) effect of changing the shape parameter σ

$$z = \frac{\ln\left(\dfrac{t - \gamma}{\psi}\right)}{\sigma} \qquad (16)$$

which is of the form of Eq. 5. From Eq. 6 then,

$$f_L(t) = \frac{f_z(z)}{\sigma\,(t-\gamma)} \qquad (17)$$

where

$$f_z(z) = \frac{1}{\sqrt{2\pi}}\,\exp\,(-z^2/2) \qquad (18)$$

From Eq. 7, the cdf of the lifetime distribution is

$$Q_L(t) = F_z(z) = \Phi(z) = \frac{1}{\sqrt{2\pi}} \int_{-\infty}^{z} \exp\ (-z^2/2)\ dz \qquad (19)$$

This **gaussian** function as it is called, which is often given the symbol $\Phi(z)$, is also not integrable but its value is easily determined from z, called the **probit**,[12] using the approximating (but quite accurate) formula,[13] given in Appendix 1. An inverse formula (also given) enables z to be obtained from $\Phi(z)$. The transformation can also be effected using "probability" graph paper as in Fig. 3 because the ordinate scale is linear in probits with the rulings showing the corresponding values of the Gaussian function.

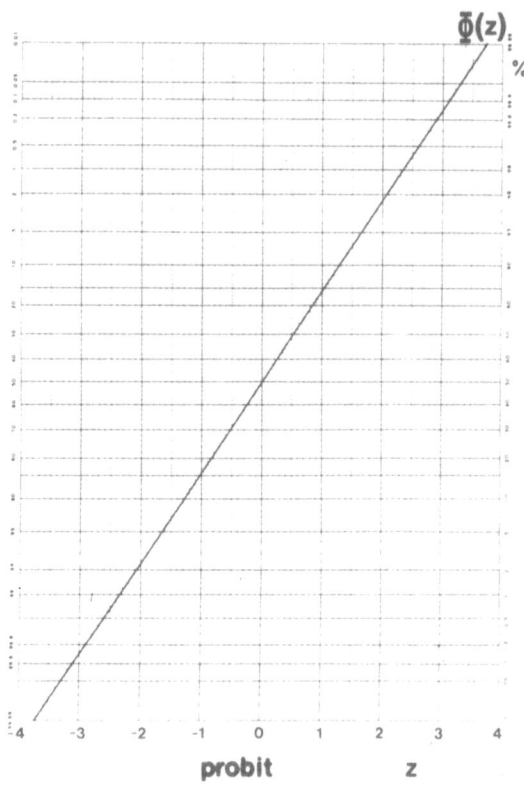

Fig. 3 - The Gaussian transformation: probit: Z, Gaussian: $\Psi(Z)$

From Eq. 16, the lifetime t_p corresponding to a cdf p can be written as

$$t_p = \gamma + \psi\ \exp\ (\sigma z_p). \qquad (20)$$

For a cdf of 0.5, z = 0 from Fig. 3 so $t_{0.5} = \gamma + \psi$ which is accordingly the **median lifetime** of the lognormal distribution. For any other cdf, t_p is known as the **p% reliable lifetime**.

Summarising the probit, Eq. 16 shows that it has a simple monotonic relationship to the lifetime t; when $t = \gamma$, $z = -\infty$ and the fdf and cdf are both zero. When $t = \gamma + \psi$, $z = 0$ and the cdf is 0.5, and when t approaches ∞, so does z with the cdf reaching unity and the fdf returning to zero.

If the logarithms in Eq. 16 are removed by writing

$$y = \ln(t - \gamma) \tag{21}$$

and

$$\mu = \ln \psi \tag{22}$$

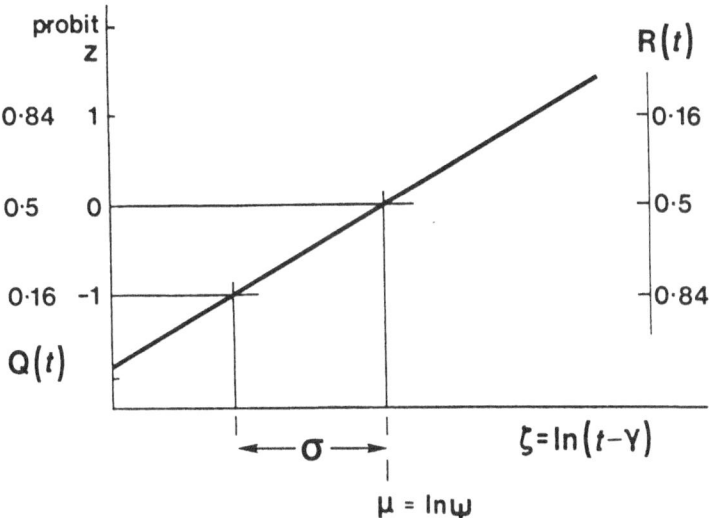

Fig. 4 - Plot of the cumulative distribution function for the lognormal distribution. For $\gamma = 0$, abscissa scale is $\ln$ (lifetime), ψ = median lifetime, $\mu = \ln$ (median lifetime).
cdf: $Q(t)$; reliability function: $R(t)$

then a plot of the cdf of the lifetime distribution against y will be linear of slope $1/\sigma$ (referred to the probit scale) and intercept μ on the abscissa as shown by Fig. 4 where the reliability function $R_L(t)$ is also marked. Using Eqs. 12, 14 and 16, a summary of all the lifetime functions in terms of the probit can be prepared as shown in Table I.

A graphical expression of the hazard function has been derived in Fig. 5 having noted, from Table I, that the product $\psi\lambda(t)$ depends upon z and σ only. Looking at the cdf scale added to the abscissa and selecting the range 0.0001% to 5% (which is of future interest) the hazard function is seen to increase steadily with z (and therefore with lifetime) over this range if σ is less than about 2, and decrease when σ exceeds about 4. From the last entry of Table I, the cumulative hazard function is, like the cdf, zero at $t = \gamma$, but, in contrast, as $t \to \infty$, $H(t)$ $\to \infty$. In the same way that an ordinate scale of z can be marked off with corresponding values of the cdf as in Fig. 3, so a cumulative hazard scale can be made by calculating the corresponding value of $H(t)$ using Eq. 14. Such "hazard graph paper" is available.

TABLE I

LOGNORMAL DISTRIBUTION OF LIFETIMES

parameters: location γ
scale η
shape σ

Lifetime Function	Symbol*	Expression in terms of
failure density function (fdf)	$f(t)$	$\dfrac{f_{/}(z)}{\sigma\psi \; \exp \; (z\sigma)}$
cumulative failure function (cdf)	$Q(t)$	$\Phi(z)$
reliability function	$R(t)$	$1 - \Phi(z)$
hazard function	$\lambda(t)$	$\dfrac{f_{/}(z)}{\sigma\psi \; (1 - \Phi(z)) \; \exp \; (z\sigma)}$
cumulative hazard function	$H(t)$	$\ell n \left\{ \dfrac{1}{1 - \Phi(z)} \right\}$

*suffix L omitted

where $z = \dfrac{\ell n \left(\frac{t - \gamma}{\psi} \right)}{\sigma}$ $\qquad\qquad f_{/}(z) = \dfrac{\exp \; (-z^2/2)}{\sqrt{2\pi}}$

median lifetime $= \psi + \gamma$ $\qquad\qquad \Phi(z) = \int_{-\infty}^{z} f_{/}(z) \; dz$
dispersion $= \sigma$

The lognormal distribution is often used in 2-parameter form, obtained by setting $\gamma = 0$. The lifetimes then range from 0 to ∞ and the scale parameter, ψ, becomes the median lifetime.

3.2 Weibull Distribution

The frequency density function for the Weibull lifetime distribution is given by

$$f_w(t) = \frac{\beta}{\eta} \left(\frac{t - \gamma}{\eta} \right)^{\beta-1} \exp \left[- \left(\frac{t - \gamma}{\eta} \right)^{\beta} \right] \quad t > \gamma \geq 0$$
$$= 0 \qquad\qquad\qquad\qquad 0 \leq t \leq \gamma$$

(23)

where σ and η are the parameters of location and scale and have the dimensions of time whilst β, the shape parameter, is dimensionless. The restrictions shown above on t again imply that

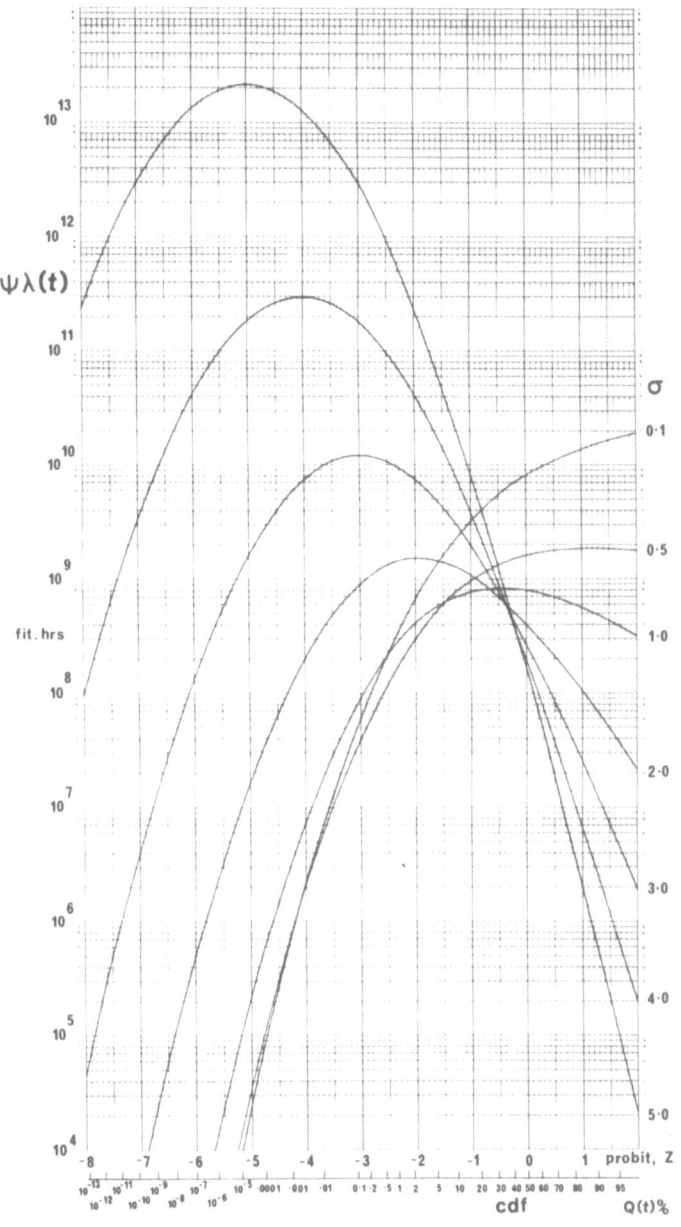

Fig. 5 - Hazard function for the lognormal distribution.
From parameters, find probit z then λ(*t*) and Q(*t*).

the range 0 to γ includes no lifetimes. Of the specimen profiles
of Fig. 6 (all drawn for $\gamma = 0$)
those of Fig. 6(b) reveal that,
unlike the lognormal case, the
fdf is not always at zero at $t =$
0. For $\beta = 1$, the fdf is actually
$1/\eta$ at $t = 0$ and it starts at $+\infty$
for all lower β values. In other
respects, $1/\beta$ influences the profile similarly to σ of the lognormal distribution whilst an increase in the scale factor η (Fig.
6(a)) broadly corresponds to increasing ψ.

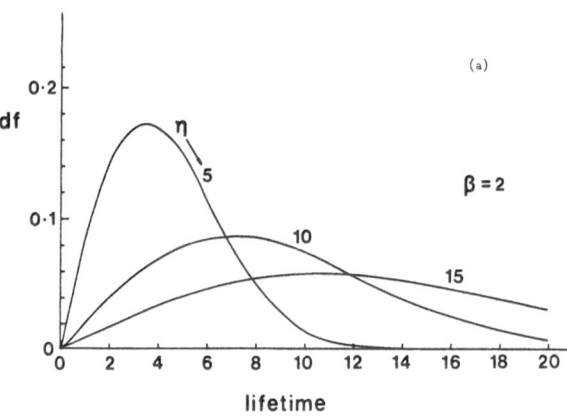

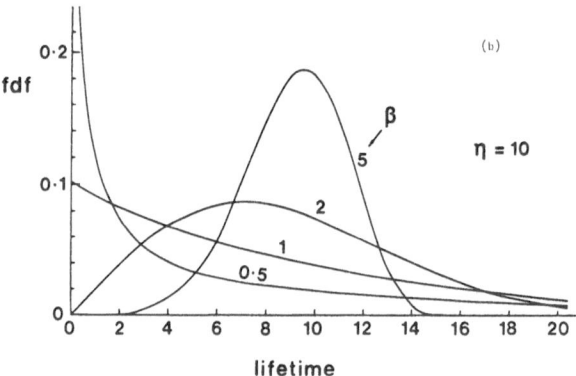

Fig. 6 - The frequency density function for the Weibull
distribution: (a) the effect of changing the scale parameter η; (b) effect of changing the shape parameter β

In contrast to the impossi-
bility of integrating the lognor-
mal fdf, Eq. 23 can be integrated directly but it will prove helpful to demonstrate the similari-
ties between the lognormal and Weibull distributions by defining, in analogy with Eq. 16, the
variable

$$z = \frac{\ell n\left(\dfrac{t - \gamma}{\eta}\right)}{1/\beta} \tag{24}$$

which is again of the form of Eq. 5. If z is distributed according to $f_z(z)$ then, from Eq. 6,

$$f_w(t) = \frac{\beta f_z(z)}{t - \gamma} \tag{25}$$

corresponding to Eq. 17 where

$$f_z(z) = \exp[z - \exp z] \qquad (26)$$

equivalent to Eq. 18. From Eq. 7, the cdf of the lifetime distribution is then

$$Q_w(t) = F_z(z) = 1 - \exp[-\exp z] \qquad (27)$$

In analogy with the probit, z will be called the **Weibit**. On Weibull graph paper, the scale is linear in Weibits with the rulings showing the corresponding values of $F_z(z)$. The p% reliable lifetime for the Weibull distribution is, from Eq. 24

$$t_p = \gamma + \eta \exp\left(\frac{z_p}{\beta}\right). \qquad (28)$$

When z = 0, corresponding to a cdf of 0.63 (Eq. 27) then $t_{0.63} = \gamma + \eta$ which is called the **characteristic lifetime** of the Weibull distribution.

Summarising the Weibit, Eq. 24 shows that it has a simple monotonic relationship to the lifetime *t*; when $t = \gamma$, z = -∞ and the cdf (but not necessarily the fdf) is zero. When $t = \gamma + \eta$, z = 0 and the cdf is 0.63, and when *t* approaches ∞, so does z with the cdf reaching unity and the fdf returning to zero.

If the logarithms in Eq. 25 are removed using Eq. 21 again and if

$$\xi = \ln \eta, \qquad (29)$$

then a plot of the cdf of the lifetime distribution against ξ will be linear of slope β (referred to the Weibit scale) and intercept ξ as shown by Fig. 7. Dropping a perpendicular on to the cdf

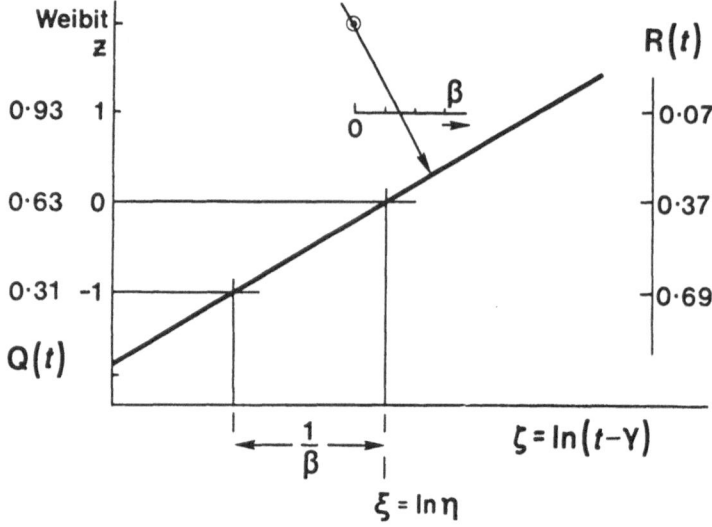

Fig. 7 - Plot of the cumulative distribution function for the Weibull distribution. For $\gamma = 0$, abscissa scale is ℓn (lifetime)m η = characteristic lifetime.
$z = \ell$n (characteristic lifetime); cdf: $Q(t)$; reliability function: $R(t)$

line as shown enables a scale of β to be added as on commercial Weibull graph papers. Using

Eqs. 12, 14 and 24, a summary of all the lifetime functions in terms of the Weibit can be

prepared as in Table II where however, the direct expressions (in terms of t) are also shown.

In order to complete the analogy, Fig. 8 is drawn in correspondence with Fig. 4 revealing that

the hazard function has no maximum; for β less than unity, the function falls with increasing z

(and therefore lifetime) whilst for β greater than unity, it always rises. When $\beta = 1$, the

hazard function is constant, which is considered further as a special case in the next section.

As for the lognormal distribution, the cumulative hazard function is zero at $t = \gamma$ whilst

as $t \to \infty$, $H(t) \to \infty$, and again similarly, Weibull cumulative hazard paper can be prepared by

marking a linear Weibit scale with values of $H(z)$ (= exp z). Plotting the cumulative hazard

on this scale against ℓn($t - \gamma$) then yields a straight line of slope β, with the characteristic

life, η, given by the intercept at unity cumulative hazard. The same result is obtained using

common logarithms as on Fig. 9.

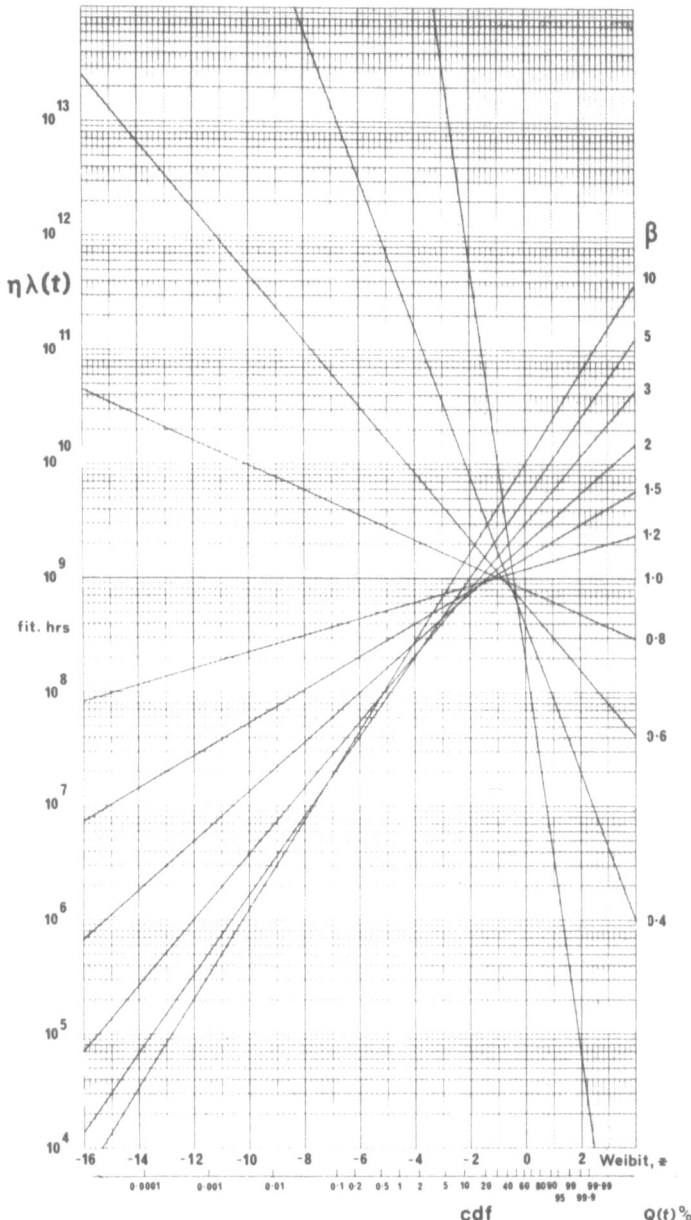

Fig. 8 - Hazard function for the Weibull distribution.
From parameters, find Weibit z and then $\lambda(t)$ and $Q(t)$.

Sometimes the Weibull distribution is expressed in terms of the parameter α or its reciprocal, where

$$\alpha = \eta^{\beta} \qquad (30)$$

The distribution is also used in its 2-parameter form with $\gamma = 0$ whereupon the scale parameter η becomes the characteristic lifetime.

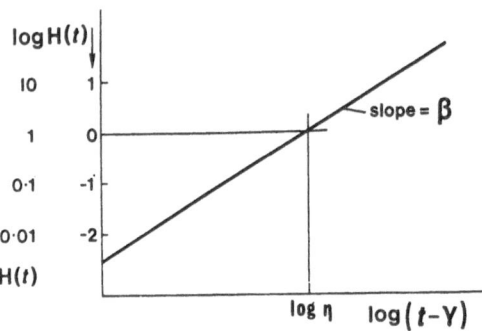

Fig. 9 - Plot of the cumulative hazard function for the Weibull distribution: for $\gamma = 0$, abscissa is log (lifetime), η = characteristic lifetime , cumulative hazard function: H(t)

TABLE II

WEIBULL DISTRIBUTION OF LIFETIMES

parameters: location γ
scale η
shape β

Lifetime Function	Symbol*	Expression in terms of	Expression in terms of
failure density function (fdf)	$f(t)$	$\frac{\beta}{\eta} \cdot \frac{f_r(z)}{\exp (z/\beta)}$	$\frac{\beta}{\eta} \left(\frac{t-\gamma}{\eta}\right)^{\beta-1} \exp\left[-\left(\frac{t-\gamma}{\eta}\right)^{\beta}\right]$
cumulative failure function (cdf)	$Q(t)$	$F_r(z)$	$1 - \exp\left[-\left(\frac{t-\gamma}{\eta}\right)^{\beta}\right]$
reliability function	$R(t)$	$1 - F_r(z)$	$\exp\left[-\left(\frac{t-\gamma}{\eta}\right)^{\beta}\right]$
hazard function	$\lambda(t)$	$\frac{\beta}{\eta} \cdot \frac{\exp z}{\exp (z/\beta)}$	$\frac{\beta}{\eta} \left(\frac{t-\gamma}{\eta}\right)^{\beta}$
cumulative hazard function	$H(t)$	$\ln\left(\frac{1}{1-f_r(z)}\right)$	$\left(\frac{t-\gamma}{\eta}\right)^{\beta}$

*Suffix W omitted

where $z = \beta \ln\left(\frac{t-\gamma}{\eta}\right)$

characteristic lifetime = $\eta + \gamma$

$f_r(z) = \exp (z - \exp z)$

$F_r(z) = 1 - \exp (- \exp z)$

3.2.1 Exponential Distribution

If the shape and location parameters of the Weibull distribution are set to $\beta = 1$ and $\gamma = 0$, then the resulting 1-parameter distribution has a significance justifying a separate identity as the exponential distribution. The fdf is thus

$$f_E(t) = \frac{1}{\eta} \exp\left(-\frac{t}{\eta}\right) \qquad (31)$$

which steadily decreases from an initial value of $1/\eta$ as already recorded on Fig. 6(b).

The cdf

$$Q_E(t) = 1 - \exp\left(-\frac{t}{\eta}\right)$$

(32)

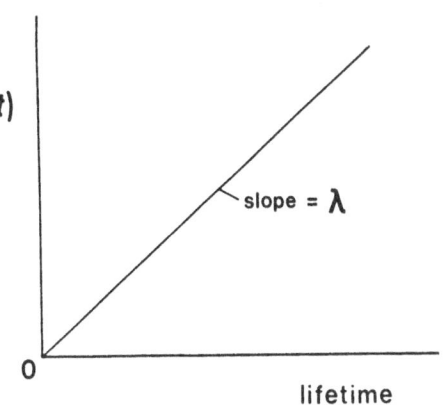

rises exponentially with lifetime towards unity. The hazard function, as already seen on Fig. 4, is constant at

$$\lambda(t) = \frac{1}{\eta}$$

(33)

Fig. 10 - Plot of the cumulative hazard function for the exponential distribution λ = hazard function (constant) cumulative hazard function: $H(t)$.

and the cumulative hazard function rises linearly with lifetime as

$$H(t) = \frac{t}{\eta}$$

(34)

illustrated in Fig. 10.

3.3 Lognormal-Weibull Parameter Relationship

It is not possible to find a quantitative relationship between the parameters of the lognormal and Weibull distributions but an approximate translation can be made by fitting a set of Weibull-distributed lifetimes to a lognormal distribution and vice-versa. The method adopted uses numerical estimating theory, yet to be described, but the result is

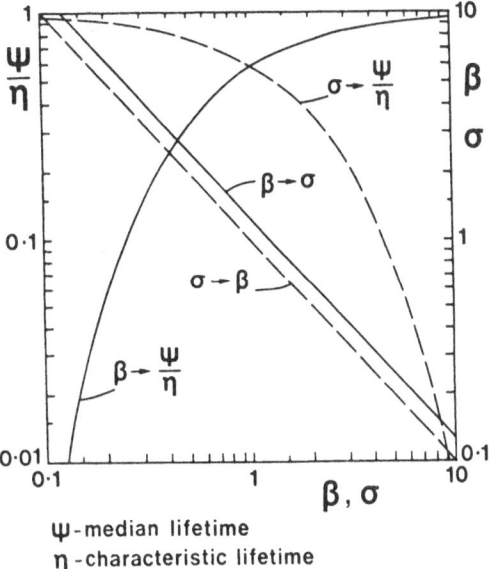

ψ - median lifetime
η - characteristic lifetime

Fig. 11 - Approximate parameter translation between lognormal and Weibull distributions.
Weibull → lognormal: enter abscissa with β and use full straight line to find σ, curved full line to find ψ/η and hence ψ.
lognormal → Weibull: enter abscissa with σ and use broken straight line to find β, curved broken line to find ψ/η and hence η.

more appropriate at this point and is shown by Fig. 11. Entering the abscissa with β or σ as given, yields the corresponding shape factor on the right-hand scale and the ratio of the median lifetime of the lognormal distribution to the characterisitic lifetime of the Weibull distribution on the left hand scale. The shape factors clearly have an approximately reciprocal relationship whilst for high β or low σ, the ψ/η ratio tends to unity.

4. MULTIMODAL LIFETIME FUNCTIONS

Reference was made in Sec. 3 to the possibility that the lifetimes of a component population may actually come from several sub-populations, each having a different set of parameter values. Suppose that there are k sub-populations of fdf, $f_i(t)$, with the size of each representing a fraction q_i of the whole population. The fraction of the population belonging to sub-population i and having lifetimes in the range t to $t+dt$ is then $q_i f_i(t)$ dt so the fdf for the whole population is

$$f(t) = \sum_{i=1}^{k} q_i f_i(t) \tag{35}$$

and the cdf is

$$Q(t) = \sum_{i=1}^{k} q_i Q_i(t) \tag{36}$$

where the $Q_i(t)$ are the cdfs for the sub-populations. The remaining lifetime functions are readily expressed using Eqs. 11, 12 and 14. Many types of multimodal distribution exist depending upon the relationship between the parameters of the sub-populations but only two are of particular interest, namely, the **bimodal**, for which k = 2, and the **staggered multimodal** where k is unlimited but the shape and scale parameters are constant.

4.1 Bimodal Distribution

Whilst the two sub-populations of a bimodal distribution could be of different type (one lognormal and the other Weibull) this possibility is never entertained. In, say, a lognormal bimodal distribution none of the corresponding parameters will, in general, be equal, so the fdf becomes, using Eq. 17,

$$f(t) = \frac{q_a\, f_z\,(z_a)}{\sigma_a\,(t - \gamma_a)} + \frac{q_b\, f_z(z_b)}{\sigma_b(t - \gamma_b)} \tag{37}$$

where

$$z_{a,b} = \frac{\ell n \left(\dfrac{t - \gamma_{a,b}}{\psi_{a,b}} \right)}{\sigma_{a,b}} \tag{38}$$

and

$$q_a + q_b = 1. \tag{39}$$

Usually, the fdf of Eq. 37 is used with $\gamma_a = \gamma_b$ and often with both the location parameters at zero. The cdf of the lognormal bimodal distribution is

$$Q(t) = q_a\, \Phi\,(z_a) + q_b\, \Phi\,(z_b) \tag{40}$$

directly from Eq. 19. The corresponding expressions for the Weibull bimodal distribution are

$$f(t) = \frac{q_a\, \beta_a\, f_z(z_a)}{t - \sigma_a} + \frac{q_b\, \beta_b\, f_z(z_b)}{t - \sigma_b} \tag{41}$$

and

$$Q(t) = q_a\, F(z_a) + q_b\, F(z_b) \tag{42}$$

where

$$z_{a,b} = \beta_{a,b}\, \ell n \left(\frac{t - \gamma_{a,b}}{\eta_{a,b}} \right) \tag{43}$$

With the lifetime functions dependent upon five parameters (assuming $\gamma_a = \gamma_b = 0$), a generalised graphical representation of the bimodal distribution is not possible. For the cdf, however, most of its parameter dependence can be appreciated from Fig. 12 plotted in the style of Fig. 4 for the lognormal distribution. For each curve, the proportion of the population in the early distribution is $q_a = 0.1$, whilst the logarithms of the median lives are 0 for the early sub-population and 10 for the main one. The ranges within which 90% of the constituent distributions fall are also marked showing, of course, that the plots are straighter the more

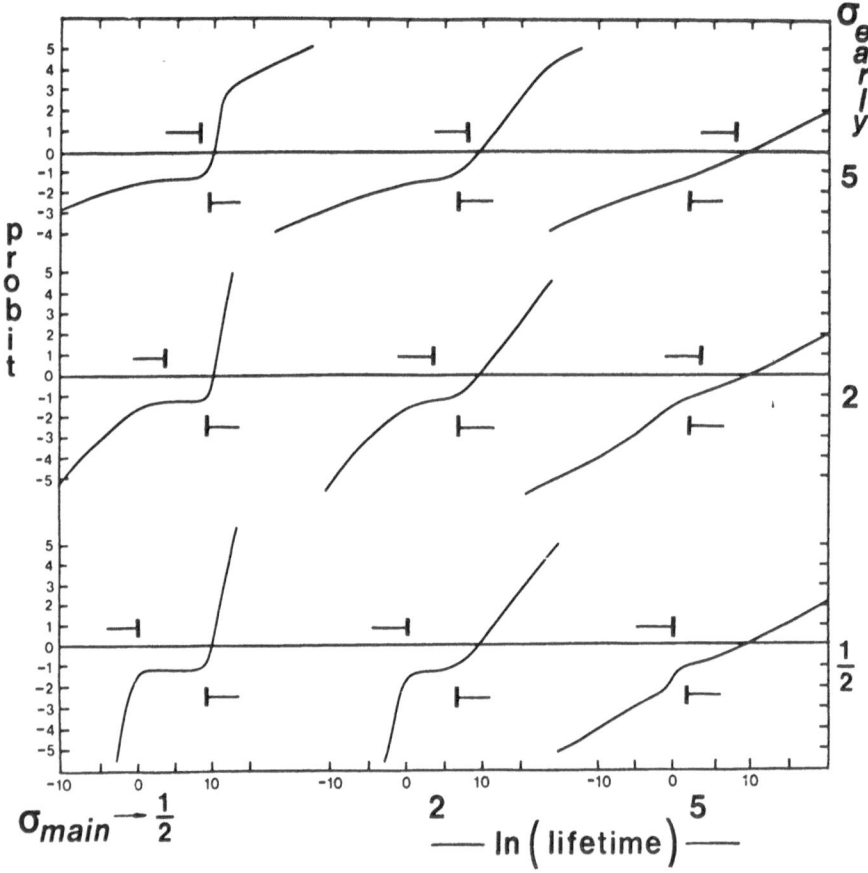

Fig. 12 - Bimodal-lognormal plots of the cumulative distribution function showing effect of early and main dispersion (σ) combinations. Proportion of early distribution = 0.1 $\mu_{main} - \mu_{early}$ = 10 units of ℓn (lifetime) scale - 90% of early distribution lies to left of upper vertical bar, 90% of main distribution lies to right of lower vertical bar.

the distributions overlap. The effect of varying q_a is easily inferred by noting that horizontal regions of the curve, where visible, occur at a cdf of about 0.1. A similar diagram can be constructed for a bimodal Weibull distribution (corresponding to Fig. 7) with β instead of σ as the variable.

4.2 Staggered Multimodal Distribution

If the shape and scale factors for all the sub-populations are the same but each has a different location parameter, the fdf is given by Eq. 35 with, for the lognormal distribution,

$$f_i(t) = \frac{1}{\sigma\sqrt{2\pi}(t - \gamma_i)} \exp\left[-\frac{1}{2}\left(\frac{\ell n\,(t - \gamma_i) - \mu}{\sigma}\right)^2\right] \tag{44}$$

or, for the Weibull

$$f_i(t) = \frac{\beta}{\eta}\left(\frac{t - \gamma_i}{\eta}\right)^\beta \exp\left[-\left(\frac{t - \gamma_i}{\eta}\right)^\beta\right] \tag{45}$$

The fdf is thus the summation of a series of similar profiles, scaled by the q_i and starting at γ_1, γ_2, ..., γ_k.

If the sub-population profiles are themselves bimodal in the manner discussed in the previous section, the the fdf becomes, for the lognormal distribution,

$$f(t) = q_a \sum_{i=1}^{k} q_i \frac{f_z(z_{ai})}{\sigma_a(t - \gamma_i)} + q_b \sum_{i=1}^{k} q_i \frac{f_z(z_{bi})}{\sigma_b(t - \gamma_i)} \tag{46}$$

and the cdf is

$$Q(t) = q_a \sum_{i=1}^{k} q_i\,\Phi(z_{ai}) + q_b \sum_{i=1}^{k} q_i\,\Phi(z_{bi}) \tag{46}$$

where

$$z_{ai,bi} = \frac{\ell n\,(t - \gamma_i) - \mu_{a,b}}{\sigma_{a,b}} \tag{48}$$

A similar expression can be derived for the Weibull distribution. It is not possible to discuss the properties of the staggered multimodal distribution in general terms.

5. PARAMETER MODIFICATION

The lifetime functions which have been considered referred, as mentioned earlier, to a fully characterized population but its parameters are not invariant; they will depend upon the conditions of usage which can be quantified as an applied stress. The nature of this dependence is the central issue of the topic of accelerated testing. Although complex stress relationships certainly exist, the simple **linear stress model** is often valid, which for the lognormal distribution has the form

$$\mu = b_o + b_1 x_1 + b_2 x_2 + \ldots \tag{49}$$

where the x's are **stress factors** and the b's are **stress coefficients**. A similar expression, for ξ, applies to the Weibull distribution. If the median lifetimes are ψ and ψ' for stresses x_1, x_2 ... and x_1', x_2' ... then from Eqs. (22) and (49),

$$\frac{\psi}{\psi'} = \exp \, b_1 \, (x_1 - x'_1) \, \exp \, b_2 \, (x_2 - x'_2)\ldots \tag{50}$$

which is the **acceleration factor** associated with the stress change from x to x'. For a single stress, the probit for a lifetime t_p becomes (from Eq. 16)

$$z_p = \frac{\ell n \, (t_p - \gamma) - b_o - b_1 x_1}{\sigma} \tag{51}$$

so a set of cdf plots for different stresses will have the form of Fig. 13 where the separation of the lines is $b_1 \Delta x$ and their slope is still $1/\sigma$. For p constant, corresponding to a fixed cdf, a plot of $\ell n \, (t_p - \gamma)$ against the stress factor, x, will be linear with slope b_1 as in Fig. 14. An exactly parallel argument is valid for the Weibull distribution.

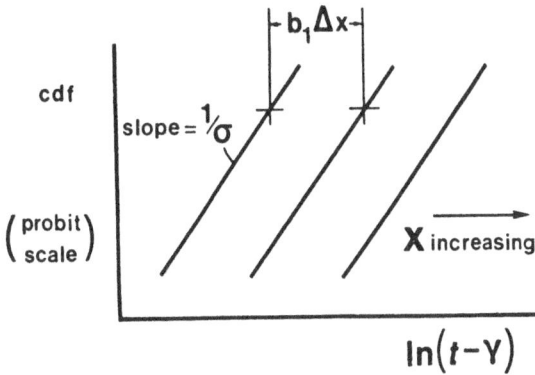

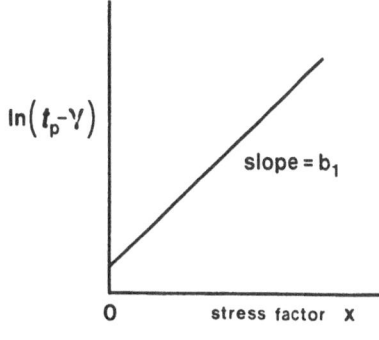

Fig. 13 - The cumulative distribution function for different stresses conforming to the linear stress model; x = stress factor; Δx = change in stress factor; b_1 = stress coefficient; γ = location parameter. For $\gamma = 0$, abscissa scale is ℓn (lifetime).

Fig. 14 - Dependence of reliable lifetime on stress for a constant value of the cdf; t_p = lifetime for cdf = p; b_1 = stress coefficient; γ = location parameter. When x = 1/temperature the plot is the Arrhenius Line with activation energy = $b_1/11606$ eV.

Known expressions for the stress factor, x, which increases as the actual stress decreases, include $\frac{1}{T}$ and $\left(\frac{const}{T} - \ell nT \right)$ where T is the temperature of some part of the component, $\ell n \frac{1}{V}$ where V is an applied voltage and -E where E is an electric field, $\ell n \frac{1}{J}$ where J is a current density and $-h^2$ where h is a relative humidity. The most widely used stress is temperature with $x = \frac{1}{T}$ whereupon Fig. 14 becomes an Arrhenius plot,[14-18] and b_1 is the activation energy which can be expressed in units of volts by dividing by the ratio, electron charge/Boltzmann's constant, $e/k = 11606°K/volt$. Expressions such as $x = \ell n \frac{1}{J}$ correspond to the inverse power model.[3,19,20] For a bimodal lifetime distribution, the stress dependence can become very complicated; the median lifetimes of the constituent distributions may for example, have different temperature sensitivities so that cdf plots for different stresses will vary in shape as well as position. It may also happen that the bimodal proportion is stress sensitive so that at a low stress only a few components belong to the early distribution whereas at large stresses nearly all do so.

It follows from Eq. 51, that another cdf plot must exist, analogous to that of Fig. 4, in which the reliable lifetime t_p is held constant whereupon the probit will depend upon the stress in the manner of Fig. 15 with the slope of the line

equal to $-b_1/\sigma$. If a set of lines were drawn for dif-

ferent reliable lifetimes, their separation parallel to

the stress scale would evidently be $\Delta \ell n t / b_i$.

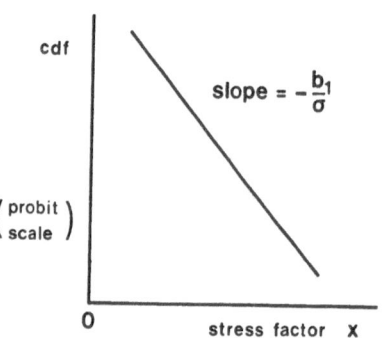

Fig. 15 - Dependence of the cumulative distribution function on stress for a constant lifetime value; b_1 = stress coefficient.

6. PARAMETER ESTIMATION

Attention will shortly be turned to the main task where a population of components exists but its lifetime parameters, and even the distribution type are unknown. The basic approach is always the same; a sample is withdrawn from the population, deployed in some way and the occurrence of failures recorded. A lifetime distribution is then postulated, — and in the present work either the lognormal or Weibull will be taken with the lognormal as the first choice — and the sample data is used to estimate the lifetime parameters. The data will, of course, comprise real-time observations whether they arise from laboratory life tests or from service. Because the interpretation of lifetime distributions in real time is not always as straightforward as it might appear, it will prove useful to consider the translation in the very simple situation where the members of the population are all deployed at the same instant — they may be said to share a common birth time which is taken as the real-time zero. The fdf of the lifetimes $f(t)$ then becomes f(t) the fraction of the population failing per unit time at time t; in other words, f(t) is the **failure rate** referred to the size of the original population. The cdf, Q(t), is simply the fraction of the population which has failed at time t whilst R(t) is the fraction surviving. The hazard function also becomes a failure rate, $\lambda(t)$, but is expressed in terms of the number of survivors (from Eq. 12) and is thus the true **instantaneous failure rate** or **hazard rate**. The product $\lambda(t)$ dt is evidently the number of components failing in an interval, dt, expressed as a fraction of the number exposed at time t and is called the **incremental hazard** or just the **hazard** in dt. The cumulative hazard function evaluated at real times t_1 and t_2 enables the **average**

hazard rate to be found over that interval from

$$\bar{\lambda}_{t_1 \rightarrow t_2} = \frac{H(t_2) - H(t_1)}{t_2 - t_1} \tag{52}$$

The average hazard rate from time zero to t is thus

$$\bar{\lambda}(t) = \frac{H(t)}{t} \tag{53}$$

which is called the **cumulative hazard rate**. For the Weibull distribution (Table II) with $\gamma = 0$,

$$\bar{\lambda}(t) = \frac{t^{\beta-1}}{\eta^{\beta}} \tag{54}$$

which is related to the hazard rate by

$$\lambda(t) = \beta \bar{\lambda}(t) \tag{55}$$

and is the basis of the **Duane**[8,21] **method** of predicting instantaneous failure rates from cumulative hazard rate observations (with the "α" used by Duane equal to 1 - β).

Because the hazard rate is usually a very small number, it is convenient to multiply it by 10^9 for practical use and refer to the unit as a **fit** (**failure unit**). Thus 1 fit is 1 failure in 10^9 component hours (100 fits = 0.01%/1000 hours).

6.1 **Lifetest Data**

In describing the ways in which lifetest data can be used for parameter estimation, real data will be exclusively used but as the emphasis is on methodology rather than on the results themselves, only essential details of the components used and their operating conditions will be given. The criterion of failure in all the component examples is the manufacturer's performance specification unless a contrary indication is given.

6.1.1 **Graphical Analysis**

Suppose that all the members of a component sample, of size n, are deployed simultane-

ously, and the failure occurrences are recorded. Preferably, each individual lifetime should be

obtained but very often it will only be possible to count the accumulated failures at arbitrary

inspection occasions; in a life test under an accelerating applied stress, for example, the stress

will usually have to be removed in order to make measurements, from which it will be

ascertained which components have failed. If some of the failures occur because a measured

quantity (such as a leakage current) has drifted outside its specified limit, then it is desirable

to record the values on each measurement occasion and estimate the lifetime by linear, or

possibly quadratic, interpolation. If this method is impracticable (as with catastrophic failures)

then the lifetimes must be determined from the mid-way point of the neighbouring inspections

or, if there are several failures, they can be assumed to be uniformly distributed between the

inspection times.

If the testing period is allowed to continue until the entire sample has failed, which could

happen in a life test under severe applied stress, then the lifetime of each member will be

known and the data is said to be **complete**. Often however, the observations will be discontin-

ued after only a fraction of the sample has failed, the data then being described as **censored**

with the running times without failure called **censoring times**.

Having determined or estimated the lifetimes, they are then set down in ascending order

$$t_1 \ \dots \ t_2 \ \dots \ t_j \ \dots \ t_r$$

where there are r observed lifetimes. A value of the cdf is then estimated from each lifetime

using a "plotting position" formula, of which several exist, but that preferred is[16]

$$Q(t_j) \ = \ \frac{j - 0.5}{n}. \tag{56}$$

If it is now postulated that the components of the population from which the sample was drawn have a 2-parameter lognormal lifetime distribution, then the values of $Q(t_j)$ are equated with the gaussian to give corresponding values of z_j the probit (using say, the $\Phi(z)$ → z formula in Appendix 1).

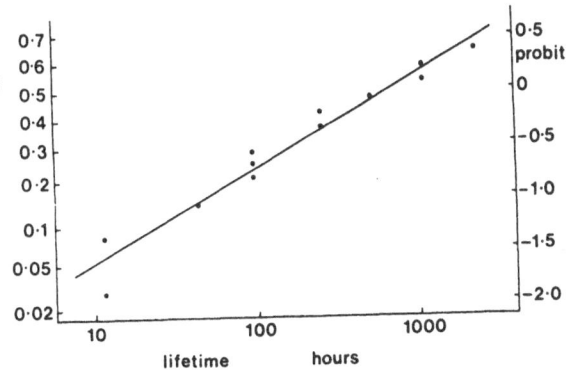

cdf

Fig. 16 - Lognormal lifetime cdf plot for LPTTL integrated circuits (type SN74L00/hermetic). Ambient temperature = 200°C, reverse bias operation; sample size = 18, ψ = 660h, σ = 2.6.

A plot of z_j against $\ell n t_j$ in the style of Fig. 4 will then yield estimates of the median lifetime and dispersion; if, as is usually more convenient, log t_j rather than $\ell n t_j$ is plotted, then the slope of the line is actually $\frac{\ell n 10}{\sigma}$. Specimen plots are shown on

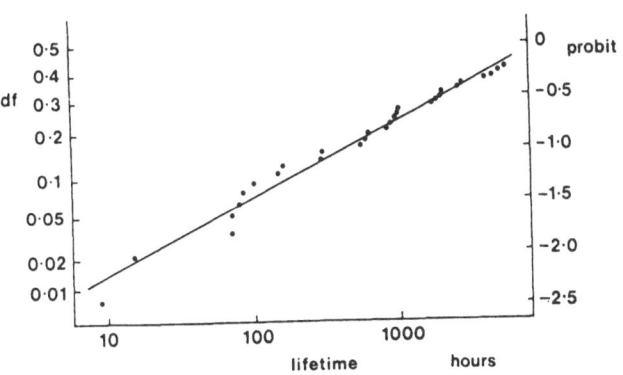

cdf

Fig. 17 - Lognormal lifetime cdf plot for operational amplifiers (type μA741/hermetic). Ambient temperature = 125°C, dissipating; die temperature = 150°C; sample size = 70, ψ = 9520h, σ = 3.1.

Fig. 16 for low power-TTL integrated circuits operated at 200°C where the lifetimes are taken to be mid-way between inspection occasions and on Fig. 17 for operational amplifiers in a 125°C ambient[22,23] where the lifetimes were obtained by interpolation. The lines of best fit were found by the method of least squares; the parameter estimates can be obtained directly from

$$\sigma = \frac{(\Sigma \ell nt_j)^2 - n \Sigma (\ell nt_j)^2}{\Sigma \ell nt_j \Sigma z_j - n \Sigma z_j \ell nt_j} \qquad (57)$$

$$\mu = (\Sigma \ell nt_j - \sigma \Sigma z_j) / n \qquad (58)$$

the summations being carried out for j = 1 to r.

In the above examples, the data was censored with all the censoring times greater than the longest lifetime but it may happen that the data comprises lifetimes and censoring times intermingled because components are prematurely removed from a test (due to a measuring accident for example).

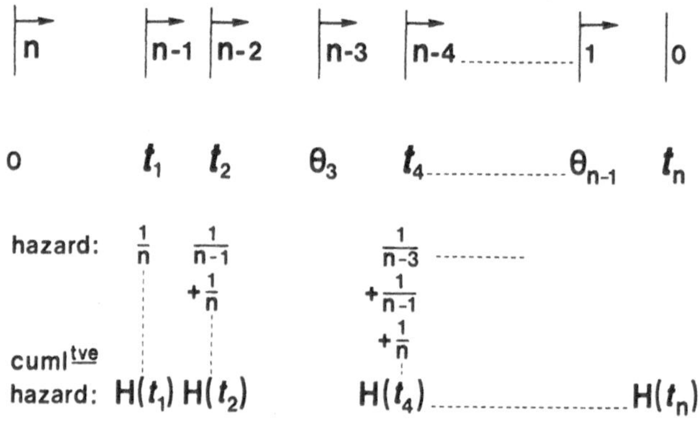

Fig. 18 - Hazard calculations for multiply-censored data; lifetime:t, censoring times: Θ, sample size: n

Alternatively, if the causes of each failure are known, it may be useful to re-analyse the data with the failures for selected causes removed which is achieved by converting each deleted lifetime t_j to a censoring time θ_j. For a crude analysis of this **multiply-censored data**,[24,25] the censoring times could just be ignored but they can quite easily be accommodated by estimating the cumulative hazard from each lifetime as illustrated by Fig. 18. Up to the first lifetime, the hazard is 1/n which is equal to the cumulative hazard $H(t_1)$. After the first lifetime, the number of survivors is n-1 so the hazard from the first to the second lifetime is 1/(n-1) and the cumulative hazard at the second lifetime is $H(t_2) = 1/n + 1/(n-1)$ and so on. At θ_3 however, there is no addition to the cumulative hazard but the influence of the censored component is exerted through the next hazard contribution being 1/(n-3) and not 1/(n-2).

The cumulative hazard can be plotted directly on hazard plotting paper,[26] to give a straight line; on Weibull paper the parameters can then be found directly as described in Sec. 3.2. An example of a plot for multiply-censored data is given on Fig. 19 for transistors used in an equipment field trial,[27] where the failure of one transistor (or any other component in the equipment) entailed the

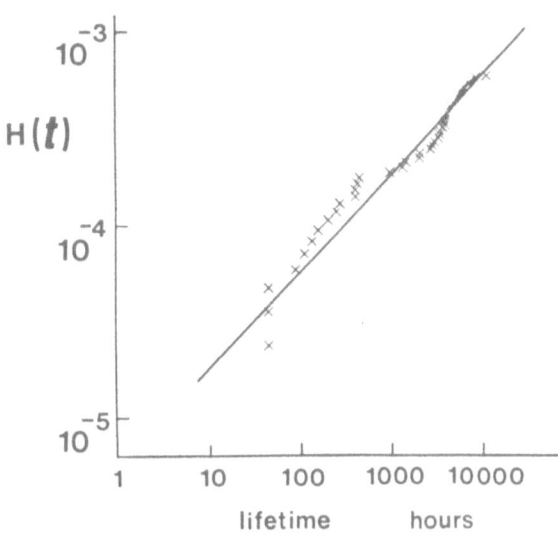

Fig. 19 - Weibull cumulative hazard plot for small-signal transistors (CV-type/hermetic). Service operation, multiply-censored data, sample size = 85124, $\eta = 3 \times 10^{10}$, $\beta = 0.51$

removal of all the transistors in that equipment.

For a lognormal distribution, the cumulative hazard can also be plotted on appropriately scaled paper but it is usually more convenient to convert each value of H(*t*) to a cdf point using Eq. 14 rewritten as

$$Q(t_j) = 1 - \exp H(t_j) \tag{59}$$

The cdf plots of Figs. 16 and 17 are considered to be acceptably linear meaning that the postulate of a lognormal lifetime distribution has been accepted. If any curvature is found, then the possibilities are summarised by Fig. 20 but in making a judgement it should be appreciated that the errors in experimental cumulative plots are, by definition, not random; an error in one point is carried over to the next. Cumulative plots often therefore display winding as seen in Fig. 17 and other plots to follow.

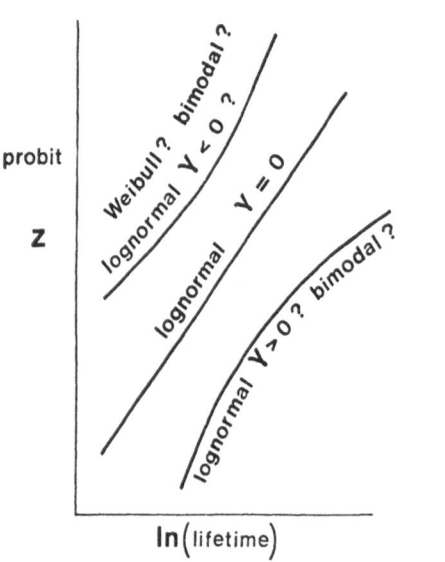

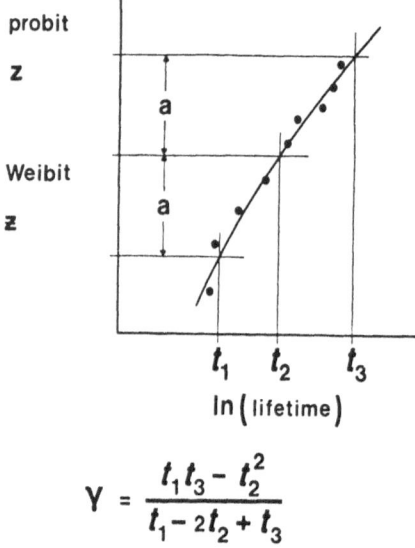

$$ \gamma = \frac{t_1 t_3 - t_2^2}{t_1 - 2t_2 + t_3} $$

Fig. 20 - Interpretations of a 2-parameter lognormal cdf plot.

Fig. 21 - Graphical estimation of the location parameter: lognormal or Weibull cdf plot.

If a non-zero location parameter is contemplated, its value can be estimated from Fig. 21[8] whereupon the foregoing methods can still be applied except that each lifetime and censoring time must be reduced by γ as a prior step. An example of lifetimes fitted to a 3-parameter lognormal distribution is given by Fig. 22 obtained on a large MOS shift register showing that it is only over the range of early lifetimes that the linearity is improved. The plot retains a bend at its other end illustrating the more general point that any non-linearity in a cdf plot is best regarded as evidence of a bimodal distribution as discussed in the next section.

6.1.1.1 Bimodal Distributions

Whereas the graphical analysis of sample data from a monomodal population is relatively straightforward, and is often quite adequate, a bimodal population is much less easily tackled in this way. Fortunately, it is often found that a small early constituent is associated with a later main distribution so that Fig. 12 can be used as a guide for effecting a dissection as will

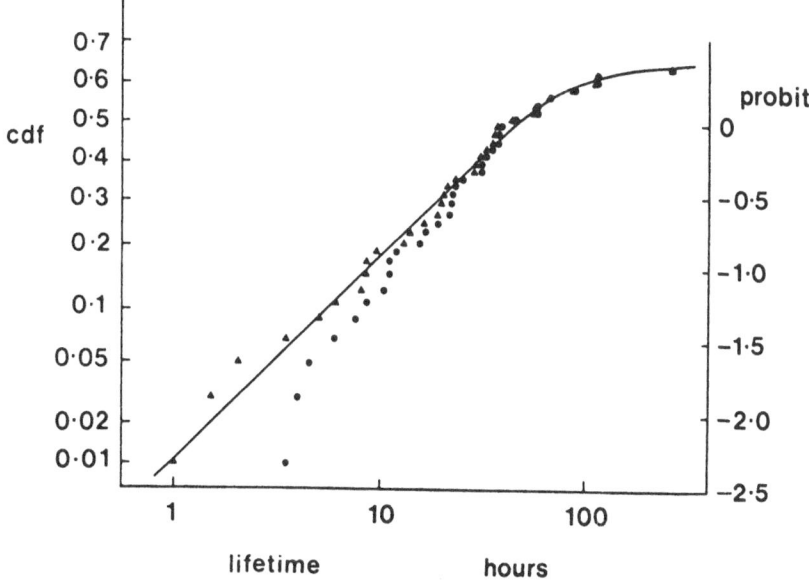

Fig. 22 - Lognormal lifetime cdf plot for MOS shift registers (quad 80 bit, metal-gate MOS, hermetic). Sample size = 50, ambient emperature = 125°C, die temperature rise = 20°C,
Location parameter: •γ = 0, $\Delta\gamma$ = 2.5h

be illustrated using the data of the cdf plot of Fig. 23 obtained on CMOS integrated circuits,[22]

stressed under bias at 200°C.

From Eq. 36 with k = 2

$$Q_a(t) = \frac{Q(t) - q_b\, Q_b(t)}{q_a} \qquad (60)$$

where the suffixes a and b are identified with the early and main distributions respectively. If the distributions do not overlap too severely, Eq. 60 can be approximated by setting $Q_b(t) = 0$ whence

$$Q_a(t) \div Q(t)\,/\,q_a \qquad (61)$$

and so from Eq. 56, the plotting positions for the early lifetimes become

$$Q_a(t) = \frac{j - 0.5}{n \, q_a} \tag{62}$$

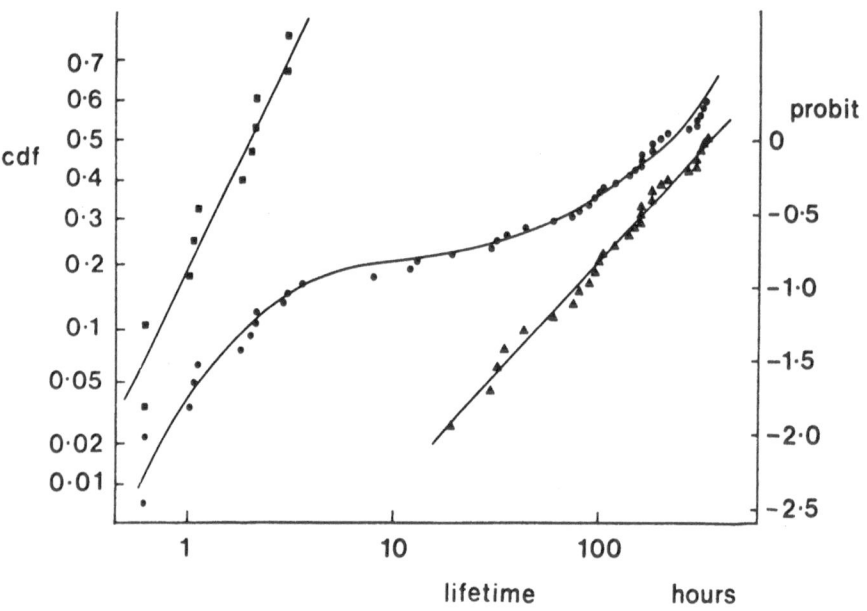

Fig. 23 - Lognormal lifetime cdf plot for CMOS integrated circuits (type 4007). Ambient temperature = die temperature = 200°C; sample size = 70, static bias; bimodal distribution early proportion = 0.2; early distribution: ψ_a = 1.9h, σ_a = 0.78, main distribution: ψ_b = 340h, σ_b = 1.5.

The comparison of Fig. 23 with Fig. 12 suggests that the early distribution comprises about 20% of the whole, so applying Eq. 61 to the first 11 experimental points gives the early cdf plot added to Fig. 23, which is acceptably linear, thus enabling the parameters to be estimated by the method of least squares.

The main distribution can be extracted in a comparable manner by approximating $Q_a(t)$ to unity to give the plotting positions,

$$Q_b(t) = \frac{j - 0.5}{n \, q_b} - \frac{q_a}{q_b} \tag{63}$$

where, of course, $q_b = 0.8$ for Fig. 23. Using this expression from the 16th point onwards, gives another linear cdf plot (with least-squares line) as shown on Fig. 23 where the complete set of five estimated parameters ψ_a, σ_a, ψ_b, σ_b and q_a is also given. The combined cdf can now be calculated using Eq. 38 with Eq. 40 which, as seen on Fig. 23 provides a very good fit to the experimental points.

6.1.1.2 Accelerated Lifetests[28,29]

In the lifetests al-
ready discussed, environ-
mental temperatures
have been used which
would justify their being
called "accelerated life
tests" but this term is
now reserved for an in-
vestigation of the accel-
eration produced by sub-
dividing the items available into sub-samples and

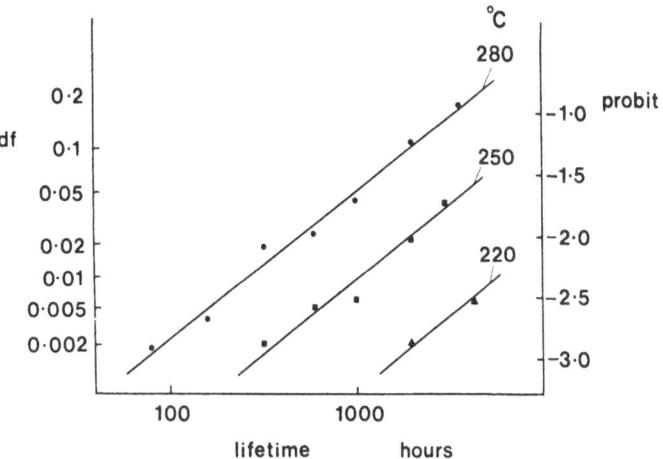

Fig. 24 - Lognormal lifetime cdf plots at different stresses for submarine cable transistors (type 4A). Sub-sample sizes = 500 (approx.): failure criterion; $h_{FE} > 10\%$ fall; $\sigma = 1.9$

stressing each at a different intensity as measured by the x's in Eq. 49. The task is then a matter of estimating the values of the coefficients b_0, b_1 ... although in practice, only one varied stress, or at most two, can be handled in one accelerated test. For a single stress, each sub-sample set of lifetimes is separately treated as just described to yield a plot corresponding to Fig. 13. In the example of Fig. 24, obtained on submarine-cable transistors,[30] the stress factor is the reciprocal of the absolute temperature of the silicon die. Recalling from Fig. 13 that the shape factor is assumed unaffected by the stress, the lines of best fit have to be drawn parallel to each other. In this example, the task is fairly easily undertaken by eye with the help of the least-squares method applied to the 280°C and 250°C plots. Plotting the 0.025%

reliable lifetime in the manner of Fig. 14 with the stress x being 1000/T where T is the die absolute temperature, then gives the Arrhenius plot of Fig. 25. Because the reliable lifetime is plotted on a scale of common logarithms, the slope must be multiplied by $\ln 10$ to give b_1. With the slopes on Fig. 24 yielding $\sigma = 1.92$, the coefficients of Eq. 49 become $b_0 = -20$ and $b_1 = 16.4$ in units of $^\circ$K/1000 with b_1 corresponding to an activation energy of 1.4 eV.

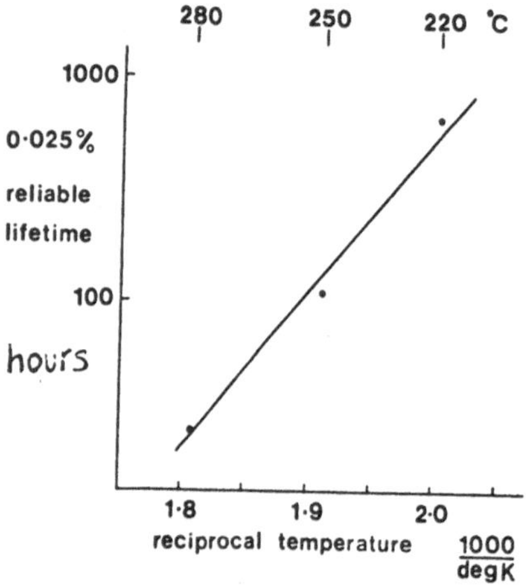

Fig. 25 - Arrhenius plot for submarine-cable transistors (type 4A); $b_0 = -20$, $b_1 = 16.^\circ$K/1000; activation energy $\phi = 1.4$eV

6.1.1.2.1 Step-Stress Tests[31]

In all the life-test procedures already described, the applied stress remained constant during any single test and this method of "steady-stress" testing, in which the lifetimes are measured directly, is generally preferred. It will be clear from Fig. 13, however, that if the occurrence of failures is highly stress sensitive, meaning that b_1 is large, then the lifetime cdf plots of Fig. 13 will tend to be widely spaced necessitating inconveniently long or unacceptably short testing times unless small stress separations are used, which is likely to introduce experimental difficulties. In this event, there are advantages in not measuring the lifetimes directly. Instead, a series of equal testing times - the period is called the **dwell time** - are interleaved with stress increments, the cumulative fraction failed being recorded after each period. From such a "step-stress" test, a cdf plot in the form of Fig. 15 can be extracted and if repeated for different dwell times, the stress coefficients and σ can be found just as from a steady-stress test. As shown in Appendix 2, the stress increments should be equal whereupon Fig. 26 gives the necessary adjustment to each stress in order to convert the cumulative failure plot to a cdf plot. A difficulty arises because the correction procedure uses the value of the stress coefficient which is not yet known, but a prior estimate of b_1, obtained by neglecting the correction, will usually suffice. The corrections evidently became smaller as the stress increment and the stress coefficient increase and very often may be neglected. An example of a corrected pair of step-stress plots obtained on transistors is given by Fig. 27. The value of b_1 can be estimated using

$$b_1 = \frac{\ell n\,(t_2/t_1)}{x_2 - x_1} \quad (64)$$

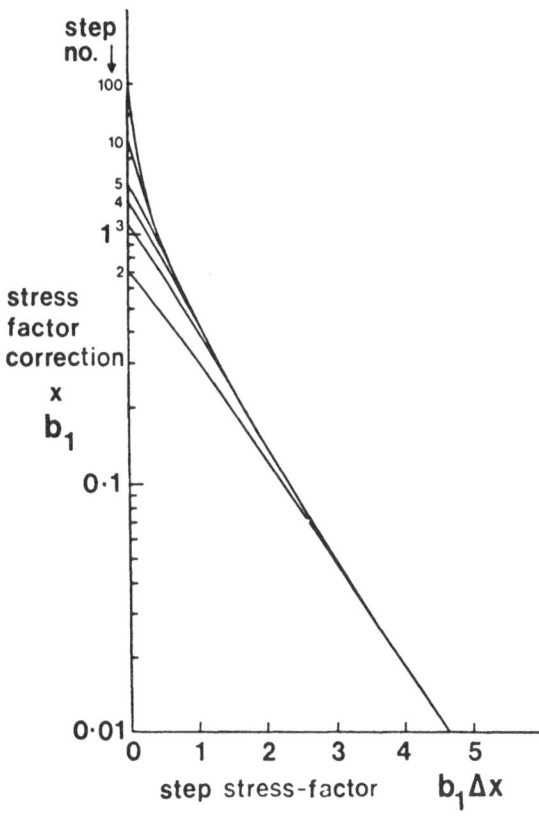

Fig. 26 - Correction chart for step-stress test; Knowing stress increment, enter abscissa using an assumed value of stress coefficient b_1. Correction for a given step number is then substrated from stress factor value at that step.

where t_1 and t_2 are the dwell times corresponding to stress factors x_1 and x_2 for the same cdf. Because the slope of the corrected plots gives b_1/σ (Fig. 15), σ can also be estimated. The effect of neglecting the correction is shown on Fig. 27; the activation energy would be over-estimated by about 10%.

6.1.1.3 **Drift Tests**

As remarked in Sec. 6.1.1, it is sometimes possible to record the change in an electrical property of a component during a lifetest and, given a failure limit, determine lifetimes by interpolation. It may, alternatively, be possible to obtain the population parameters directly from the measured properties of the component. Thus suppose that the property is denoted by v and that it is normally distributed over the sample with variance σ_v and mean $\bar{v}$. The frequency density function for v is thus

$$f(v) = \frac{1}{\sigma_v \sqrt{2\pi}} \exp\left[-\frac{1}{2} \left(\frac{v - \bar{v}}{\sigma_v} \right)^2 \right] \tag{65}$$

Let is be further assumed that the mean, $\bar{v}$, is dependent upon the stress factor x applied to the sample and upon the testing time, t, according to

$$\bar{v} = a_0 + a_1 x + c_0 \ell n\, t \tag{66}$$

where a_0, a_1 and c_0 are constants, and σ_v is assumed constant. If failure is deemed to occur when v and v_c, then the integral of Eq. 65 from v_c to ∞ is the fraction of the sample which has failed at time t, namely, Q(t) which gives

$$Q(t) = \Phi \left(\frac{\ell n t - (v_c - a_0 - a_1 x) / \sigma}{\sigma_v / c_0} \right) \tag{67}$$

and by comparison with Eq. 19 is seen to represent a lognormal lifetime distribution having

$$\mu = b_0 + b_1 x \tag{68}$$

where

$$b_0 = \frac{v_c - a_0}{c_0} \qquad b_1 = -\frac{a_1}{c_0} \tag{69}$$

and

$$\sigma = \sigma_v / c_0 \tag{70}$$

A plot of $\bar{\nu}$ against ℓnt at a steady stress will, from Eq. 66, yield a straight line of slope c_0 and if repeated for several stresses, the times for a constant value of $\bar{\nu}$ can be plotted against the stress to yield a_0 and a_1. Knowing ν_c, the desired stress coefficients b_0 and b_1, can then be found. Examples of these plots are shown by Figs. 28 and 29 obtained on early p-channel metal-gate MOS transistors[32,33] in which ν was taken as the natural logar-

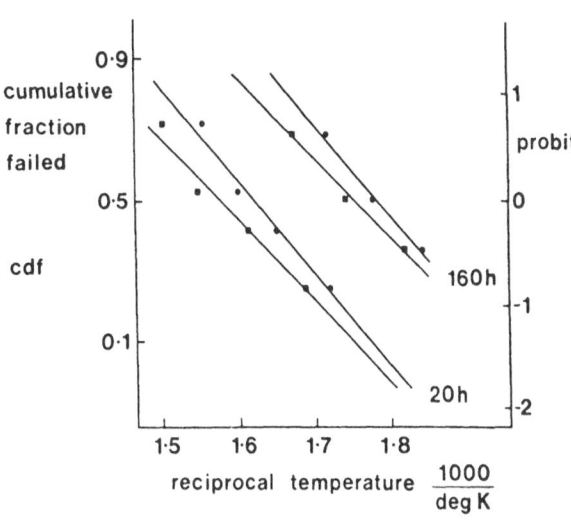

Fig. 27 - Lognormal step-stress plot for submarine-cable transistors. Sub-sample sizes = 30 (approx.); 20 and 160h; failure criterion, h_{FE} > 10% fall. cumulative fraction failed gives uncorrected estimates; $\sigma = 0.85$, $\phi = 1.15$eV cdf plots give corrected estimates; $\sigma = 0.87$, $\phi = 1.02$eV.

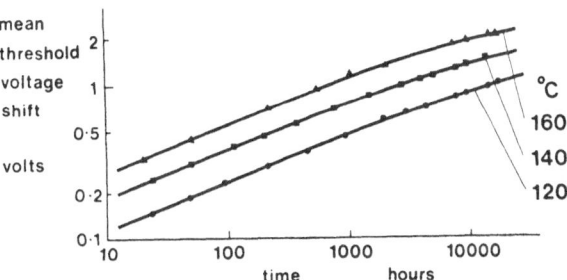

Fig. 28 - Mean threshold voltage shift at different stresses in p-channel, metal-gate MOS transistors. Sample sizes = 20-25V, negative gate bias.

ithm of the shift of the threshold voltage stimulated by temperature stress. The value of c_0 from Fig. 28 is 0.3 and with σ_ν known to be sensibly constant at 0.14, $\sigma = 0.47$. With the limiting shift taken as 0.3 volts, b_0 and b_1 were then found to have the values given on Fig. 29, the activation energy being very close to 1 eV.

6.1.2 <u>Maximum Likelihood Analysis</u>

A graphical analysis of life test data will very often suffice and even the use of a least-squares analysis to draw a line of best fit may be rather academic even though easily done.   In general however, the less well-ordered the data - and lifetest data is often subject to consider-able scatter - and the more com-plex the analy-

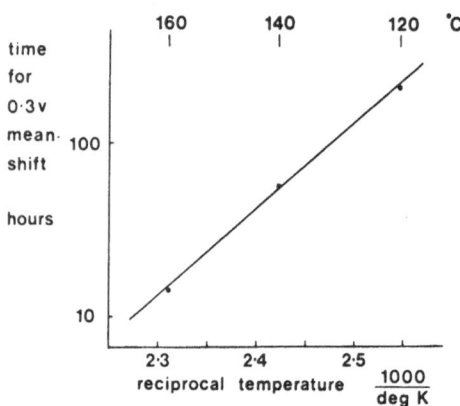

Fig. 29 - Temperature sensitivity of thresh-old voltage shift in p-channel, metal-gate transistors; b_0 =   -23.76, b_1 = $11.4°K/1000$, activation energy = 1.0eV.

sis problem, the greater are the advantages of a numerical me-thod and espe-cially the maxi-mum likelihood The   method entails a con-siderable "ov-

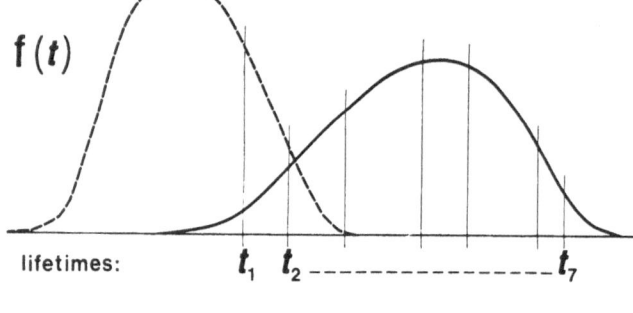

Fig. 30 - Basis of the maximum likelihood method of parame-ter estimation. Broken line; L low: full line, L high.

erhead" in programming but once done, a given task is easily excuted. The method has also the advantage that it enables confidence limits to be placed on the estimated parameters and on any lifetime functions calculated from them. The principle of the method is easily grasped with the help of Fig. 30 which shows a set of lifetimes t_1 to t_7. Suppose that trial population values are used to yield the distribution profile of the broken curve. Multiplying the ordinates together gives the likelihood, L, which would then clearly be very small due to the several lifetimes in the tail of the distribution. For another proposed set of parameters, however, yielding the full curve, the product would be much larger meaning that those parameter values would be more likely. The task then, is one of finding that unique set of parameters which maximises the likelihood. If the data is censored, the likelihood function is modified to

$$L = \prod_{j=1}^{r} f(t_j) \prod_{j=1}^{c} \{1 - Q(\theta_j)\} \tag{71}$$

for r lifetimes t_j and c censoring times θ_j. If a lognormal lifetime distribution is postulated, then $f_L(t_j)$ and $Q_L(\theta_j)$ are found from Eqs. 16, 17 and 19.

The method is also applicable to accelerated life-test data, the likelihood function being

$$L = \prod_{i=1}^{k} \prod_{j=1}^{r_i} f_i(t_{ij}) \prod_{i=1}^{k} \prod_{j=1}^{c_i} \{1 - Q_i(\theta_{ij})\} \tag{72}$$

where there are k sub-samples each with r_i lifetimes and c_i censoring times. For a 2-parameter lognormal lifetime distribution (say), the fdf and cdf are found from

$$f_i(t_{ij}) = \frac{f(z_{ij})}{\sigma t_{ij}} \tag{73}$$

and
$$Q(\theta_{ij}) = \Phi(z_{ij}) \tag{74}$$

where

$$z_{ij} = \frac{\ln t_{ij} - b_o - b_1 x_{1i} - b_2 x_{2i}}{\sigma}, \tag{75}$$

x_{1i} and x_{2i} being the stress factors for the ith sub-sample. The method of adjusting the parameters in order to maximise the likelihood is described in Appendix 3 but the basis of the procedure is easily appreciated by considering one parameter only, say σ in a lognormal lifetime distribution. For a range of trial values of σ, the likelihood will pass through a maximum when $\frac{dL}{d\sigma} = 0$. It is virtually never possible to solve this equation for σ explicitly but by Newton's method,[38] σ' is a better approximation where

$$\sigma' = \sigma - \left(\frac{dL}{d\sigma} \bigg/ \frac{d^2L}{d\sigma^2} \right) \tag{76}$$

and the procedure can be iterated until σ has any desired accuracy. The labour involved in the method is largely the initial task of obtaining the derivatives of L with respect to each parameter which is however, made much easier by taking the logarithm of the likelihood and maximising ℓnL.

The second derivative has an additional significance because when the maximum in the likelihood function is sharp (say) implying that the uncertainty in the estimate is small, the second derivative will be highly negative. The reciprocal of the negated second derivative is thus a measure of the uncertainty and can actually be shown to be equal to the variance of the parameter estimate which, in turn, can be used to place confidence limits on the parameter value. As seen in Appendix 3, the method of maximising the likelihood, which is called the Newton-Raphson[39,40] procedure when more than one parameter is adjusted, involves the manipulation of vectors and matrices of dimensions determined by the number of parameters being adjusted ranging from 2 for a simple 2-parameter lifetime distribution to 5 for a bimodal assumption (with 2-parameter constituent distributions). The variance/covariance matrix, which is obtained from the solution, can be used to place confidence limits not only on the parameters but also on any function of them, such as the hazard rate. Sufficient detail is given in Appendix 3 for programming the maximisation of the likelihood of Eq. 71 for a bimodal lifetime distribution and of Eq. 72 for two stress coefficients, b_1 and b_2, both solutions assuming 2-parameter lognormal distributions.

For the straightforward analysis of a set of lifetime data, postulated to be mono-modally distributed, the ML method has but little advantage as shown by the plot of Fig. 31 where the data actually comprised 16 lifetimes and 4 censoring times, the latter arising because testing was halted at 13,000 hours. The confidence bands do, however, enable the goodness of the fit to be assessed rather more quantitatively. The plot of Fig. 31 also shows how the ML solution is modified if censoring had actually occurred immediately after the last observed failure.

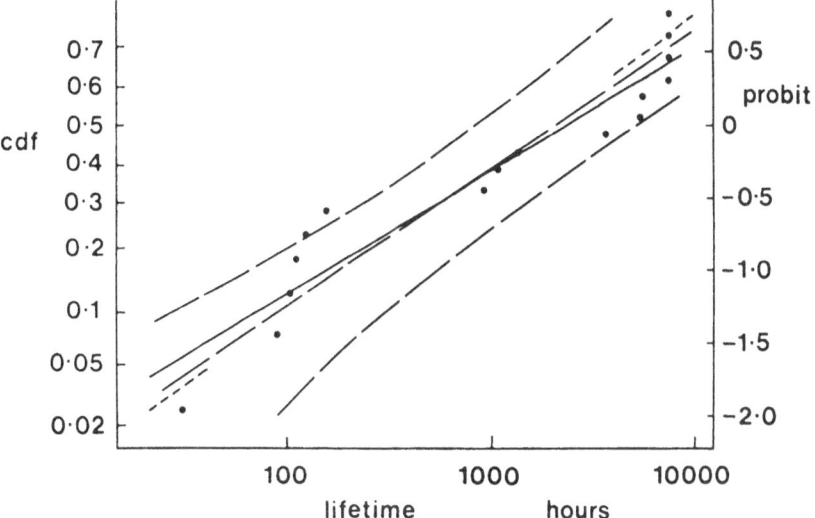

Fig. 31 - Comparison of maximum likelihood and least-squares parameter estimates from lognormal lifetime cdf plot. Sample size = 20. Full line, least squares, ψ = 2343h, σ = 2.72; broken lines, ML with 90% confidence limits, ψ = 2070h, σ = 2.41; fine broken lines, ML showing effect of censoring immediately after last recorded lifetime, ψ = 1848h, σ = 2.27.

An illustration of the ML method applied to some accelerated-test data obtained on operational amplifiers is shown by Fig. 32. The plotted points are quite scattered making it very difficult to draw parallel lines and prepare a graphical Arrhenius plot. From the results of the ML analysis, however, Fig. 33 can be drawn where the Arrhenius lines are shown both for the median lifetime and the 2% reliable lifetime together with their 90% confidence limits. At 50°C, for example, the reliable lifetime is about 20 years but in 10 samples out of every 100, it would be a mere 6 months.

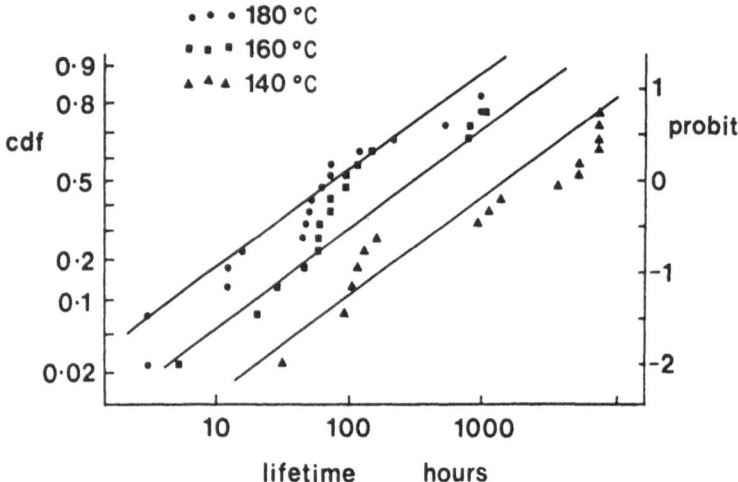

Fig. 32 - Lognormal lifetime cdf plots at different stresses for
operational amplifiers (type μA741). Sub-sample sizes = 20;
failure criterion, any of 15 measurements.

6.1.3 Constant Hazard Rate

If the failures observed in a life test are very few in number, even zero, it will not

be possible to postulate a lifetime distribution characterised by more than one parame-

ter and expect to find meaningful estimates of them. The only avenue then is to

assume an exponential distribution and find the single parameter η of Eq. 31 which is

equal (Eq. 33) to the reciprocal of the constant hazard rate λ. It is then unnecessary

to know any individual lifetimes; dividing the accumulated component hours by the

total number of observed failures gives an estimate of λ directly. Confidence limits

can moreover be placed on this estimate by an argument which starts from the obser-

vation that if a series of samples are withdrawn from a population of known λ and

tested for a certain time, then the total number of observed failures will satisfy a

Poisson distribution which may be approximated by a χ^2 distribution.[41,42] Usually,

only the upper confidence limit is of interest, conventionally expressed by $(1-\alpha)$ where

$1-\alpha$ lies between 0 and 1. The hazard rate at the upper $(1-\alpha)\%$ confidence level is then

$$\lambda_{1-\alpha} = \frac{1}{2C} \chi^2_{1-\alpha} \{2 (r + 1) \}$$ (77)

$\chi^2_p(\nu)$ being the ordinate of the χ^2 distribution[41,42] with ν degrees of freedom at which its integral has the value p. A tabulation for the evaluation of the upper limit, given the component hours C and the number of failures observed, r, is given in Appendix 4.

7. FAILURE MECHANSIMS

Examples of life tests applied to transistors and integrated circuits have already been given but the analysis methods described are valid also for tests designed to expose failure by a single mechanism. It is sometimes possible to use real components for this purpose as in the example of Fig. 28 where a very small integrated circuit was used to investigate the slow hole trapping phenomenon. In general,

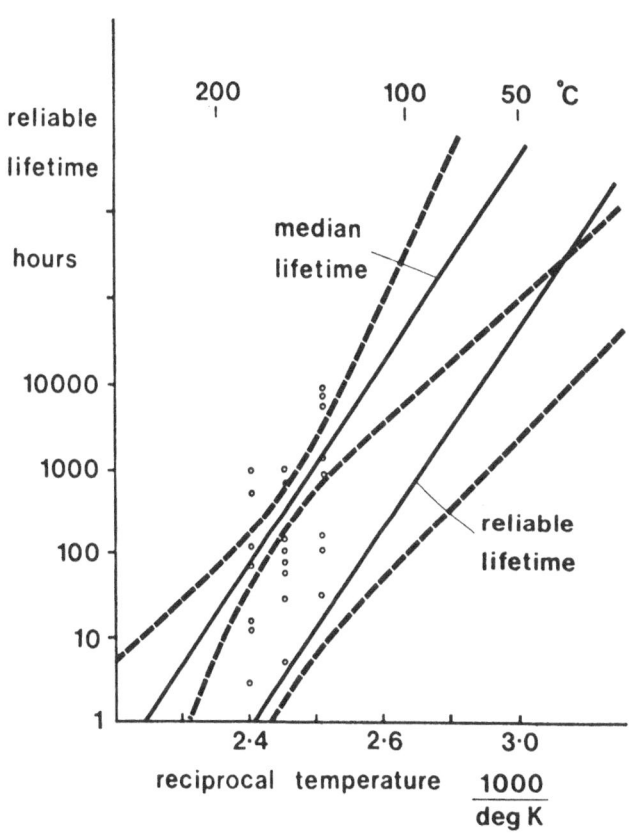

Fig. 33 - Arrhenius plots for operational amplifiers (type μA741); o observed lifetimes; reliable lifetime at 2%, 90% confidence bands, ML estimates: σ = 1.95, activation energy = 1.15 ± 0.42eV.

however, special test dice have to be fabricated incorporating integrated circuit fragments or elements such as capacitors or single transistors. The characterisitcs of encapsulations can be similarly assessed using appropriately designed enclosures. Some examples of failure-mechanism stress sensitivities follow, using the analysis methods already described.

7.1 <u>Electromigration</u>[43]

The phenomenon of electromigration in aluminum conductor tracks depends upon both track temperature and current density, so the model of Eq. 49 could be postulated with two stresses. For the tests to be described,[44] two different samples were used so for one the temperature stress was varied at a constant current density whilst

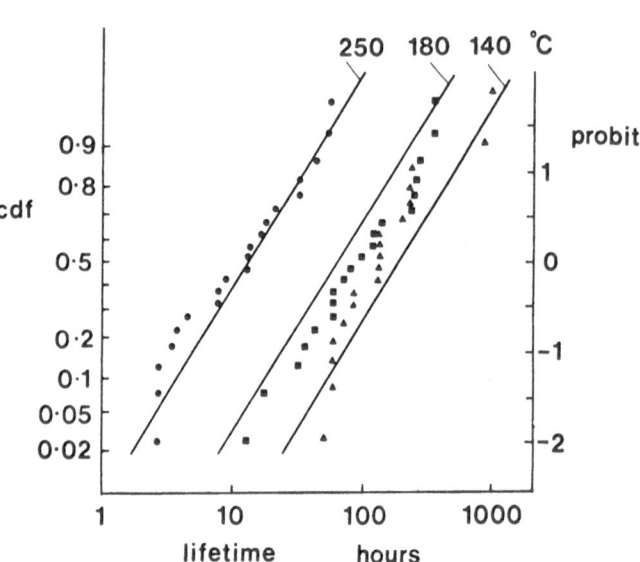

Fig. 34 - Temperature acceleration of electromigration; Aluminum track current density = 2×10^6 amps/sq cm; sub-sample sizes, 20, 20, 18 (complete data); failure criterion, open circuit; activation energy = 0.44 ± 0.09eV.

for the other, the temperature was held constant and a different current density was used for each sub-sample. For a set of three different temperatures with $x = \dfrac{1000}{T}$, the lifetime plots of Fig. 34 can be considered to satisfy the linear stress model although it would be difficult to draw the set of lines of best fit by eye. The maximum likelihood method readily solves the problem to yield the drawn lines and the parameters b_0, b_1, and σ which for the median lifetime expressed in hours gives

$$\mu = -7.12 + \frac{5.075}{T}$$

with $\sigma = 0.96$. The b_1 term (Eq. 49) corresponds to an activation energy of 0.44 eV with lower and upper confidence limits of 0.35 and 0.53 eV.

For the second sample, five different current densities were used with $x = \log(10^8/J)$ where J, the current density is in amps/sq cm. From the ML results of Fig. 35, the median lifetime is expressed in hours where

$$\mu = 29.6 + \frac{1.865}{\ell \, nJ}$$

with $\sigma = 0.30$ and 90% confidence limits on the stress coefficient of 1.80 and 1.93. The narrowness of the limits reflects the low scatter in the cdf plots. The acceleration factor for electromigration, as defined by Eq. 50, is then

$$\frac{\psi}{\psi'} = \left(\frac{J'}{J}\right)^{1.865} \exp\left\{\frac{0.44e}{kT}\left(\frac{1}{T} - \frac{1}{T'}\right)\right\}$$

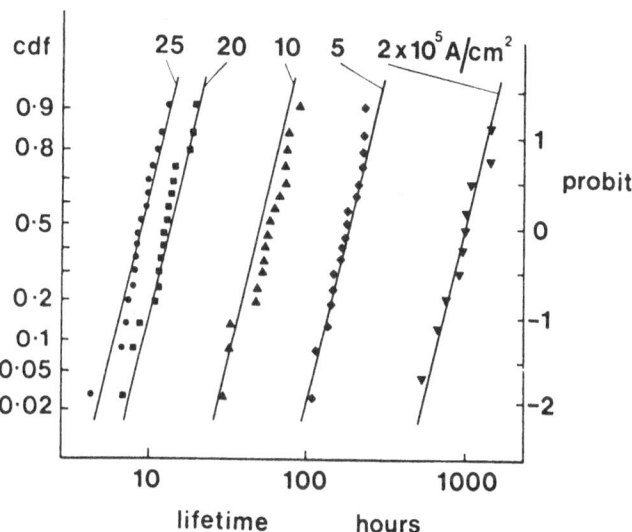

Fig. 35 - Current acceleration of electromigration. Aluminum track temperature = 180°C; sub-sample sizes, 4×18, 11, failure criterion; open circuit; median lifetime dependence = $J^{-1.865 \pm 0.065}$

Although the linear stress model with temperature and current-density stress has been frequently verified, a wide range of values of the stress coefficients has been reported, the differences often being ascribable probably to experimental difficulties. The best estimate for the current density index is believed to be 2 and for the activation energy 0.45 eV.

7.2 **Dielectric Breakdown**

It is generally accepted that the breakdown of integrated circuit capacitors is accelerated by both temperature and electric field.
The set of temperature-stress plots at constant field on Fig. 36 obtained on silicon dioxide capacitors[45] shows excellent linearity
but from the Arrhenius plot for a cdf of 0.1
on Fig. 38, it has to be concluded that the
linear stress model is not very definitely vali-
dated. The drawn line corresponds to an activation energy of 0.4 eV.

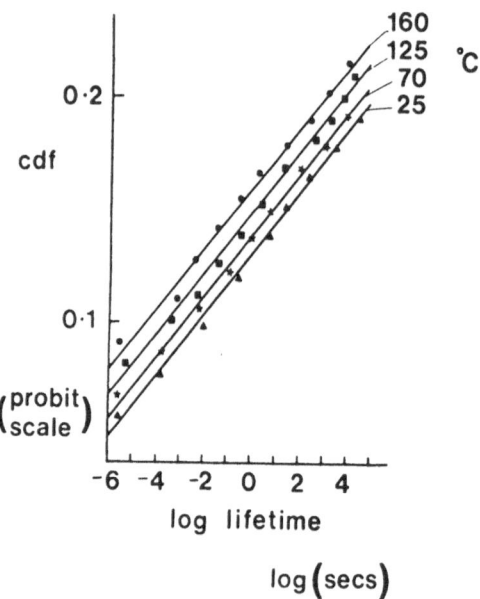

Fig. 36 - Temperature acceleration of dielectric breakdown. Silicon dioxide capacitors with n-type polysilicon electrodes; electric field 2MV/cm.

On another sample, with the temperature
held constant and dielectric field regarded as
the stress variable, good linear plots were
again obtained as shown by
Fig. 37. Equating the stress
x with -E where E is the
dielectric field, then enables
the field acceleration plot of
Fig. 39 to be drawn for a
cdf of 0.1. The linear stress
model is now clearly satisfied. The acceleration fac-

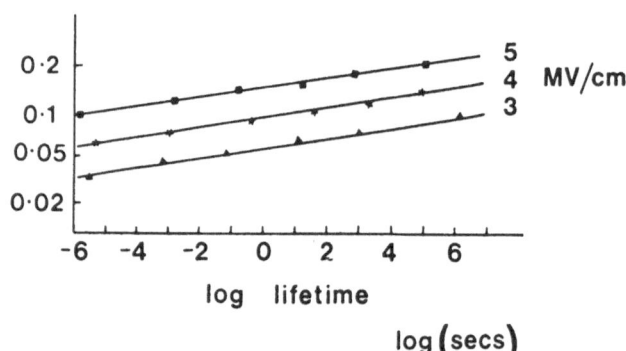

Fig. 37 - Field acceleration of dielectric breakdown.
Silicon dioxide capacitors with n-type polysilicon electrodes; temperature = 25°C.

tor for dielectric breakdown is thus

$$\frac{\psi}{\psi'} = \exp\{14.6\ (E' - E)\}\ \ \exp\{\frac{0.4e}{k}\left(\frac{1}{T} - \frac{1}{T'}\right)\}$$

where E is in MV/cm.

The linear stress model is not always verified in dielectric breakdown tests. Threshold effects are, for example, known which cause lifetime cdf plots to appear bimodal because a fraction of the sample has infinite lifetimes, this fraction becoming smaller at higher fields.

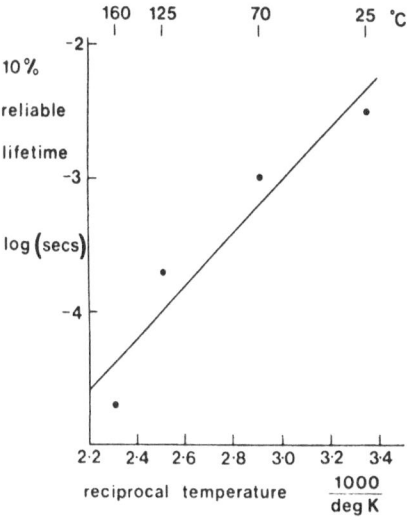

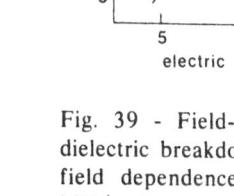

Fig. 38 - Arrhenius plot for dielectric breakdown. Activation energy = 0.4eV

Fig. 39 - Field-acceleration plot for dielectric breakdown. Median lifetime field dependence = 1/exp (14.6 × MV/cm)

7.3 Aluminum Corrosion in Plastic Encapsulation[46,47]

Another example of a degradation phenomenon dependent upon two stresses is provided by the corrosion of aluminum metallisation tracks under bias when plastic encapsulated and operated in a humid environment. The stress factor associated with the die temperature, T, was taken as $x_1 = \frac{1000}{T}$ and for the fractional relative humidity over the die, $x_2 = -h^2$.

A total of 6 cdf plots were obtained as shown by Figs. 40 and 41 for which the humidity and temperature were respectively held constant. Because all the test specimens could be considered as

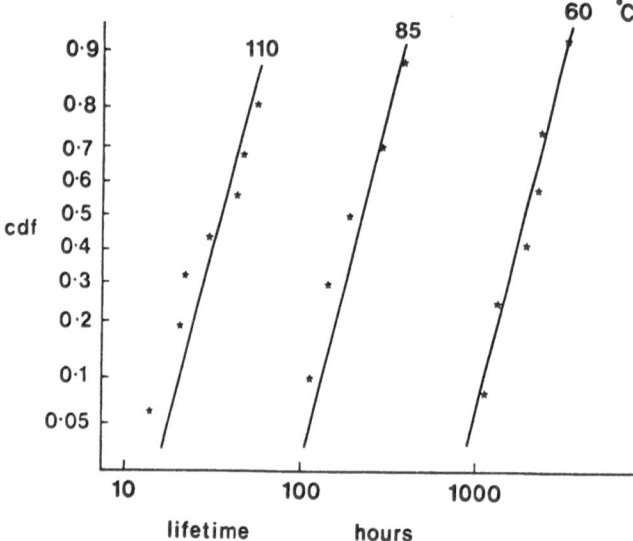

Fig. 40 - Temperature acceleration of aluminum corrosion; unpassivated interdigitated metal pattern on silicon dioxide; dil-encapsulated in epoxy-novolac resin; rh = 90%, bias 30 volts; failure criterion: open circuit.

members of one sample, a single maximum likelihood analysis was performed yielding directly, for the median lifetime in hours,

$$\mu = -16.1 + 10.2 \left(\frac{1000}{T} \right) - 8.57h^2$$

where $\sigma = 0.4$.

The ML lines of best fit on Figs. 40 and 41 show good conformance to the linear stress model with two stress coefficients. The activation energy in the temperature term in 0.87 eV, the 90% confidence limits being 0.8 and 0.95 eV, whilst the limits on the humifity stress coefficient are 8.0 and 9.2. The acceleration factor for corrosion in a plastic encapsulation is thus

$$\frac{\psi}{\psi'} = \exp \left\{ 8.57 \, (h'^2 - h^2) \right\} \cdot \text{xp} \left\{ \frac{0.87e}{kT} \left(\frac{1}{T} - \frac{1}{T'} \right) \right\}$$

It may be noted that holding one stress factor constant is essential for a graphical analysis but it is not necessary when an ML analysis is performed.

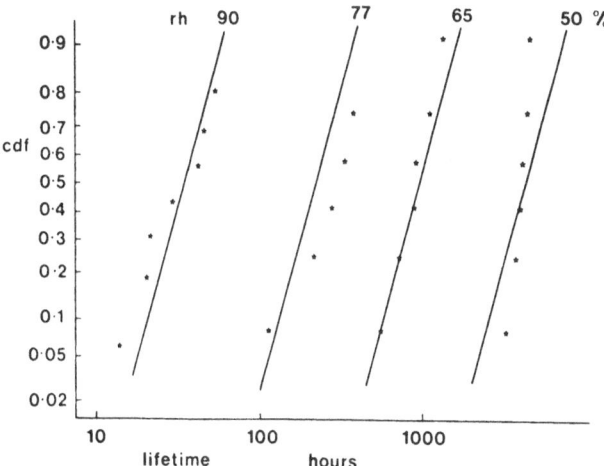

Fig. 41 - Humidity acceleration of aluminum corrosion. Unpassivated interdigitated metal pattern on silicon dioxide; dil-encapsulated in epoxy-novolac resin; temperature = 110°C, 30 volts; failure criterion, open circuit.

8. LIFE TESTING

The examples of cdf plots already given show that component lifetime data can often be satisfactorily described by a lognormal distribution, either monomodal or bimodal. Very often, a reasonable fit to the data would also have been obtained by postulating a Weibull distribution. The results show, however, that the best results tend to be obtained on functionally simple items like transistors and circuit elements. On integrated circuits, the cdf plot can be quite poor as in Fig. 31. A similar conclusion can be drawn in respect of the validity of the linear stress model; excellent results were seen in Fig. 24 for transistors by comparison with the scattered data of Fig. 32 on operational amplifiers. There are two related reasons for this trend. The first concerns the number of failure mechanisms at work. If only one mechanism exists, the prospects for a linear cdf plot and a valid linear stress model are good. Even if two mechanisms exist, it may be possible to separate them by a bimodal dissection although the estimation of the activation energies, if different, necessitates a set of bimodal accelerated-test plots which has only rarely been achieved. If, of course, the failure mechansims can be identified directly, by, say, a physical examination of the failed specimens, then the prospects for a meaningful statistical analysis improve again, assuming that the classified lifetimes are sufficiently numerous. The other advantage of simple components stems from that very simplicity; when a transistor, say, is thermally stressed, its internal electrical state can be fairly confidently controlled by the applied electrical and environmental conditions. The popular "reverse bias" applied to a transistor can, for example, be relied upon to bias the junctions as intended and could well stimulate the major failure mechanism such as loss of current gain. This control also permits high stresses to be used, even up to 300°C, as for the transistors of Fig. 24. On small-scale integrated circuits, "static bias" may also determine most of the internal node potentials but less assuredly and certainly for a more limited stress range. An example of an obviously invalid accelerated test is given by Fig. 42 where the cdf plot at 200°C is probably valid but the cdf at 260°C is actually lower than at 230°C. The origin of this behaviour in this illustration was traced to a reduction in the voltage on the output transistor which occurred above 220°C, as seen in Fig. 43, thus indicating a stress "validity limit" of about 220°C for these integrated circuits.

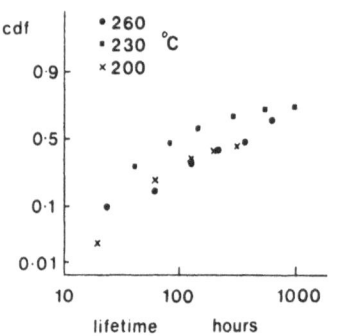

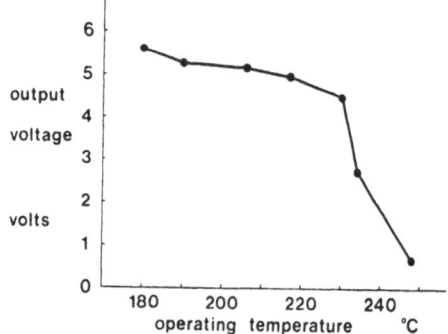

Fig. 42 - Invalid temperature-accelerated life test cdf plots. Low power TTL integrated circuits.

Fig. 43 - Identification of validity limit; reduction of electrical stress at elevated temperatures. Low power TTL integrated circuits.

For LSI circuits, accelerated testing is predictably still more difficult from the greater impact of both multiple failure mechanisms and the electrical inaccessibility of most of the circuit die. Before reviewing what seems to be experimentally possible, however, a closer examination of the implications of multiple failure mechanisms will be made.

8.1 Multiple Failure Mechanisms

The earlier discussion of bimodal lifetime distributions referred specifically to a **mixed population**; that is to say, a part of the population was at risk from one cause of failure only with the other part vulnerable only to another failure mechanism. The causes could be, say, bond wire detachment and oxide breakdown, and assuming that they were stress-accelerable, their stress coefficients (activation energies for thermal stress) would be different. With, as already discussed, $f_a(t)$ and $f_b(t)$ describing the distribution of lifetimes in the two parts, the fdf for the whole population is given by Eq. 35 (with k = 2). Knowing the lifetime distribution as a result of a test together with the activation energies for the two mechanisms, then allows the service

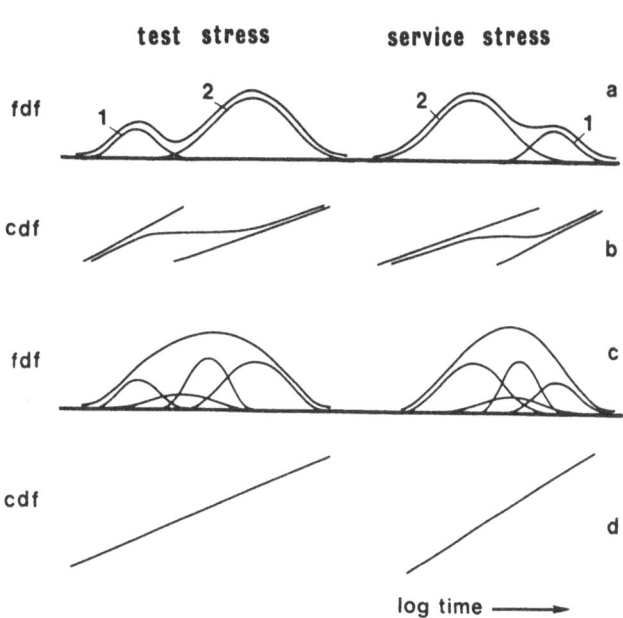

Fig. 44 - Mixed populations; test and service lifetime distributions: (a) fdf for bimodal lifetime distribution showing scaled fdfs for lifetimes determined by mechanisms 1 and 2, activation energy 1 > activation energy 2 with (b) cdf for bimodal lifetime distribution with lines for mechanisms 1 and 2, (c) fdf for multimodal distribution showing scaled fdfs for constituent distributions, (d) apparent cdf plots for multimodal lifetime distributions.

lifetime distribution to be predicted as summarised schematically on Fig. 44 (a) and (b). This diagram emphasises the importance of exposing both distributions during a life test; a prediction based on mechanism 1 alone would not detect the early onset of failures in service from mechanism 2.

Reynolds

If the mixed population comprises several parts, then the dissection of the cdf observed under high stress is virtually impossible experimentally; the plot might well appear to be lognormal again as illustrated in Fig. 44(b) and (c). From the corresponding cdf plots there shown, it is clear that unless all the mechanisms are known and characterised by their μ and σ values and stress coefficients, the service lifetime distribution cannot be predicted with much confidence; even the apparent dispersion may change. The mixed population is however, only one manner in which multiple failure mechanisms can occur.

Suppose, for the purpose of

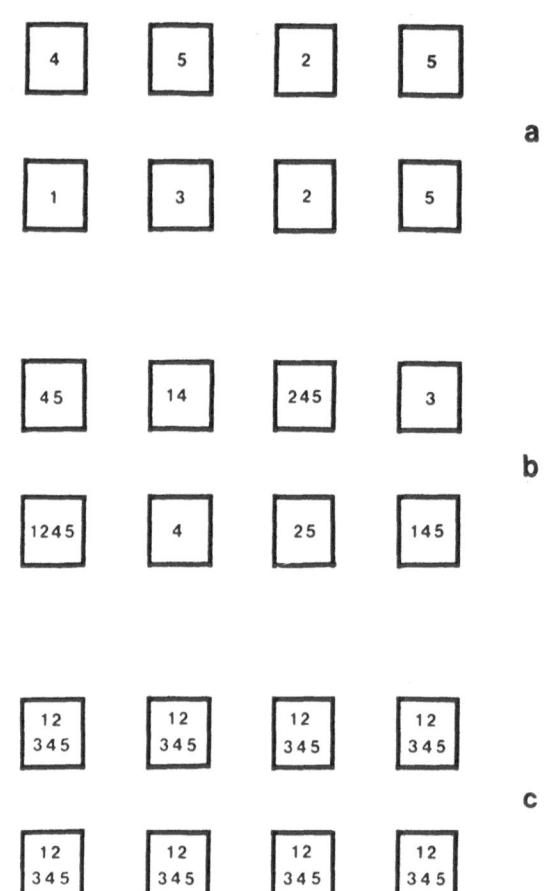

Fig. 45 - Multiple failure mechanisms with each component subject to, (a) one of five possible mechanisms (mixed population), (b) some (or all) of five possible mechanisms, (c) all five mechanisms.

illustration, that an integrated circuit is vulnerable to five different failure mechanisms. The ways in which they might exist are then shown in Fig. 45 where the first, (a) is the **mixed** population already discussed. In Fig. 45(c), each component is at risk from all five mechanisms so failure depends upon the **competing risks** to which they are exposed. The intermediate, and doubtless often real situation, is portrayed in Fig. 45(b) where each component is subject to some of the (competing) risks. The cdf's for the three cases are

(a)
$$Q(t) = \sum_{i=1}^{k} q_i \, Q_i(t), \quad (\Sigma \, q_i = 1) \tag{78}$$

(b)
$$Q(t) = 1 - \prod_{i=1}^{k} [1 - q_i \, Q_i(t)], \quad (\Sigma \, q_i > 1) \tag{79}$$

(c)
$$Q(t) = 1 - \prod_{i=1}^{k} [1 - Q_i(t)] \tag{80}$$

where $Q_i(t)$ = lifetime cdf for failures caused by mechanism i, $Q(t)$ = population lifetime cdf (and k = 5 in Fig. 45).

Desirably, expressions for these three cases would be obtained with $Q_i(t)$ expressed for the lognormal and Weibull distributions and combined with the linear stress model, but unfortunately only a few solutions are possible of which the following is one.

The competing risk model is more conveniently expressed in terms of survival functions rather than cdf's. Thus for k competing mechanisms causing lifetime distributions described by $R_i(t)$, the reliability function for the population is

$$R(t) = \prod_{i=1}^{k} R_i(t) \tag{81}$$

from which it is easily shown that the hazard function for the population is

$$\lambda(t) = \sum_{i=1}^{k} \lambda_i(t) \tag{82}$$

where $\lambda_i(t)$ is the hazard function due to mechanism i. If the lifetime distributions due to mechanisms i are exponentially distributed, then from Eqs. 11 and 32

$$R_i(t) = \exp\left(-\frac{t}{\eta_i}\right) \tag{83}$$

and if the mechanism conforms to the linear stress model (Eq. 49),

$$\eta_i = \exp (b_{oi} + b_{1i}x). \qquad (84)$$

If the proportion of the population failing for mechanism i at a stress x is p_i, it can be shown that the (constant) hazard rate λ' for a stress factor x' is related to the hazard rate λ at a stress factor x according to

$$\lambda' = \lambda \sum_{i=1}^{k} p_i \exp \{b_{1i} (x - x') \} \qquad (85)$$

This formula can be applied in two ways. If, at a high stress, sufficient failures are accumulated for the proportions p_i to be estimated and the mechanisms identified, then x becomes the test stress factor. Estimating the hazard rate λ at stress x from Eq. 77, then enables the hazard rate λ' to be found at a lower service stress of factor x', and the ratio $\dfrac{\lambda}{\lambda'}$, (>1) becomes a **deceleration** factor. If, alternatively, sufficient field data is available for p_i to be estimated in service, the x and x' correspond respectively, to service and test whereupon $\dfrac{\lambda'}{\lambda}$ is greater than unity and becomes an **acceleration** factor. Used in this

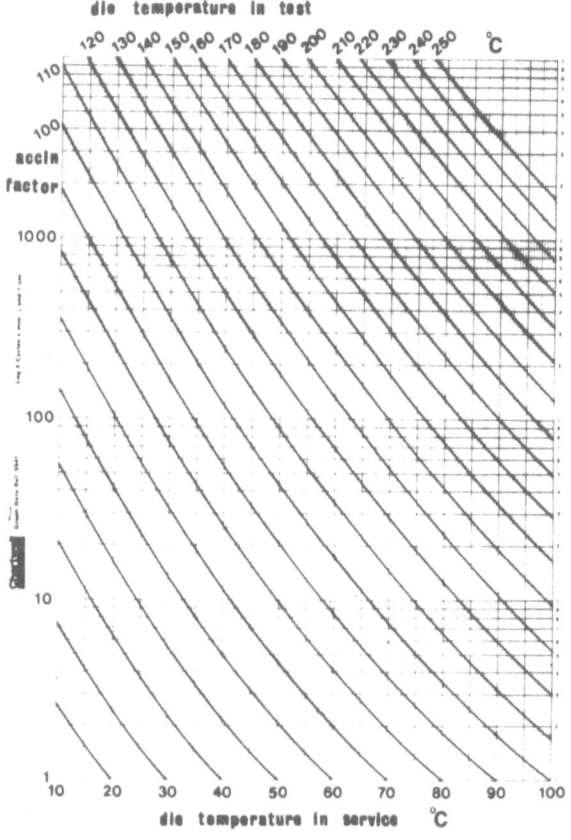

Fig. 46 - Acceleration factor for MOS integrated circuits in hermetic encapsulations.

latter manner, the formula enables the permitted number of
failures in a lifetest to be estimated given a desired service
hazard rate. Two proposed
sets of data for this purpose,
based on field-failure mechanism experience[48] and stress
coefficient determinations of
the kind given in Sec. 7, are
shown in Tables III and IV, the
former applying to MOS circuits in hermetic encapsulations
where the ambient temperature
provides the stress. The temperature used in the formula is,
however, the die temperature
which must be estimated as

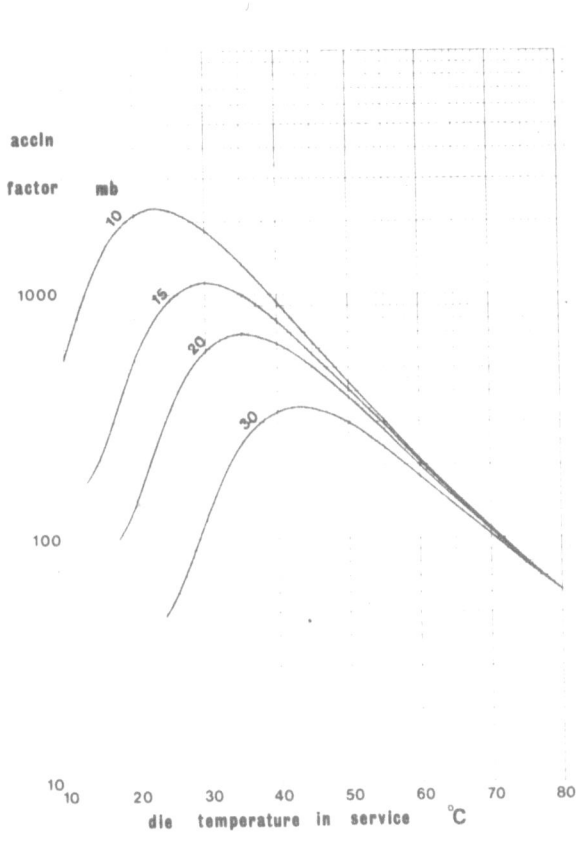

Fig. 47 - Acceleration factor for MOS integrated circuits in plastic encapsulations for ambient partial pressures of water vapour in millibars. Test ambient; 85°C, 85% rh; die temperature rise on test = 0°C.

discussed in Sec. 8.3. The data of Table IV applies to MOS circuits in plastic encapsulations

where both elevated ambient temperature and relative humidity are the stresses with, for

humidity, $x = -h^2$, as introduced in Sec. 7.3. The relative humidity, like the temperature,

refers to that obtaining at the die surface, the method of calculation being given in Appendix

5. When (as for Table IV) two stresses are applied, the acceleration factors are multiplied

together to give the overall factor. Curves of the acceleration factors from the data of Tables

III and IV are given in Figs. 46 and 47; those for a humid ambient are calculated for the

widely-used 85°C, 85% rh test condition and a figure must be assumed for the partial pressure of water in the service atmosphere.

TABLE III
Acceleration-factor data for temperature stress testing testing of MOS integrated circuits in hermetic encapsulations

Service Failure Mechanism	Temperature Stress Coefficient $b_1/11606$ eV	Proportion %
bulk	1.0	3
surface/oxide instability	0.9	40
aluminum corrosion	0.85	10
intermetallic bond formation	0.7	8
electromigration	0.45	5
oxide breakdown	0.3	20

TABLE IV
Acceleration-factor data for temperature/humidity stress testing of MOS integrated circuits in plastic encapsulations

Service Failure Mechanism	Temperature Stress Coefficient $b_{11}/11606$ eV	Humidity Stress Coefficient b_{12}	Proportion
bulk	1.0		3
aluminum corrosion	0.85	8.1	10
surface/oxide instability	0.7	4.4	40
galvanic bond-pad corrosion	0.6	4.4	15
oxide breakdown	0.3		12

8.2 LSI Circuit Life-Tests

Against the rather complex background of multiple failure mechanisms, the interpretation of life-test results obtained on complex integrated circuits necessarily requires each case to be taken on its merits as will be seen from the examples which follow.

The cdf plots of Fig. 48 were obtained on a sample of 100 p-channel MOS shift registers,[49] dynamically life-tested in two equal-size subsamples at ambient temperatures of 80

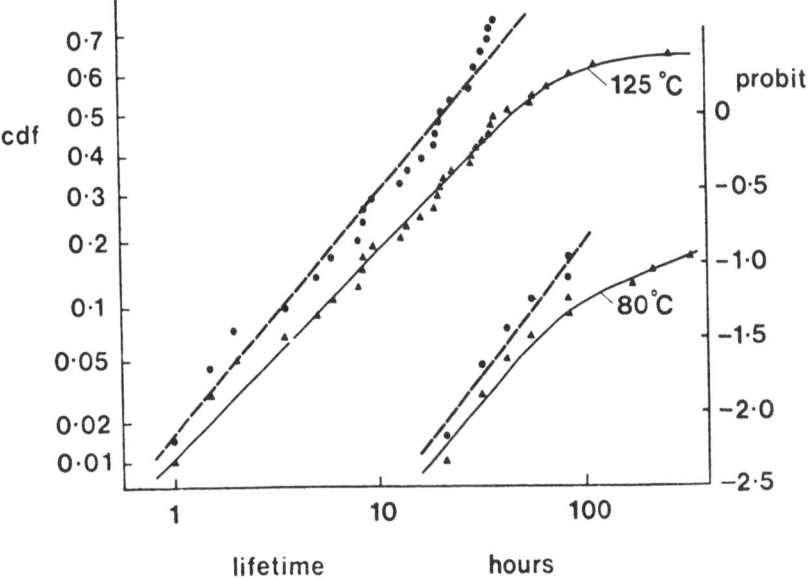

Fig. 48 - Lognormal lifetime cdf plots for MOS shift registers at different stresses (quad 80-bit, metal-gate PMOS hermetic). Die temperature rise = 20°C; dynamic operation (~ 40mW dissipation); sub-
sample sizes = 50; broken lines, early distribution, proportion 0.65; at
125°C ambient ψ = 20.3 h, σ = 1.40; at 80°C ambient ψ = 278 h, σ
= 1.25.

and 125°C. The operating conditions were calculated to produce a die temperature rise of

about 20°C above the ambient temperature. The lifetimes were plotted after adjustment for a

location parameter of 2.5 hours as described in Sec. 6.1.1. The population is apparently

bimodal with the early-distribution proportion estimated as approximately 0.65 from the

125°C plot. Performing dissections as described in Sec. 6.1.1.1 then yields reasonably linear

plots for the early constituent distribution with its parameters added to Fig. 48. The σ-values

are reasonably consistent and the median-lifetime ratio, with die temperatures of 100 and

145°C, implies an activation energy of 0.78 eV. If the service ambient temperature is taken

as 30°C, the service median lifetime is approximately 12,000 hours with the 2.5% reliable

lifetime (in the whole population) estimated (using Eq. 20) as 1160 hours, indicating a

defective product.

It is sometimes possible to isolate a single failure mechanism in an LSI circuit as in the UV-erasable programmable memories of Fig. 49. After programming each component, the sample was stressed at 250°C with no electrical potential applied. (On the measurement occasions, the data

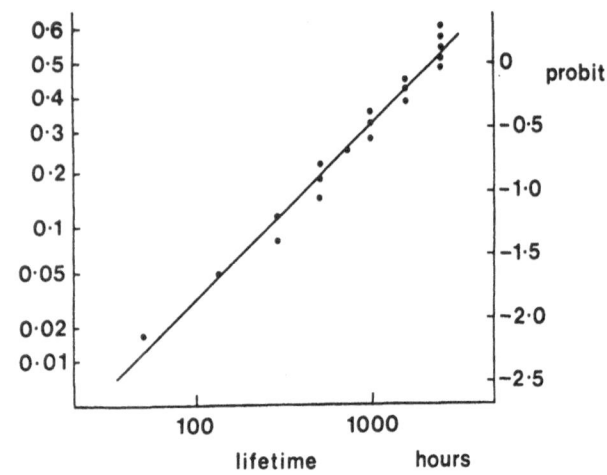

Fig. 49 - Lognormal lifetime cdf plot for 16K UV-erasable proms. Storage at 250°C; sample size = 30; failure criterion; stored data error; ψ = 2180 h, σ = 1.67.

was simply read out and checked, any incorrect bit signaling failure.) Because the mechanism of failure in these memories has been fairly well characterised, it was deemed sufficient to use the generally-accepted minimum activation energy of 0.75 eV for the mechanism. Estimating the parameters graphically from the cdf plot (given on Fig. 49), the median lifetime at an assumed die service temperature of 70°C is then 10^7 hours, the 2.5% reliable lifetime being 50 years which is evidently satisfactory. Predicting the hazard rate from Fig. 5 or Table I reveals, by Fig. 50, that these items will eventually "wear out" after an acceptable service life.

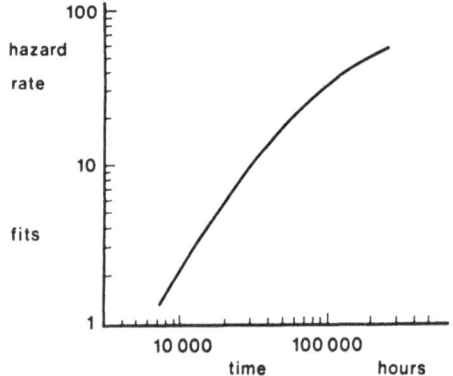

Fig. 50 - Expected hazard rate for eproms failing by data loss; median lifetime ψ = 1.3×10^7 hours; σ = 1.67.

An example of an accelerated lifetest which seems to follow the classic pattern is shown by the cdf plots of Fig. 51 obtained on 4K dynamic random-access memories.[50] Parallel lines of best fit drawn by eye give an activation energy estimate of 0.93 eV which, translating to a service temperature of 50°C, predicts a median lifetime of 2×10^{11} hours. The high value of the dispersion seen on Fig. 51 ($\sigma = 6.0$) means that the hazard rate will steadily fall as shown on Fig. 52, the level, 100 fits after 1 year, probably being acceptable for many applications.

A much more difficult interpretive problem is presented by the results of Fig. 53 obtained on samples of 16K random-access memories dynamically tested at an ambient temperature of 160°C, a temperature carefully chosen to be within the validity limit discussed in Sec. 8. The

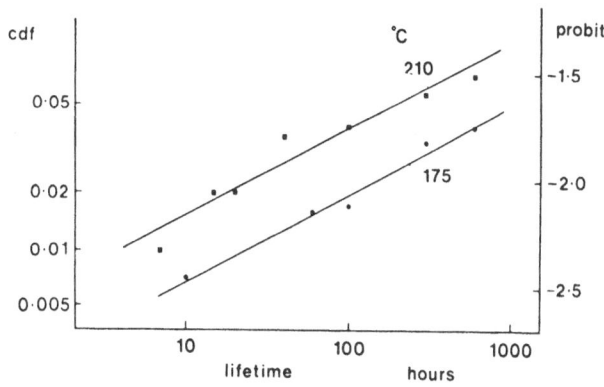

Fig. 51 - Lifetime cdf plots at two different temperatures for 4K dynamic memories. Sample sizes, 64 at 175°C, 184 at 210°C; dynamic write/read operation; activation energy = 0.93eV, $\sigma = 6.0$.

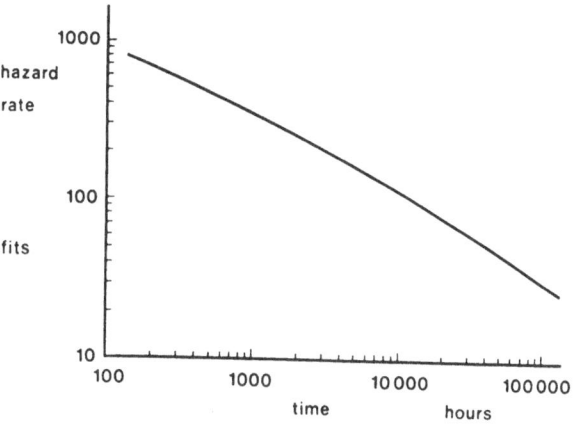

Fig. 52 - Expected hazard rate for 4K dynamic memories. Median lifetime $\psi = 2 \times 10^{11}$ h, $\sigma = 6.0$.

die temperature rise above the ambient is approximately 15°C. The plot has an evident bimodal character with the early distribution effectively having a median lifetime of 2 hours with negligible dispersion. A rather high activation energy, over 1.3 eV, would have to apply if numerous failures were to be avoided in a long life application at an ambient temperature of say, 50°C so this sample would disqualify the lot, at least, from which it came. If the early failures were screened out by a suitable burn-in treatment, (24 hours at 125°C would probably suffice) then the other 3 failures could be handled by resorting to the exponential model. Calculating the failure rate from Eq. 73 at 60% confidence on the basis of some 15000 hours of testing (at 160°C) and using an acceleration factor of 1000, having entered Fig. 46 with an expected service die temperature of 65°C, (for a 50°C ambient) a field hazard rate of about 250 fits is predicted, which would be reasonable for these components.

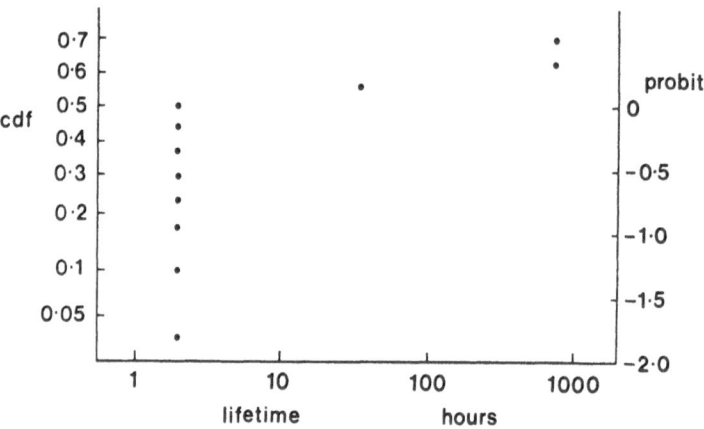

Fig. 53 - Lifetime cdf plot for 16K dynamic memories. Sample size = 15, dynamic operation, ambient temperature = 160°C; die rise = 15°C.

8.3 Testing Semiconductor-Components at Elevated Temperature

In steady-stress life tests, a choice is available between three different modes of operation, namely, storage, static bias and dynamic bias. For all three options, the validity limit must not be exceeded. On transistors and simple integrated circuits, it will probably be possible to identify the limit by measurements similar to those of Fig. 43 but for complex circuits, it will be necessary to apply a functional checking routine to the component (using, say, a computer-controlled measurement system) with the temperature slowly raised. The limit will usually be quite obvious, 150°C being typical.

Storage testing has the advantage of simplicity and, as seen in the memories of Fig. 49, may have a specific purpose but, in general, the method is not considered to be very searching. When static bias is used, there is obviously a choice in the way the fixed potentials are applied; the supply rails will usually be connected normally but it is generally difficult to show that any particular method of connection of the other pins induces more stress than does another, assuming that the manufacturer's limits are not exceeded. For dynamic bias, which has received more consideration with the advent of LSI circuits, an elementary function is continuously performed such as writing a systematic pattern into a memory and then reading it out. Appropriate biasing methods for microprocessor parts are still at a very early stage but the use of a random instruction sequence operating on random data is being investigated. In general, dynamic bias is claimed to induce the most failures, both in quantity and in type, but static bias reveals particular weaknesses like oxide and surface instabilities, more efficiently.

Whichever powering method is adopted, the effect of dissipation on the die temperature must be assessed either from the thermal resistance of the encapsulation or by measurements on companion parts using an infrared microradiometer or thermo-sensitive liquid crystals. If a significant temperature distribution over the die exists, the maximum temperature should be chosen for use in the statistical analysis. Thermal resistances vary widely from one type of encapsulation to another; measured from the component amounting (say a printed circuit board) to the component die, the resistance may be as low at $30°C/watt$ for a large die in a ceramic (hermetic) enclosure, up to $500°C/watt$ for a small plastic-encapsulated transistor.

In some life-testing equipment, it may be possible to monitor the parts under test whilst they are at the elevated temperature but in order to assess whether they have truly failed, the temperature must be reduced to a normal operating temperature prior to taking a set of measurements which should as far as possible encompass the operating specification. For these inspection occasions, the necessary temperature reduction should be completed without changing the applied electrical conditions, otherwise annealing effects may occur which will reverse changes induced by the stress. Even at the measurement temperature, it is desirable to preserve those conditions until the measurements are made because some drift effects are known which occur at room temperature. It is obviously desirable that the temperature transitions should be completed in times which are small by comparison with the stress periods.

The foregoing remarks apply also to thermal step-stress tests but still greater care is required because the changing stress may also disturb the electrical and thermal conditions of the die; the die temperature may, for example, require a different dissipation correction at each step. For this reason, coupled with the need for the correcting procedure discussed in Sec. 6.1.1.2.1, temperature step-stress test results are often not formally analysed but are used to determine suitable stress levels for a subsequent set of steady-stress tests.

9. SERVICE-DATA ANALYSIS

The ultimate component reliablity question will now be addressed, namely, the analysis of real service data. Components in service again constitute a sample drawn from a population of unknown parameters. If the service records are adequately comprehensive and the failures are sufficiently numerous, it may well be possible to extract the lifetimes and censoring times

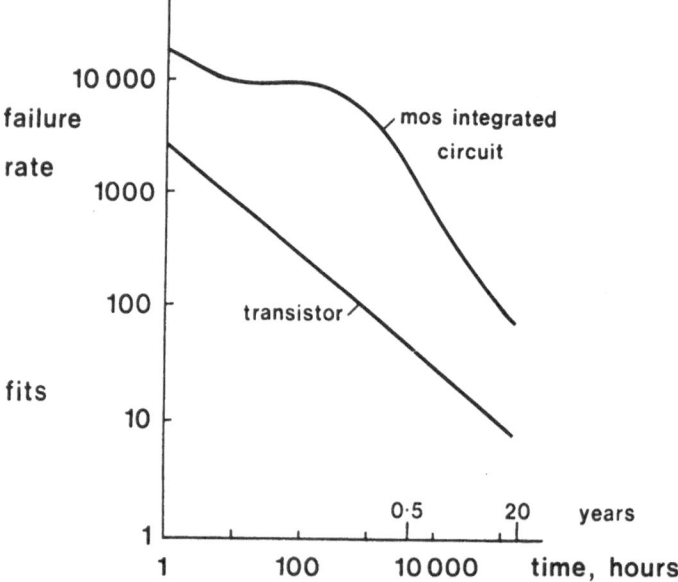

Fig. 54 - Predicted component hazard rates in service. Metal-gate p-channel integrated circuit and small-signal transistors in telecommuncations equipment, determined from lifetime obser-vations over 24,000 hours.

for the sample, whereupon the methods already described can be used again. Unfortunately, field data is rarely available in such a convenient form because the sample members do not enter service simultaneously; recording the real time at which failures occur (the failure times) or when removals from service are made (the cessation times) will not yield lifetimes and censoring times unless the full life history of every component is individually documented. Such an approach was actually adopted for the transistors and MOS circuits of Fig. 54[27] but the cost involved is usually unacceptable. If, however, the real-time build-up of the sample is known — the service profile is known, as illustrated in Fig. 55. —

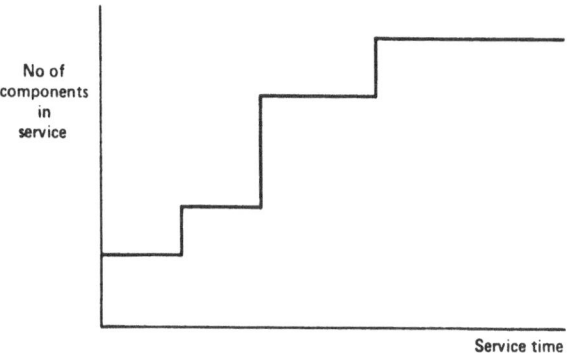

No of components in service

Service time

Fig. 55 - A service profile.

and the failure times are both known and are sufficiently numerous, then it is possible to adapt the ML method in order to estimate the population lifetime parameters.[51] The method is also applicable if the failure times are known only to lie between two (real) inspection times and, rather as a bonus, the effect of replacing failed components can be accommodated. It must be recognised that if the accumulated failures total no more than about 10, the method is not really suitable, leaving no choice but to find the accumulated component hours (the area under the service profile), postulate an exponential lifetime distribution and calculate the hazard rate by the method of Sec. 6.1.3.

9.1 **Analysis**

In Sec. 4.2, it was shown how an fdf and cdf could be obtained to express the distribution of lifetimes in a multimodal population comprising a set of sub-populations each with a different value of the location parameter γ. If the members of each sub-division of the service profile are considered to come from such a sub-population, then the entire sample can be considered to have been deployed at the same **conceptual** birth time t = 0 by setting γ_i (i = 1 to k) equal to the real birth time less the conceptual birth time. The failure time fdf, f(t), and cdf, Q(t), are then given by Eqs. 46 and 47 (with Eq. 48) with t replaced by t. As all the γ_i are known, the task is one of finding the five lifetime parameters μ_a, σ_a, μ_b, σ_b and q_a. In the event that a monomodal lifetime population is actually postulated, the problem simplifies to finding μ $(=\mu_a)$ and σ $(=\sigma_a)$.

The formation of the likelihood function follows the principles outlined in Sec. 6.1.2 but if, as often occurs with field data, the failures are only observed at inspection occasions, Eq. 71 should be modified to read

$$L = \prod_{j=1}^{r} \{Q(\tau_r) - Q(\tau_{r-1})\}^{n_r} \prod_{j=1}^{c} \{1 - Q(\theta)\} \tag{86}$$

where n_r is the number of failures occurring between τ_{r-1} and τ_r.

As discussed in Sec. 6.1.2, the maximisation of the likelihood necessitates the determination of the first and second derivatives of L with respect to the parameters. The ML procedure for lifetimes and censoring times is, however, only a special case of the present problem which occurs when all the γ-values are zero; the failure times become the lifetimes and the service profile is just a vertical line. Only one set of derivatives need therefore be calculated for both Eqs. 71 and 86.

Unlike the simple lifetime/censoring time solution, initial parameter values for the ML procedures are more difficult to obtain and if they are too far from the optimum set, the Newton-Raphson procedure will fail to converge. Other algorithms can, however, be used to obtain better initial values. In a simple incremental algorithm for example, each parameter is increased in turn by a small percentage, the new value being adopted if the likelihood increases but rejected in favour of a similarly decremented value if the likelihood falls. It is desirable to monitor the log-likelihood after each correction operation but the actual value of the maximum has no absolute significance.

When a component fails in a real system, it is usually replaced by a new item but if the accumulated failures represent only a small fraction of the total population, the errors in the estimated parameters will be negligible. If desired however, these errors can be eliminated by simply adding a step of one component to the service profile at each failure time. Two examples of the application of the foregoing method follow.

9.1.1 Small-Scale Integrated Circuits

The service profile for nearly 92000 small-scale low power TTL integrated circuits located in electromechanical telephone exchanges is shown by Fig. 56 with the recorded failure times and a single cessation time for the remainder of the sample.

With as many as 11 steps in the service profile, suitable initial parameters for the staggered multimodal analysis were sought by preparing a lognormal cumulative failure plot on the untrue assumption that all the components in the sample entered service at the time of the first group. The visually plotted line of best fit, seen

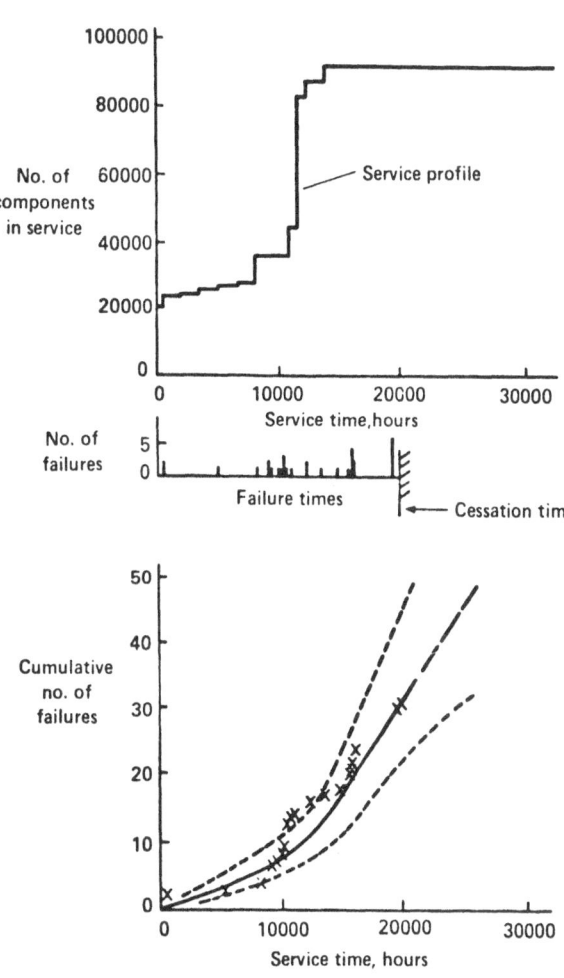

Fig. 56 - Service data for low-power TTL integrated circuits; (a) service profile and failure times with one common cessation time; (b) cumulative failures, x = observed, — = calculated from monomodal ML solution, -- = 90% confidence limits.

on Fig. 57, suggests a median lifetime of about 10^7 hours and a dispersion of 2. With these initial values, followed by a single incremental correction cycle of 5%, about 6 cycles of the Newton-Raphson procedure sufficed for the likelihood maximum, the median lifetime being 10^9 hours and the dispersion 3.34. The fact that some of the experimental points fall outside the 90% limits on Fig. 56 suggests that the lognormal distribution is only marginally satisfactory, although, as mentioned earlier, the winding effect on cdf plots can be misleading. The corresponding hazard plot of Fig. 58 predicts a substantially constant

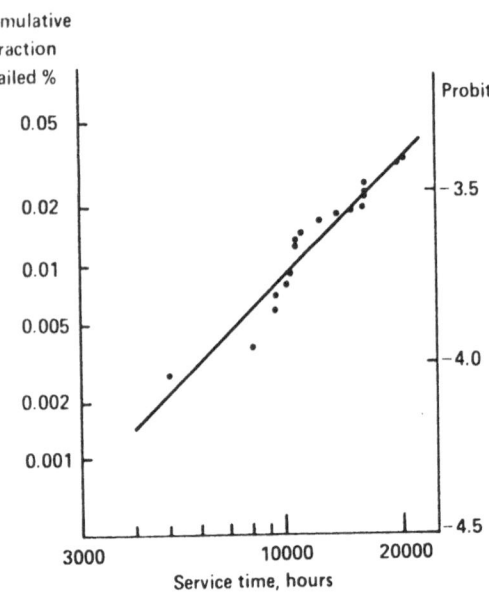

Fig. 57 - Low-power TTL integrated circuits. Estimation of initial parameters for maximum likelihood analysis assuming common birth time; $\psi = 10^7$ h, $\sigma = 2$.

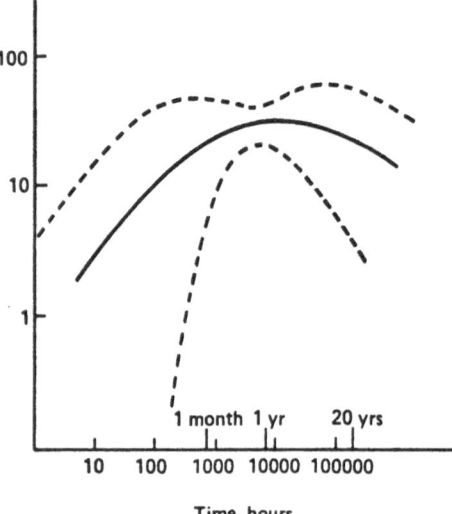

Fig. 58 - Hazard plot for low-power TTL integrated circuits in service. Calculated from monomodal ML solution, ---- 90% confidence bands.

failure rate over most of the proper service life of these components, at a value of about 30 fits.

9.1.2 Dynamic Random-Access Memories

The service profile of over 20,000 1024-bit dynaic random-access memories also from telephone exchanges, is shown in Fig. 59 together with the number of failures observed at monthly inspections.[52]

As before, a crude lognormal cumulative failure plot, shown as Fig. 60, was prepared on the assumption of a common birth time. The graph suggests that the lifetime distribution might be bimodal so a

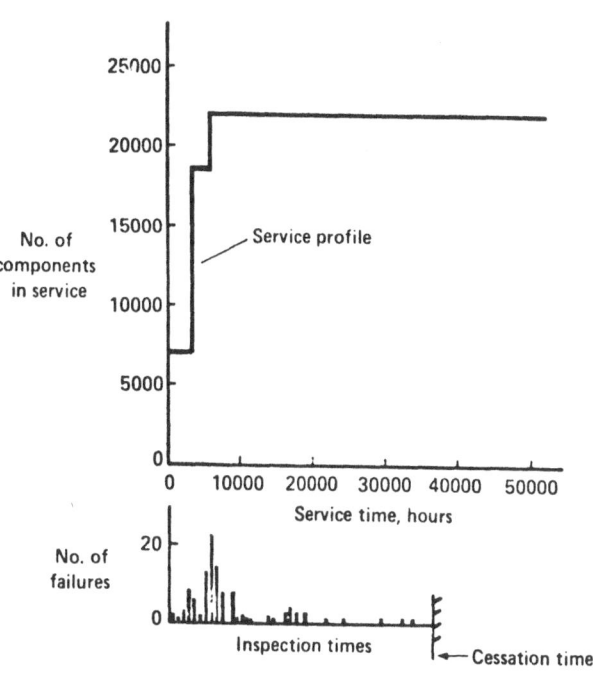

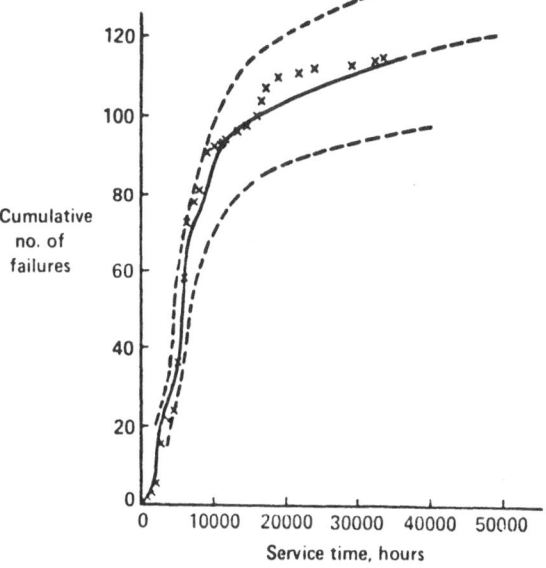

Fig. 59 - Service data for 1K dynamic memories; (a) service profile and failure times with one common cessation time, (b) cumulative failures, x = observed, — = calculated from bimodal ML solution, -- = 90% confidence limits.

graphical dissection could be made as described in Sec. 6.1.1.1 but this labour is

unnecessary. By first

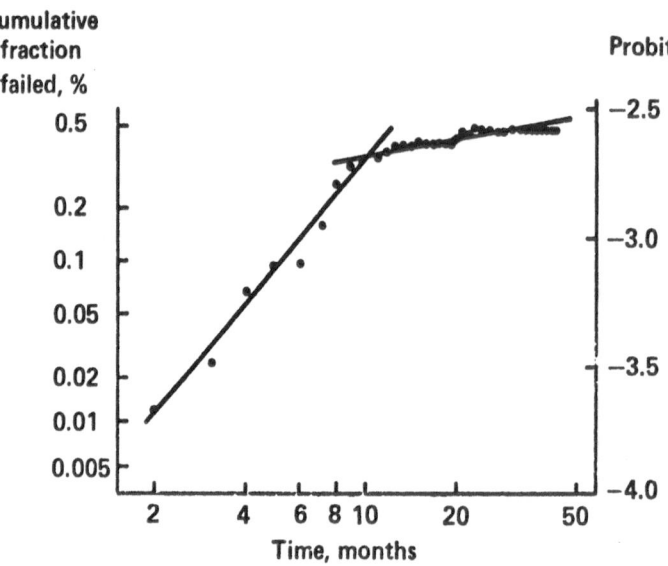

Fig. 60 - Dynamic memories; estimation of initial parameters for
maximum likelihood analysis assuming common birth time.

postulating a monomodal solution, for which crude guesses suffice, the results cas be

used an initial parameter estimates for the main distribution. Comparing then Fig. 60

with Fig. 12, the easily distribution proportion might be about 0.4%. Equally roughly,

the median lifetime for the early distribution is, say 10 months, with the dispersion

perhaps 1.6. With these initial figures, the full bimodal solution is easily obtained as

summarised in Table V. The likelihood for the bimodal solution is evidently higher

than that for the monomodal analysis thus confirming that the population is more

likely bimodal. The comparison between the actual and calculated cumulative failures

on Fig. 59 is acceptable with no points going outside the 90% limits. The correspond-

ing hazard plot of Fig. 61 clearly shows the effect of the early distribution which could

have been usefully removed by a suitable burn-in procedure if its presence had been

known before the parts were passed into service. Thereafter, the failure rate is

predicted to fall steadily over most of the intended lifespan of the equipment.

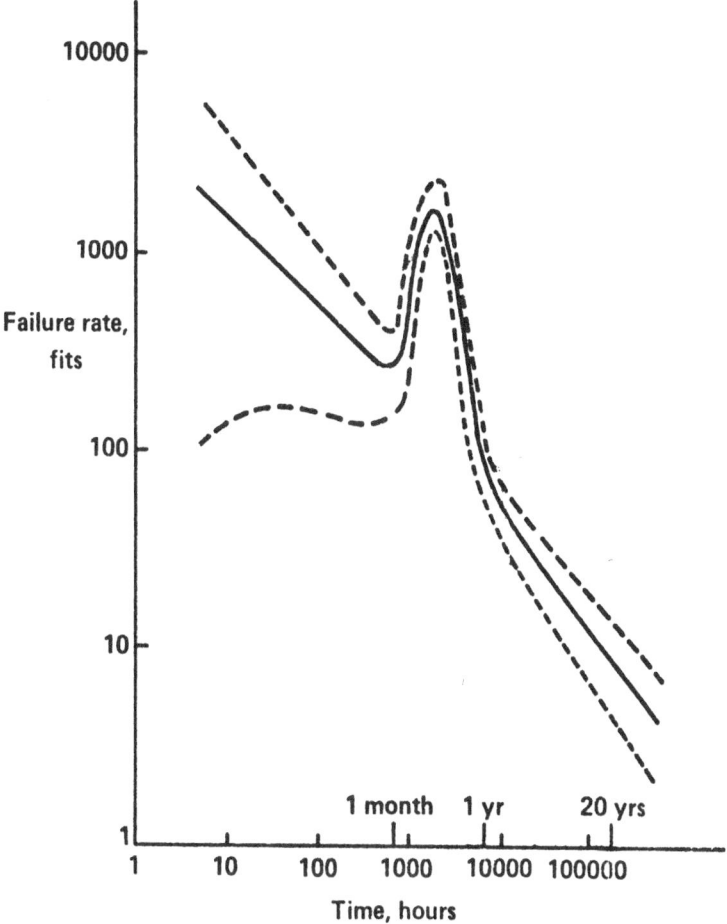

Fig. 61 - Hazard plot for 1K dynamic memories in service; -- = 90% confidence bands.

TABLE V

Parameters					Steps in the Analysis of Memory Data			
Early		Main		Prop^n				
$\mu_\perp$	$\sigma_\perp$	μ_h	σ_h	$q_\perp$	Log Likelihood	No of Iterations		Analysis
					Iteration	Incremental	Newton-Raphson	
		32	11		1141 initial	0	6	Monomodal
		22.7	7.4		1134 final			
2.3	1.6	22.7	7.4	0.004	1137 initial	3	6	Bimodal
1.14	0.40	24.3	7.04	0.0034	1082 final			

10. CONCLUSIONS

Having reviewed accelerated life testing and data analysis methods for semiconductor components, it is concluded that:

(i)   The lifetime distribution for simple components and component elements often satisfies a lognormal or Weibull distribution (or both) together with the linear stress model using temperature and other stresses. Component lifetime distributions are often, however, bimodal.

(ii)   The validity of the above conclusion diminishes as components become physically larger and functionally more complex due to the impact of multiple failure mechanisms.

(iii)   It is still profitable to subject LSI circuits to life tests at elevated stresses but the data obtained needs individual interpretation and the outcome may often be a good/bad indication rather than a quantitative life prediction. An investigation of contributory failure mechanisms, with an estimation of their stress sensitivites and the frequencies of the failures caused, is seen as a better alternative to formal accelerated tests.

(iv)   Graphical methods of data analysis are, for many purposes, sufficient, but the maximum likelihood method is a desirable option and may sometimes provide the only justifiable analysis procedure.

(v)   There is ample evidence that the common assumption of a steady failure rate for semiconductor parts is not generally valid; it is much more likely, the failure rate will continue to fall steadily with time.

11. ACKNOWLEDGEMENTS

The author is grateful for the assistance of his colleagues in the Component Reliability Division of the British Telecom Research Laboratories and, particularly, for the help given by Mr. C. Johnston. Data supplied by Ellemtel, Sweden, and the Plessey Co., U.K. is appreciated. Acknowledgement is also made to the Director of Research, British Telecom for permission to publish this paper.

12. APPENDICES

APPENDIX 1

Approximation to the probit transformation:

$$z = \text{probit} \qquad \Phi(z) = \frac{1}{\sqrt{2\pi}} \int_{-\infty}^{z} \exp\left(-z^2 / 2\right) dz \qquad (A1)$$

$\underline{z \rightarrow \Phi(z)}$

$$\text{Let } \mu = \frac{1}{1 + 0.23164192} \qquad (A2)$$

and

$$P = (0.3193815u - 0.3565638u^2 + 1.781478u^3 - 1.821256u^4 +$$

$$1.330274u^5) \; \frac{\exp\left(-\dfrac{z^2}{2}\right)}{\sqrt{2\pi}} \qquad (A3)$$

then for:

$$z < 0, \quad \Phi(z) = p$$

$$z > 0, \quad \Phi(z) = 1 - p$$

$\underline{\Phi(z) \rightarrow z}$

For $\quad 0 < \Phi(z) < 0.5, \quad$ let $p = \Phi(z)$

$\quad\quad 0.5 < \Phi(z) < 1 \qquad\qquad p = 1 - \Phi(z)$

$$\text{Let } u = \sqrt{\ell n \left(1 / p^2\right)} \qquad (A4)$$

and

$$\eta = \frac{2.515517 + 0.802853u + 0.010328u^2}{1 + 1.432788u + 0.189269u^2 + 0.001308u^3} - u \qquad (A5)$$

then for $\quad 0 < \Phi(z) < 0.5, \quad z = \eta$

$\quad\quad\quad\quad 0.5 < \Phi(z) < 1, \quad\quad z = -\eta$

APPENDIX 2

Step-Stress Correction

If a sample of components is operated with a stress factor* x for a time t, then the fraction failed for a lognormal distribution is given by $\Phi(z)$ where the probit z is

$$z = \frac{\ell n\,t - b_0 - b_1 x}{\sigma} \qquad (A6)$$

from Eqs. 16 and 49 of the main text.

If the stress applied at the mth step of a step-stress test is represented by x_m and the dwell times at each step are all equal to t_d, the fraction failed after step 1, given by z_1, could then have been obtained from operation for a shorter time t_1 at stress x_2 where

$$t_1 = \exp\,(z_1\sigma + b_0 + b_1 x) \qquad (A7)$$

and substituting for z_1,

$$t_1 = t_d \,\exp\,\{-b_1\,(x_1 - x_2)\,\}.$$

The fraction failed after the second step is thus determined by an effective stress period at stress x_2 of $t_d + t_1$, the probit being

$$z_2 = \frac{\ell n\,(t_d + t_1) - b_0 - b_1 x_2}{\sigma}. \qquad (A9)$$

This fraction failed is the same as would have been obtained after a single stress period t_d at a

---

* The subscripts used in Eq. 49 are now dropped to allow different values of the stress factor x to be subscripted 1, 2,.....

higher stress of factor x_2', whereupon

$$x'_2 = x_2 - \frac{1}{b_1} \ell n \left(1 + \frac{t_1}{t_d} \right). \qquad \text{(A10)}$$

By progressively applying the foregoing argument, and assuming that all the stress increments are equal to Δx (which is positive) or

$$\Delta x = x - x_2 = x_2 - x_3 = \ldots \qquad \text{(A11)}$$

then

$$x'_m = x_m - \frac{1}{b_1} \ell n \left\{ \frac{1 - \exp(-m \, b_1 \, \Delta x}{1 - \exp(-b_1 \, \Delta x)} \right\} \qquad \text{(A12)}$$

The right-hand term is thus the correction to be subtracted from the stress factor for the mth step in order to give the observed cumulative failure fraction using a fresh sample for that step. The logarithmic term is the ordinate of Fig. 26 of the main text. The same result is obtained for a Weibull lifetime distribution.

APPENDIX 3

Maximising the Likelihood Function

Suppose that the likelihood contains m parameters p_i to be adjusted. Making P the vector of initial parameters define the matrices,

$$R = \frac{\partial L}{\partial P} \quad \text{(the Jacobian)} \qquad \text{(A13)}$$

and

$$C = \frac{\partial^2 L}{\partial \bar{P} \partial P} \quad \text{(the Hessian)} \qquad \text{(A14)}$$

The matrix analogue of Eq. 76 of the main text is the Newton-Raphson formula yielding the corrected parameter matrix

$$P_1 = P - C^{-1} R \qquad (A15)$$

In order to find R and C, the matrices H and F and the scalar quantities are determined from

the likelihood function using functions, called ϕ- functions, which have derivatives of the fdf

and cdf and are special to the problem under analysis. For each data item (such as a failure

time) H, F and s are evaluated and used to give

$$R1 = (s) \cdot H \qquad (A16)$$

$$C1 = (s) \cdot F - (s^2) \cdot (H \times H') \qquad (A17)$$

where the prime signifies transposition. Summing the R1 and C1 matrices for every data item

then gives R and C. The variance/covariance matrix V is given by

$$V = (-C)^{-1} \qquad (A18)$$

Bimodal 2-Parameter Lognormal Lifetime Distribution with Service Profile

For failure times t, the ϕ-functions are

$$\phi_{xi}(t) = \sum_{i=1}^{k} \frac{q_i \, z^m \, f_z(z)}{t - \gamma_{xi}} \qquad (A19)$$

for both $x \equiv a$ and $x \equiv b$, m being an integer, whilst for inspection times τ_r and

cessation times θ,

$$\phi_{xi}'(\tau_r, \theta) = \sum_{i=1}^{k} q_i \, z^m \, f_z(z) \qquad (A20)$$

where

$$z_{xi} = \frac{\ell n \, (t, \tau_r, \theta - \gamma) - \mu_x}{\sigma_x} \qquad (A21)$$

using whichever time is appropriate. For inspection and cessation times, it is also necessary to find

$$\omega = \sum_{i=1}^{k} q_i \, \Phi \, (z_{xi}) \tag{A22}$$

The scalar, s, is for,

      failure times s $= 1/f(t)$

      inspection times $s = 1/\{Q(\tau_r) - Q(\tau_{r-1})\}$

$$\tag{A23}$$

      cessation times s $= -1/\{1 - Q(\theta)\}$

The elements of the H and F matrices are then obtained from Table VII except that for each inspection time the H and F elements are the differences between the listed elements and the corresponding elements for the preceding inspection time.

Monomodal 2-Parameter Lognormal Lifetime Distribution with Two Stresses

For failure times t, the ϕ-functions are

$$\phi_m(t) = \frac{z^m \, f_z(z)}{t} \tag{A24}$$

and for cessation times

$$\phi'_m(\theta) = z^m \, f_z(z) \tag{A25}$$

where

$$z = \frac{\ell n(t,\theta) - b_o - b_1 x_1 - b_1 x_2}{\sigma} \tag{A26}$$

The scalar s, is for,

      lifetimes s $= 1 / f(t)$

(A27)

cessation times $s = -1/\{ \quad - \quad Q(\theta)\}$

The elements of the H and F matrices are then obtained from Table VIII.

TABLE VII
φ-Functions for Bimodal 2-Parameter Lognormal Lifetime Distribution Analysis

Element	Failure Times	Inspection Times and Cessation Times
H(1,1)	$\frac{p}{\sigma_a^2}\,\phi_{a1}$	$-\frac{p}{\sigma_a}\,\phi'_{a0}$
H(2,1)	$\frac{p}{\sigma_a^2}\,(\phi_{a2}-\phi_{a0})$	$-\frac{p}{\sigma_a}\,\phi'_{a1}$
H(3,1)	$\frac{(1-p)}{\sigma_b^2}\,\phi_{b1}$	$-\frac{(1-p)}{\sigma_b}\,\phi'_{b0}$
H(4,1)	$\frac{(1-p)}{\sigma_b^2}\,(\phi_{b2}-\phi_{b0})$	$-\frac{(1-p)}{\sigma_b}\,\phi'_{b1}$
H(5,1)	$\frac{\phi_{a0}}{\sigma_a}-\frac{\phi_{b0}}{\sigma_b}$	$\omega_a-\omega_b$
F(1,1)	$\frac{p}{\sigma_a^3}\,(\phi_{a2}-\phi_{a0})$	$-\frac{p}{\sigma_a^2}\,\phi'_{a1}$
F(1,2)	$\frac{p}{\sigma_a^3}\,(\phi_{a3}-3\phi_{a1})$	$-\frac{p}{\sigma_a^2}\,(\phi'_{a2}-\phi'_{a0})$
F(1,5)	$\frac{\phi_{a1}}{\sigma_a^2}$	$-\frac{\phi_{a0}}{\sigma_a}$
F(2,2)	$\frac{p}{\sigma_a^3}\,(\phi_{a4}-5\phi_{a2}+2\phi_{a0})$	$-\frac{p}{\sigma_a^2}\,(\phi'_{a3}-2\phi'_{a1})$
F(2,5)	$\frac{\phi_{a2}-\phi_{a0}}{\sigma_a^2}$	$-\frac{\phi_{a1}}{\sigma_a}$
F(3,3)	$\frac{(1-p)}{\sigma_b^3}\,(\phi_{b2}-\phi_{b0})$	$-\frac{(1-p)}{\sigma_b^2}\,\phi'_{b1}$
F(3,4)	$\frac{(1-p)}{\sigma_b^3}\,(\phi_{b3}-3\phi_{b1})$	$-\frac{(1-p)}{\sigma_b^2}\,(\phi'_{b2}-\phi'_{b0})$
F(3,5)	$-\frac{\phi_{b1}}{\sigma_b^2}$	$\frac{\phi'_{b0}}{\sigma_b}$
F(4,4)	$\frac{(1-p)}{\sigma_b^3}\,(\phi_{b4}-5\phi_{b2}+2\phi_{b0})$	$-\frac{(1-p)}{\sigma_b^2}\,(\phi_{b3}-2\phi_{b1})$
F(4,5)	$-\frac{\phi_{b2}-\phi_{b0}}{\sigma_b^2}$	$\frac{\phi'_{b1}}{\sigma_b}$

The matrices H and F are symmetrical with all other elements zero. The arguments of the φ-functions are omitted for simplicity. The symbol p is used in place of q_a of the main text.

TABLE VIII
φ-Functions for Monomodal 2-Parameter Lognormal Lifetime Analysis

Element	Failure Times	Cessation Times
H(1, 1)	$\frac{\phi_1}{\sigma^2}$	$-\frac{\phi_0'}{\sigma}$
H(2, 1)	$\frac{x_1\phi_1}{\sigma^2}$	$-\frac{x_1\phi_0'}{\sigma}$
H(3, 1)	$\frac{x_2\phi_2}{\sigma^2}$	$-\frac{x_2\phi_0'}{\sigma}$
H(4, 1)	$\frac{\phi_2-\phi_0}{\sigma^2}$	$-\frac{\phi_1'}{\sigma}$
F(1, 1)	$\frac{\phi_2-\phi_0}{\sigma^3}$	$-\frac{\phi_1'}{\sigma^2}$
F(1, 2)	$\frac{x_1}{\sigma^3}(\phi_2-\phi_0)$	$-\frac{x_1\phi_1'}{\sigma^2}$
F(1, 3)	$\frac{x_2}{\sigma^3}(\phi_2-\phi_0)$	$-\frac{x_2\phi_1'}{\sigma^2}$
F(1, 4)	$\frac{\phi_3-3\phi_1}{\sigma^3}$	$-\frac{\phi_2'-\phi_0'}{\sigma^2}$
F(2, 2)	$\frac{x_1^2}{\sigma^3}(\phi_2-\phi_0)$	$-\frac{x_1^2\phi_1'}{\sigma^2}$
F(2, 3)	$\frac{x_1x_2}{\sigma^3}(\phi_2-\phi_0)$	$-\frac{x_1x_2\phi_1'}{\sigma^2}$
F(2, 4)	$\frac{x_1}{\sigma^3}(\phi_3-3\phi_1)$	$-\frac{x_1}{\sigma^2}(\phi_2'-\phi_0')$
F(3, 3)	$\frac{x_2^2}{\sigma^3}(\phi_2-\phi_0)$	$-\frac{x_2^2\phi_1'}{\sigma^2}$
F(3, 4)	$\frac{x_2}{\sigma^3}(\phi_3-3\phi_1)$	$-\frac{x_2}{\sigma^2}(\phi_2'-\phi_0')$
F(4, 4)	$\frac{\phi_4-5\phi_2+2\phi_1}{\sigma^3}$	$-\frac{\phi_3'-2\phi_1'}{\sigma^2}$

H and F are symmetrical matrices
The arguments of the φ-functions are omitted for simplicity

APPENDIX 4

Hazard Rate for an Exponential Lifetime Distribution

Hazard rate at an upper confidence limit $= \frac{X}{2C}$ where:

C = accumulated component hours

X is given in Table VI

TABLE VI

No of failures	X for confidence level:		
	60%	90%	99%
0	1.833	4.605	9.210
1	4.045	7.779	13.277
2	6.211	10.645	16.812
3	8.351	13.362	20.090
4	10.473	15.988	23.209
5	12.584	18.549	26.217
6	14.685	21.064	29.141
7	16.780	23.542	32.000
8	18.868	25.989	34.805
9	20.951	28.412	37.566
10	23.031	30.813	40.289
11	25.106	33.196	42.980
12	27.179	35.563	45.642
13	29.249	37.916	48.278
14	31.316	40.256	50.892
15	33.381	42.585	53.486

APPENDIX 5

Relative Humidity at the Surface of a Semiconductor Die

The saturated vapour pressure (svp) P_s of water at temperature T is given by[53]

$$\ell n \, P_s \;=\; 48.5334 \,-\, 4.0843 \, \ell nT \,-\, \frac{6505.72}{T} \tag{A28}$$

where T is in °K and P_s is in millibars.

Given a test ambient condition expressed as a temperature and relative humidity, the rh at the die surface is obtained by finding, in turn, the:

(i)   svp (ambient), from the above equation using the ambient temperature,

(ii)   partial-pressure of water vapour = svp (ambient) × rh (ambient),

(iii) svp (die), from the above equation using the die temperature,

(iv)   rh (die) = partial-pressure of water vapour/svp (die).

For the service ambient condition, the same sequence is used by starting with the fixed partial pressure of water vapour as given on Fig. 47 of the main text.

REFERENCES

1.   I. Bazovsky, Reliability Theory and Practice, Prentice Hall, NJ (1961).

2.   D. R. Byrkit, Elements of Statistics, Van Nostrand, Reinhold (1972).

3.   N. R. Mann, R. E. Schafer and N. D. Singpurwalla, Methods for Statistical Analysis of Reliability and Life Data, Wiley, N.Y. (1974).

4.   A. Hald, Statistical Theory with Engineering Applications, Wiley, N.Y. (1962).

5.   Myers, K. L. Wong and Gordy, Reliability Engineering for Electronic Systems - Reliability Mathematics (Wong), Wiley, N.Y. (1964).

6.   B. Winer, Statistical Principles for Experimental Design, McGraw Hill, N.Y. (1962).

7.   N. A. J. Hastings and J. B. Peacock, Statistical Distributions, Butterworths, London (1975).

8.   P. D. T. O'Connor, Practical Reliability Engineering, Heyden, London (1981).

9.   F. H. Reynolds, Accelerated-Test Procedures for Semiconductor Components, IEEE Annual Proc. Reliability Physics, p. 166-178, (1977).

10.   F. H. Reynolds, Thermally Accelerated Aging of Semiconductor Components, Proc. IEEE **62**, 2 p. 212-222 (1974).

11.   F. H. Reynolds, Non-Destructive Evaluation of Semiconductor Materials and Devices, Nato ASI (Ed. J. N. Zemel) Chapter 14, Plenum Press (1979).

12.   D. J. Finney, Probit Analysis, Cambridge University Press, 3rd Edition, Cambridge (1971).

13.   M. Abramowitz and I. A. Stegun, Handbook of Mathematical Functions, Eqs. 26.2.17 and 26.2.23, Dover Publications, N.Y. Nov. (1970).

14.   G. A. Dodson and B. T. Howard, High Stress Aging to Failure of Semiconductor Devices, Proc. 7th Nat. Symp. on Reliability and Quality Control in Electronics, IEEE, p. 262-272, Jan. (1961).

15.   D. S. Peck, Semiconductor Reliability Predictions from Life Distribution Data: in Semiconductor Reliability, J. E. Shwop and H. J. Sullivan, Eds. Van Nostrand Reinhold, (1961).

16.   W. Nelson, Analysis of Accelerated Life Test Data - Part I: The Arrhenius Model and Graphical Methods, IEEE Trans. on Electrical Insulation **E1-6**, 4 p. 165-181 (1971).

17.   G. J. Hahn and W. Nelson, Graphical Analysis of Incomplete Accelerated Life Test Data Insulation/Circuits, p. 79-84, Sept. (1971).

18.   G. J. Hahn and W. Nelson, "A Comparison of Methods for Analysing Censored Life Data to Estimate Relationships Between Stress and Product Life, IEEE Trans. on Reliability **R-23**, 1 (1974)

19. W. Nelson, Graphical Analysis of Accelerated Life Test Data with the Inverse Power Law Model, IEEE Trans. on Reliability **R-21**, 1 p. 2-11 (1972).

20. W. Nelson, Analysis of Accelerated Life Test Data - Least Squares Methods for the Inverse Power Law Model, IEEE Trans. on Reliability **R-24**, 2 p. 103-107 (1975).

21. E. O. Codier, Reliability Growth in Real Life, IEEE Reliability Symposium, (1968).

22. G. M. Johnson, Evaluation of Microcircuit Accelerated Test Techniques, McDonnell Douglas Report to RADC, (1975).

23. M. Stitch, G. M. Johnson, B. P. Kirk and J. B. Brauer, Microcircuit Accelerated Testing Using High-Temperature Operating Tests, IEEE Trans. of Reliability **R-24**, 4, p. 238-250 (1975).

24. W. Nelson, Hazard Plotting Methods for Analysis of Life Data with Different Failure Modes, J. Quality Technol. **2**, 3 p. 126-147 (1970).

25. W. Nelson, Graphical Analysis of Accelerated Lifetest Data with a Mix of Failure Modes, IEEE Trans. on Reliability **R-25**, 4 p. 230-287 (1975).

26. W. Nelson, Theory and Applications of Hazard Plotting for Censored Failure Data Technometrics **14**, 4 p. 945-966 (1972).

27. F. H. Reynolds and J. W. Stevens, Semiconductor Component Reliability in an Equipment Operating in Electromechanical Telephone Exchanges, IEEE 16th Annual Proc. Reliability Physics, p. 7-13 (1978).

28. D. S. Peck, The Analysis of Data From Accelerated Stress Tests, 9th Annual Proc. Reliability Physics, p. 68-83 (1971).

29. D. S. Peck, New Concerns About Integrated Circuit Reliability, IEEE 16th Annual Proc. Reliability Physics, p. 1-6 (1978).

30. H. W. Rouhof, Accelerated Lifetest Results on Submarine Cable Transistors, IEEE Transactions on Reliability **R-24**, 4 p. 226-227 (1975).

31. W. Nelson, Accelerated Life Testing Step Stress Models and Data Analysis, IEEE Trans. on Reliability **R-29**, 4 p. 103 (1975).

32. F. H. Reynolds, R. W. Parrott and D. Braithwaite, Use of Tests at Elevated Temperatures to Accelerate the Life of an MOS Integrated Circuit, Proc. IEEE **118**, p. 475-485, March-April (1971).

33. F. H. Reynolds, The Response of the Threshold Voltages of the Transistors in Simple MOS Circuits to Tests at Elevated Temperatures, IEEE 9th Annual Proc. Reliability Physics, p. 46-56, (1971).

34. A. C. Cohen, Maximum Likelihood Estimation in the Weibull Distribution Based on Complete and on Censored Samples, Technometrics **7**, 4 p. 579-588, (1965).

35. W. Nelson and W. Q. Meeker, Weibull Percentile Estimates and Confidence Limits from Singly Censored Data by Maximum Likelihood, IEEE Trans. on Reliability **R-25**, 1 p. 20-24 (1976).

36. M. L. Shooman, Probabilistic Reliability; An Engineering Approach, McGraw Hill, N.Y. (1968).

37. P. R. Stopher, Survey Sampling and Multivariate Analysis for Social Scientists and Engineers, Heath, Lexington, Mass. (1979).

38. L. Toft and A. D. D. McKay, Practical Mathematics, Pitman, London (1942).

39. N. R. Draper and H. Smith, Applied Regression Analysis, Wiley, N.Y. (1967).

40. A. Ralston and H. S. Wilf, Statistical Methods for Digital Computers, Wiley, N.Y. (1977).

41. S. R. Calabro, Reliability Principles and Practices, McGraw Hill, N.Y. (1962).

42. B. Epstein, Estimation From Life-Test Data, IRE Trans. on Reliability and Control, p. 104-107, April (1960).

43. J. R. Black, Electromigration - A Brief Survey and Some Recent Recent Results, IEEE Trans. Electron Devices **ED-16**, (1969).

44. S. P. Sim, Procurement Specification Requirements for Protection Against Electromigration Failures in Aluminum Metallisations, Microelectronics and Reliability **19**, p. 207-218 (1979).

45. D. L. Crook, Method of Determining Reliability Screens for True Dependent Dielectric Breakdown, IEEE, 17th Annual Proc. Reliability Physics, p. 1-7 , (1979).

46. S. P. Sim and R. W. Lawson, The Influence of Plastic Encapsulants and Passivation Layers on the Corrosion of Thin Aluminum Films Subjected to Humidity Stress, IEEE, 17th Annual Proc. Reliability Physics, p. 103-112, (1979).

47. R. W. Lawson, The Accelerated Testing of Plastic Encapsulated Semiconductor Components, IEEE, 12th Annual Proc. Reliability Physics, p. 243-147, (1974).

48. H. C. Rickers, LSI/Microprocessor Reliability Prediction Model Development, Rome Air Development Center, RADC-TR-79-97, New York.

49. T. Cunningham and S. Lincoln, Private communication.

50. H. A. Batdorf, D. H. Hensler and R. D. Wasson, Reliability Evaluation Program and Results for a 4K Dynamic RAM, IEEE, 16th Annual Proc. Reliability Physics, p. 14-18, (1978).

51. F. H. Reynolds, C. Johnston, Component Field-Failure Data Analysis, IEEE Proc. 31st Electronic Components Conf., p. 506-514, (1981).

52. Ö. Hallberg, Private communication.

53. E.A. Moelwyn-Hughes, Physical Chemistry, Pergamon Press, London, (1961).

BASIC INTEGRATED CIRCUIT FAILURE MECHANISMS

C.M. Bailey

Bell Telephone Laboratories
Allentown, Pennsylvania   18103

I. INTRODUCTION

This paper reviews basic failure mechanisms which are or can be experienced in today's technology integrated circuits. Perspectives are also discussed for the future.

It is useful to describe failure rates in terms of the historical bathtub curves (Fig. I-1). The regions in this curve include infant mortality, "constant" or steady state life and wearout. Considering that, in most cases, semiconductor devices do not exhibit wearout within their useful lifetime, the regions of interest for the failure mechanisms discussed are infant mortality and steady state life.

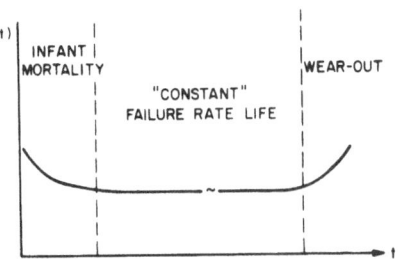

Fig. I-1 - Bathtub curve

Failure   mechanisms   are sometimes broken down into categories of package, chip, and use problems. For example, package problems can be caused by corroded leads and leaking seals, while chip problems can result from dielectric breakdown of MOS devices or surface inversion. Use problems can result in electrostatic discharge failures. In truth,

TABLE I-1.   NORMALIZED DISTRIBUTIONS OF
MALFUNCTIONS FOR MOS DEVICES (1)

HERMETIC PACKAGES

Defect Category	PMOS	NMOS	CMOS
Surface	35%	42%	52%
Bulk	1	3	1
Oxide	25	26	18
Diffusion	12	5	3
Metallization	27	24	26

NON-HERMETIC PACKAGES

Defect	PMOS	NMOS	CMOS
Surface	61%	70%	74%
Bulk	1	2	--
Oxide	12	12	10
Diffusion	8	4	1
Metallization	18	12	15

package, chip and use problems are interactive in many cases. Table I-1 shows a breakdown of failure causes observed in hermetic and plastic packages.[1] These data show the same general failure causes for hermetic and plastic devices. The difference is in the incidence.

DEVICE ASSOCIATION	PROCESS	RELEVANT FACTORS	ACCELERATING FACTORS	ACCELERATION (EA APPARENT ACTIVATION ENERGY)
SILICON OXIDE AND SILICON-SILICON OXIDE INTERFACE	SURFACE CHARGE ACCUMULATION	MOBILE IONS, V, T	T	BIPOLAR: E_A=1.0 - 1.05 eV / MOS: E_A=1.2 - 1.35 eV
	DIELECTRIC BREAKDOWN	E, T	E	
	CHARGE INJECTION	E, T, QSS	E, T	EA = 1.3 eV(SLOW TRAPPING)
METALLIZATION	ELECTRO MIGRATION	T, j, A, GRADIENTS OF T AND j, GRAIN SIZE	T, j	EA = 0.5 - 1.2 Ev / j^2 TO j^4
	CORROSION CHEMICAL GALVANIC ELECTROLYTIC	CONTAMINATION, HUMIDITY (H) V, T,	H, V, T	STRONG H EFFECT / EA ≈ 0.3 - 0.6 eV(FOR ELECTROLYSIS) / V MAY HAVE THRESHOLDS
	CONTACT DEGRADATION	T, METALS, IMPURITIES	VARIED	
BONDS AND OTHER MECHANICAL INTERFACES	INTERMETALLIC GROWTH	T, IMPURITIES, BOND STRENGTH	T	AL-AU: EA=1.0 - 1.05 eV
	FATIGUE	TEMPERATURE CYCLING, BOND STRENGTH	T EXTREMES IN CYCLING	
HERMETICITY	SEAL LEAKS	PRESSURE DIFFERENTIAL, ATMOSPHERE	PRESSURE	

V – VOLTAGE   E – ELECTRIC FIELD   A – AREA
T – TEMPERATURE   j – CURRENT DENSITY   H – HUMIDITY

Fig. I-2 - Time dependent failure mechanisms in silicon semiconductor devices.

The relative importance of the failure mechanisms depends on the device. For example, LSI devices have more chip related problems than SSI devices, MOS has more oxide problems than bipolar, and low power/high voltage devices have more corrosion problems.

As an introduction to some of the time dependent failure mechanisms, an overview is given in Fig. I-2 of the relevant factors, accelerating factors and associated activation energies.

II. CORROSION

Corrosion is a common problem because the accelerating factors of electric field, ionin contamination, and moisture are so often present. It can affect every metal part of the device. Table II-1 lists some examples. (A notable exception is polysilicon MOS gate metalli-

```
TABLE II-1.   METAL CORROSION

Al, Au, Ni, Cr
Kovar, Lead Frame
Ag, Sn/Pb, Lead Frame Plate
```

zation.) Susceptibility is a function of size and spacing of chip elements. As spacing goes down, the electric field goes up, accelerating the corrosion processes. Plastic packages are obviously more susceptible because of the relatively easy ingress of moisture. On the other

hand, devices in hermetic packages are not immune due to sealed-in moisture, leaks, etc.

There are three types of corrosion cells of interest including electrolytic, galvanic and

concentration. The two most commonly observed types, electrolytic and galvanic are illustrat-

ed in Fig. II-1. In galvanic corrosion the two electrodes are two dissimilar metals connected in

an electrolyte and the driving force is the difference in the electrochemical potentials. In the

electrolytic cell the electrodes are similar and the driving force is an applied potential between

them. One example of a galvanic corrosion is that where aluminum bond pads are corroded

away around the gold ball in an Al-Au wire bond system. Another example concerns the

atmospheric corrosion of Au-plated Kovar leads. The principal for this is illustrated in Fig.

II-2.[2] In the presence of moisture, corrosion of the Kovar begins at pinholes in the Au plating.

Examples of electrolytic corrosion, to be discussed later, include anodic corrosion of aluminum

in the presence of chloride, and cathodic corrosion in the presence of p-glass.

1. ELECTROLYTIC CORROSION
   APPLIED BIAS CAUSES CORROSION

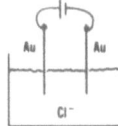

2. GALVANIC CORROSION
   DISSIMILAR METALS CAUSE CORROSION

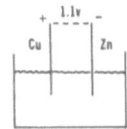

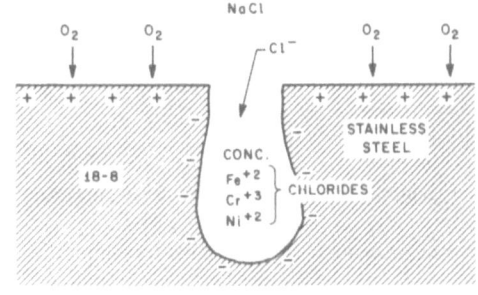

Fig. II-2 - Chloride ion pitting.

Fig. II-1 - Commonly observed corrosion cells.

Corrosion reactions, like all chemical reactions, are considered from two aspects, namely,

thermodynamics and kinetics. Thermodynamic considerations yield relative corrosion tenden-

cies. However, these tendencies may have little relation to the rate at which corrosion

proceeds, i.e., the reaction kinetics.

Thermodynamic considerations involve the Nernst equation

$$E = E^\circ - \frac{RT}{2.3} nF \log \frac{A_C A_D}{A_A A_B}$$

for the reaction $A + B \rightarrow C + D$.

E is the cell EMF, E° is the standard oxidation po-

tential, A's are activities, R is the gas constant

(8.314 joule/mol°K), T the absolute temperature, n

is the number of chemical equivalents, and F is the

Faraday constant (96,500 coulombs/equivalent).

The quantity (RT/2.3F) equals 0.059 volt at 25°C.

For electrolytic corrosion, if the potential on the

metal line is greater than E, then the reaction will

$$Au + 4Cl^- \rightarrow [AuCl_4]^- + 3e$$

$$E = 1.0 - \frac{.059}{3} \; LOG \; \frac{[AuCl4]^-}{[Cl-]4}$$

FOR SAY, $[Cl^-]$, 1 MOLAR: $AuCl_4^-$, 10^{-1} MOLAR

$$E = 1.0 - 0.3 \approx 0.7$$

I^2L DEVICE WORKING AT 650mV MAX. VOLTAGE IS SAFE.

Fig. II-3 - Low voltage operation example.

occur; if it is less than E then the reaction is thermodynamically impossible. An illustration of

such a condition is given in Fig. II-3 for an I^2L device metallized with gold and using a supply

voltage of 0.65V. A lifetime in excess of 1000 hours was observed when the device was

operated in salt water.[3]

Kinetics are dominant in most corrosion processes. The important factors are passivity

and environment. Aluminum for example, is a very active metal. On thermodynamic grounds

it might be expected to corrode readily. However the formation in air of a thin, stable oxide

effectively passivates it. If kept clean and free of ionic contamination, aluminum films will

survive for long periods. The important environmental factors are humidity, electrode

microstructure, substrate surface, etc. (Low device operating temperature presents a problem

with plastic packages.) Another factor is polarization, i.e., the change in electrode potentials

from their equlibrium value as a result of current flow in the circuit. However, this is

generally not important as the corrosion   currents are low.

Aluminum Metallization Corrosion

Aluminum is susceptible to corrosion attack by a variety of materials and environments. Because of the specific nature of aluminum corrosion and the variable environmental and contamination factors few generalizations can be made. However, two frequent causes of corrosion are chlorine and phosphorus.

In the presence of ionic contamination the passive aluminum oxide skin may be breached, allowing rapid corrosion to occur due to the reaction between the exposed aluminum and moisture in the atmosphere. Chloride contamination is partic-

$$Al(OH)_3 + Cl^- \rightarrow Al(OH)_2\,Cl + OH^-$$

$$Al + 4Cl^- \rightarrow Al(Cl)_4^- + 3e^-$$

$$2AlCl_4^- + 6H_2O \rightarrow 2Al(OH)_3\,6H^+ + 8Cl^-$$

Fig. II-4 - Chloride corrosion of aluminum.

ularly effective. The total reaction is shown in Fig. II-4.[4] With electrical bias the chloride ions migrate to the anode causing corrosion. Fig. II-5 is an example of voltage specific corrosion. There has been some concern about the possible corrosion effects of the addition of Cu to Al. However Al/Cu devices have been shown to have excellent reliability under severe test conditions.[5] The Cu is not incorporated in the Al oxide, and, in the absence of chloride, the cor-

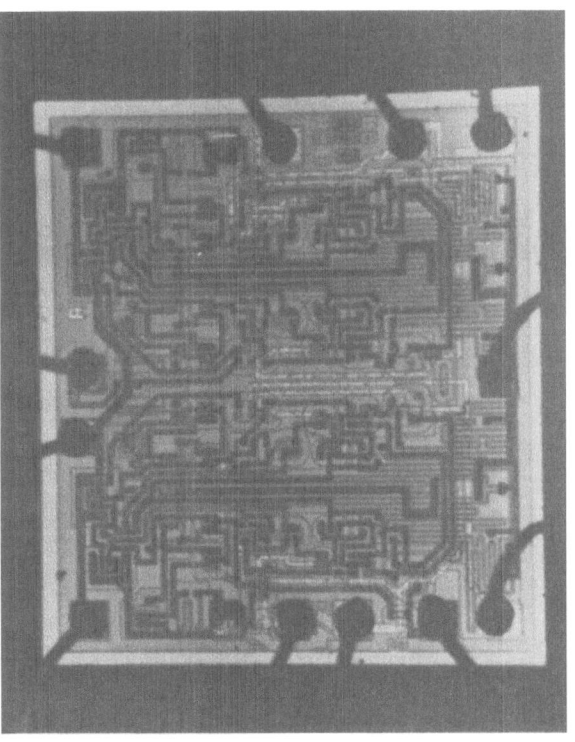

Fig. II-5 - Example of voltage specific corrosion of aluminum.

rosion properties of Al/Cu are similar to those of pure aluminum.

Phosphorus (as contained in phosphosilicate passivation glasses) is an inherent component in integrated circuits - serving as an Na^+ ion getter. Aluminum corrosion will occur if the phosphorus concentration is above some threshold. Cathodic corrosion occurs in the presence of phosphorus. Koelmans[6] postulated that OH⁻ ions, generated at the cathode, attack the aluminum. A scheme of reactions is suggested in Fig. II-6. The corrosion is usually grain boundary attack. Figure II-7 illustrates the effect. Figure II-8[5] shows the effects of phosphorus additions to dielectric layers.

REACTION	NATURE OF PROCESS
ANODIC REACTIONS	
$H_2O \rightarrow 2H^+ + \tfrac{1}{2}O_2 + 2e^-$	OXYGEN EVOLUTION
$2Al + 3H_2O \rightarrow Al_2O_3 + 6H^+ + 6e^-$	ALUMINUM PASSIVATION
CATHODIC REACTIONS	
$H_2O + e^- \rightarrow \tfrac{1}{2}H_2 + (OH)^-$	HYDROGEN EVOLUTION
$Al + 3(OH)^- \rightarrow CORROSION\ PRODUCT + H_2$	ALUMINUM CORROSION

Fig. II-6 - Electrode reactions in phosphorus-related corrosion of aluminum metallization.

Gold Metallization Corrosion

Electrolytic corrosion is observed in most gold metallization systems to some degree (Ti/Pd/Au, Ni/Cr/Au, Ti/W/Au, Ti/Pt/Au). Gold corrosion produces electrical shorts caused by the migration of gold between two biased conductors resulting in a metal

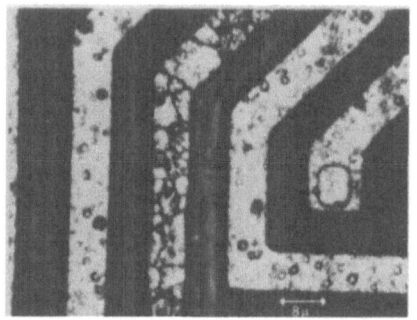

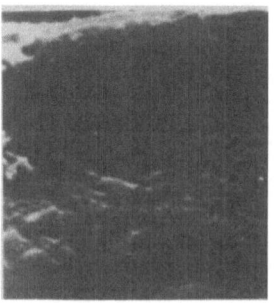

Fig. II-7 - Example of phosphorus-related corrosion of aluminum metallization.

bridge between them. This is commonly referred to as dendritic growth. Dissolution (corrosion) of the gold occurs at the anode, followed by migration of the soluble corrosion product to the cathode where it is reduced and electrodeposited in the familiar dendritic pattern. The current density increases rapidly near the tip of the dendrite encouraging preferential growth of the spike. The most likely reaction is

$$Au + 4Cl \rightleftharpoons AuCl_4^- + 3e^-, \quad E_o = 1.02$$

Figure II-9 is an example of electrolytic gold corrosion. As with aluminum corrosion, the acclerating factors are temperature, humidity, and voltage.

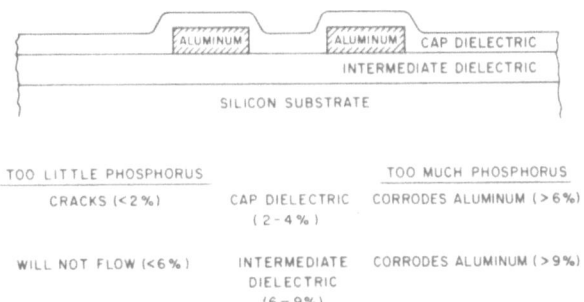

Fig. II-8 - Effects of phosphorus.

III. ELECTROMIGRATION

Electromigration is the transport of metal ions throuuh a conductor resulting from the passage of direct current. It is caused by a modification of the normally random diffusion process to a directional one by the

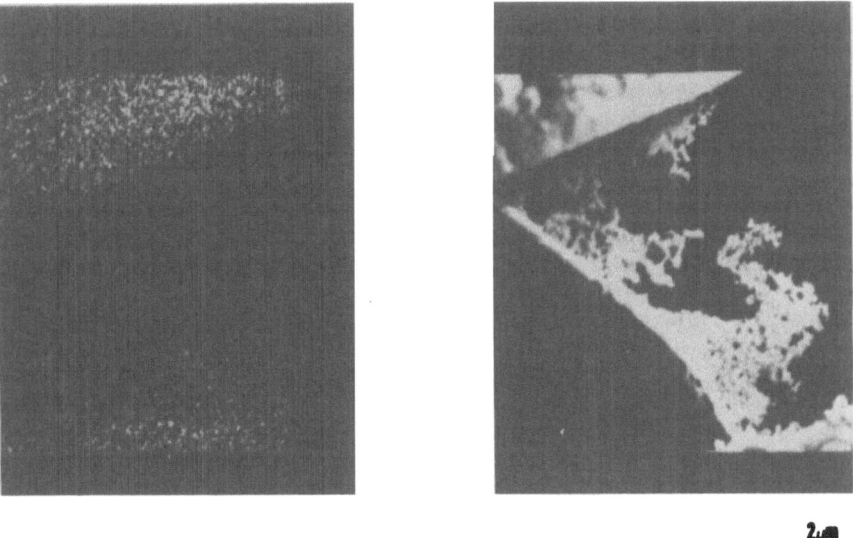

Au X-ray   SEM

Fig. II-9 - Example of electrolytic corrosion of gold.

charge carrier "electron wind." This directional effect caus-

es ions to migrate or diffuse downstream in terms of elec-

tron wind direction, and vacancies to move upstream.

The physics of the interaction is not well understood,

but it appears to be due to non-uniformities in the ion flux.

The causes of these non-uniformities have been attributed

to temperature gradients, structural gradients and compos-

itional gradients.   Structural details are quite important.

Metal films consist of an agglomeration of single crystal

grains where the crystal structure of each grain is oriented

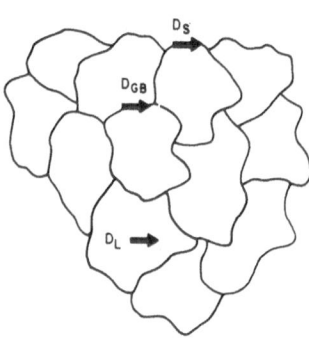

ALUMINUM CRYSTALLITIES
D_S SURFACE DIFFUSION
D_{GB} GRAIN BOUNDRY DIFFUSION
D_L LATTICE DIFFUSION

Fig. III-1 - Lattice, grain
boundary and surface diffusion.

in a different direction (Fig. III-1). Grain boundaries provide easy paths for self-diffusion and

electromigration as compared to diffusion and electromigration through the lattice.   The

surface of the metal also provides an easy path for the diffusion. Hence grain boundary and

surface diffusion are the more significant factors. Electric current parallel to a temperature

gradient leads to accumulation or depletion of metal, depending on the sign of the gradient.

Depletion causes voids which eventually may grow to form an open circuit. Accumulations

result in the formation of hillocks or whiskers and short circuits.

The basic factors affecting electromigration failure rates are summarized in Table III-1.

The dependence of time-to-failure on current density is indicated as being in the range of J^2 to

J^4, but at densities above 10^6 A cm^{-2} where joule heating starts to become significant even

higher powers can result.

TABLE III-1.   FACTORS AFFECTING ELECTROMIGRATION RATES

Current Density:	Life Time $\propto \dfrac{1}{J^N}$	N = 2-4
Temperature:	Life Time $\propto \dfrac{1}{\exp{-E_A/kt}}$	E_A = 1/2-1.2eV
Crystal Structure:	D_s and $D_{gb} \gg D_l$	

Much higher current density would appear to be usable under pulsed d.c. at low duty

factors. For long pulses (10^{-3} to 10^{+3} Hz) electromigration dominates by vacancy diffusion

during the "off time." For short pulses ($>10^{+3}$ Hz), failure now involves pulse heating leading

to "thermal-fatigue. Current densities up to 10^7 A cm^{-2} and above can be used for low duty

cycles. Data obtained by English *et al.*[7] on the effects of pulsed d.c. on AU film conductors

are shown in Fig. III-2.

Grain size and its homogeneity is an important

factor affecting electromigration phenomena. Large

grain size appears to help by limiting the number of

diffusion paths - a single crystal film of aluminum is

virtually indestructible.[8] Black[9] showed the effect on

time to failure of some aluminum alloys as a function

of current density, temperature, and cross sectional

area (Fig. III-3). Small grained crystals have a high

density of grain boundary paths and exhibit a

relatively short life at lower temperatures, and

an activation energy of 0.48 eV. Large

grained crystals provide electromigration paths

mainly at the surface and exhibit moderate life-

times at the lower temperature, and an activa-

tion energy 0.8 eV. Large grained conductors

with CVD glass exhibit the greatest life - ap-

parently due to reduced grain boundary and

surface diffusion. The activation energy is 1.2

eV and approaches that of bulk aluminum at

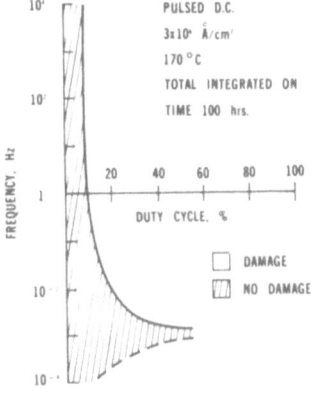

Fig. III-2 - Electromigration effects of pulsed DC of Au film
conductors.

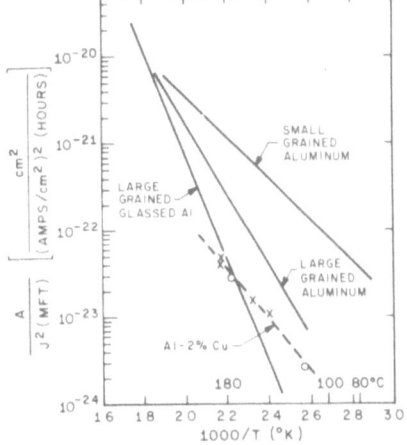

Fig. III-3 - Median time to failure of
aluminum and al Al-2% Cu alloy as a
function of current density j, temperature T, and cross-sectional area A.

1.48 eV. It is noted, however, that the role of a confining glass is controversial because

electromigration can produce cracks in overlying passivation glass layers, particularly for thick narrow stripes. From Fig. III-3 it appears that an Al-2% Cu alloy is useful in inhibiting

electromigration at high temperatures.

In 1977, Howard *et al.*[10] reported that greatly increased life was obtained in Al

"sandwich" stripes with a thin middle layer of Cr, Ti, or other metal that forms a low diffusiv-

ity intermetallic phase with Al upon heating. Recently, Jaspal and Dolal[11] reported an

improvement in electromigration behavior using an underlying $Cr-Cr_2O_3$ material with Al-Cu

metallization. They reported an improvement of electromigration lifetime of 10X, an activa-

tion energy of 0.7 eV and a three-fold increase in the current carrying capability of the Al-Cu.

This improvement was related to the structural changes and diffusions that occur between the

underlying layer and the Al-Cu metallization.

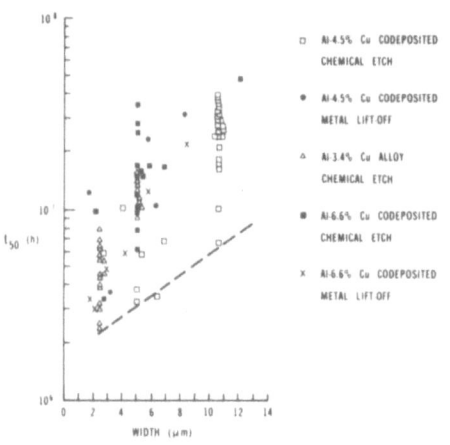

Fig. III-4 - Width dependence of t_{50} for many lots of Al-Cu samples.

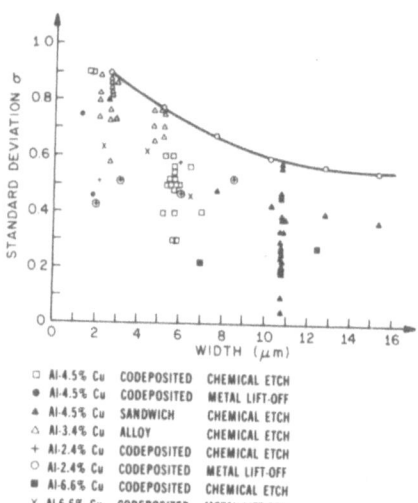

Fig. III-5 - Width dependence of σ for many lots of Al-Cu samples.

Line width is another structural factor in electromigration-induced failures. Studies by

Agarwala *et al.*[12] have shown that the lifetime of Al conductors decreases when strip width is

decreased. Scoggin *et al.*[13] did some work to determine the dependence of the median lifetime

and of σ on Al-Cu, Al-Cu-Si, and Cr-Ag-Cr stripes of various widths. Figure III-4 is a plot of

median life vs. stripe widths of their data on different lots of Al-Cu samples, and Fig. III-5 is

a plot of σ vs. line width. These plots show a decrease of median life with decreasing stripe

width, and a general increase of σ with increasing width. The dashed line in Fig. III-4 is the worst case lower bound for the width dependence of median life, and the solid curve in Fig. III-5 is the worst case upper bound for the width dependence of σ. They postulated the line width dependence as being due to a grain size/line width relationship (Fig. III-6). Figure III-7 shows the effect of median life and σ width dependence on failure rate.

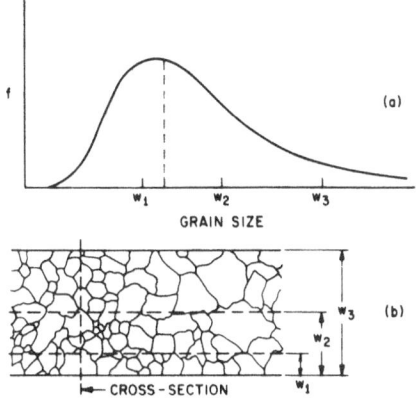

Fig. III-6 - Effects of grain size in metal stripes.

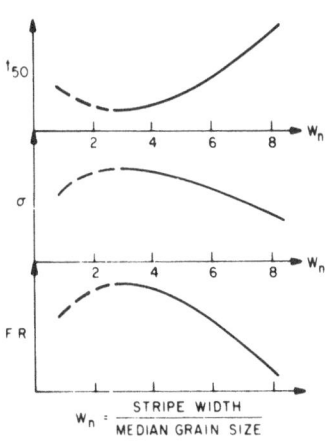

Fig. III-7 - Effect of t_{50} and σ width dependences on F.R.

The lifetime might be expected to decrease with increasing length of the line because of an increasing probability of a fatal structure defect in longer lines. The combination of long lines (>1 cm) with narrow widths (<3 μm) would appear to present a formidable obstacle to the use of fine-line Al for VLSI devices. However, Vaidya *et al.*[14] determined the electromigration lifetimes for a combination of long lines (up to

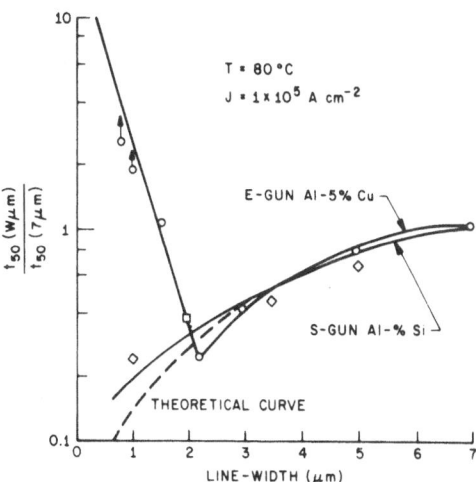

Fig. III-8 - Effect of linewidth on electromigration lifetimes normalized to that for 7μm material.

3 cm) and narrow linewidths (down to 1 μm) of evaporated and sputter-source deposited Al-Cu-Si films. The lifetimes of the sputtered films were significantly smaller than those for

e-beam evaporated films. The latter displayed an unusually large improvement in the lifetime for finer linewidths (Fig. III-8).

Integrated circuit designs that avoid high current (CCDs, I^2L) minimize the risk of electromigration failure. Small, high operating temperature devices, such as microwave transistors and GaAs FETs, are the most susceptible. In general, it can be expected that, as device geometries shrink, problems with electromigration will increase.

IV. INTERDIFFUSION

One failure mechanism due to interdiffusion is that resulting from the formation of ntermetallic phases in the commonly used goldaluminum wire bond system. (Figure IV-1 is an example of intermetallic formation.) Five phases can possibly form from this interdiffusion: $AuAl_2$, $AuAl$, Au_2Al, Au_5Al_2, and Au_4Al. If there is a sufficient supply of gold and aluminum, and the bond sees temperatures above $250°C$ for a long enough time, all five phases will be present. Generally, the majority of the intermetallic formation will be Au_5Al_2 regardless of time or temperature. If, however, there is a shortage of gold or aluminum the bond will contain only aluminum-rich or gold-rich intermetallics after a reasonable time at temperature. In themselves, all of these phases are stronger than aluminum. However, problems such as weak or open bonds sometimes result. These problems have been related to

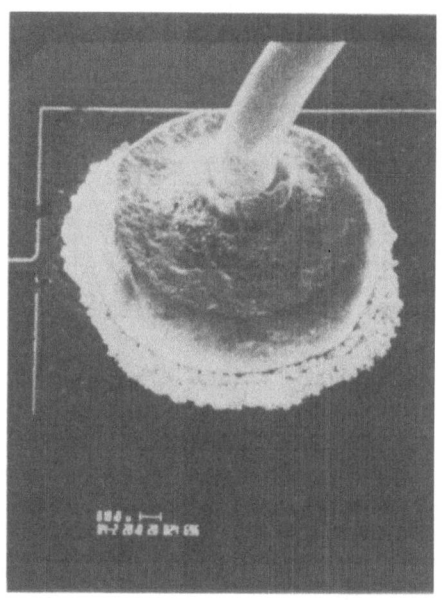

Fig. IV-1 - **Example of intermetallic formation.**

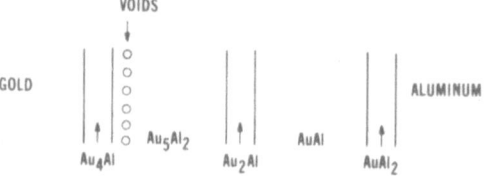

Fig. IV-2 - Effect of Kirkendall voids.

the formation of Kirkendall voids in intermetallic phases (Fig. IV-2), which can form as a result of the differing diffusion rates in the phases. An excess vacancy flux is generated on the side from which the faster species diffuses and hence voids can nucleate. Nucleation is fastest in the predominant phase, Au_5Al_2 (although it has also been shown to occur in the $AuAl_2$ phase).

Philofsky[15] measured the time for voiding to become continuous in Au_5Al_2 as a function of temperature (Fig. IV-3). This served as the basis for design limits for temperature excursions during and subsequent to bonding.

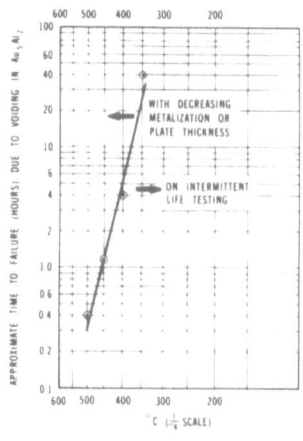

Fig. IV-3 - Time for failure caused by continuous voiding in Au_5Al_2 vs. temperature.

It was also shown that these times were substantially reduced by as much as an order of magnitude by thermal cycling. The solution is to keep time/temperature excursions relatively low.

Aluminum/Silicon Interdiffusion

Interdiffusion at aluminum-silicon contacts is another possible source of device failure. The diffusivity of silicon in thin film aluminum is much higher than in bulk samples.[16] Also, the activation energy is appreciably lower, presumbaly due to a much stronger grain boundary contribution. When an aluminum-silicon metallization is annealed at elevated temperature, interface reactions can occur. As deposited, the silicon is in a supersaturated solution which below 500°C is not thermodynamically stable. Silicon precipiation is diffusion controlled. Long anneals or slow cooling

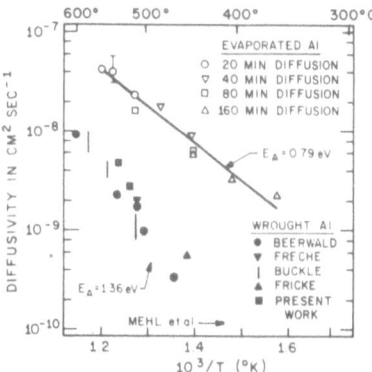

Fig. IV-4 - The diffusivity of Si in solid Al. Conventional Al specimens, e.g., drawn wires and rolled sheets, are indicated by filled symbols. The evaporated Al of the present integrated-circuit specimens are indicated by open symbols (16).

give large precipitation. Fast cooling gives small precipitates and super saturation (see Fig. IV-4).

Thermal interdiffusion and solubility can result in emitter shorts due to aluminum penetration. This is a particular risk for very shallow junction diffusions. Figure IV-5 shows alloy penetration pits after sintering for 1 hour at 400°C. Another effect is the degradation of Schottky barrier contacts (Fig. IV-6). Here the Al doped silicon is p-type and this thin (100Å) epitaxial layer raises the barrier height, but subsequent heat treatment tends to precipitate the Al atoms onto electrically inactive sites so the barrier drops. Figure IV-7 is an example of epitaxial mesas formed by this process.

Electromigration at aluminum-silicon contacts is an additional interdiffusion effect. This is a major problem now in power devices but may become more prevalent as shallow emitters and small contacts come into use. The net effect is that hillocks form at the negative contact, and voids occur at the silicon-aluminum interface at the positive

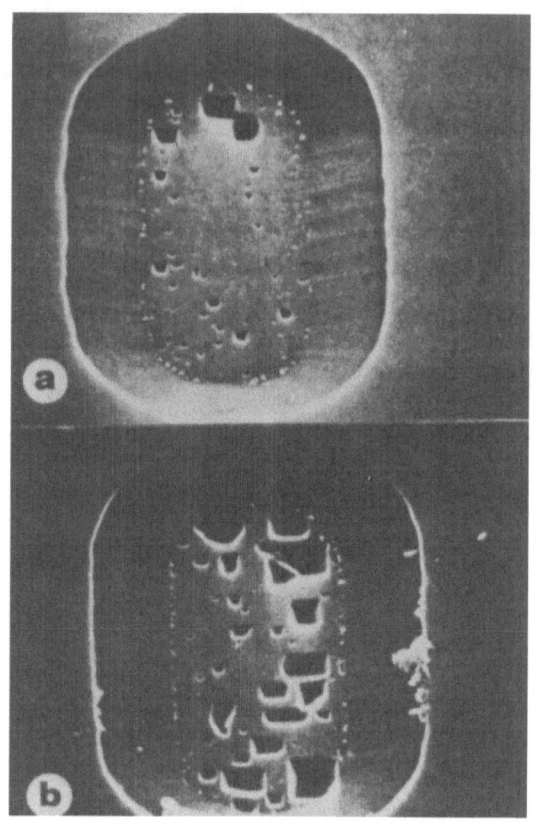

Fig. IV-5 - Alloy penetration pits formed by aluminum/silicon interdiffusion.

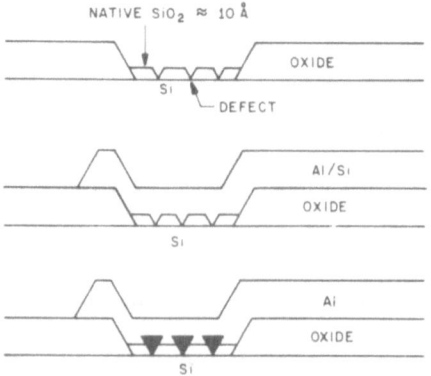

Fig. IV-6 - Epitaxial precipitation of silicon in SBD.

contact. Figure IV-8 is an example of these electromigration effects. Prokop and Joseph[17] have investigated this effect using the configuration shown in Fig. IV-9. They monitored mean time to failure (open circuit) at constant current density as a function of contact geometry. They found a dependence on contact width (Fig. IV-10) where current crowding causes open circuits to develop at the leading edge of the contact and sweep across the interface causing an open circuit. Results for a constant current density (unspecified but probably in the 1×10^3 A cm^{-2} range) are shown in Fig. IV-11. Dependence on contact

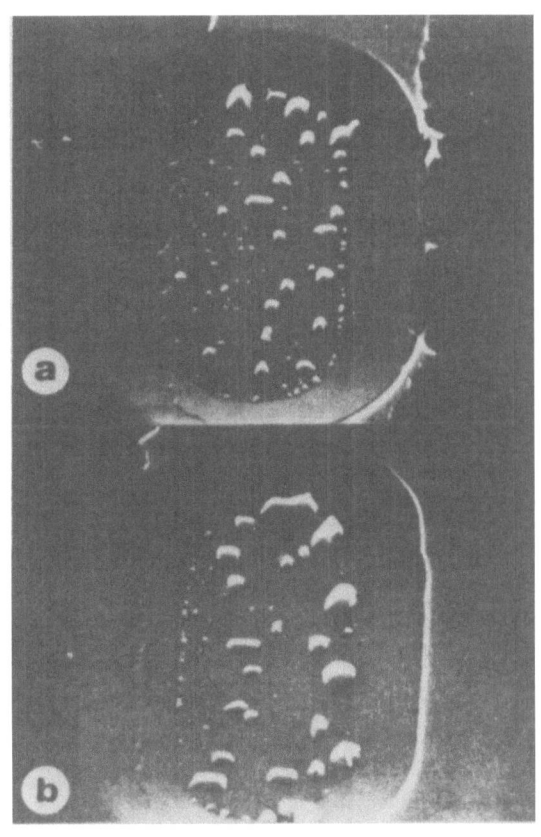

Fig. IV-7 - Epitaxial mesas formed by aluminum/silicon interdiffusion.

length, again at a constant current density, is shown in Fig. IV-12. The dependence was thought to be related to contact/grain size ratio. (Small contacts have few aluminum grain boundaries intersecting the silicon interface.) This does not apply to the width dependence because

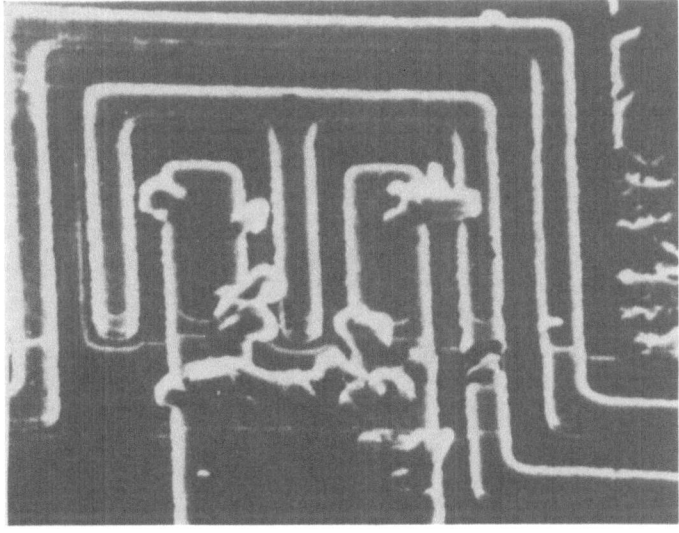

Fig. IV-8 - Example of electromigration effects at contact windows.

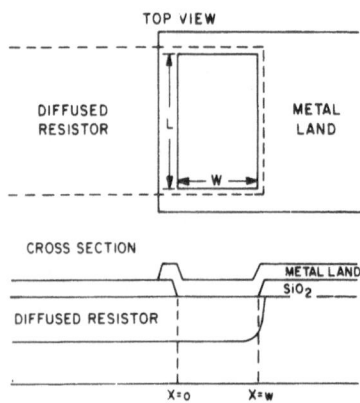

Fig. IV-9 - Drawing of a typical metal-semiconductor contact.

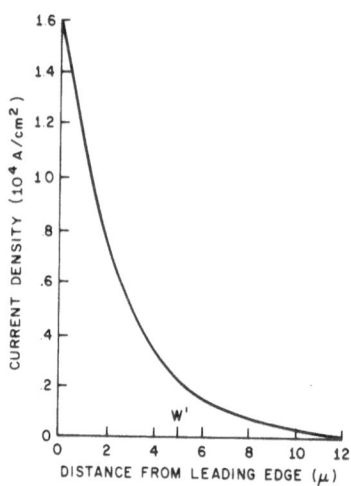

Fig. IV-10 - Current density at contacts calculated from TL model.

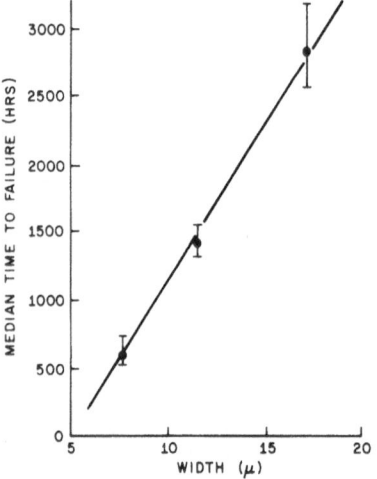

Fig. IV-11 - Dependence of medial time to failure on width of contact for a constant J_0.

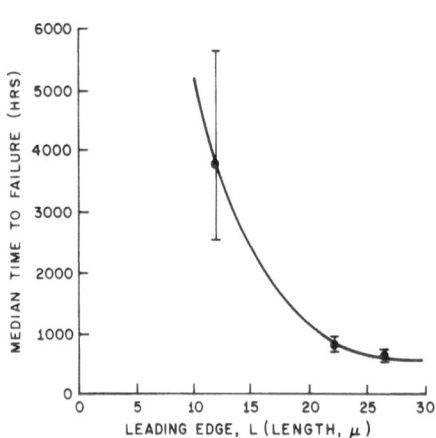

Fig. IV-12 - Dependence of median time to failure on contact leading edge for a constant $J_0 = I/L$.

current crowding breaks up the contact into a series of "unit-width" contacts. Prokop and Joseph provided the empirical equation

$$t_{50} = A J^{-1} W \exp \left(\frac{\alpha}{L} + \frac{E_A}{kT} \right)$$

where A and α are constants depending on process conditions. E_A is probably around 0.85 eV for silicon diffusion in aluminum.

V. SURFACE INVERSION

One failure mechanism which often limits the reliability of semiconductor devices is surface inversion. Figure V-1 illustrates how this can occur. Here, the presence of positive charge in (or on) the oxide inverts the p-type base material due to induced negative charge, making it an n-type material. Similarly the presence of negative charges can invert the n-type collector material to p-type. These inversions result in increases in leakage current (shunt) paths, decreases in breakdown voltages, and, hence failure of the device.

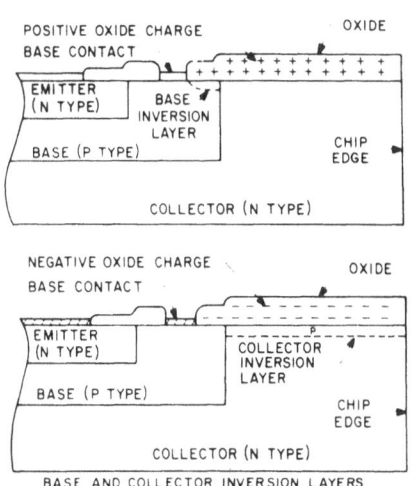

Fig. V-1 - Illustration of surface inversion due to ionic contamination.

Sodium is probably the worst culprit because: (1) It is mobile in SiO_2 and will diffuse to the SiO_2/Si interface; (2) It is a prevalent element; (3) It readily dissociates into a positive ion; (4) Silicon is a blocking contact to sodium.

Low surface dopant concentrations are particularly susceptible - especially at high temperature and humidity, which increases the surface mobility of ions. Surface inversion can be reduced by using silicon nitride or p-glass passivation. Other solutions, which are not without their penalties, include: hermetic encapsulation at low relative humidity (expensive), thick field oxides (step coverage problems), chan-stops (increased chip size), uniform increase in surface doping (decreases breakdown voltage), and the use of metal field plates over implanted resistors (less area for metal routing).

The activation energy for surface inversion is generally accepted as approximately 1.0 eV. The rate of reaction is given by the Arrhenius equation

$$K = A e^{-E_A/RT}$$

where

K = rate of reaction

A = pre-exponential factor

E_A = activation energy

R = Boltzmann Constant, 8.6×10^{-5} eV/°/K if E_a is in eV

T = temperature in degrees absolute

If a reaction rate is known at one temperature and also the activation energy, it is possible to calculate the rate at a second temperature

$$\log \frac{K_2}{K_1} = \frac{E_A}{2.303R} \frac{T_2 - T_1}{T_1 T_2}$$

For example, for surface inversion, if the median time to failure at 300°C is 60 hours, it will be 3.15×10^6 hours at 100°C.

VI. MOS FAILURE MECHANISMS

The basic causes of MOS failures are those due to oxide states, dielectric breakdown, and hot electron injection. Each of these mechanisms is discussed separately. Another mechanism, soft errors in RAMs due to alpha particles is discussed in another section.

Sodium Ion Drift

Sodium, introduced during processing into the
SiO$_2$ gate insulator of MOS devices, is easily ionized
and can drift readily through the oxide under the influence of an electric field causing considerable
changes in parameters such as threshold voltage. Figure VI-1 illustrates simplistically the effect. The clos-

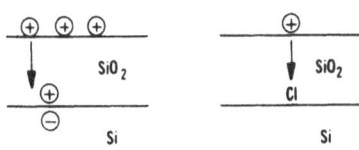

Fig. VI-1 - Sodium ion drift through gate oxide.

er the ions come to the SiO$_2$/Si interface, under the influence of the gate voltage, the larger
the effect is on the silicon. The ions retain their charge even when they approach very close
to the interface.

Kriegler[18] provides the following expression for the change in the threshold voltage

$$\Delta V_t = \frac{Q_o}{C_o}$$

where Q_o and C_o are the total mobile charge and oxide capacitance per unit oxide area
respectively. This equation provides a qualitative illustration of the sensitivity of MOS
structures to sodium ion drift. In a layer 1000Å thick, the migration of 10^{11} ions/cm^2 could
result in a threshold voltage shift of 0.5V. In a
smaller (10^{-5} cm^2) gate area configuration,
only 10^6 ions would be required to produce
this shift. The forward drift activation energy
has been measured as 1.1 eV. The recovery
may be faster with an activation energy of
about 0.75.

Phosphosilicate glass is widely used to
immobilize the ions. The use of an O$_2$/HCl
oxidizing ambient has been reported.[18] This

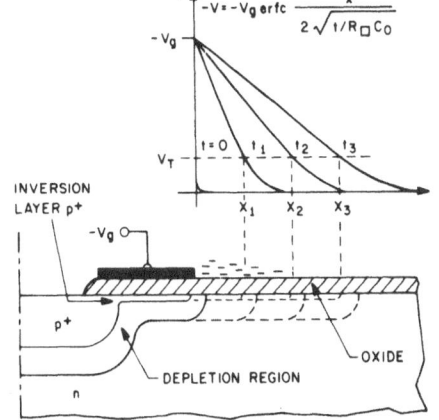

Fig. VI-2 - Field-induced junction by mobile surface ions.

acts to reduce sodium problems by "unblocking" the silicon surface to sodium.

In addition to the oxide bulk ion drift described above, there is the possibility of ion movement in lateral electric fields.[19] The drift occurs on the outer surface of the insulator, or along insulator interfaces. The insulator surface conduction is illustrated in Fig. VI-2. Lateral charge spreading can result in silicon surface inversion and channel formation. The activation energy is estimated at 1 eV. The effect is strongly influenced by protective layers.

Other Oxide Charge States

Two types of electrically active sites in the Si/SiO_2 interface region cause performance and reliability problems in MOS devices.[20] These are surface state oxide charges (Q_{ss}) and interface states (N_{ss}).

Surface state oxide charge is intrinsic, and is generally considered to be located very near the interface. Its density is independent of band bending or applied bias. It is immobile and its charge state cannot be changed by varying gate bias. Its polarity is always positive. Its magnitude depends on silicon surface orientation. It is practically independent of impurity type and concentration in the silicon, and independent of SiO_2 thickness for a specific preparation condition. The value of Q_{ss} is related to oxidation conditions. Control is a function of oxidation conditions such as temperature, wet or dry ambients, and silicon substrate orientation. Q_{ss} density is two or three times larger on a (111) surface than on a (100) surface,[20] and is an inverse function of oxidation temperature. Oxidation in dry oxygen is preferred for minimum fixed charge.

Interface states are related to the termination of the silicon lattice at the Si/SiO_2 interface, and have been described as centers which have energy levels in the silicon band gap and can exchange charge with the silicon.[20] The traps are most likely caused by defects at the Si/SiO_2 interface of distorted or broken silicon-oxygen bonds. Interface traps can be either charged or discharged by varying gate voltages. Interface trap density has the same dependence on orientation as fixed charge. They most likely occur at the Si/SiO_2 interace. Interface trap density can be reduced by annealing in hydrogen.

Q_{ss} and N_{ss} can change with time during devuce life - resulting in stability problems. One effect which can change the densities is slow trapping. Temperature-bias aging accelerates the effect, producing an irreversible change in the densities. Evidence exists, however,[20] that the slow trapping effect is negligible in many practical cases. Another way in which Q_{ss} and N_{ss} densities can change at low temperatures is transfer of hydrogen in or out of the oxide, e.g., from moisture in the device package. Protective layers, such as silicon nitride, can provide protection. However, it has been reported[21] that n-channel MOS transistors, encapsulated with plasma deposited silicon nitride, exhibited drift in saturation mode threshold when operated with high drain voltage. The effect was attributed to hydrogen, present in the nitride, diffusing to the Si/SiO_2 interface, creating interface states near the drain diffusion.

Dielectric Breakdown

LSI MOS devices are characterized by large areas of thin gate oxides. Dielectric breakdown is now the major failure mechanism for MOS devices, accounting for about 50% of all failures. It is characterized by a low thermal activation energy but is sharply dependent on voltage as shown in Fig. VI-3.[22]

As oxides are further thinned down with scaling, dielectric breakdown will become even more of a problem. Figure VI-4 is an illustration of the fine line impact on dielectric breakdown.

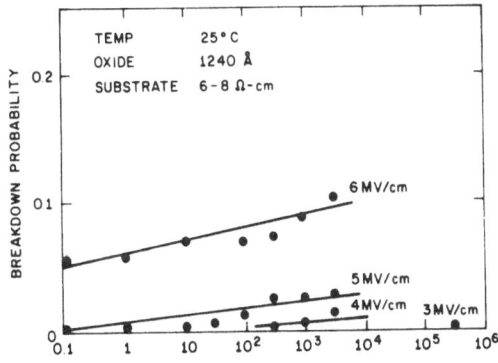

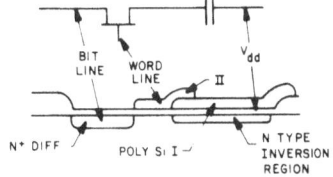

MEMORY ERROR RATE IS PROPORTIONAL TO SIGNAL/NOISE RATIO

S/N PROPORTIONAL TO CAPACITANCE
                      BIT LINE RESISTANCE

$$\text{CAPACITANCE} = \frac{\text{AREA} \times \text{DIELECTRIC CONSTANT}}{\text{THICKNESS}}$$

$$\frac{\text{BIT LINE}}{\text{RESISTANCE}} = \frac{\text{LENGTH} \times \text{THICKNESS} \times \text{RESISTIVITY}}{\text{WIDTH}}$$

Fig. VI-3 - Oxide breakdown probability as a function of field and time.

Fig. VI-4 - Fine line impact on dielectric breakdown.

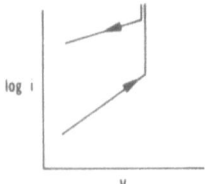

BREAKDOWN SCENARIO

1. INITIATION PROCESS
2. INSTABILITY AND CURRENT RUNAWAY
3. VOLTAGE COLLAPSE AND DISCHARGE OF ELECTRO STATIC ENERGY IN CAPACITOR.
4. A) ESTABLISHMENT OF LOW VOLTAGE STATE   (FATAL)

            OR .

   B) QUENCHING OF BREAKDOWN CHANNEL } (HEALING)
      VOLTAGE RECOVERY VIA RC TIME CONSTANT }

Fig. VI-5 - Electrical definition of oxide breakdown.

Fig. VI-6 - Dielectric breakdown scenario.

There are three general breakdown categories: intrinsic, defect related, and field fatigue.

We can define breakdown from various points of view: (1) Materials - a localized change in dielectric structure or composition; (2) Electrical - an irreversible change in electrical characteristics of the dielectric (see Fig. VI-5); (3) Device - the device does not function, e.g., leakage current becomes so large that storage time becomes less than refresh time.

A breakdown scenario can be summarized in terms of Fig. VI-6.

Figure VI-7[5] shows a histogram of breakdown voltages in $Al/SiO_2/Si$ capacitors with a relatively thin dielectric. The low-voltage failures are presumably due to dielectric defects. The high-voltage failures represent a skewed Gaussian distribution and are believed due to intrinsic breakdown. For intrinsic breakdown of SiO_2, time dependence appears to be present, but very steep,[23] as shown in Fig. VI-8. The intrinsic breakdown mechanism is belived to be due to impact ionization:

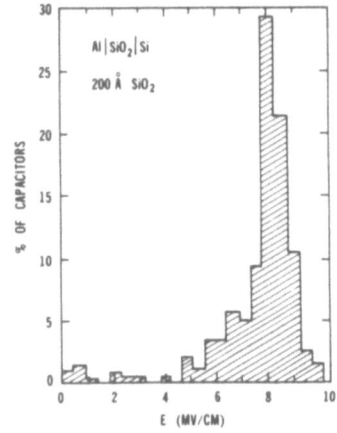

Fig. VI-7 - Distribution of breakdown voltages.

A. Electrons tunnel into the SiO_2 conduction band from the cathode by Fowler-Nordheim tunneling.

B. These electrons are accelerated by the SiO_2 field and some gain sufficient energy to cause electron-hole pairs by impact ionization.

C. These secondary electrons are rapidly swept out of the oxide, but the holes, being relatively immobile, remain and increase the field and electron tunneling in the oxide.

D. The process is balanced by drift and recombination of holes. Defect related oxide breakdown is the major cause of poor reliability and low yield in MOS memories. There are several causes; including particulates (and pinholes), aluminum migration, and sodium contamination. The yield and reliability is a function of the electric field and time. Screening procedures use cell stress voltage to remove "weak" devices - generally limited to V + 50%.

Interdiffusion can result in lower dielectric breakdown voltages. Aluminum polysilicon interdiffusion kinetics have been measured by Nakemura *et al.*[24] Sixty minutes at 400°C is usually sufficient to cause significant interdiffusion.

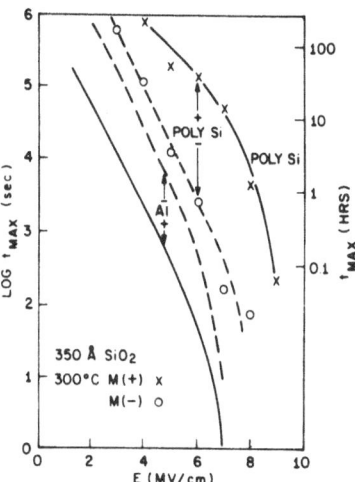

Fig. VI-8 - Maximum time to failure as a function field at 300°C for poly-Si and Al electrodes.

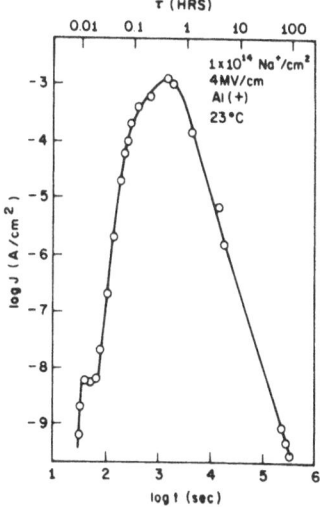

Fig. VI-9 - Current through a Na^+ contaminated, 200Å SiO_2 films as a function of time.

Interdiffusion can be reduced by doping polysilicon with phosphorus and by annealing in Ar at 1000-1100°C for 30 minutes.

Sodium related breakdown is important in time dependent breakdown of positive biased gates. Mobile sodium ions drifting through the SiO_2 film increase the field strength at the Si-SiO_2 interface. When the field becomes as large as the breakdown strength of the oxide, breakdown occurs. The time required was found to approximate Peek's law, i.e., the electric field, E, α $t^{-1/4}$. The leakage current is also time dependent as shown in Fig. VI-9.[25] In the rising region the field due to Na^+ ions increases Fowler-Nordheim emission to silicon. In the falling region, Na^+ has drifted to the Si-SiO_2 interface.

Temperature stress is a factor in dielectric breakdown. Intrinsic breakdown strength increases with temperature; conversely field fatigue time decreases with temperature. For intrinsic breakdown, the activation energy is approximately 1.4 eV for polysilicon. For defect breakdown, the activation energy for polysilicon is in the order of 0.3 to 0.4 eV, with aluminum somewhat higher.

More important is voltage stress. Crook[26] provides the following equation for the electric field acceleration factor

$$A_V \; + \; \frac{V^1 - V}{6.2 \times 10^{-4} \; t_{ox}}$$

where

V^1 = stress voltage

V = operating voltage

t_{ox} = thickness of oxide in Å

Dielectric breakdown is dependent on processing variables and ion implantation species as shown in Table VI-1 and VI-2.

Hot Electron Injection

Electrons traversing the gate channel of an n-channel device can be scattered into the gate oxide if they have
sufficient energy (see Fig. VI-10).[27] In
saturation there is a large field in the
drain depletion region.   Channel current electrons are accelerated by the
field.   Impact ionizing and phonon
scattering direct some of these electrons to the Si-SiO$_2$ interface. If their
energy is sufficient to surmount the interface barrier they will be injected
into the oxide, and the trapped charge
will alter the threshold voltage.

Hot electron injection is increased
with reduced channel lengths and oxide
thicknesses, shallower diffusions, and
increased diffusion doping. All of the
these will occur with device scaling.

The effect is dependent on the effective   electron   energy   or
"temperature," and the oxide trapping
efficiency. The injection current

TABLE VI-1. PROCESSING DEPENDENCE OF DIELECTRIC BREAKDOWN		
Parameter Increased	Maximum Break-Down Field	Effect on Defect Density
Oxidation temperature	X	Increases
Ambient (wet-dry)	Constant	X
N$_2$ anneal (extended)	X	Increases
HCl	X	Decreases
PSG (~100Å)	Increases	Decreases

(X - Not well established or inconclusive)

DIELECTRIC BREAKDOWN

TABLE VI-2. EFFECT OF ION IMPLANTATION ON DIELECTRIC BREAKDOWN				
Species	Range (Å)	Fluence (cm^{-2})	Unan- nealed	Annealed
He	100	10^{16}	Increases	Increases
P	~1000	10^{15}	Increases	Constant
N$_2$	~1000	10^{15}	Increases	Constant
N$_2$	250	10^{15}	X	Decreases
B	280	10^{15}	Increases	Decreases
P	220-800	$>10^{15}$	Increases	Increases (Al$^-$)
As	220-800	$>10^{15}$	Increases	Increases
Ar,N	<50	$>10^{13}$	Decreases	X

(X - Not well established or inconclusive)

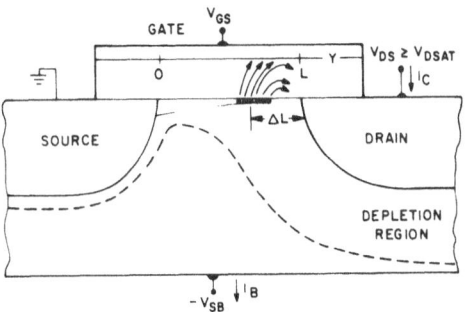

Fig. VI-10 - MOS transistor in saturation.

$$I_E \; \alpha \; \exp \; (-E_B/kT_E)$$

where the interface barrier potential is 3.1 eV, and T_E is the "effective electron temperature" which is related to the electric field, E_{SI}, in the channel.

$$k \, T_E \; = \; \frac{q(E_{SI}\lambda)^2}{2E_R}$$

where

λ = electron mean free path

E_R = energy loss per phonon collision

E_{SI} increases with increasing channel doping. It increases with decreasing channel length, insulator thickness, source/drain junction depth, and temperature. E_{SI} is greatest near the drain and this is where most injection occurs. The trapping efficiency, T_E, is thought to be related to density, location of trap sites, the electric field of conductive metallization over the oxide. Generally, SiO_2 is better than SiO_2 - Si_3N_4 for the gate dielectric.

VII. ALPHA PARTICLES

Naturally ocurring alpha particles, emitted during radioactive decay of uranium and thorium, contained in minute proportions within packaging material, cause soft errors in integrated circuit memories.[28] The term "soft error" refers to a random single-bit failure not related to a physically defective device. Replacement of the device before the failure would not necessarily have prevented the failure, nor does the occurrence of this type of failure increase the probability of another occurrence with that bit. To give an idea of the seriousness of the problem, a typical soft-error rate may be of the order of one bit per 1000 hours in a system containing 1000 16K memory devices. This rate corresponds to a device rate of one failure per million hours of 1000 FITs. Furthermore, this high rate does not improve with time, since the radioactive uranium and thorium compounds have effectively infinite half lives. These compounds have been found by quantitative analysis methods to exist in most materials used in semiconductor packaging. Table VII-1 lists the results of quantitative analysis of

TABLE VII-1. RESULTS OF
MATERIALS ANALYSIS (28)

Material	ppmU [1]	ppmTh [2]	α/cm^2-hour [3]	% Zr [4]
Alumina - A	2.5	0.6	0.6	1
Alumina - B	-	-	0.3	-
Alumina - C	-	-	0.5	-
Glass - A	12	6	29	17.5
Glass - B	2.5	3	5.2	3
Glass - C	17	6	45	25
Glass - D	12	6	18	6
Glass - E	-	-	32	20
Epoxy	-	-	1.7	-
Silicone	-	-	1.3	-
Au-Plated Lids	-	-	.04-1.0 [5]	-

(1),(2) Mass spectrographic analysis.
(3)   Alpha scintillation counting.
(4)   Emission spectrographic analysis.
(5)   Activity varied widely.

various materials.[28] Even the lowest level of .04 α/cm^2-hour would be enough to produce a high failure rate for some dynamic memories.

In dynamic RAM's, data is stored as the presence or absence of minority carrier charge in potential wells in p-type silicon under positively charged polysilicon gate electrodes. Refresh is needed to maintain the charge. A ZERO is represented by stored charge in the range, depending on the design, of several hundred thousand electrons to a few millions. A one is represented by an empty potential well. A factor, called the critical charge, Q_{crit}, differentiates between a One and a ZERO in terms of the number of electrons. Electron-hole pairs are generated by the alpha particles and are collected by depletion layers such that electrons end up in the potential wells. If collection of electrons is such as to exceed Q_{crit} then a ONE can be changed into a ZERO. Collection does not change a ZERO into a ONE. Figure VII-1 shows the stages

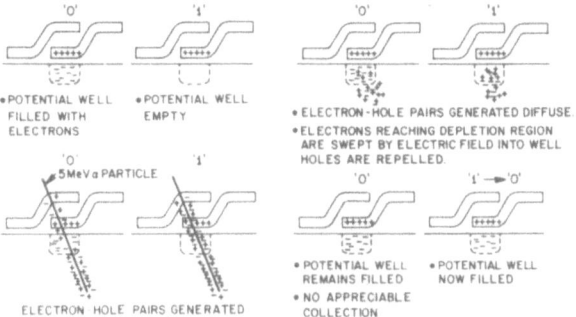

Fig. VII-1 - Effects of alpha particles.

624

Bailey

of the production of a soft error.[28] The error rate is directly proportional to the alpha flux as shown by experiments conducted by May and Woods. Figure VII-2[28] shows the error rate to be directly proportional to alpha flux for several decodes of source intensity. In addition to the flux dependence, the actual soft error rate of a specific device will depend on such factors as target area, collection efficiency, critical charge, geometry of the cell, and others. Obviously these factors are all interrelated as a function of the device design and technology and the package type and composition.

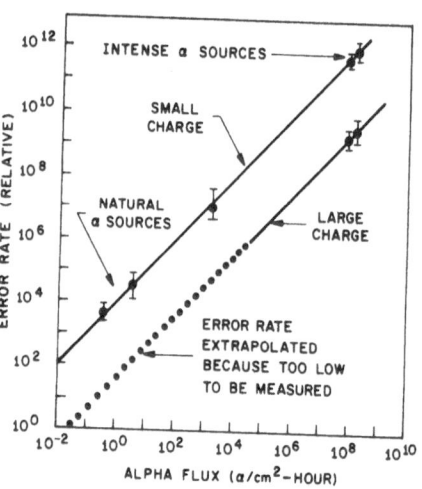

Fig. VII-2 - Error rate vs. alpha flux for two test devices of different critical charges.

Critical charge has been called "the most important factor" in determining device sensitivity to soft errors.[28] As an example of Q_{crit} in a commercial dynamic RAM, the value for a 16K design is about one million electrons. May and Woods measured soft error rates versus Q_{crit} for a number of dynamic RAMs and CCDs from several manufacturers. (See Fig. VII-3).[28] In other experiments, they showed that the critical

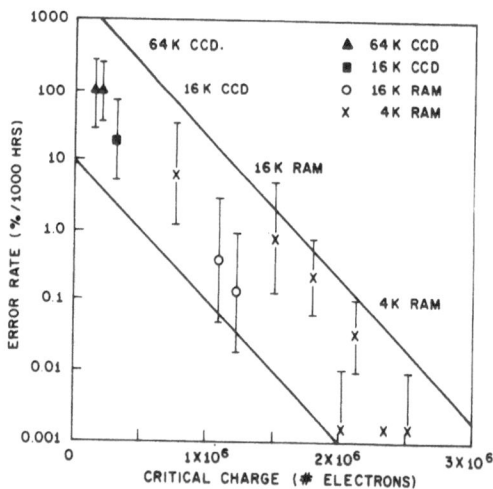

Fig. VII-3 - Error rate vs. critical charge for devices from several manufacturers.

charge stored can be varied by changing cell geometry and hence storage capacitance, or by changing voltages. (See Figs. VII-4a and VII-4b[28])

Other error rate factors and their interrelating effects are summarized in Fig. VII-5. At least one, the angle at which alpha particles strike the device surface bears special attention.

It affects attenuation, penetration depth, collection efficiency, and effective target area.
Figure VI-6[28] shows the results of one angle of incidence experiment.

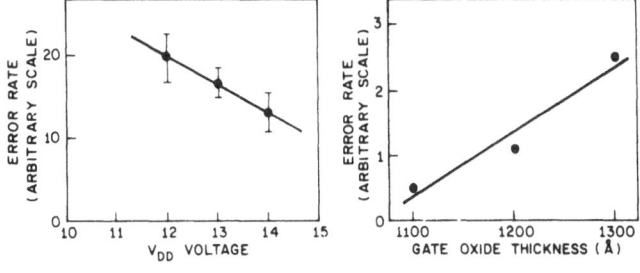

Fig. VII-4 - Effects of critical charge on soft error rate. a)
critical charge proportional to operational voltage; b) critical
charge proportional to oxide capacitance[28].

ALPHA FLUX

ALPHA ENERGY

COLLECTION EFFICIENCY

CELL GEOMETRY

PASSIVATION LAYERS

ANGLE OF INCIDENCE

Fig. VII-5 -
Other alpha
particle error
rate factors.

Estimation of error rate for a given device is difficult to do. One needs to know the chip layout and details of its dopant profiles, in order to estimate the critical charge and the collection efficiencies. Then one can estimate the incident alpha flux if one knows the package part materials and their radio-active properties. An estimation of error rate can then be made.

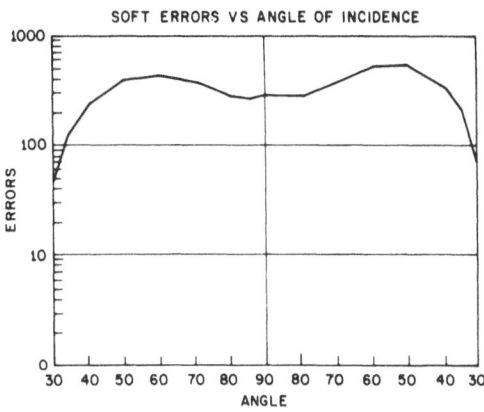

Fig. VII-6 - Effect of angle of incidence on soft error rate.

The practical prospects for characterization of a specific device for alpha particle soft errors are somewhat uncertain. Accelerated testing will undoubtedly be necessary. For example, consider a system with 1000 RAMs, and assume that we can tolerate one system error per six weeks. (Many cannot tolerate such a high rate.) One error per 1000 RAMs per six weeks means an error rate/RAM of 0.1% per thousand hours. To verify this maximum allowable device error rate requires millions of hours of testing - an unlikely prospect. The present alternative, accelerated testing, involved the replacement of the normal alumina device lids with a "hot" glass lid and a foil of thorium. The claim for this technique[28] is that the time acceleration is increased by a factor of 1×10^6.

There are some prospects for reduction of the alpha flux at the device surface by "cleaning-up" the materials. Sealing glasses, which have high fluxes, are easiest because the most radioactive constituent of the glass can be eliminated. The selection of the proper alumina from those presently available might achieve a four-fold reduction in error rate. It is estimated that a factor of ten reduction in alpha flux may eventually be achieved. For hermetic packages, an eventual lower limit of the order of .001 to .01 α/cm^{-2} - hour may be possible. This is about the limit of measurability.

Various design and technology changes have been proposed to reduce alpha particle sensitivity. These are summarized in Fig. VII-7. McPartland *et al.* reported on the design of a 64K dynamic RAM, having a soft error of 100 FITs achieved by designing alpha immunity into the circuit.[29] One possible method of improvement involves shielding of the chip surface with a material which will stop the alpha particles. A promising technique was recently reported by White *et al.*[30]

FOLDED BIT SENSE LINES

DECREASED FRONT-END GATE OXIDES

TIMING CHANGES

CHARGE TRANSFER

NOVEL CELL DESIGNS

DECREASED DIFFUSION LENGTH

Fig. VII-7 - Alpha particles - some proposed design and technology changes.

They found that the use of a thick (2 mil minimum) RTV rubber coating over the chip surface of 4K static RAM's, packaged in a side-brazed multilayer ceramic package, effectively shielded the chip from alpha radiation. A soft error rate of less than 0.2%/1000 hours was observed with this coating, and the coating did not introduce any other undesirable effects.

As device trends to higher density continue, smaller values of Q_{crit} and changes in collection efficiency will be encountered. On the other hand there will certainly be lower values of alpha flux in next generation packages.

Nevertheless, the error rate is still likely to be high and error detection and correction (EDAC) will be necessary. This is accomplished by reserving some bits for reconstruction of the data following an error. In a dynamic RAM reconstruction is done when refreshing. Table VII-1[28] shows that EDAC becomes more efficient as word length increases.

TABLE VII-2. ERROR DETECTION AND CORRECTION - BITS NEEDED (28)		
Bits/Word	EDAC Bits	Overhead
8	5	63%
16	6	38%
32	7	22%
64	8	13%

VIII. ELECTROSTATIC DISCHARGE

While not an intrinsic failure mechanism, electrostatic discharge (ESD) has become recognized as a significant cause of device failures at all stages of device and equipment production, assembly, test installation and field use. It will become even more significant with newer devices incorporating smaller geometries and design rules.

Many models have been suggested for ESD damage, mostly variations of the human body model. Unger[31] has proposed two additional models (besides a human body model): 1) Charged device model where the device itself acting as one plate of a capacitor can store charge. When the device is grounded, the discharge pulse can create damage. 2) An electrostatic field is always associated with charged objected. Under certain circumstances, a device inserted in the field can have a potential induced across an oxide that creates breakdown.

Figure VIII-1 is an equivalent circuit of Unger's human body model. A body capacitance of 100 to 250 pf and a body resistance of 1000 ohms is generally used. It is possible to develop body potentials (thousands of volts) that far exceed the damage potential of devices. With a charge of only 2000 volts the human body stores approximately 0.4 microjoules of energy. With the human body equivalent circuit shown in Fig. VIII-1, this energy is released

628 Bailey

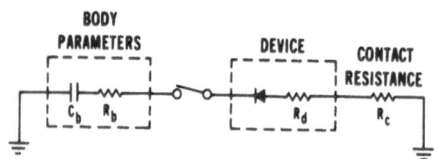

Fig. VIII-1 - Human body model

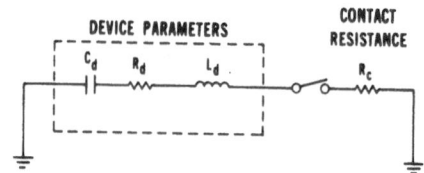

Fig. VIII-2 - Charged device model (bipolar representation).

with time constants in tenths of microseconds, providing average powers up to several kilowatts sufficient to melt metal and crater the chip surface.[31]

Unger postulated two equivalent circuits for the charged device model - one generally for bipolar devices (Fig. VIII-2) and one applicable to MOS (Fig. VIII-3). With bipolar device capacitance in the 1 to 20 pf range, a device can store up to 100 microjoules of energy. With low resistance to ground and a

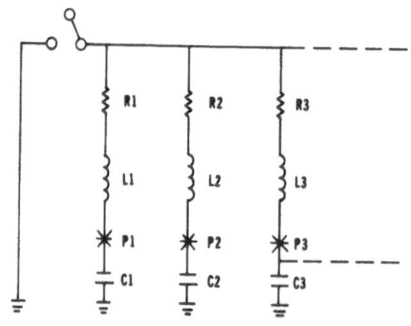

$R_i \equiv$ RESISTANCE IN i^{th} DISCHARGE PATH
$L_i \equiv$ INDUCTANCE IN i^{th} DISCHARGE PATH
$C_i \equiv$ CAPACITANCE OF ith DISCHARGE PATH
P_i AND P_j ARE THE POINTS BETWEEN WHICH
THE POTENTIAL IS CALCULATED AS A FUNCTION OF TIME.

Fig. VIII-3 - Charged device model (multipath discharge representation).

lead frame inductance of 10 nanohenries, this energy results in nanosecond discharge pulses with average power ranging from several hundred to several thousands of watts - enough to destroy metal. Figure VIII-4 shows damage to an I^2L device that was charged to 1000 volts and then discharged to ground.

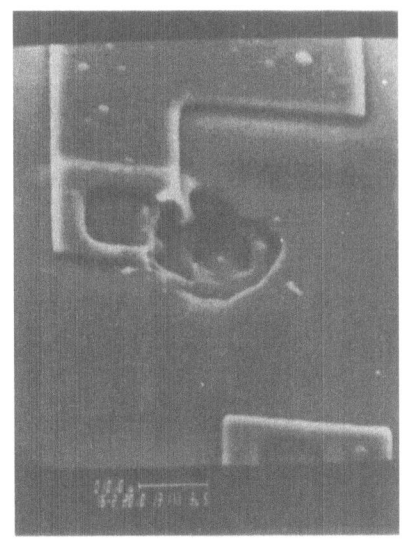

400X **1600X**

Fig. VIII-4 - Charged device breakdown of bipolar device.

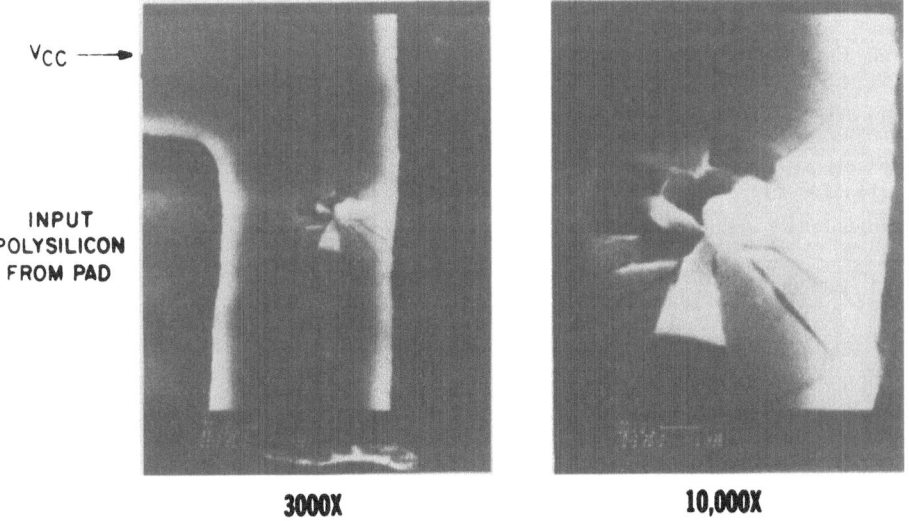

3000X **10,000X**

Fig. VIII-5 - Charged device breakdown of MOS oxide.

630   *Bailey*

In the MOS equivalent circuit, multiple paths are represented with lumped elements. Grounding of one pin results in potential differences between paths sufficient to cause oxide breakdown to occur. Figure VIII-5 is an example.

ESD is a cause of failure in all technologies and devices. The difference is in the degree of susceptibility. MOS devices are the most susceptible due to breakdown of the thin gate oxide. In addition to oxide breakdown, or as a consequence of it, junction leakage may be increased in MOS devices by ESD stress, causing a circuit to fail specifications even though it is still functionally operational.[31] DeChiaro[32] demonstrated an ESD induced electro-thermomigration of aluminum metallization through contacts into the substrate of NMOS LSI devices. Bipolar devices, especially linear devices, fail due to P-N junction damage. Minear and Dodson[33] showed this effect on linear bipolar integrated circuits. ESD induced silicon melting was caused by localized overheating in the depletion region of a PN junction (see Fig. VIII-6). After the current pulse has dissipated the molten silicon solidifies into a polycrystalline structure resulting in redistribution of dopant and recombination centers. This causes increased base current and reduced common emitter current gain of NPN transistor. In

general, ESD failures are event dependent but an ESD may also initiate a device weakness or fault that degrades and causes failure with continued use - an event/time dependent failure.

Several investigators have determined the damage threshold for different device technologies. The results of one investigation is shown in Table VIII-1.

Many models have been proposed for device ESD protection techniques.

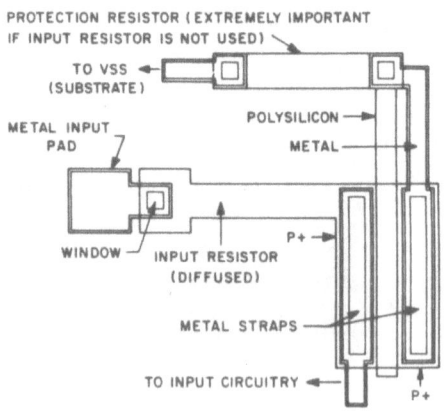

Fig. VIII-7 - Gated punch through protection network.

For MOS, such techniques are input resistors, input diodes, resistor/diode combinations, field plate diode, thick-oxide MOS transistor, punch-through diode, spark-gap device, and the gated punch-through device. Of these the most effective appears to be the gate punch-through device (see Fig. VIII-7). A technique for bipolar linear devices is the "phantom-emitter" transistor structure illustrated in Fig. VIII-8.[33] This adds a second emitter diffusion shorted to the base contact. Also,

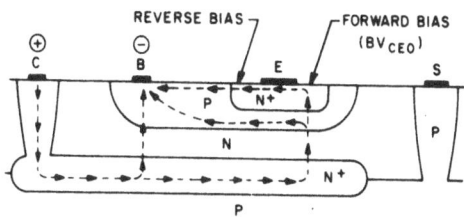

Fig. VIII-6 - ESD current paths in NPN transistor from C^+B^- pulse.

TABLE VIII-1 (31)

SUSCEPTIBILITY RANGES OF VARIOUS DEVICES
EXPOSED TO ELECTROSTATIC DISCHARGE

Device Type	Range of ESD Susceptibility (Volts)
MOSFET	100 - 200
JFET	140 - 10,000
CMOS	250 - 2,000
SCHOTTKY Diodes, TTL	300 - 2,500
Bipolar Transistors	380 - 7,000
ECL (For Hybrid use, PC Board Level)	500 -
SCR	680 - 1,000

there is a deliberate separation of base contact from the normal emitter diffusion. In normal operation the diffusion does nothing. But under ESD it completes a lower breakdown voltage path between the buried collector and the base contact. This transient clamping action, with the help of the extrinsic base resistance limits the ESD pulse current through the vulnerable emitter sidewall, effectively protecting the regular emitter from damage.

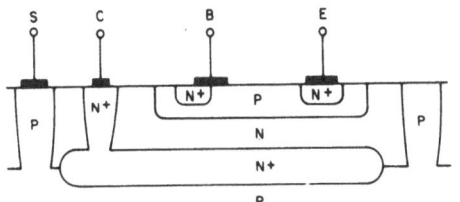

Fig. VIII-8 - "Phantom-emitter" transistor structure.

In the final analysis the best protection is prevention (grounding, static free environments, etc.).

IX. INFANT MORTALITY

The term "infant mortality" as used here refers to failures which occur early in time. We have defined infant mortality to include both dead-on-arrivals (DOA's) and device-operating-failures (DOF's). DOA's refer to devices that fail when initially tested after shipment or incorporation in the next assembly level. For example, incoming inspection failures are DOAs, first circuit pack test are also DOA failures, etc. Device operating-failures refer to failures which occur after some period of operating time. The total of DOAs and DOFs constitutes the early failure or infant mortality population. Estimates of infant mortality DOF percentages are included in Table IX-1.

TABLE IX-1. ESTIMATES OF INFANT MORTALITY DOF PERCENTAGES	
Device Type	Typical Range of Cumulative % DOF (1 Year)
Transistors	.02 - .04
Diodes	.01 - .02
Bipolar TTL IC's (*)	.05 - .07
Linear IC's	.10 - .18
Digital CMOS IC's (*)	.05 - .07
MOS Memory IC's	.07 - .40

*SSI/MSI

The failures are generally due to manufacturing defects of a gross nature. They tend to be such things as oxide pinholes, photoresist or etching defects resulting in near opens or shorts, conductive debris on the chip, scratches, weak bonds, partially cracked chips or ceramics, etc. Some appear to result from surface inversion problems, probably due to gross contamination or gross passivation defects. Some obviously result from electrical overstress. Table IX-2 shows a breakdown of infant mortality failure mechanisms observed by the author in returns from the field. However, infant mortality mechanisms may change in type and degree from product-to-product and from lot-to-lot within a product.

TABLE IX-2. BREAKDOWN OF INFANT-MORTALITY FAILURE MECHANISMS (PERCENT OF FAILURES ANALYZED)					
	Commercial			Western Electric	
	T^2L	CMOS	Memory	T^2L^*	Memory**
Overstress	4	60	17	35	9
Oxide Defects	2	1	51	-	53
Surface Defects	18	-	24	-	-
Bonds, Beams	37	5	7	39	27
Metallization	30	34	-	4	2
Misc.	9	-	1	22	9

*Beam-Lead
**Wire-Bonded

The emphasis on mechanical failures and those largely due to voltage would suggest a limited dependence on temperature. Figure IX-1 shows a comparison between: 1) the temperature acceleration of infant mortality failures and 2) the expected activation energy of about 1 eV for ionic contamination failures of commercial bipolar devices.[34] The data points, collected

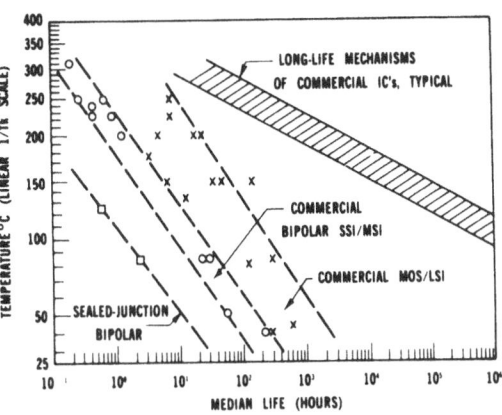

Fig. IX-1 - An Arrhenius comparison of infant-mortality data with expected long-term failures.

from many sources, suggests an activation energy in the range of 0.37 to 0.42 eV. It would appear, from these data, that an activation energy of 0.4 eV would be appropriate for infant mortality in general.

X. CONCLUSIONS

A review has been presented of the basic failure mechanisms intrinsic in today's integrated circuit technologies. The nature, cause, and effects of these mechanisms are in many cases interactive. Perspectives as to the incidence of these mechanisms with increasing scale of integration have been mentioned. Many of the mechanisms discussed are likely to be exacerbated as feature spacings diminish. Technological innovations will be needed to offset the possible increased risks.

XI. ACKNOWLEDGEMENTS

The author wishes to acknowledge the direct contributions of many colleagues at Bell Laboratories, those others who contributed to the literature, and C. M. Melliar-Smith for his helpful and critical review.

REFERENCES

1. "LSI/Microprocessor Reliability Prediction Model Development", Rome Air Development Center Report RADC-TR-79-97, March, 1979.

2. S. C. Kolesar, "Principles of Corrosion", 12th Annual Proceedings, 1974 Reliability Physics Symposium.

3. F. W. Hewlett, Jr. and R. A. Pederson, "The Reliability of Integrated Injection Logic Circuits for the Bell System", 14th Annual Proceedings, 1976 Reliability Physics Symposium.

4. W. M. Paulson and R. P. Lorrigan, "The Effect of Impurities on the Corrosion of Aluminum Metallization", 14th Annual Proceedings, 1976 Relaibility Physics Symposium.

5. A. T. English and C. M. Melliar-Smith, "Reliability and Failure Mechanisms of Electronic Materials", Ann. Rev. Mater. Sci. 8, 459 (1978).

6. H. Koelmans, "Metallization Corrosion in Silicon Devices by Mositure-Induced Electrolysis", 12th Annual Proceedings, 1974 Reliability Physics Symposium.

7. English, *et al.*, "Electromigration in Conductor Stripes Under Pulsed DC Powering", Appl. Phys. Lett. 21, 8 (1972).

8. F. M. de'Heurle and I. Ames, Appl. Phys. Lett. 16, 80 (1970).

9. J. R. Black, "Physics of Electromigration", 12th Annual Proceedings, 1974 Reliability Physics Symposium.

10. J. K. Howard, *et al.*, 1977 Electrochemical Society Fall Meeting, Atlanta, Georgia, Ext. Abstract Nr. 178:480-81.

11. J. S. Jaspal and H. M. Dabal, "A Three-Fold Increase in Current Carrying Capability of Al-Cu Metallurgy by Predepositing a Suitable Underlay Material", 19th Reliability Physics Symposium (1981).

12. B. N. Agarwala, *et al.*, J. Appl. Phys. 41, 10 3945 (9170).

13. G. A. Scoggin, "Wdith Dependence of Electromigration Life in Al-Cu, Al-Cu-Si and Ag Conductors", 13th Annual Proceedings, 1975 Reliability Physics Symposium.

14. S. Vaidya, *et al.*, "Electromigration Resistance of Fine-Line Al for VLSI Applications", 18th Annual Proceedings, 1980 Reliability Physics Symposium.

15. E. Philofsky, "Design Limits When Using Gold-Aluminum Bonds", 9th Annual Proceedings, 1971 Reliability Physics Symposium.

16. J. O. McCaldin and H. Sankur, "Diffusivity and Solubility of Si in the Al Metallization of Integrated Circuits", Appl. Phys. Lett. 19, 12 (1971).

17. G. S. Prokop and R. R. Joseph, "Electromigration Failure at Aluminum-Silicon Contacts", J. Appl. Phys. 43, 6 (1972).

18. R. J. Kriegler, "Ion Instabilities in MOS Structures", 12th Annual Proceedings, 1974 Reliability Physics Symposium.

19. W. H. Schroen, "Process Testing for Relaibility Control", 16th Annual Proceedings, 1978 Reliability Physics Symposium.

20. E. H. Nicollian, "Interface Instabilities", 12th Annual Proceedings, 1974 Reliability Physics Symposium.

21. R. C. Sun, *et al.*, "Effects of Silicon Nitride Encapsulation on MOS Device Stability", 18th Annual Proceedings, 1980 Reliability Physics Symposium.

22. C. R. Barrett and R. C. Smith, "Failure Modes and Reliability of Dynamic RAMS", 14th IEEE Computer Society International Conference, San Francisco, Feb-March, 1977.

23. C. M. Osburn and E. Bassous, "Improved Dielectric Reliability of SiO_2 Films with Polycrystalline Silicon Electrodes", J. Electrochem. Soc.; Solid State Sci. Technol. (1975).

24. K. Nakamura, *et al.*, "Interaction of Al Layers with Polycrystalline Si", J. Appl. Phys. 46, 11 (1975).

25. C. M. Osburn and S. I. Raider, "The Effect of Mobile Sodium Ions on Field Enhancement Dielectric Breakdown in SiO_2 Films on Silicon", J. Electrochem. Soc.; Solid State Sci. Technol. (1973).

26. D. L. Crook, "Method of Determining Reliability Screens for Time Dependent Dielectric Breakdown", 17th Annual Proceedings, 1979 Reliability Physics Symposium.

27. B. Euzant, "Hot Electron Efficiency in IGFET Structures", 15th Annual Proceedings, 1977 Reliability Physics Symposium.

28. T. C. May and M. H. Woods, "A New Physical Mechanism for Soft Errors in Dynamic Memories", 16th Annual Proceedings, 1978 Reliability Physics Symposium.

29. R. J. McPartland, *et al.*, "Alpha-Particle Induced Soft Errors and 64K Dynamic RAM Design Interaction", 18th Annual Proceedings, 1980 Reliability Physics Symposium.

30. M. White, *et al.*, "The Use of Silicone RTV for Alpha-Particle Protection on Silicon Integrated Circuits", 1981 Reliability Physics Symposium.

31. B. A. Unger, "Electrostatic Failures of Semiconductor Devices", 1981 Reliability Physics Symposium.

32. L. F. DeChiaro, "Electro-Thermomigration in NMOS LSI Devices", 1981 Reliability Physics Symposium.

33. R. L. Minear and G. A. Dodson, "Effects of Electrostatic Discharge on Linear Bipolar Integrated Circuits", 15th Annual Proceedings, 1977 Reliability Physics Symposium.

34. D. S. Peck, "New Concerns About Integrated Circuit Reliability", 16th Annual Proceedings, 1978 Reliability Physics Symposium.

CHAPTER VII

FUTURE TRENDS

MINIATURIZATION LIMITS FOR MOS TECHNOLOGY

Robert H. Dennard

IBM Thomas J. Watson Research Center
Yorktown Heights, NY   10598

I. INTRODUCTION

The progress in miniaturization of MOSFET devices and integrated circuits has continued steadily.   Even while development work using 1μm dimensions is building up, exploratory efforts in submicron devices has begun.   Government funding has served to stimulate this effort, a notable step in this direction being the VHSIC program which aims to establish feasibility of $1/2\mu$m MOSFET technology in the next few years.[1] In this context the subject of limits to miniaturization has more than remote academic interest; indeed, reducing dimensions further by only a factor of two will require dealing with severe technical barriers which eventually will limit the size of MOSFET devices.   None of these barriers is precisely and sharply defined, so that many design trade offs and compromises remain to be made before the ultimate dimensional limit will be known.

In this section, the approach will be to examine the properties of a MOSFET device for $1/4\mu$m channel length produced by scaling down a 1μm design.   The principles of scaling and the initial 1μm design are given in the chapter on Advanced Silicon MOS Devices and Related Problems.   By examining where scaling begins to fail, and where the underlying assumptions become invalid, one may see the problem areas which limit miniaturization.   In some cases, use of new materials and structures are identified which can help solve the problems.

II. ILLUSTRATIVE SCALED $1/4\mu$m MOSFET

The cross-section of a design for a $1/4\mu$m MOSFET is shown in Fig. 1, where the design is produced by scaling - as far as possible - from a 1μm technology.[2] An inherent assumption in this procedure is that the manufacturing tolerances used for such a device can also be scaled down, so that a channel length of 0.25μm will have a reasonable variation of say $\pm0.10\mu$m due to use of very high-resolution patterning and etching tools.   The insulator thickness, the

junction depth, and the
operating voltage have
all been reduced by a
factor of 4, while the
substrate doping has
been increased by 4
times.

In fundamental
terms the reduction of
the insulator thickness

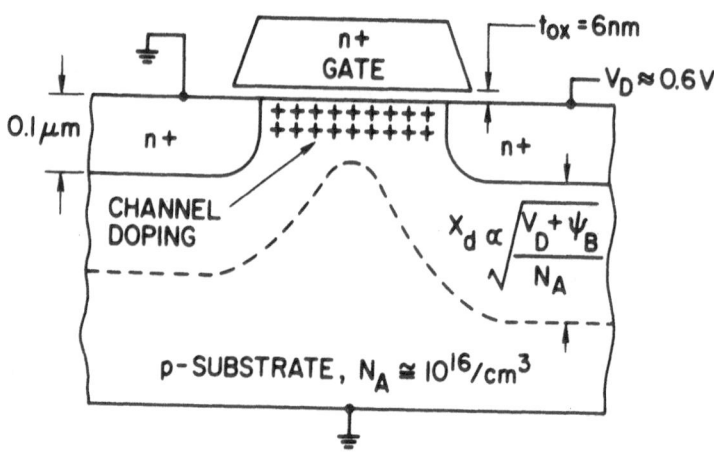

Fig. 1 - Cross-section of FET with $1/4\mu m$ channel length, scaled from $1\mu m$ technology.

to 6 nm does not represent a concern. Direct tunneling through the gate oxide is not a problem, and the applied voltage is so low that lowering of the barrier due to Fowler-Nordheim tunneling is insignificant. However, further reduction of the insulator thickness by less than a factor of two would cause severe direct tunneling. Therefore, this poses a limit to much further miniaturization if insulator thickness has to be scaled. In terms of material problems, there is no indication that thin insulators cause a basic difficulty. Good breakdown voltages have been reported in thin insulators, and reliability has been shown to be unaffected as long as electric fields are not increased.[3] It should be noted that very small devices have been built without scaling the insulator thickness.[4] However, if one is to work at low voltage to conserve power, thin oxides are necessary to get high transconductance in a reasonable area and to make the device threshold less susceptible to oxide charge variations. Thin oxides are also very important in controlling threshold shifts with applied drain voltage - the so called "short-channel effects." The only other way to control short-channel effects is by high substrate doping, which is not viable for circuits where the source is not grounded.

The junction depth of $0.1\mu m$ can be readily achieved with ion implantation, but solid solubility of the implanted arsenic dopant will limit the sheet resistance to about 80-100 ohms/square which is very difficult to use in circuit layouts. Beam annealing could possibly help with this problem by increasing the activated arsenic. A surface layer of metal silicide across the n^+ regions could also potentially lower the resistance.

The operating voltage from scaling considerations is about 0.6 volts for logic applications. As indicated in Fig. 1 by the expression for depletion depth, one of the reasons for reducing the applied voltage is to reduce the width of the depletion region around the drain junction. However, the built-in junction potential, ψ_b, is larger than the applied voltage so that the depletion region does not scale as desired, even though the back bias on the substrate has been reduced to zero. The result is not too severe for the $1/4\mu m$ device and can be compensated with a somewhat heavier substrate doping, N_A, than indicated by direct scaling. There is a price for this in terms of increased "substrate sensitivity" of devices where the source is not always grounded, such as transfer devices widely used in structured logic designs. The whole problem of operating at such low voltages and the alternative of using higher voltages will be discussed in the next section, after some of the other physical limitations are considered here.

The boron doping impurities introduced into the channel region in a thin layer under the gate insulator, as indicated in Fig. 1, give rise to several problems. As discussed in the chapter on Advanced Silicon MOS Devices and Related Problems, the threshold voltage does not scale perfectly with oxide thickness which requires an increased boron dose. The work function of the n^+ gate material disposes the device to be normally conductive (depletion mode) except for the bulk charge due to the substrate doping. The non-scaling term in the threshold equation, $\Delta W_f + \psi_s$, which can be thought of as the difference in work function between the gate material and the electron inversion layer, amounts to about -0.15 volts at room temperature. The boron implant dose required to overcome this effect and obtain an enhancement threshold in the proper proportion to the scaled power supply voltage is larger because of the thin gate insulator. The larger bulk charge increases the electric field at the semiconductor surface. The increased electric field causes a reduced mobility due to surface scattering.[5,6] The thickness of the implanted layer is also reduced by scaling, raising the peak concentration to about 2×10^{17} atoms / cm^3. This is not believed sufficient to cause significant impurity scattering. However, even at this level, there are only about 600 doping atoms within a $1/4\mu$m-long channel. The spacing of atoms in Fig. 1 is approximately drawn to scale. The problem of statistical fluctuations in this small number of doping atoms has been raised and studied.[7] A simple model considering the standard deviation of the doping atoms shows that within a silicon chip containing say 100,000 devices, at least a few are likely to have threshold shifts of 20% due to this effect. This is a bound on a more accurate solution which considers non-uniform spacing of the dopant atoms over the channel region. This problem behaves roughly inversely with channel length, and thus could represent a serious limit for even smaller devices. These problems involved with channel doping are not insurmountable since it is possible to have a gate material whose work function gives nearly the desired turn-on threshold voltage without a channel implant. The choice of materials is limited, however, and the candidates are not necessarily comparable with the thin gate insulator (without stresses or instabilities) or with all the processing requirements. Without a channel implant, the bulk charge in the background doping would still be a factor in the threshold voltage and would also vary randomly, but the effect on threshold voltage would be smaller.

III. PROBLEMS ASSOCIATED WITH LOW-VOLTAGE OPERATION

Low-voltage operation is desirable in integrated circuits in order to get the best performance at a given power level. It is necessary, however, to maintain adequate margins to allow for fluctuations in the parameters such as the threshold voltage. Tolerances in threshold are reduced by scaling down the gate insulator thickness. However, some terms may not scale, such as any variation in the gate work function relative to the channel. Also, change of the threshold with temperature is fundamental, due to the Fermi-level shift in the substrate and due to the change in the source-channel barrier height for a given conduction level. This causes a change in threshold voltage somewhat less than 1mv/°C for the $1/4\mu$m device under discussion. This produces some difficulty in designing for fluctuations in ambient temperatures, if the supply voltage is only 0.6 volts, though it can be done for a reasonably controlled environment.

Thermal noise is not a problem, and internally generated noise (due to coupling between lines) scales down with the power supply. Noise due to ionizing radiation such as alpha or cosmic particles could easily cause circuit malfunction due to the scaling down of the charge levels involved in the circuits. These malfunctions would have to be controlled by minimizing the ionization sources and by error recovery techniques (using redundancy, for example, though this can seriously increase the complexity of logic circuitry).

Another problem for low-voltage operation is in the small-signal device behavior. The turn-on characteristic of an FET becomes relatively poorer in scaled-down devices as illustrated in Fig. 2. Under the indicated measurement condition, the drain current normally varies linearly with applied gate voltage after a strong inversion layer is produced. The

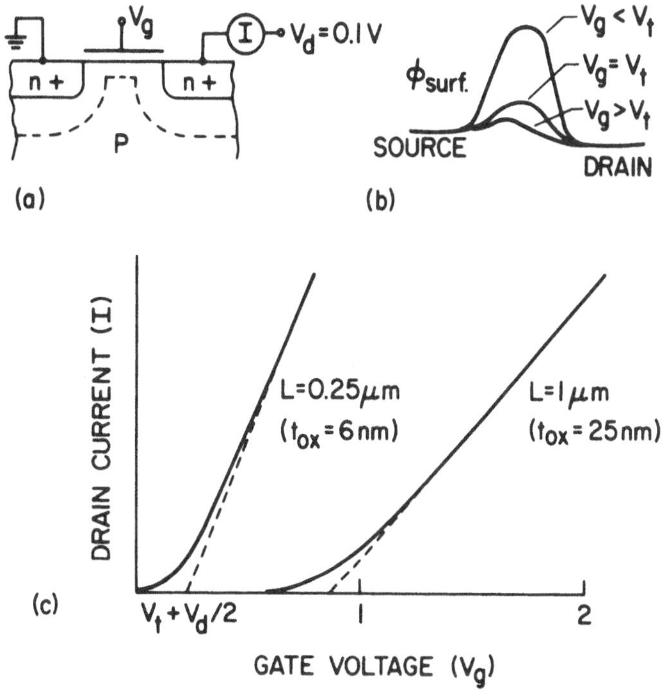

Fig. 2 - Turn-on behavior in small dimension scaled device.

projection of this linear region to zero current gives the threshold voltage V_T plus a small offset $\frac{V_D}{2}$. With the gate voltage below V_T, the current is exponentially related to the gate voltage, and this characteristic is unaffected by scaling so that it becomes more significant in scaled-down devices where V_T is smaller. This is the well-known subthreshold scaling problem.[8] With the gate voltage above $V_T + \frac{V_D}{2}$, however, there is also a problem since the drain current approaches the ideal linear function of gate voltage only at large gate voltages.

This is because the potential at the silicon surface, as indicated in Fig. 2(b), is not constant as commonly assumed for these conditions, but actually changes by a small amount to allow the increased flow of channel current across the potential barrier at the edge of the source.[9] Thus, there is a loss of transconductance at low gate voltaie which affects small devices more severely since they operate solely in this region. As a consequence, the $1/4\mu m$ device can lose something like 25% of its transconductance, depending on the exact operating condition. Combined with the reduction in mobility described previously due to higher electric field, the $1/4\mu m$ device can be severely hampered.

A part of the problem of degraded transconductance can be related to the thickness of the inversion layer. An illustration of the distribution of carriers in the inversion layer is shown in Fig. 3, which gives both the classical field solution as well as the solution which accounts for quantum effects due to the constraints at the surface.[10] The average distance of the carriers below the surface, allowing for the dielectric constant, is equivalent to having a 1 nm thicker gate insulator. Fig. 3 is only for illustration, since the inversion layer in the $1/4\mu m$ device would be closer to

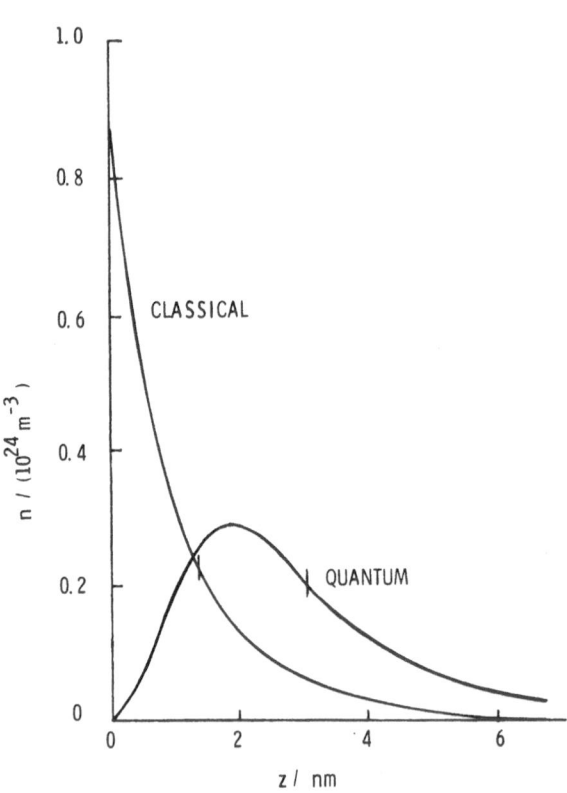

Fig. 3 - Inversion layer density versus distance from surface. $T=150°K$, $N_A = 1.5 \times 10^{22}$ m^{-3}, total electron concentration $= 10^{16}$ m^{-2}, <100> surface; (from Stern).

the surface due to the higher doping (and thus higher electric field). However, the depth of the inversion layer does increase for low inversion layer density, which accounts for some of the reduced transconductance at low gate drive in the scaled device of Fig. 2.

More careful analysis needs to be done in order to resolve the design problems in 1/4 micron devices. For example, a change in the gate work function to reduce the electric field in the silicon will improve the mobility due to reduced surface scattering; however, the same change will increase the inversion layer depth below the surface, which is counterproductive. Lower operating temperature has been shown to improve the sharpness of the turn-on characteristic below threshold,[11] and it also clearly will improve the behavior just above threshold by minimizing the surface potential variation and reducing the inversion layer depth. For such small devices, low temperature appears to be a necessity for dynamic RAM's because of the requirement for very small device leakage with the gate near zero voltage. It would also greatly benefit devices used for general logic applications.

Raising the supply voltage is another way to help the problems with turn-on sharpness of scaled devices, as well as to allow for threshold spreads due to temperature changes, etc. Increasing the 0.6 volt supply to, say 1.0 volt for the 1/4 micron device should not pose any fundamental problems. Device breakdown problems, source-drain avalanching, and hot-electron injection could not be a problem at such low voltage levels even though the electric fields are increased. Gate-insulator reliability could be affected, but little is known for such voltage levels. Threshold lowering due to short-channel effects is known to be only linearly related to drain voltage,[12] so that the problem can be contained. Thus, higher operating voltage appears to be practical up to the point that punchthrough becomes a factor. The increased current and power would be accompanied by some increase in speed. For a modest voltage increase, tradeoffs in device size can generally be made to handle the power problem at the expense of performance.

646

Dennard

IV. CONCLUSION

With all these design considerations, it is difficult to establish an exact limit for miniaturization of MOS devices. Many problem areas have been identified in scaling existing technology. Structural and material changes were suggested to deal with source/drain resistance and with problems related to channel doping. Non-scaling of certain potential terms was shown to hinder depletion layer scaling and to cause poor turn-on behavior. Somewhat larger operating voltage can help with those problems in the $1/4\mu$m device design. It seems clear that even shorter device lengths can be handled and that workable design points can be found for some applications. Low temperature operation is advantageous with respect to some of the problems and can help extend the limits.

REFERENCES

1. D. F. Barbe, Tech. Digest of Int. Elec. Dev. Meeting, p. 20 (Dec. 1980).

2. R. H. Dennard, F. H. Gaensslen, E. J. Walker, and P. W. Cook, IEEE J. Solid-State Circuits SC-14, 247 (1979).

3. P. Solomon, J. Vac. Sci. Technol. 14, 1122 (1977).

4. W. Fichtner, E. N. Fuls, R. L. Johnson, T. T. Sheng and R. K. Watts, Tech. Digest of Elec. Dev. Meeting, p. 24 (Dec. 1980).

5. A. G. Sabnis and J. T. Clemens, Tech. Dig. of Int. Elec. Dev. Meeting, p. 18, Dec., (1979).

6. S. C. Sun and J. D. Plummer, IEEE Tran(on Elec. Dev. ED-27, 1497 (1980).

7. R. W. Keyes, IEEE J. Solid-State Circuits SC-10, 245 (1975).

8. R. H. Dennard, F. H. Gaensslen, H. N. Yu, V. L. Rideout, E. Bassous, and A. R. LeBlanc, IEEE J. Solid-State Circuits SC-9, 256 (1974).

9. Y. Hayashi, Electronics Letters 11, 618 (1975).

10. F. Stern, CRC Critical Reviews in Solid-State Science 4, 499 (1974).

11. F. H. Gaensslen, V. L. Rideout, E. J. Walker and J. J. Walker, IEEE Tran. on Elec. Dev. ED-24, 218, (1977).

12. R. R. Troutman, IEEE Tran. on Elec. Dev. ED-26, p. 461 (1979).

PHYSICS OF SUBMICRON DEVICES

Giorgio Baccarani

Istituto di Electtronica
Facolta di Ingegneria
Universita di Bologna
Bologna, Italy

1.   INTRODUCTION

During the last decade we have been witnessing a tremendous increase in the number of elementary electron devices contained within a monolitic integrated circuit. Such an increase is partially due to the feasibility of larger chips but, most importantly, is the result of a continuous effort toward reduced device dimensions, the driving force being improved performance, and lower cost per function.

In view of this rapidly evolving scene, several studies have been carried out in the last few years,[1-3] aiming at the definition of the physical limits for device dimensions and electrical performances. Such an evaluation, however, is often made on the basis of the well-known physical model including continuity, transport and Poisson's equations, which has historically proved to be adequate for the analysis of large devices.

One question is therefore in order: Is our present knowledge of important physical effects occurring in submicron devices adequate for predicting their performance? In this lecture we are examining a few physical phenomena which can be relevant when the spatial and temporal scales over which device behaviour must be considered are approaching the carrier mean-free path and the mean time between collisions, respectively, and will show how the traditional model fails in interpreting these effects.

Electron devices can be classified according to the required description of carrier transport as follows:[4-6]

i)   Large devices. These devices have an active length larger than 1 μm, and have already been manufactured even at their lower edge, in several industrial laboratories.[7-10] For modelling purposes, the classical drift-diffusion equations are used in conjunction with continuity and Poisson's equation, and the boundary conditions at ohmic contacts or Schottky-barrier junctions are usually simplified, assuming equilibrium or quasi-equilibrium for the carrier densities. Velocity saturation effects are accounted for by considering drift velocity as a function of the local field, or better, of the quasi-Fermi level gradient. Surface scattering in FET's is taken into account making use of a normal-field dependent carrier mobility, according to empirical relationships. Analytical models based upon a partitioning of the whole device in space-charge regions and quasi-neutral ones can sometimes be used except, perhaps, at the lowest dimensional size. Such models, although containing a number of fitting parameters, are currently being used in circuit simulation programs.

ii) Medium-small devices (MSD's). These devices have typical dimensions in the $0.1 \div 0.5$ μm range, and are being studied and manufactured in a few industrial laboratories. The classical transport equations, which result from an approximate, local solution of the Boltzmann equation, are not expected to provide reliable results, due to the small spatial, and fast temporal scales over which the electric field is changing. Finally, the active length and the depletion width can be quite comparable to the Debye length, thus preventing the use of the depletion approximation.

iii) Very small devices (VSD's). These devices are characterized by active lengths in the $10 \div 50$ nm range, and have not been designed and manufactured yet. Moreover, the physical principles on which their behavior is based, is not clearly identified. At these extremely small dimensions, one expects that the concepts of effective mass and band structure fail. Due to the large fields typical of these devices, the collision process cannot be considered instantaneous; rather, the energy gained from the field during collisions is comparable with that of a "free flight". Thus, even the BTE is not applicable for VSD's. An additional feature is that these devices are strongly coupled with their environment,[5] i.e., the boundary conditions and scattering at the interface appreciably alter the transport properties, which require a fully quantum-mechanical approach. This problem has been treated by Barker and Ferry[5] who have attempted

to lay a conceptual framewok for the study of VSD's, but these efforts are still in a

very preliminary stage and, as already mentioned, the physical effects to be profitably

used in VSD's are not identified yet.

This paper will survey the basic transport phenomena in semiconductors, identifying the

fundamental limitations of the drift-diffusion approach. Next, the Monte Carlo method of

solution of the BTE is analyzed, and the most important, non-conventional physical effects

occurring in MSD's, such as ballistic transport, velocity over and undershoot, complex

mobility, etc., are described.   Carrier transport in thin inversion layers is then treated and,

finally, the evolution of device modelling is discussed.

2.1 The Boltzmann Transport Equation. Physical Background and Limitations

Transport phenomena in semiconductors are usually described by the Boltzmann equation

(BTE), which can be expressed as follows:

$$\frac{df}{dt} = \frac{\partial f}{\partial t} + \vec{u}_g \cdot \vec{\Delta}_r f - \frac{q}{\hbar}\vec{E} \cdot \vec{\Delta}_k f =$$

$$\tag{1}$$

$$= \int_{k'} [S(\vec{r}, \vec{k}', k)\, f\, (\vec{r}, \vec{k}', t) - S(\vec{r}, \vec{k}, k')\, f\, (\vec{r}, \vec{k}, t)] d^3 k'$$

where $f(\vec{r}, \vec{k}, t)d^3r\, d^3k$ represents the number of particles in the elementary volume $d^3r\, d^3k$ of

the phase space at time t, $\vec{u}_g = \frac{1}{\hbar}\vec{\Delta}_r\, \varepsilon(\vec{k})$ represents the group velocity and $\vec{E} = \vec{E}(\vec{r})$ is the

electric field. Finally, $S(\vec{r}, \vec{k}, \vec{k}')\, d^3k'$ represents the scattering probability per unit time from

the state $\vec{k}$ in $\vec{r}$, to any state $\vec{k}'$ contained in d^3k'. Equation (1) essentially represents a

continuity equation in the phase space $(\vec{r}, \vec{k})$ and implies the following assumptions:

i)   The band theory, and the effective mass theorem apply to the material under consider-

       ation.

ii)   The electric field is nearly constant over a scale length comparable to the physical

       dimensions of the wave packet describing the motion of the particle. This imples the applicability of the Ehrenfest theorem, i.e. the classical description of electron dynam-

ics.

iii)   Collisions are instantaneous or, at least, the collision duration is much shorter than the

average time of flight. As a consequence, scattering is a spatially localized event, and

the kinetic energy gained by the electron from the field during collisions is negligible.

iv)   Electron-electron interaction is neglected. The inclusion of this effect within the

scattering integral would render Eq. (1) non-linear in f.

v)   The scattering probability is independent of the electric field.

The last limitation is perhaps more practical than fundamental, and is related to the

difficulty of including such an effect within the collision integral. Attempts have been made to

account for it in Monte Carlo simulations, with a limited success.

In essence Eq. (1) represents a semi-classical approach to the transport problem, and

relies upon the concept of a single carrier distribution function which can be used for evaluat-

ing statistical averages and expectation values of the basic physical quantities.

By defining, as in Ref. 11

$$\tau^{-1}(\vec{r}, \vec{k}) = \int_{\vec{k}'} S(\vec{r}, \vec{k}, \vec{k}') \, d^3k' \tag{2a}$$

$$\tilde{f}(\vec{r}, \vec{k}, t) = \tau(\vec{r}, \vec{k}) \int_{\vec{k}'} S(\vec{r}, \vec{k}', \vec{k}) \, f(\vec{r}, \vec{k}', t) d^3k' \tag{2b}$$

Eq. (1) takes the form:

$$\frac{df}{dt} = \frac{f - \tilde{f}}{\tau} \tag{3}$$

τ representing the mean time between collisions.

2.2 Iterative Solution of the BTE. The Classical Transport Equation

From the mathematical point of view, the BTE is an integro-differential equation which, unfortunately, does not allow for a closed form solution but, rather, involves the use of iterative procedures which, moreover, are scarcely suitable even for numerical approaches, due to the multi-dimensional nature of the phase space.

Along these lines one might, in principle, determine the sequence of functions $f^{(o)}$, $f^{(1)}$, ..., $f^{(n)}$, ... such that[11]

$$\frac{df^{(n)}}{dt} + \frac{f^{(n)}}{\tau} = \frac{\tilde{f}^{(n-1)}}{\tau} \tag{4}$$

where $f^{(o)}$ is the zero-order approximate solution. The reason for using a lower-order approximation within the integral (2b) is due to the smoothing effect of the integration, when the function to be integrated is locally incorrect. Equation (4) is a purely differential equation if $f^{(n-1)}$ is known, and, for a given initial condition

$$f^{(n)}[\vec{r}(0), \vec{k}(0), 0] = f[\vec{r}(0), \vec{k}(0), 0] \tag{5}$$

its solution is

$$f^{(n)}(\vec{r}, \vec{k}, t) = \tilde{f}^{(n-1)}(\vec{r}, \vec{k}, t) + [f(\vec{r}, \vec{k}, 0) - \tilde{f}^{(n-1)}(\vec{r}, \vec{k}, 0)]$$

$$\exp\left\{ -\int_0^t \frac{dt'}{\tau(\vec{r}, \vec{k})} \right\} - \int_0^t \frac{d\tilde{f}^{(n-1)}}{dt'} e^{-\int_{t'}^t \frac{dt''}{\tau(\vec{r}, \vec{k})}} dt' , \tag{6}$$

where the integrals must be evaluated along the classical trajectories of the particles in the absence of collisions, $\vec{r} = \vec{R}(t)$, $\vec{k} = \vec{K}(t)$.

If we let the initial condition recede to the infinite past, Eq. (6) simplifies to

$$f^{(n)} = \tilde{f}^{(n-1)} - \int_{-\infty}^t \frac{d\tilde{f}^{(n-1)}}{dt'} e^{-\int_{t'}^t \frac{dt''}{\tau}} dt'. \tag{7}$$

By assuming, as a zero-order solution, the quasi-thermal equilibrium distribution function

$$f^{(o)} = f_o(\vec{r}, \vec{k}, t) = \frac{1}{4\pi^3}$$

$$\left\{1 + \exp\left[E_c(\vec{r}, t) + \varepsilon(\vec{k}) - E_{Fn}(\vec{r}, t)\right]/k_B T_e(\vec{r}, t)\right\}^{-1} \tag{8}$$

T_e being an effective temperature of the carriers, the first-order solution becomes

$$f^{(1)} = f_o(\vec{r}, \vec{k}, t) - \int_{-\infty}^{t} \frac{df_o}{dt} e^{-\int_{t'}^{t} \frac{dt''}{\tau}} dt' \tag{9}$$

which is known as Chambers integral.

Integrating iteratively by parts on the right-hand side of (9) we find,[12]

$$f^{(1)} = \sum_{n=0}^{m} f_n^{(1)}(\vec{r}, \vec{k}, t) - \int_{-\infty}^{t} \frac{df_m^{(1)}}{dt} e^{-\int_{t'}^{t} \frac{dt''}{\tau}} dt' \tag{10}$$

where

$$f_n^{(1)} = -\tau(\vec{r}, \vec{k}) \frac{df_{n-1}^{(1)}}{dt} \tag{11}$$

If $\lim_{n \to \infty} f_n^{(1)} = 0$, which is certainly true if the electron gas is slighly displaced from equilibrium, we may write

$$f^{(1)} = \sum_{n=0}^{\infty} f_n^{(1)}(\vec{r}, \vec{k}, t) \tag{12}$$

To our knowledge, no general proof exists of the convergence of (12). However, provided the electron gas is slightly displaced from equilibrium, one expects the following inequality to hold

$$\frac{f_n^{(1)}}{f_{n-1}^{(1)}} = -\frac{\tau}{f_{n-1}^{(1)}} \frac{df_{n-1}^{(1)}}{dt} << 1 \tag{13}$$

in which case (12) would be rapidly converging.

Let us limit ourselves to the consideration of the first-order term in (12). By remembering (8) one finds

$$
\begin{aligned}
f_1^{(1)} &= -\tau(\vec{r},\,\vec{k})\left(\frac{\partial f_o}{\partial t} + \vec{u}_g \cdot \vec{\Delta}_r f_o - \frac{q}{\hbar}\,\vec{E}\cdot\vec{\Delta}_k f_o\right) \\
&\approx -\tau(\vec{r},\,\vec{k})\, f_o\,(\vec{r},\,k,t)[1-4\pi^3\, f_o(\vec{r},\,k,\,t)]\frac{1}{k_B T_e}\vec{u}_g\cdot\vec{\Delta}_r E_{Fn}(\vec{r},\,t)
\end{aligned}
\tag{14}
$$

and, therefore,

$$
f^{(1)} \approx f_o(\vec{r},\,\vec{k},\,t)\, -\tau(\vec{r},\,\vec{k}) f_o(\vec{r},\,\vec{k},\,t)[1-4\pi^3 f_o(\vec{r},\,\vec{k},\,t)]\,\frac{1}{k_B T_e}\,\vec{u}_g\cdot\vec{\Delta}_r E_{Fn}
\tag{15}
$$

Equation (15) shows that, to first order, the electron transport driving force is the quasi-Fermi level gradient, and not the electric field. In deriving (14), however, the partial derivative of f_o has been neglected, as well as the space variation of carrier temperature $\vec{\Delta}_r T_e$. The current density $\vec{J}_n$ can now be derived

$$
\begin{aligned}
\vec{J}_n &= -q\int_k \vec{u}_g f^{(1)}\, d^3k = \hat{\mu}_n\, n\, \vec{\Delta}_r\, E_{Fn} \\
&= q\hat{\mu}_n\, n\, \vec{E} + q\, \hat{D}_n\, \vec{\Delta}_r\, n
\end{aligned}
\tag{16}
$$

where

$$
\hat{\mu}_n = \frac{q}{\hat{m}^*}\,<\tau>
\tag{17}
$$

$\hat{m}^*$ being the effective mass operator, and $<\tau>$ a suitable average of the mean time between collisions. If more than one valley contributes to carrier transport, several contributions like (16) must be considered, leading, for cubic crystals, to a scalar mobility.

In order to derive (16), several simplifying assumptions have been performed:

i)   Having neglected higher order terms in Eq. (12), we have dropped the second and higher order derivatives of the quasi-Fermi potential $E_{Fn}(\vec{r},t)$. This means that we transform a non-local solution of the BTE into an approximate one depending only upon the local value of $\vec{\Delta}_r\, E_{Fn}$.

ii)   An additional consequence of the previous simplifying assumption is that non-linear

terms in $\vec{\Delta}_r E_{Fn}$ are neglected, i.e. the dependence of the distribution function upon

the quasi-Fermi level gradient is linearized. Such a simplification is reasonable if

the following condition is fulfilled

$$\frac{1}{k_B T_e} \tau \vec{u}_g \cdot \vec{\Delta}_r E_{Fn} < < 1 \tag{18}$$

which amounts to say that the scale of length over which E_{Fn} varies by $k_B T_e$ must

be large compared with the carrier mean-free path.

iii)   Higher order terms in Eq. (12) contain explicitly the electric field, due to the $\vec{k}$

dependence of $\tau(r,k)$. Thus, only to first order the electron transport driving force

is the quasi-Fermi level gradient. Far from equilbirium, the electric field can play a

role.

iv)   The spatial gradient of $\tau(\vec{r}, \vec{k})$ is neglected. This implies a slowly-varying impurity

concentration over a carrier mean-free path.

v)   The partial time derivative of f_o has been neglected in (14), which holds true only

if the time scale over which external excitations are changing is large compared

with the average time between collisions.

vi)   The time and spatial variation of carrier temperature is neglected which, again, is

related to the assumption of a nearly uniform, and slowly-varying quasi-Fermi level

gradient.

vii)   Parabolic energy bands are assumed.

viii)   Finally, we have assumed an infinite semiconductor. In a real device additional

problems arise due to the boundary conditions, which can force highly irregular

distribution functions. If a metal semiconductor junction behaves as an absorbing

boundary, the distribution function is completely asymmetrical there, because there

are no electrons entering the semiconductor. In such a situation, electron transport

in the vicinity of the junction is heavily altered, and we expect that the drift-diffusion equation fails within a few carrier mean-free paths from the boundary itself.

2.3 The Monte Carlo (MC) Method

An alternative method for solving the BTE consists in simulating the motion of one or more electrons at microscopic level.[13,14] This motion results from a sequence of drifts within the electric field, followed by a scattering event. The drifting time, the type of scattering process, and the final state are random quantities which can be expressed in terms of the transition rates due to the various processes, and the spatial distribution of the electric field. Indicating with $\lambda(\vec{r}, \vec{k}) = \tau^{-1}(\vec{r}, \vec{k})$ the cumulative probability of a scattering event per unit-time, the probability that an electron drifts for a time t before colliding is given by

$$P(t) = \lambda(\vec{r}, \vec{k}) \exp \left\{ -\int_0^t \lambda(\vec{r}, \vec{k}) \, dt' \right\} \tag{19}$$

where the integral on the RHS of (19) must be evaluated along the classical trajectory within the phase space, resulting from integration of the motion equations.

We must therefore generate a sequence of random numbers characterized by the probability distribution (19), in order to determine the successive locations of the scattering events. On the other hand, it is straightforward to generate by computer random numbers with uniform probability between 0 and 1; thus, we must convert such a sequence s_1, s_2, ..., s_i, into a sequence of times of flight t_1, t_2, ... t_i,... satisfying (19). This can be achieved by writing

$$p(s) \, ds = P(t) \, dt \tag{20}$$

and therefore

$$s = \int_0^t P(t') dt' = 1 - \exp \left[-\int_o^t \lambda(\vec{r}, \vec{k}) \, dt' \right] \tag{21}$$

Due to the complex nature of the $\lambda(\vec{r}, \vec{k})$ function, shown, in simplified form, in Fig. 1, inverting Eq. (21) may be a difficult task. It is therefore convenient to consider an additional, fictitious scattering process which leaves unmodified the state of the particle, and with a probability distribution $\lambda_o(\vec{r}, \vec{k})$ such that

$$\lambda_o(\vec{r}, \vec{k}) \; + \; \lambda(\vec{r}, \vec{k}) \; = \; \Gamma \qquad\qquad (22)$$

where Γ is a constant which must be chosen large enough so as to prevent λ_o to be negative. So doing, Eq. (21) becomes

$$s \; = \; 1 - e^{-\Gamma t} \qquad (23)$$

which can easily be inverted to yield

$$t \; = \; \frac{1}{\Gamma} \, \ell \, \frac{1}{1-s} \qquad (24)$$

Having determined the time of flight and, therefore, the scattering location within the phase-space, in order to define the collision type

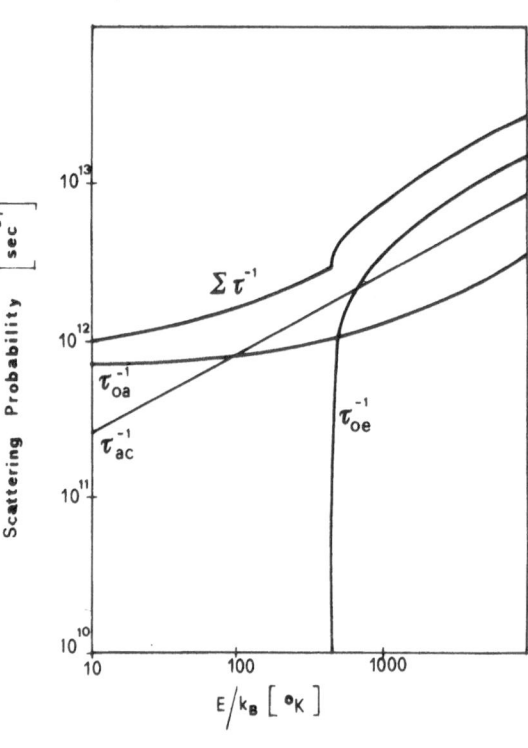

Fig. 1 - Simplified scattering probability in silicon.

occurring at that location, it is only necessary to generate a random number with uniform probability between 0 and Γ and test the inequality

$$s \; < \; \sum_{q=0}^{m} \lambda_q(\vec{r}, \vec{k}) \qquad m \; = \; 0,1,2,3,..,M$$

M being the number of "true" scattering processes. This provides the final energy of the electron. In case of a completely randomizing collision, the final state is easily determined since all momenta on a given energy surface are equally probable. On the other hand, with non-randomizing collisions such as impurity scattering or polar-optical scattering, the angle between $\vec{k}$ and $\vec{k}'$ must be determined accounting for the differential cross section of the scattering process, whereas the azimuthal angle has a uniform probability between 0 and 2π. The final state of the electron after collision represents the initial condition for the next flight within the phase-space.

In order to determine the distribution function, histograms are set up by subdividing the phase space in subcells and recording the time each electron spends in every subcell. If we assume the ergodic theorem to hold, such a cumulative time is proportional to the distribution function. In case of a semiconductor with non-equivalent minima, separate histograms are set up for different valleys, and the electron switches from one histogram to the other when an intervalley scattering takes place.

Once the distribution function has been calculated, the physical observables of interest such as drift velocity, mean energy, etc. can in principle be obtained from numerical integration. However, in order to avoid having to use a fine histogram mesh to achieve the desired accuracy, it is more convenient to calculate these quantities directly.[13]

It can be demonstrated that the distribution function resulting from a MC simulation satisfies the BTE.[13,15] Thus the MC method represents a useful tool for the study of transport properties in bulk material, since it allows riddance of analytical approximations, such as that of the relaxation time, or those based on "a priori" choice of a distribution function. Its most important limitation, however, is that the electric field shape must practically be known, because coupling Poisson's equation with a MC simulation for one or two species of carriers would lead to a prohibitive CPU time.

3.1 Phenomena Due to the Fast Temporal Variation of the Electric Field: Velocity Overshoot, Complex Mobility

The study of electron dynamics in silicon, subject to a step-varying electric field, has been carried out by Nougier *et al*[16] using a MC analysis method. Their result is summarized in Fig. 2, showing the temporal variation of the carrier drift velocity v_d and average energy $<\varepsilon>$

following the application of an electric field of 50 KV/cm. The most striking feature of such a result is the strong velocity overshoot which mostly occurs within 0.5 psec from the electric field step, while the electron average energy increases more slowly and saturates after 1.0 psec. The maximum velocity slightly exceeds 2×10^7 cm.sec^{-1}, i.e. twice the scattering-limited velocity, which is a typical non-equilibrium phenomenon due to the sudden variation of the electric field.

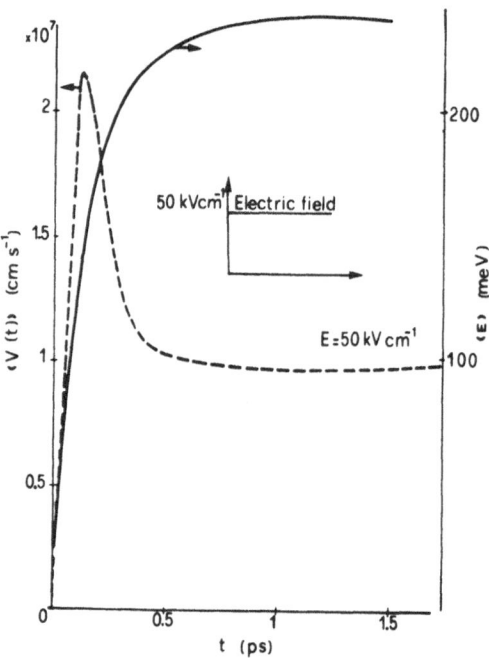

Fig. 2 - Drift velocity and average electron energy in silicon following a step-varying electric field. After Nougier *et al.* (16).

In a homogeneous semiconductor subject to a uniform and constant field, a dynamic equilibrium is reached by the carriers when momentum and energy gained from the field are exactly balanced by momentum and energy losses due to randomizing collisions. The resulting drift velocity is constant, and so is the average carrier energy, both values depending upon the electric field strength. Such an equilibrium can be reached because the collision probability is an increasing function of carrier energy. As the electric field increases, both carrier energy and scattering rate increase, thus rendering more efficient the energy dissipation process, until a new equilibrium situation is reached. As a consequence, drift velocity is a sub-linear function of the field.

When, however, the latter is suddenly applied to a semiconductor, carriers are rapidly accelerated to a velocity in excess of 10^7 cm/sec before randomizing collisions take place, resulting in the illustrated velocity overshoot. Correspondingly, the distribution function is heavily skewed by the field, while carrier energy is well below its asymptotic value. As time progresses, however, carrier energy increases, and drift velocity decreases to its equilibrium value with a time constant essentially corresponding to the energy relaxation time.

Even more complex phenomena occur in multivalley semiconductors such as GaAs, due to the different values of the electron effective mass in (000) and (100) minima. The drift velocity following the application of an electric field pulse[17] is shown in Fig. 3. Due to the smaller value of the effective mass in the (000) val-

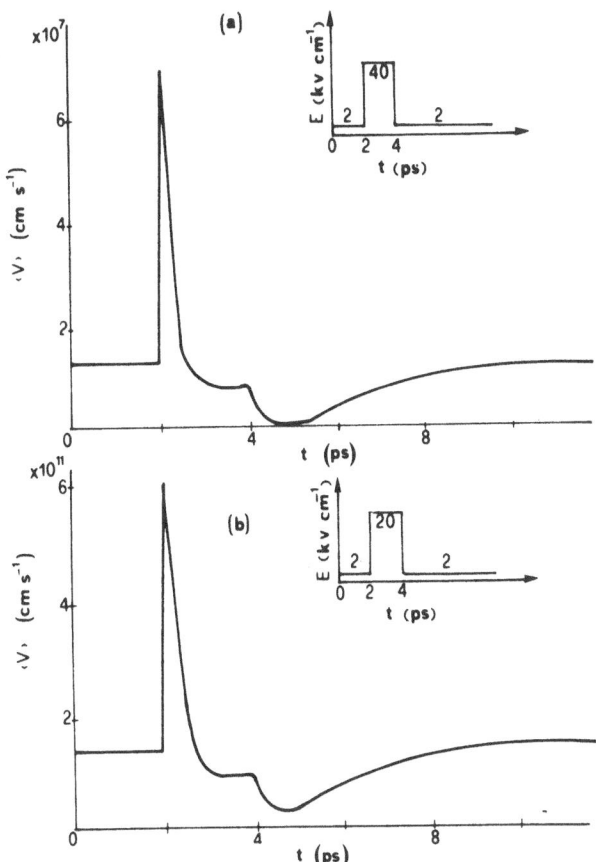

Fig. 3 - Electron dirft velocity in GaAs following a pulse-varying electric field. (a) $E_{max}=40$ kV/cm; (b) $E_{max}=20$ kV/cm. After Carnez *et al.* (17).

ley, the overshoot effect is more pronounced in GaAs than in Si, and the drift velocity reaches a value of $6 \div 7 \times 10^7$ cm/sec. As the carrier energy increases, however, intervalley scattering leads to a larger population of the (100) valleys, and drift velocity decays to its asymptotic value. Furthermore, when the electric field drops to the value of 2 KV/cm, the relative population of the (000) and (100) valleys is not in equilibrium with the new field, leading to an undershoot effect, until finally the equilibrium velocity is reached.

An additional effect related to the finite response time of the carriers is a complex differential mobility resulting at large operating frequencies. If the latter is small compared with τ^{-1}, the carriers are in equilibrium with the differential field superimposed to the constant one, and the differential mobility is simply given by the derivative of the velocity-field curve.

If, however, the signal frequency is comparable with τ^{-1}, electrons are accelerated and decelerated by the incremental field between collisions, and a velocity modulation becomes effective even in the saturated velocity range.[18] Such a result is shown in Fig. 4 where the real and imaginary parts of the complex differential mobility of electrons in silicon are represented against frequency. The real part of the differential mobility increases in the 100 ÷

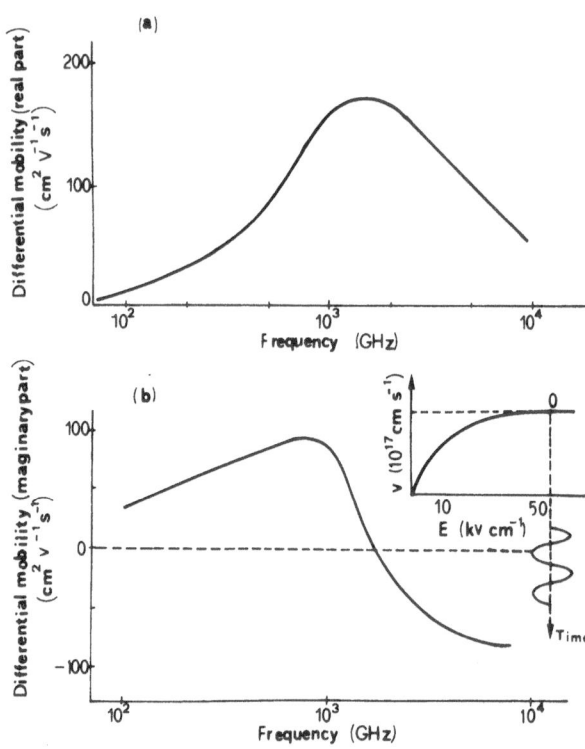

Fig. 4 - Real and imaginary parts of electron differential mobility in silicon against frequency. The static field is assumed to be 50 kV/cm. After Zimmermann *et al.* (18).

2000 GHz range, where electron collisions cannot maintain equilibrium with the electric field, and decreases at higher frequencies where electron inertia does not allow them to follow the signal variations. Correspondingly, the imaginary part varies from positive to negative values, indicating that a resonance effect occurs in the neighborhood of 2000 GHz.

3.2 Phenomena Due to the Spatial Variation of the Electric field

This case is particularly interesting for submicron devices, which are intrinsically characterized by large variations of the electric field over short distances. Figures 5 and 6 show the spatial variation of electron concentration and drift velocity, respectively, within the base and collector space charge region of a bipolar transistor having a base width of 200 nm. Results of a MC simulation are compared with a classical model, taking velocity saturation into account. The electric field has been assumed constant, and equal to 1 KV/cm within the base,

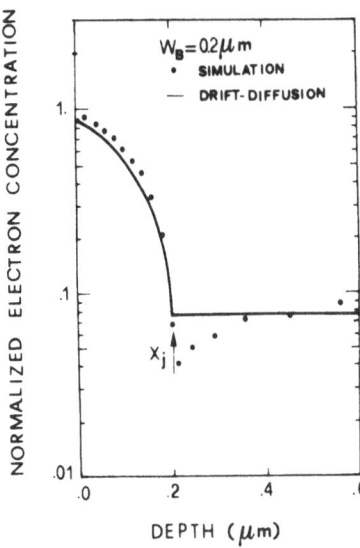

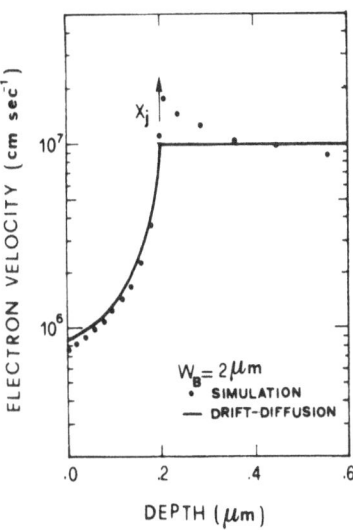

Fig. 5 - Normalized electron distribution in an idealized bipolar transistor. The solid line represents the drift-diffusion approach, whereas the dots represent the M.C. simulation. After Baccarani et al. (15).

Fig. 6 - Electron velocity distribution in an idealized bipolar transistor. After Baccarani et al. (15).

and linearly-varying within the collector space-charge region, with a maximum value at the interface of 10^5 V/cm. The main features of this result are the following:[15]

i)   The carrier velocity resulting from the MC simulation turns out to be smaller within the base than the classical one, the difference, however, being smaller than 12%.

ii)   A velocity overshoot of 80% occurs at the beginning of the collector space-charge region, whose spatial extent in about $0.1 \mu m$.

iii) The carrier drift velocity at the junction is $v_d(x_j) = 1.15 \times 10^7$ cm/sec., i.e. 15% larger than both the scattering-limited velocity and the thermionic-emission velocity, which are remarkably similar in silicon, at room temperature.

The electron distribution function at $x = x_j$ is shown in Fig. 7 and, due to the field discontinuity, appears to be strongly asymmetrical (only 8% of the injected electrons are backscattered into the base). The carrier temperature, however, is remarkably close to the

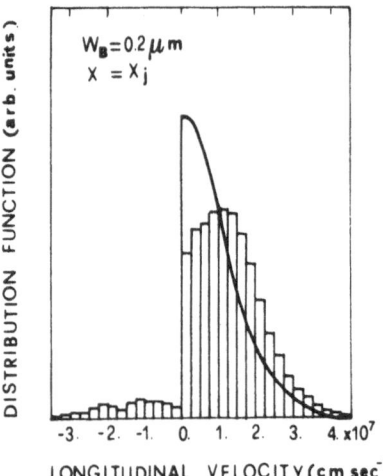

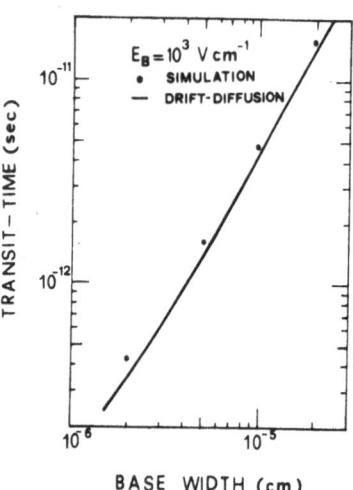

Fig. 7 - Electron distribution function at the base-collector junction of a bipolar transistor. After Baccarani *et al.* (15).

Fig. 8 - Base transit time against base width in a bipolar transistor. After Baccarani *et al.*

lattice temperature, so that the large drift velocity comes out as a consequence of the asymmetrical distribution function, rather than from carrier heating.

In spite of the roughness of the drift-diffusion approach which, as already mentioned, neglects the influence of the neighboring points on the shape of the distribution function, as well as heat conduction and thermoelectronic current, the resulting integral quantities such as the base transit time are remarkably close to the classical predictions, as shown in Fig. 8 where the latter is plotted against base width.

3.3 Carrier Transport in Thin Inversion Layers

Carrier transport in thin inversion layers has been the subject of many investigations, both theoretical and experimental. Early analyses were carried out on the assumption of a classical distribution of carriers, and accounted for diffuse or partially diffuse scattering at the interface.[19-21] Subsequently, sophisticated quantum mechanical models were developed, based on a simultaneous solution of the coupled Poisson's and Schrödinger equations.[22-25] Being the inversion layer a thin potential well, the motion normal to the interface is quantized, and therefore the bulk conduction band is split in several sub-bands near the surface, each of which is a two dimensional continuum associated with one of the quantized energy levels. At low temperatures, only the first subband is occupied, and the shape of the inversion layer is substantially altered with respect to the classical model. At higher temperatures, however, electrons spread among the various subbands, and their energy distribution becomes closer to the classical one. At room temperature it has been estimated[24] that the average distance from the interface of the inversion-layer electrons only differs by about 10% in the quantum-mechanical and classical models. A treatment of the transport properties of a two-dimensional electron gas falls beyond the scope of this paper; however, the main experimental features of carrier mobility in inversion layers is summarized below:

i) At low temperature, the main scattering mechanism in weakly inverted surfaces are due to oxide charges and impurities, leading to a conductivity mobility proportional to T/N_I. The effect of surface charges and impurities is less important in moderately and strong inverted layers because of screening; in this case, however, surface scattering comes into play. As a consequence, mobility becomes relatively independent of temperature, whereas the variation with the normal field exhibits a distinct maximum, followed by a rapidly decreasing behavior. Surprisingly enough, at low temperature even channel conductance decreases with the normal field.[26]

ii) At room and higher temperatures, the main scattering mechanisms are acoustic and intervalley scattering in weakly inverted layers, and surface scattering in strong inversion. Thus, carrier mobility is monotonically decreasing with the normal field and temperature, with a low-field value about one half of the bulk mobility.

iii) Anisotropy effects are observed due to the electron band structure in silicon, especially at large applied fields. In particular (100) oriented surfaces exhibit a larger mobility than (111) and (011).

iv) As the parallel field increases beyond $10^4 V/cm$ velocity saturation effects are observed for both electrons and holes. The scattering-limited velocities, however, are significantly smaller than the bulk ones, resulting in 5×10^6 cm/sec and 2×10^6 cm/sec., respectively, at room temperature.[28]

A recent MC simulation performed by Zimmermann *et al.*[29] of electron transport in silicon inversion layes, assuming either diffuse or specular scattering at the interface, has led to velocity-field curves ip surprisingly good agreement with experimental data by Cohen and Muller,[28] as shown in Fig. 9. The authors claim that the mobility loss in the low field region is due to the combined effect of surface scattering and transverse field. However, at least for the case of specular scattering, by which the parallel component of carrier momentum is conserved, it is not obvious how this mobility loss can take place, since the normal field cannot heat the carriers.

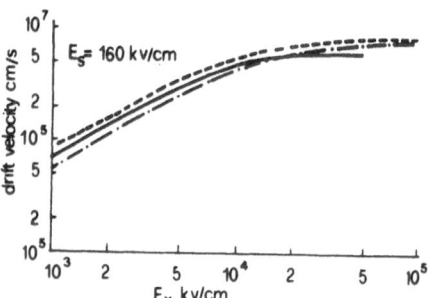

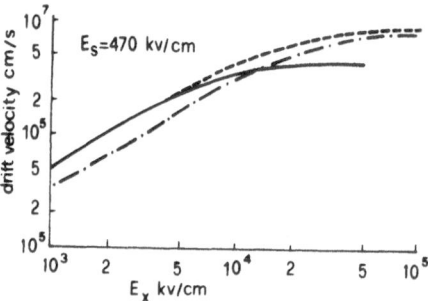

Fig. 9 - Electron drift velocity against parallel field in a silicon inversion layer. (a) Normal field $E_s = 160kV/cm$; (b) $E_s = 470$ kV/cm. After Zimmermann *et al.* (29).

Referring now more specifically to submicron MOSFET's, velocity saturation has already proved to play a role, leading ultimately to a drain current in saturation proportional to $(V_G - V_T)$ instead of $(V_G - V_T)^2$. An additional effect which probably deserves some consideration is the change of the inversion layer shape induced by carrier heating. As the electron temperature increases, their spatial distribution broadens and, for a given gate bias, the surface concentration decreases, leading to a reduced transconductance.

Finally ballistic transport can take place in submicron MOSFET's and a reduced carrier transit time is expected.

4.1 The Problem of Simulating MSD's

So far we have been considering the study of carrier dynamics in highly idealized, mostly one-dimensional, electric field distributions. So doing, non-equilibrium phenomena due to the non-local solution of the BTE have been emphasized. In order to become a useful, predictive tool of device performance for any given physical and geometrical structure, however, the MC simulation must be coupled with a two or even three-dimensional solution of Poisson's equation, and requires very long computation times due to the random fluctuations of the calculated parameters, the resulting accuracy being proportional to the square root of the number of simulated particles. In addition, if we want to study time-dependent effects occurring on a time scale much larger than the mean time between collisions, such as trapping-detrapping phenomena, this would require extremely long computation times.

A realistic simulation of a GaAs MESFET using a multiparticle MC method was first carried out by Hockney *et al.*[30]. This method consists in storing the coordinates and momenta of a large number of electrons contained in a given device. Poisson's equation is then solved, accounting for the carrier spatial distributions and imposed boundary conditions. Subsequently, the motion of each electron is simulated for a given time Δt, leading to a new electron distribution. This procedure is iterated, until steady state is reached. Although this method is certainly appealing, it is not clear how well it performs in terms of external device properties, such as static characteristics, switching time, etc.

Due to the difficulties associated to a complete MC simulation, it is desirable to identify a simplified approach which, while accounting for the fundamental effects included in the BTE, nevertheless leads to a tractable problem. This result can be achieved by assuming that the distribution function be a drifted maxwellian, with drift velocity and electron temperature treated as non-local physical quantities.

From the BTE within the relaxation-time approximation

$$\frac{\partial f}{\partial t} + \vec{u}_g \cdot \vec{\Delta}_r f - \frac{q}{\hbar} \vec{E} \cdot \vec{\Delta}_k f = -\frac{f - f_o}{\tau} \tag{25}$$

multiplying each term by $\hat{m} \vec{u}_g$ and integrating over the first Brillouin zone, we find

$$\frac{\partial}{\partial t}(n\, m^* \vec{v}_d) + \vec{\Delta}_r(nk_BT_e) + q\, n\, \vec{E} = -n\, \frac{m^* \vec{v}_d}{\tau_p} \tag{26}$$

where the momentum relaxation time τ_p is given by

$$\frac{1}{\tau_p} = (m \vec{v}_d)^{-1} \int_k \frac{1}{\tau} \hat{m}^* \vec{u}_g f \, d^3k .$$ (27)

Multiplying each term of (26) by $\mu_n = \dfrac{q}{m^*}\tau_p$, we find a generalized expression of the transport equation, namely

$$\vec{J}_n + \tau_p \frac{\partial \vec{J}_n}{\partial t} = q \mu_n n \vec{E} + k_B \mu_n \vec{\Delta}_r(T_e n)$$ (28)

which, in steady-state, reduces to the well-known drift-diffusion equation; μ_n however, is a function of T_e.

Multiplying instead each term of (25) by $\varepsilon(\vec{k})$ an integrating, one finds

$$\frac{\partial W}{\partial t} + \vec{\Delta S} - \vec{E} \cdot \vec{J}_n = - \frac{W - W_0}{\tau_\varepsilon}$$ (29)

where W represents the electron energy per unit volume, $\vec{S}$ is the energy flux, and τ_ε is the energy relaxation time, defined as

$$\frac{1}{\tau_\varepsilon} = <\varepsilon>^{-1} \int_k \frac{1}{\tau} \varepsilon(k) f \, d^3k$$ (30)

Equation (29) expresses energy conservation: In fact it states that the energy increase per unit time and per unit volume results from the algebraic sum of three contributions; namely, energy flux through the surface limiting the volume, energy supplied by the electric field, and energy dissipated by exciting lattice vibrations.

For the analysis of unipolar devices, where only one type of carrier is really involved, the mathematical model results from the system of (28-29), Poisson's and current continuity equations. The unknown functions are: electric potential ψ, electron concentration n, average electron energy $<\varepsilon>$ or, equivalently, electron temperature T_e, and current density. The energy flux is related to the current density by a relationship which reduces to $\vec{S} = \frac{5}{3}(<\varepsilon>/q)\vec{J}_n$ for small values of the drift velocity.

In order to solve such a system, the energy dependence of τ_p, τ_ε (and eventually m*, if the band is not assumed parabolic) must be known. Such a dependence can be determined by numerical integration of the total scattering probability $\lambda(\vec{t},\vec{k})$, with the necessary weighting functions. Alternatively, it can be obtained from the static $v_d(E)$, and $T_e(E)$ curves, as resulting from a MC simulation.[31] An approximate, although similar approach, has been used in Ref. 30 in simulating the static characteristics of a GaAs MESFET: the diffusive term has in fact been neglected in (26) and (29), while being added to the drift velocity in the expression of current density. For a 0.2 μm gate length, the drift velocity turns out to be about twice as large as the static $v_d(E)$ characteristics, thus considerably improving the expected device transconductance.

4.2 Conclusions

In this paper, the theoretical background underlying current transport in semiconductors has been reviewed, in order to identify the physical origin of new phenomena such as ballistic transport, velocity overshoot, etc., which can somehow affect the final performances of medium-small devices. Several examples have been illustrated, including temporal and spatial non-equilibrium effects, and conduction phenomena in thin inversion layers.

The MC simulation procedure has been illustrated, and its main limitations pointed out with reference to both single carrier, and multi-carrier approaches.

Finally, a mathematical model is developed, which makes use of a generalized transport equation, and includes an additional equation expressing energy conservation. Such a model allows us to get rid of the local-mobility assumption, and renders the treatment of ballistic effects feasible, even in two dimensions, at a reasonable cost.

REFERENCES

1.   B. Hoenisen and C. A. Mead, Solid-State Electron, 15, 819 (1972).

2.   R. W. Keyes, Proc. IEEE, 63, 740 (1975).

3.   K. N. Ratnakumar and J. D. Meindl, ISSCC Digest, 72 (1980).

4.   J. R. Barker and D. K. Ferry, Solid-State. Electron. 23, 519 (1980).

5.   J. R. Barker and D. K. Ferry, Solid-State Electron., 23, 531 (1980).

6.   D. K. Ferry and J. R. Barker, Solid-State Electron., 223, 545 (1980).

7.   R. H. Dennard, F. H. Gaensslen, IEEE J. Solid State Circuits, SC-14, 247 (1979).

8.   P. I. Suciu, E. N. Fuls and H. J. Boll, IEEE Electron Dev. Letters, EDL-1, 10 (1980).

9.   K. Ohta, K. Yamada, M. Saitoh, K. Shimizu and Y. Tarui, IEEE J. Solid-State Circuits, SC-15, 417 (1980).

10.   T. Ito, T. Nozaki, H. Ishikawa and Y. Fukukawa, ISSCC Digest, 74 (1980).

11.   E. De Castro, "Fondamenti di Elettronica" Cap. VIII, UTET (1975).

12.   G. Baccarani, J. of Appl. Phys., 47, 4122 (1976).

13.   W. Fawcett, A. D. Boardman and S. Swain, J. Phys. Chem. Solids, 31, 1963 (1970).

14.   C. Canali, C. Jacoboni, F. Nava, G. Ottaviani and A. Alberigi Quaranta, Phys. Rev. B, 12, 2265 (1975).

15.   G. Baccarani, C. Jacoboni and A. M. Mazzone, Solid-State Electron, 20, 5 (1977).

16.   J. P. Nougier, J. C. Vaissiere, D. Gasquet, J. Zimmermann and E. Constant, J. of Appl. Phys., 51, (1980).

17.   B. Carnez, A. Cappy, A. Kaszynski, E. Constant and G. Salmer, J. of Appl. Phys., 51, 784 (1980).

18.   J. Zimmermann, Y. Leroy and E. Constant, J. of Appl. Phys., 49, 3378 (1978).

19.   J. R. Schrieffler, Phys. Rev., 97, 641 (1955).

20.   R. F. Greene, D. R. Frankl and J. Zemel, Phys. Rev., 118, 967 (1960).

21.   R. F. Pierret and C. T. Sah, Solid-State Electron, 11, 279 (1968).

22.   F. Stern and W. E. Howard, Phys. Rev. 163, 816 (1967).

23.   E. D. Siggia and P. C. Kwok, Phys. Rev. B, 2, 1024 (1970).

24.   J. A. Pals, Phys. Rev. B, 5, 4028 (1972).

25.   C. T. Sah, T. H. Ning and L. L. Tschopp, Surface Sci., 32, 561 (1972).

26.   F. F. Fang and A. B. Fowler, Phys. Rev., 169, 619 (1968).

27.   S. C. Sun and J. D. Plummer, IEEE J. Solid-State Circuits, SC-15, 562 (1980).

28.   R. W. Coen and R. S. Muller, Solid-State Electron., 23, 35 (1980).

29.   J. Zimmermann, R. Fauquembergue, M. Charet and E. Constant, Electron. Lett., 16, 17 (1980).

30.   B. Carnez, A. Cappy, A. Kaszynski, E. Constant and G. Salmer, J. of Appl. Phys., 51, 784 (1980).

31.   M. Shur, Electron. Lett., 12, 615 (1976).

32.   E. Constant, Inst. Phys. Conf. Ser. No. 57, 141 (1981).

THE ROLE OF GaAs IN HIGH SPEED INTEGRATED CIRCUITS[*]

Richard C. Eden[t]
Enhanced Energy Systems
Newbury Park, CA

and

Bryant M. Welch
Rockwell International
Thousand Oaks, CA

I. INTRODUCTION

The achievement of ultra high speed VLSI, that is, of integrated circuit chips with complexities of $N_g > 10^4$ equivalent logic gates with propagation delays of $\tau_d \sim$ 100ps or less, would make possible computational powers or chip throughput rates hard to imagine by todays standards. The realization of such potential is very difficult, of course, since it necessitates achieving in one circuit and fabrication technology: 1) ultra high speed (very low τ_d); 2) low power per gate (P_D); 3) extremely low dynamics switching energies (power-delay products, $P_D\tau_d$); 4) very high gate densities; and 5) very high process yields (sufficient to allow economic fabrication of such complex parts).[1,2] The problem here is the requirement for improvements on both sides of what are classical tradeoffs such as speed-power or lithographic resolution-yield. For example, in silicon MOS the speed may improve by increasing the supply voltage and logic voltage swings in order to increase the average device transconductances or current gain-bandwidth products (f_τ's). Increasing V_{DD}, however, while reducing τ_d, sharply increases gate power dissipation, P_D, and switching energies, $P_D\tau_d$, which would lead to unacceptable power levels in $> 10^4$ gate VLSI chips.[1,2] Reducing geometry by pressing

[*] The planar LSI SDFL GaAs integrated circuit work at Rockwell International was supported in part by the Advanced Research Projects Agency of the Department of Defense and was monitored by the Air Force Office of Scientific Research under Contract F49620-77-C-0087.

[t] Consultant with Rockwell International, and full-time employee while most of the SDFL GaAs LSI research described was in progress. Present address: GigaBit Logic, Culver City, CA.

lithographic resolution is an approach to coping with this speed-power tradeoff, and it improves density as well, but this approach can easily result in unacceptable reduction in yield if pressed too far.

Some of the aspects of these tradeoffs are illustrated in Table I, which reviews the circuit requirements for such ultra high speed, low power logic, and their consequences in terms of device requirements if such circuits are to be realized. Clearly, since logic gates classically must have current and power gains greater than their fanout loadings, the f_τ and f_{max} (current and power gain-bandwidth products) of the active devices must be very high. Perhaps slightly less obviously, the devices must have a high degree of nonlinearity, i.e., they must develop their high transconductance (or f_τ or f_{max}) at control voltages only a small logic voltage swing above threshold. This is necessitated by the need to drastically reduce dynamic switching energies ($P_D\tau_d$) for ultra high speed VLSI. The dynamic switching energy must exceed the stored energy on the switched capacitance, C,

TABLE I
CIRCUIT AND CONSEQUENT DEVICE REQUIREMENTS
FOR VERY HIGH SPEED, LOW POWER LOGIC CIRCUITS

CIRCUIT REQUIREMENT	CONSEQUENT DEVICE REQUIREMENTS
1. LOW LOGIC VOLTAGE SWINGS (FOR LOW POWER ~ ½CΔV^2)	VERY UNIFORM THRESHOLD VOLTAGES FOR ACTIVE DEVICES (PARTICULARLY FOR VLSI)
2. LOW CAPACITANCES (BOTH DEVICE AND PARASITIC)	LOW INPUT CAPACITANCE DEVICES HIGH CIRCUIT DENSITIES SEMI-INSULATING SUBSTRATE FOR LOW CIRCUIT PARASITICS
3. HIGH SWITCHING SPEEDS WITH REASONABLE FANOUT LOADINGS AT LOW SWITCHING VOLTAGES	VERY HIGH f_τ (CURRENT GAIN-BANDWIDTH) VERY HIGH f_{max} (POWER GAIN-BANDWIDTH) VERY RAPID INCREASE IN TRANSCONDUCTANCE ABOVE THRESHOLD (STRONG NONLINEARITY)
4. HIGH CIRCUIT COMPLEXITIES WITH ACCEPTABLE COST (HIGH YIELDS)	EASILY FABRICABLE DEVICE STRUCTURES HIGH YIELD PROCESS THRESHOLD VOLTAGES AND OTHER CRITICAL DEVICE PARAMETERS INSENSITIVE TO REASONABLE PROCESSING VARIATIONS TIGHT, PREDICTABLE PARAMETER STATISTICS
5. DELIVERY ASAP	ESTABLISHED FABRICATION TECHNOLOGY

$$P_D \tau_d > \frac{1}{2} C (\Delta V_L)^2 \qquad (1)$$

where C is the sum of the input capacitances of the fanout of loading gates plus the parasitic capacitance, and ΔV_L is the logic voltage swing. (More precisely[1,2] the $(\Delta V_L)^2$ term is the product of the supply voltage, V_{DD}, times ΔV_L). Hence, the power supply and logic swing voltages must be kept small for low power-delay products, and the devices must not only be fast, but must be fast with small logic voltage swings. A further consequence of this requirement for small logic voltage swings is that the active devices have threshold voltages that are very precisely controlled, i.e., with standard deviation of threshold voltage a small fraction (preferably <5% or better) of the logic voltage swing.

II. GaAs DEVICE APPROACHES FOR ULTRA HIGH SPEED VLSI

As illustrated in Table I, the circuit constraints for very high speed, low power logic for VLSI generate a number of requirements for the characteristics of the active devices needed to implement such circuits. These required device characteristics also have numerous implications for the general physical characteristics needed for the devices themselves. These considerations can be very useful in selecting promising device approaches for ultra high speed VLSI. Table II reviews some of the desired device characteristics from Table I with focus on their consequences in terms of the physical parameters required for the device (e.g., structures, geometries, semiconductor parameters, etc.).

A. Importance of Mobility for VLSI

As indicated in Table II, a key consequence of the requirement for strong nonlinearity, that is, high transconductance, g_m, and f_τ at control voltages only small logic swings above threshold is the need for high carrier mobilities and/or very short carrier transit path lengths in the devices. This is easily illustrated for an FET structure,[1,2] where, for gate voltages, V_{gs}, near threshold, V_p, the saturated drain current I_{ds}, is given by

TABLE II
DEVICE CHARACTERISTICS DESIRED FOR OPTIMAL
HIGH-SPEED LOW-POWER SWITCHING WITH THE CONSEQUENT
DEVICE PHYSICAL PARAMETERS AND STRUCTURAL CHARACTERISTICS

DESIRED ELECTRICAL CHARACTERISTIC	CONSEQUENT PHYSICAL PARAMETERS
HIGH TRANSCONDUCTANCE AT CONTROL VOLTAGES A SMALL SWING ABOVE THRESHOLD	HIGH CARRIER MOBILITIES EXTREMELY SHORT CONTROL TRANSIT PATH (e.g., GATE LENGTH OR BASE WIDTH)
VERY UNIFORM THRESHOLD VOLTAGES, V_T (TO ALLOW LOW VOLTAGE SWING LOGIC)	EXTREMELY LOW V_T SENSITIVITY TO <u>HORIZONTAL</u> GEOMETRY VARIATIONS LOW V_T SENSITIVITY TO REASONABLE <u>VERTICAL</u> GEOMETRY VARIATIONS LOW V_T SENSITIVITY TO <u>DOPING</u> VARIATIONS
VERY LOW INPUT CAPACITANCES	SMALL GEOMETRIES (e.g., GATE LENGTH AND WIDTH IN FETs OR EMITTER AREA IN BIPOLARS)
HIGH f_τ AND f_{max} GAIN-BANDWIDTH PRODUCTS	HIGH CARRIER MOBILITIES AND SATURATION VELOCITIES, SMALL GEOMETRIES (SHORT TRANSIT LENGTHS; e.g., GATE LENGTH, BASE WIDTH OR EMITTER STRIPE WIDTH)
HIGH YIELD	SIMPLE, EASILY-FABRICATED STRUCTURE

$$I_{ds} = K(V_{gs} - V_p)^2 = \frac{\epsilon \mu_n W}{2aL_g} (V_{gs} - V_p)^2 \tag{2}$$

where W is the channel width, L_g the gate length, a the distance between gate and channel, μ_n the electron mobility (assuming n-channel), and ϵ is the dielectric constant of the semiconductor in a MESFET or JFET or of the gate insulator in an IGFET (MOSFET). The transconductance, g_m, for this Shockley FET model is, then

$$g_m = \frac{\epsilon \mu_n W}{aL_g} (V_{gs} - V_p) \tag{3}$$

and the current gain-bandwidth product in this near-threshold, small $V_{gs} - V_p$, region is given by

$$f_\tau = \frac{g_m}{2\pi C_{gs}} = \frac{\mu_n}{2\pi L_g^2} (V_{gs} - V_p). \tag{4}$$

Equation (3) illustrates that the strength of the nonlineatity, that is, the increase of transconductance with small gate voltages above threshold increased as μ_n/L_g, while the f_τ expression of Eq. (4), which ignored fringing capacitances, gains an additional $1/L_g$ dependence due to the assumed capacitance variation. (In real short-channel devices the capacitance variation is much slower than $1/L_g$ due to fringing effects).

Clearly, to maximize the transconductance at any given $V_{gs} - V_p$ voltage swing one can reduce the gate length L_g to shorter and shorter values. Unfortunately, taking this course alone places unreasonable pressures on the required lithographic precision and will ultimately prove disastrous to yield. Note, however, that we can obtain an equivalent improvement in near-threshold transconductance by go-

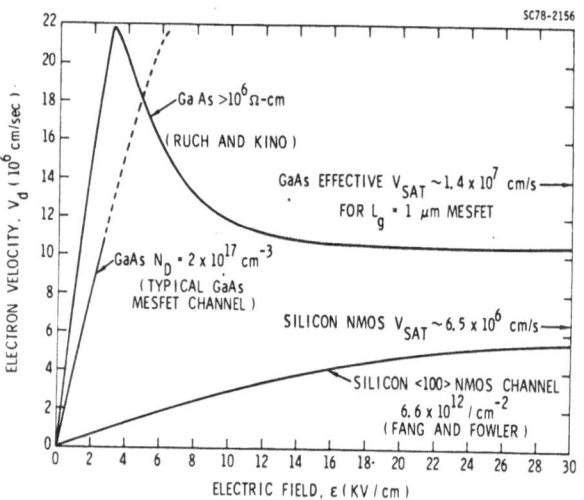

Fig. 1 - Comparison of the equilibrium velocity-field characteristics for electrons in silicon MOS channels and GaAs at 300°K. The high resistivity GaAs exhibits larger mobilities than typical GaAs MESFET channels (the curve for which is simply extrapolated from the low-field mobility).

ing to a semiconductor with a higher channel mobility, μ_n. This is, of course, the prime motivation for using GaAs, rather than silicon in high speed FETs and integrated circuits. Electron mobilities in GaAs of several hundred thousand have been obtained in undoped material at 77°K (8500 cm^2/Vs at room temperature), while in typical MESFET n-channel dopings (N ~ 10^{17} cm^{-3}) electron mobilities in the μ_n ~ 4000 to 5000 cm^2/V_s range are obtained (see Fig. 1). These GaAs electron mobilities are about 6 times higher than those for correspondingly-doped bulk silicon, and nearly an order of magnitude larger than typical silicon MOS n-channel mobilities under strong inversion conditions. As a consequence, very high f_τ's and switching speeds can be obtained in GaAs short channel devices even with small voltage swings, ΔV_L. For example, at room temperature, L_g = 0.6 μm E-beam fabricated

GaAs E-MESFET ring oscillators have given τ_d = 30 ps delays at $\Delta V_L \sim 0.5$ V, while at 77°K (where the electron mobility is even higher), τ_d = 17 ps speeds were measured in the FI = FO = 1 ring oscillators.[3]

B. Super-Mobility Devices: The HEMT

While these 77°K E-MESFET results are impressive, the electron mobilities in the channel at 77°K are not too much higher than at room temperature due to electron Rutherford scattering off of ionized donor impurities (typically $N_D^+ \sim 10^{17}$ in MESFET channels). Special GaAs/Ga$_{1-x}$Al$_x$As FET structures variously referred to as modulation-doped FET's or high electron mobility transistors (HEMTs)[4] should greatly improve this 77°K channel mobility. The key in these structures is to place the donor atoms in a wider-bandgap Ga$_{1-x}$Al$_x$As layer adjacent to an undoped GaAs channel layer, which in fact receives the free electrons from the ionized donors. Electron channel mobilities of μ_e = 49,300 cm^2/Vs at 77°K have been reported for HEMT structures.[5] From Eqs. (3) and (4), such FETs, fabricated with reasonably short gate lengths, should achieve high g_m and f_τ with logic swings of only $\Delta V \sim 100$ mV or so.

C. Saturation and Ballistic Transport Effects

It is important to note the limitations of Eq. (2)-(4), particularly when applied to very short channel devices operating at substantial voltage swings. If we naively used Eq. (4) to estimate the f_τ for a L_g = 0.5 μm HEMT with μ_n = 5×10^4 cm^2/Vs at V_{gs} - V_p = 5 volts we would calculate an f_τ = 1.6×10^{13} Hz (16,000 GHz). However, by these same assumptions, the electron velocity in the channel would be

$$v_d = \mu_n \mathscr{E}_c \simeq \mu_n(V_{gs} - V_p)/L_g \qquad (5)$$

or v_d = 5×10^9 cm/s for the conditions assumed. Inasmuch as the velocity of light in GaAs is only 9×10^9 cm/s, this is, to say the least, suspiciously high (as is, of course, the calculated f_τ). The problem with our naive calculation, was, of course, that we have assumed ohmic,

$v_d = \mu_n \mathscr{E}_c$ behavior at any $\mathscr{E}_c$, while in fact the equilibrium velocity-field curves, as shown in Fig. 1, show velocity saturation behavior at high field strengths. One must hasten to point out, however, that it would be just as naive to assume that in very short channel, low voltage MESFETs or HEMTs that the equilibrium GaAs velocity-field characteristics of Fig. 1 would rigorously apply. If we took, for example, the previous case of an $L_g = 0.5$ μm HEMT with $\mu_n = 5 \times 10^4$ cm^2/Vs, only now operating at a more reasonable $V_{gs} - V_p = 250$ mV, we would have $\mathscr{E}_c = (V_{gs} - V_p)/L_g = 5000$ V/cm. This would give from Eq. (5), $v_d = 2.5 \times 10^8$ cm/s, as based on the ohmic assumption, which is clearly too high, since a purely ballistic (kinetic energy = electrostatic energy) electron of mass m_e dropping though a potential of $V_{gs} = V_p = 250$ mV would have a velocity of only

$$v_B = \sqrt{\frac{2(K.E.)}{m_e}} = \sqrt{\frac{2q(V_{gs}-V_p)}{m_e}} \tag{6}$$

or $v_B = 1.1 \times 10^8$ cm/s for $m_e = 0.072$ times the free electron mass (actually v_B would be slightly less than this, as the GaAs electron effective mass increases somewhat away from the conduction band minimum).

While the infinite electron mass puts an upper limit of about $v_d = 10^8$ cm/s on the electron velocity in our $V_{gs} - V_p = 250$ mV, $L_g = 0.5$ μm HEMT example, the equilibrium static velocity-field curve for GaAs at $\mathscr{E}_c = 5000$ V/cm gives a velocity of only $v_d = 1.8 \times 10^7$ cm/s (this would drop even lower at higher field strengths, approaching $v_s \sim 1 \times 10^7$ cm/s). That this equilibrium velocity-field projection is completely unreasonable for this example is clear when one considers <u>why</u> the GaAs velocity-field curve shows the peak and negative differential mobility behavior (finally leading to the $v_s \sim 1 \times 10^7$ cm/s saturation) that it does. The abrupt reduction in mobility and electron velocity in GaAs at $\mathscr{E}_c \sim 3200$ V/cm is caused by transfer of electrons from the very narrow low m_e conduction band minimum at Γ (K = 000) in GaAs to a plurality of high mass, silicon-like electron minima lying about $\Delta E = 0.3$ eV above the Γ minimum. Electrons in the higher valleys have high mass and strong internal scattering and so exhibit very low mobility (like conduction electrons in silicon). However, note that since these minima lie $\Delta E = 0.3$ eV above the Γ minimum, and we have assumed only a $V_{gs} - V_p = 0.25$ volt maximum potential across the channel for this example, it is physically impossible for electrons to transfer to these low-mobility valleys. Hence, the equilibrium v-$\mathscr{E}$ curve of Fig. 1, measured for long devices with tens of volts across them, is completely irrelevant for this low voltage, short channel HEMT case. In fact, even for $V_{gs} - V_p > 0.3$ V operation in very short ($L_g \sim 0.5$ μm) channel GaAs FETs, the electron transit times through the channel are so short (approaching 1 ps) that electron transfer to higher valleys is not completed during transit. These "velocity overshoot," or "ballistic" effects are predicted to improve the performance of all short transit path GaAs devices, particularly when the operating voltages are low.

D. Effect of Velocity Saturation on FET Performance

Based on the above arguments, we expect that in out $L_g = 0.5$ μm short-channel 77°K HEMT example, electron velocities well above the $v_d = 1.8 \times 10^7$ cm/s equilibrium v-$\mathscr{E}$ curve (Fig. 1) prediction, but definitely below the $v_B = 1.1 \times 10^8$ cm/s ballistic limit for $V_{gs} - V_p = 0.25$ V (and certainly far below the $v_d = 2.5 \times 10^8$ cm/s predicted from $v_d = \mu_n \mathscr{E}$, ignoring the effect of electron mass). Reasonable estaimates from calculations and diode measurements[6] of the transient electron velocity in GaAs under conditions similar to this example are around $v_d \sim 5 \times 10^7$ cm/s. This is as compared to saturation velocities of $v_{sat} \sim 6.5 \times 10^6$ cm/s in bulk silicon (at very high $\mathscr{E}$-fields, approaching 10^5 V/cm). If we assume saturated electron velocity, $v_d = v_{sat}$, (rather than $v_d = \mu_n \mathscr{E}_c$), Eqs. (3) and (4) become

$$g_m = \frac{\epsilon v_{sat} W}{a} \quad \text{(for } v_d = v_{sat}) \tag{7}$$

for transconductance in this velocity-saturation region, and

$$f_\tau = \frac{v_{sat}}{2\pi L_g} \quad \text{(for } v_d = v_{sat}) \tag{8}$$

for the current gain-bandwidth product. By this calculation, the 0.5 μm HEMT with a = 0.1

μm at V_{gs} - V_p < 0.3 V could have g_m/W values approaching 400 mS/mm and f_τ values from

Eq. (8) near 100 GHz for effective v_{sat} values approaching 4×10^7 cm/s (as compared to

rather unrealistic f_τ = 800 GHz and g_m/W = 5300 mS/mm values from Eqs. (4) and (3)

assuming $v_d = \mu_n \mathscr{E}_c$). More realistic capacitance estimates, including fringing, would probably

roughly halve the 100GHz f_τ estimate for the (μ_n = 50,000) HEMT to perhaps 50 GHz. This

is not a great deal higher (perhaps 2X) than would be expected for a normal (μ_n = 4000) L_g

= 0.5 μm, 300°K GaAs MESFET, but the normal FET would require 5 to 10 times higher

voltage swings to achieve full f_τ. As a consequence, the HEMT could in principal achieve

somewhat higher switching speeds in 77°K operation than could a normal GaAs MESFET of

similar geometry, but at far lower logic swings so that the dynamic switching energies might be

one or two orders of magnitude less. This would, of course, be very attractive for VLSI.

E. Limitations to Very Low Logic Voltage Swing Operation

Achieving these exceedingly low $P_D\tau_d$ products in very short channel HEMTs (or other

devices) by operating with very low logic swings ($\Delta V \sim$ 100 mV or less), assumes that such

operation is in fact practical from a circuit standpoint. Given that the device nonlinearities are

strong, as is the case for the HEMT, the limitations on logic swing are that ΔV_o must be large

in comparison to circuit noise levels and large in comparison to key parameter variations, such

as threshold voltage. The thermal noise leVel, $v_n = kT/q$, is only 6.64 mV at 77°K, and logic

swings of 10 kT/q or higher are adequate, so that < 100 mV swings are thermodynamically

acceptable at 77°K. A more serious "noise" problem is the "hash" or transient noise spikes

from high speed switching operations electromagnetically coupled onto the circuit power

supply and ground conductors, as well as directly to other digital signal lines. While, in principle this "hash" can be made arbitrarily small by careful electromagnetic design of the integrated circuit layouts, packages and interconnects, in fact, as switching speeds increase, the L (dI/dt) or M (dI/dt) and C (dV/dt) terms all tend to make this problem worse. It is not unreasonable to expect that switching noise or "hash" levels could become a limiting factor on how low the logic swings may be reduced in practical ultra high speed VLSI circuits.

A much more severe limitation to the reduction of logic voltage swing in GaAs MESFET or HEMT circuits, is the minimum variation of threshold voltage, V_p, of the FETs over a circuit or wafer achievable in the fabrication process. As discussed later, in order to accommodate the statistical variations in V_p without unacceptable degradation of dynamic switching characteristics, the logic voltage swing ΔV_L must be at least ~ 20 times the standard deviation of V_p (preferably higher). This would mean that for the HEMT example of $\Delta V_L = 100$ mV, a $\sigma_{V_p} < 5$ mV uniformity would be required for the FETs. As discussed later, the best FET V_p uniformities achieved over a GaAs wafer are $\sigma_{V_p} = 34$ mV, with $\sigma_{V_p} = 50$ mV more reasonable[21] (see Fig. 10) - much too high for this super low logic swing operation. Hence, attaining ultra-low dynamic switching energies through very low logic swing circuit operation will necessitate either major improvements in materials and processing techniques to allow exceedingly tight control of FET pinch-off voltage, or else the use of a device approach in which the threshold voltage is inherently extremely uniform.

F. The GaAs Heterojunction Bipolar Transistor for Super-Speed VLSI

The realization of very low logic voltage swings requires a high degree of nonlinearity (rapid increase in transconductance for small voltage swings, ΔV_L, above the threshold) combined with extremely uniform threshold voltage devices, which probably means device structures in which the threshold voltage is highly insensitive to normal processing parameter variations (horizontal dimensions, vertical dimensions, doping level variations, etc.). While the GaAs MESFET pinch-off voltage is insensitive to horizontal geometry variations (e.g., L_g), V_p is sensitive to both the thickness (vertical geometry) and doping level in the channel layer. In comparison, an almost ideal device from the standpoint of threshold voltage variations is the bipolar transistor. In a bipolar transistor, V_{BE} varies linearly with temperature, T (which can presumable be kept quite uniform over a chip), but only logarithmically with basewidth, W_B, base doping level, N_A, mobility, collector current, I_c, etc. Threshold uniformities ($\sigma_{V_{BE}}$) of a few millivolts over a circuit or wafer are fairly easily achievable in silicon bipolar transistors.

In spite of the inherent threshold uniformity of bipolar and the great success of silicon (homojunction) bipolar transistors in high speed logic (such as ECL), the homojunction bipolar transistor involves some very unfortunate inherent design tradeoffs which drastically limit performance and make it quite unattractive for GaAs implementation. These design compromises center on the fact that to maintain emitter injection efficiency, the base doping level, N_A, must be limited to a fraction of a percent of the emitter doping. Further, the active (transiting) carrier is the minority carrier in the base (e.g., electrons in an NPN bipolar) and it is up to the rather modest density, N_A, of holes in the base to establish electrical contact from the control electrode to the active region with acceptably small lateral base resistance, R_B. The inherent performance of an NPN bipolar improves strongly as the electron transit time through the base is reduced by lowering the basewidth, W_B, but this leads immediately to unacceptably high levels of base resistance, R_B, unless the emitter stripe width, W_E, is exceedingly small (e.g., $<< 1$ μm). Further, if we lower W_B too far (e.g., < 1000Å), with the modest base doping levels involved in homojunction bipolars, the number of doping atoms in a cube of dimesion W_B on each side gets so small ($\sim 10^2$) that simply the Gaussian statistical variations in this number can lead to emitter-collector punchthrough in a statistically signifi-cant (for VLSI) number of transistors. So far as GaAs is concerned, the fact that the homojunction bipolar transistor required both high electron mobility, μ_n, for short transit time and high hole mobility, μ_p, for low R_B, is a problem because the hole mobility in GaAs is even less than that in silicon. Hence there is little reason to expect that the performance of a homojunction bipolar transistor in GaAs would be much better than that in silicon.

While homojunction bipolar transistors in GaAs offer little apparant advantages over silicon, the availability of high-quality $GaAl_xAs_{1-x}$ to GaAs heterojunctions, as used in the

HEMT FETs for example, makes it possible to fabricate heterojunction bipolar transistors (HJBT) which avoid most of the disadvantageous tradeoffs of homojunction bipolars.[7,8,23] The heterojunction bipolar transistor makes use of a wider-bandgap emitter (typically 10 kT/q larger than that of the base, or $\sim$ 0.25 eV for room temperature use), such as an n-type $GaAs_xAs_{1-x}$ emitter on a heavily-doped p^{++} GaAs base in an NPN HJBT. The larger emitter bandgap makes hole injection from the base into the emitter virtually impossible, even when the base doping level is much higher than that in the emitter. Hence in the HJBT there is no tradeoff between maintaining β or emitter injection efficiency and increasing base doping, N_A, as required to reduce basewidth, W_B without increasing the lateral base resistance, R_B. In addition, the high N_A minimizes base depletion (Early) and punchthrough effects from the collector bias and the lighter emitter doping level reduces the emitter-base capacitance. Like the homojunction bipolar transistor, the current flow is vertical, hence the f_τ - limiting transit path distance (the basewidth, W_B) is limited only by our ability to make thin semiconductor layers, and molecular-beam epitaxial (MBE) layers of a few tens of angstroms in thickness have been demonstrated. Unlike homojunction bipolars, though, the heavy base dopings in HJBTs allow practical use of base thicknesses of a few hundreds of angstroms.

In non-saturating (ECL-type) logic circuits the HJBT may be used with a GaAs homojunction collector(However, if this structure is allowed to go into saturation (excessive I_B with strongly forward biased collector-base junction), hole injection from the base into the collector region will results in a large minority carrier stored charged in the collector which will slow turn-off unacceptably, just as in silicon bipolar transistors operated in saturated switching. Fortunately, in the HJBT, it should be possible to use a wide bandgap collector region (along with emitter) to confine the excess stored charge in saturated operation to the very thin base region only.[23] This should make it possible to use for very high speed VLSI, such simple, very low power, high-density saturated logic approaches as I^2L. Of course, higher speeds would be expected for non-saturated logic approaches (ECL, CSL, etc.), but at somewhat higher power levels and lower gate densities.

The heterojunction bipolar transistor (HJBT), from both an ultimate performance and a parameter statistics standpoint, appears to be ideal for ultra high speed VLSI. Even with comparatively modest (1 μm) emitter stripe widths and comfortable (for MBE) 500Å base widths, f_τ values of the order of 100 to 200 GHz (or intrinsic transistor switching speeds of ~ 1 ps), should be attainable at reasonable emitter current densities, and this is at room temperature. Because, as with the homojunction bipolar transistor, the "threshold voltage" (V_{BE} for some specific I_c) is determined mainly by the energy gaps of the base and emitter regions, rather than geometrical and doping factors, the threshold uniformity should be outstanding ($\sigma_{V_{BE}}$ ~ few mV). Consequently, the logic swings, ΔV, may be reducable to the basic thermal noise or switching "hash" limits for ultra low dynamic switching energy operation. In the ECL-type circuit approaches, the balanced current operation tends, with proper layout and design rules, to minimize the switching noise or "hash", to that very low (ΔV ~ 10 kT/q) logic swing operation may prove practical in HJBT ECL circuits).

If f_τ's of 200 GHz indeed prove achievable in 1 μm emitter geometry heterojunction bipolar transistors with low parasitics (i.e., semi-insulating GaAs substrate), then an FO = FI = 1 HJBT ring oscillator could show logic delays of under 2 ps, or more reasonably loaded circuits propagation delays in the τ_d = 5 ps to 10 ps range. At these propagation delays, on-chip interconnect delays (~ 10 ps/mm) could be quite significant and interconnect lines longer than a few millimeters would probably have to be driven and terminated in their characteristic impedance to avoid reflections. Of course, the high transconductances in the HJBTs should readily facilitate achieving typical line impedances in source follower drivers or active terminations in minimally-sized devices.

In summary, this type of GaAs heterojunction bipolar transistor circuit technology could achieve, at room temperature or 77°K, logic speeds as high or higher than those of Josephson junction circuits at 4°K. Unfortunately, very little work relevant to ultra high speed VLSI has been done on HJBTs. While some work on GaAs/GaAl$_x$As$_{1-x}$ heterojunction bipolars for microwave amplifier applications has been published,[9] most of the very thin base, high $-f_\tau$,

MBE or MOCVD-fabricated HJBT work has been oriented toward phototransistor applications (see references cited in 9 and 23). A complete fabrication technology will have to be devised before this highly attractive device approach can be brought to bear on the achievement of ultra high performance logic for VLSI.

G. Comparison of Prospective GaAs Device Approaches for VLSI

Table I outlined some of the circuit requirements for ultra high speed, low power logic for VLSI and interpreted these in terms on desirable material qualities, such as semi-insulating substrate for low parasitics and device requirements such a very high f_τ at low voltage swings and very uniform threshold voltages. Table II reviewed these device requirements for ultra high speed VLSI in terms of the device structural features and semiconductor parameters (e.g., electron mobilities and velocities) necessary to realize them. The previous five sections have illustrated these device requirements by focussing on specific examples in which their roles are

TABLE III
COMPARISON OF GaAs DEVICE APPROACHES
FOR HIGH SPEED – LOW POWER LOGIC CIRCUITS

DEVICE TYPE	NONLINEARITY HI g_m @ L_o V_C - V_T	THRESHOLD VOLTAGE UNIFORMITY				f_τ	FABRICABILITY, YIELD
		COMBINED	HORIZONTAL SENSITIVITY	VERTICAL SENSITIVITY	DOPING SENSITIVITY		
HETERO-JUNCTION BIPOLAR TRANSISTOR ($W_E \sim 1\,\mu m$, $W_B < 0.1\,\mu m$)	EXCELLENT (KT LIMITED) (g_m @ V_{BE} - V_T $\sim 100\,mV$)	EXCELLENT (ENERGY GAP) $\sigma_{V_T} \sim 1\,mV$	EXC.	EXC.	EXC.	EXCELLENT (100-200 GHz)	UNKNOWN – NO PROCESS DEVELOPED
GaAs MESFET ($L_g = 1\,\mu m$, $a \gtrsim 0.1\,\mu m$)	GOOD (g_m @ V_{gs} - Vp $\sim 1V$)	OK† (DOPING PROFILE) $\sigma_{V_T} \sim 40\,mV$	EXC.	FAIR (ADEQUATE WITH I.I.)	FAIR	GOOD ($\sim 15\,GHz$)	EXCELLENT, SIMPLE, HIGH YIELD
GaAs MESFET ($L_g = 0.5\,\mu m$, $a \gtrsim 0.1\,\mu m$)	GOOD – VERY GOOD (g_m @ V_{gs} - Vp $\sim 0.5V$)	OK† (DOPING PROFILE)	EXC.	FAIR (ADEQUATE WITH I.I.)	FAIR	VERY GOOD ($\sim 30\,GHz$)	DIFFICULT LITHOGRAPHY
HETERO-JUNCTION FET (HIGH MOBILITY) ($L_g = 1\,\mu m$, $a \lesssim 0.1\,\mu m$, $T = 77°K$)	EXCELLENT WHEN COOLED (g_m @ V_{gs} - Vp $\sim 100\,mV$)	OK? (DOPING PROFILE)	EXC.	FAIR (ADEQUATE WITH MBE?)	FAIR	VERY GOOD ($\sim 20\,GHz$ OR 50 @ 0.5 μm?)	MBE FABRICA-TION, NEW PROCESS
"PERMEABLE BASE" TRANSISTOR, PRESENT METAL FINGER SIZE (0.16 μm)	VERY GOOD (g_m @ V_{gs} - Vp $\sim 0.3V$)	VERY POOR (HIGHLY GEOMETRY SENSITIVE)	POOR (FINGER SPACING)	FAIR	POOR	GOOD (17 GHz f_{max} MEASURED WITH 37 GHz f_T)	DIFFICULT EPI PROCESS AND FINGER LITHOGRAPHY
FUTURE "PERMEABLE BASE" TRANSISTOR @ 500Å FINGER SIZE	VERY GOOD	VERY POOR	POOR	FAIR	POOR	EXCELLENT (200 GHz f_τ CALCULATED)	NO KNOWN PROCESS FOR THIS FINGER SIZE

†MESFET THRESHOLD UNIFORMITIES SATISFACTORY FOR D-MESFET LOGIC WITH $\sim$ 1 VOLT LOGIC SWINGS, VERY MARGINAL FOR 0.5V SWING E-MESFET LOGIC.

clear. Table III summarizes this discussion and extends it to a comparison of most of the major contenders in GaAs device technologies for ultra high speed, low power logic for VLSI, from the "plain vanilla" (e.g., $L_g = 1$ μm MESFET) to the "far out" (e.g., 5 ps - 10ps switching heterojunction bipolar transistors). Reviewed in Table III are performance factors such as the ultimate speed capabilities (as limited by f_τ or f_{max}) and the ultimate capability to achieve those high speeds with very low dynamic switching energies (low logic voltage swings as determined by the nonlinearity or voltage swing necessary to achieve high g_m or f_τ).

On a more practical basis, however, the attainment of low voltage swing operation requires a very tight threshold voltage distribution for the switching devices. Hence in Table III, the device approaches are rated according to their inherent sensitivities to horizontal device dimension variations (the worst, since these are caused directly by the lithographic process and are most difficult to control), to vertical device dimension variations (e.g., implant depth, epitaxial layer thicknesses, etc.), and to doping density variations in the active regions of the devices.

Also shown in Table III is a "combined" overall threshold voltage uniformity rating estimated for each device type. These should be considered as tentative, inasmuch as only the horizontal (lithographic) variations are fundamental and common to all device types, and most of devices are insensitive to these horizontal variations. The vertical geometry and doping density variations are highly fabrication approach and fabrication technology dependent. For example, GaAs D-MESFETs fabricated by liquid phase epitaxy (LPE) in the early to mid 1970's had standard deviations of V_p of the order of 1 volt, principally due to epitaxial active layer thickness variations. In our planar ion-implanted GaAs IC process,[1,2,24] we have obtained D-MESFET with standard deviations over a wafer as low as 34 mV (Fig. 10) by careful control of implant range and dose and the compensating impurity level in the substrate material in the region of the active device channels. MBE is capable of fabricating very uniform thickness layers and hence should also be capable of uniform V_p's. Hence, either changes of the method by which these devices are fabricated or improvements in the particular fabrication technologies involved can produce improvements in the threshold voltage uniformities.

These threshold voltage uniformity projections can, in spite of their technology-dependent nature, be useful in weighing prospective device approaches for ultra high speed VLSI. For example, the heterojunction bipolar transistor shows outstanding f_τ and nonlinearity, in combination with inherent outstanding threshold voltage uniformity. The permeable-base transistor,[10,25] a vertical-geometry structure like an NPN bipolar transistor, but with the p-type base replaced with a very fine metal grid (similar to Shockley's analog triode structure), also should, with small enough grid finger geometries ($\sim$ 500Å lines and spaces), be able to achieve 200 GHz cutoff frequencies[10] and have very good nonlinearity, with predicted $\tau_d \sim$ 2ps propagation delays.[25] Unfortunately, in this device structure the pinch-off voltage is not only dependent on the vertical dimensions (epitaxial layer thicknesses) and donor concentrations, but it also varies strongly with a horizontal (lithographically-generated) dimension, the spacing between fingers. Inasmuch as the finger lithography (1600Å lines and spaces, fabricated by laser interference in current f_τ = 37 GHz, f_{max} = 17 GHz experimental devices) is at the edge of the state of the art, it seems unreasonable to expect that tight control of threshold voltages in these devices will be realized. Hence, their use in low power (low logic voltage swing), ultra high speed circuits, particularly at the VLSI complexity level, appears highly unlikely. These comments also apply to other, similar gridded strucutres dimensioned to take advantage of ballistic transport (transient velocity overshoot) effects in GaAs.

An even newer GaAs transistor structure, not included in Table III, which is designed to utilize ballistic electron transport effects to achieve extremely high speeds is the planar-doped barrier transistor.[26] This is another vertical current flow device geometry (like the heterostructure bipolar and pseudobase transistors) which is fabricated by MBE with an n (emitter)-i-p^+ (sheet)-i-n (base)-i-p^+ (sheet)-i-n structures form barriers $\sim$0.3 eV in height so that electrons are injected into the thin ($\sim$1000 Å) base at the optimum energy for high electron velocity (4-9 x 10^7 cm/s) and long mean-free path for energy loss so they can cross the base with sufficient energy and momentum to surmount the collector barrier (while the majority carrier electrons in the base remain confined). One advantage of this hot electron transistor structure over a heterojunction bipolar transistor structure with the same basewidth is that the higher electron velocities should give shorter transit times, e.g., for a $W_B = 500$ Å basewidth, the planar doped barrier transistor would have a base transit time of $t_{TR} < 0.1$ ps, as compared to $t_{TR} \sim 0.25$ ps for the HJBT. However, since the performance for both devices will be limited in the main by parasitic capacitances, this difference is probably not important. A more substantial advantage, in principal, for the planar doped barrier transistors is that they could operate with very low logic voltage swings (at least at 77K where 10 kT/q is low enough) and very low supply voltages ($V_{CC} \sim 0.5$ V) for extremeley low dynamic switcing energies. This would, in principal, make them very attractive for ultra high-speed VLSI. Unfortunately, the inherent threshold voltage uniformity of the heterojunction bipolar transistors the makes their use in very low logic swing VLSI so exciting is not shared by the planar doped barrier transistors. While the planar doped barrier transistor structure is quite immune to horizontal geometry variations, the barrier heights, and hence the threshold voltages, are critically dependent both on vertical geometry (MBE layer dimensions) and the exact number of acceptors/cm^2 in the two p$^+$ doping sheets. Hence, it would appear that while this structure offers performance attractions, its application to VLSI where very precise threshold control is needed will require significant improvements in MBE fabrication technology.

The last column in Table III, while perhaps mundane from a device physics point of view, is probably the most important of all for VLSI: fabricability and yield potential. The enormous numbers of semiconductor devices in VLSI circuits ($\sim 10^5$/chip) virtually demands that the devices themselves have very high yields and that the process for IC fabrication be as simple as possible (as few steps as possible) and high yield in every step. While the realization of high yield is quite dependent on the state of development of the fabrication technology, certain device approaches clearly are more easily fabricated and promise higher yields in the context of the technology known at a given time. Hence, while from fundamental device point of view, the heterojunction bipolar transistor appears the device of choice for ultra high speed, low power logic, major fabrication technology problems need be solved before the device can even seriously be evaluated for logic ICs, let alone actually be used for VLSI circuits.

From a practical fabrication/yield standpoint, the only GaAs device approach having demonstrated potential for LSI/VLSI circuits is L_g = 1μm D-MESFETs (with 1000-gate circuits demonstrated with Schottky diode-FET logic). While these 1μm gate length MES-FETs (at τ_d ~ 50 ps to 150 ps) are an order of magnitude slower than what heterojunction bipolar circuits may one day achieve, they exist today and they are still nearly an order of

magnitude faster than available silicon homojunction bipolar circuits
and take a small fraction of the
power. Figure 2 shows the structure for a planar, double-implanted
GaAs MESFET frabricated by
localized implantations into semi-

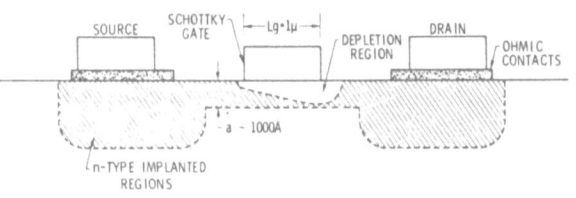

Fig. 2 - Cross-section of the planar GaAs Schottky-gate field effect transistor (MESFET) structure used in SDFL LSI circuits. The source-drain ohmic contact spacing is 4 μm and the deep n^+ contact implant spacing is 2.5 μm in this L_g = 1 μm geometry.

insulating (~ 10^8 Ωcm) GaAs.[1,2] The basic structure is very simple, comprising a thin (a ~ 1000Å-2000Å) n-type (typically N_d ~ 2×10^{17} cm^{-3}) active region joining two ohmic contacts (separated typically by a distance of L_{ch} = 4-5μm) with a narrow (L_g ~ 1μm) metal Schottky barrier gate separating the source and drain. The GaAs FET operation is similar to that of a normal JFET, with the conducting n-channel confined between the gate depletion region and the semi-insulating GaAs substrate. By varying the thickness and doping level of the active region, the pinch-off voltage, V_p, and saturated drain current (at V_{gs} = 0), I_{dss}, for the GaAs MESFET can be varied, even down to the point where the normal built-in voltage of the Schottky barrier gate can pinch-off the FET (enhancement-mode MESFET operation). Note in Fig. 2 that one narrow (L_g = 1μm) metal line is required, but that the alignment accuracy (~ ±1μm) required is not too stringent for conventional projection or contact photolitho-graphic methods. For this reason, the device fabrication for the L_g = 1μm GaAs MESFET technology is readily accomplished with commercial step and repeat projection aligners, including, presumably, those capable of large production volume levels.

We have reviewed in this section, and summarized in Table III, what should be achievable with various GaAs device approaches in future ultra high speed VLSI. Rather than concentrate further on what could be done with different devices if you could make VLSI complexity circuits of them, we will shift our focus to whether it is reasonable to expect that circuits of such complexity can be fabricated in GaAs at all. The most practical approach to demonstrating that GaAs VLSI is indeed reasonable is to describe the process approaches used at Rockwell in obtaining our 1000-gate LSI GaAs ICs and yield and performance results obtained from this work.

III. YIELD ANALYSIS AND IMPLICATIONS FOR VLSI DESIGN

The largest fully working GaAs integrated circuits fabricated to date, the Rockwell SDFL 1008-gate 8×8 bit latched parallel multipliers, contain 6000 semiconductor devices (3000 D-MESFETs and active loads and 3000 Schottky barrier diodes), 12,000 circuit overcrossings and 7000 VIA interconnections between the 1st and 2nd level metallizations, on a chip area of 6.7 mm^2. If we consider even a "lower VLSI" part such as a 1ns access time, 4K bit static RAM, the chip area would be only slightly larger ($\sim$ 8.7 mm^2 est.), but there could be 60,000 or more semiconductor devices (30,000 FETs and active loads and 30,000 diodes), with nearly 50,000 VIAs and 75,000 overcrossings. To achieve fully working circuit operation, it is necessary that none of these devices or circuit features be fatally defective (e.g., a shorted overcrossing, open VIA, or a broken gate on a MESFET). It further requires that all of the semiconductor devices (MESFETs and diodes) not only must be functional, but must be functional with their key device parameters (e.g., FET pinch-off voltage or diode forward voltage drop) lying within a range that will allow proper circuit functioning (both static and dynamic). These two yield-limiting factors, which we will refer to as "fatal defects" and "statistical parameter variations," respectively, are best treated independently, for clarity in understanding the VLSI yield. In this view, the overall yield for an IC chip, Y_{CHIP}, is given as the product of two yield factors,

$$Y_{CHIP} = Y_D \, Y_S \qquad (9)$$

where Y_D is the yield associated with the finite density of "fatal defects" and Y_S is the yield associated with the finite probability that the parameters of one or more of the devices making up the circuit may be out of the range required for proper static or dynamic performance.

A. Yield Reduction from Fatal Defects Density

It would be, in principal, possible to consider the yield due to such "fatal defects" as overcrossing shorts on the basis of the probability of one overcrossing being good, taken to the power of the number of overcrossings. These probabilities are so close to unity as to lack physical meaning (aside from statistical insight) and device features such as overcrossings (unlike the semiconductor device parameters) have no physical basis for showing an inherent statistical nature which would explain non-unity feature yields. Useful insight into these fatal defect yield loss mechanisms has been made on the basis of associating the occurrence of these features or gross device failures with an assumed finite area density for such defects. This basic approach has proven very useful in explaining silicon LSI and VLSI circuit yields as a function of chip area and chip scaling. In this model, the defect-related chip yield is associated with the probability of including fatal defects in a given chip area. If the density of these fatal defects is D_D (per cm^2) and the area of the IC chips is A_C (cm^2), then the mean number of defects per chip on the wafer will be

$$\overline{N}_D = D_D \, A_C \qquad (10)$$

Considerable effort has gone into developing yield models based on various assumptions about the nature of these area defect distributions (in particular, the degree of correlation between defect locations). The simplest model is to assume the defects are totally uncorrelated, so that Poisson statistics apply. This model is known to give excessively pessimistic results for LSI chip yields as area is increased, but it will serve adequately to illustrate the point here. Assuming uncorrelated defects, the probability of finding any number (positive integer), N_D, of these defects in a chip is given by the Poisson distribution as

$$P_D(N_D) = e^{-\overline{N}_D} \frac{\overline{N}_D^{N_D}}{N_D} \quad \text{for } N_D = 0,1,2,3... \tag{11}$$

In most integrated circuit chip architectures, even one fatal defect is too many for the circuit to function. Hence, in this "zero allowable defects" IC approach, the chip yield is given by

$$Y_{D0} = P_D(0) = e^{-\overline{N}_D} = e^{-D_D A_C} \tag{12}$$

Typical values of defect density-chip area products and their associated "zero defect" yields, Y_{D0}, are shown in the column labelled Y_0 in Table IV. From Eq. (12), we would predict, for

TABLE IV
YIELD IMPROVEMENT EFFECT FROM MEMORY ARRAY REDUNDANCY

CHIP AREA x DEFECT DENSITY $\overline{N}_D = A_C \times D_D$	NUMBER OF REDUNDANT ROWS, N_{RR}						
	0	1	2	3	4	5	6
	Y_0	Y_1	Y_2	Y_3	Y_4	Y_5	Y_6
0.6931	50%	84.66%	96.67%	99.44%	99.92%	99.99%	99.999%
1.6094	20%	52.19%	78.09%	91.99%	97.58%	99.38%	99.86%
2.3026	10%	33.03%	59.54%	79.88%	91.59%	96.99%	99.06%
2.9957	5%	19.98%	42.41%	64.82%	81.60%	91.65%	96.67%
3.9120	2%	9.82%	25.13%	45.08%	64.60%	79.87%	89.83%
4.6052	1%	5.61%	16.21%	32.48%	51.23%	68.49%	81.73%
5.2983	0.5%	3.15%	10.17%	22.56%	38.98%	56.38%	71.74%
6.2146	0.2%	1.44%	5.31%	13.31%	25.74%	41.19%	57.19%
6.9078	0.1%	0.79%	3.18%	8.67%	18.16%	31.26%	46.35%
7.6009	0.05%	0.43%	1.87%	5.53%	12.49%	23.06%	36.45%
8.5172	0.02%	0.19%	0.92%	2.98%	7.36%	14.83%	25.44%
9.2103	0.01%	0.10%	0.53%	1.83%	4.83%	10.35%	18.83%
9.9035	0.005%	0.05%	0.30%	1.11%	3.11%	7.08%	13.63%
10.8198	0.002%	0.02%	0.14%	0.56%	1.70%	4.18%	8.63%
11.5129	0.001%	0.01%	0.08%	0.33%	1.07%	2.75%	5.99%

example, that if the 1000-gate 8×8 bit multiplier chip with A_C 6.7 mm^2 gave a 20% yield

$(D_D = 1.6094/0.067$ cm$^2 = 24$ cm$^2)$ then a 4K static RAM chip requiring $A_C = 9$ mm^2

would give a yield of exp $(-24 \times 0.09) = 11.5\%$.

In fact, for parts such as memory chips which are very regular in structure, we needn't

limit ourselves to accepting the stringent "zero allowable defect" criterion for yield. If we can

provide some number of redundant rows (or columns), N_{RR}, in the memory array to be

reprogrammed into use at wafer probe to replace rows containing defective cells, then we can

accept up to N_{RR} defects, rather than zero defects, on the chip and still obtain functional parts

(assuming all defects are in the array itself). For a circuit tolerating up to N_{RR} defects (rather

than 0) defects, we have for the chip yield

$$Y_{DN_{RR}} = \sum_{N_D=0}^{N_{RR}} P_D(N_D) \tag{13}$$

where $P_D(N_D)$ is found from Eq. 11. In general, $Y_{DN_{RR}}$ is much higher, of course, than Y_{DO},

as illustrated in Table IV for various defect density-chip area products and N_{RR} values from 0

to 6.

To illustrate the effect of this redundancy on yield, consider the simple case when the

average number of defects/chip is $\overline{N}_D = A_C D_D = 1$. Here the yield of perfect parts is, Y_O

$= 36.8\%$, while adding only one redundant row would double it to $Y_1 = 73.6\%$. For

$\overline{N}_D = 3$, a higher defect density would drop the yield of perfect parts to $Y_O = 5\%$. while

even one redundant row would quadruple that yield to $Y_1 = 20\%$ and more rows would given

even better results ($Y_2 = 42.3\%$, $Y_3 = 64.7\%$ and $Y_4 = 81.5\%$). As shown in Table IV,

more dramatic improvements are obtained in cases when the defect density is so high as to be

a total "wipeout" for non-redundant parts. For example, with a mean defect count of

$\overline{N}_D = A_C D_D = 6.908$, the yield of "perfect" parts would be $Y_O = 0.1\%$ (unuseably low),

but only 4 redundant rows would raise this to $Y_4 = 18.2\%$ and 6 redundant rows would give a

very reasonable $Y_6 = 46.4\%$ from this otherwise useless wafer. Even providing 6 redundant rows would increase the memory array by less than 10%. Reprogramming the replacement rows would add significant cost to the test procedure, but if this correction process were reasonably automated the costs should be more than amply repaid by the increased yield, at least until the "perfect part" yield gets up into the $Y_O = 50\%$ to 80% region.

B. Yield Reduction from Statistical Variations of Device Parameters

While the chip area increase from the 1000-gate GaAs SDFL 8×8 bit multiplier chip (6.7 mm^2) to that estimated for a 1 ns 4K static RAM ($\sim 8.7\ mm^2$) is not too large, the increase in the number of GaAs MESFET (from 3000 to 30,000) and diodes (same X10 increase) is quite dramatic. VLSI parts with $10,000^+$ gates would have nearly 10^5 semiconductor devices on each chip. This causes yield problems over and above the "fatal defect density" problem discussed in the previous section.

The achievement of ultra high speed VLSI requires low power circuits having very low dynamic switching energies, which is made possible by operation with reduced logic voltage swings, ΔV_L (as discussed in Secs. I and II). As ΔV_L is reduced, however, the uncertainties in gate or FET threshold voltage may no longer be negligible in comparison to the logic voltage swing. This means, for example, that there is a finite statistical probability that one of the FETs will have a threshold voltage so far from the mean that the logic gate will not function properly (either statically or dynamically) in the circuit. These statistical variations were, in the main, unimportant in older $V_{DD} = 5$ volt supply silicon NMOS LSI chips, since the several-volt logic swings were very large in comparison to typical standard deviations in threshold voltage of the MOSFETs. However, as high-performance VLSI with logic swings approaching 1 volt emerges, this issue will be important for silicon NMOS, just as it is today in the ultra high speed, low power GaAs circuits. This problem also becomes drastically worse as the number of devices per chip is increased. In fact, it might best be called the "stone wall" effect because of the sudden, drastic dropoff of yield in parts with large numbers of devices when the ratio, X, of the allowable limit for a parameter ΔV_A (measured from the mean), to the standard deviation of the parameter, σ_v, or

$$X = \frac{V_{UL} - \overline{V}}{\sigma_V} = \frac{\Delta V_A}{\sigma_V} , \qquad (14)$$

gets too large. If we assume Gaussian statistics for the device parameter, then the probability that a given device will be beyond the acceptable excursion limit of ΔV_A from the mean will be $P(\Delta V_A / \sigma_V) = P(X)$, where $P(X)$ is the well known error integral.[11] If this is the only statistical parameter limiting yield and there are N such devices with the same limit, V_A, and statistics, σ_V, on the chip, then the statistical chip yield, Y_s, in Eq. (9) will be given by

$$Y_S = [P(X)]^N \quad \text{(single limit case).} \quad (15)$$

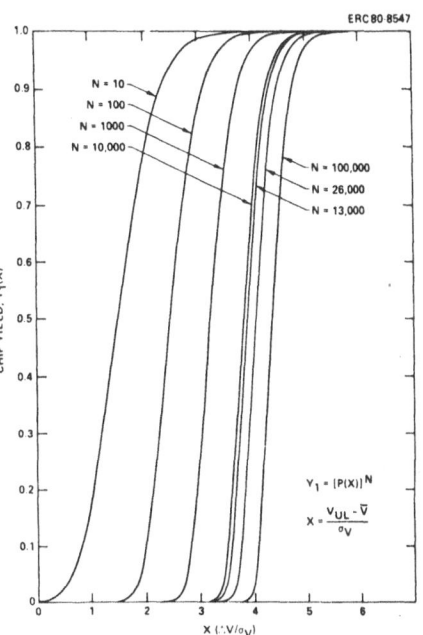

This function is graphed in Fig. 3 for various numbers of devices on a chip, N. Note that for N = 13,000, taking a parameter limit from $\Delta V_A = 4.5$ α_V down to $\Delta V_A = 3.5 \ \sigma_V$ drops the yield from over 95% to under 5%, and going to $\Delta V_A = 3 \ \sigma_V$ reduces the yield to 2.4×10^{-8}. Actually, in general, there are both upper and lower limits on a parameter, typically with, for best yield, the nominal mean located in the center, so that the yield expression for Eq. (9) should be

$$Y_S = [P(X^+) \ (P(X^-)]^N \quad \text{(two − limit case)} \quad (16)$$

where $X^+ = (V_{UL} - \overline{V})/\sigma_V$ and $X^- = (\overline{V} - V_{LL})/\sigma_V$, with V_{UL} the upper limit of the parameter, V_{LL} the lower limit, and $\overline{V}$ the mean of the

Fig. 3 - Chip yield due to statistical parameter variations, calculated assuming Gaussian statistical variations of a parameter (like FET V_p) and a single (e.g., upper) limit on the value of that parameter for circuit operability. Yield, Y_s, is plotted for various numbers of devices in the chip, N, as a function of X, the ratio of the difference between the limit and the mean, ΔV_A, and the standard deviation σ_V.

parameter over the wafer. For the simple example of an optimum wafer mean, $\overline{V} = \frac{1}{2}$ ($V_{u1} + V_{LL}$), then taking $\Delta V = \frac{1}{2} (V_{UL} - V_{LL})$, Eq. (15) and Fig. 3 still apply except that we must replace N by 2N. (i.e., a 13,000 device chip would use the N = 26,000 curve in Fig. 3; this is even steeper, dropping from 91.5% at 4.5 σ_v to 0.24% at 3.5 σ_v).

In real integrated circuits, the situation is even worse than illustrated in Fig. 3 because of the fact that the <u>mean itself</u>, $\overline{V}_p$, varies from wafer to wafer, usually with a standard deviation, σ_w, exceeding that of the parameter variation over a given wafer (σ_v). This distribution of means, $\overline{V}_p$, for an assemblage of wafers has a degrading effect on average yield which can be calculated in integrating Eq. (16) over the distribution of wafer means with the appropriate normalized probability weighting to give the average. A computer program for this analysis has been written for the case when the "mean of means" is optimum ($\overline{V}_p = \frac{1}{2} (V_{UL} + V_{LL})$) and the distribution of $\overline{V}_p$ is Gaussian with a standard deviation of σ_W. For example, if we consider an E-MESFET RAM with a pinch-off voltage limit separation of $V_{UL} - V_{LL} = 300$ mV for circuit function ($\Delta V_A = 150$ mV), we find that if we take (a very optimistic value of) $\sigma_v = 35$ mV for the standard deviation of pinch-off voltage, from Fig. 15 (with X = 4.286 on the N = 26,000 curve), a theoretical maximum yield for N = 13,000 FET RAMs of 78.9% for a wafer with precisely optimum $\overline{V}_p$. Among a group of wafers with an average value of $\overline{V}_p$ at optimum, but with a standard deviation $\sigma_W = 70$ mV (also very optimistic); the average theoretical maximum yield would fall to less than 15%, while for a more realistic $\sigma_W = 150$ mV, to less than 7%. In fact, a very good standard deviation of GaAs MESFET pinch-offs over one wafer is $\sigma_v = 45$ mV (from some of our planar IC wafers), which would give, for $V_{UL} - V_{LL} = 0.3$ V, a theoretical maximum yield of 1.4×10^{-5} (with $\sigma_W = 0$; 5.5×10^{-7} with $\sigma_W = 100$ mV).

This example can also serve to illustrate the extreme danger of extrapolating SSI yield data to VLSI parts. While the $\sigma_v = 45$ mV E-MESFET example for the N = 13,000-FET RAM gave an unusable $\overline{Y}_S = 5.5 \times 10^{-7}$ with $\sigma_W = 100$ mV, this same wafer collection would give an average yield of $\overline{Y}_S = 65\%$ on an N = 5-FET part or $\overline{Y}_S = 18\%$ for an N

= 10-FET SSI chip. Even if σ_v and σ_W were both as high as 100 mV, N = 10-FET SSI parts would yield up to 10.5% (or 27% for N = 5), so that it would be easy to feel, on the basis of SSI results that FET pinch-off control was "no problem," when in fact yield would hit the "stone wall" at N = 30 ($\overline{Y}_S$ = 0.09%). Yield for N = 13,000 FETs would be, of course, non-existent.

The conclusion to be drawn from this examination of the relation of statistical parameter variations to yield is that it would be <u>extremely</u> unwise to make a commitment to a device technology approach for VLSI, whatever the apparent attractions, if the peak to peak parameter variation limits (2 V_A) allowable in the circuit approach (for proper static and dynamic operation) is not at least 10 times the standard deviation for the device parameter (e.g., σ_{v_p}) which can reasonably be expected to be obtained with the fabrication process. This will typically mean that the logic voltage swing, ΔV_L, must be at least 20 (preferably more) times the standard deviation of FET pinch-off voltage in FET logic to avoid dynamic switching speed and noise margin degradation. With the present "good" σ_{v_p}'s ~ 50 mV for the best MESFET process and the logic swings of GaAs E-MESFET logic limited to ΔV_L ~ 500 mV, it would appear that this technology is not quite ready to attack VLSI. (The somewhat larger voltage swings allowable in E-JFET circuits would, for the same σ_{v_p}'s, help this problem.)

In our depletion-mode SDFL logic circuits (see Fig. 4(a)), the typical voltage noise margins are of the order of 0.6 V or more on either a "high" or "low". In comparison to the observed σ_{v_p} ~ 50 to 100 mV wafer uniformities (let alone the σ_{v_p} ~ 17-30 mV short range uniformities), the statistical probability of exceeding the voltage noise margin is negligible. This is a consequence, of course, of the fact that the logic voltage swings and noise margins have been kept fairly large in the SDFL circuit designs in order to cope with the significant switching transient induced ("hash") noise levels that are commonly present in very high speed digital circuits and systems. The <u>current</u> noise margins, or relationship between the current drive capability of a gate and the input current requirements of its fanout of loading gates, is a different issue, however. In general, optimum switching speed performance is obtained when the source drive capabilities and loading gate input current requirements are closely matched. This is always the case for transient currents in dynamic measurements on FET logic circuits but it is even more obvious in SDFL where this is also reflected in a static fanout limitations because of the DC input current requirements of the gates (Fig. 4(a)). The static fanout, and hence the current noise margin, may be increased in the SDFL circuits by increasing the ratio of the output pullup, (PU) to the input pulldown (PD) active load current ratio (by adjusting device widths). On the other hand, increasing this ratio much beyond that required to meet the particular fanout loading conditions for the gate involves some sacrifice in switching speed. For this reason, most of out SDFL circuits are fabricated with a current noise margin of only about 30%. With the observed long-range standard deviations in device saturation currents, σ_{Idss}, in the 6% to 8% range, little influence on yield would be expected, even for > 1000 gate LSI circuits. If σ_{Idss} were 10%, significant yield reductions (to about 26% on 1000 gate circuits) would be expected, so that it would be necessary to raise the current noise margin slightly. For VLSI designs it will be important to watch these active load statistics carefully, and to optimize the performance/yield tradeoff according to the statistical device parameter control available in the process.

IV. THE SCHOTTKY DIODE-FET LOGIC CIRCUIT APPROACH FOR ULTRA HIGH SPEED GaAs VLSI

The achievement of LSI or VLSI requires high gate densities, as attained in part in this work by the use of high resolution (1μm) lithographic techniques. Another important factor defining the number of logic gates attainable per unit chip area is the circuit design used. Whereas in all previously published GaAs FET digital IC work the GaAs FETs themselves are used as the nonlinear logic elements, with this planar, multiple localized implantation approach, optimized GaAs Schottky barrier switching diodes can be fabricated. Inasmuch as GaAs Schottky barrier diodes are very strongly nonlinear and are among the fastest-switching semiconductor devices existent, and their switching energies ($\sim 10^{-15}$ joules), required chip areas (1μm × 2μm active area) and capacitances ($\sim$ 2 fF) are very low, their use as logic elements in GaAs digital ICs seems very desirable. We have developed a logic circuit approach[22] using such high speed GaAs Schottky barrier switching diodes[12,14] for most logic functions, with, in its simplest form, GaAs Schottky gate FETs used primarily for inversion and gain, as shown in Fig. 4(a).

Figure 4(a) shows the circuit diagram for a simple 4-input Schottky diode-FET logic NOR gate.[22] The logical OR function (prior to inversion) is accomplished by the very small high conductance GaAs Schottky barrier diodes. In addition to their logic function, these diodes provide ~ 0.75 V of level shifting between the positive logic levels from previous stages and the negative gate voltages required by the normally-on, depletion mode Schottky

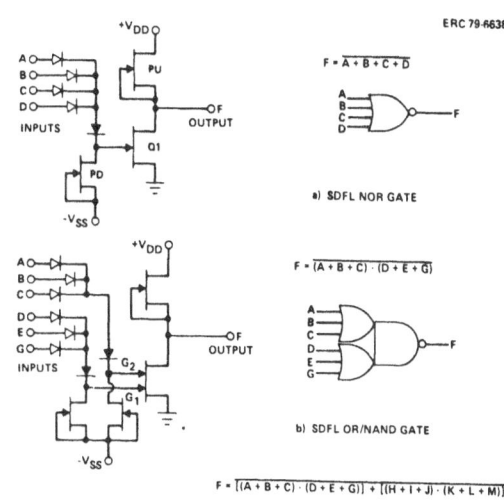

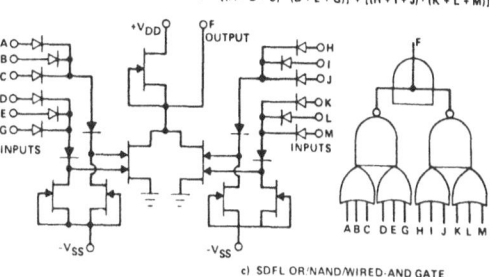

Fig. 4 - Schottky diode-FET logic (SDFL) gate configurations for 1-, 2- and 3-logic level gates. All FETs are depletion-mode typically -1.5 V < V_p Schottky diodes, while shaded diodes are larger area, higher capacitance voltage shifting diodes.

gate GaAs FETs. For very low pinch-off voltage GaAs FETs ($V_p \approx$ -0.5 V), the switching diodes (D_A - D_D) alone provide adequate level shifting for very low voltage and very low power ICs. By adding the level shift diode, D_s, in Fig. 4, GaAs FETs with slightly higher pinch-offs ($V_p \sim$ -0.7 to -1.4 V) may be used with higher supply voltages and logic swings; the data presented here are for gates with this shift diode.

The pulldown to -V_{SS}, PD, sinks the bias current for the diodes and provides most of the turn-off current for the gate of Q1. PD is kept small (typically < 20% of the width of Q1). However, since in this current-sourcing logic the dc fanout is limited to the current ratio between PU and PD (typically 3 or 4), larger fanouts are achieved, where necessary, by adding a source-follower or other output buffer configuration to the gate. Because of the very small size of the logic diodes, and because of the fact that the diodes, which are two-terminal devices, require fewer overcrossings and vias than circuits using 3-terminal devices like FETs, multi-input SDFL NOR gates require only a fraction of the chip area needed by FET-logic approaches.

In the interests of efficiency and speed it is often desirable to implement logic functions with multi-level gates. In FET logic, the use of combinations of series (NAND) or parallel (NOR) FET configurations to achieve two-level logic functions is common. For example, in earlier GaAs work, Van Tuyl, *et al.*[13] demonstrated depletion-mode GaAs MESFET buffered-FET logic (BFL) NAND/NOR (or NAND/WIRED-AND) gates, with up to two NAND input terms (series or dual-gate FETs) and up to two of these NAND functions drain dotted together, having propagation delays as low as $\tau_d = 110$ ps. Four of these gates were utilized to implement a fast ($f_c \sim 1/(2\tau_d)$ complementary-clocked $\div$ 2 frequency divider stage which gave toggle frequencies up to 4.5 GHz, for $\tau_d \sim 111$ ps effective gate delays, at a $P_D \sim 40$ mW/gate power level.[13] Here, the advantage of a two-level gate is clear, since the fastest NOR-gate implemented complementary-clocked divider circuit uses 8 gates and has a maximum speed of $f_c = 1/(4 \ \tau_d)$. In SDFL, we have fabricated more easily used, single-clocked T-connected D-flip flops using 6 NOR gates, achieving maximum toggle frequencies of $f_c = 1.9$ GHz at $P_D = 2.5$ mW/gate, which corresponds to $\tau_d = 110$ ps ($f_c = 1/(4.85 \ \tau_d)$ in this configuration[1,2,17]). Hence, while the SDFL NOR gates gave the same propagation delays as the BFL gates (and at much lower power levels), the architecture of the NAND/NOR dividers allowed much higher toggle frequencies to be reached; this has been (inaccurately) cited as a significant limitation of the SDFL circuit approach.

In fact, such multi-level logic gate configurations may indeed be realized in SDFL[15] with up to 3-level gates with many (10-20) inputs; not being restricted to 4-input 2-level gates as in BFL. Thus, the Schottky diode-FET logic subsumes normal FET logic, to which is added the additional level of logic provided by the ultra high speed switching diodes.

In our earlier work[12,14,16] the FET logic function utilized was principally the inverter, so that, with the diode-OR, a NOR gate function was realized; Fig. 4(a). On the other hand, by using such diode-OR clusters on each gate of a dual gate FET (or series-FET connection), an OR/NAND 2-level gate is achieved - Fig. 4(b) - which is the complement of the 2-level NAND/NOR gate realized in BFL, except that the number of first level terms in the SDFL version is no longer restricted to two. Further, drain dotting can be used between two of these SDFL OR/NAND gates to achieve a 3-level OR/NANA/WIRED-AND gate configuration (Fig. 1(c)), achieving the (positive) logic function

$$F = \overline{[(A + B + C) \cdot (D + E + G)] + [(H + I + J) \cdot (K + L + M)]} \quad (17)$$

or equivalently

$$F = \overline{[(A + B + C) \cdot (D + E + G)]} \cdot \overline{[(H + I + J) \cdot (K + L + M)]} \quad (18)$$

While the application of the two-level SDFL OR/NAND gates of Fig. 1(b) is obvious because of its complementary relationship to the widely-used NAND/NOR configuration,[13] the utility of the three-level SDFL OR/NAND/WIRED-AND gates - Fig. 1(c) - is less clear and best illustrated with an example. One of the major accomplishments of our GaAs IC effort has been the demonstration of a 1008 (NOR) gate 8×8 bit latched parallel multiplier. The parallel multiplier array in this part consists of 760 NOR gates, principally used in the 48 full adder cells which (with 8 half adders) make up the array. A 12-gate, 3 τ_d to sum and $2\tau_d$ to carry full adder cell configuration was selected for this design. The speed of the multiplier could be improved by reducing the number of logic gate propagation delays, τ_d, required to accomplish the full-adder function. Even with NOR gates a 2 τ_d delay for both sum and carry can be achieved, but at the (prohibitive) expense of an 18-gate cell complexity. By using 2-level and 3-level SDFL gates, a full adder circuit (Fig. 3 of Ref. 15) can achieve a 1 τ_d to carry (C or $\overline{C}$) and 2 τ_d to sum (S or $\overline{S}$) delay with only 6 gates (4 OR/NAND and 2

OR/NAND/WIRED-AND). This would reduce the delay for an 8×8 bit parallel multiply from the 35 τ_d of the NOR implementation to 21 τ_d, a 40% reduction. The power shouldbe greatly reduced, since only half as many gates are required, although the yield would not be expected to be improved since similar numbers of logic FETs are required. The essential speed advantage of the 3-level OR/NAND/WIRED-AND gate in generating $\overline{C}_{OUT}$ from C or C_{OUT} from $\overline{C}$ in only 1 τ_d can be utilized in fast ripple-adders without generating all four (S, $\overline{S}$, C_{OUT}, $\overline{C}_{OUT}$) terms by changing from propagating C_{OUT} to $\overline{C}_{OUT}$ in alternate

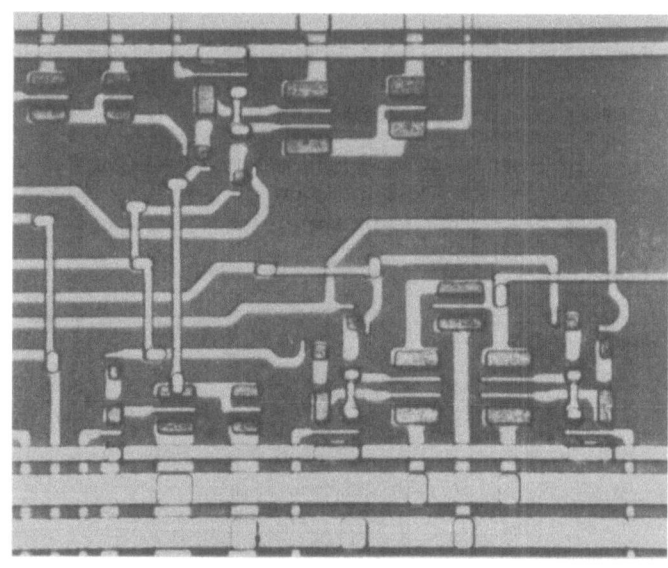

Fig. 5 - Microscope photograph or portion of planar SDFL circuit (Fig. 4 of Ref. 15) showing 1-, 2- and 3-level SDFL gates. At the lower left is an inverter (1-input NOR gate), at the top center is a 2-level OR/NAND gate, while at the lower right is a 3-level OR/NAND/WIRED-AND gate.

stages. This approach (illustrated in Fig. 4 of Ref. 15) has been demonstrated in fast ripple carry 4-bit adder chip (Table VII). Figure 5 shows a microscope photograph of a section of this adder circuit showing 1-, 2- and 3-level SDFL gates.

V. PLANAR GaAs LSI/VLSI FABRICATION TECHNOLOGY

During the past several years the development of GaAs integrated circuits has been making rapid progress.[1] Currently, several GaAs IC technologies have achieved MSI status and are expected to impact near term high speed signal processing applications.[2] However, as the trend of modern electronics suggests, for GaAs IC technology to make a serious impact on future electronic systems, it must be capable of LSI/VLSI complexities. GaAs fabrication approaches and processing techniques much be selected that are extendable and fully compatible with the yield and reliability demands of high density complex LSI/VLSI.[18]

This section described Rockwell's fabrication technology which was specifically designed and developed to satisfy the goal of LSI/VLSI.[18,20] This goal provided strong motivation for developing a planar fabrication process using the low power Schottky diode FET logic (SDFL) circuit approach shown in Fig. 4. In the following we will discuss a planar GaAs integrated circuit fabrication technology which has been demonstrated at the LSI level and is potentially capable of achieving VLSI. This technology combines advanced planar device and multi-level interconnect structures with state-of-the-art LSI processing techniques including multiple localized ion implantations, reduction photolithography, plasma etching, reactive ion etching and ion milling. Using this process technology, LSI (1008 gate) level GaAs ICs have been successfully fabricated and tested.[24]

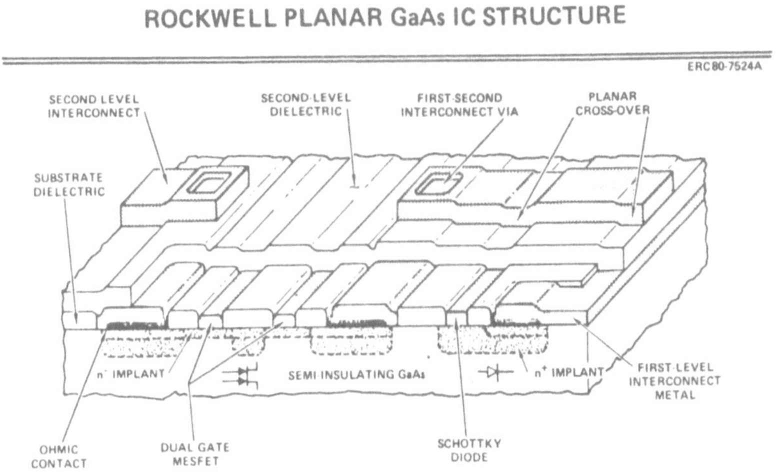

ROCKWELL PLANAR GaAs IC STRUCTURE

Fig. 6 - Cutaway view of a planar GaAs IC showing a dual-gate FET, a diode, and interconnects.

A. Planar Ion Implanted Fabrication Process

The development of an ion implantation technology in GaAs has made it possible to conceive new fabrication approaches not viable within the limitations of epitaxial layers. GaAs implantation advancements, coupled with the observation that the rapid development of

Si LSI was greatly facilitated by planar fabrication techniques using ion implantation and dielectric passivation, suggested that similar developments in GaAs would enhance the development of LSI/VLSI. The schematic of Fig. 6 represents a cutaway view of the Rockwell planar IC approach. The planar circuits are fabricated by using multiple localized ion implants directly into semi-insulating GaAs substrates. Hence, individual devices can be optimized by using different implants with the unimplanted GaAs substrate providing isolation between adjacent devices. This fabrication method ideally compliments the Schottky diode-FET logic (SDFL) circuit approach (shown in Fig. 4) since SDFL requires the MESFETs and Schottky diodes to have different implanted active layers. The sophistication of this process technology is illustrated (Fig. 6) by the use of a 1 μm wide n^+ implant region placed between dual 1 μm MESFET gates in order to lower the channel resistance between the dual gates. Another important aspect of this approach is the use of dielectrics. Dielectric regions are utilized for post implantation annealing, protecting the GaAs surfaces during processing, and passivating the surface. Also, this fabrication approach is compatible with any number of implantation steps. Therefore, this process technology has, in principle, the capabilities for producing mixed device types, such as E-MESFET (or E-JFET) and D-MESFET on the same GaAs IC chip.

The planar implantation and lithography process steps used in this work are shown in Fig. 7. It is a totally dry, (no wet chemical process steps) simple, (only six mask levels) high yield LSI/VLSI compatible process. The material processes are characterized by multiple, localized implantations, (non-epitaxial) providing planar device structures

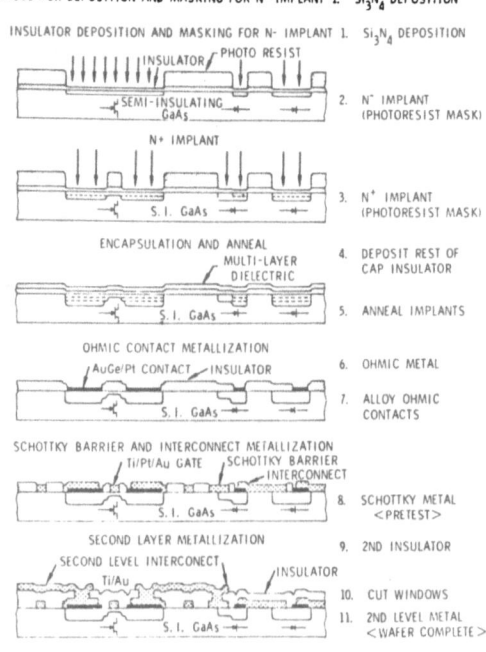

Fig. 7 - Planar SDFL GaAs IC fabrication steps.

isolated by high resistivity semi-insulating GaAs substrates. The use of any number of localized Se, S, or Si implants or combinations of implants provides the process flexibility for optimizing various LSI/VLSI circuits. All of the dielectrics and metals are deposited or grown using conventional processing methods including rf diode and magnetron sputtering, E-Beam evaporation and plasma enhanced CVD techniques.

Resist delineation is accomplished by using reduction projection photolithography for defect free, micron resolved photoresist patterns. Pattern replication is accomplished without any chemical etching. A unique high yield replication process has been developed by Rockwell specifically for planar GaAs ICs (see Sec. B). This fabrication process incorporates plasma etching in conjunction with an intermediate lift-off technique providing extremely high yield self-aligned metal/dielectric patterns. Second level metal interconnects are replicated using a dry etched ion milling technique. As will be discussed in Sec. C, the resulting planar structures are used to great advantage in the fabrication of high yield complex multi-level GaAs ICs.

The implant process steps used in this work are shown in Fig. 7 steps 2-5. Initially, the GaAs is coated with a thin layer of Si_3N_4 which remains on the wafer throughout all of the subsequent processing steps. The first process steps are the two localized implantations carried out through the thin Si_3N_4 layer using thick photoresist as the ion beam mask. Following the implants, additional dielectric (SiO_2) is added prior to the post implantation annealing shown in step 5.

Recently, the continued development of planar localized implant processes for LSI performance and yield optimization has led to the investigation of Si as an alternate ion implant specie for both the Se and S implants normally used in our planar GaAs process.[20] Silicon appears to be a suitable ion specie for use a "universal" n-type implant ion in GaAs. The insert table and profiles shown in Fig. 8 illustrate how Si implanted profiles can satisfy both the D-MESFET and Schottky switching diode requirements.

Typical implantation parameters and the resulting electron concentration doping profiles for both IC implants conducted through the Si_3N_4 cap are shown in Fig. 8. The shallow 400 KeV Se or 160 KeV Si implantation profiles are peaked near the GaAs surface due to the energy absorbed in penetrating the thin Si_3N_4 layer. In contrast the S or Si (n^+) implants provide much deeper ($\sim$ 4000Å) active layers than the shallow ($\sim$ 1500Å) low threshold ($V_p \sim$ -1 V) MESFET channel (n^-) implant. These deeper implanted layers are ideally suited jor the high speed switching diodes required in SDFL. Both n^- and n^+ implants are used for the level shifting diodes and for enhancing the doping under ohmic contact regions. As demonstrated in Fig. 8, the appropriate choice of Si implant energy and dose results in profiles similar to the typical Se and S profiles normally used. However, merely substituting Se and S implants with Si implants is not the important issue. In the case of D-MESFET channel profiles, and to a lesser

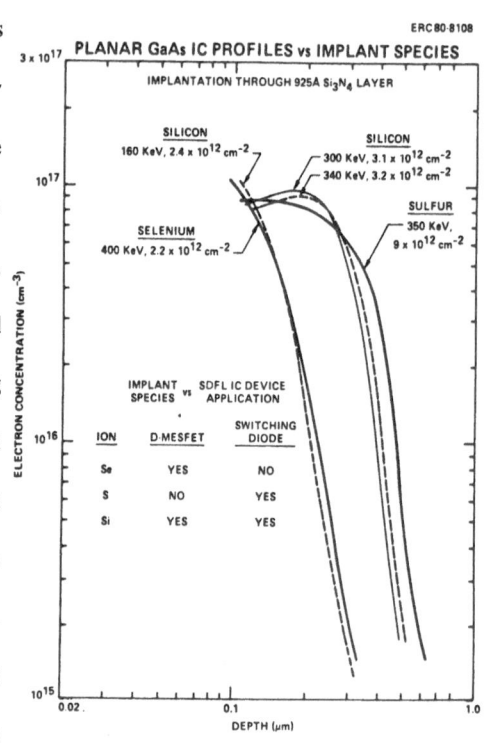

Fig. 8 - Carrier concentration profiles for Se, Si and S implants.

extent the Schottky diode profiles, the reproducibility uniformity and IC device performance are the real issues relevant to the optimal implant species experimentation. Silicon active layer profile reproducibility and uniformity investigations have been initially carried out by comparing the dc performance of D-MESFETs and Schottky diodes fabricated with various Si, Se and S variations.

The emphasis of implant specie comparison experiments was placed on evaluating the shallow lightly-doped high resistance n^- MESFET channel layer. This active layer directly controls the pinch-off voltage, V_p, the key parameter in the optimization of power-delay product of the logic gates. The uniformity of MESFET pinch-off voltages for four implant specie combinations is shown in Fig. 9. As shown in Fig. 8 the n⁻ FET channel is implanted

with either Se or Si while the n⁺ ohmic contact region is implanted with either S or Si. In Fig.

9 the average pinch-off voltage, $\overline{V}_p$, measured for 72 FETs equally distributed across the

wafer lies within a narrow 0.974 to 1.184 range indicating adequate control for all the dopant

variations studied. Uniformity of

V_p is also quite good as indicated

from the standard deviation values

ranging from a low of 68 mV

(6.8%) to a high of 101 mV

(9.4%). The results shown in Fig.

9 are average threshold voltage val-

ues (83 mV, 8.0%) routinely with

the implant processes described.

This implant optimization study has

not yet singled out any obvious ad-

vantage for choosing one specie

combination over another. To date,

this data supports the conclusion

that Se, Si and S are all viable GaAs

n-type implant dopants when uti-

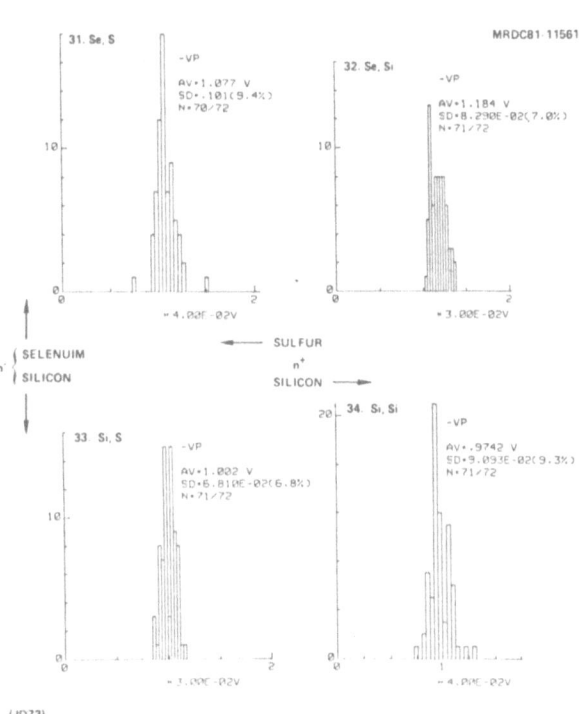

Fig. 9 - Histograms of pinch-off voltages of test FETs comparing various implant specie combinations of (Se, S), (Se, Si), (Si, S), and (Si, Si) n⁻ and n⁺ implanted regions respectively.

lized in conjunction with a mature material, ion implantation and encapsulation processes.

Recently, we have experienced 60 mV pinch-off voltage standard deviations for full

wafers averaged over many GaAs IC wafer lots. The constant uniformity improvements that

we have observed with time is indicative of the advancements made in both the semi-insulating

GaAs substrates and the fabrications processes. State-of-the-art results for D-MESFET

threshold uniformities are shown in Fig. 10 for both long range (full wafer) and short range

(FET array) distributions.[21] Full wafer pinch-off voltage standard deviations of 35 mV have

been monitored on numerous wafers with the best short range data being 17 mV (1.2%). The
constantly improving wafer device uniformity strongly supports the viability of this technology
for reaching VLSI levels.

B. Advanced Microstructure Pattern Replication Techniques

The first level circuit lithography processes encompass
the most crucial steps in the fabrication of planar GaAs ICs.
Since GaAs ICs are designed
with 1 μm features to achieve
speed goals, a dry pattern replication process capable of extremely high LSI/VLSI compatible yields has been developed.

The following summarizes

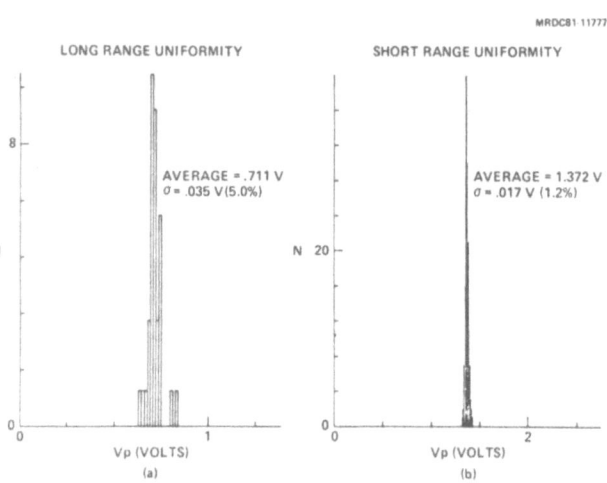

Fig. 10 - Histograms of pinch-off voltages of test FETs
comparing the uniformity for MESFETs uniformly distributed over an entire wafer area (long range) with that for
81 MESFETs packed in the area of a small chip (~ 300
μm x 300 μm - short range).

the first level circuit lithography steps used to define ohmic contacts, Schottky barriers, gates
and first-level interconnects. At this point in the process the entire surface of the GaAs wafer
is covered with dielectric. Subsequent ohmic contact and Schottky metallization layers are
deposited within windows in this dielectric (as shown in Fig. 7, steps 6-8). This process
represents a significant departure from typical microwave GaAs MESFET fabrication approaches and more closely resembles Si planar IC processing.

Common to all of the GaAs IC process steps is the delineation of the fine line resist
patterns required for the fabrication of these circuits. Since 1977, the photolithography steps
for our GaAs ICs have been accomplished by using a Cannon FPA 141 4X projection mask
aligner. Circuit pattern replication is accomplished using projection photolithography in
conjunction with photoresist lift-off techniques.

Photoresist lift-off techniques have been commonly used in fine line ($\sim$ 1 μm) lithography

applications for many years. However, when the direct photoresist lift-off technique is used

for defining metal patterns, excellent edge acuity of the photoresist profile is required.

Normally,

in practice, excellent
edge acuity (vertical
side walls) is not al-
ways attainable and
limited process yield is
commonly experienced.
This is particularly true
using reduction projec-
tion photolithography
techniques where resist
images and profiles are
often difficult to con-
trol due to the small
depth of focus

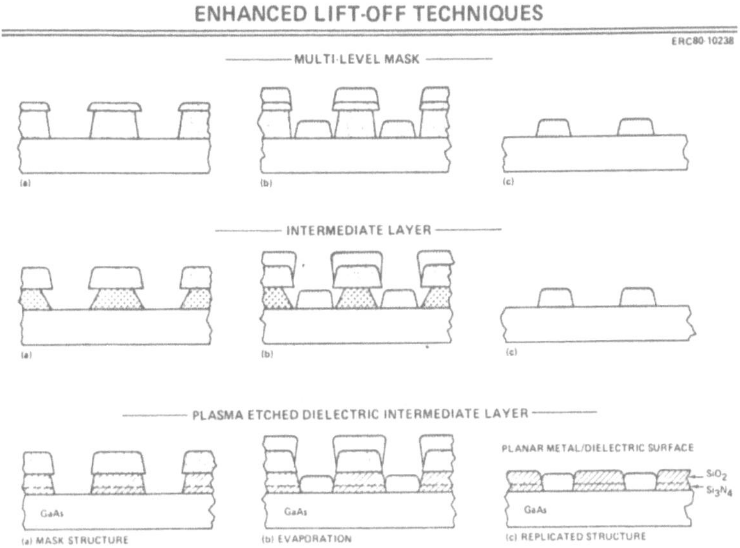

ENHANCED LIFT-OFF TECHNIQUES

Fig. 11 - Schematics of two commonly used enhanced lift-off methods; the multi-level mask and intermediate layer approaches, in contrast with the plasma etched dielectric intermediate layer technique described in this work.

(typically $\pm 1.5 \mu$m for 1 μm resolution) of this optical technique combined with large wafer

flatness variations normally experienced. Therefore, in order to insure high yields from lift-off

processes, enhanced lift-off techniques are commonly used.[19] Two examples of enhanced

lift-off techniques are shown in the upper portion of Fig. 11 along with the plasma etched

dielectric intermediate layer method used in this work shown at the bottom of Fig. 11. The

key principle used in both the multi-level mask or intermediate layer methods is use of

evaporation shadowing. Mask structures are formed with resist ledge (multi-mask techniques)

or under-cut resist/dielectric structures (intermediate layer) that allow convenient shadowing

of the evaporated metal, thus insuring an extremely high yield lift-off process.

The enhanced lift-off method described here was developed specifically for GaAs IC applications. This process approach has taken into account the inherent differences between processing on GaAs versus Si. For example, in this work the substrate is semi-insulating (the Si analogy would be insulating sapphire substrates in SOS technology), therefore, the first level interconnects may be fabricated directly on the substrate without the usual constraints associated with isolation and stray capacitance common to Si. Additionally, the commonly used freon based plasma etching techniques used for etching Si_3N_4 and SiO_2 do not etch GaAs. These two unique features of GaAs have motivated the development of the plasma etched dielectric intermediate layer enhanced lift-off technique shown in Fig. 11. The structural advantage of the intermediate layer enhanced lift-off method can be seen in the

scanning electron micrograph of Fig. 12(a). The intermediate dielectric layer is first plasma etched, then the appropriate metal is evaporated to a thickness less than or equal to the dielectric thickness. The resulting microstructure automatically provides a separation between the metal deposited in the

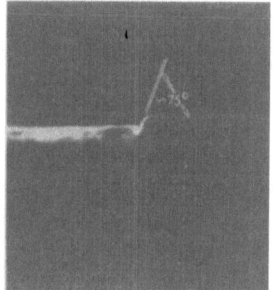

Fig. 12 - (a) Scanning electron micrograph showing the intermediate layer enhanced lift-off microstructure prior to lift-off. (b) Scanning electron micrograph showing the planar self-aligned metla/dielectric structure resulting from use of the enhanced lift-off method described in this paper.

dielectric window and the metal deposited on the top and side wall of the photoresist pattern. This structure allows solvents to easily penetrate and readily dissolve the photoresist. This process is easy to implement, very reproducible and extremely high yield.

The key difference between this approach and other intermediate layer lift-off approaches is that plasma etching techniques have been utilized for replicating micron resolved dielectric windows with minimum resist degradation which results in a lift-off process for providing the precisely aligned metal/dielectric structure shown in Fig. 12(b). Optimization of the plasma etching parameters for near ($\sim 75°$ slope) anisotropic dielectric profiles allows the metal to actually seal the dielectric opening (see Fig. 12(b)). This is advantageous for GaAs IC processing since critical, lightly doped, thin GaAs MESFET channel active layers are sealed and isolated from subsequent process steps. This unique first level lithography pattern replication technique has continuously and successfully been used in the development of GaAs LSI.

C. High Yield Planar Multi-level Interconnect

The Rockwell multi-level interconnect fabrication approach is illustrated in the planar multi-level GaAs IC structure shown in Fig. 13. As illustrated in the upper portion of Fig. 13, this fabrication approach hinges on maintaining a smooth metal/dielectric surface after the

ROCKWELL PLANAR
MULTI-LEVEL INTERCONNECT STRUCTURE

ERC 80-8111

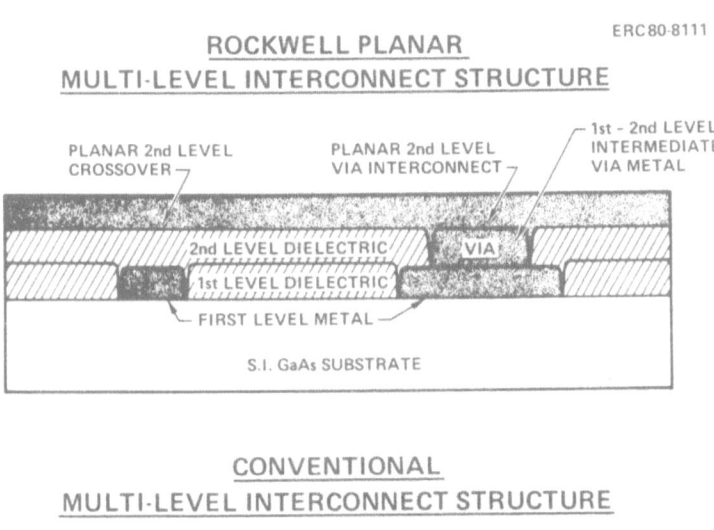

CONVENTIONAL
MULTI-LEVEL INTERCONNECT STRUCTURE

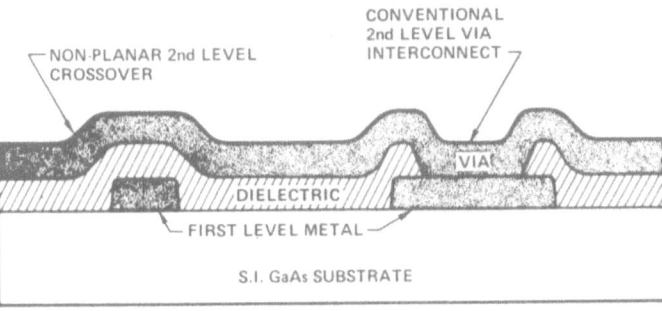

Fig. 13 - Schematic of the Rockwell planar multi-level interconnect structure compared to conventional fabrication structures.

first level metallization steps, leading to the planar multi-level interconnect crossover structure shown. Planar crossovers greatly enhance the yield of multi-level interconnects as compared to conventional crossover structures shown in the lower portion of Fig. 13. Also shown in Fig. 13 is the use of "filled" vias interconnecting the first level metal to the second level metal. Since there is no step coverage consideration for filled vias, higher yield, denser multi-level interconnect structures are possible.

The planar multi-level interconnect process is accomplished by fabricating the first-level metal within dielectric windows, and maintaining the metal thickness close to the dielectric thickness resulting in a smooth planar surface greatly facilitating the fabrication of complex multilayer interconnects. Figure 14(a) shows an actual cross section of the planar crossover structure resulting from these process approaches. Figure 14(b) shows a top view of a similar structure. The process used to fabricate the planar multi-layer structure is described as follows: an ~ 5000 Å plasma enhanced CVD silicon nitride (plasma nitride) layer is deposited on the ICs. Via windows are then reactively ion etched (RIE) through the nitride in order to interconnect the first-level metal to the second-level metal. A second-level metal composed of Ti/Au is deposited over the second-level dielectric and subsequently defined by ion milling, completing the process.

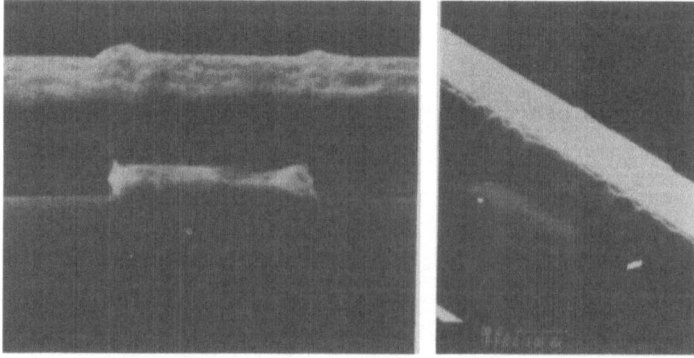

Fig. 14 - (a) SEM cross section showing the planar multi-level inter-connect structure illustrated in Fig. 13; (b) SEM photo showing a planar second level interconnect overcrossing, isolated from the first level interconnect by plasma nitride.

Planar crossovers eliminate potential crossover problems such as shorts between first- and second-level interconnects, and high-resistances or open interconnects resulting from poor step coverage. Another very important advantage is that these planar structures allow both thinner metal and dielectric films to be used resulting in less film stress, positively impacting yield and reliability. We also believe that these planar approaches have real potential for extension to three levels of interconnects which will greatly impact the optimization of future complex VLSI.

More recently, implemented into our process, is the planar second-level via interconnect shown in the schematic of Fig. 13. A similar process to the enhanced lift-off method already described is used to fill the via windows with intermediate metal between the first and second level interconnects. Figure 15 shows SEM cross sections of a conventional via structure with its inherent step coverage limitations in contrast with the planar "filled" via structure described here.

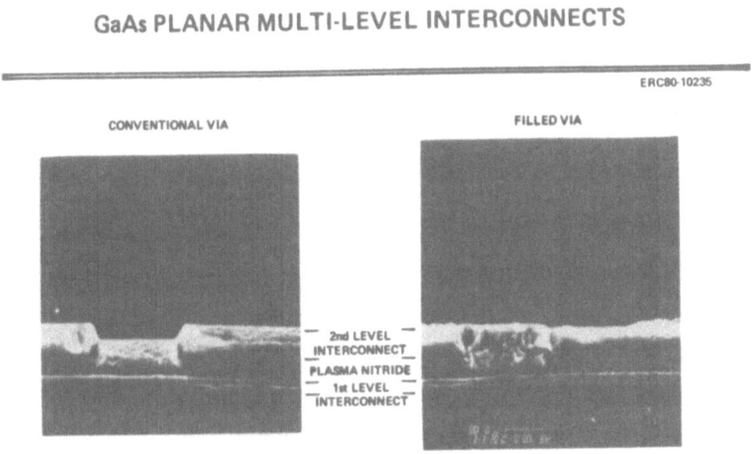

GaAs PLANAR MULTI-LEVEL INTERCONNECTS

Fig. 15 - SEM cross section comparing conventional via structures with the planar second level "filled" via structure.

In this process reactive ion etching is used to etch the via window with an anisotropic profile followed by evaporated metal in order to fill the window. Any small voids or gaps ($\sim$ 500 Å) between the intermediate via metal and the dielectric are easily filled during the deposition of the second-level metal. Figure 15(b) shows a SEM cross section of a filled via window along with the planar second-level interconnect. The motivation for developing this process is illustrated in Fig. 16 which contrasts two different via interconnect layouts, both having a common via hole cross-sectional area (1.5 μm x 1.5 μm). Conventional processing and layout designs call for isotropic via side-walls which require larger geometries for the second level interconnects and tight alignments for insuring adequate step coverage. However, filled vias do not require any step coverage; therefore, smaller second-level interconnects with relaxed alignment tolerances may be utilized. The capability of achieving higher densities with increased process yield is evident. These via approaches are expected to positively impact future VLSI requirements.

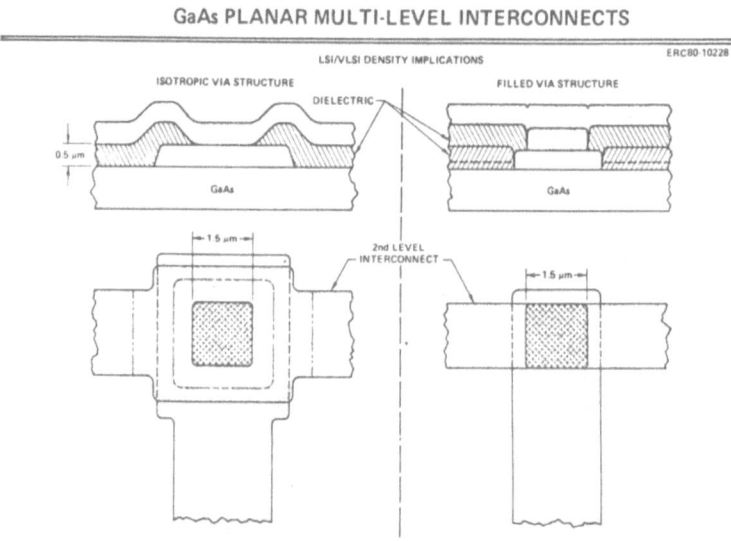

Fig. 16 - Comparison of a conventional IC via layout vs. a filled via layout design, illustrating the impact on density and alignment tolerance using these different approaches.

D. GaAs LSI/VLSI Process Status

The integrated circuit performance advantages which can be obtained by virtue of gallium arsenide's high electron mobility and aided by the availability of semi-insulating GaAs substrates has motivated the research in GaAs IC processing technology presented here. Results of this research have provided significant advances in GaAs processing techniques, allowing the full development of a planar "silicon-like" LSI fabrication technology.

The planar implantation microstructure pattern replication techniques described here have been developed specifically for LSI/VLSI requirements. A key goal for this process development was to demonstrate the successful operation of a GaAs LSI circuit. This goal has been achieved through the successful fabrication and functional operation of the GaAs 8 x 8 bit I/O latched parallel multiplier containing over 1000 gates shown in Fig. 17. Testing of this circuit has verified full functionality for all 64,000 possible multiplier combinations with gate delays of 150 ps having also been measured. This gate delay corresponds to a ~ 5 ns multiplication time to obtain a 16-bit product. These speeds are significantly higher than any currently available SI ICs (see Section VI).

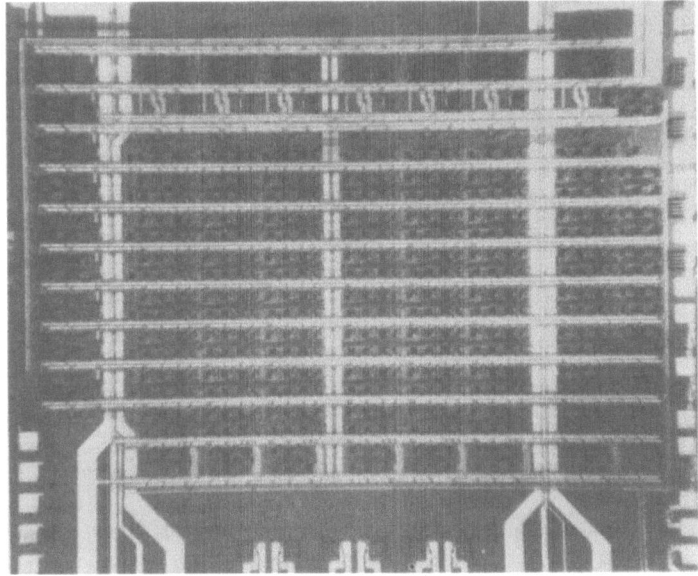

Fig. 17 - Photograph of a planar GaAs LSI (1008 gate) 8×8 bit parallel multiplier.

Process capabilities have been clearly demonstrated at the LSI level as indicated by the component and feature count for this multiplier circuit shown in Table V. The gate count of 1008 corresponds to over 6000 individual devices, ~ 12,000 crossovers, and ~ 7000 vias. This advanced planar GaAs 8 x 8 multiplier LSI circuit has been made possible through the use of high yield planar microstructure fabrication techniques. Demonstration of this LSI part provides us with excellent prospects for achieving future VLSI.

In summary, we have pres-
ented a GaAs fabrication process which is well on the way to
meeting the goals of future
GaAs VLSI. The excellent
prospects of GaAs VLSI are
enhanced by the fact that the
basic process steps used in fa-

TABLE V
COMPONENT COUNT FOR GaAs LSI
8 x 8 BIT PARALLEL MULTIPLIER

ERC80-10579

COMPONENT	LOGIC CELLS (ADDERS)	PERIPHERAL (LATCHES, I/O)	TOTAL
LOGIC GATES	728	280	1,008
DIODES	2184	840	3,024
FETs	1456	560	2,016
ACTIVE LOADS	728	280	1,008
TOTAL DEVICES	4368	1680	6,048
OVERCROSSINGS	7400	4600	12,000
VIAS	4900	2100	7,000

bricating GaAs ICs are generally fewer than those used in Si ICs. This is possible mainly

because of the availability of semi-insulating substrates, providing minimal parasitic capaci-

tances and direct device isolation without any additional complicated processes as found in Si

NMOS, CMOS, or CMOS/SOS. These factors should have a significant favorable impact on

the ultimate yield attainable for GaAs VLSI applications.

VI. EXPERIMENTAL PERFORMANCE RESULTS FOR PLANAR SDFL LSI GaAs ICs

As indicated in Table VII, we have fabricated many different circuits, from SSI to LSI, to

demonstrate the capability of this planar SDFL GaAs IC technology. The simplest demonstra-

tion circuits, very useful for establishing baseline gate propagation and dynamic switching

energy values, are ring oscillators. Ring oscillators are formed of an odd number, N_R, of

inverting gates (e.g., NAND, NOR or inverter gates).[14] Hence, with the proper logic state

applied to inputs not in the ring signal path, no stable state exists for the ring and the outputs

must oscillate at a frequency determined by the gate delay, τ_d ($f_{osc} = 1/(2N_R\tau_d)$). Some of

the ring oscillator results from our $L_g = 1$ μm gate length planar SDFL gates are shown in

Table VI. The highest speed obtained in low power ($P_D = 1.1$ mW) gates was

$\tau_d = 62$ ps at $P_D\tau_d = 68$ fJ, while the lowest speed-power products were measured in

very small geometry, $W = 3$ μm inverter gates with $P_D\tau_d = 16$ fJ at $\tau_d = 136$ ps. Of particu-

TABLE VI
COMPARISON OF GaAs IC RING OSCILLATOR RESULTS
FOR L_g = 1 μm SDFL OR/NAND AND SDFL NOR GATES

ERC80-8912

	GATE POWER		MODE	f_{OSC} (MHz)	τ_d (ps)	$P_D\tau_d$ (fJ)
W = 10 μm OR/NAND	P_D = 1.7 mW		G_1	640	111.6	201.1
GATE 7-STAGE RO	V_{DD} = 2.34V @ 625 μA		$G_1 + G_2$	550	129.9	216.9
FI = 4, FO = 2	V_{SS} = -0.86V @ -250 μA		G_2	680	105.0	179.6
W = 10 μm NOR GATE	P_D = 865 μW					
9-STAGE RING OSC	V_{DD} = 1.675V @ 460 μA			740	75	65
FI = 2, FO = 1	V_{SS} = -0.95V @ -100 μA					
W = 5 μm OR/NAND	P_D = 475 μW		G_1	405	176.4	83.9
GATE 7-STAGE RO	$V_{DD} \simeq$ 1.9V @ 200 μA		$G_1 + G_2$	348	205.2	93.7
FI = 4, FO = 2	$V_{SS} \simeq$ -1V @ ~ -95 μA		G_2	410	174.2	94.4
W = 5 μm NOR GATE	P_D = 372 μW					
9-STAGE RING OSC	V_{DD} = 1.57V @ 218 μA			502	110.7	41.1
FI = 2, FO = 1	V_{SS} = -0.61V @ -48 μA					

(ALL OF THE ABOVE WERE MEASURED ON THE SAME CHIP OF THE SAME WAFER FOR COMPARISON)

	GATE POWER		MODE	f_{OSC} (MHz)	τ_d (ps)	$P_D\tau_d$ (fJ)
HIGHEST SPEED:						
W = 10 μm NOR GATE	P_D = 1.10 mW			896	62	68
9-STAGE FI = 2, FO = 1	V_{DD} = 1.7V @ 546 μA					
LOWEST $P_D\tau_d$:						
W = 3 μm INVERTER	P_D = 110 μW			817	136	16
9-STAGE FI = 1, FO = 1						

lar interest in Table VI is the comparison between the simple SDFL NOR gate (Fig. 4(a)) and the 2-level SDFL OR/NAND gate Fig. 4(b) results for L_g = 1 μm ring oscillators of W = 10 μm and W = 5 μm sizes measured on the same chip of the same IC wafer. These OR/NAND ring oscillators[15] were fabricated with FI = 4 gates (as Fig. 4(b) but with diode inputs B and G missing) with the output connected to both G_1 and G_2 inputs of the following stage (i.e., F connected to C and D in Fig. 4(b)). Hence, if a positive control line signal is applied to the common "A" inputs in Fig. 4(b), oscillation through G_1 (the FET gate closest to the source) is obtained, while if the "A" inputs are low and the common "E" inputs are high, oscillation through the G_2 gates is obtained. If both the "A input and "E" input control lines are low, then simultaneous propagation through G_1 and G_2 is required (the worst case since it simulates a 2 μm gate length FET). As indicated in Table VI, the speed penalty for going to the 2-level SDFL OR/NAND gates is quite modest, particularly for the W = 10 μm gate sizes.

Hence, for circuit applications favoring the use of the multi-level gate implementations, these configurations should be extremely attractive. For example, extrapolating the $\tau_d = 105$ to 130 ps ring oscillator speeds to a fast complementary-clocked divider ($F_c \sim 1/2 \ \tau_d$) implemented with OR/NAND gates would give a maximum clocking frequency on the order of $F_c \sim 4$ GHz. As mentioned in Section IV, the use of 3-level SDFL gates has been demonstrated for 1 τ_d carry delay ripple adder circuits and would be attractive for the parallel multiplier as well. Of course, for many applications it would be desirable to eliminate the fan-in restrictions at the second and third level of logic (typically FI = 2 for high circuit speeds) inherent in FET logic implementations, but with the unlimited diode fan-in available at the first logic level in SDFL many very useful functions can be achieved.

While ring oscillators are useful devices for measurement of the basic speed-power properties of logic gates, a more realistic index of performance is the operation of gates in real sequential or combinational logic circuits. These circuits generally require use of gates with several inputs and fanouts of two or more. For this study, circuits using D flip-flops were chosen as representative examples of sequential logic circuits; a parallel multiplier, implemented with half and full adders, was selected as a combinatorial logic circuit example. Data multiplexer and demultiplexer circuits with on-chip synchronous address generators, representing a mixture of sequential combinatorial functioning, were also fabricated.[1,16,17]

Circuits in the MSI to 1000±-gate LSI range of complexity have been designed and fabricated using the planar SDFL approach (Table VII). All high speed measurements required for testing the circuits have been made at wafer probe level, although many

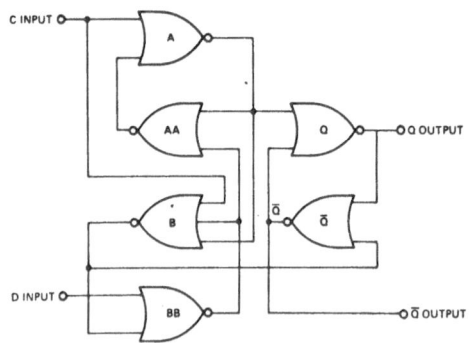

ERC78-2590A

Fig. 18 - Logic diagram for the NOR-implemented type D flip flop used in various shift register and divider circuits in this work. In the ripple divider applications the D-FF is T-connected as a ÷2 by connecting the Qbae output to the D (data) input. In shift register operation, the Q output is connected to the D input of the following stage.

TABLE VII
REVIEW OF ROCKWELL PLANAR
Gaas IC DEMONSTRATION CIRCUITS ERC80-8911

FULLY WORKING, SUCCESSFUL TESTED CIRCUITS

5 x 5 PARALLEL MULTIPLIER	260 GATES
8-STAGE SHIFT REGISTER/217 BIT P/N CODE GENERATOR	96 GATES
3 x 3 PARALLEL MULTIPLIER	75 GATES
4-BIT FAST RIPPLE CARRY ADDER USING 2 AND 3-LEVEL GATES	43 (63 EQUIV)
÷ 10/11 OR ÷ 20/22 VAR. MOD. DIVIDER FOR FREQ. SYNTHESIZER	37 GATES
8:1 DATA MULTIPLEXER WITH ADDRESS GENERATOR	64 GATES
1:8 DATA DEMULTIPLEXER WITH ADDRESS GENERATOR	60 GATES
1 GHz ANALOG SAMPLE/HOLD OR TRACK/HOLD CIRCUIT	25 COMPONENTS
DIGITAL FM DEMODULATOR CIRCUIT	65 COMPONENTS
÷ 8 FREQUENCY PRESCALER (D-FLIP FLOP RIPPLE DIVIDER)	25 GATES
÷ 8 SYNCHRONOUS COUNTER IMPLEMENTED WITH D-FFs	33 GATES
NOR GATE RING OSCILLATORS IN W = 3, 5, 10 AND 20 μm WIDTHS	9 GATES
OR/NAND GATE, 3 PROPAGATION MODE RING OSCILLATORS, 5 & 10 μm	7 GATES

CIRCUITS AT FABRICATION/INITIAL TESTING STAGE

2 x 32 SHIFT REGISTER/DUAL 4.3 x 10^9 BIT P/N CODE GEN.	550 GATES
8 x 8 PARALLEL MULTIPLIER	1000 GATES

packaged devices have also been tested, frequently with slightly better performance results. Output buffers of various types (source follower, open drain, etc.) are located on-chip to drive the output package capacitances or transmission line interconnections, as required, to prevent loading of the internal circuitry. Binary ripple counters or frequency dividers have been made by using the D Flip-Flop (D-FF) implemented with 6 NOR gates as shown in Fig. 18, as a basic building block. By connecting the D (data) input to the $\overline{Q}$ output, the clock input will produce an output transition for every full clock cycle (÷2; this is called a toggle or T-connected D-FF). Logic simulation of this D-FF configuration indicates that correct operation should be obtained up to a clock frequency of $1/(4.58 \ \tau_d)$, where τ_d is the propagation delay for the logic gates in the circuit.

A simple multi-stage binary frequency divider configuration is the ripple divider, in which the Q output of the first T-connected D-FF is connected to the clock input of the second, etc. We have fabricated three stage ripple dividers (÷8) containing 25 SDFL NOR gates and have operated these up to a clock frequency of 1.9

GHz. Figure 19(a) shows "low frequency" (f_c

= 100 MHz) $\div 2$, $\div 4$ and $\div 8$ outputs, while

Fig. 19(b) shows the $\div 8$ output at 237 MHz

from this divider when the clock input is at f_c

= 1.9 GHz. This corresponds to an equivalent

propagation delay of 110 ps, in good agree-

ment with ring oscillator data for 10 μm SDFL

NOR gates from the same wafer. The ob-

served dynamic switching energy varied over

the 0.25 to 0.45 pJ/gate range, depending on

the bias conditions and wafer pinch-off volt-

age. Power dissipations were as low as 45 mW

for the three stage dividers.[1,2,17]

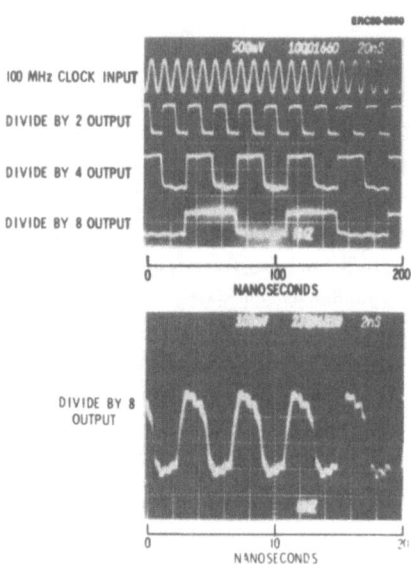

Fig. 19 - SDFL D-FF binary ripple divider results.[1,16,17] (a) Low frequency
operation showing 100 MHz sinewave
claock and $\div 2$, $\div 4$ and $\div 8$ circuit outputs. (b) $\div 8$ output of this ripple divider operating at a clock frequency of
1.9 GHz.

In addition, larger MSI circuits including an 8-input data multiplexer containing 64 gates

have been evaluated (logic diagram shown in Fig. 3 of Ref. 17). This circuit would be useful,

for example, for parallel-to-serial conversion or a high speed data transmission link. A 1 input

to 8 output data demultiplexer containing 60 gates was also fabricated and evaluated. Both

circuits utilize a presettable 3-stage synchronous counter[17] to generate the address for the

multiplexer or demultiplexer gate arrays. (A ripple counter could not be used here since the

outputs from the 3 stages are not "valid" at the same time; in the synchronous counter where

all of the D-FFs are clocked simultaneously, all of the Q and $\overline{Q}$ outputs form a valid address

for the multiplexer.) A SEM photograph of a multiplexer chip is shown in Fig. 10.4 of Ref. 1

or Fig. 4 of Ref. 17; the chip size, excluding probe pads, is only 0.77 mm x 0.54 mm.

Operation of both the multiplexer and demultiplexer chips has been demonstrated at a clock

frequency of 1.1 GHz. Power dissipation of the multiplexer circuits varied from 75 mW to

375 mW for wafers with pinch-off voltages of V_p = -0.5 and -1.45 V, respectively.

In addition to the binary ripple and synchronous counters, we have fabricated and tested at clock rates up to 1.4 GHz variable-modulus dividers ($\div 5/6$; $\div 10/11$, $\div 40/41$, $\div 10/12$, $\div 20/22$ and $\div 80/82$ ratios) of the type used in frequency synthesizers. We have also fabricated larger sequential logic circuits of the type used for code generation in, for example, spread-spectrum communications. As noted in table VII, we have fabricated and successfully tested a 96-gate 8-bit shift register/217 bit P/N code generator, implemented with eight D-FFs of the type shown in Fig. 18. In the shift register, the Q-output of a stage is connected to the D-input of the succeeding stage and all flip flops are clocked together. This chip is set up for various modes of shift register operation (serial load, parallel preset, recirculate, etc.). In addition, an exclusive-or gate with inputs taken from the 5th and 8th stages, and with its output connected to the serial input at the 1st stage, makes the circuit into a pseudo-noise (P/N) code generator. Depending on the initial state set in with the parallel preset inputs, P/N codes of 217 bit, 31 bit, and 7 bit lengths may be obtained. Proper operation in the shorter code lengths can be verified by directly examining the P/N code generator output on an oscilloscope, but this is rather marginally practical for the 217 bit case. It has proven easier for thse high speed P/N code generator circuits to verify operation by examining the output on a spectrum analyzer (a Tektronix 7L13, 0 - 1860 MHz was used) for the proper overall amplitude envelope (as calculated from the Fourier transform of the P/N code) and the proper spacing of the frequency lines in the output spectrum (e.g., $f_c/217$, $f_c/31$, or $f_c/7$.).

A particularly attractive candidate for a representative combinatorial logic part is a parallel multiplier, since multiplication frequently represents a bottleneck in signal processing and computer systems. Typical sizes for a high-speed parallel multiplier would be 8 x 8 bits to 16 x 16 bits, with larger products formed of combinations of these. A straight parallel multiplier, N x N bits, without carry lookahead and not using more complex (e.g., Wallace tree) approaches, requires (N-2) full adders and N half adders in its implementation and requires a total of (N-1) sum delays plus (N-1) carry delays to obtain the product. A goal of our ARPA GaAs LSI program was to construct an ~ 1000 gate 8 x 8 parallel multiplier chip

(760 gate multiplier plus I-O latching and buffering). As a first step toward this goal, a

75-gate 3 x 3 bit parallel multiplier was fabricated, followed by a 260-gate 5 x 5 bit version.

A fully latched 1008-gate 8 x 8 bit parallel multiplier was then designed and fabricated, with

the first fully working devices achieved in late 1980.

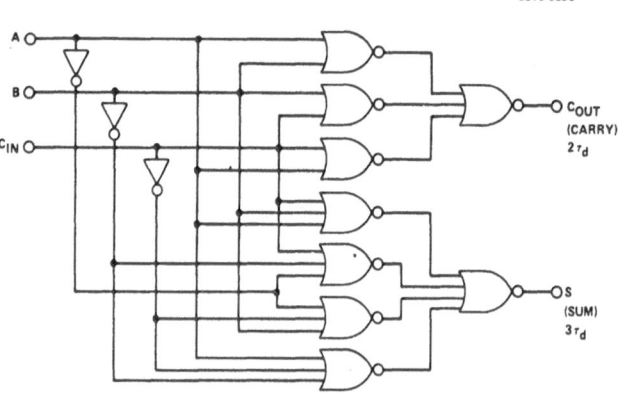

SC79-5650

Since the bulk of a large multiplier is full adder cells, the speed of these cells dictates the multiply time. Figure 20 shows the logic diagram for a NOR-implemented full adder cell. By using a full minterm expansion for both sum and carry, the propagation delays are kept

Fig. 20 - Full adder cell used for array multipliers, implemented with 12 SDFL NOR gates. Carry delay is $2\tau_d$ and sum delay is $3\tau_d$ where τ_d is the basic NOR gate propagation delay.

to $2\tau_d$ for the carry and 3 τ_d for the sum (where τ_d is the NOR gate propagation delay). A

total of 12 gates (9 NORs and 3 inverters) are required per full adder cell. Figure 21 shows

the logic diagram for the 5 x 5 parallel multiplier chip (FA stands for a full-adder cell and HA

for a 5-gate half-adder cell). For this N = 5 case, 5 half adders and 15 full adders are used.

A microscope photograph of this multiplier GaAs IC chip is shown in Fig. 20 of Ref. 2. These

260-gate 5 x 5 bit parallel multiplier circuits have given good chip yields on GaAs substrate

wafers grown by both the horizontal Bridgman and liquid encapsulated Czochralski (LEC)

methods (the LEC wafers were 3 in. diameter), with chip yields up to 20% obtained.

The 8 x 8 bit latched parallel multiplier chip consists of a 760-gate parallel multiplier

array plus input and output latches and drivers (248 gates). The 8 x 8 bit parallel multiplier

array itself is the same as shown for the 5 x 5 bit multiplier in Fig. 21, except that it is

expanded from 5 rows x 4 column to 8 rows x 7 columns. Unlike the earlier 3 x 3 and 5 x 5

multiplier chips which consisted of the parallel multiplier array only, full input and output latching was incorporated on the 8 x 8 bit multiplier chip. This makes it possible to use the multiplier in a pipelined mode with slower (silicon) systems, since a multiply can be in progress using the latched input operands while the outside system is disposing of the old latched 16-bit product and getting the new operands for the next multiply. The overall architecture for this GaAs SDFL 8 x 8 bit parallel multiplier chip is shown in Fig. 22. Note that the input operand and/or output product latches may be defeated for direct parallel multiplier array operation. Figure 17 shows a photomicrograph of the

LOGIC DIAGRAM OF 5 x 5 ARRAY MULTIPLIER CHIP

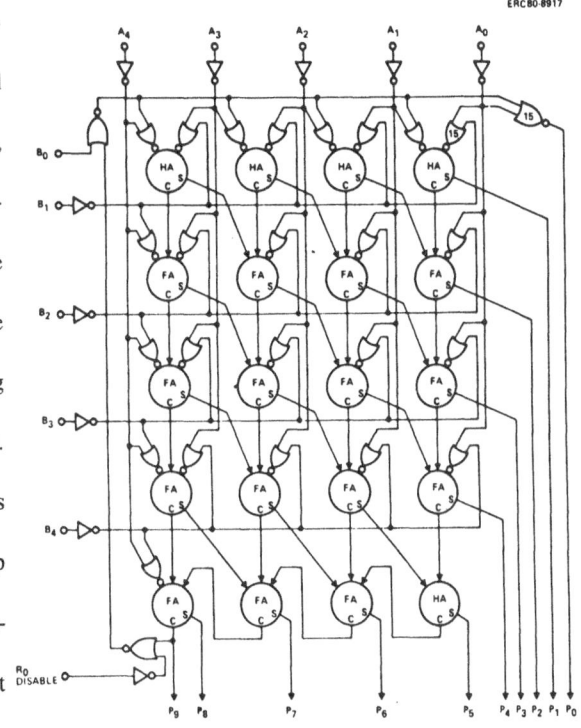

Fig. 21 - Logic diagram of the 5×5 bit parallel multiplier chip. Note the "ring oscillator" mode feedback path from P_9 to B_0 for testing propagation delay through the multiplier.

1008-gate 8 x 8-bit multiplier chip. The orientation of the photograph is similar to the block diagram in Fig. 22, with input latches at the top and left, parallel multiplier array in the center, and output latches and drivers at the bottom and right. The overall chip size is 2.457 mm x 2.7 mm. While the bulk of the chip area is consumed in power supply busses on signal interconnection lines, even including these, the average gate density in the parallel multiplier array is 370 gates/mm^2 ($\sim$ 4 mil^2/gate as compared to an actual gate area of $\sim$ 1 mil.^{2})

The wafer probe test procedure for these multiplier chips allows for both static and dynamic performance measurements. Initial probe measurements for gross functionality involve the application of a specific input code with a square-wave signal on one of the inputs

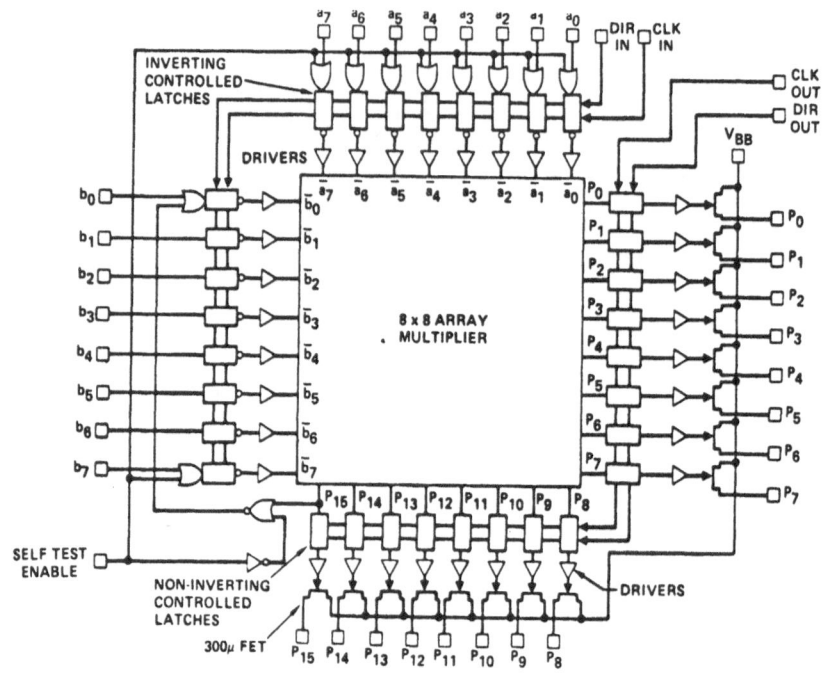

Fig. 22 - Block diagram of the latched input/output 8×8 bit parallel
multiplier chip. The DIR IN and DIR OUT pins make the input and
output latches transparent for direct parallel multiplier operation. The
SDFL TEST ENABLE pin enables the $\overline{P}_{15}$ to B_O feedback path for
"ring oscillator" mode multiply time measurements.

which should produce a square-wave output, either in-phase or inverted, on all of the product

bits (e.g., for the 3×3, 111 × 10S = $\overline{SSSSSS}$ where S is the square wave). If proper gross

functionality is observed, complete low-speed functional testing is carried out for all possible

combinations of operand inputs, using a **MACSYM II** computer system to generate inputs and

measure and verify outputs (for all 65,536 combinations for the case of the 8×8 bit multipli-

er). The problem of measuring multiply speed at wafer probe on such high-speed circuits was

dealt with by adding a few extra gates to the multiplier chip to make an on-chip feedback

path. For the 5×5 bit multiplier (see Fig. 21), the invert of the most significant product bit,

P_9, may be applied back to the least significant multiplier bit, B_O. The input operands are

selected as A = 11111 and B = 1000 B_O, which gives the product $B_O\overline{B}_O\overline{B}_O\overline{B}_O\overline{B}_O\overline{B}_O$

$B_OB_OB_OB_O$; here the signal from the B_O input to the $P_9 = B_O$ output (diagonally from upper left to lower right in Fig. 21 and then down and from right to left along the bottom row) represents the worst case multiple time (3 sums + 5 carriers) for the multiplier. Since after a multiply time is completed, $P_9 = B_O$, there can be no stable product when $\overline{P}_9$ is fed back to B_O. Hence in this feedback mode the whole chip behaves like a very large ring oscillator, oscillating at essentially a frequency of $1/(2T_M)$, where T_M is the worst-case multiple time. (More precisely, $f_{osc} = 1/(2N\tau_d)$, where N is the total number of gate delays, including those in the feedback path; $N = 22$ for the 5×5 circuit of $f_{osc} = 1/44\ \tau_d$, whereas $T_M = 21\ \tau_d$). The feedback path for the 8×8 bit latched parallel multiplier (Fig. 22) is similar ($\overline{P}_{15}$ to B_O), but the feedback circuit goes through the input latch structure (which must be in its direct or "transparent" mode for proper operation). The SDFL NOR gate delays, τ_d, measured on the 3×3 multiplier chips were $\tau_d = 172$ ps at $P_D = 750\ \mu W/gate$ ($P_D\tau_d = 129$ fJ), or $\tau_d = 225$ ps at $P_D = 420\ \mu W/gate$ for an excellent $P_D\tau_d = 95$ fJ dynamic switching energy, with similar results on the 5×5's. The layout was improved somewhat on the 8×8 bit multiplier chip, with gate delays as low as $\tau_d = 150$ ps measured, for a parallel multiply time ($T_M = 35\ \tau_d$) of $T_M = 5.25$ ns at chip dissipation of about 2 watts. At lower power levels (~ 1 watt for the chip), the gate delays are typically $\tau_d \sim 200$ ps ($T_M \sim 7$ns).

The $T_M = 5.25$ ns multiply time, corresponding to 190 M multiplies/s, obtained with our GaAs SDFL 8×8 bit parallel multiplier chips is nearly an order of magnitude faster than commercial silicon bipolar parallel multipliers, at a fraction of the power dissipation. (A new ECL 8×8 bit multiplier with $T_M = 19$ ns at $P_D = 4.4$ watts has been announced, but this chip uses a more exotic architecture and cannot, on an IC technology basis, be directly compared with the straightforward parallel multiplier array results). As discussed previously, the use of multilevel SDFL gates would reduce the full adder propagation delay time from $3\ \tau_d$ sum, $2\ \tau_d$ carry to $2\ \tau_d$ sum, $1\ \tau_d$ carry so that the 8×8 parallel multiply time (6 sums and 8 carries plus τ_d) is reduced by 40% (from $35\ \tau_d$ to $21\ \tau_d$). This alone would increase, for the same $\tau_d = 150$ ps gate delay value, the multiply rate from 190 million multiplies/s to 317 million multiplies/second ($T_M = 3.15$ ns). It is felt that with modest further improvements in gate delay and improved multiplier architecture, 8×8 bit multiply times of 2 ns should be attainable (0.5 G multiplies/s).

VII. SUMMARY

In summary, the future prospects for using GaAs to achieve ultra high speed VLSI looks very bright. GaAs logic circuits have shown ring oscillator gate delays as low as 30 ps at room temperature of 17.5 ns at 77°K using L_g = 0.6 μm MESFETs. Other GaAs device approaches, such as the heterojunction bipolar transistor, might reduce these gate delays by an order of magnitude as the new processing technologies to fabricate these devices are mastered. As an indication of the present state of GaAs technology, we have developed in our ARPA/Rockwell GaAs LSI program, a super planar fabrication process capable of optimizing two or more types of devices as required for sophisticated LSI/VLSI circuit approaches. This process has demonstrated the capability for achieving tight parameter control on low power D-MESFETs and high speed switching diodes, and for achieving good yields on lower LSI complexity (hundreds of gates) circuits (up to 20% on 260-gate 5×5 bit parallel multipliers). The Schottky diode-FET logic circuit approach used in this work has demonstrated high speeds (τ_d as low as 62 ps), low power ($P_D \sim$ 100 μW to 2 mW/gate), high densities (NOR gate areas as low as 600 μm^2), and outstanding $P_D\tau_d$ products (as low as 16 fJ in ring oscillators, or 95 fJ even in large MSI parts). The simplicity, high density, and large device parameter margins of SDFL, along with its speed and power characterisitcs, make it ideal for ultra high speed VLSI. We have demonstrated scores of different SDFL circuits, from SSI to LSI; some of these listed in Table VII. We have fabricated 1008-gate LSI parts, 8 bit x 8 bit latched parallel multiplier, and have obtained fully working samples of these parts both on the smaller horizontal Bridgman-grown GaAs substrates and the new 3 in. diameter LEC GaAs substrate material. We have demonstrated that even in these LSI parts, the inherent GaAs is maintained, with SDFL NOR gate delays as low as 150 ps obtained in the 8 bit x 8 bit parallel multiplier chips (corresponding to a 5.25 ns multiply time).

Based on the results to date, extension of this planar GaAs IC technology to the capability for 10,000-gate VLSI circuits seems a reasonable goal, and efforts to achieve this goal are in progress.

ACKNOWLEDGEMENT

The authors are indebted to the many hard working individuals contributing to the GaAs integrated circuit programs at Rockwell. In particular we wish to thank P. Asbeck, C. Kirkpatrick, and Y.D. Shen for their contributions in material studies, ion implantation and processing, and S. Long, R. Zucca, G. Kaelin, F. Lee and C.P. Lee for circuit design, testing and device analysis. F. Eisen has made invaluable contributions in the development of GaAs ion implantation technology and along with A. Firsternberg has provided valuable guidance and support to the development of GaAs integrated circuit technology. Finally, the GaAs IC program at Rockwell is indebted for the encouragement, support and technical guidance provided by R. Reynolds and S. Roosild through the Defense Advanced Research Project Agency.

REFERENCES

1. R. C. Eden, B. M. Welch, R. Zucca and S. I. Long, "The Prospects for Ultrahigh-Speed VLSI GaAs Digital Logic", IEEE Trans. on Electron Devices, ED-26, 4 (April 1979), p. 299-317, and IEEE J. Solid State Circuits SC-14, 2 (April 1977), p. 221-239.

2. R. C. Eden and R. M. Welch, "GaAs Digital Integrated Circuits for Ultra High Speed LSI/VLSI," Chapter 5 of Springer Series in Electrophysics Volume 5: Very Large Scale Integration (VLSI) Fundamentals and Applications, D. F. Barbe, Ed., Springer-Verlag, Berlin (1980).

3. Takashi Mizutani, "Gigabit Logic Operation with Enhancement-Mode GaAs MESFET IC's," presented at IEEE MTT-S workshop on Gigabit Logic for Microwave Systems, May 1980. Data presented for L_g = 0.6 μm 15-stage inverter ring oscillators with room temperature speeds up to τ_d = 30 ps at P_D = 1.9 mW/gate or τ_d = 63 ps at P_D = 100 μW/gate and 77°K speeds up to τ_d = 17.5 ps at P_D = 9.2 mW/gate. Earlier, L_g = 0.8 μm data is presented in the paper: T. Mizutani, N. Kato, M. Ida and M. Ohmori, "High-Speed Enhancement-Mode GaAs MESFET Logic", IEEE Trans. on Microwave Theory and Tech. MTT-28, 5 (May 1980), p. 479-483.

4. T. Mimura, S. Hiyamizu and K. Nanbu, "A New Field-Effect Transistor with Selectively Doped GaAs/n-Al$_x$Ga$_{1-x}$As Heterojunctions", Jap. J. Appl. Phys. Lett. 19, 5 (1980), p. 225-227.

5. T. Minura, S. Hiyamizu, H. Ishikawa and T. Misugi, "An Enhancement-Mode High Electron Mobility Transistor for VLSI", High Speed Digital Technologies Conference, Paper III-5, San Diego, Jan. 14, 1981.

6. L. Eastman, R. Stall, D. Woodard, C. Wood, N. Dandekar, M. Shur and K. Board, "Ballistic Electron Transport in Thin Layers of GaAs", presented at 1980 Device Research Conference, Cornell U., Ithaca, NY, June 1980; Abstract published IEEE Trans. on Electron Devices ED-27, 11 (Nov. 1980), p. 219.

7. W. Shockley, U.S. Patent, 2,569,347, 1951.

8. H. Kroemer, "Theory of a Wide-Gap Emitter for Transistors", Proc. IRE 45, (1957), p. 1535-1537.

9. J. P. Bailbe, A. Mary, P. H. Hiep and G. E. Ray, "Design and Fabrication of High-Speed GaAlAs/GaAs Heterojunction Transistors," IEEE Trans. on Electron Devices Ed-27, 6 (June 1980), p. 1160-1164 and references cited therein.

10. C. O. Bozler and G. D. Alley, "Fabrication and Numerical Simulation of the Permeable Base Transistor," IEEE Trans. on Electron Devices ED-27, 6 (June 1980), p. 1128-1141.

11. M. Abramowitz and I. A. Stegun, "Handbook of Mathematical Functions", National Bureau of Standards Applied Mathematics Series 55, U.S. Government Printing Office, Washington, DC, p. 931 and p. 966-977.

12. R. C. Eden, B. M. Welch and R. Zucca, "Low Power GaAs Digital ICs Using Schottky Diode-FET Logic", 1978 Int. Solid State Circuits Conference, Digest of Tech. Papers, p. 68-69, (Feb. 1978).

13. R. F. VanTuyl, C. A. Leichti, R. E. Lee and E. Gowen, "GaAs MESFET Logic with 4 GHz Clock Rate", IEEE J. Solid State Circuits SC-12, (Oct. 1977), p. 485-496.

14. R. C. Eden, B. M. Welch and R. Zucca, "Planar GaAs IC Technology: Applications for Digital LSI", IEEE J. Solid State Circuits SC-13, (Aug. 1978), p. 419-426.

15. R. C. Eden, F. S. Lee, S. I. Long, B. M. Welch and R. Zucca, "Multi-Level Logic Gate Implementation in GaAs ICs Using Schottky Diode-FET Logic", ISSCC Digest of Technical Papers, (Feb. 1980), p. 122-123 and 265-266.

16. R. C. Eden, "GaAs Integrated Circuits, MSI Status and VLSI Prospects", 1978 Int. Electron Devices Mtg., Tech, Digest, (Dec. 1978), p. 6-11.

17. S. I. Long, F. S. Lee, R. Zucca, B. M. Welch and R. C. Eden, "MSI High Speed Low Planar GaAs ICs Using Schottky Diode FET Logic", IEEE Trans. on Microwave Theory and Techniques MTT-28, (May 1980).

18. B. M. Welch, Y. D. Shen, R. Zucca, R. C. Eden and S. I. Long, "LSI Processing Technology for Planar GaAs Integrated Circuits", IEEE Trans. Electron Devices ED-27, 6 (June 1980).

19. J. Moran, "Multiple Resist Layers and Their Applications", in Microcircuit Engr. 80, (Sept. 1980), Amsterdam, The Netherlands.

20. B. M. Welch and R. C. Eden, "Planar GaAs Integrated Circuits Fabricated by Ion Implantation", Int. Electron Devices Meeting, Tech. Digest, (Dec. 1977), pp. 205-208.

21. R. Zucca, B. M. Welch, C. P. Lee, R. C. Eden and S. I. Long, "Process Evaluation Test Structures and Measurement Techniques for a Planar GaAs Digital IC Technology", IEEE Trans. Electron Devices ED-27, 12 (Dec. 1980).

22. Richard C. Eden, U.S. Patent 4,300,064, 1981.

23. H. Kroemer, "Heterostructure Bipolar Transistors and Integrated Circuits", Proc. of IEEE, 70, 1 13 (Jan. 1982).

24. S.I. Long, B.M. Welch, R. Zucca, P.M. Asbeckm C.-P. Lee, C.G. Kirkpatrick, F.S. Lee, G.R. Kaelin and R.C. Eden, "High Speed GaAs Integrated Circuits", Proc. of IEEE, 70, 1 (Jan. 1982) pp. 35-45.

25. C.O. Bozler and G.D. Alley, "The Permeable Base Transistor and Its Application to Logic Circuits", Proc. of IEEE, 70, 1 (Jan. 1982) pp. 46-52.

26. R.J. Malik, M.A. Hollis, L.F. Eastman, D.W. Woodard, C.E.C. Wood and T.R. AuCoin, "GaAs Planar Doped-Barrier Transistors", IEEE Trans. Electron Devices, ED-28, 10 (Oct. 1981), p. 1246.

JOSEPHSON INTEGRATED CIRCUITS

Ernest J. Van Derveer

IBM Zurich Research Laboratory
8803 Rüschlikon
Switzerland

1. INTRODUCTION

Nineteen years ago, B. D. Josephson[1] predicted a supercurrent at zero voltage between two superconducting metals, separated by a thin insulator. Today, 20-25 laboratories around the world are actively studying the possible exploitation of devices based on the Josephson effect. Although the promise of a very fast device (picosecond) at low power levels (microwatt) has yet to be realized in a computing system, usage of Josephson junctions as a voltage standard and an extremely sensitive magnetic-field detector is widespread. The attractions of Josephson devices are summarized in Table 1. The combination of a fast, low power circuit plus superconducting transmission lines offers the possibility of packing and interconnecting hundreds of thousands of circuits and hundreds of millions of memory bits into a volume very comparable to that occupied by the human brain;[2] a feat unachievable with a high-performance semiconductor technology.

This achievement of high component packing density is required to improve the performance of large systems simply because electrical signals can travel only 15 centimeters at best in one nanosecond (30 centimeters in free space) in a computer environment. To impact existing computer technology, the cycle-time objective for a Josephson computer should be in the one to two nanosecond range, line delays alone can easily amount to two or more nanoseconds without incurring any of the usual circuit and timing delays.

2. SUPERCONDUCTIVITY

Over 25 elements when cooled below a critical temperature (T_c) assume an electronic ordered state and exhibit superconductivity properties.[3] The transition to a superconductive state is abrupt. Above T_c, the elements behave as normal metals with resistivity depending on

impurity defect and phonon scattering. Below T_c, a superconductor acts as if it had no

resistivity. Currents once established in superconducting loops have persisted undiminished

for as long as two and a half years. The range of critical temperatures is from a few millide-

grees Kelvin to over $20°K$ for certain alloys. If a large magnetic field is applied, superconduc-

tivity can be destroyed. Also the presence of a magnetic field lowers T_c. There is also a

current value called the critical current above which, due to self-fields, the superconductor

goes normal (Silsbee).

Provided the frequency of an AC signal is below the gap energy divided by $\hbar$, (approx.

10^{12} Hz for lead) a superconductor is dissipationless.[4] A superconductor will expel an applied

magnetic field (Meissner-Ochsenfeld), if the field is below a critical value H_c. A current flows

on the surface of the superconductor that exactly cancels the applied magnetic field in the

interior. The depth to which currents and magnetic fields penetrate a superconductor is

known as the London penetration depth λ. A field or current decreases as $e^{-x/\lambda}$ from the

surface. Values of λ for typical materials used in Josephson technology range from 500 -

2000 Å, and λ increases considerably as the temperature approaches T_c. An additional

magnetic phenomenon that will prove useful in device considerations is the quantization of

magnetic flux. A superconductor ring encloses flux only in integer values of the quantity $\psi_0 =$

$h/2e = 2.0679 \times 10^{-15}$ Vs, a quantity known as fluxoid or flux quantum. ψ_0 is small the flux

enclosed by the cross section of a human hair in the earth's magnetic field is several flux

quanta. Another unusual effect is the appearance of a gap in the energy levels centered at the

Fermi level of a superconductor. This gap is temperature dependent increasing to a maximum

at temperatures close to zero degrees Kelvin. Gap voltages for lead and niobium are approxi-

mately 2.8 millivolts. The concept of a gap voltage is useful to explain the I-V curves of a

Josephson junction. Giaever was the first to demonstrate experimentally the existence of an

energy gap.[5]

A theory on the microscopic level that is fundamental in understanding superconductivity was developed by Bardeen, Cooper and Schrieffer.[6] This is commonly called the BCS theory. Particularly striking is the agreement between BCS theory and measurement of the gap voltage vs. temperature curve.

3. TUNNELING

The behavior of electrons in a solid can be described in quantum mechanics by a wave. This representation leads to the possibility of an electron appearing on the other side of a barrier despite the fact the electron had insufficient energy to surmount the barrier. The probability of tunneling is an exponential function of the barrier thickness, meaning a barrier must be extremely thin to pass an appreciable number of electrons to the other side. Tunneling phenomena in solids were studied mostly by physicists for theoretical understanding until 1957 when Esaki[7] invented a practical device, the tunnel diode. A heavily doped semiconductor diode resulted in a sufficiently thin junction barrier that permitted electrons to tunnel through the junction at low values of junction voltage.

In addition to the need for a thin barrier, an electron must find an available energy state on the other side of the barrier into which it can tunnel for current to flow.

One essential ingredient of the BCS theory is the formation of a loosely coupled pair of electrons called Cooper pairs. The energy required to break up a Cooper pair is called the gap energy. It is the movement of these coupled pairs through the atomic lattice without collision that explains the infinite conductivity of a superconductor.

4. JOSEPHSON PREDICTIONS

In 1962, a graduate student at Cambridge University, Brian Josephson predicted that under certain conditions, an insulator sandwiched between two superconductors could behave like a superconductor. Cooper pairs could tunnel through a thin oxide with no applied voltage.

Since thin oxides ($\sim$ 50 Å) can have holes, experimental verification of Josephson tunneling seems extremely difficult since insulator shorts would behave as tunneling Cooper pairs. Josephson also predicted that the maximum current that flows at zero applied voltage varies as sin H/H if an external magnetic field H is applied. The zero points in the maximum current would occur whenever an integral number on magnetic flux quanta links the barrier. The experimental verification of this magnetic-field dependence was the key to general acceptance of Josephson's predictions which won him the Nobel Prize in 1973.

As in superconductors, the magnetic fields and tunneling currents of a junction can penetrate only to a certain depth λ_J (Josephson Penetration Depth) which is approximately one hundred times larger than the London Penetration depth.

So far, only the effects at zero barrier voltage have been discussed. When a DC voltage is applied, an AC supercurrent flows. The supercurrent varies sinusoidally with time with a frequency of $\omega_0 = 2eV/h$. It is the interaction of this oscillation with the resonant cavity formed by the junction that causes peaks in the IV characteristics.[8] Because of the extensive research on superconductors being conducted in the 1960's, experimental verification of Josephson's predictions came quickly.[9,10]. Early demonstration of sub-nanosecond switching speeds by Matisoo[11] increased the interest in digital application of Josephson-type devices.

5. JOSEPHSON DEVICES

Practical Josephson devices come in many forms. The one that has received the most attention over the years is the magnetically controlled device often called SQUID (Superconducting Quantum Interference Device). A SQUID contains one or more superconducting loops including Josephson junctions. Two superconducting layers are isolated from a third layer, control line, that generates a field to lower the maximum Josephson current (Fig. 1). Isolation of input and output is achieved and because the control-line current is smaller than the device current, device gain is realized. The threshold characteristic, device or gate current vs. control current can be tailored by combining two or more junctions in parallel separated by inductances[12] (Fig. 2.). Different junction geometries give another device design dimension. Devices made with sine-shaped junction areas result in a transfer characteristic in

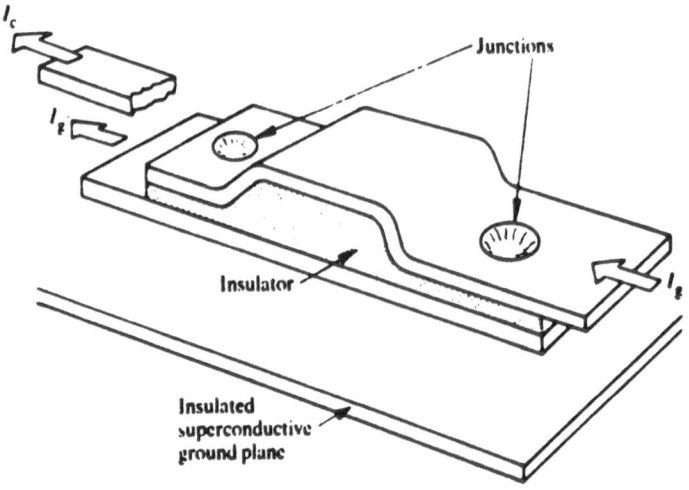

Fig. 1 - Josephson Device (SQUID) with two junctions. (After [32])

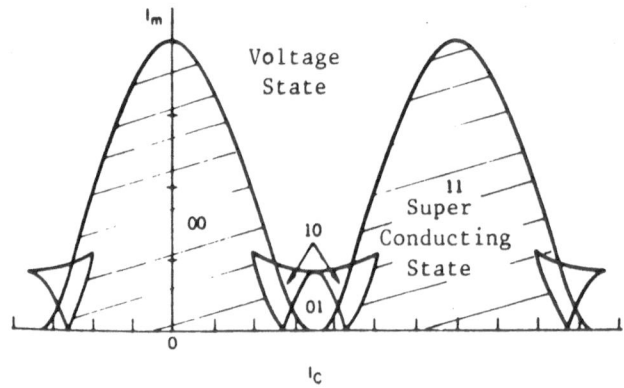

which the secondary maxima (side lobes) are greatly reduced compared to those of a rectangular junction.[13] These have proved useful in memory circuits.

Fig. 2 - Threshold characteristic for a three-junction SQUID. (After [2])

Direct injection of current into a two-Josephson-junction SQUID has been developed and is called CID (Current Injection Device)[14] (Fig. 3). However, the necessary isolation does not naturally occur and must be provided in some other fashion. The CID device

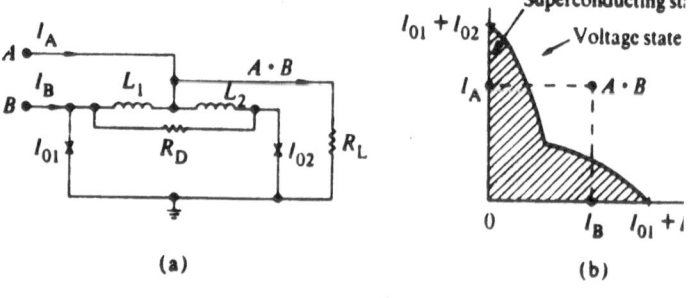

Fig. 3 - a) Current-Injection-Device equivalent circuit. b) Threshold characteristic for the AND junction of A and B. (After [14])

is attractive since the area occupied is one third that of the magnetic coupled devices. A Josephson device model first suggested by Stewart and McCumber[15,16] (Fig. 4) has proved useful in circuit simulations of point junctions.

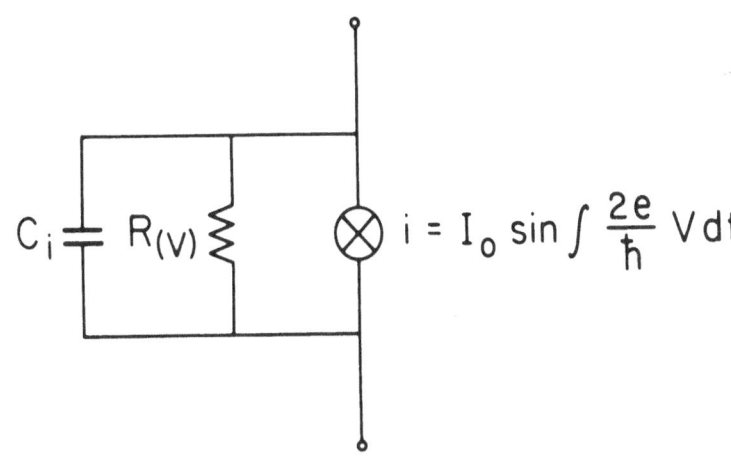

$$i = I_0 \sin \int \frac{2e}{\hbar} V dt$$

Fig. 4 - Stewart-McCumber model of a point Josephson junction. $R_{(v)}$ is a voltage-dependent resistor.

6. DEVICE/CIRCUIT GAIN

The gain of a SQUID is inherently low compared to semiconductor devices, and when the realizable control of critical parameters such as the maximum Josephson current are considered, high device gain is required. One method to increase gain of magnetic coupled devices is to increase the number of turns the control winding links the device, which increases the device area. Another is to use multiple-junction devices which increases gain due to coupling currents from the multiple SQUID loops. A third obvious

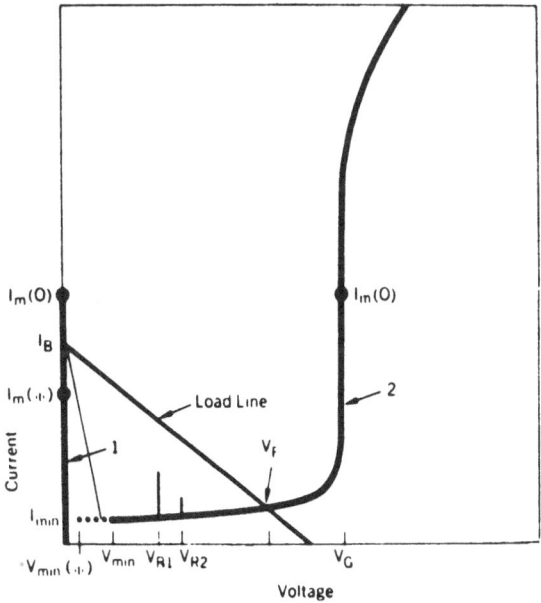

Fig. 5 - Josephson current-voltage curve with a latching and resetting load line. Spikes in curve at V_{R1} and V_{R2} are caused by junction acting as resonant microwave cavity. Switching occurs either by increasing I_B above $I_m (0)$ or applying a magnetic field to lower I_m. (After [2])

method is to add an additional buffer stage. A significant speed penalty must be paid due to the additional stage or stages.

7. LATCHING AND NON-LATCHING CIRCUITS

To understand a unique characteristic of a simple logic circuit, terminate a Josephson device with a load resistor R_L. Application of signal input control current large enough to exceed the Josephson current, will cause the device to go from a superconducting state to the resistive or voltage state. Depending on the internal resistance, capacitance and Josephson current of the device, the value of R_L, the device will either latch up in the voltage state or return to the superconductivity state when the input signal is removed (Fig. 5). For the practical reasons of keeping L/R time constants low and the width of matching transmission lines to reasonable values, most circuits are of the latching variety. Latching circuits present a new set of problems to a computer system, since every time the state of a logic gate has changed, a means must be provided to reset the latched gates. AC power supplies are used to accomplish the reset function.[17] Josephson junctions clip the AC supply providing a clipped alternating supply current.

Two complications arise due to an AC power supply. The first is the need to store the required information in a latch that is not affected by the AC power supply.[18] The second is a little more subtle in that, depending on the speed with which the AC supply goes through the zero point, the device may fail to reset (or punch through) (Fig. 6). Low device capacitance is necessary to prevent punch though from becoming a limitation to device performance.[19]

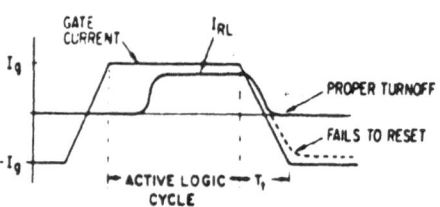

Fig. 6 - Alternating power (I_g) used to power up and reset the latching Josephson logic circuits. Punch-through occurs when the gate fails to reset and the output remains "ON" into the next cycle. (After [14])

A positive consequence of an AC power supply is the elimination of a separate system clock since the latches are updated every change of the AC polarity. A method to achieve non-latching operation developed by Baechtold *et al.*[20] is to reset one Josephson device by a second device, which switches to the voltage state.

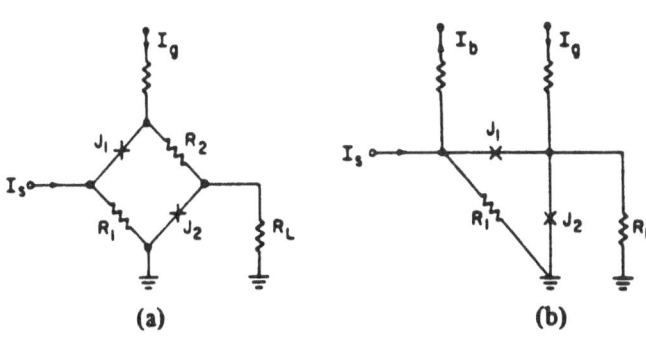

(a)   (b)

Fig. 7 - a) Direct-Coupled-Isolation (DCI) device. b) Josephson Atto Weber switch (JAWS) device. In both circuits, input current switches J_1 to voltage state for isolation, then J_2 is switched to divert I_g to output. (After [25])

8. DIRECT-COUPLED DEVICES

The large area required for a magnetically-coupled SQUID can be reduced by using a device structure in which current is direct-coupled or injected into the device. The main disadvantage of these devices is their lack of isolation. Additional Josephson junctions have been added to effect isolation. Gheewala and Mukherjee[21] have proposed a Direct-Coupled Isolation

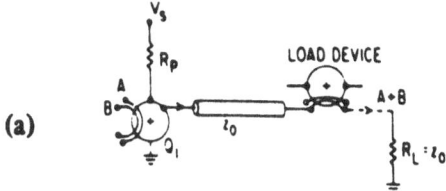

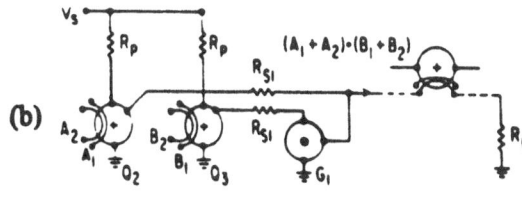

Fig. 8 - Josephson Current-Injection-Logic (CIL) circuits. a) Two-input OR. b) Two-input AND gate. (After [25])

device and Fulton *et al.*[22] the Josephson Otto Weber Switch (Fig. 7). A third device, Current Injection Device, utilizes flux quantization. Two Josephson junctions whose maximum currents differ from each other by a factor of 3 to 5 are interconnected with inductances L_1 and L_2 to form the CID device. A set of circuits satisfactory for general system logic implementation of both the CIL (Fig. 8) and DCL (Fig. 9) variety is given by Gheewala.[23]

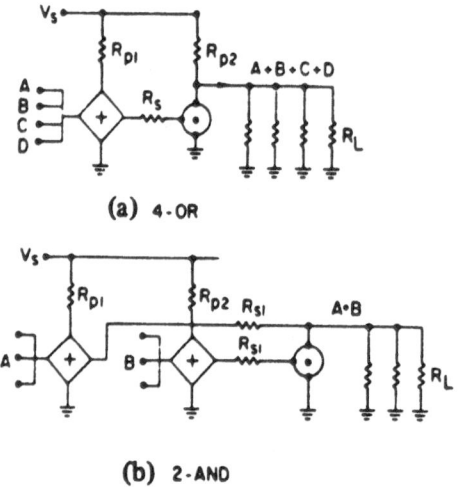

(a) 4-OR

(b) 2-AND

Fig. 9 - Josephson Direct-Coupled
Logic (DCL) circuits. a) Four-input
OR. b) Two-input AND. (After [25])

9. SWITCHING DELAYS OF CIRCUITS

There are three delay components in a Joseph-

son circuit. The application of an input pulse to a

Josephson circuit results in a turn-on delay or time to

establish the proper phase relations in the junction.[24]

Depending on the amount of overdrive, the turn-on

delay can be either very long or of the order of se-

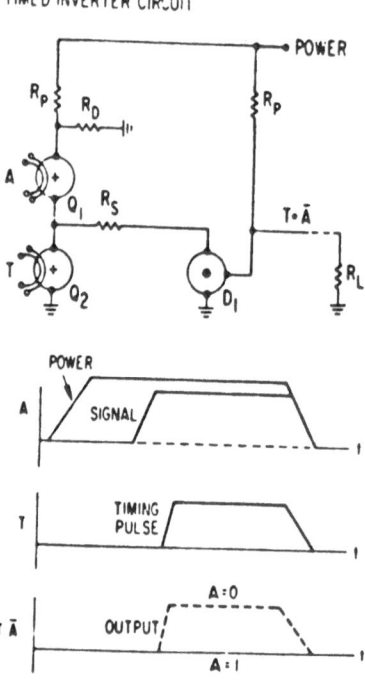

TIMED INVERTER CIRCUIT

Fig. 10 - A Timed-Inverter circuit to generate the complement of signal A using latching
logic circuits. The timing pulse
T is designed to arrive after A.
(After [25])

veral picoseconds. A second delay component is the ZC time constant associated with

charging of the device capacitance C in parallel with the output transmission line which has a

characteristic impedance Z. The finite speed of signal transmission down the superconducting

line will add more delay depending on the distance to the next input point. In Josephson

Technology, the transmission line is usually terminated in its characteristic impedance which

results in the quickest transfer of signal (first incident switching). When magnetically-coupled

devices are used, with a certain conductance of the control line, an additional delay occurs

because of the inductive discontinuity in the output transmission line. Fan-out for a magnetically-coupled device can be very large depending only on the delay skew variation that can be tolerated in the first-to-last switched device.

10. SIGNAL INVERSION

Realization of an inversion circuit in a latching scheme is not straightforward since the output is independent of input once latched. One solution is to use a delayed enable signal that enables the logic gate after all input signals have arrived[25] (Fig. 10). A second possibility is to run a dual rail where both the true and complement are generated at each stage. At first glance, this would seem to double the gate count but in actual cases studied the penalty is usually much less.

11. LOGIC EXPERIMENTAL RESULTS

Gheewala has done extensive experimental verification of ASTAP[26] circuit simulations. An experimental two-input OR gave a 130 psec delay over ten lightly loaded stages. A complete experimental evaluation of a circuit family of OR's and AND with an average fan-in of 4.5 and fan-out of 2.5 with a power dissipation of 4.8 μ/gate gave an average delay of 45 psec.

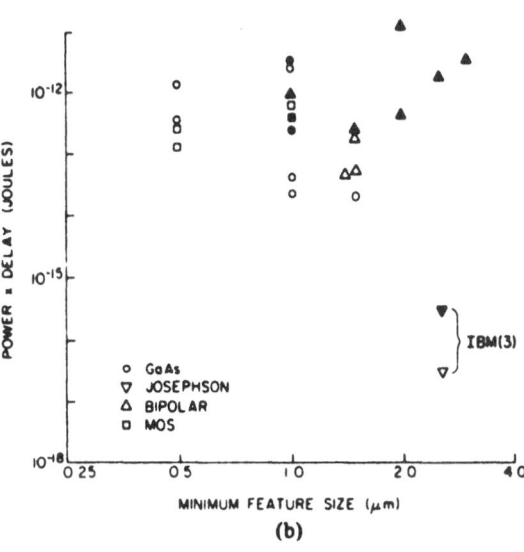

Fig. 11 - Power-delay product for logic circuits in current technologies. (After [25])

The basic fast switching speed, of 9-10 psec, of individual devices has been verified by Hamilton[27] and Tuckerman[28] by unique on-chip sampling techniques.

A comparison of competing technologies is given in Fig. 11. Josephson circuits are clearly faster and dissipate far less power by several orders of magnitude when compared to other technologies. It should be noted that the maximum heat-removal capability of freon at 300°K to liquid helium at 4.6°K is approximately 30:1.

12. NON-DESTRUCTIVE READ-OUT MEMORY CELLS (NDRO)

Memory cells can be ei-
ther destructive or non-
destructive when their contents
are read out.   Henkels and
Zappe[29-31] have done extensive
work on the design of fast
NDRO (Non-Destructive

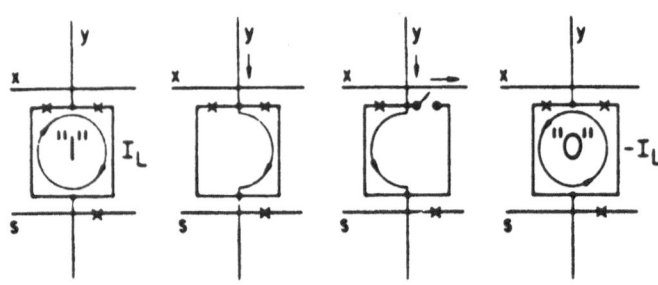

Fig. 12 - NDRO cell-storage operation. (After Zappe, IEEE Trans. Elect. Dev., <u>ED-27</u>, 10 (1980).)

Read-Out) cells suitable for cache-memory applications.   Assuming the original cell state has

no circulatory current (zero state), application of a current I_T causes $I_T/2$ to flow in the drive

gate (Fig. 12).   Application of control currents I_x and I_y' cause the drive to switch causing the

right branch to carry the full I_T.   Removal of the external I_x I_y' and I_y sources establishes a

circulatory $I_y(1)$.   The sense gate is designed such that $I_T/2$ does not cause it to be switched

but I_T does.   Read out is therefore non-destructive.   To remove a one, I_x and I_y' but no I_y are

applied, causing the driver to switch and destroy the superconducting current storage loop.

This cell is the basis for a 4K x 1 bit array chip design with a 0.5 nsec access time and 6 mW

power dissipation.[32]

13. DESTRUCTIVE READ-OUT MEMORY
CELLS (DRO)

A different approach to memory-cell design
has been explored by Gueret.[33,34] This cell uses a
destructive read-out method (Fig. 13).   If a two-
junction interferometer is biased by I_B to the inter-
section of the n = 0 and n = 1 modes and two
additional positive currents I_x and I_D are applied
and then a device current I_y is supplied, a 1 is writ-

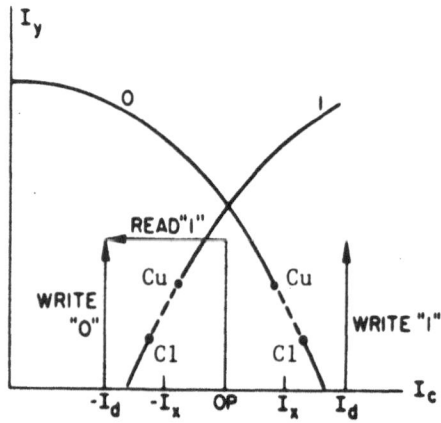

Fig. 13 - Operation of a DRO single flux-quantum memory cell.

ten. Control currents of $-I_x$ and $-I_D$ causes a "0" or zero flux quantum to be written or stored. Read out is accomplished by first applying device current I_y and then $-I_D$ and $-I_x$. If the cell were in a "1" state, the mode boundary n = 1 to n = 0 is crossed above the critical point, the cell is switched to a voltage state and a current

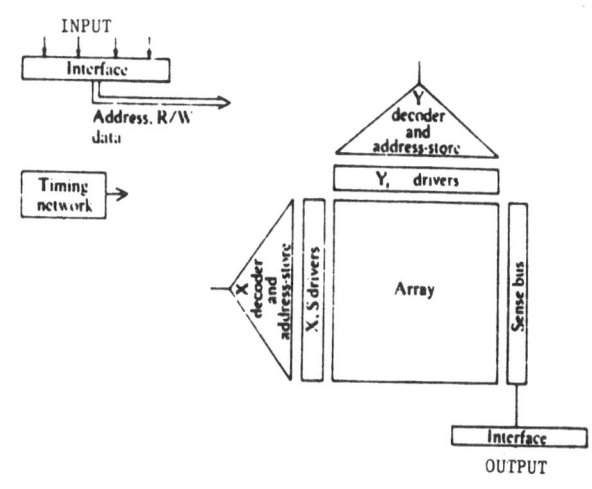

Fig. 14 - Block diagram of memory array (After [32])

pulse fed to a sense amplifier. Had the cell contained a "0", no pulse would have been sent ot the sense amplifier. One additional design constraint is that the gray zone noted by C_L and C_U must be avoided during writing and reading. Crossing the boundary in the zone results in an indeterminate cell state. Beha[35] has shown NDRO operation of a SFQ cell by use of an asymmetrical two-junction interferometer design. By causing the zero lobe to read "0", it is possible for the cell to reset to zero after switching.

A 2 K bit cross section of a DRO SFQ memory chip has been experimentally tested with good results.[36] This has established the base of a 16 K bit memory chip with an access time of 15 nsec and 40 μw power at full repetition rate. Stand-by power is zero.[37]

The block diagram of a Josephson memory chip looks like any other technology memory chip, with input latches, decoders, array drivers, memory cells and sense amplifiers (Fig. 14). As is the case with the logic, the performance and power required are in general one or more orders of magnitude superior to any other present-day technology. Figure 15 shows a comparison of performance.

14. COOLING

Estimates have been made of the amount of power dissipated at 4.2°K for a large system application to be approximately 10 W. Experiments of Guernsey indicated that a three-stage refrigeration unit with reasonable cost and maintenance schedule can be achieved.[38] It should be noted that 2000 times more power is required to be supplied at room temperature to the compressor. For the small system when the refrigerator load is 25 to 50 mW, a simple reliable low-cost 4°K cryocooler is only in the proposal stage.[39, 40] A diagram of a proposed cooler for a large system is given in Fig. 16.

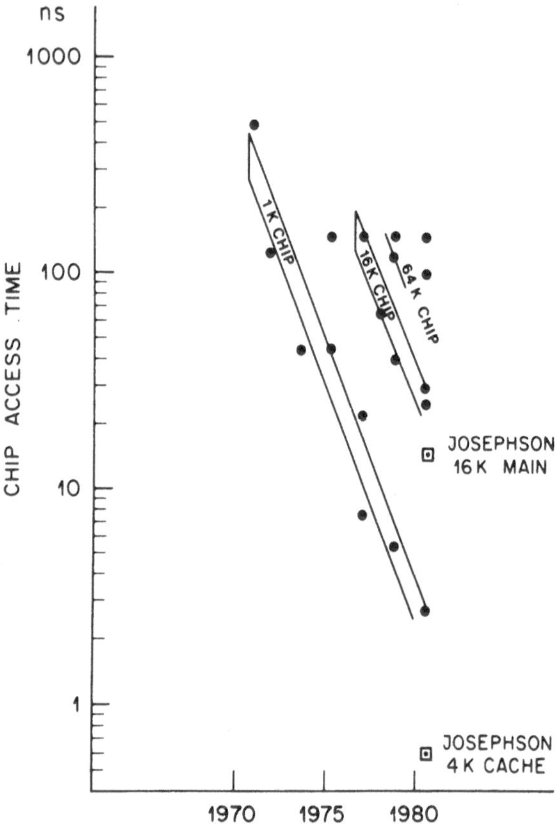

Fig. 15 - Memory-chip access time for several technologies.

15. FABRICATION

Although not within the framework of this paper, fabrication of actual devices involves a series of lithographic thin-film metal and insulator steps. One of the most demanding is the formation of the thin tunnel barrier oxide. Greiner et al[41] give details on the control possible with a lead technology, while Broom et al.[42] report on the possibilities of niobium technology. Although the basic device is a simple diode, by the time ground planes, damping resistors, control windings, load resistors and additional levels of wiring are added, the total mask count equals that of the more complex bipolar semiconductor chips.

16. PACKAGING

High-performance chips must be efficiently interconnected to realize a total computer system. Superconducting transmission lines permit the propagation of rapid rise-time pulses with little distortion. Small lines over ground planes permit close packing of lines in a single wiring channel with little interaction. Interconnections should be as short as possible to limit signal delay due to the relatively slow propagation velocity in low impedance lines.

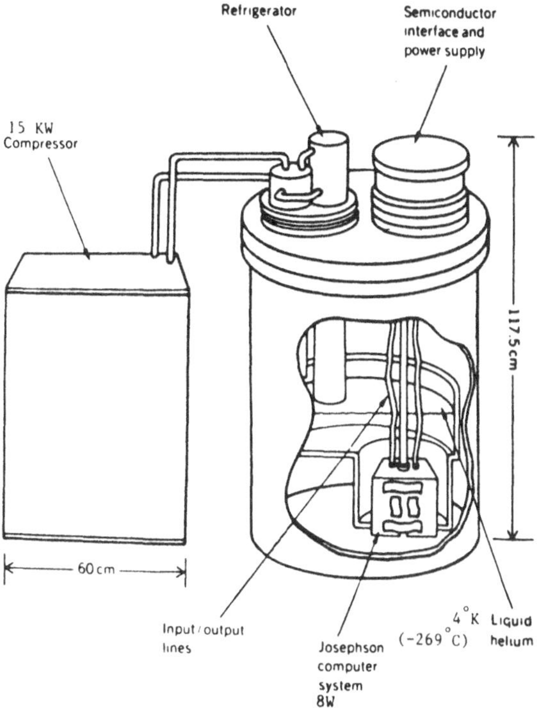

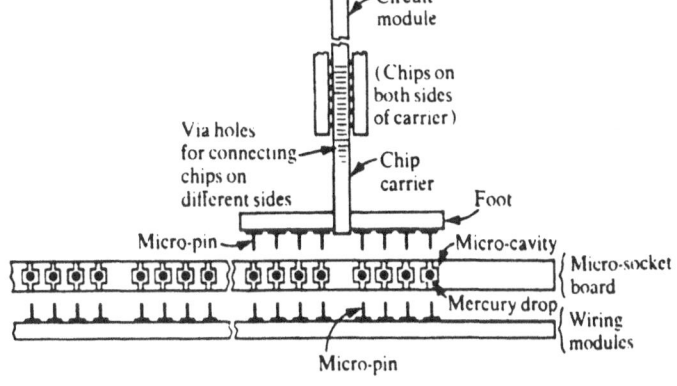

Fig. 16 - Josephson cryogenic cooler for large system. (After [2])

Rapid removal of system components is essential for engineering change capability for both initial system configuration and upgrading. Another requirement is the need to remove the heat dissipated during circuit operation. Although liquid helium is not as conductive as water, the power dissipation of Josephson circuits is several orders of magnitude lower than that of silicon technology thereby permitting closer circuit packing density.

A proposal of Anacker to forma microsockets filled with mercury appears rather attractive (Fig. 17). Assembly with

Fig. 17 - Packaging concept for Josephson computer. (After Tsui, IBM J. Res. Develop., 24, 2 (1980).)

essentially zero insertion force is realizable since the mercury is liquid at room temperature. Cooling to liquid-helium temperature freezes the inserted pin mechanically and electrically causes a low impedance contact.

The 300°C temperature excursion from room to liquid helium forces the expansion coefficient of large package pieces to be equal. Silicon has proved to be an excellent base on which to make Josephson circuit chips. It is also the material used for the other levels of packaging. Figure 18 shows how the modules are plugged into a board to form a small system test vehicle. See A. Brown's[43] paper for additional details.

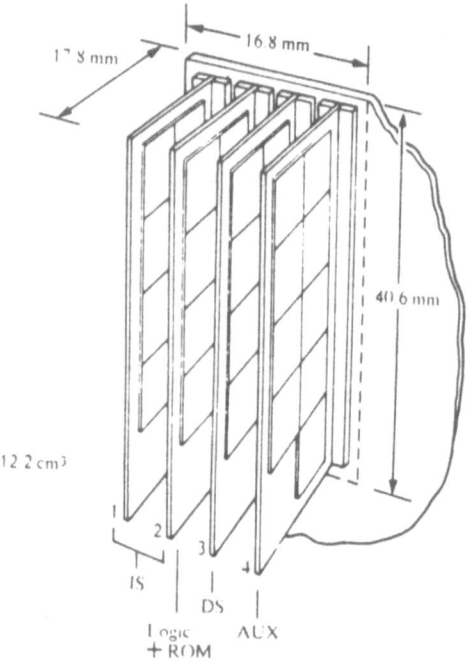

Fig. 18 - Small system packaging modules plug into board to form system. (After Tsui, IBM J. Res. Develop., 24, 2 (1980).)

17. CHALLENGES OF JOSEPHSON TECHNOLOGY

Although Josephson-type devices offer many unique advantages over other technologies, there are several challenges that may limit full commercial exploitation. The one characteristic that has proved most troublesome is that the Josephson junction is basically a threshold device. Design of LSI chips requires extremely tight control of several device parameters, the most notable one being the Josephson supercurrent. Control of the oxide barrier to better than 0.1 Å appears to be necessary to fulfill circuit requirements.[44] A second limitation is the use of inductances in devices to obtain desired properties. These inductances require the total circuit area to be much larger than the actual junction. A third constraint is the inability to run interconnecting wiring over devices, thereby limiting circuit density. A final challenge occurs due to the sensitivity of Josephson devices to magnetic fields. The earth's magnetic field is strong enough to cause problems requiring extensive shielding.

18. FUTURE

Anacker[45] has outlined the requirements for a Josephson computer to process information at the rate of 250 million instructions per second. This achievement will require the entire computer to occupy only a 10 cm box an have a basic machine cycle time of 1 nanosecond. Clearly, these numbers are far beyond the projections of silicon technology. It is difficult to tell whether projections of the researchers will ever be realized in a commercial product.

The fabrication of edge junctions in niobium by Broom *et al.*[46] presents the opportunity to fabricate stable and low capacitance junctions. Designers need no longer be constrained by the lead-alloy junction.

Van Duzer,[47] using a trend analysis curve of machine cycle time vs. year of introduction, has projected that in the 1990's we can expect a Josephson prototype machine.

In essentially all the previously discussed work, nominal or center values of device and circuit parameters have been used. For Josephson circuits to become useful in the VLSI era, actual distributions of parameters must be established and simulations of circuits made to determine an operating window, see Baechtold.[48]

A superconducting transistor has been proposed and fabricated by Gray.[49] This development is interesting because of the similarity to the semiconductor junction transistor.

Although the main emphasis in this paper has been high-speed digital applications, signal processor usage has equally unique potential.

As more researchers enter the field, we can expect an ever quickening pace of new devices and applications based on the Josephson junction.

REFERENCES

1. B. D. Josephson, (July 1962), Possible New Effects in Superconductive Tunnelling. Phys. Rev. Lett., vol. 1, no. 7., p. 251.

2. W. Anacker, (May 1979), Computing at 4 Degrees Kelvin. IEEE Spectrum, vol. 16, no. 5, p. 26.

3. E. A. Lynton, (1967), Superconductivity, Wiley, New York.

4.   R. L. Kautz, (Jan. 1978), Picosecond Pulses on Superconducting Striplines. J. Appl. Phys., vol. 49, no. 1, pp. 308-314.

5.   I. Giaever, (Aug. 1960), Energy Gap in Superconductors Measured by Electron Tunnelling. Phys. Rev. Lett., vol. 5, p. 147.

6.   J. Badeen L. M. Cooper and J. R. Schrieffer, (April 1957), Microscopic Theory of Superconductivity. Phys. Rev., vol. 106, p. 162; and (December 1957), Theory of Superconductivity. Phys. Rev., vol. 108, p. 1175.

7.   L. Esaki, (Jan 1958), New Phenomenon in Narrow Germanium p-n Junctions. Phys. Rev., vol. 109, p. 603.

8.   H. H. Zappe and B. S. Landman, (Jan. 1978), Analysis of Resonance Phenomenon in Josephson Quantum Interference Devices. J. Appl. Phys., vol. 49, p. 344. H. H. Zappe and B. S. Landman, (July 1978), Experimental Investigation of Resonances in Low-Q Josephson Interference Devices, J. Appl. Phys., vol. 49, p. 4149.

9.   P. W. Anderson and J. M. Rowell, (1963), Probable Observation of the Josephson Superconducting Tunnelling Effect. Phys. Rev. Lett., vol. 10, p. 230.

10.   J. M.Rowell, (1963), Magnetic Field Dependencies of the Josephson Tunnel Current. Phys. Rev. Lett., vol. 11, p. 200.

11.   J. Matisoo, (1966), Subnanosecond Pair Tunnelling to Single-Particle Tunnelling Transition in Josephson Junctions. Appl. Phys. Lett., vol. 9, p. 167.

12.   B. S. Landman, (1977), Calculations of Threshold Curves for Josephson Quantum Interference Devices. IEEE Trans. Mag., vol. MAG-13, pp. 871. Also, E. O. Schultz-DuBois and P. Wolf, (Oct. 1967), Static Characteristics of Josephson Interferometers. Presented at the Int. Conf. on Superconducting Quantum Devices.

13.   R. F. Broom, W. Kotyczka and A Moser, (March 1980), Modeling of Characteristics for Josephson Junctions Having Non-Uniform Width or Josephson Current Density. IBM J. Res. Develop., vol. 24, no. 2, p. 178.

14.   T. Gheewala, (March 1980), Design of 2.5 Micrometer Josephson Current Injection Logic. IBM J. Res. Develop., vol. 24, no. 2, p. 130.

15.   W. C. Stewart, (April 1968), Current-Voltage Characteristics of Josephson Junctions. Appl. Phys. Lett., vol. 12, p. 277.

16.   D. E. McCumber, (June 1968), Effect of AC Impedence on DC Voltage-Current Characteristics of Superconducting Weak-Link Junctions. J. Appl. Phys., vol. 19, p. 3113.

17.   P. C. Arnett and D. H. Herrell, (1979), Regulated AC Power for Josephson Interferometers for Latching Logic Circuits. IEEE Trans. Magn., vol. MAG-15, p. 544.

18.   A. Davidson, (Oct. 1978), A Josephson Latch. IEEE J. Solid-State Circuits, vol. SC-13, p. 583.

19.   T. A. Fulton and R. C. Dynes, (June 15, 1971), Switching to Zero Voltage in Josephson Tunnel Junctions. Solid-State Commun., vol. 9, p. 1069.

20. W. Baechtold, T. Forster, W. Heuberger and Th.O. Mohr, (May 15, 9175), Complementary Josephson-Junction Circuit: A Fast Flip-Flop and Logic Gate. Electron Lett., vol. 11, no. 10, p. 203.

21. T. Gheewala and A. Mukherjee, (Washington DC, Dec. 3-5, 1979), Josephson Direct-Coupled Logic (DCL). IEDM Tech. Dig., p. 482.

22. T. A. Fulton, S. S. Pei and L. N. Dunkleberger, (May 15, 1979), A Simple High Performance Current Switched Josephson Logic. Appl. Phys. Lett., vol. 34, no. 10, p. 709.

23. T. Gheewala, (Oct. 1979), 30-picosecond Josephson Current Injection Logic (CIL). IEEE J. Solid-State Circuits, vol. SC-14, no. 5, p. 787.

24. E. P. Harris, (Jan. 1979), Turn-On Delay of Josephson Interferometer Logic Devices. IEEE Trans. Magn., vol. MAG-15, p. 562. Also, D. G. McDonald, R. L. Patterson, C. A. Hamilton, R. E. Harris and R. L. Kautz, (Oct. 1980), Picosecond Applications of Josephson Junctions, IEEE Trans. Electron Devices, vol. ED-27, no. 10, p. 1945.

25. T. Gheewala, (Oct. 1980), Josephson-Logic Devices and Circuits. IEEE Trans. Electron Devices, vol. 27, no. 10, p. 1857.

26. IBM Advanced Statistical Analysis Program, IBM Publication No. SH20-1118-0, available through IBM Branch Offices.

27. C. A. Hamilton, F. L. Lloyd, R. L. Peterson and J. R. Andrews, A Superconducting Sampler for Josephson Logic Circuits. Appl. Phys. Lett., vol. 35, p. 718. (Nov. 1, 1979).

28. D. Tuckerman, (June 15, 1980), A Josephson Ultra-High Resolution Sampling System. Appl. Phys. Lett., vol. 36, p. 1008.

29. W. H. Henkels nd H. H. Zappe, (Oct. 1978), An Experimental 64-bit Decoded Josephson NDRO Random Memory. IEEE J. Solid-State Circuits, vol. SC-13, no. 5, p. 591.

30. H. H. Zappe, (Feb. 1975), A Subnanosecond Josephson Tunnelling Memory Cell with Non-Destructive Readout. IEEE J. Solid-State Circuits, vol. SC-10, no. 1, p. 12.

31. H. H. Zappe, Josephson Quantum Interference Computer Devices. IEEE Trans. Magn., vol. MAG-13, p. 41, Jan. 1977.

32. S. M. Faris, W. H. Henkels., E. A. Valsamakis and H. H. Zappe, (Mar. 1980), Basic Design of a Josephson Cache Memory. IBM J. Res. Develop. (Special Issue on Josephson Technology), vol. 24, no. 2. p. 143.

33. P. Gueret, (Mar. 1975), Storage and Detection of a Single Flux Quantum in Josephson Junction Devices. IEEE Trans. Magn., vol. MAG-11, no. 2, p. 751.

34. P. Gueret, Th.O. Mohr and P. Wolf. (Jan. 1977), Single Flux-Quantum Memory Cells. IEEE Trans. Magn., vol. MAG-13, no. 1. p. 52.

35. H. Beha, (Sept. 1977), Two-Josephson-Junction Interferometer Memory Cell for NDRO. Electron. Lett., vol. 13, no. 20, p. 596.

36. R. F. Broom, P. Geuret, W. Kotyczka, Th.O. Mohr, A. Moser, A. Oosenbrug and P. Wolf, (Feb. 1977), Model for a 15 ns 16 K RAM with Josephson Junctions. IEEE Int. Solid-State Circuits Conf., Dig. Tech. Papers, p. 60.

37. P. Gueret, A. Moser and P. Wolf, (Mar. 1980), Investigations for a Josephson Computer Main Memory with Single Flux Quantum Cells. IBM J. Res. Develop. (Special Issue on Josephson Technology), vol. 24, no. 2, p. 155.

38. R. Guernsey and E. Flint, (May 1981), Refrigeration Requirement for Superconducting Computers. NBS special refrigeration for cryogenic sensors and electronic systems.

39. J. E. Zimmermann and D. B. Sullivan, (1979), A Milliwatt Sterling Cycle Cryocooler for Temperatures below 4°K, Cryogenics, vol. 19, p. 170.

40. D. B. Sullivan and J. E. Zimmermann, (1979), Very Low Power Sterling Cryocooler Using Plastic and Composite Materials. Int. J. Refrig., vol. 2, p. 211.

41. J. H. Greiner, *et al.* (Mar. 1980), Fabrication Process for Josephson Integrated Circuits. IBM J. Res. Develop. (Special Issue on Josephson Technology), vol. 24, no. 2, p. 195.

42. R. F. Broom, S. I. Raider, A. Oosenbrug, R. E. Drake and W. Walter, (Oct. 1980), Niobium Oxide Barrier Tunnel Junction. IEEE Trans. Electron Devices, vol. ED-27, no. 10, p. 1998.

43. A. V. Brown, (March 1980), An Overview of Josephson Packaging. IBM J. Res. Develop., vol. 21, no. 2, p. 167.

44. R. F. Broom and Th.O. Mohr, (May/June 1978), Studies on Arrays of Josephson Tunnel Junction Interferometers. J. Vac. Sci. Technol., vol. 15, p. 1166.

45. W. Anacker, (March 1980), Josephson Computer Technology: An IBM Research Project. IBM J. Res. Develop., vol. 24, no. 2 p. 107.

46. R. F. Broom, A. Oosenbrug and W. Walter, (1980), Josephson Junctions of Small Area Formed on the Edge of Niobium Films. Appl. Phys. Lett., vol. 37, p. 237.

47. T. Van Duzer (May 1980), Proceedings of Second International Conference on Superconductivity Quantum Devices, West Berlin.

48. W. Baechtold, (1980), Josephson High-Performance Logic. Proc. IEEE Int. Conf. on Circuits and Computers, ICCC 80, vol.2, p. 879.

49. K. E. Gray, (Mar. 15, 1978), A Superconducting Transistor. Appl. Phys. Lett., vol. 36, no. 6, p. 392.